U0172652

# nature

## The Living Record of Science

## 《自然》百年科学经典

英汉对照版（平装本）

第十卷（上）

总顾问：李政道（Tsung-Dao Lee）

英方主编：Sir John Maddox
Sir Philip Campbell

中方主编：路甬祥

2002-2007

外语教学与研究出版社 · 麦克米伦教育 · 自然科研
FOREIGN LANGUAGE TEACHING AND RESEARCH PRESS · MACMILLAN EDUCATION · NATURE RESEARCH
北京 BEIJING

**图书在版编目（CIP）数据**

《自然》百年科学经典. 第十卷. 上，2002–2007 ：英汉对照 /（英）约翰 · 马多克斯（John Maddox），（英）菲利普 · 坎贝尔（Philip Campbell），路甬祥主编. -- 北京 ：外语教学与研究出版社，2019.12
ISBN 978-7-5213-1471-7

Ⅰ. ①自… Ⅱ. ①约… ②菲… ③路… Ⅲ. ①自然科学－文集－英、汉 Ⅳ. ①N53

中国版本图书馆 CIP 数据核字 (2020) 第 021652 号

地图审图号：GS（2019）3264 号

出 版 人　徐建忠
项目统筹　章思英
项目负责　刘晓楠　黄小斌
责任编辑　黄小斌
责任校对　王丽霞
封面设计　孙莉明　曹志远
版式设计　孙莉明
出版发行　外语教学与研究出版社
社　　址　北京市西三环北路 19 号（100089）
网　　址　http://www.fltrp.com
印　　刷　北京华联印刷有限公司
开　　本　787 × 1092　1/16
印　　张　34
版　　次　2020 年 4 月第 1 版　2020 年 4 月第 1 次印刷
书　　号　ISBN 978-7-5213-1471-7
定　　价　168.00 元

购书咨询：（010）88819926　电子邮箱：club@fltrp.com
外研书店：https://waiyants.tmall.com
凡印刷、装订质量问题，请联系我社印制部
联系电话：（010）61207896　电子邮箱：zhijian@fltrp.com
凡侵权、盗版书籍线索，请联系我社法律事务部
举报电话：（010）88817519　电子邮箱：banquan@fltrp.com
物料号：314710001

# 《自然》百年科学经典（英汉对照版）

# 编译委员会

# Contents
# 目录

# Volume X

# (2002-2007)

# Estimating the Human Health Risk from Possible BSE Infection of the British Sheep Flock

N. M. Ferguson *et al.*

## Editor's Note

**The neurodegenerative disease bovine spongiform encephalopathy (BSE) was first recognized in British cattle in the 1980s. The government's response was to monitor herds closely and restrict beef sales to ensure that organs known to harbour BSE were not sold. Nevertheless, an outbreak occurred in the late 1980s, leading to the culling of British cattle and a European ban on British beef exports. BSE is believed to trigger a similar, fatal disease in humans called Creutzfeldt-Jakob disease, of which more than 160 people have died following the 1980s BSE outbreak. Here epidemiologist Roy Anderson and colleagues model the likely course of the human epidemic and predict future deaths to be in the range 50–50,000.**

---

Following the controversial failure of a recent study[1] and the small numbers of animals yet screened for infection[2], it remains uncertain whether bovine spongiform encephalopathy (BSE) was transmitted to sheep in the past via feed supplements and whether it is still present. Well grounded mathematical and statistical models are therefore essential to integrate the limited and disparate data, to explore uncertainty, and to define data-collection priorities. We analysed the implications of different scenarios of BSE spread in sheep for relative human exposure levels and variant Creutzfeldt–Jakob disease (vCJD) incidence. Here we show that, if BSE entered the sheep population and a degree of transmission occurred, then ongoing public health risks from ovine BSE are likely to be greater than those from cattle, but that any such risk could be reduced by up to 90% through additional restrictions on sheep products entering the food supply. Extending the analysis to consider absolute risk, we estimate the 95% confidence interval for future vCJD mortality to be 50 to 50,000 human deaths considering exposure to bovine BSE alone, with the upper bound increasing to 150,000 once we include exposure from the worst-case ovine BSE scenario examined.

---

THE aim of this study was not to evaluate the probability that BSE has entered the sheep flock, but rather, given the pessimistic assumption that infection has occurred, to explore its potential extent and pattern of spread. In this, we used epidemiological parameter estimates from experimental BSE infections of sheep, and, where data are unavailable, assumed (given the observed similarities in BSE and scrapie pathogenesis in sheep) that other aspects of disease epidemiology resemble those of scrapie. Analyses were constrained to be consistent with the failure to detect the BSE agent in a small sample of 180 brains[2] collected between 1996 and 2000 from sheep diagnosed with scrapie (giving

# 英国羊群可能的BSE感染带来的人类健康风险评估

弗格森等

编者按

神经退行性疾病牛海绵状脑病(BSE)于20世纪80年代首次在英国牛中发现。英国政府的回应是密切监控牧群，限制牛肉销售，以确保患有BSE的牛的器官不被出售。然而，20世纪80年代末爆发了一场疫情，导致英国牲畜遭到扑杀，欧洲禁止英国牛肉出口。据信，BSE在人类中引发了一种类似的致命疾病(称为克雅氏病)。自20世纪80年代BSE爆发以来，已有160多人死于这种疾病。流行病学家罗伊·安德森和他的同事们模拟了人类流行病的可能进程，并预测未来因这种疾病死亡的人数将在50~50,000之间。

---

随着最近一项有争议的研究的失败[1]加上少数动物已经接受过感染筛查[2]，但牛海绵状脑病(BSE)过去是否是通过饲料添加剂传染到羊身上以及现在是否仍然存在都还不能确定。因此，有充分依据的数学和统计学模型对于整合有限的不同数据、探索不确定性以及确定数据收集的重点都是十分必要的。我们分析了BSE在羊群中传播的不同方式与相关的人类暴露水平以及变异型克雅氏病(vCJD)患病率之间的关系，我们发现，如果BSE进入了羊群并且发生了一定程度的传播，那么羊BSE导致的不断增加的公共卫生风险很可能比牛的大，但是通过附加限制羊产品进入食物供应，可以将任何这类风险减少高达90%。将这个分析延伸到绝对风险，在95%置信区间内，我们估计单纯暴露于牛BSE的未来vCJD死亡率是50到50,000人，而一旦暴露于最坏方式传播的羊BSE，则上限增加到150,000人。

---

本研究的目的不是评估BSE进入羊群的可能性，而是悲观地假设感染已经发生，以此来研究感染传播的潜在程度和方式。在本文中，我们使用了实验羊的BSE感染得出的流行病学参数估计，对于那些无法获得的数据，就假定该病流行病学的这些数据与瘙痒病类似(考虑到羊身上BSE和瘙痒病发病机理的相似性)。此次分析的是从1996年到2000年之间患有瘙痒病的羊中收集的180份脑标本[2]，要求在这些脑标本中不能检测到BSE病原体的存在(考虑到在明显患有瘙痒病的绵羊

an upper bound for BSE prevalence within apparently scrapie-affected sheep of 2%; Fig. 1a), and are also broadly consistent with an assessment of historical exposure of the ovine population to meat and bonemeal (MBM)[3].

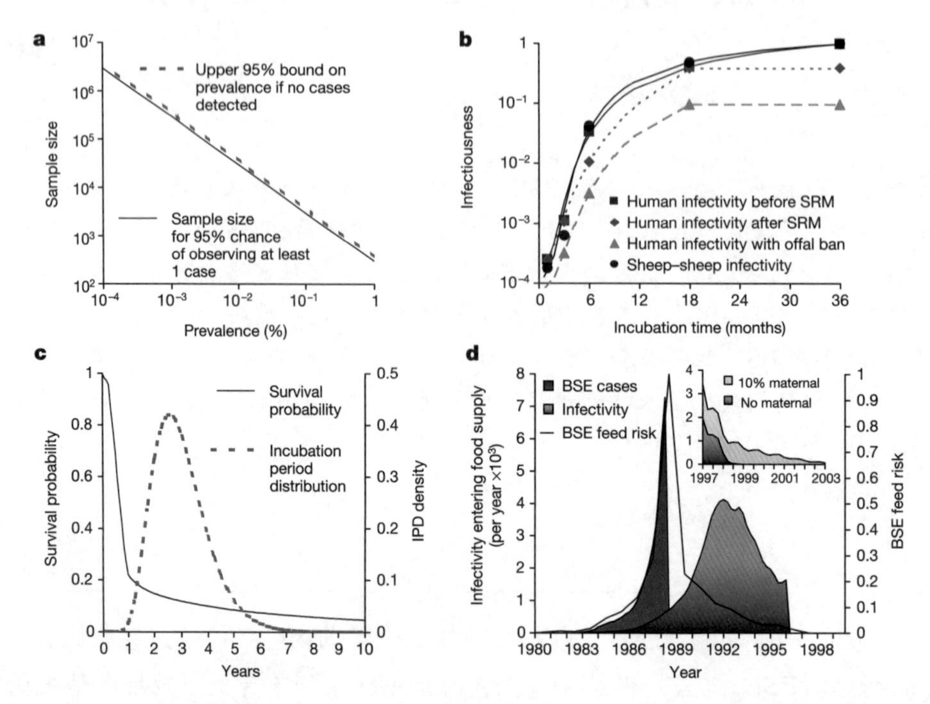


Fig. 1. Epidemiological inputs to transmission model. **a**, Relationship between sample size and detectable prevalence in screening studies. **b**, Infectiousness of sheep as a function of time from infection (see Methods). Sheep and human profiles are separately normalized to give maxima of 1. **c**, Survival probability of sheep as a function of age, and assumed incubation period distribution (IPD) of BSE in sheep. The survivorship function was estimated from annual data from the June census, slaughter, export and disappearance statistics, and data on the seasonality of lamb slaughter. **d**, Before mid-1988, both reported and unreported clinical BSE cases could be used for food (red-shaded curve). After that time, BSE was made notifiable and cases were destroyed, so we assume none entered food. The blue-shaded curve represents the rate of slaughter (per year) of pre-clinical infected cattle weighted by infectiousness relative to disease onset, assuming infectiousness grows at the exponential rate of 4 per year. The over-30-month scheme ended most bovine exposure in 1996, with estimates of residual levels (inset) being dependent on the extent of maternal transmission of BSE. The solid black curve represents the estimated infection hazard to cattle and sheep posed by infectivity in contaminated feed, relative to the maximum level reached in 1988.

Key to our analysis is estimates of the infectivity in animal tissues during disease incubation. Data are limited for BSE in sheep acquired by oral challenge, but using new experimental results and published data from studies of both scrapie[4-7] and sheep BSE[8,9] pathogenesis, we constructed an infectiousness profile. This profile was based on temporal changes in the density of the agent in different tissues, weighted by the proportion of such tissue in the host's body (Fig. 1b). This profile differs from that of BSE in cattle[10,11], with a more rapid rise in overall infectivity early in the incubation period in a wide range of tissues (for example, spleen and lymph nodes). The sheep–human infectiousness profiles (Fig. 1b) adjust for tissue-specific usage in food[12] and the effect of the 1997 specified risk materials (SRM) ban in sheep.

中 BSE 的患病率上限是 2%；图 1a)，并且要求与对这些绵羊群曾经与肉类和骨粉(MBM)接触的历史评估大致一样[3]。

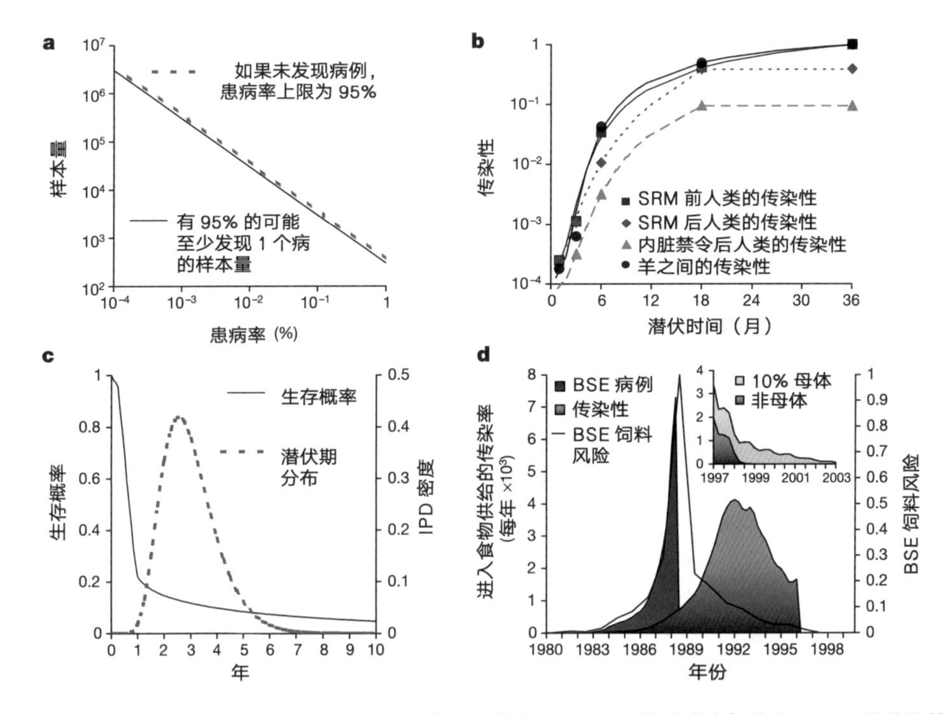


图 1. 传播模型中的流行病学数据。**a**，在筛选研究中样本量和检测到的患病率间的关系。**b**，羊传染性随感染时间的变化(见方法)。羊和人的患病率概况分别归一化到最大值为 1。**c**，羊随着年龄增长的生存概率，以及假设的羊群中 BSE 的潜伏期分布状况(IPD)。生存函数是根据每年 6 月份进行的羊群普查、屠宰、出口和失踪的数据以及羊羔屠宰的季节性数据进行估计的。**d**，1988 年中期之前，报告的和未报告的 BSE 患病性畜都可以做成食品(红色阴影曲线)。此后，BSE 病例都要上报，并且患畜都被销毁，因此我们假定没有患畜做成食品。蓝色阴影曲线表示临床感染前期牛的屠宰率(每年)，以相对于疾病发病的传染性来表示，假设传染性以每年 4 的速度指数式增长。这个超过 30 个月的方案在 1996 年结束了大部分牛群的暴露，而残存水平的估计(插入图)取决于 BSE 母体传播的程度。黑色的实线表示相对于 1988 年达到的最高水平，在受污染饲料中的传染性对牛羊造成的感染风险的估计。

我们分析的关键是评估疾病潜伏期内动物组织的传染性。通过进食感染的 BSE 羊群的数据有限，但是根据对瘙痒病[4-7]和羊 BSE[8,9]发病机理研究所获得的新的研究结果和公布的数据，我们构建出了一个传染性的概况。这个概况基于不同组织中病原体密度的暂时性改变，以该组织在宿主体内所占的比例来表示(图 1b)。此概况与牛中的 BSE 感染情况不同[10,11]，表现为在不同组织(例如脾和淋巴结)的潜伏期早期，总体传染性增加更快。羊–人传染的概况(图 1b)根据食物中特殊组织的含量[12]以及 1997 年针对羊的高风险食品(SRM)禁令的效果进行调整。

The distribution of the BSE incubation period in sheep is not well characterized, but on the basis of the limited available data, we used an offset gamma distribution with a mean of 3 years (Fig. 1c) and substantial variance (intended to capture variation caused by dose dependency and host genotype) both for sheep infected by feed and those infected horizontally. Pathogenesis and susceptibility are dependent on the genetic background of the host[13-15], and genotype frequencies of the key polymorphisms vary considerably within and between flocks of different breeds[16,17]. Collation of limited available data suggests that roughly one-third of sheep in Great Britain (England, Scotland and Wales) have BSE-susceptible genotypes (see Supplementary Information). Exposure of the sheep flock to the BSE agent via contaminated feed (Fig. 1d) is assumed to mirror (albeit at a much lower level) that of British cattle, as estimated in back-calculation analyses of the BSE epidemic[18,19]. Exposure is likely to have occurred as far back as the early 1980s, before the disease was identified in cattle[18,19]. As for BSE in cattle[20], host survivorship is important given the long incubation period of the disease. Best estimates (see Supplementary Information) are presented in Fig. 1c (giving a mean life expectancy of 1.5 years), although more precise data are urgently required, perhaps based on a sheep equivalent of the British Cattle Tracing System (http://www.bcms.gov.uk).

Capturing the information in Fig. 1 requires a framework that integrates the temporal evolution of BSE pathogenesis in the individual host (incorporating age-dependent susceptibility and exposure) into a mathematical model of transmission within (including seeding and spread) and between flocks. This model builds on previous analyses of within- and between-flock transmission of scrapie[21,22], and consists of a set of nonlinear partial differential equations detailing the transmission dynamics of the agent and the demography of the sheep flock under time-dependent exposure to BSE-contaminated feed. The dynamics of disease transmission within the sheep population are determined by the magnitude of the respective basic reproduction numbers of the agent within a flock ($R_0^A$) and between flocks ($R_0^F$). The reproduction numbers define the average number of secondary cases or flocks generated by one primary case or infected flock in a susceptible flock or population of flocks. We considered three representative scenarios: (I) $R_0^A > 1$, $R_0^F < 1$—self-sustaining transmission within a flock but not between flocks; (II) $R_0^A > 1$, $R_0^F > 1$—the worst-case scenario for future spread, with spread within and between flocks inducing an expanding epidemic; and (III) $R_0^A < 1$, $R_0^F < 1$—the best-case scenario, with non-self-sustaining transmission both within and between flocks (see Fig. 2 for precise parameter values).

For each scenario, the level of flock infection due to MBM exposure was adjusted to give BSE prevalence below 2% of scrapie prevalence at present (consistent with the results of ongoing studies screening sheep brains). Judgements of scenario consistency therefore depend on the estimated prevalence of scrapie in British sheep, with such estimates being based on limited data from detailed surveys of specific flocks (using clinical criteria for diagnosis) and a postal survey of farmers intended to characterize national historical patterns of scrapie incidence[23,24]. The uncertainties in interpreting these data (giving estimates of infection prevalence anywhere between 0.1% and 1%, see Supplementary

羊群中BSE潜伏期的分布没有很明显的特征，但是根据有限的可用数据，我们对饲料感染以及水平感染的羊均使用平均数为3年的偏移γ分布（图1c）以及显著的差异（试图发现由剂量依赖性和宿主基因型引起的变化）。发病机制和易感性由宿主的遗传背景决定[13-15]，重要多态性的基因型频率在不同品种的羊群内部和羊群之间的差别相当大[16,17]。对有限可用数据的整理表明，英国（英格兰、苏格兰和威尔士）大约三分之一的羊具有BSE易感基因型（见补充信息）。正如对BSE流行病的反算法分析所估计的那样，这些羊群通过污染的饲料暴露于BSE病原体中（图1d），这与英国中的情况差不多（尽管水平要低得多）[18,19]。暴露很可能在20世纪80年代早期就已经存在，那时尚未在牛群中发现此病[18,19]。对于牛的BSE[20]，由于潜伏期很长，宿主的生存概率就显得很重要。尽管迫切需要更加精确的数据，但图1c已经显示了最佳估计值（见补充信息）（平均预期寿命是1.5年），这可能基于在羊群中使用的英国牛群追踪系统（http://www.bcms.gov.uk）。

获得图1所示的信息需要一个模型，而这个模型能够将个体宿主中BSE发病机制的时间演变（包括年龄依赖的易感性和暴露）整合到在羊群内部（包括播种和传播）和羊群之间传播的数学模型中。该模型基于先前对羊群内部和羊群之间瘙痒病传播的分析[21,22]，并且由一组非线性偏微分方程组成，详细描述了病原体传播的动力学和暴露在受BSE污染饲料下的具有时间依赖性的羊群的种群统计学。羊群中疾病传播的动力学是由群内部（$R_0^A$）和群之间（$R_0^F$）病原体各自的基础繁殖数量的大小决定的。繁殖数目定义了以下四种情况的平均数量，或者次要病例，或者由一个主要病例产生的羊群，或者易感群体中受感染的群体，或者种群。我们考虑了三个代表性情况：（I）$R_0^A > 1$，$R_0^F < 1$——群内部是自我维持的传播，而群之间不是；（II）$R_0^A > 1$，$R_0^F > 1$——未来最差的情况，在群内部和群之间传播，导致流行病不断扩大；（III）$R_0^A < 1$，$R_0^F < 1$——最好的情况，群内部和群之间都是非自我维持的传播（精确的参数值见图2）。

对于每一种情况，由于MBM暴露导致的群体感染的水平都调整到使得BSE的患病率低于目前2%的瘙痒病患病率（与正在进行的羊脑筛查的研究结果一致）。因此，对情况一致性的判断就取决于英国羊群中估计的瘙痒病患病率，而这个估计是基于对特定羊群（使用临床诊断标准进行诊断）的详细调查以及对农民的邮政调查所得到的有限数据，目的是描述全国瘙痒病发病的历史模式[23,24]。解释这些数据的不确定性（感染的患病率估计值在0.1%～1%之间，见补充信息）使得大规模（图1a）

Information) make large-scale (Fig. 1a) screening of the national flock for transmissible spongiform encephalopathies (TSEs) a priority. In constructing the scenarios, we assumed a scrapie prevalence of about 0.3%, with scenarios II and III then corresponding to a BSE prevalence of 0.5% that of scrapie, and scenario I to a prevalence of about 2% that of scrapie. Scenario I was thus intended to represent something of a worst case in terms of the numbers of animals infected to date, although we cannot exclude the possibility of even larger epidemics given the limited data currently available.

Figure 2 displays the epidemiological characteristics of these scenarios in terms of within-flock and overall prevalence (Fig. 2a–c), and their implications for human exposure via food (Fig. 2d–f). The estimates of exposure incorporate data on human consumption of ovine material (see Supplementary Information), which indicate that 67% of lambs and 83% of sheep older than 12 months slaughtered for consumption in 1999 were consumed domestically in the UK. Very few live sheep are imported into Great Britain, and most imported lamb meat originates from New Zealand, which has never detected signs of BSE or scrapie infection in either its bovine or ovine populations. Thus the potential risk from BSE-infected sheep arises from home-bred animals.

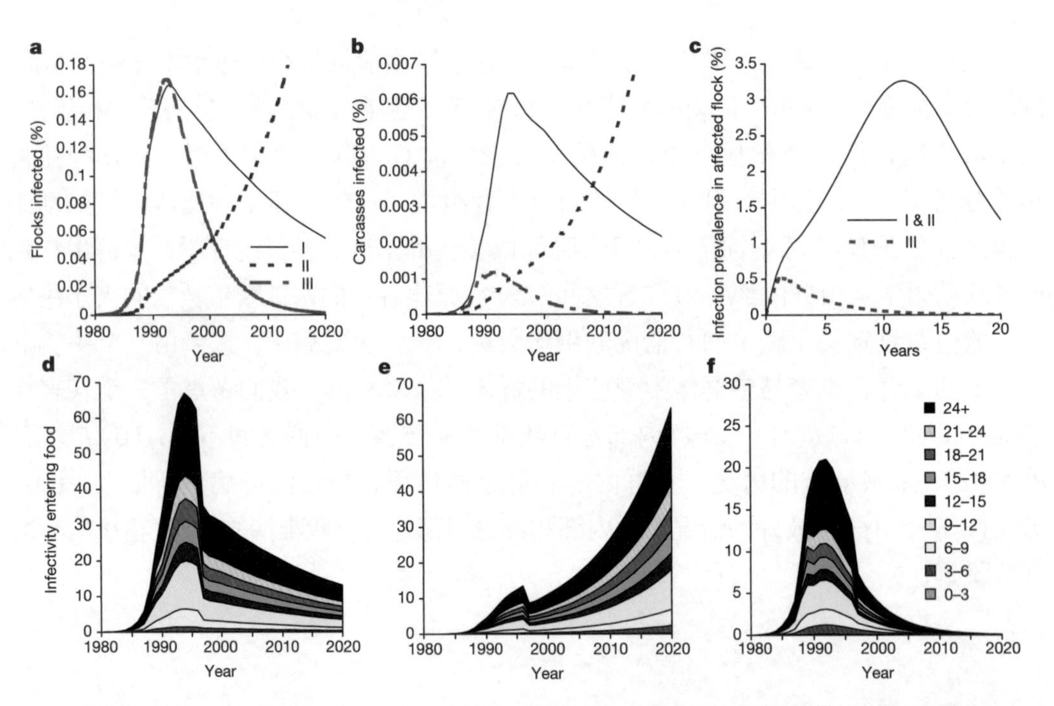


Fig. 2. Epidemiological characteristics of BSE transmission scenarios in sheep. $R_0^A = 2$, $R_0^F = 0.8$ and $\beta_B = 0.2\%$ per year for scenario I; $R_0^A = 2$, $R_0^F = 1.5$ and $\beta_B = 0.025\%$ per year for scenario II; $R_0^A = 0.8$, $R_0^F = 0.5$ and $\beta_B = 0.2\%$ per year for scenario III; where $\beta_B$ is assumed flock infection incidence rate per unit of feed risk profile shown in Fig. 1c. By comparison, $\beta_B$ values of 0.2% and 0.025% represent per-animal infection hazards about 50- and 400-fold less than that experienced by cattle. **a**, Proportion of flocks affected through time for scenarios I–III. **b**, Proportion of carcasses entering the food supply infected (at any incubation stage) with BSE. **c**, Prevalence in affected flock as a function of time since initial entry of infection into the flock. **d–f**, Estimated infectivity (in units of maximally infectious carcasses) entering the food supply under scenarios I, II and III, respectively, derived from the infectiousness profiles shown in

筛查全国羊群中的传染性海绵状脑病(TSE)成为优先事项。在构建上述情况时，我们假定瘙痒病的患病率大约是0.3%，情况II和III相对应的BSE的患病率是瘙痒病的0.5%，而情况I对应的BSE患病率是瘙痒病的2%。因此，情况I预期能够代表目前为止感染动物数量的最差状况，尽管我们不能排除由于目前可用的数据有限而发生更大范围流行病的可能性。

图2显示了这些情况在群内部和总体患病率方面的流行病学特征(图2a～2c)，以及它们在人类通过食物暴露时的影响(图2d～2f)。暴露估计包括人类对绵羊身体组成部分进行消费的数据(见补充信息)，这表明1999年屠宰的67%的羊羔和83%的超过12个月龄的绵羊都是在英国国内消费的。很少有活羊进口到英国，大部分进口的羔羊肉来自新西兰，而新西兰从未在羊群或者牛群中检测到BSE或者瘙痒病感染的迹象。因此，BSE感染的羊的潜在风险都来自于家养动物。

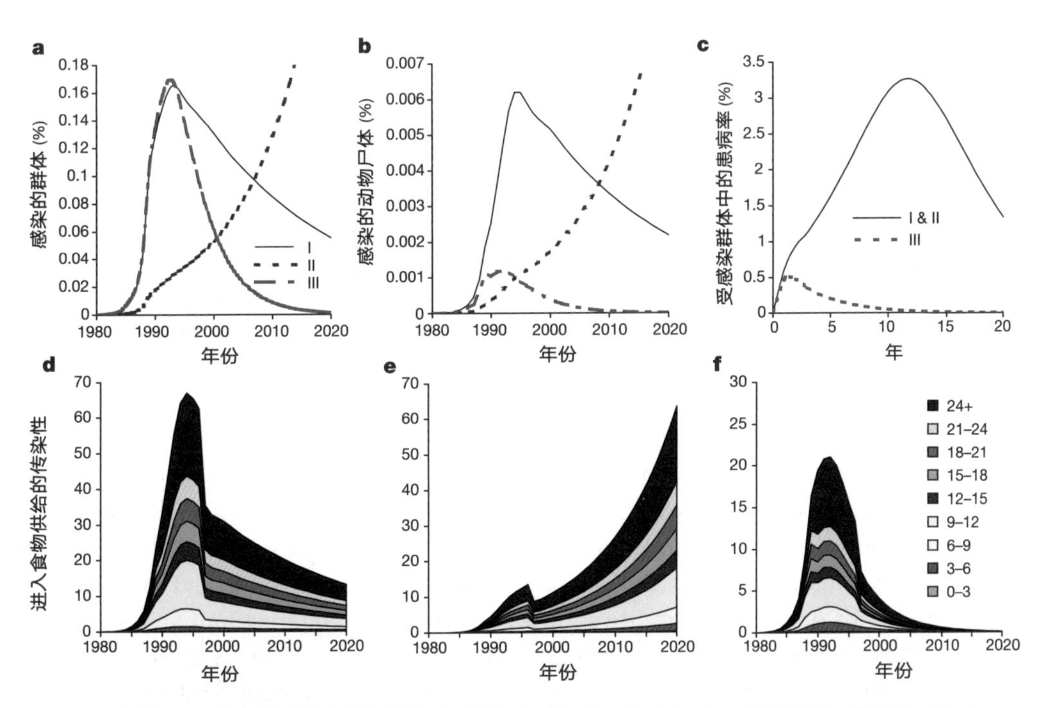


图2. 羊群中BSE传播情况的流行病学特征。情况I，$R_0^A=2$，$R_0^F=0.8$，$\beta_B=0.2\%$ 每年；情况II，$R_0^A=2$，$R_0^F=1.5$，$\beta_B=0.025\%$ 每年；情况III，$R_0^A=0.8$，$R_0^F=0.5$，$\beta_B=0.2\%$ 每年；其中 $\beta_B$ 是每单位饲料风险分布的假定群体感染发病率，如图1c所示。相比之下，$\beta_B$ 值为0.2%和0.025%分别表示每个动物感染的风险是牛感染的风险的五十分之一和四百分之一。**a**，三种情况下，随着时间的推移，受感染群体的比例。**b**，感染了BSE的动物尸体进入食物供给的比例(包括各种潜伏阶段)。**c**，从感染进入群体开始，受感染群体的患病率随时间的变化的函数。**d~f**，根据图1a所示的传染性概况，在I、II、III三种情况下进入食物供给的估计传染性(以最大传染性尸体为单位)。改变受BSE感染的羊群的不同比例

Fig. 1a. Varying the proportion of flocks affected with BSE (shown in **a**, and determined by $\beta_B$) scales the exposure curves in **d–f** proportionately.

The sudden drop in human exposure in the late 1990s (Fig. 2d–f) is due to the SRM ban for sheep (banning the consumption of the skull, tonsils and spinal column of animals older than 12 months and the spleen of all animals). In the worst-case scenario (II), the degree of risk is still rising steeply, whereas for the other scenarios risk is falling, approaching very low levels in the best case (III). Critically, even in the worst-case scenario, the great majority of past exposure of the human population to BSE arose from cattle (Fig. 1d). However, for all three scenarios considered (and therefore assuming BSE did enter the sheep flock), the number of maximally infectious sheep entering the food supply in Great Britain at present is estimated to be greater than the number of maximally infectious cattle[18,19] (Fig. 1d, inset). This comparison equates the risk from one maximally infectious sheep with that of a cow, despite the difference in body mass, an assumption motivated by the evidence of more widespread distribution of infectivity in sheep.

Potential risk-reduction strategies include restrictions on the age of sheep slaughtered for consumption and enhanced tissue-based controls to reduce the amount of infectivity entering the food supply; additional measures based on flock history or ram genotype might also be possible but are not considered here. Combined tissue- and age-based restrictions are estimated to reduce current and future risk by at least 80% for all three sheep BSE scenarios considered (Fig. 3a). This is encouraging, but the precise values shown depend on the accuracy of the data on the development of infectivity in different tissues of BSE-infected sheep. Furthermore, verifying the age of sheep is difficult without an identification scheme, so in the short term it would be feasible only to exploit seasonal birthing patterns and dental indicators to impose approximate age restrictions. How the impact of 12-month age restrictions might be reduced by compensatory changes in slaughter patterns is shown in Fig. 3b.

Translation of the patterns of relative exposure through time into measures of absolute risk requires estimation of potential vCJD mortality in the human population of Great Britain. Given the uncertainty in the parameters determining such predictions (in particular, the incubation period distribution and human susceptibility), analyses must be based on the best available estimates of temporal changes in human exposure to infected material. The confidence bounds on vCJD mortality shown in Table 1 characterize human exposure by infectivity-weighted estimates of the numbers of BSE-infected cattle (Fig. 1d) and sheep (Fig. 2) slaughtered for consumption through time (correcting for early under-reporting of cattle BSE incidence). We modelled the vCJD epidemic solely in the 40% of the population that are methionine homozygous at prion protein (PrP) codon 129, assuming no other genetic variation in susceptibility. Currently no case data exist with which to constrain epidemic scenarios for other genotypes, but if future cases are diagnosed then upper bounds on epidemic size will increase. Compared with our previous estimates[25], upper bounds on epidemic size in the absence of BSE in sheep have reduced (largely as a result of a change in statistical methods), being in the range 50,000–100,000 depending on the assumptions

（如图 **a** 所示，由 $\beta_B$ 决定）按比例缩放 **d~f** 的暴露曲线。

20 世纪 90 年代末人类暴露的骤降（图 2d ~ 2f）是因为人们发布了针对羊的 SRM 禁令（禁止食用 12 个月龄以上动物的头骨、扁桃体和脊柱以及所有动物的脾脏）。风险程度在最坏的情况下（II）仍然急剧升高，而在其他情况下都下降了，在最好的情况下（III）达到最低。准确地说，即便是在最坏的情况下，人类既往暴露于 BSE 的大部分情况来源于牛（图 1d）。但是，对所有三种情况来说（假设 BSE 确实进入了羊群），目前英国进入食物供给的受传染的羊的最大数目估计超过了感染的牛的数量 [18,19]（图 1d，插图）。这个比较将一只传染性最强的羊的风险与一头奶牛的风险等同起来，尽管羊和牛的体重差别很大，但是做这一比较是因为有证据表明羊的传染性分布更为广泛。

可能的降低风险的措施包括限制供消费而屠宰的羊的年龄以及加强基于动物组织的控制，以减少进入食物供给的传染性，基于羊群历史以及公羊基因型的其他措施可能也起作用，但是这里不作考虑。对于羊 BSE 三种情况的考虑，结合基于组织和年龄的限制，预计能够将目前和未来的风险降低至少 80%（图 3a）。这很令人鼓舞，但是显示的精确的数值取决于 BSE 感染羊中不同组织传染性发展的数据的准确性。此外，在没有鉴定方案的情况下确定羊的年龄是很困难的，因此在短期内可行的仅仅是根据季节性生育模式以及牙齿指标来施加近似年龄的限制。图 3b 中显示了 12 个月龄限制的效果是如何被屠宰模式的补偿性改变所降低的。

将相对暴露随时间改变的模式转换成绝对风险值需要估计英国人群中潜在的 vCJD 死亡率。考虑到决定这些预测的参数的不确定性（尤其是潜伏期分布以及人类易感性），我们的分析必须基于人类暴露于感染物质的时间变化的最佳现有估计。表 1 显示的 vCJD 死亡率的置信区间以不同时间下屠宰用于消费的 BSE 感染的牛（图 1d）和羊（图 2）数量的传染力加权估计值（对早期牛 BSE 发病率的漏报进行了更正）特征性地描述了人类暴露的情况。我们仅在 40% 的朊蛋白（PrP）密码子第 129 位是蛋氨酸纯合子的人群中模拟 vCJD 流行病，假设这些人在易感性方面没有其他遗传变异。目前，还没有病例数据来限制其他基因型的流行情况，但是如果将来的病例得到诊断，那么流行范围的上限将会增加。与我们之前的估计相比 [25]，在没有疯牛病的情况下，羊疫情的规模上限已经降低（主要是由于统计方法的改变）。根据所作的假设，其范围在 50,000 ~ 100,000 之间。这些数值远远超过了最近发表的估计，它是根据人类先前暴露于 BSE 传染性趋势的粗略描述得出的 [26,27]。2001 ~ 2080 年的

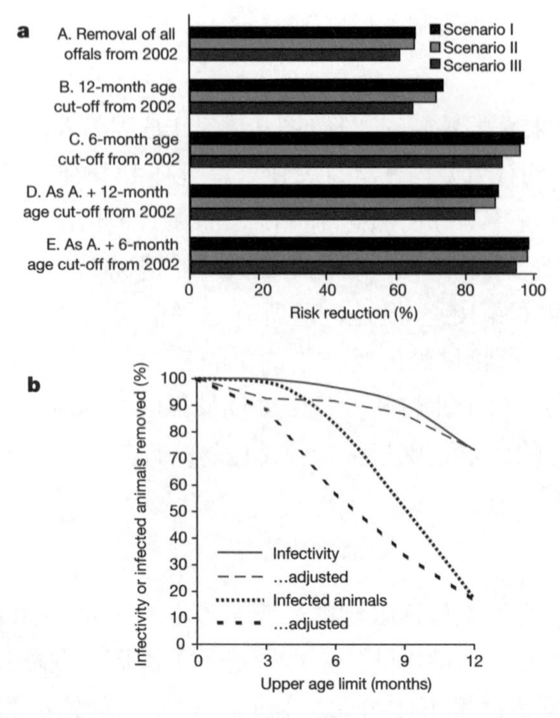


Fig. 3. Impact of risk-reduction measures. **a**, Estimated relative impact of various measures on current exposure to BSE infectivity in sheep food products for scenarios I–III. "Removal of all offals" corresponds to a ban on all internal organs and central nervous system tissue of sheep entering the human food supply, but does not assume removal of lymph nodes. Estimates presented were calculated assuming that age restrictions would not change slaughter patterns. **b**, Estimated impact of imposing age restrictions as a function of the upper age limit imposed for scenario I. Effects on infectivity and total number of infected animals entering the food supply are shown, with dashed curves showing how the impact is reduced if slaughter patterns are adjusted after such a measure, such that all animals currently slaughtered under 12 months old are then slaughtered under the upper age limit imposed.

made. These values are substantially greater than recently published estimates derived assuming a cruder representation of past trends in human exposure to BSE infectivity[26,27]. Best-fit estimates associated with 2001–2080 confidence bounds generally lie in the range 100–1,000, but the fits of the model vary little for fewer than 10,000 deaths. It should be noted (Table 1, E) that large epidemics are still possible even if the mean incubation period is less than 60 years. In the presence of BSE in sheep, the upper bound is substantially increased only if BSE is capable of becoming endemic in the national flock (scenario II).

Table 1. 95% confidence intervals for future vCJD deaths

| Cattle only* | 2000–2002 | 2001–2005 | 2001–2010 | 2001–2020 | 2001–2040 | 2001–2080 |
|---|---|---|---|---|---|---|
| A | 20–100 | 40–400 | 40–1,200 | 40–5,000 | 40–20,000 | 50–50,000 |
| B | 20–100 | 40–400 | 40–1,200 | 40–5,000 | 40–20,000 | 40–40,000 |
| C | 20–100 | 30–400 | 40–1,400 | 40–7,000 | 40–40,000 | 40–90,000 |
| D | 20–100 | 30–400 | 30–1,300 | 40–7,000 | 40–40,000 | 40–100,000 |
| E | 20–100 | 40–350 | 40–1,100 | 40–5,000 | 40–20,000 | 50–35,000 |

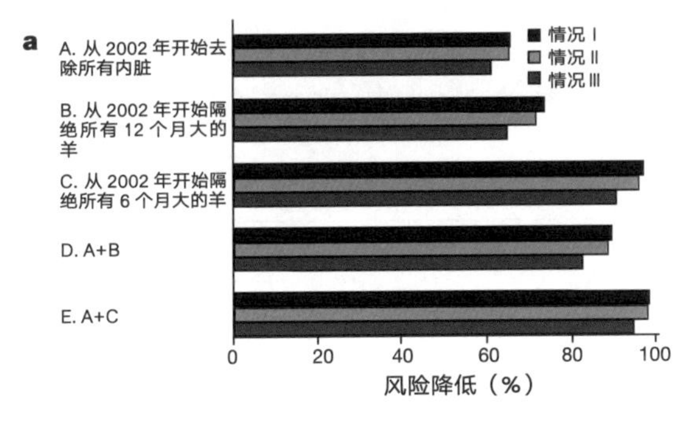


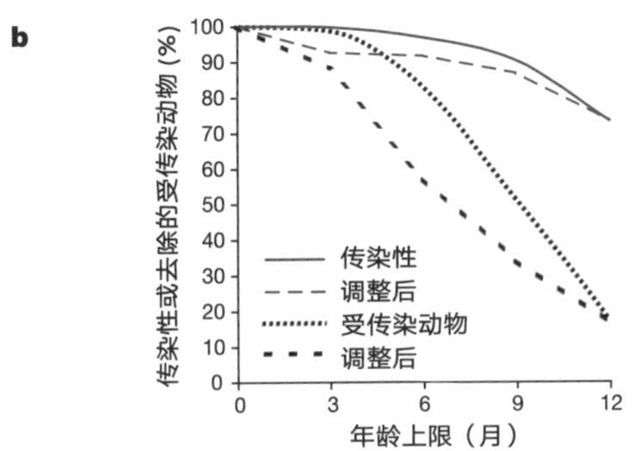


图 3. 降低风险的措施的效果。**a**，三种情况下，估计各种措施对于目前羊类食品暴露于 BSE 传染性的相对影响。"去除所有内脏"表示禁止羊的内脏和中枢神经系统组织进入人类的食物供给，但并不意味着要去除淋巴结。显示出来的估计效果是假设年龄限制不会改变屠宰状况的情况下得到的。**b**，对于情况 I，将年龄限制强行作为年龄上限的函数的估计效果。图中显示了传染性的影响以及进入食物供应的受感染动物总数，而虚线显示的是假如屠宰状况根据下面的措施进行调整后这种影响是如何降低的。该措施是将目前在 12 个月龄以下屠宰的所有动物限定到年龄上限以下屠宰。

置信区间相关的最佳估计在 100～1,000 之间，但对于少于 10,000 的死亡数，这个模型的最佳值变化不大。需要注意的是（表 1，E），大规模的流行仍然是可能的，即便平均潜伏期小于 60 年。羊群中 BSE 存在的情况下，只有 BSE 能够在全国的羊群中流行时，其上限才会大幅度地增加（情况 II）。

表 1. 未来 vCJD 死亡数的 95% 置信区间

| 仅仅牛 * | 2001～2002 | 2001～2005 | 2001～2010 | 2001～2020 | 2001～2040 | 2001～2080 |
|---|---|---|---|---|---|---|
| A | 20～100 | 40～400 | 40～1,200 | 40～5,000 | 40～20,000 | 50～50,000 |
| B | 20～100 | 40～400 | 40～1,200 | 40～5,000 | 40～20,000 | 40～40,000 |
| C | 20～100 | 30～400 | 40～1,400 | 40～7,000 | 40～40,000 | 40～90,000 |
| D | 20～100 | 30～400 | 30～1,300 | 40～7,000 | 40～40,000 | 40～100,000 |
| E | 20～100 | 40～350 | 40～1,100 | 40～5,000 | 40～20,000 | 50～35,000 |

*Continued*

| Cattle and sheep† | Sheep:cattle infectivity ratio | 2001–2020 | 2001–2040 | 2001–2080 |
|---|---|---|---|---|
| Scenario I | 1:1 | 40–5,000 | 40–20,000 | 40–60,000 |
| | 10:1 | 40–5,000 | 50–30,000 | 50–70,000 |
| Scenario II | 1:1 | 20–5,000 | 20–20,000 | 20–70,000 |
| | 10:1 | 40–5,000 | 70–30,000 | 110–150,000 |
| Scenario III | 1:1 | 40–5,000 | 40–20,000 | 40–50,000 |
| | 10:1 | 40–5,000 | 50–30,000 | 40–50,000 |

* Cattle-only calculations assume: A, BSE exposure in humans was proportional to the BSE cases before the mid-1988 in addition to infected animal slaughter rates shown in Fig. 1c; B, as A but excluding exposure to reported BSE cases; C, as A but additionally fitting to at least three vCJD deaths reported before 2001 being infected before 1986, as suggested by analysis of the Queniborough cluster; D, as A but assuming the 1989 specified bovine offal ban reduced human exposure by at least 80%; E, as A but restricting the mean incubation period to be no longer than 60 years.

† Calculations for cattle and sheep assume exposure as A from cattle plus exposure from sheep as in Fig. 2d–f from scenarios I–III, assuming sheep are equally or 10 times as infectious as cattle. Short-term predictions for these scenarios are similar to those excluding sheep.

Although the risk analysis presented here incorporates a wide variety of available information into a single integrated framework, its reliability depends on the quality and volume of data available for parameter estimation. The limited data highlight the need for further studies to measure: (1) scrapie and BSE prevalence in sheep (stratified by age), employing sample sizes sufficiently large to detect low prevalence; (2) sheep survivorship more precisely; (3) BSE infectivity in sheep quantitatively, by stage of incubation, tissue and sheep genotype; (4) age-dependent susceptibility to TSE infection in a variety of species, including sheep and cattle; and (5) historical trends in bovine and ovine tissue consumption. Molecular typing methods giving rapid results[28,29] are clearly valuable for prevalence screening and strain typing (whether for BSE or scrapie), but study design should take account of test sensitivity and therefore consider which tissue should be tested (brain not necessarily being optimal). Given the uncertainty regarding the presence of BSE in the British sheep flock, such large-scale testing is a priority. In the interim, this analysis informs prevalence survey design and policy consideration of the potential benefits of additional risk-reduction measures.

## Methods

Additional detail is provided as Supplementary Information.

### Modelling transmission of BSE in sheep

Infectiousness of a BSE-infected sheep as a function of time $\tau$ since infection (relative to the incubation period $T$) was estimated by fitting the parametric form $\rho(\frac{\tau}{T}) = \exp\,(a(\frac{\tau}{T})^b/((\frac{\tau}{T})^b+c))$

续表

| 牛和羊† | 羊：牛传染性比例 | 2001 ~ 2020 | 2001 ~ 2040 | 2001 ~ 2080 |
|---|---|---|---|---|
| 情况 I | 1∶1 | 40 ~ 5,000 | 40 ~ 20,000 | 40 ~ 60,000 |
| | 10∶1 | 40 ~ 5,000 | 50 ~ 30,000 | 50 ~ 70,000 |
| 情况 II | 1∶1 | 20 ~ 5,000 | 20 ~ 20,000 | 20 ~ 70,000 |
| | 10∶1 | 40 ~ 5,000 | 70 ~ 30,000 | 110 ~ 150,000 |
| 情况 III | 1∶1 | 40 ~ 5,000 | 40 ~ 20,000 | 40 ~ 50,000 |
| | 10∶1 | 40 ~ 5,000 | 40 ~ 20,000 | 40 ~ 50,000 |

* 只计算牛的假设：A，除了图 1c 中所示的受感染动物的屠宰率之外，人类中的 BSE 暴露与 1988 年中期以前的 BSE 病例成比例；B，和 A 一样，但是排除了暴露于已经报告过的 BSE 病例；C，和 A 一样，但是额外地加入了至少 3 例 2001 年前报告的并且在 1986 年前感染的 vCJD 死亡病例，正如对 Queniborough 群体的分析所提示的一样；D，和 A 一样，但是假设 1989 年规定的牛内脏禁令将人类的暴露减少了至少 80%；E，和 A 一样，但是限制平均潜伏期不超过 60 年。

† 计算牛和羊：假设来自牛的暴露和 A 一样，来自羊的暴露如图 2d~2f 所示的三种情况。假定羊的传染性与牛相同或是牛的 10 倍。这些情况下的短期估计值和不包括羊的情况类似。

尽管这里所描述的风险分析将很多可用信息整合到了单个的整体框架中，但是其可靠性依赖于用于参数估计的数据的质量和数量。有限的数据强调了进一步测量研究的必要性：(1) 羊的瘙痒病和 BSE 患病率（根据年龄分层），需要足够大的样本量来检测低患病率；(2) 羊的更加精确的生存期；(3) 根据潜伏期的阶段、组织和羊基因型，定量分析羊 BSE 的传染性；(4) 不同物种对于 TSE 感染的年龄依赖的易感性，包括牛和羊；以及 (5) 对牛和羊组织消费的历史趋势。分子分型方法能够快速得出结果[28,29]，对于患病率筛查和种类分型都非常有价值（无论是 BSE 还是瘙痒病），但是研究设计需要考虑测试的敏感性，因此需要考虑用什么组织进行测试（脑组织并不是最佳选择）。由于英国羊群中是否存在 BSE 仍不确定，这种大规模的测试是需要优先考虑的。在此期间，这个分析为患病率调查的设计以及政策考量提供了依据，说明了额外减少风险措施的潜在好处。

## 方 法

补充信息中提供了详细内容。

### BSE 在羊中传播的模型

通过将参数式 $\rho(\frac{\tau}{T}) = \exp(a(\frac{\tau}{T})^b/((\frac{\tau}{T})^b+c))$ 与羊 BSE[8,9] 和瘙痒病[4-7] 发病机制研究的组织特异性的传染力数据进行组合，估算感染后（相对于潜伏期 $T$）羊 BSE 传染性随时间 $\tau$

to tissue-specific infectivity data from studies of pathogenesis of ovine BSE[8,9] and scrapie[4-7]. The data were weighted by total tissue mass when estimating sheep-to-sheep infectiousness, and by the proportion of tissue mass entering food when estimating human exposure to infectivity (Fig. 1b). This approach captures the impact on overall infectiousness of between-tissue variation in PrP accumulation during pathogenesis of ovine BSE—namely, that some tissues (for example, lymph nodes) rapidly develop detectable infectivity, the growth of which later slows and/or saturates, whereas others (for example, brain) only exhibit high levels of infectivity close to clinical onset.

Because mass-action models cannot capture the observed clustering of cases[20], we developed a model of TSE transmission in sheep that incorporated three relevant tiers: (1) the individual animal—capturing age-dependent susceptibility and exposure, and pathogenesis of disease; (2) the individual flock—capturing within-flock sheep-to-sheep transmission; and (3) the national population of flocks—capturing between-flock transmission and exposure to contaminated feed.

Assuming homogenous mixing within a flock, the deterministic susceptible-infected model is:

$$\frac{\partial x}{\partial t}+\frac{\partial x}{\partial a}=-(\Lambda(t)\kappa(a)+\mu(a))x \qquad x(t,0)=B \qquad \mu(a)=-\frac{1}{S}\frac{\mathrm{d}S}{\mathrm{d}a}$$

$$\frac{\partial y}{\partial t}+\frac{\partial y}{\partial a}+\frac{\partial y}{\partial \tau}=-(F_A(\tau)+\mu(a))y \qquad F_A(\tau)=\frac{f_A(\tau)}{1-\int_0^{\tau} f_A(\tau')\mathrm{d}\tau'}$$

$$y(t,a,0)=\Lambda(t)\kappa(a)x(t,a)$$

$$\Lambda(t)=\beta_A\int_{\tau=0}^{t}\int_{T=\tau}^{\infty}F_A(T)\,\rho(\tau/T)y(t,a,\tau)\mathrm{d}T\,\mathrm{d}\tau \qquad y(0,a,0)=\frac{\kappa(a)\,Y_0}{\int\kappa(a)\,\mathrm{d}a}$$

where we denote the densities of susceptible and infected animals of age $a$ at time $t$ by $x(t, a)$ and $y(t, a, \tau)$; time from infection by $\tau$; the incubation period distribution by $f_A(\tau)$ (see Fig. 1c); the force of infection by $\Lambda(t)$; the transmission coefficient for horizontal spread by $\beta_A$; the fixed birth rate by $B$; the mortality rate by $\mu(a)$; and age-dependent susceptibility by $\kappa(a)$ (conservatively assumed to be constant for the first 12 months of life and zero thereafter). The number of animals infected at the start of a flock outbreak is $Y_0$. Assuming that the infectiousness at time $t$ of a flock infected at $t = 0$ to other flocks is proportional to the within-flock force of infection $\Lambda(t)$, we modelled transmission within a population of 100,000 flocks of 400 animals each. Genetic heterogeneity in susceptibility to infection was modelled at the flock level, with 33,000 flocks assumed to have all animals susceptible and the remainder completely resistant. The introduction of infection into a flock from an external source (contaminated feed or other sheep) was modelled as a rare event, initially infecting 1% of animals.

The infection dynamics of flocks were modelled using an SIRS framework, whereby flocks are initially "susceptible" to infection, enter an "infected" state of extended duration, "recover" and

的变化。估算羊之间的传染性时，数据通过组织总质量进行加权，而在估计人类暴露的传染性时则通过组织进入食物供给的比例进行加权（图 1b）。这种方法捕捉到羊 BSE 发病过程中不同组织间 PrP 聚集的不同对总体传染性的影响——也就是说，一些组织（比如淋巴结）很快就可以检测到传染性，其增长随后放缓和（或）达到饱和，而另一些组织（比如脑）仅仅显示出高水平的传染，接近于临床发作。

由于质量作用模型不能捕捉观察到的病例群[20]，我们建立了一个 TSE 在羊中传播的模型，包含了三个相关的层次：(1) 个体动物——得到依赖年龄的易感性和暴露以及疾病的发病机制；(2) 独立的羊群——捕捉群内羊之间的传播；(3) 全国的羊群——捕捉群间传播以及对污染饲料的暴露。

假定群内的混合是均匀的，那么确定的易感染模型是：

$$\frac{\partial x}{\partial t}+\frac{\partial x}{\partial a}=-(\Lambda(t)\kappa(a)+\mu(a))x \qquad x(t,0)=B \quad \mu(a)=-\frac{1}{S}\frac{\mathrm{d}S}{\mathrm{d}a}$$

$$\frac{\partial y}{\partial t}+\frac{\partial y}{\partial a}+\frac{\partial y}{\partial \tau}=-(F_A(\tau)+\mu(a))y \qquad F_A(\tau)=\frac{f_A(\tau)}{1-\int_0^{\tau}f_A(\tau')\mathrm{d}\tau'}$$

$$y(t,a,0)=\Lambda(t)\kappa(a)x(t,a)$$

$$\Lambda(t)=\beta_A\int_{\tau=0}^{t}\int_{T=\tau}^{\infty}F_A(T)\,\rho(\tau/T)y(t,a,\tau)\mathrm{d}T\,\mathrm{d}\tau \qquad y(0,a,0)=\frac{\kappa(a)\,Y_0}{\int\kappa(a)\,\mathrm{d}a}$$

其中，我们用 $x(t, a)$ 和 $y(t, a, \tau)$ 表示年龄为 $a$ 的易感动物和已感染的动物在 $t$ 时刻的密度；感染后的时间是 $\tau$；潜伏期分布是 $f_A(\tau)$( 见图 1c)；感染的强度是 $\Lambda(t)$；水平传播系数是 $\beta_A$；固定生产率是 $B$；死亡率是 $\mu(a)$；依赖年龄的易感性是 $\kappa(a)$(保守地假定生命的前 12 个月为常数，之后是 0)。羊群疫情爆发开始时感染动物的数量是 $Y_0$。假设一个羊群在 $t=0$ 时受感染，在 $t$ 时刻对其他羊群的传染性与群内传染性 $\Lambda(t)$ 成正比。我们在含有 100,000 个羊群的群体中建立传播模型，每个羊群中包含 400 只羊。对感染易感性的遗传异质性在羊群水平进行建模，假定 33,000 个羊群中所有的动物都是易感的，而剩余的羊群都是完全耐药的。通过外部来源将感染引入羊群（污染的食物或者其他羊）被建模为小概率事件，最初感染 1% 的动物。

使用 SIRS 框架对羊群的感染动态进行建模，其中羊群最初对感染是“易感状态”，然后进入长时间的“感染”状态，“恢复”，并对再次感染具有耐药性（大约持续 20 年），然后恢复

become resistant to further infection (for a period of 20 years), then revert to the "susceptible" state. The flock-level "recovery" process approximates flock outbreak extinction mechanisms, such as demographic stochasticity (likely to be critical[30] given the low reported prevalence of scrapie), control measures and selection for scrapie-resistant genotypes.

Denoting the number of susceptible flocks at time $t$ by $s(t)$, flocks infected time $\theta$ ago by $h(t, \theta)$ (with infectivity $\Lambda(\theta)$ from the within-flock model) and recovered/resistant flocks by $r(t)$, model dynamics are described by:

$$\frac{\mathrm{d}s}{\mathrm{d}t} = \xi r - \Omega(t)s \qquad \Omega(t) = \beta_F \int_{\theta=0}^{\infty} \Lambda(\theta)h(t, \theta)\mathrm{d}\theta + \beta_B \phi(t)$$

$$\frac{\partial h}{\partial t} + \frac{\partial h}{\partial \theta} = -F_F(\theta)h \qquad F_F(\theta) = \frac{f_F(\theta)}{1 - \int_0^{\theta} f_F(\theta')\mathrm{d}\theta'}$$

$$\frac{\mathrm{d}r}{\mathrm{d}t} = \int F_F(\theta)h(t, \theta)\mathrm{d}\theta - \xi r \qquad s(0) = N \quad h(t, 0) = \Omega(t)x \quad h(0, \theta) = 0$$

Here, $f_F(\theta)$ is the "incubation period" distribution for flocks (gamma distributed with mean 6 years; see Supplementary Information); $\Omega(t)$ the force of infection for flocks at time $t$; $\beta_F$ and $\beta_B$ the coefficients for between-flock transmission and exposure of flocks to contaminated feed; $\phi(t)$ the relative risk from contaminated feed at time $t$ (estimates from refs 18, 19); and $\xi$ ( = 0.05 per year) the rate at which recovered flocks re-enter the susceptible pool. The values of $\beta_A$ and $\beta_F$ corresponding to required values of $R_0^A$ and $R_0^F$ were determined numerically.

Exposure of the human population to BSE infectivity in sheep was represented by the number of infected animals slaughtered for food (stratified by age and incubation stage, weighted by infectivity; Fig. 2d–f). The effectiveness of risk-reduction measures was evaluated by examining the distribution of exposure as a function of animal age, and by comparing results obtained from the infectivity profiles corresponding to current SRM controls and a ban on all offal.

It should be noted that this model framework reproduces observed within-flock and overall scrapie incidence patterns well if run to endemicity (results not shown).

## Prediction of vCJD incidence

As in earlier work[25], the probability density that an individual develops clinical disease at time $t$ and age $a$ is

$$p(t, a) = S_H(t, a) \int_{t-a}^{t} f(t-u)\, I(u, a-t+u)\exp\left[ -\int_0^{u} I(u', a-t+u')\mathrm{d}u' \right] \mathrm{d}u$$

where $S_H(t, a)$ is the probability that someone born at time $t - a$ will survive to age $a$ (derived from

到“易感”状态。羊群水平“恢复”的过程类似于羊群爆发灭绝的机制，比如人口统计学上的随机性（由于报道的瘙痒病的患病率很低，因此这可能很关键[30]）、控制措施和选择耐瘙痒病基因型。

用 $s(t)$ 来表示 $t$ 时刻易感羊群的数量，用 $h(t,\theta)$ 来表示 $\theta$ 时刻之前感染的羊群数量（根据羊群内的模型，传染性是 $\Lambda(\theta)$），用 $r(t)$ 来表示恢复的/耐药性的羊群数量，动态模型表示为：

$$\frac{\mathrm{d}s}{\mathrm{d}t}=\xi r-\Omega(t)s \qquad \Omega(t)=\beta_{\mathrm{F}}\int_{\theta=0}^{\infty}\Lambda(\theta)h(t,\theta)\mathrm{d}\theta+\beta_{\mathrm{B}}\phi(t)$$

$$\frac{\partial h}{\partial t}+\frac{\partial h}{\partial \theta}=-F_{\mathrm{F}}(\theta)h \qquad F_{\mathrm{F}}(\theta)=\frac{f_{\mathrm{F}}(\theta)}{1-\int_{0}^{\theta}f_{\mathrm{F}}(\theta')\mathrm{d}\theta'}$$

$$\frac{\mathrm{d}r}{\mathrm{d}t}=\int F_{\mathrm{F}}(\theta)h(t,\theta)\mathrm{d}\theta-\xi r \qquad s(0)=N \quad h(t,0)=\Omega(t)x \quad h(0,\theta)=0$$

这里，$f_{\mathrm{F}}(\theta)$ 是羊群的“潜伏期”分布（平均 6 年的 $\gamma$ 分布，见补充信息）；$\Omega(t)$ 是 $t$ 时刻的羊群感染强度；$\beta_{\mathrm{F}}$ 和 $\beta_{\mathrm{B}}$ 是羊群间传播和羊群暴露于污染食物的系数；$\phi(t)$ 是 $t$ 时刻污染食物的相对风险（从文献 18 和 19 预估）；$\xi$(= 每年 0.05) 是恢复的羊群重新进入易感群体的速度。与 $R_0^{\mathrm{A}}$ 和 $R_0^{\mathrm{F}}$ 的需求值对应的 $\beta_{\mathrm{A}}$ 和 $\beta_{\mathrm{F}}$ 的数值用数字来确定。

人群对羊群 BSE 传染性的暴露程度由屠宰后作为食物的感染动物数量来表示（根据年龄和潜伏阶段分层，用传染性加权，图 2d ~ 2f）。降低风险措施的有效性通过以下两种方式评估：检查以动物年龄为函数的暴露分布；将目前 SRM 控制和禁止所有内脏相对应的传染性结果进行比较。

需要注意的是，如果遇到区域性流行，这个模型框架可以很好地复制观察到的羊群内以及总体的瘙痒病发病率模式（结果没有显示）。

## vCJD 发病率的预测

如之前的工作所述[25]，个体在 $t$ 时间和 $a$ 年龄发生临床疾病的概率密度是

$$p(t,a)=S_{\mathrm{H}}(t,a)\int_{t-a}^{t}f(t-u)\,I(u,a-t+u)\exp\left[-\int_{0}^{u}I(u',a-t+u')\mathrm{d}u'\right]\mathrm{d}u$$

其中 $S_{\mathrm{H}}(t,a)$ 是某个生于时间 $t-a$ 的个体能够活到 $a$ 年龄的可能性（来自于人口普查数据），

census data) and $f(t-u)$ is the incubation period distribution (modified lambda distribution). The infection hazard is given by

$$I(t, a) = \beta g(a) \left( v_c(t) \int \Omega_c(z)\omega_c(z, t)\mathrm{d}z + v_s(t) \int \Omega_s(z)\omega_s(z, t)\mathrm{d}z \right)$$

where $\beta$ is the transmission coefficient and $g(a)$ an age-dependent susceptibility/exposure function (uniform with gamma-distributed tails). For each infectious species $i$ (C for cattle, S for sheep), $\Omega_i(z)$ is the relative infectiousness of an animal time $z$ from disease onset, $v_i(t)$ is the time-dependent effectiveness of risk-reduction measures (for cattle, this is parameterized as a step reduction in 1989 due to the specified bovine offal ban), and $\omega_i(z, t)$ is the number of animals slaughtered for consumption stratified by time and incubation stage. (Values for sheep were obtained from the above model, and those for cattle used updated back-calculation estimates[18,19], with $\int \Omega_c(z)\omega_c(z, t)\mathrm{d}z$ plotted in Fig. 1d.)

The relative infectiousness of cattle by incubation stage was assumed to increase exponentially from a baseline level to a maximum value before onset of clinical signs[25]. We restricted analysis to the 40% of the population that are methionine homozygous at PrP codon 129, assuming no other genetic variation in susceptibility.

Through numerical solution of the inverse problem (see Supplementary Information or further details), $\beta$ was calculated as a function of case incidence, allowing incidence of vCJD deaths in any time interval to be treated as a model parameter. This enabled nonlinear optimization techniques to be used to obtain likelihood profiles by fitting the model to the joint age- and time-stratified mortality data to the end of 2000. We obtained 95% confidence bounds from the one-dimensional likelihood profiles. Infection prevalence is poorly constrained by the observed incidence data, and can range from being equal to mortality to 100-fold larger, with little effect on model fit.

(**415**, 420-424; 2002)

**N. M. Ferguson, A. C. Ghani, C. A. Donnelly, T. J. Hagenaars & R. M. Anderson**
Department of Infectious Disease Epidemiology, Faculty of Medicine, Imperial College of Science, Technology and Medicine, St Mary's Campus, Norfolk Place, London W2 1PG, UK

Received 21 November; accepted 12 December 2001. Published online 9 January 2002, DOI 10.1038/nature709.

---

References:

1. Frankish, H. Samples blunder renders sheep-BSE study useless. *Lancer* **358**, 1436 (2001).
2. Beckett, M. *House of Commons Hansard Debates, 22 October 2001* 373 (2001) ⟨http://www.parliament.the-stationery-office.co.uk/pa/cm200102/cmhansrd/cm011022/debindx/11022-x.htm⟩ .
3. Det Norske Veritas *Assessment of Exposure to BSE Infectivity in the UK Sheep Flock* Report no. C782506 (The Meat and Livestock Commission, London, 1998).
4. Hadlow, W. J., Kennedy, R. C. & Race, R. E. Natural infection of Suffolk sheep with scrapie virus. *J. Infect. Dis.* **146**, 652-664 (1982).
5. van Keulen, L. J. M., Schreuder, B. E. C., Vromans, M. E. W., Langeveld, J. P. M. & Smits, M. A. Scrapie-associated prion protein in the gastrointestinal tract of sheep with natural scrapie. *J. Comp. Pathol.* **121**, 55-63 (1999).
6. van Keulen, L. J. M., Schreuder, B. E. C., Vromans, M. E. W., Langeveld, J. P. M. & Smits, M. A. Pathogenesis of natural scrapie in sheep. *Arch. Virol. Suppl.* **16**, 52-21 (2000).

$f(t-u)$ 是潜伏期分布（修改后的 $\lambda$ 分布）。感染风险计算如下

$$I(t, a)=\beta g(a)\left(v_c(t)\int\Omega_c(z)\omega_c(z, t)\mathrm{d}z+v_s(t)\int\Omega_s(z)\omega_s(z, t)\mathrm{d}z\right)$$

其中 $\beta$ 是传输系数，$g(a)$ 是依赖年龄的易感性/暴露函数（与 $\gamma$ 分布一致）。对于每个感染性物种 $i$（C 表示牛，S 表示羊），$\Omega_i(z)$ 是疾病出现后 $z$ 时刻某个动物的相对传染性，$v_i(t)$ 是降低风险措施的时间依赖有效性（对牛来说，由于特定的牛内脏禁令，因此在 1989 年发病率骤减），$\omega_i(z, t)$ 是根据时间和潜伏阶段分层的用于消费的动物屠宰数量。（羊的数据从以上的模型得出，而牛的数据用更新后的反算法估计 [18,19]，$\int\Omega_c(z)\omega_c(z, t)\mathrm{d}z$ 在图 1d 中绘出。）

在潜伏期，牛的相对传染性假定从基线水平指数式地增长到临床症状出现前的最高水平 [25]。我们将分析限制在那些 PrP 密码子第 129 位是蛋氨酸纯合的 40% 的人群中，这些人被认为在易感性方面没有其他遗传变异。

通过对反问题的数值解（详情请参阅补充资料），$\beta$ 以病例发病率的函数来计算，将任意时间段 vCJD 死亡发生率作为模型参数。这使得非线性优化技术通过将截至 2000 年底的年龄和时间分层的死亡率数据套入到该模型中来获得似然分布。我们从单维似然分布中获得了 95% 置信区间。感染的患病率几乎不受观察到的发病率数据的限制，其范围从与死亡率相等到是死亡率的 100 多倍，对模型的拟合影响不大。

（毛晨晖 翻译；肖景发 审稿）

7. Andreoletti, O. *et al.* Early accumulation of $PrP^{Sc}$ in gut-associated lymphoid and nervous tissues of susceptible sheep from a Romanov flock with natural scrapie. *J. Gen. Virol.* **81**, 3115-3126 (2000).

8. Foster, J. D., Parnham, D. W., Hunter, N. & Bruce, M. Distribution of the prion protein in sheep terminally infected with BSE following experimental oral transmission. *J. Gen. Virol.* **82**, 2319-2326 (2001).

9. Jeffrey, M. *et al.* Oral inoculation of sheep with the agent of bovine spongiform encephalopathy (BSE). 1. Onset and distribution of disease-specific PrP accumulation in brain and viscera. *J. Comp. Pathol.* **124**, 280-289 (2001).

10. Fraser, H., Bruce, M. E., Chree, A., McConnell, I. & Wells, G. A. H. Transmission of bovine spongiform encephalopathy and scrapie to mice. *J. Gen. Virol.* **73**, 1891-1897 (1992).

11. Wells, G. A. H. *et al.* Infectivity in the ileum of cattle challenged orally with bovine spongiform encephalopathy. *Ver. Rec.* **135**, 40-41 (1994).

12. Hart, R. J., Church, P. N., Kempster, A. J. & Matthews, K. R. *Audit of Bovine and Ovine Slaughter and By-products Sector (Ruminant Products Audit)* (Leatherhead Food Research Association, London, 1997).

13. Hunter, N. PrP genetics in sheep and the implications for scrapie and BSE. *Trends Microbiol.* **5**, 331-334 (1997).

14. Hunter, N., Goldmann, W., Marshall, E. & O'Neill, G. Sheep and goats: natural and experimental TSEs and factors influencing incidence of disease. *Arch. Virol. Suppl.* **16**, 181-188 (2000).

15. Jeffrey, M. *et al.* Frequency and tissue distribution of infection-specific PrP in tissues of clinical scrapie and cull sheep obtained from scrapie affected farms in Shetland. *J. Comp. Pathol.* (submitted).

16. Baylis, M., Houston, F., Goldmann, W., Hunter, N. & McLean, A. R. The signature of scrapie: differences in the PrP genotype profile of scrapie-affected and scrapie-free UK sheep flocks. *Proc. R. Soc. Lond.* B **267**, 2029-2035 (2000).

17. Hunter, N. *et al.* Is scrapie solely a genetic disease? *Nature* **386**, 137 (1997).

18. Anderson, R. M. *et al.* Transmission dynamics and epidemiology of BSE in British cattle. *Nature* **382**, 779-788 (1996).

19. Ferguson, N. M., Donnelly, C. A., Woolhouse, M. E. J. & Anderson, R. M. The epidemiology of BSE in cattle herds in Great Britain II: Model construction and analysis of transmission dynamics. *Phil. Trans. R. Soc. Lond.* B **352**, 803-838 (1997).

20. Donnelly, C. A. & Ferguson, N. M. *Statistical Aspects of BSE and vCJD—Models for Epidemics* (Chapman & Hall and CRC, London, 2000).

21. Woolhouse, M. E. J. *et al.* Population dynamics of scrapie in a sheep flock. *Phil. Trans. R. Soc. Lond. B* **354**, 751-756 (1999).

22. Gravenor, M. B., Cox, D. R., Hoinville, L. J., Hoek, A. & McLean, A. R. The flock-to-flock force of infection for scrapie in Britain. *Proc. R. Soc. Lond. B* **268**, 587-592 (2001).

23. Hoinville, L. J. A review of the epidemiology of scrapie in sheep. *Rev. Sci. Techn. Office Int. Epizoories* **15**, 827-852 (1996).

24. Hoinville, L. J., Hoek, A., Gravenor, M. B. & McLean, A. R. Descriptive epidemiology of scrapie in Great Britain: results of a postal survey. *Vet. Rec.* **146**, 455-461 (2000).

25. Ghani, A. C., Ferguson, N. M., Donnelly, C. A. & Anderson, R. M. Predicted vCJD mortality in Great Britain. *Nature* **406**, 583-584 (2000).

26. d'Aignaux, J. N. H., Cousens, S. N. & Smith, P. G. Predictability of the UK variant Creutzfeldt–Jakob disease epidemic. *Science* **294**, 1729-1731; published online 25 October 2001 (10.1126/ science.1064748).

27. Valleron, A.-J., Boelle, P.-Y., Will, R. & Cesbron, J.-Y. Estimation of epidemic size and incubation time based on age characteristics of vCJD in the United Kingdom. *Science* **294**, 1726-1728(2001).

28. Hill, A. F. *et al.* Molecular screening of sheep for bovine spongiform encephalopathy. *Neurosci. Lett.* **255**, 159-162 (1998).

29. Maissen, M., Roeckl, F., Glatzel, M., Goldmann, W. & Aguzzi, A. Plasminogen binds to disease-associated prion protein of multiple species. *Lancer* **357**, 2026-2028 (2001).

30. Hagenaars, T. J., Ferguson, N. M., Donnelly, C. A. & Anderson, R. M. Persistence patterns of scrapie in a sheep flock. *Epidemiol. Infect.* **127**, 157-167 (2001).

**Supplementary Information** accompanies the paper on *Nature*'s website (http://www.nature.com).

**Acknowledgements.** This work was funded by the Food Standards Agency. N.M.F. and A.C.G. acknowledge funding from The Royal Society. C.A.D., T.J.H. and R.M.A. acknowledge funding from the Wellcome Trust. We are very grateful to M. Bruce, S. Bellworthy and M. Jeffrey for access to pre-publication data on infectivity. We also thank L. Hoinville, J. Wilesmith and R. Will for access to data. We thank N. Hunter, L. Green, J. Anderson, A. James, A. Bromley, H. Mason, P. Comer and members of the Spongiform Encephalopathy Advisory Committee for discussions.

**Competing interests statement.** The authors declare that they have no competing financial interests.

Correspondence and requests for materials should be addressed to N.M.F. (e-mail: neil.ferguson@ic.ac.uk).

# The Role of the Thermohaline Circulation in Abrupt Climate Change

P. U. Clark *et al.*

## Editor's Note

**The discovery of rapid shifts in climate, some over just a few decades, during the past ice age and the preceding interglacial period has transformed our view of global climate, forcing us to regard it as something potentially unstable that has been unusually quiescent during the present interglacial. The possible implications for human-induced global warming are obvious. The most likely mechanism of these sudden changes involves patterns of water circulation in the North Atlantic Ocean, perhaps caused by changes in the global hydrological cycle. This review article by Peter Clark and coworkers not only supplied an overview of this emerging understanding but also helped to establish the centrality of ocean circulation to the entire process of global climate change.**

---

The possibility of a reduced Atlantic thermohaline circulation in response to increases in greenhouse-gas concentrations has been demonstrated in a number of simulations with general circulation models of the coupled ocean–atmosphere system. But it remains difficult to assess the likelihood of future changes in the thermohaline circulation, mainly owing to poorly constrained model parameterizations and uncertainties in the response of the climate system to greenhouse warming. Analyses of past abrupt climate changes help to solve these problems. Data and models both suggest that abrupt climate change during the last glaciation originated through changes in the Atlantic thermohaline circulation in response to small changes in the hydrological cycle. Atmospheric and oceanic responses to these changes were then transmitted globally through a number of feedbacks. The palaeoclimate data and the model results also indicate that the stability of the thermohaline circulation depends on the mean climate state.

---

THE ocean affects climate through its high heat capacity relative to the surrounding land, thereby moderating daily, seasonal and interannual temperature fluctuations, and through its ability to transport heat from one location to another. In the North Atlantic, differential solar heating between high and low latitudes tends to accelerate surface waters polewards whereas freshwater input to high latitudes together with low-latitude evaporation tend to brake this flow. Today, the former thermal forcing dominates the latter haline (freshwater) forcing and the meridional overturning in the Atlantic drives surface waters northward, while deep water that forms in the Nordic Seas flows southward as North Atlantic Deep Water (NADW). This thermohaline circulation (THC) is responsible for

# 温盐环流在气候突变中的作用

克拉克等

## 编者按

在末次冰期和之前的间冰期期间，气候快速变化的发现已改变了我们对全球气候的看法，其中一些变化仅发生在短短几十年内，这迫使我们将气候变化视为一种潜在的不稳定因素，而这种不稳定因素在目前的间冰期异常沉寂。人类引起的全球变暖产生了显而易见的影响。产生这些突变最可能的机理造成了北大西洋的水循环模式，这个模式也可能是由全球水文循环的变化引起的。彼得·克拉克及其同事的这篇评论文章不仅概述了这一新认识，而且帮助建立了海洋环流在全球气候变化整个过程的中心作用。

---

许多海洋–大气耦合环流模式的模拟结果指出，大西洋温盐环流的减弱可能是温室气体浓度的增加引起的。但是，要估计未来温盐环流的变化仍有一定困难，这主要是因为模式参数化难以确切，以及气候系统对温室效应响应机制的不确定性。分析过去气候的变化有助于解决这一问题。资料和模式的结果都指出，末次冰期中，气候的突然变化源于大西洋温盐环流的变化对水文循环微小变化的响应。此后，大气和海洋对这些变化的响应通过一系列反馈机制传到全球。古气候资料和模式的结果都指出，温盐环流的稳定性依赖于气候平均态。

---

海洋通过以下两种途径影响气候：一是通过海洋相对于周围陆地的更大的热容量调节每日的、季节的和年际的温度振荡，二是通过海洋把热量从一地输送到另一地的能力。在北大西洋，高低纬度之间太阳辐射加热的差异使海面的水加速向极地流去，而高纬度的淡水输入和低纬度的蒸腾作用又使水流减速。现在，前者的热力强迫作用超过后者的高盐水（淡水）的强迫作用，大西洋的经向翻转作用使海洋表层水向北流动，而形成于北欧海域的深层水（北大西洋深层水，NADW）则向南流动。温盐环流（THC）负责把大部分大西洋的热量输向极地，输送量峰值在北纬 24° 达到

much of the total oceanic poleward heat transport in the Atlantic, peaking at about $1.2 \pm 0.3$ PW (1 PW equals $10^{15}$ watts) at 24°N (ref. 1).

No such deep overturning occurs in the North Pacific, where surface waters are too fresh to sink[2]. The lack of a meridional land barrier in the Southern Ocean precludes the existence of strong east–west pressure gradients needed to balance a southward geostrophic surface flow, so that poleward heat transport associated with the THC is small there. Deep-water formation in the Southern Ocean occurs along the Antarctic continental shelf in the Weddell and Ross Seas either through intense evaporation or, more typically, through brine rejection that produces dense water that sinks down and along the slope[3]. In addition, supercooled water may be formed at the base of the thick floating ice shelves during freezing or melting and this dense water may in turn flow downslope[4].

The idea that the Atlantic THC may have many speeds is now a century old[2], but not until the 1960s did a quantitative, albeit idealized, framework emerge to explain the physics behind the potential existence of these multiple equilibria[5]. Subsequently, ocean[6] and coupled atmosphere–ocean general circulation models (GCMs) (ref. 7) were shown to support multiple equilibria. Such studies have revealed that multiple equilibria exist because the atmosphere responds to anomalies of sea surface temperature, but not salinity[8]. They have further shown that transitions between different states are often abrupt and can be induced through small perturbations to the hydrological cycle.

The concept of multiple equilibria of the THC and the transitions between these states is now commonly invoked as a mechanism to explain the abrupt climate changes that were characteristic of the last glaciation[9,10]. Here we refer to an abrupt change as a persistent transition of climate (over subcontinental scale) that occurs on the timescale of decades. Although understanding the mechanisms behind abrupt climate transitions in the past is interesting in its own right, there is a pressing need to gain insight into the likelihood of their future occurrence[11,12]. Most, but not all, coupled GCM projections of the twenty-first century climate show a reduction in the strength of the Atlantic overturning circulation with increasing concentrations of greenhouse gases[13]—if the warming is strong enough and sustained long enough, a complete collapse cannot be excluded[14,15]. The successful simulation of past abrupt events that are found in the palaeoclimate record is the only test of model fidelity in estimating the possibility of large ocean–atmosphere reorganizations when projecting future climate change.

## Past Changes in the Thermohaline Circulation

Considerable progress has been made in linking past abrupt changes in North Atlantic surface-ocean and atmospheric temperature with changes in deep ocean circulation, confirming the important role that major reorganizations of the Atlantic THC have played in abrupt climate change. Although a continuum of possible modes of NADW formation may exist[16], palaeoceanographic records suggest that the largest changes in North

$1.2 \pm 0.3$ PW（1 PW = $10^{15}$ W）（参考文献 1）。

北太平洋没有这种翻转作用，那里海洋表层水很淡，不会下沉[2]。在南大洋，经向没有陆地阻隔，因而不存在和向南的地转流相平衡的东西向强气压梯度，因此这里由 THC 引起的向极输送热量较少。在南大洋，深层水形成于威德尔海和罗斯海的南极大陆架附近，它是通过强的蒸发作用或者盐析作用形成的，后一种过程形成的密度大的海水沿着斜坡下沉流动，该过程在南大洋更典型一些[3]。另外，在厚的浮冰大陆架的底部，冰在冻结或融化时可以形成过冷水，这种密度大的水会沿着斜坡向下流动[4]。

大西洋 THC 存在多种流速的想法至今已有一个世纪之久了[2]。但直至 20 世纪 60 年代才出现一个定量化、理想化的模式来解释多平衡态的物理原因[5]。此后，海洋环流模式[6]和海气耦合环流模式（GCMs）（参考文献 7）说明这一多平衡态确实存在。这些研究揭示了，多平衡态是存在的，因为大气是响应于海面温度异常现象，而不是海水盐度[8]。这些研究进一步指出，不同状态之间的转换常常突然发生，而且可能是由水文循环的小扰动引起的。

THC 的多平衡态和不同状态之间的转换现在常常用来解释气候突变的机制，这种突变是末次冰期气候的特征[9,10]。此处，我们把发生在几十年尺度的持续的气候（超越次大陆范围）转换叫做气候突变。虽然以往人们把兴趣集中于了解气候突变的机制，但更需要了解的是未来发生气候突变的可能性[11,12]。大部分（并非全部）关于 21 世纪气候的 GCM 耦合模式的结果指出，随着温室气体浓度的增加，大西洋经向翻转流环流的强度将减少[13]，如果增暖足够强，且持续的时间足够长，则不能排除这一环流有完全停滞的可能性[14,15]。当用模式来预测未来气候变化情况下大尺度的海洋–大气重组的可能性时，对古气候记录中的突发事件的成功模拟，是模式逼真度的唯一检验手段。

## 以往温盐环流的变化

将北大西洋洋面和大气温度过去的突变与深海环流的变化联系在一起的研究已取得了很大进展，这些方面的研究证明了大西洋 THC 的重组在气候突变中起了重要作用。虽然可能存在一系列 NADW 模态[16]，但是古海洋记录指出，北大西洋气候的最大变化与三种可能模态之间的转换有关，即现代模态、冰期模态和海因里希模

Atlantic climate were associated with transitions among three possible modes: a modern mode, a glacial mode, and a Heinrich mode[9]. The modern mode is characterized by the formation of deep water in the Nordic Seas and its subsequent flow over the Greenland–Scotland ridge[17]. The newly formed NADW flows into the Labrador Sea where it entrains recirculating, relatively cold and fresh Labrador Sea intermediate waters[17] that are largely confined to the subpolar gyre of the North Atlantic[18] before progressing southward. During the glacial mode, NADW probably formed through open-ocean convection in the subpolar North Atlantic, sinking to depths of less than 2,500 m (ref. 19). Buoyancy loss by brine rejection under sea ice in the Nordic Seas may have provided an additional source of glacial-mode NADW[20]. Finally, during the Heinrich mode, Antarctic-derived waters filled the North Atlantic basin to depths as shallow as 1,000 m (ref. 21).

Changes in the location, depth and volume of newly formed NADW associated with mode shifts are now relatively well constrained, but the magnitude of related changes in the rate of the overturning circulation and deep ocean ventilation remains controversial[22,23]. In this regard, atmospheric radiocarbon (expressed as $\Delta^{14}C_{atm}$, the per mil deviation from a $^{14}C$ standard after correction for radioactive decay and fractionation) offers great promise for identifying past changes in the globally integrated THC. $\Delta^{14}C_{atm}$ is a function of the production rate of $^{14}C$ in the upper atmosphere and the sizes of and exchange rates between the major carbon reservoirs. Considerable work yet remains before production-rate effects are quantified so as to yield a residual $\Delta^{14}C_{atm}$ signal that uniquely reflects changes in ocean ventilation back to 22.0 kyr before present (BP), or the period in which existing records show coherent changes in $\Delta^{14}C_{atm}$ (Fig. 1a). As a first approximation, we account for the long-term decrease in $^{14}C$ production rate that resulted primarily from a gradual increase in the geomagnetic field intensity over the last 22 kyr (ref. 24), isolating changes in $\Delta^{14}C_{atm}$ that, on shorter timescales ($\leqslant 10^3$ yr), are largely due to changes in cosmic radiation or ocean ventilation.

Polar ice core records of the cosmogenic radionuclide $^{10}Be$ can be used to estimate past changes in cosmic radiation because $^{10}Be$ is rapidly (1–2 yr) removed from the atmosphere to the ice surface by precipitation. On this basis, the $^{10}Be$ record from the GISP2 ice core[25] suggests that changes in cosmogenic production rates are an unlikely cause of first-order changes in $\Delta^{14}C_{atm}$ between 15.0 and 22.0 kyr (Fig. 1b). Between 11.5 and 15.0 kyr BP, the $^{10}Be$ signal in the GISP2 ice core is more strongly influenced by climate, but a detailed analysis suggests that a geomagnetic influence may have been important in explaining the decrease in $\Delta^{14}C_{atm}$ during the Younger Dryas cold event[26,27] (Fig. 1a). The two large increases in $^{10}Be$ flux at 10.1 and 11.1 kyr BP, which may reflect solar variability[26], correspond to two increases in $\Delta^{14}C_{atm}$ of about 30‰. (The small age offsets indicate that the GISP2 timescale is too old in this interval by about 60 years[26].) In contrast, the fluctuations in $\Delta^{14}C_{atm}$ before 14.6 kyr BP with amplitudes of over 80‰ do not have corresponding fluctuations in $^{10}Be$ flux, and thus probably reflect changes in the THC, although a geomagnetic influence cannot yet be ruled out for explaining some of the signal.

态[9]。现代模态的特征是北欧海域有深水区形成以及格陵兰–苏格兰脊地区有相应的流场[17]。这个新形成的NADW流入拉布拉多海，并产生回流，即相对较冷并且较淡的拉布拉多海中层水[17]，这种水团在向南流之前受到北大西洋副极地环流[18]的很大限制。在冰期模态时，NADW可能是由副极地北大西洋的开阔海洋中的对流形成的，它下沉到不到2,500米的深处(参考文献19)。北欧海域海冰下面的盐析作用引起的浮力减小是形成冰期模态NADW的另一个原因[20]。最后，在海因里希模态时，来自南极的水填满了北大西洋海盆较浅的1,000米处(参考文献21)。

新形成的NADW的位置、深度和体积的变化与模态转换相联系，这种联系现在已经了解得相当清楚了，不过它的翻转环流速率和大洋深部的流动性(即深部通风)的变化量级仍有争论[22,23]。在这方面，大气放射性碳(用$\Delta^{14}C_{atm}$表示，以经过放射性衰减和分馏订正后对$^{14}C$标准值的偏差来表征，以千分之一计)对弄清全球THC过去的变化提供了很大的可能性。$\Delta^{14}C_{atm}$是上层大气$^{14}C$产生率、主要碳库的大小和碳库之间交换率的函数。在使产生率的影响定量化以便得到一个剩余的$\Delta^{14}C_{atm}$信号方面，我们还需要做大量工作。剩余的$\Delta^{14}C_{atm}$信号是唯一能够反映大洋通风在距今2.2万年以来变化的量，或者反映这一时期的记录存在$\Delta^{14}C_{atm}$的一致变化量(图1a)。作一个初步近似，我们给出$^{14}C$产生率的长期减少趋势，这一结果主要来自2.2万年以来地磁场强度的逐渐增加(参考文献24)，而后分离出较短的时间尺度($\leq$1,000年)的$\Delta^{14}C_{atm}$变化，这一部分反映了宇宙辐射或者大洋通风的变化。

极地冰芯记录的宇生放射性核素$^{10}Be$可以用于估计以往宇宙辐射的变化，因为$^{10}Be$在发生降水时很快(1～2年)由大气进入冰面。在此基础上，由GISP2冰芯[25]得出的$^{10}Be$记录表明，宇生产生率的变化似乎不是1.5万到2.2万年期间$\Delta^{14}C_{atm}$的一阶变化的原因(图1b)。距今1.15万到2.2万年，GISP2冰芯中$^{10}Be$的信号受到气候的更大影响，但是详细的分析得出，地磁产生的影响在解释新仙女木冷期事件$\Delta^{14}C_{atm}$的减少也是重要的[26,27](图1a)。距今1.01万到1.11万年，$^{10}Be$通量的两次大的增加，可以反映太阳辐射的变率[26]，其与两次$\Delta^{14}C_{atm}$约30‰的增加相对应。(小的年代偏离说明GISP2时间尺度对于这个时期来讲，相对老了约60年的时间[26])。相反，距今1.46万年振幅超过80‰的振荡并未对应$^{10}Be$通量的波动，它可能反映的是THC的变化，虽然在解释某些信号时地磁的影响不能完全排除。

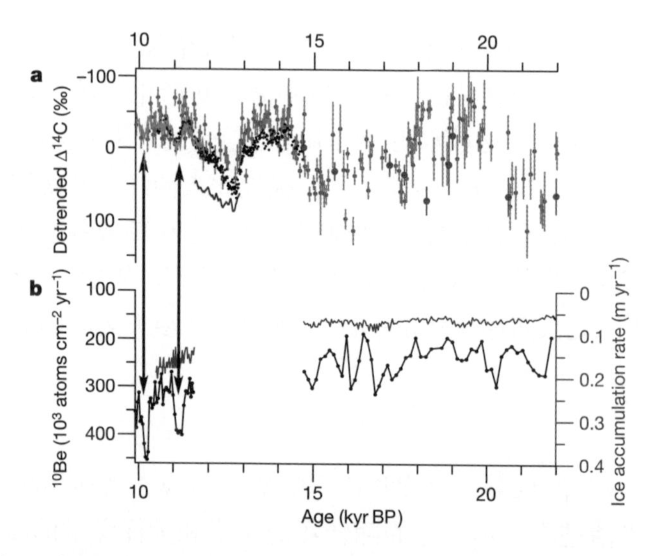


Fig. 1. Comparison of record of linearly detrended $\Delta^{14}C_{atm}$ to $^{10}Be$ flux data from the GISP2 ice core. **a,** Linearly detrended records of $\Delta^{14}C_{atm}$ used to infer changes in the global thermohaline circulation between 9.9 and 22.0 kyr BP. Records include tree rings (solid green line)[72], varves from the Cariaco basin (blue circles)[33], varves from Lake Suigetsu, Japan (grey circles with error bars)[73], and U/Th-dated corals (red circles with error bars)[74]. The thin red line is an estimate of production-rate decrease in $\Delta^{14}C_{atm}$ between 11.5 and 12.9 kyr BP (from ref. 26 with age model modified to be on GISP2 timescale); vertical scale is offset by 80‰. We linearly detrended the combined $\Delta^{14}C_{atm}$ records for the interval 5 to 25 kyr in order to account for the long-term decrease in $^{14}C$ production rate that resulted primarily from a nearly linear increase in the geomagnetic field intensity over this time[24]. Short-term variations in the geomagnetic field intensity may also contribute, but existing geomagnetic records showing variability during this interval are not coherent. **b,** Records of $^{10}Be$ flux (filled circles) and ice accumulation rate (blue line) from GISP2 ice core. $^{10}Be$ flux is the product of the $^{10}Be$ concentration[25] and the ice accumulation rate[75] averaged over the sample interval of the $^{10}Be$ measurements. We do not show the accumulation rate and $^{10}Be$ flux records from 14.6 to 11.5 kyr BP because the $^{10}Be$ flux variations in this interval are dominated by the large variations in snow accumulation rate, thus obscuring production-rate controls on $^{10}Be$ during this time. Accumulation-rate variability was small before 14.6 kyr BP, thus indicating that there is little potential influence of changes in snowfall rate on $^{10}Be$ flux.

NADW is presently the major source of $^{14}C$ to the deep sea[28], and changes in the strength of this water mass (and its preformed properties) probably dominate the variations in $\Delta^{14}C_{atm}$ associated with the global carbon cycle over the last 22.0 kyr. These changes can be modulated by changes in the THC elsewhere, but surface waters in the Southern Ocean and North Pacific are now old[28], and may have been significantly older during glacial periods[29], suggesting that any increase in formation of deep or intermediate waters in these regions will have a lesser impact on the $\Delta^{14}C_{atm}$ budget than changes in NADW.

The hypothesis that changes in $\Delta^{14}C_{atm}$ largely reflect variations in the Atlantic THC is supported by the relation between first-order changes in $\Delta^{14}C_{atm}$ and in proxies that record changes in the volume of NADW and North Atlantic climate (Fig. 2). High $\Delta^{14}C_{atm}$ during the Last Glacial Maximum (LGM) (Fig. 2c) is associated with the glacial mode of NADW[19] and cold North Atlantic atmospheric[30] (Fig. 2d) and sea surface[31] (Fig. 2e) temperatures. The subsequent decrease in $\Delta^{14}C_{atm}$ to essentially interglacial levels was associated with

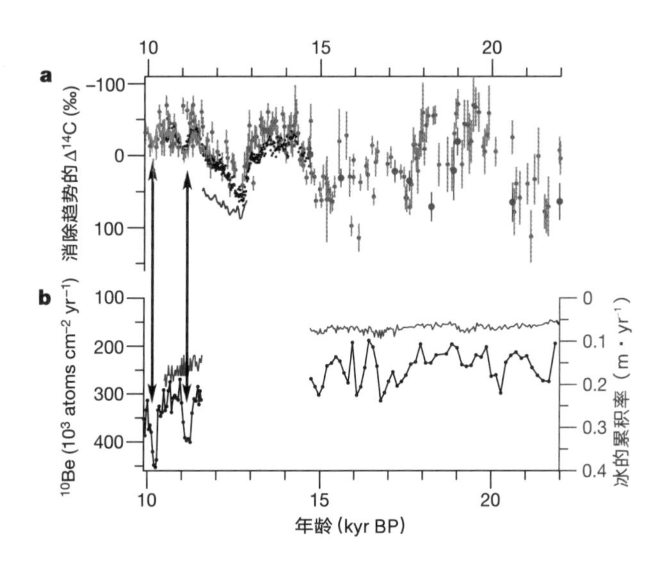


图 1. 消除线性趋势后的 $\Delta^{14}C_{atm}$ 记录和 GISP2（格陵兰冰盖计划）冰芯 $^{10}Be$ 通量资料的比较。a 图为消除线性趋势后的 $\Delta^{14}C_{atm}$，用于推断距今 0.99 万到 2.20 万年期间全球温盐环流的变化，其记录包括树木年轮（绿实线）[72]、卡里亚科盆地纹泥（蓝点）[33]，日本水月湖纹泥（带有误差棒的灰点）[73]，以及 U/Th 定年的珊瑚（带有误差棒的红点）[74]。细红线表示距今 1.15 万到 1.29 万年 $\Delta^{14}C_{atm}$ 产生率减少的估量（该指标源于参考文献 26，修改其年代模式以吻合 GISP2 冰芯的时间尺度）；垂直尺度偏移了 80‰。对距今 0.5 万到 2.5 万年期间的综合 $\Delta^{14}C_{atm}$ 记录消除线性趋势是为了估计长期 $^{14}C$ 产生率的减少，产生率的减少是由这一时间尺度地磁场强度的线性增加造成的 [24]。地磁场强度的短期变化对此也有影响，不过已有的地磁场记录指出在这一时期变化不一致。b 图为 GISP2 冰芯的 $^{10}Be$ 通量记录（实心圈）和冰累积率（蓝线）。$^{10}Be$ 通量由 $^{10}Be$ 浓度求得 [25]，冰的累积率是样本时段 $^{10}Be$ 的平均观测值 [75]。我们没有给出距今 1.46 万到 1.15 万年期间的累积率和 $^{10}Be$ 通量的记录，因为这一时期 $^{10}Be$ 通量的变化主要由雪的累积率决定，而掩盖了冰的产生率对 $^{10}Be$ 的影响。在距今 1.46 万年以前，累积率的变化小，因此降雪率对 $^{10}Be$ 通量的影响小。

NADW 是现今深海 $^{14}C$ 的主要来源 [28]，水团强度的变化（以及其此前形成的性质）可能主导了 2.2 万年以来与全球碳循环相联系的 $\Delta^{14}C_{atm}$ 的变化。这些变化在其他地方受 THC 调制，但在南大洋和北太平洋表层水现在是古老的 [28]，而在冰期就更加古老了 [29]，这说明在这些区域，深层水或中层水的形成的任何增加对 $\Delta^{14}C_{atm}$ 的影响都没有 NADW 变化的影响大。

$\Delta^{14}C_{atm}$ 的变化很大程度上反映大西洋 THC 的变化这一假说，得到了 $\Delta^{14}C_{atm}$ 的一阶变化与记载着 NADW 体积和北大西洋气候变化的代用资料之间关系的支持（图 2）。在末次冰盛期（LGM）（图 2c），$\Delta^{14}C_{atm}$ 的高值与 NADW 冰期模态 [19]，北大西洋冷大气的 [30]（图 2d）和海面的 [31]（图 2e）温度相联系。此后，$\Delta^{14}C_{atm}$ 基本上下降到间冰期水平，这和北大西洋地区增温有关，而随后在距今 1.90 万到 1.47 万年期间，长

warming in the North Atlantic region, while a subsequent long-term increase between 19.0 kyr BP to 14.7 kyr BP coincides with the Oldest Dryas cooling, during which Heinrich event 1 (H1) occurred (~17.5 kyr BP) (Fig. 2b). During H1, $\Delta^{14}C_{atm}$ rapidly increased to levels similar to the LGM (Fig. 2c), whereas the volume of NADW during the Heinrich mode (as inferred from geochemical proxies) was < 50% of that during the LGM[21]. Relative to the glacial mode, these relations suggest that the Heinrich mode reflects some combination of further shallowing of NADW to intermediate depths while maintaining similar rates of overturning, and a compensatory increase in deep-water formation elsewhere.

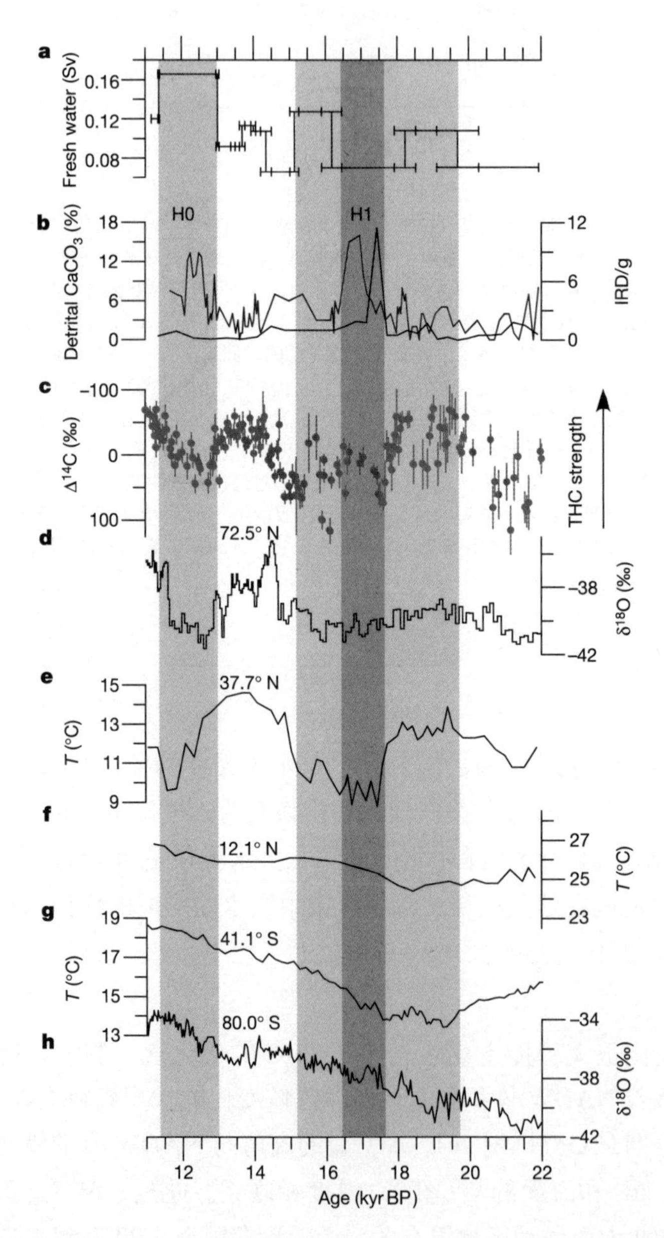


Fig. 2. Comparison of reconstructed forcings and responses in the Atlantic basin during the last

期的增长则和老仙女木冷期同时发生，在这一时期（距今约 1.75 万年）海因里希 1（H1）事件发生（图 2b）。在海因里希模态时期，$\Delta^{14}C_{atm}$ 迅速上升到类似 LGM（图 2c）的水平，而 NADW 的体积（据地球化学代用资料）小于 LGM 时期的 50%[21]。相对于冰期模态，这些关系说明海因里希模态反映的是 NADW 进一步变浅到中层水深度，同时保持了类似的翻转速度，这种变化得到别处深水形成增加的补偿。

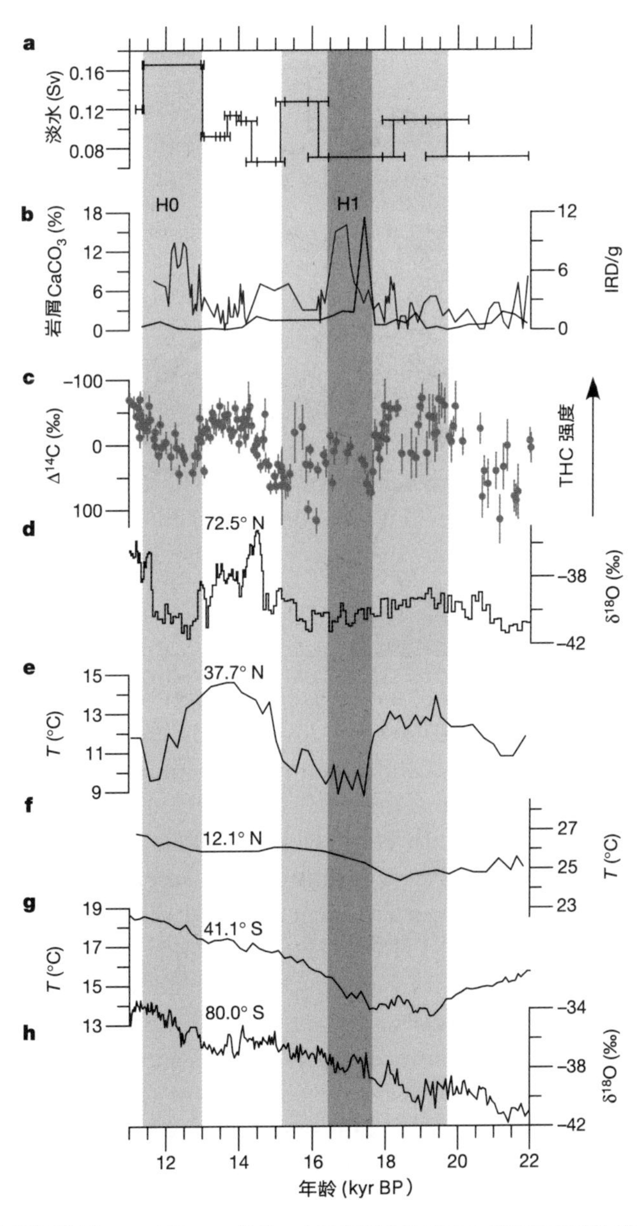


图 2. 末次冰川消退期（距今 1.1 万到 2.2 万年），大西洋海盆的强迫和响应的重建比较。淡水强迫变化

deglaciation (11–22 kyr BP). Time series of changes in freshwater forcing (**a, b**) and response of the THC as inferred by first-order changes in $\Delta^{14}C_{atm}$ (**c**). Panels **d–h** show atmospheric and sea surface temperature changes during the last deglaciation along a north–south transect of the Atlantic basin (72.5° N to 80° S), illustrating the operation of the bipolar seesaw in response to large changes in the Atlantic THC. **a**, Changes in freshwater runoff from North America through the Hudson and St Lawrence Rivers[38]. Two vertical light-grey bars identify intervals of enhanced freshwater flux to the North Atlantic from these sources as well as from icebergs inferred from records (not shown) of total ice-rafted debris (IRD)[37]. **b,** Changes in the amount of IRD containing detrital carbonate lithologies, with large increases identifying times of Heinrich events H1 (identified by darker-grey vertical bar) and H0. The IRD records are from North Atlantic cores VM23-081 (blue curve)[37] and SU8118 (black curve)[31]. **c,** Detrended record of $\Delta^{14}C_{atm}$ from Lake Suigetsu, Japan[73], with first-order changes indicating changes in the meridional overturning of the Atlantic THC (see text); a stronger THC is indicated by more negative values of $\Delta^{14}C_{atm}$. **d,** The GISP2 ice-core oxygen-isotope record[76,77]. Results of calibrating the isotopic palaeothermometer by various methods show a temperature depression of 20 °C at the Last Glacial Maximum ~21.0 kyr BP (ref. 30), a warming of 10 °C at the onset of the Bølling at ~14.6 kyr BP (ref. 32), and a warming of 15 °C at the end of the Younger Dryas ~11.5 kyr BP (ref. 41). **e,** Record of sea surface temperatures (SSTs) derived from alkenones measured in North Atlantic core SU8118 (ref. 31). **f,** Record of SSTs derived from alkenones measured in tropical North Atlantic core M35003-4 (ref. 78). **g,** Record of SSTs derived from alkenones measured in South Atlantic core TN057-21-PC2 (ref. 79). The timescale for this record is provisional (J. Sachs, personal communication). **h,** The Byrd ice-core oxygen-isotope record[50].

A large decrease in $\Delta^{14}C_{atm}$ to interglacial levels coincides with the onset of the Bølling–Allerød warm interval at around 14.7 kyr BP; the abrupt warming recorded in the GISP2 ice core[32] (Fig. 2d) appears to be a nonlinear response to the more gradual increase in the THC (Fig. 2c). A subsequent increase in $\Delta^{14}C_{atm}$ began precisely at the onset of the Younger Dryas cold interval at 13.0 kyr BP (Fig. 2c)[33]. The decrease in $\Delta^{14}C_{atm}$ during the remainder of the Younger Dryas may be the result of an increase in the production rate of $^{14}C$ (refs 26, 27) (Fig. 1a), suggesting that the THC remained weakened until the end of the Younger Dryas at 11.5 kyr BP when it abruptly switched back to the modern mode[27].

## Mechanisms of Past Abrupt Climate Changes

The freshwater budget in the North Atlantic is one of the major components that governs the strength of the Atlantic THC, and dynamical ocean models show that the THC is sensitive to freshwater perturbations on order of 0.1 Sv (1 Sv = $10^6$ $m^3$ $s^{-1}$) (refs 34–36). A clear goal for understanding past changes in the THC, therefore, is to identify the mechanisms that influenced the North Atlantic freshwater budget. Reconstructions of changes in the freshwater flux from ice sheets around the North Atlantic reveal good agreement with past changes in the THC and North Atlantic climate (Fig. 2a–e), identifying these processes as being important in causing abrupt climate change[37,38]. Paradoxically, although the THC in current models responds to freshwater forcings without delay, the largest deglacial meltwater event on record, referred to as meltwater pulse 1A (MWP-1A), occurs more than 1,000 years before the next significant change in the THC associated with the Younger Dryas cold interval[39]. This paradox may be resolved, however, if MWP-1A originated largely from the Antarctic Ice Sheet[40], where its impact on the Atlantic THC would be substantially reduced.

的时间序列（a 图和 b 图）和 $\Delta^{14}C_{atm}$ 一阶变化所表示的 THC 的响应（c 图）。d ~ h 图表示末次冰川消退期，沿大西洋海盆的一个南北剖面（北纬 72.5 度到南纬 80 度）上大气和海面温度的变化，说明了在大西洋 THC 大变化响应方面的两极跷跷板作用。**a** 图为经过哈得孙河和圣劳伦斯河的北美淡水径流的变化[38]。两个垂直的浅灰色柱表示这些源头或者冰山流向北大西洋增加的淡水输入的时间段，冰山流根据冰筏碎屑（IRD）总体的记录[37]（未给出）推断而来。**b** 图为岩屑碳酸盐中 IRD 含量的变化，其增值明显的时期为海因里希事件 H1（用深灰色柱表示）和 H0。IRD 的记录来自北大西洋冰芯 VM23-081（蓝色曲线）[37] 和 SU8118（黑色曲线）[31]。**c** 图为来自日本水月湖、消除了趋势线的 $\Delta^{14}C_{atm}$ 记录[73]，$\Delta^{14}C_{atm}$ 一阶变化表示北大西洋 THC 经向翻转的变化（见正文）；其中一个较强的 THC 用 $\Delta^{14}C_{atm}$ 的负值标出。**d** 图为 GISP2 冰芯的氧同位素记录[76,77]。在末次冰盛期距今约 2.10 万年（参考文献 30）用不同方法算出的同位素古温度有一个 20 ℃的下降，在距今约 1.46 万年的波令事件开始时增温达到 10 ℃（参考文献 32），在新仙女木事件末期距今约 1.15 万年增温达到 15 ℃（参考文献 41）。**e** 图为根据在北大西洋 SU8118 冰芯中测得的烯酮推断的海面温度记录（SST）（参考文献 31）。**f** 图为根据在热带北大西洋 M35003-4 冰芯中测得的烯酮推断的 SST（参考文献 78）。**g** 图为根据在南大西洋 TN057-21-PC2 冰芯中测得的烯酮推断的 SST（参考文献 79）。该记录的时间尺度是临时性的（萨赫斯，个人交流）。**h** 图为伯德站冰芯的氧同位素记录[50]。

$\Delta^{14}C_{atm}$ 大量减少至达到间冰期水平发生在距今 1.47 万年左右的波令–阿勒罗德暖期；这个 GISP2 冰芯记录的一个突然升温[32]（图 2d）看来是对 THC 逐渐增强的非线性响应（图 2c）。其后 $\Delta^{14}C_{atm}$ 的增加与距今 1.30 万年新仙女木冷期的开始精准地一致（图 2c）[33]。新仙女木后期 $\Delta^{14}C_{atm}$ 的减少看来是 $^{14}C$ 产生率增加的结果（参考文献 26 和 27）（图 1a），这说明 THC 直到距今 1.15 万年新仙女木冷期结束时都很弱，而后它又突然转换回现代模态[27]。

## 过去气候突变的机制

北大西洋淡水注入的多寡是控制大西洋 THC 强度的主要因子之一，动力学海洋模式指出 THC 对量级为 0.1 Sv（$1\ Sv = 10^6\ m^3 \cdot s^{-1}$）的淡水扰动敏感（参考文献 34 ~ 36）。因此，了解过去 THC 的变化是为了弄清影响北大西洋淡水注入多寡的机制。用大西洋周围冰川资料重建的淡水通量的变化与过去 THC 和北大西洋气候的变化相当一致（图 2a ~ e），这说明这些过程是导致气候突变的重要因素[37,38]。自相矛盾的是，虽然 THC 的现代模式对淡水强迫没有延迟，然而在记录中显示的最大的冰期融水事件，被称为融水脉动 1A（MWP-1A），却比而后在新仙女木冷期发生的 THC 显著变化超前了一千多年[39]。不过这一悖论可以解释为：MWP-1A 很大程度上源于南极冰川[40]，它对大西洋 THC 的影响是相当有限的。

Millennial-scale ($10^3$ yr) changes in the Atlantic THC that involved transitions from modern to glacial modes of NADW are manifested as dramatic fluctuations of North Atlantic climate referred to as Dansgaard–Oeschger (D–O) events, each having a characteristic pattern of abrupt (years to decades) warming followed by gradual (centuries) cooling. The largest changes ( > 15 °C over Greenland)[41] occurred during times of intermediate global ice volume, $CO_2$, or insolation[42]. In Greenland ice core records, abrupt warmings exhibit a preferred waiting time of 1,500 ± 500 yr (ref. 43). This spacing is similar to a 1,000–2,000-yr climate cycle identified from marine records during interglacial as well as glacial periods, leading some to speculate that D–O events may be amplified expressions of an ongoing persistent and stable climate cycle[37,43]. However, this timescale of variability does not constrain one mechanism over another. Several mechanisms may have operated jointly, and the relative contribution of potential mechanisms probably changed in response to the large-scale changes in global boundary conditions that accompanied the last deglaciation.

Atmospheric transmission of the D–O signal beyond the North Atlantic region is suggested by high-resolution climate records that display the same structure of change that, within dating uncertainties, is synchronous with the North Atlantic signal. It is also suggested by simulations using atmospheric GCMs[44], although these models have yet to include all potential feedbacks that may be important in transmitting the signal to regions far from the North Atlantic. Greenland ice-core records of methane and $\delta^{18}O$ strongly support the hypothesis of North Atlantic forcing in showing a nearly instantaneous response of the tropical water balance to changes in high-latitude temperature[32,41]. Palaeoclimate records identify additional atmospheric responses to North Atlantic climate that would potentially further amplify and transmit the D–O signal. These include changes in (1) the strength of trade winds and associated oceanic upwelling[45], (2) the position of the Intertropical Convergence Zone, with attendant effects on water vapour transport for the Atlantic to the Pacific basin[46], (3) the strength of the Asian monsoon[47], (4) sea surface temperatures of the Pacific warm pool[48] and (5) ventilation of the North Pacific[49].

A large reduction in the THC during the Heinrich mode (Fig. 2c) causes additional cooling in the North Atlantic[31] (Fig. 2e), while contemporaneous warming observed in some regions of the Southern Hemisphere[9,50] indicates that the large reduction in NADW formation decreased meridional heat transport from the South Atlantic (Fig. 2f–h). A similar "bipolar seesaw" effect may hold for the Younger Dryas event[51] (Fig. 2), but the absence of a comparable thermally antiphased southern signal associated with many of the older D–O events[50] suggests that any changes in oceanic heat transport accompanying the changes from modern to glacial circulation modes were either too small or too short to be clearly registered in existing climate records.

We thus expect to see two mechanisms of variability in climate related to changes in the THC: one associated with atmospheric transmission and one associated with an oceanic seesaw, although these are not independent as an oceanic change necessarily implies an atmospheric response and vice versa. We use empirical orthogonal function (EOF) analysis

NADW 从现代模态到冰期模态的转换中，称为丹斯果–厄施格尔（D–O）事件的北大西洋气候的突然变动，显示了大西洋 THC 千年尺度的变化，其中每一次都有典型的突然（年际到年代际）增温和随之而来的逐渐（世纪尺度）降温。最大的温度变化（在格陵兰 ＞ 15℃）[41] 发生在中等大小的全球冰量、$CO_2$ 浓度和太阳辐照量的时期 [42]。在格陵兰冰芯记录中，突然增温通常有一个 1,500 ± 500 年的酝酿期（参考文献 43）。这段间隔期与源自冰期和间冰期期间的海洋记录中 1,000 ~ 2,000 年的气候循环相似，这使有些人推测 D–O 事件可能是一个持续发展并且稳定的气候循环的放大表现 [37,43]。但是，这一时间尺度变化并不限定于某种特定的机制，而是由若干机制协同控制，每种潜在机制的相对贡献可能随末次冰消期时全球边界条件的大尺度变化而改变。

在北大西洋以外地区，D–O 信号的大气传输，可从高分辨率的气候记录中找到。在一定的定年误差范围内，这些气候记录显示与北大西洋信号同步调的一致变化。这些信号也可由大气环流模式模拟出来 [44]，虽然这些模式尚未能完全包括所有能把信号传到离北大西洋以外很远的地方的重要反馈过程。格陵兰冰芯记录的甲烷和 $\delta^{18}O$ 强烈地支持北大西洋强迫导致热带水的平衡对高纬度温度变化有近乎同时的响应这一假设 [32,41]。古气候记录指出，大气对北大西洋气候另外的响应，说明它能相当大地放大和传输 D–O 信号。这些响应包括以下几种变化：(1) 信风和与之相联系的海洋上升流的强度 [45]，(2) 热带辐合带的位置，影响着大西洋到太平洋的海盆尺度的水汽输送 [46]，(3) 亚洲季风的强度 [47]，(4) 太平洋暖池的海面温度 [48]，(5) 北太平洋的通风作用 [49]。

海因里希模态期间 THC 的大减弱（图 2c）引起北大西洋的进一步降温 [31]（图 2e），而南半球某些地区同时观测到的增温 [9,50] 说明 NADW 大幅度地减弱了来自南大西洋的经向热量输送（图 2f ~ h）。类似的“两极跷跷板”效应在新仙女木事件时也有 [51]（图 2），但是没有与许多较古老的 D–O 事件联系的从南面来的反相位的高温信号 [50]，这说明伴随着从现代模态向冰期模态的变化而产生的任何海洋热传输的变化，或者强度太小，或者时间太短，以至在已有的气候记录中找不到。

因此，我们看到两种与 THC 有关的气候变化的机制：一种和大气传输有关，另一种和海洋“跷跷板”效应有关，虽然两者并不是相互独立的，因为海洋的变化意味着大气的响应，或者反过来大气的变化意味着海洋的响应。我们应用经验正交函数

to provide an objective characterization of these modes of climate variability during the last deglaciation by reducing 18 well-dated time series of climate change (Fig. 3) into spatially coherent, orthogonal eigenvectors. Our analysis indicates that the last deglaciation was dominated by two climate responses. The first EOF (68% of variance) captures the global warming from glacial to interglacial conditions that evolved over a timescale of about 10 kyr. Included in this EOF is the interruption of the warming trend by the Younger Dryas. The second EOF (15% of the variance) quantifies the spatial and temporal expression of millennial changes centred at 16 and 12 kyr BP (Fig. 3a). The spatial pattern of this EOF, with negative scores over Antarctica (except Taylor Dome) and in the South Atlantic and positive scores at all other sites (Fig. 4b), is consistent with an atmospheric transmission of the North Atlantic signal except for those areas in the Southern Hemisphere where operation of the seesaw during the last deglaciation produced an antiphased response as predicted by a large change in the THC[52-54]. We find that a similar spatial pattern holds for the interval between 26 and 50 kyr BP, indicating that the seesaw operated during this time as well[50].

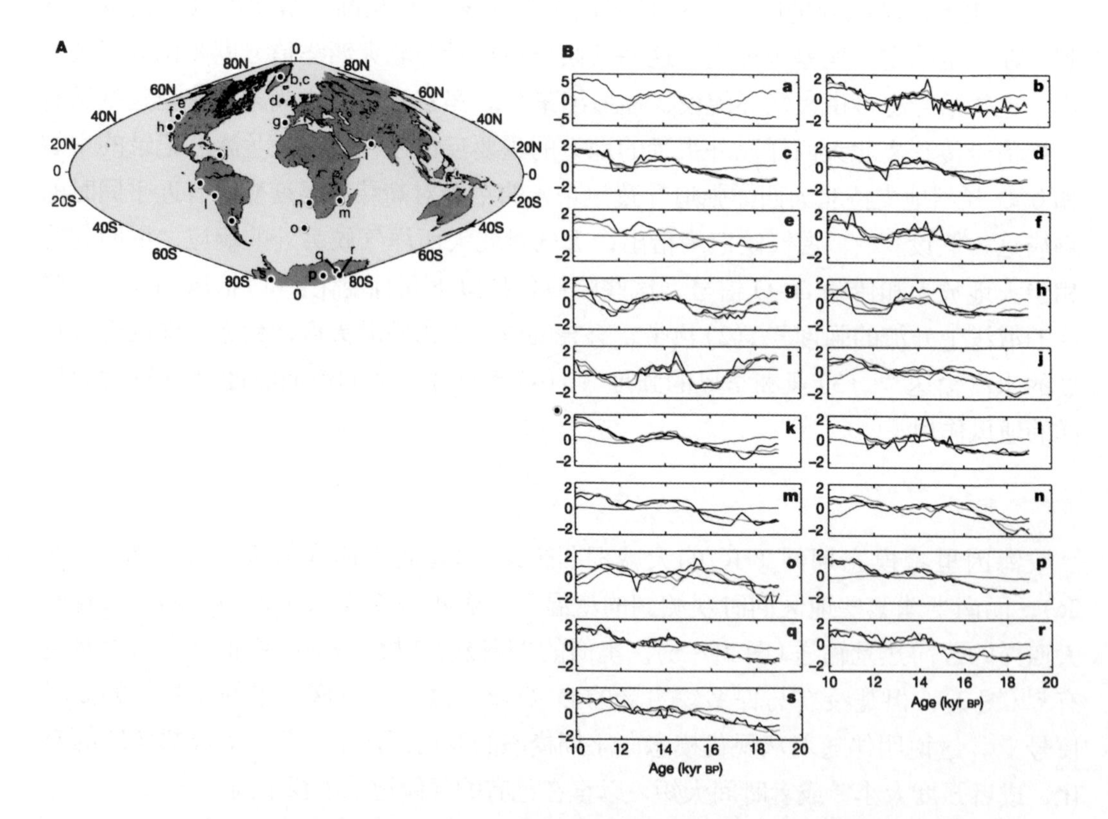


Fig. 3. Time-history of the first two empirical orthogonal functions (EOFs) determined from 18 time series. **A,** Data sampling locations. **B,** Time series. These data sets provide time series of climate change with temporal sampling resolution of about 250 years or less. All time series were interpolated to a constant sampling interval of 125 years. Numerical experiments demonstrate that the results presented here are insensitive to factor-of-2 changes in the selected constant sampling interval. All data sets were then transformed to mean zero and standard deviation of one. We note that where the second EOF is negative, time series are characterized by early warming but a weaker Younger Dryas; where the second EOF is

(EOF)方法将 18 个记录完好的表示气候变化的历史资料时间序列(图 3)加以降维，形成空间上相互关联的正交特征向量，得到了在末次冰消期气候变化模态的客观表征。我们的分析指出，末次冰消期主要有两种气候响应。第一个 EOF(可解释 68% 的方差)捕获了冰期到间冰期 1 万年以上时间尺度的全球增温，其中包括在新仙女木期增暖趋势的中断。第二个 EOF(可解释 15% 的方差)量化了距今 1.6 万到 1.2 万年期间以千年尺度变化的空间和时间的变化(图 3a)。EOF 空间分布在南极洲(泰勒冰穹除外)和南大西洋为负，在其他地方为正(图 4b)。这种分布和北大西洋信号的大气传输方向相同，但南半球的一些地区除外，在这些地区，末次冰消期跷跷板的作用产生了一个反相位的响应，这是由 THC 的大幅度变化预示的[52-54]。我们还发现了距今 2.6 万到 5.0 万年期间类似的空间分布，这说明在这一时期跷跷板的作用也很明显[50]。

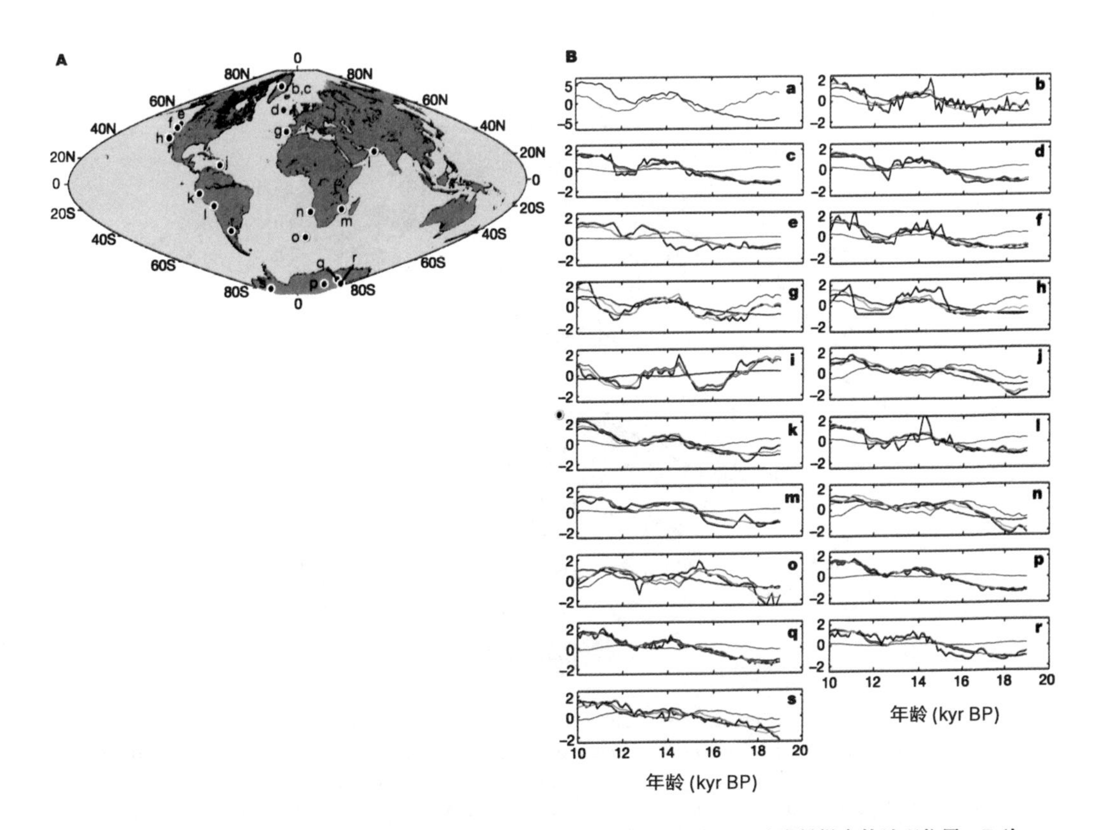


图 3. 由 18 个时间序列求出的前两个经验正交函数(EOF)的时间演变。**A** 为资料样本的地理位置。**B** 为时间序列。这些资料集给出了具有 250 年左右或更少的采样分辨率的气候变化的时间序列。所有的时间序列都以 125 年等距时间间隔内插。数值实验说明，此处的结果对选定的等距时间间隔 2 倍采样率是不敏感的。所有的资料都已换算成平均值为 0、标准差为 1 的序列。我们注意到，在第二个 EOF 数值为负的地方，其时间序列以早期的增温而后是一个弱的新仙女木期为其特征；在 EOF 数值为正的地方，新仙女木期趋向一个更强的事件。我们强调，第二个 EOF 的空间分布强烈地依赖时间尺度的性质和对

positive, the Younger Dryas tends to be a more significant event. We emphasize that the pattern of the second EOF depends strongly on the quality of the timescale and the precision of the synchronization between these palaeoclimatic records (for example refs 80, 81). **a,** The time series for the first two EOFs, where the blue line is the first EOF and the red line is second EOF. The first EOF accounts for 68% of the data variance while the second EOF accounts for 15%. **b–s,** Plots of the original time series and the results of the EOF analysis. The normalized time series from each location (black line), the weighted EOFs using the weights from **a** (blue line is first EOF, red line is second EOF), and the sum of the first two EOFs (green line). **b,** Oxygen-isotope record from the GISP2 ice core[76,77]. **c,** Methane record from the GISP2 ice core[50]. **d,** Percentages of *Neogloboquabrina pachyberma* (s.) from North Atlantic core VM23-81 (ref. 37). **e,** Relative abundance of a radiolarian assemblage identified in northeastern Pacific core EW9504-17PC (ref. 82). **f,** Percentages of *N. pachyberma* (s.) from ODP Site 1019 in the northeastern Pacific[83]. **g,** Alkenone-derived sea surface temperatures from the subtropical northeast Atlantic[31]. **h,** Changes in the bioturbation index from ODP Site 893 in the Santa Barbara basin[49]. **i,** Total organic carbon from Arabian Sea sediments[47]. **j,** Alkenone-derived sea surface temperatures from the tropical North Atlantic[78]. **k,** Oxygen-isotope record from the Huascaran ice core, Peru[84]. **l,** Oxygen isotope record from the Sajama ice core, Bolivia[85]. **m,** Alkenone-derived sea surface temperatures from the tropical Indian Ocean[86]. **n,** Percentages of *N. pachyberma* (s.) from the South Atlantic[87]. **o,** Residual oxygen-isotope record measured on planktonic foraminifera for the South Atlantic[88]. **p,** The Vostok ice-core deuterium record[89]. **q,** Deuterium record from the Dome C ice core[90]. **r,** Oxygen-isotope record from the Taylor Dome ice core[80]. **s,** Oxygen-isotope record from the Byrd ice core[50].

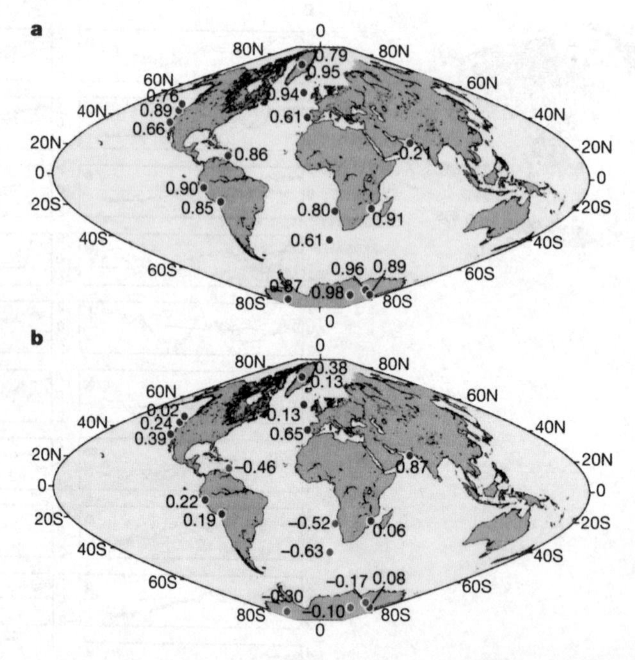


Fig. 4. The spatial pattern of the first two empirical orthogonal functions extracted from the data set (see Fig. 3a). Numbers shown are loadings of the first two EOFs for 18 deglacial time series. Loadings indicate the importance of each EOF in explaining the variations in each time series. To reconstruct the normalized (mean zero and standard deviation of one) time series at any one location, we take the loadings shown in Fig. 3a, multiply them by the time series of the EOFs and add them together. The two loadings next to the Greenland site correspond to the GISP2 oxygen isotope (upper) and methane (lower) records. **a,** First EOF; **b,** second EOF.

这些古气候记录做同时化处理的精度(比如参考文献 80 和 81)。**a** 图为前两个 EOF 的时间序列，蓝线是第一个 EOF 的，红线是第二个 EOF 的。第一个 EOF 可解释 68% 的方差，而第二个 EOF 可解释 15% 的方差。**b** ~ **s** 图为原始时间序列图和 EOF 分析的结果。每一个测站归一化的时间序列(黑线)，应用 **a** 图中的权重求得的加权结果(蓝线是第一个 EOF 的，红线是第二个 EOF 的)和前两个 EOF 的合成结果(绿线)。**b** 图为 GISP2 冰芯的氧同位素记录 [76,77]。**c** 图为 GISP2 冰芯的甲烷记录 [50]。**d** 图为由北大西洋冰芯 VM23-81 获得的左旋厚壁新方球虫百分率(参考文献 37)。**e** 图为由东北太平洋冰芯 EW9504-17PC 获得的放射虫的相对富集丰度(参考文献 82)。**f** 图为东北太平洋大洋钻探计划(ODP) 1019 位点获得的左旋厚壁新方球虫百分率 [83]。**g** 图为由烯酮推断的副热带东北大西洋海面温度 [31]。**h** 图为位于圣巴巴拉盆地的大洋钻探计划 893 位点获得的生物扰动指数的变化 [49]。**i** 图为阿拉伯海沉积物中的总有机碳 [47]。**j** 图为由烯酮推断的热带北大西洋海面温度 [78]。**k** 图为秘鲁瓦斯卡兰冰芯的氧同位素记录 [84]。**l** 图为玻利维亚萨哈马冰芯的氧同位素记录 [85]。**m** 图为由烯酮推断的热带印度洋海面温度 [86]。**n** 图为南大西洋左旋厚壁新方球虫百分率 [87]。**o** 图为南大西洋浮游有孔虫上残余的氧同位素记录 [88]。**p** 图为东方站冰芯中氘的记录 [89]。**q** 图为冰穹 C 冰芯中氘的记录 [90]。**r** 图为泰勒冰穹冰芯中氧同位素记录 [80]。**s** 图为伯德站冰芯中氧同位素记录 [50]。

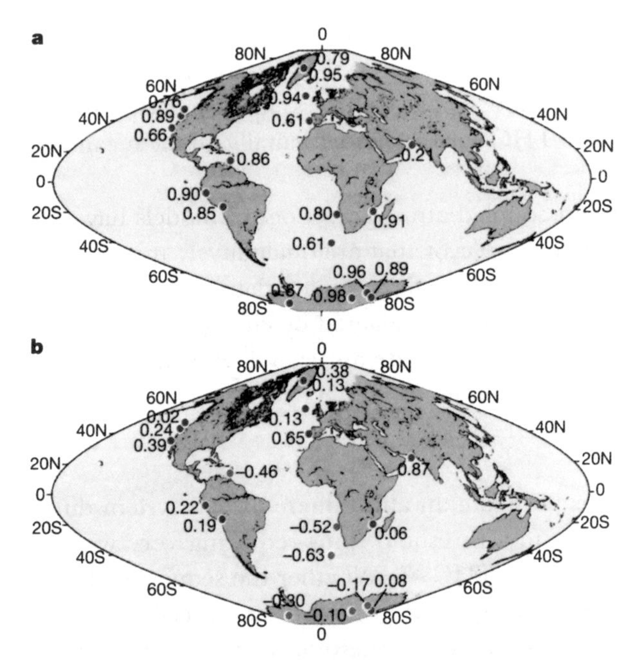


图 4. 由资料集(见图 3a)降维得出前两个经验正交函数的空间分布。图上数字是 18 个间冰期时间序列的前两个 EOF 的载荷。载荷表示每个 EOF 对各个时间序列在解释方差方面的重要程度。为了在任一测站重建归一化(平均值为 0，标准差为 1)的时间序列，我们用图 3a 中测站的载荷乘以 EOF 的时间函数，然后相加。格陵兰测站附近的两个载荷是 GISP2 冰芯的氧同位素记录(上)和甲烷记录(下)。**a** 为第一个 EOF。**b** 为第二个 EOF。

## Modelling Abrupt Climate Change

Modelling abrupt climate change faces several challenges: (1) a disparity in timescales between transient states (lasting many centuries) and mode changes (occurring in a few years to decades); (2) a compromise in model complexity due to long integrations and sufficient model resolution to capture abruptness; and (3) uncertainties in initial conditions due to patchy coverage as well as calibration problems in palaeoclimatic proxy data. Although at present there exists no self-consistent model that simulates, without prescribed forcing, changes that resemble the palaeoclimatic record, important progress has been made.

Abrupt change manifests itself in two different ways in climate models: an abrupt transition across a threshold to a new equilibrium state, or a response to a fast forcing. Although multiple equilibria are not necessary for abrupt change[35], the palaeoclimatic records suggest that the ocean–atmosphere system may have preferred modes of operation, that is, the system operates like a "flip-flop" mechanism[55]. This implies the existence of hysteresis of the THC[16,34,56], which appears to be common in coupled climate models. However, the shape and structure of the hysteresis loop strongly depends on model parameters and therefore is still a tunable feature[57] so that the exact location of thresholds cannot yet be determined with current models. Nevertheless, it is clear that thresholds and hence the stability properties of the THC depend fundamentally on the mean climate state[8,58].

A number of ocean and coupled atmosphere–ocean models have found the existence of three distinct modes of the conveyor that are qualitatively the same as the modern, glacial and Heinrich modes identified by palaeoclimate data[16,59,60]. These and other models have also revealed that transitions can be triggered by changes in the freshwater balance of the North Atlantic and changes in modes are associated with changes in the heat budget of the Atlantic basin. For large reductions in NADW formation, this leads to the same "seesaw effect" seen in the data[34,52,53] (Figs 2, 4).

Recent modelling ideas postulate an atmosphere–ocean system during the last glaciation that was extremely close to a threshold, thus requiring very weak freshwater forcing to trigger abrupt changes of the THC[16,60]. Whether the sequence of abrupt events originates from unknown periodic forcing[43,60] or instabilities and feedbacks associated with circum-Atlantic ice sheets[16,38] remains an open question. Furthermore, qualitative consistence with the palaeoclimatic records remains very sensitive to parameter choices in these models. The basic question about the origin of abrupt change is therefore not solved. However, all model simulations until now point towards the key role of the freshwater balance in the Atlantic Ocean. More realistic models and improved reconstructions of the various components of the hydrological cycle (precipitation, run-off, iceberg discharge, sea ice) are urgently needed.

Progress towards a mechanistic understanding of abrupt climate change can be expected from more complete models. These are coupled models with higher resolution, models that no longer require flux adjustments, and models that include biogeochemical cycles.

## 气候突变的模拟

模拟气候突变面临如下挑战：(1) 暂态变化（几个世纪尺度）和模态变化（发生在几年到几十年间）之间时间尺度的差异；(2) 如何在长期积分法需要的模式复杂性和为了模拟突变特征而采取精细的模式分辨率之间进行取舍；(3) 数据融合涵盖范围和古气候代用资料的校准问题引起的初始条件的不确定性。虽然现在尚无自协调的模式可以在没有事先给定强迫的情况下模拟类似于古气候记录的变化，但研究已经取得了重要的进展。

气候突变本身在气候模式中表现为两种不同的方式：一种是跨过一个阈值时突然转到一个新的平衡态，另一种是对快速强迫的响应。虽然多平衡态对气候突变来说并不是必需的[35]，但是古气候记录指出海气系统的运行有一定倾向性，即常出现“双稳态”的机制[55]。这就意味着 THC 的滞后现象[16,34,56]的存在，这种情况在耦合气候模式中经常出现。但是这种滞回曲线的形态和结构与模式参数很有关系，因而仍是一种可调特征[57]，因此阈值的准确位置在当前的模式中仍然是难以确定的。不过，阈值以及由此而来的 THC 的状态稳定性基本上依赖于气候的平均状态[8,58]。

大量海洋和海气耦合模式已经发现大洋传送带存在着三种不同模态，即与从古气候资料中分析出来的现代模态、冰期模态和海因里希模态[16,59,60]大致一样的模态。这些模式和其他模式都指出模态之间的转换由北大西洋淡水平衡的变化启动，而模态内部的变化则和大西洋洋盆的热量收支有关。当 NADW 大量减少时，从记录中看到的“跷跷板效应”就会产生[34,52,53]（图 2 和 4）。

近期的模拟思路假定了末次冰期期间的大气–海洋系统非常接近一个阈值，因此需要非常弱的淡水强迫来启动 THC 的突然变化[16,60]。这些突然事件的连续发生究竟是由未知周期的强迫引起[43,60]还是由与大西洋周围的冰川有关的不稳定性和反馈作用引起[16,38]，仍是尚未弄清的问题。再则，模式结果和古气候记录的大致一样，仍然依托于模式中参数的选择。因此，关于突然变化的起因这个基本问题仍然是尚未解决的。但是直到现在，所有模式的模拟结果都指向大西洋淡水平衡这一关键作用上。设计更多的接近实际的模式和改进水文循环中不同成分（降水、径流、冰山冰量的增减、海冰）的重建方法是急需的。

设计更完善的模式有望在了解气候突然变化的机制方面取得进展。这包括更高分辨率的耦合模式、不再要求通量调整的模式和包含生物化学循环的模式。后者能

The latter will enable a direct and quantitative comparison with palaeoclimatic data (for example, the stable isotopes of water and carbon), and significantly facilitate model-based hypothesis testing. The tracer $\Delta^{14}C_{atm}$ affords additional constraints and helps to estimate how much of the observed changes (Fig. 1a) can be associated with changes in the THC[27]. Results based on simplified ocean models show that a shutdown of the Atlantic THC produces a significant increase in $\Delta^{14}C_{atm}$ but the amplitudes are only about half of those observed[27]. Further model simulations are needed to fully explore the dependence of $\Delta^{14}C_{atm}$ changes on model parameters.

Tropical Pacific variability represents the dominant mode of modern climate variability with its effects felt across the globe. Although a basic understanding of the physics of ENSO has been achieved, it is still not clear how ENSO responds in colder and warmer mean climates. For example, some models suggest that tropical Pacific variability will remain similar to the present, whereas others suggest stronger and more frequent warm events could be in store in a warmer future climate. Nevertheless, it is apparent that Atlantic-to-Pacific moisture transport is sensitive to the phase of ENSO[61], so that a persistent trend towards an enhanced frequency of warm events[62] could lead to an increase in net export of moisture from the Atlantic across the Isthmus of Panama, thereby possibly affecting the THC.

## Discussion

Publication of the results from the first ice core from Summit, Greenland, detailing abrupt climate change during the last glaciation[63] motivated an intensive investigation for evidence of similar climate instability occurring elsewhere on the globe and its causes. The resulting palaeoclimate records clearly reveal the global extent of millennial-scale climate variability, with varying responses that are consistent with atmospheric and oceanic changes associated with changes in the Atlantic THC (Figs 3, 4). Moreover, the relation between times of increased freshwater flux to the North Atlantic and corresponding decreases in the THC (Fig. 2) supports modelling results showing that the THC was sensitive to small changes in the hydrological cycle (order of 0.1 Sv) during the last glaciation[35,36,60].

With respect to cause and effect, however, our understanding of abrupt climate change remains incomplete. Insofar as the palaeoclimate record provides the fundamental basis for evaluating the ability of models to correctly simulate behaviour of the THC, additional information is needed to address these issues. In particular, several areas that require immediate attention include: (1) an increase in the distribution of sites in the Pacific and Southern Oceans; (2) development of new tools to synchronize palaeoclimatic records and constrain phasing relations; (3) improvements in calibrating the climate signal from proxy records; and (4) further analysis of the $\Delta^{14}C_{atm}$ record for the last 50 kyr BP.

A variety of simulations from coupled models of varying complexities are beginning to simulate the temporal evolution and global signature of millennial-scale change revealed by

够和古气候资料（例如水和碳的稳定同位素）进行直接的定量比较，也能很方便地进行关于模式的假设检验。示踪物 $\Delta^{14}C_{atm}$ 需要更多的约束，有助于估计观测到的变化量（图 1a）与 THC 的变化 [27] 的联系到底有多大。基于简化海洋模式的结果说明，北大西洋 THC 的停滞将导致 $\Delta^{14}C_{atm}$ 的显著增加，但其模拟到的振幅只是观测结果的一半 [27]。为了弄清 $\Delta^{14}C_{atm}$ 的变化对模式参数的依赖性，进一步的模拟是需要的。

热带太平洋的变化代表现代气候变化的主要模态，其对全球都产生影响。我们虽然已对 ENSO（厄尔尼诺–南方涛动）的物理性质有了一个基本的了解，但是仍然不清楚 ENSO 在较冷和较暖的平均气候下会产生什么样的响应。例如，某些模式表明在未来较暖的气候状态下热带太平洋的变化与现代的情况相似，而另一些模式则认为会有更强、更频繁的暖事件发生。但是，大西洋到太平洋的水汽传输对 ENSO 位相的响应非常敏感 [61]，这是很清楚的。因此，暖事件频率持续增大的趋势 [62] 会导致从大西洋穿过巴拿马地峡的水汽净输出增加，因而可能影响 THC 的状态。

## 讨　论

格陵兰萨米特站的第一个冰芯研究结果的发表，详尽揭示了末次冰期期间气候的突然变化 [63]，推动了对发生在全球其他地方类似的气候不稳定性及其原因的广泛研究。已有的古气候记录清楚地说明，全球范围千年尺度气候变率及其不同的响应，与大西洋 THC 变化相联系的大气和海洋的变化有一致性（图 3 和 4）。此外，北大西洋的淡水输入流量增加的时间和相应的 THC 减少的时间的关系（图 2）支持了模拟结果，这些结果说明末次冰期期间 THC 对水文循环的小变化（量级为 0.1 Sv）是敏感的 [35,36,60]。

但是，在因果关系方面，我们对气候突变的了解仍然是不完全的。在某种程度上，古气候记录为评估模式是否能够正确模拟 THC 的行为方面提供了重要基础，尚需要增加信息以便弄清这一重要问题。如下几个方面尤其需要重视：(1) 增加太平洋和南大洋测站的数目；(2) 发展研究新方法以解决古气候记录同时性和确定位相关系；(3) 改进用代用资料来校正气候信号的方法；(4) 进一步分析距今 5.0 万年以来的 $\Delta^{14}C_{atm}$ 记录。

各种各样的复杂耦合模式得到的不同模拟结果，已经能够初步模拟古气候记录所揭示的千年尺度的时间演变和全球特征，并对这种变化的机制提供了启示。特别

the palaeoclimate record, and provide important insights into the mechanisms of change. In particular, palaeoclimate records and modelling experiments are providing a framework for the possible magnitude of future warming and the response of the interconnected Earth system to such a warming. Moreover, coupled GCM experiments incorporating geologic data (for example, continental runoff history)[36,64] provide new constraints on the mechanisms of abrupt climate change and will lead to model improvements that are essential for achieving the ability to simulate future climate. Nevertheless, modelling past abrupt climate change remains one of the greatest challenges for palaeoclimate modellers. Further progress will probably be realized as fully interactive and non-flux adjusted coupled Earth system models are developed that treat the full range of climate feedbacks[16].

Some modelling experiments find that during the next few centuries, the THC moves to an "off" state in response to increasing greenhouse gases[14,15,65]. A reduction of the meridional heat transport into the circum-Atlantic region would partially compensate the warming due to increasing greenhouse gases, although such a change could have serious climatic consequences for the climate in the circum-Atlantic region through modifying long-established regional air–sea temperature contrasts, seasonal variations in the direction and strength of wind patterns[66] and the location of convective areas[67]. The implication of such changes on regional climate remains largely unexplored. Reorganizations in the THC would also change the distribution of water masses and hence the density in the world ocean. A warmer and more stratified North Atlantic would also take up less anthropogenic $CO_2$ (ref. 68). On the other hand, other experiments suggest little or no reduction of the THC to the same greenhouse gas forcing[13]. This indicates the possible dominance of negative feedback mechanisms such as changes in the amplitude and frequency of ENSO[69], or modifications of atmospheric variability patterns in the Northern Hemisphere[70].

The fate of the THC in the coming century largely depends on the response of air–sea heat and freshwater fluxes to the increased load of greenhouse gases. Uncertainties in modelled responses are particularly large for the latter[13]. Moreover, the threshold for the occurrence of an abrupt change in a particular climate model depends on poorly constrained parameterizations of sub-grid-scale ocean mixing[57]. Because a complete THC shutdown is a threshold phenomenon, the assessment of the likelihood of such an event must involve ensemble model simulations[71], as well as continued efforts to simulate past abrupt climate changes that so remarkably affected the global climate system.

(**415**, 863-869; 2002)

**Peter U. Clark*, Nicklas G. Pisias†, Thomas F. Stocker‡ & Andrew J. Weawer§**

* Department of Geosciences, Oregon State University, Corvallis, Oregon 97331, USA

† College of Oceanic and Atmospheric Sciences, Oregon State University, Corvallis, Oregon 97331, USA

‡ Climate and Environmental Physics, University of Bern, Physics Institute, Sidlerstrasse 5, 3012 Bern, Switzerland

§ School of Earth and Ocean Sciences, University of Victoria, PO Box 3055, Victoria, British Columbia V8W 3P6, Canada

是，古气候记录和模拟实验提供了未来增温的可能程度，以及相互联系的地球系统对这种增暖的响应。此外，包含地质资料（如大陆径流的历史演变）的耦合 GCM 实验 [36,64] 给出了关于气候突变机制的新约束，并将使模式在模拟未来气候的能力方面有本质的改进。不过，模拟过去气候突变对古气候模拟工作者来说仍然是一个很大的挑战。只有考虑完全相互作用和非通量调整的耦合地球系统模式以处理气候反馈的大多数方面，进一步的进展才可能实现 [16]。

一些模拟实验得出，在未来的几个世纪，THC 将转向“关闭”的状态以响应温室气体的增加 [14,15,65]。沿经线向环大西洋地区热量输送的减少将部分抵消由于温室气体增加引起的增温，然而这种变化将通过下列因素对环大西洋地区的气候造成严重的后果，这些因素是：改变长时间建立的区域性大气–海洋温度差，及风系的方向、强度 [66] 和对流区位置 [67] 的季节变化。这些变化对区域气候变化的影响意义仍然有待研究。THC 的重构将改变水团的分布，因而引起世界海水密度的变化。一个更暖和分层更多的北大西洋会吸收更少的人类活动产生的 $CO_2$(参考文献 68)。但是，也有一些实验指出同样强度的温室气体强迫对 THC 的减少作用不大甚至没有作用 [13]。这说明，诸如 ENSO 振幅和频率的变化 [69]，或者北半球大气变率分布的变化 [70]，这些负反馈机制可能是主要的 [70]。

下一世纪 THC 的命运很大程度上取决于大气–海洋中的热量和淡水通量对温室气体浓度增加的响应。模式模拟给出的淡水通量响应的不确定性特别大 [13]。而且，在特定的气候模式中发生气候突变的阈值受不够确切的次网格海洋混合参数化的影响 [57]。由于 THC 的完全消失是一个临界点的现象，对这一事件的评估需要组合模式的模拟 [71]，并需要持续努力以对全球气候系统有显著影响的过去气候突变进行模拟。

（周家斌 翻译；郑旭峰 审稿）

References:

1. Ganachaud, A. & Wunsch, C. Improved estimates of global ocean circulation, heat transport and mixing from hydrographic data. *Nature* **408**, 453-457 (2000).

2. Weaver, A. J., Bitz, C. M., Fanning, A. F. & Holland, M. M. Thermohaline circulation: High latitude phenomena and the difference between the Pacific and Atlantic. *Annu. Rev. Earth Planet. Sci.* **27**, 231-285 (1999).

3. Killworth, P. D. Deep convection in the world ocean. *Rev. Geophys. Space Phys.* **21**, 1-26 (1983).

4. Grumbine, R. W. A model of the formation of high-salinity shelf water on polar continental shelves. *J. Geophys. Res.* **96**, 22049-22062 (1991).

5. Stommel, H. Thermohaline convection with two stable regimes of flow. *Tellus* **13**, 224-2S0 (1961).

6. Bryan, F. High-latitude salinity effects and interhemispheric thermohaline circulations. *Nature* **323**, 301-304 (1986).

7. Manabe, S. & Stouffer, R. J. Two stable equilibria of a coupled ocean-atmosphere model. *J. Clim.* **1**, 841-866 (1988).

8. Weaver, A. J. in *Natural Climate Variability on Decade-to-Century Time Scales* (eds Martinson, D. G. *et al.*) 365-381 (National Research Council, National Academy Press, Washington DC, 1995).

9. Alley, R. B. & Clark, P. U. The glaciation of the northern hemisphere: a global perspective. *Annu. Rev. Earth Planet. Sci.* **27**, 149-182 (1999).

10. Stocker, T. F. Past and future reorganization in the climate system. *Quat. Sci. Rev.* **19**, 301-319 (2000).

11. Broecker, W. S. Thermohaline circulation, the Achilles heel of our climate system: Will man-made $CO_2$ upset the current balance? *Science* **278**, 1582-1588 (1997).

12. Alley, R. B. *et al. Abrupt Climate Change: Inevitable Surprises* (National Research Council, National Academy Press, Washington DC, in the press).

13. Cubasch, U. *et al.* in *Climate Change 2001—The Scientific Basis: Contribution of Working Group I to the Third Assessment Report of the Intergovernmental Panel on Climate Change* (eds Houghton, J. T. *et al.*) 525-582 (Cambridge Univ. Press, Cambridge, 2001).

14. Manabe, S. & Stouffer, R. J. Century-scale effects of increased atmospheric $CO_2$ on the ocean–atmosphere system. *Nature* **364**, 215-218 (1993).

15. Stocker, T. F. & Schmittner, A. Influence of $CO_2$ emission rates on the stability of the thermohaline circulation. *Nature* **388**, 862-865 (1997).

16. Schmittner, A., Yoshimori, M. & Weaver, A. J. Instability of glacial climate in an earth system model. *Science* (in the press).

17. Dickson, R. R. & Brown, J. The production of North Atlantic deep water: Sources, sinks and pathways. *J. Geophys. Res.* **99**, 12319-12341 (1994).

18. McCartney, M. S. Recirculating components to the deep boundary current of the northern North Atlantic. *Prog. Oceanogr.* **29**, 283-383 (1992).

19. Labeyrie, L. *et al.* Changes in vertical structure of the North Atlantic Ocean between glacial and modern times. *Quat. Sci. Rev.* **11**, 401-413 (1992).

20. Vidal, L., Labeyrie, L. & van Weering, T. C. E. Benthic $\delta^{18}O$ records in the North Atlantic over the last glacial period (60–10 kyr): Evidence for brine formation. *Paleoceanography* **13**, 245-251 (1998).

21. Sarnthein, M. *et al.* Changes in east Atlantic deepwater circulation over the last 30,000 years: Eight time slice reconstructions. *Paleoceanography* **9**, 209-262 (1994).

22. Yu, E. F., Francois, R. & Bacon, M. P. Similar rates of modern and last-glacial ocean thermohaline circulation inferred from radiochemical data. *Nature* **379**, 689-694 (1996).

23. Rutberg, R. L., Hemming, S. R. & Goldstein, S. L. Reduced North Atlantic Deep Water flux to the glacial Southern Ocean inferred from neodymium isotope ratios. *Nature* **405**, 935-938 (2000).

24. Frank, M. Comparison of cosmogenic radionuclide production and geomagnetic field intensity over the last 200,000 years. *Phil. Trans. R. Soc. Lond. A* **358**, 1089-1102 (2000).

25. Finkel, R. C. & Nishiizumi, K. Beryllium 10 concentrations in the Greenland Ice Sheet Project 2 ice core from 3–40 ka. *J. Geophys. Res.* **102**, 26699-26706 (1997).

26. Muscheler, R., Beer, J., Wagner, G. & Finkel, R. C. Changes in deep-water formation during the Younger Dryas event inferred from $^{10}Be$ and $^{14}C$ records. *Nature* **408**, 562-520 (2000).

27. Marchal, O., Stocker, T. F. & Muscheler, R. Atmospheric radiocarbon during the Younger Dryas: production, ventilation, or both? *Earth Planet Sci. Lett.* **185**, 383-395 (2001).

28. Broecker, W. S., Mix, A., Andree, M. & Oeschger, H. Radiocarbon measurements on coexisting benthic and planktic foraminifera shells: potential for reconstructing ocean ventilation times over the past 20000 years. *Nucl. Instrum. Methods Phys. Res.* **B5**, 331-339 (1984).

29. Sikes, E. L., Samson, C. R., Guilderson, T. P. & Howard, W. R. Old radiocarbon ages in the southwest Pacific Ocean during the last glacial period and deglaciation. *Nature* **405**, 555-559 (2000).

30. Cuffey, K. M. *et al.* Large Arctic temperature change at the Wisconsin-Holocene glacial transition. *Science* **270**, 455-458 (1995).

31. Bard, E., Rostek, F., Turon, J.-L. & Gendreau, S. Hydrological impact of Heinrich events in the subtropical northeast Atlantic. *Science* **289**, 1321-1324 (1999).

32. Severinghaus, J. P. & Brook, E. J. Abrupt climate change at the end of the last glacial period inferred from trapped air in polar ice. *Science* **286**, 930-934 (1999).

33. Hughen, K. A., Southon, J. R., Lehman, S. J. & Overpeck, J. T. Synchronous radiocarbon and climate shifts during the last deglaciation. *Science* **290**, 1951-1954 (2000).

34. Stocker, T. F. & Wright, D. G. Rapid transitions of the ocean's deep circulation induced by changes in surface water fluxes. *Nature* **351**, 729-732 (1991).

35. Manabe, S. & Stouffer, R. J. Coupled ocean-atmosphere model response to freshwater input: comparison to Younger Dryas event. *Paleoceanography* **12**, 321-336 (1997).

36. Fanning, A. F. & Weaver, A. J. Temporal-geographical meltwater influences on the North Atlantic Conveyor: Implications for the Younger Dryas. *Paleoceanography* **12**, 307-320 (1997).

37. Bond, G. C. *et al.* in *Mechanisms of Global Climate Change at Millennial Timescales* (eds Clark, P. U., Webb, R. S. & Keigwin, L. D.) 35-58 (Geophysical Monograph 112, American Geophysical Union, Washington DC, 1999).

38. Clark, P. U. *et al.* Freshwater forcing of abrupt climate change during the last glaciation. *Science* **293**, 283-287 (2001).

39. Fairbanks, R. G. A 17,000-year glacio-eustatic sea level record: influence of glacial melting rates on the Younger Dryas event and deep ocean circulation. *Nature* **342**, 637-642 (1989).

40. Clark, P. U., Mitrovica, J. X., Milne, G. A. & Tamisiea, M. Sea-level fingerprinting as a direct test for the source of global meltwater pulse 1A. *Science* (submitted).

41. Severinghaus, J. P., Sowers, T., Brook, E. J., Alley, R. B. & Bender, M. L. Timing of abrupt climate change at the end of the Younger Dryas interval from thermally fractionated gases in polar ice. *Nature* **391**, 141-146 (1998).

42. McManus, J. F., Oppo, D. W. & Cullen, J. L. A 0.5-million-year record of millennial-scale climate variability in the North Atlantic. *Science* **283**, 971-975 (1999).

43. Alley, R. B., Anandakrishnan, S. & Jung, P. Stochastic resonance in the North Atlantic. *Paleoceanography* **16**, 190-198 (2001).

44. Hostetler, S. W., Clark, P. U., Bartlein, P. J., Mix, A. C. & Pisias, N. Atmospheric transmission of North Atlantic Heinrich events. *J. Geophys. Res.* **104**, 3947-3952 (1999).

45. Hughen, K. A., Overpeck, J. T., Peterson, L. C. & Trumbore, S. Rapid climate changes in the tropical Atlantic region during the last deglaciation. *Nature* **380**, 51-54 (1996).

46. Peterson, L. C., Haug, G. H., Hughen, K. A. & Rohl, U. Rapid changes in the hydrologic cycle of the tropical Atlantic during the last glacial. *Science* **290**, 1947-1951 (2000).

47. Schulz, H., von Rad, U. & Erlenkeuser, H. Correlations between Arabian Sea and Greenland climate oscillations of the past 110,000 years. *Nature* **393**, 54-57 (1998).

48. Kienast, M., Steinke, S., Stattegger, K. & Calvert, S. E. Synchronous tropical South China Sea SST change and Greenland warming during deglaciation. *Science* **291**, 2132-2134 (2001).

49. Behl, R. J. & Kennett, J. P. Brief interstadial events in the Santa Barbara basin, NE Pacific, during the past 60 kyr. *Nature* **379**, 243-246 (1996).

50. Blunier, T. & Brook, E. J. Timing of millennial-scale climate change in Antarctica and Greenland during the last glacial period. *Science* **291**, 109-112 (2001).

51. Broecker, W. S. Paleocean circulation during the last deglaciation: A bipolar seesaw? Paleoceanography **13**, 119-121 (1998).

52. Crowley, T. J. North Atlantic deep water cools the southern hemisphere. *Paleoceanography* **7**, 489-497 (1992).

53. Schiller, A., Mikolajewicz, U. & Voss, R. The stability of the thermohaline circulation in a coupled ocean-atmosphere general circulation model. *Clim. Dyn.* **13**, 325-347 (1997).

54. Stocker, T. F. The seesaw effect. *Science* **282**, 61-62 (1998).

55. Oeschger, H. *et al.* in *Climate Processes and Climate Sensitivity* (eds Hansen, J. E. & Takahashi, T.) 299–306 (Geophysical Monograph 29, American Geophysical Union, Washington DC, 1984).

56. Mikolajewicz, U. & Maier-Reimer, E. Mixed boundary conditions in ocean general circulation models and their influence on the stability of the model's conveyor belt. *J. Geophys. Res.* **99**, 22633-22644 (1994).

57. Schmittner, A. & Weaver, A. J. Dependence of multiple climate states on ocean mixing parameters. *Geophys. Res. Lett.* **28**, 1027-1030 (2001).

58. Tziperman, E. Inherently unstable climate behaviour due to weak thermohaline ocean circulation. *Nature* **386**, 592-595 (1997).

59. Mikolajewicz, U., Maier-Reimer, E., Crowley, T. J. & Kim, K.-Y. Effect of Drake and Panamanian gateways on the circulation of an ocean model. *Paleoceanography* **8**, 409-426 (1993).

60. Ganopolski, A. & Rahmstorf, S. Rapid changes of glacial climate simulated in a coupled climate model. *Nature* **409**, 153-158 (2001).

61. Schmittner, A., Appenzeller, C. & Stocker, T. F. Enhanced Atlantic freshwater export during El Niño. *Geophys. Res. Lett.* **27**, 1163-1166 (2000).

62. Timmermann, A. *et al.* Increased El Niño frequency in a climate model forced by future greenhouse warming. *Nature* **398**, 694-696 (1999).

63. Johnsen, S. J. *et al.* Irregular glacial interstadials recorded in a new Greenland ice core. *Nature* **359**, 311-313 (1992).

64. Rind, D. *et al.* Effects of glacial meltwater in the GISS coupled atmosphere-ocean model: Part I: North Atlantic Deep Water response. *J. Geophys. Res.* **106**, 27335-27354 (2001).

65. Rahmstorf, S. & Ganopolski, A. Long-term global warming scenarios computed with an efficient coupled climate model. *Clim. Change* **43**, 353-367 (1999).

66. Mikolajewicz, U. & Voss, R. The role of the individual air-sea flux components in $CO_2$-induced changes of the ocean's circulation and climate. *Clim. Dyn.* **16**, 627-642 (2000).

67. Wood, R. A., Keen, A. B., Mitchell, J. F. B. & Gregory, J. M. Changing spatial structure of the thermohaline circulation in response to atmospheric $CO_2$ forcing in a climate model. *Nature* **399**, 572-575 (1999).

68. Joos, F., Plattner, G.-K., Stocker, T. F., Marchal, O. & Schmittner, A. Global warming and marine carbon cycle feedbacks on future atmospheric $CO_2$. *Science* **284**, 464-467 (1999).

69. Latif, M., Roeckner, E., Mikolajewicz, U. & Voss, R. Tropical stabilization of the thermohaline circulation in a greenhouse warming simulation. *J. Clim.* **13**, 1809-1813 (2000).

70. Delworth, T. L. & Dixon, K. W. Implications of the recent trend in the Arctic/North Atlantic Oscillation for the North Atlantic thermohaline circulation. *J. Clim.* **13**, 3721-3727 (2000).

71. Knutti, R. & Stocker, T. F. Limited predictability of the future thermohaline circulation close to an instability threshold. *J. Clim.* **15**, 179-186 (2002).

72. Stuiver, M. *et al.* INTCAL98 radiocarbon age calibration, 24,000 cal BP. *Radiocarbon* **40**, 1041-1083 (1998).

73. Kitigawa, H. & van der Plicht, J. Atmospheric radiocarbon calibration beyond 11,900 cal BP from Lake Suigetsu laminated sediments. *Radiocarbon* **42**, 369-380 (2000).

74. Bard, E., Arnold, M., Hamelin, B., Tisnerat-Laborde, N. & Cabioch, G. Radiocarbon calibration by means of mass spectrometric $^{230}Th/^{234}U$ and $^{14}C$ ages of

corals: an updated database including samples from Barbados, Mururoa and Tahiti. *Radiocarbon* **40**, 1085-1092 (1998).

75. Cuffey, K. M. & Clow, G. D. Temperature, accumulation, and ice sheet elevation in central Greenland through the last deglacial transition. *J. Geophys. Res.* **102**, 26383-26396 (1992).
76. Grootes, P. M., Stuiver, M., White, J. W. C., Johnsen, S. J. & Jouzel, J. Comparison of oxygen isotope records from the GISP2 and GRIP Greenland ice cores. *Nature* **366**, 552-554 (1993).
77. Meese, D. A. *et al.* The Greenland Ice Sheet Project 2 depth-age scale: Methods and results. *J. Geophys. Res.* **102**, 26411-26423 (1997).
78. Rühlemann, C., Mulitza, S., Muller, P. J., Wefer, G. & Zahn, R. Warming of the tropical Atlantic Ocean and slowdown of thermohaline circulation during the last deglaciation. *Nature* **402**, 511-514 (1999).
79. Sachs, J. P., Anderson, R. F. & Lehman, S. J. Glacial surface temperatures of the southeast Atlantic Ocean. *Science* **293**, 2077-2079 (2001).
80. Steig, E. J. *et al.* Synchronous climate changes in Antarctica and the North Atlantic. *Science* **282**, 92-95 (1998).
81. Mulvaney, R. *et al.* The transition from the last glacial period in inland and near coastal Antarctica. *Geophys. Res. Lett.* **27**, 2673-2676 (2000).
82. Pisias, N. G., Mix, A. C. & Heusser, L. Millennial scale climate variability of the northeast Pacific Ocean and northwest North America based on radiolaria and pollen. *Quat. Sci. Rew.* **20**, 1561-1576 (2001).
83. Mix, A. C. *et al.* in *Mechanisms of Global Climate Change at Millennial Timescales* (eds Clark, P. U., Webb, R. S. & Keigwin, L. D.) 122-148 (Geophysical Monograph 112, American Geophysical Union, Washington DC, 1999).
84. Thompson, L. G. *et al.* Late glacial stage and Holocene tropical ice core records from Huascaran, Peru. *Science* **269**, 46-50 (1995).
85. Thompson, L. G. *et al.* A 25,000-year tropical climate history from Bolivian ice cores. *Science* **282**, 1858-1864 (1998).
86. Bard, E., Rostek, F. & Sonzogni, C. Interhemispheric synchrony of the last deglaciation inferred from alkenone paleothermometry. *Nature* **385**, 707-710 (1997).
87. Little, M. G. *et al.* Trade wind forcing of upwelling seasonality, and Heinrich events as a response to sub-Milankovitch climate variability. *Paleoceanography* **12**, 568-576 (1997).
88. Charles, C. D., Lynch-Stieglitz, J., Ninnemann, U. S. & Fairbanks, R. G. Climate connections between the hemispheres revealed by deep sea sediment core/ice core correlations. *Earth Planet. Sci. Lett.* **142**, 19-22 (1996).
89. Petit, J. R. *et al.* Climate and atmospheric history of the past 420,000 years from the Vostok ice core, Antarctica. *Nature* **399**, 429-436 (1999).
90. Jouzel, J. *et al.* A new 22 ky high resolution East Antarctic climate record. *Geophys. Res. Lett.* **28**, 3199-3202 (2001).

**Acknowledgements.** We thank G. Bond, J. Jouzel, A. Mix, R. Muscheler, J. Sachs and J. van der Plicht for providing data, the NOAA-NGDC Paleoclimate Program for their data repository, and S. Hostetler, A. Mix, A. Schmittner and R. Stouffer for comments. This work was supported by grants from the Earth System History Program of the US NSF (P.U.C. and N.G.P.), the Swiss NSF (T.F.S.) and the Canadian NSERC (A.J.W.).

Correspondence and requests for materials should be addressed to P.U.C. (e-mail: clarkp@ucs.orst.edu).

# Detecting Recent Positive Selection in the Human Genome from Haplotype Structure

P. C. Sabeti *et al.*

## Editor's Note

**In this paper, geneticist Eric Lander and colleagues present a method for detecting the genetic imprint of natural selection. The technique focuses on haplotypes, sets of closely linked genetic markers present on one chromosome that tend to be inherited together. Knowing that variants of two key genes confer malarial resistance in some individuals, the team looked for evidence that a particular haplotype associated with the resistance variant arose faster in certain populations than would have happened by chance. This "signature" of natural selection was found in both the *G6PD* and CD40 ligand genes, and the team suggest that the test could be used to scan the entire human genome for evidence of recent positive selection.**

---

The ability to detect recent natural selection in the human population would have profound implications for the study of human history and for medicine. Here, we introduce a framework for detecting the genetic imprint of recent positive selection by analysing long-range haplotypes in human populations. We first identify haplotypes at a locus of interest (core haplotypes). We then assess the age of each core haplotype by the decay of its association to alleles at various distances from the locus, as measured by extended haplotype homozygosity (EHH). Core haplotypes that have unusually high EHH and a high population frequency indicate the presence of a mutation that rose to prominence in the human gene pool faster than expected under neutral evolution. We applied this approach to investigate selection at two genes carrying common variants implicated in resistance to malaria: *G6PD*[1] and CD40 ligand[2]. At both loci, the core haplotypes carrying the proposed protective mutation stand out and show significant evidence of selection. More generally, the method could be used to scan the entire genome for evidence of recent positive selection.

---

THE recent history of the human population is characterized by great environmental change and emergent selective agents[3]. The domestication of plants and animals at the start of the Neolithic, roughly 10,000 years ago, yielded an increase in human population density. Humans were confronted with the spread of new infectious diseases, new food sources and new cultural environments. The last 10,000 years have thus been some of the most interesting times in human biological history, and may be when many important genetic adaptations and disease resistances arose.

We sought to design a powerful approach for detecting recent selection. Our method relies

# 从单倍型结构检测人类基因组近期的正选择

萨贝提等

## 编者按

*在这篇文章中，遗传学家埃里克·兰德和他的同事们提出了一种检测自然选择的遗传印记的方法。这项技术关注的是单倍型，即存在于一条染色体上的一系列紧密相连的遗传标记，这些标记往往是一起遗传的。研究小组了解到两个关键基因的变异型在某些个体中产生了疟疾抗药性，因此他们寻找证据来证明在某些人群中，与抗药性变异相关的一种特定单倍型出现的速度比偶然出现的速度更快。在G6PD和CD40配体基因中都发现了这种自然选择的“特征”，研究小组认为，该测试可以用来扫描整个人类基因组，以寻找近期正选择的证据。*

---


检测人群中最近发生的自然选择对于研究人类历史和医学具有深远的意义。这里我们介绍一个通过分析人群中长范围的单倍型来检测最近正选择的基因印记的框架。首先我们在感兴趣的基因座识别单倍型（核心单倍型）。然后我们通过其与该基因座不同距离的等位基因联系的衰减评估每个核心单倍型的年龄，即通过扩展单倍型纯合性（EHH）进行测定。具有异常高的EHH和高人群频率的核心单倍型表明，在自然中性进化下，一种突变在人类基因库中上升的速度快于预期。我们应用这种方法研究了两个携带涉及抗疟疾能力的寻常变异型的基因上的选择，*G6PD*[1] 和CD40配体 [2]。在这两个基因座上，携带假设的保护性突变的核心单倍型凸现出来并显示出重要的选择证据。一般来说，该方法可以用于扫描整个基因组，以寻找近期正选择的证据。


---

人类近代史的特点是巨大的环境变化和涌现的选择性媒介 [3]。大约10,000年前新石器时代开始时，人类培育植物和驯化动物导致人群密度的增长。人类面临着新的传染病、新的食物来源和新的文化环境的传播。因此，过去的10,000年成为了人类生物学史上最有意义的一段时间，也可能是许多重要的基因适应和疾病抵抗力出现的时间。

我们试图设计一种强大的方法来检测最近的选择。我们的方法主要基于等位基

on the relationship between an allele's frequency and the extent of linkage disequilibrium (LD) surrounding it. (LD often refers to association between two alleles. Here, we use it to measure the association between a single allele at one locus with multiple loci at various distances.) Under neutral evolution, new variants require a long time to reach high frequency in the population, and LD around the variants will decay substantially during this period owing to recombination[4,5]. As a result, common alleles will typically be old and will have only short-range LD. Rare alleles may be either young or old and thus may have long- or short-range LD. The key characteristic of positive selection, however, is that it causes an unusually rapid rise in allele frequency, occurring over a short enough time that recombination does not substantially break down the haplotype on which the selected mutation occurs. A signature of positive natural selection is thus an allele having unusually long-range LD given its population frequency. The decay of LD, and therefore the relative scale of "short"- and "long"-range LD, is dependent on local recombination rates. A general test for selection on the basis of these principles must therefore control for local variation in recombination rates.

We developed an experimental design to detect positive selection at a locus using the breakdown of LD as a clock for estimating the ages of alleles. We began by genotyping a collection of single nucleotide polymorphisms (SNPs) in a small "core region" to identify the "core haplotypes". We selected SNPs of sufficient density, so that recombination between them would be extremely rare and the core haplotypes could be explained in terms of a single gene genealogy (Supplementary Fig. 1). Zones of very low historical recombination were identified by looking for clusters of SNPs where Hudson's $R_M$ was 0 and $|D'|$ was one[6,7] (see Supplementary Fig. 1).

We then added increasingly distant SNPs to study the decay of LD from each core haplotype. To visualize this process, we generated haplotype bifurcation diagrams that branch to reflect the creation of new, extended haplotypes by historical recombination proximal and distal to the core region. We measured LD at a distance $x$ from the core region by calculating the extended haplotype homozygosity (EHH). EHH is defined as the probability that two randomly chosen chromosomes carrying the core haplotype of interest are identical by descent (as assayed by homozygosity at all SNPs[8]) for the entire interval from the core region to the point $x$. EHH thus detects the transmission of an extended haplotype without recombination. Our test for positive selection involves finding a core haplotype with a combination of high frequency and high EHH, as compared with other core haplotypes at the locus. An attractive aspect of this approach is that the various core haplotypes at a locus serve as internal controls for one another, adjusting for any unevenness in the local recombination rate.

We applied our approach to two genes that have been implicated in resistance to the malaria parasite *Plasmodium falciparum*. Glucose-6-phosphate dehydrogenase (*G6PD*) is a classical example of a gene where variants can confer malaria resistance[9]. Evidence over the past 40 years has shown that the common variant *G6PD*-202A confers partial protection against malaria, with a case-control study estimating a reduction in disease risk

因的频率与其附近连锁不平衡（LD）程度之间的关系。（LD 通常指两个等位基因间的联系。这里我们用它来测量一个基因座的单个等位基因与不同距离处的多个基因座等位基因之间的联系。）在中性进化下，新的变异型需要一段很长的时间才能达到人群中较高的频率，而且在此期间，变异型附近的 LD 会由于重组大幅度地衰减[4,5]。结果，常见等位基因通常是老的并只有短范围的 LD。很少的等位基因可能是年轻的也可能是老的，因此可能具有长或者短范围的 LD。但是正选择的关键特征是它能引起等位基因频率在足够短的时间内异常快速的增长，以至于重组并未能充分打破这些选择突变发生处的单倍型。因此一个正自然选择的标志就是一个等位基因在其人类频率下具有异常长范围的 LD。LD 的衰减以及因此相对的短和长范围 LD 规模，取决于局部的重组率。根据这个原理设计的选择试验必须控制重组率的局部变异。

我们开发了一种实验设计，利用 LD 的分解作为时钟来估算等位基因的年龄，以检测一个基因座的正选择。我们首先在一个小的“核心区域”对一组单核苷酸多态性（SNP）进行基因分型，以识别“核心单倍型”。我们选取具有足够密度的 SNP，这样它们之间的重组会非常少，而且核心单倍型可以用单个基因的谱系来表示（见补充图 1）。通过寻找那些赫德森 $R_M = 0$ 且 $|D'| = 1$ 的 SNP 簇来找出历史重组率非常低的区域[6,7]（见补充图 1）。

然后我们加入逐渐增加距离的 SNP 研究每个核心单倍型 LD 的衰减。为了可视化这个过程，我们设计了单倍型分支图，用分支的方式反映核心区域近端和远端通过历史重组产生的新的扩展单倍型。我们通过计算扩展单倍型纯合性（EHH）来测量距离核心区域 $x$ 的 LD。EHH 可以被定义为从核心区域到距离 $x$ 的整个区间内，随机选择的两条携带感兴趣的核心单倍型的染色体在下降过程（通过所有 SNP 的纯合性来测定[8]）中的相同概率。因此，EHH 能够检测不含有重组的扩展单倍型的传播。我们对正选择的检测包含了找到一个与该基因座的其他核心单倍型相比具有高频率和高 EHH 组合的核心单倍型。这种方法令人满意的方面在于一个基因座的多个核心单倍型可以作为相互之间的内部对照，对任何局部重组率上的不均衡进行调节。

我们将方法用于与疟疾寄生虫恶性疟原虫的抗药性有关的两个基因上。葡萄糖–6–磷酸脱氢酶（*G6PD*）是一个基因变异型能够产生疟疾抗性的经典例子[9]。过去 40 年的证据表明常见变异型 *G6PD*-202A 对疟疾具有部分抗性，其中一项病例对照研究估计可将疾病风险降低约 50%（文献 1）。CD40 配体基因（*TNFSF5*）编码一种对

of about 50% (ref. 1). The CD40 ligand gene (*TNFSF5*) encodes a protein with a critical role in immune response to infectious agents. One case-control study suggested that a common variant in the promoter region, *TNFSF5*-726C, is associated with a similar degree of protection against malaria[2].

We first studied *G6PD* (Fig. 1). We defined a core region of 15 kilobases (kb) at *G6PD* and genotyped 11 SNPs in 3 African and 2 non-African populations. The SNPs defined 9 core haplotypes (Table 1a) (denoted *G6PD*-CH1 to 9, for core haplotypes 1 to 9). The *G6PD*-202A allele, which has been associated with protection from malaria, was carried on only one core haplotype, *G6PD*-CH8. Notably, *G6PD*-CH8 is common in Africa (18%), where malaria is endemic, but is absent outside of Africa. For carrying out our test for selection, we focused on the three African populations, which did not differ significantly with respect to core haplotype frequencies (by Fisher's exact test[10]) and hence were pooled for the main analysis. (Analyses were also performed separately for each population, yielding qualitatively similar results; see below.)

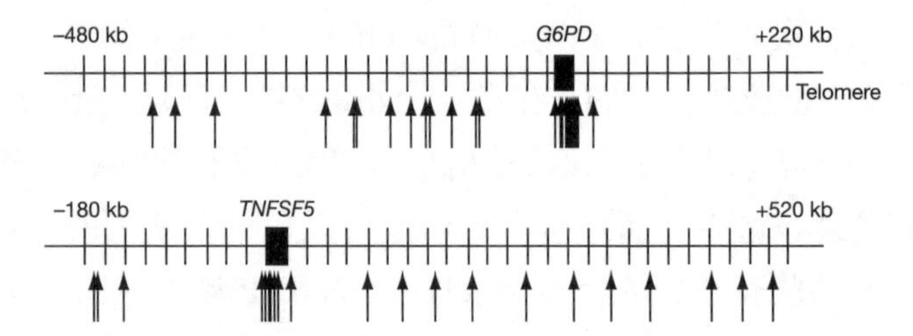


Fig. 1. Experimental design of core and long-range SNPs for *G6PD* and *TNFSF5*. The core region is highlighted by a cluster of densely spaced SNPs (arrows) at the gene. Additional, widely separated flanking SNPs, used to examine the decay of LD from each core haplotype, are also shown. Markers distal to *G6PD* were within repetitive subtelomeric sequence and could not be genotyped.

Table 1. Core haplotype frequencies in six populations

| (**a**) *G6PD* Core haplotype | Core SNP alleles (kb) | | | | | | | | | | | Core haplotype frequencies in six populations | | | | | | |
|---|---|---|---|---|---|---|---|---|---|---|---|---|---|---|---|---|---|---|
| | −10 | −2 | −2 | −1 | 0 | 1 | 1 | 3 | 3 | 4 | 4 | Total | Beni | Yoruba | Shona | African American | European American | Asian |
| 1 | C | G | C | G | G | A | C | C | G | C | C | 0.13 (54) | 0.15 (9) | 0.21 (18) | 0.13 (11) | 0.13 (12) | 0.03 (2) | 0.07 (2) |
| 2 | – | – | – | – | – | – | – | – | – | – | T | 0.16 (67) | 0 (0) | 0 (0) | 0 (0) | 0.07 (6) | 0.57 (37) | 0.80 (24) |
| 3 | – | – | – | – | – | – | – | – | – | T | – | 0.25 (102) | 0.23 (14) | 0.24 (21) | 0.45 (37) | 0.19 (17) | 0.17 (11) | 0.07 (2) |
| 4 | – | – | – | – | – | – | – | T | – | – | – | 0.01 (5) | 0.02 (1) | 0 (0) | 0.04 (3) | 0.01 (1) | 0 (0) | 0 (0) |
| 5 | – | – | – | – | – | – | – | T | A | – | – | 0.10 (41) | 0.22 (13) | 0.11 (10) | 0.06 (5) | 0.14 (13) | 0 (0) | 0 (0) |
| 6 | – | – | – | – | – | G | – | – | – | – | – | 0.12 (48) | 0.17 (10) | 0.10 (9) | 0.11 (9) | 0.22 (20) | 0 (0) | 0 (0) |
| 7 | – | – | T | A | – | G | G | – | – | – | – | 0.04 (17) | 0.05 (3) | 0.07 (6) | 0.06 (5) | 0.03 (3) | 0 (0) | 0 (0) |
| 8 | – | A* | T | A | A | G | G | – | – | – | – | 0.13 (53) | 0.17 (10) | 0.23 (20) | 0.13 (11) | 0.13 (12) | 0 (0) | 0 (0) |
| 9 | T | – | – | – | – | – | – | – | – | – | T | 0.07 (28) | 0 (0) | 0.03 (3) | 0.02 (2) | 0.07 (6) | 0.23 (15) | 0.07 (2) |
| | | | | | | | | | | | *N* | 415 | 60 | 87 | 83 | 90 | 65 | 30 |

感染性病原体的免疫应答起关键作用的蛋白。一个病例对照研究表明启动子区域的一个常见变异型 *TNFSF5*-726C 对疟疾具有类似程度的抗性[2]。

我们首先研究了 *G6PD*(图 1)。在 *G6PD* 上定义了一个 15 千碱基(kb)的核心区域，并在 3 个非洲人群和 2 个非非洲人群中对 11 个 SNP 进行了基因分型。这些 SNP 定义了 9 个核心单倍型(表 1a)(分别表示为 *G6PD*-CH1 到 *G6PD*-CH9，指代核心单倍型 1 到 9)。只有一个核心单倍型(*G6PD*-CH8)携带了 *G6PD*-202A 等位基因，该等位基因与疟疾抗病性相关。显然，*G6PD*-CH8 在疟疾流行的非洲很常见(18%)，而在非洲以外地区不存在。为了实施选择测试，我们将重点放在三个非洲人群上，他们在核心单倍型频率上没有显著差异(通过 Fisher 精确测试[10])，因此被集中在一起进行主要分析。(对每个人群也分别进行了分析，得到了定性上相似的结果；见下文)。

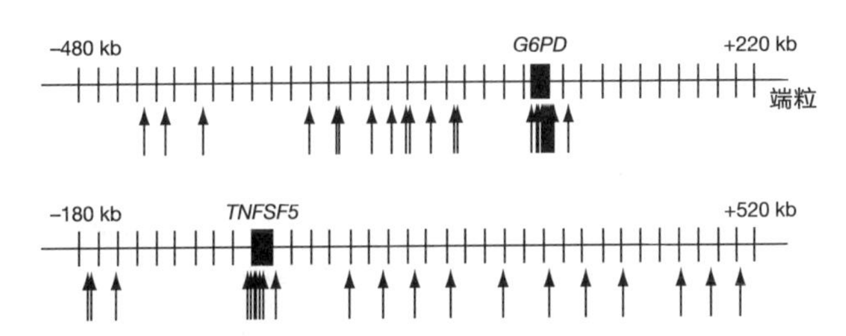


图 1. *G6PD* 和 *TNFSF5* 的核心 SNP 和长范围 SNP 的实验设计。基因的核心区域被一群密集排列的 SNP(箭头)突出显示。此外，图中还显示了高度分离的侧翼 SNP，用于检验每个核心单倍型的 LD 衰减。*G6PD* 远端的标记位于重复的亚端粒序列中，不能进行基因分型。

表 1. 六个群体中的核心单倍型频率

| (a) | 核心 SNP 等位基因(kb) | | | | | | | | | | | 六个群体中的核心单倍型频率 | | | | | | |
|---|---|---|---|---|---|---|---|---|---|---|---|---|---|---|---|---|---|---|
| *G6PD* 核心单倍型 | −10 | −2 | −2 | −1 | 0 | 1 | 1 | 3 | 3 | 4 | 4 | 总计 | 贝尼人 | 约鲁巴人 | 绍纳人 | 非裔美国人 | 欧裔美国人 | 亚洲人 |
| 1 | C | G | C | G | G | A | C | C | G | C | C | 0.13(54) | 0.15(9) | 0.21(18) | 0.13(11) | 0.13(12) | 0.03(2) | 0.07(2) |
| 2 | – | – | – | – | – | – | – | – | – | – | T | 0.16(67) | 0(0) | 0(0) | 0(0) | 0.07(6) | 0.57(37) | 0.80(24) |
| 3 | – | – | – | – | – | – | – | – | – | T | – | 0.25(102) | 0.23(14) | 0.24(21) | 0.45(37) | 0.19(17) | 0.17(11) | 0.07(2) |
| 4 | – | – | – | – | – | – | – | T | – | – | – | 0.01(5) | 0.02(1) | 0(0) | 0.04(3) | 0.01(1) | 0(0) | 0(0) |
| 5 | – | – | – | – | – | – | – | T | A | – | – | 0.10(41) | 0.22(13) | 0.11(10) | 0.06(5) | 0.14(13) | 0(0) | 0(0) |
| 6 | – | – | – | – | – | G | – | – | – | – | – | 0.12(48) | 0.17(10) | 0.10(9) | 0.11(9) | 0.22(20) | 0(0) | 0(0) |
| 7 | – | – | T | A | – | G | G | – | – | – | – | 0.04(17) | 0.05(3) | 0.07(6) | 0.06(5) | 0.03(3) | 0(0) | 0(0) |
| 8 | – | A* | T | A | A | G | G | – | – | – | – | 0.13(53) | 0.17(10) | 0.23(20) | 0.13(11) | 0.13(12) | 0(0) | 0(0) |
| 9 | T | – | – | – | – | – | – | – | – | – | T | 0.07(28) | 0(0) | 0.03(3) | 0.02(2) | 0.07(6) | 0.23(15) | 0.07(2) |
| | | | | | | | | | | | *N* | 415 | 60 | 87 | 83 | 90 | 65 | 30 |

*Continued*

| (**b**) *TNFSF5* Core haplotype | Core SNP alleles (kb) | | | | | Core haplotype frequencies in six populations | | | | | | |
|---|---|---|---|---|---|---|---|---|---|---|---|---|
| | −6 | 0 | 1 | 3 | 4 | Total | Beni | Yoruba | Shona | African American | European American | Asian |
| 1 | T | T | T | T | G | 0.03 (12) | 0.06 (4) | 0.03 (3) | 0.02 (2) | 0.04 (3) | 0 (0) | 0 (0) |
| 2 | C | – | – | – | – | 0.50 (200) | 0.32 (20) | 0.38 (33) | 0.46 (38) | 0.40 (32) | 0.78 (49) | 1.00 (28) |
| 3 | C | – | – | C | – | 0.08 (32) | 0.10 (6) | 0.12 (10) | 0.06 (5) | 0.14 (11) | 0 (0) | 0 (0) |
| 4 | – | C* | – | – | – | 0.25 (100) | 0.38 (24) | 0.38 (33) | 0.26 (21) | 0.27 (22) | 0 (0) | 0 (0) |
| 5 | – | – | C | – | – | 0.12 (47) | 0.14 (9) | 0.07 (6) | 0.18 (15) | 0.14 (11) | 0.10 (6) | 0 (0) |
| 6 | – | – | C | – | A | 0.02 (10) | 0 (0) | 0 (0) | 0.01 (1) | 0.01 (1) | 0.13 (8) | 0 (0) |
| 7 | – | C* | C | – | A | 0 (1) | 0 (0) | 0.01 (1) | 0 (0) | 0 (0) | 0 (0) | 0 (0) |
| 8 | C | – | C | – | – | 0 (1) | 0 (0) | 0 (0) | 0 (0) | 0.01 (1) | 0 (0) | 0 (0) |
| | | | | | *N* | 403 | 63 | 86 | 82 | 81 | 63 | 28 |

Observed core haplotypes at *G6PD* and *TNFSF5* in six populations of African, European and Asian descent. Relative distances of core SNP alleles from the putative malaria resistance mutations are given in kb. Frequencies for haplotypes (and numbers of observations) are given for all populations. There are no apparent recombinants among the *G6PD* core haplotypes, and $R_M$ is 0. There are 2 recombinant haplotypes among the 403 *TNFSF5* chromosomes, and $R_M$ would also be 0 if the 2 haplotypes appearing only once were removed from the analysis[6].
*Both proposed mutations associated with malaria resistance (*G6PD*-202A and *TNFSF5*-726C) are observed only in Africans and occur on *G6PD*-CH8 and on *TNFSF5*-CH4 and *TNFSF5*-CH7, respectively.

*G6PD*-CH8 demonstrates clear long-range LD (as seen by the predominance of one thick branch in the haplotype bifurcation diagram (Fig. 2a)) and has correspondingly high EHH. The EHH is 0.38 at the largest distance tested; that is, 413 kb (Fig. 2c). For each core haplotype we calculated the relative EHH; specifically, the factor by which EHH decays on the tested core haplotype compared with the decay of EHH on all other core haplotypes combined (Methods).

To test formally for selection, for each core haplotype, we compared the allele frequency to the relative EHH at various distances (Fig. 2e shows the comparison at 413 kb proximal to *G6PD*). *G6PD*-CH8 has a much higher relative EHH than other haplotypes of comparable frequency, but is this statistically significant? To obtain a sense of how unusual our observation is, we simulated haplotypes using a coalescent process (see Methods and Fig. 2e)[11]. The deviation from the simulation results is highly significant and becomes progressively more marked with increasing distance (Fig. 2g) ($P$-values at 413 kb proximal are: constant-sized population, $P < 0.0008$; expansion, $P < 0.0006$; bottleneck, $P < 0.0008$; population structure, $P < 0.0008$; see Methods and Supplementary Table 2 for details of the demographic models we considered). The frequency and LD properties of *G6PD*-CH8 are incompatible with what is expected under a model of neutral evolution for a wide range of demographies. Furthermore, when the three African populations comprising the pooled sample were considered separately, a signal of selection was identified independently in each population (Yoruba, $P < 0.0012$; Beni, $P < 0.0440$; and Shona, $P < 0.0030$, based on

续表

| (b) | 核心 SNP 等位基因(kb) | | | | | 六个群体中的核心单倍型频率 | | | | | | |
|---|---|---|---|---|---|---|---|---|---|---|---|---|
| *TNFSF5* 核心单倍型 | −6 | 0 | 1 | 3 | 4 | 总计 | 贝尼人 | 约鲁巴人 | 绍纳人 | 非裔美国人 | 欧裔美国人 | 亚洲人 |
| 1 | T | T | T | T | G | 0.03(12) | 0.06(4) | 0.03(3) | 0.02(2) | 0.04(3) | 0(0) | 0(0) |
| 2 | C | – | – | – | – | 0.50(200) | 0.32(20) | 0.38(33) | 0.46(38) | 0.40(32) | 0.78(49) | 1.00(28) |
| 3 | C | – | – | C | – | 0.08(32) | 0.10(6) | 0.12(10) | 0.06(5) | 0.14(11) | 0(0) | 0(0) |
| 4 | – | C* | – | – | – | 0.25(100) | 0.38(24) | 0.38(33) | 0.26(21) | 0.27(22) | 0(0) | 0(0) |
| 5 | – | – | C | – | – | 0.12(47) | 0.14(9) | 0.07(6) | 0.18(15) | 0.14(11) | 0.10(6) | 0(0) |
| 6 | – | – | C | – | A | 0.02(10) | 0(0) | 0(0) | 0.01(1) | 0.01(1) | 0.13(8) | 0(0) |
| 7 | – | C* | C | – | A | 0(1) | 0(0) | 0.01(1) | 0(0) | 0(0) | 0(0) | 0(0) |
| 8 | C | – | C | – | – | 0(1) | 0(0) | 0(0) | 0(0) | 0.01(1) | 0(0) | 0(0) |
| | | | | | *N* | 403 | 63 | 86 | 82 | 81 | 63 | 28 |

在非洲、欧洲和亚洲后代的 6 个群体中观察到 *G6PD* 和 *TNFSF5* 的核心单倍型。核心 SNP 等位基因与假定的疟疾抗性突变之间的相对距离用 kb 表示。所有群体的单倍型频率(以及观察次数)也显示出来。*G6PD* 核心单倍型中没有明显的重组，$R_M$ 是 0。在 403 个 *TNFSF5* 染色体中有两个重组的单倍型，但是如果将仅出现一次的这两个单倍型从分析中去除，$R_M$ 仍然是 0[6]。

* 两种与疟疾抗性相关的假定突变(*G6PD*-202A 和 *TNFSF5*-726C)仅在非洲人中观察到，分别出现在 *G6PD*-CH8、*TNFSF5*-CH4 和 *TNFSF5*-CH7 中。

*G6PD*-CH8 具有清晰的长范围 LD(在单倍型分支图中显示为明显的一个粗分支优势型(图 2a))和相应的高 EHH。EHH 在最大的检测距离下是 0.38，也就是 413 kb (图 2c)。对每一个核心单倍型，我们计算了相对的 EHH。具体来说，就是将检测的核心单倍型中 EHH 衰减的因素与所有其他核心单倍型组合起来的 EHH 衰减进行了对比(见方法)。

为了正式地检测选择，对每一个核心单倍型在不同距离下的等位基因频率与相对 EHH 进行了比较(图 2e 显示了 *G6PD* 近端 413 kb 的对比)。*G6PD*-CH8 与其他具有类似频率的单倍型相比有更高的相对 EHH，但其是否有统计学显著意义？为了了解我们的发现是多么得不寻常，我们用联合的过程模拟了单倍型(见方法和图 2e)[11]。模拟结果得出的偏离是高度显著的，并且随着距离的增加变得越来越显著(图 2g)(近端 413 kb 的 *P* 值：数量稳定的人群 $P < 0.0008$；扩张的人群 $P < 0.0006$；瓶颈效应 $P < 0.0008$；人群结构 $P < 0.0008$；我们使用的人口统计学模型的详细信息见方法和补充表 2)。*G6PD*-CH8 的频率和 LD 特性与广泛人口分布中的中性进化模型的预期不一致。此外，当把组成融合样本的三个非洲群体分开考虑时，每个群体中各自识别出了一个选择信号(约鲁巴人 $P < 0.0012$；贝尼人 $P < 0.0440$；绍纳人 $P < 0.0030$，基于数量稳定的群体的模型)，证明选择信号不是将三个群体样本融合

simulation of a constant-sized population), demonstrating that the signal of selection is not an artefact of pooling the three population samples.

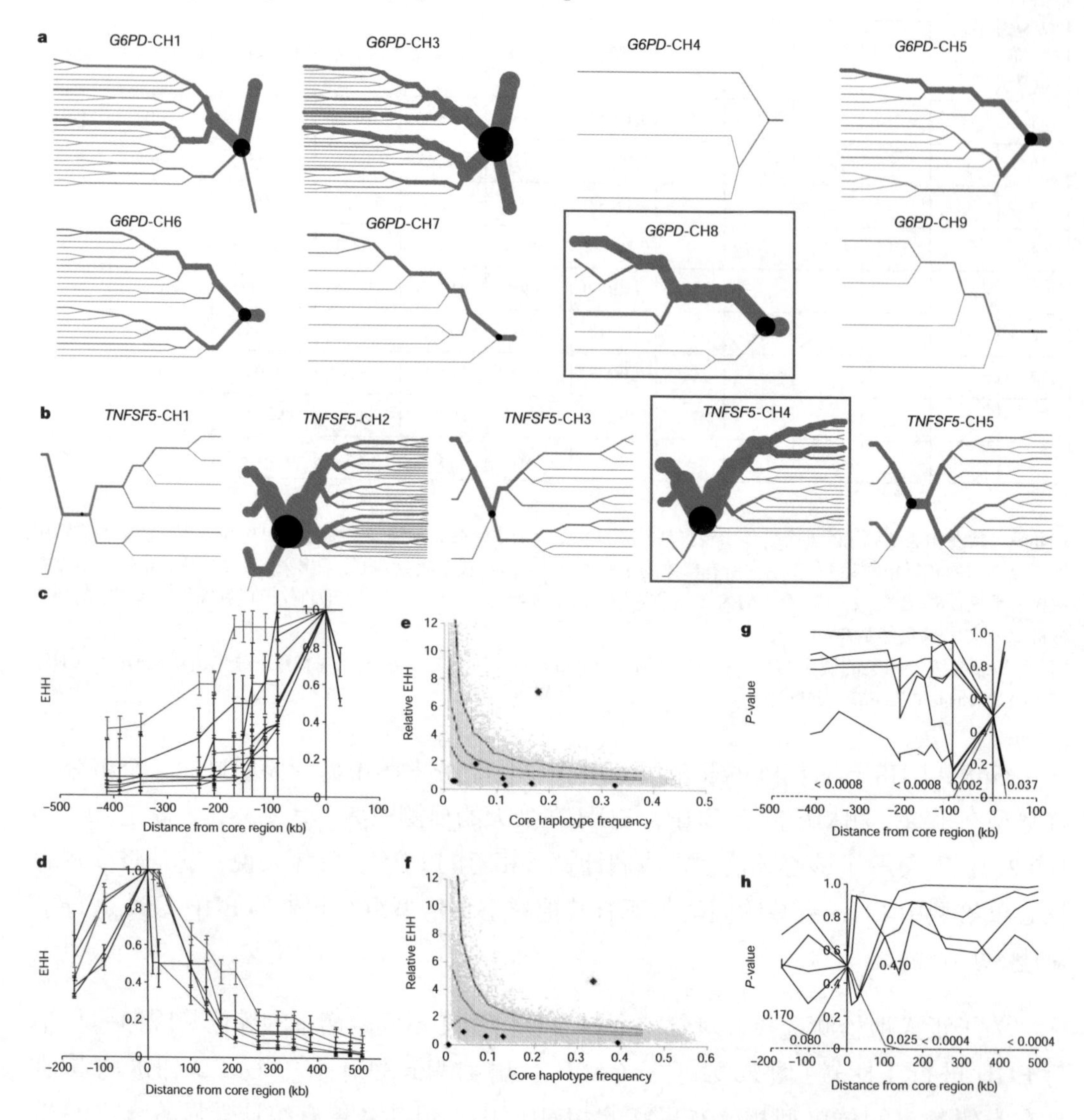


Fig. 2. Core haplotype frequency and relative EHH of *G6PD* and *TNFSF5*. **a**, **b**, Haplotype bifurcation diagrams (see Methods) for each core haplotype at *G6PD* (**a**) and *TNFSF5* (**b**) in pooled African populations demonstrate that *G6PD*-CH8 and *TNFSF5*-CH4 (boxed or labelled in red) have long-range homozygosity that is unusual given their frequency. **c**, **d**, The EHH at varying distances from the core region on each core haplotype at *G6PD* (**c**) and *TNFSF5* (**d**) demonstrates that *G6PD*-CH8 and *TNFSF5*-CH4 have persistent, high EHH values. **e**, **f**, At the most distant SNP from *G6PD* (**e**) and *TNFSF5* (**f**) core regions, the relative EHH plotted against the core haplotype frequency is presented and compared with the distribution of simulated core haplotypes (on the basis of simulation of 5,000 data sets; represented by grey dots and given with 95th, 75th and 50th percentiles). The observed non-selected core haplotypes in our data are represented by black diamonds. **g**, **h**, We calculated the statistical significance of the departure of the observed data from the simulated distribution at each distance from the core. *G6PD*-CH8 (**g**) and *TNFSF5*-CH4 (**h**) demonstrate increasing deviation from a model of neutral drift at further distances from the core region in both directions.

在一起的产物。

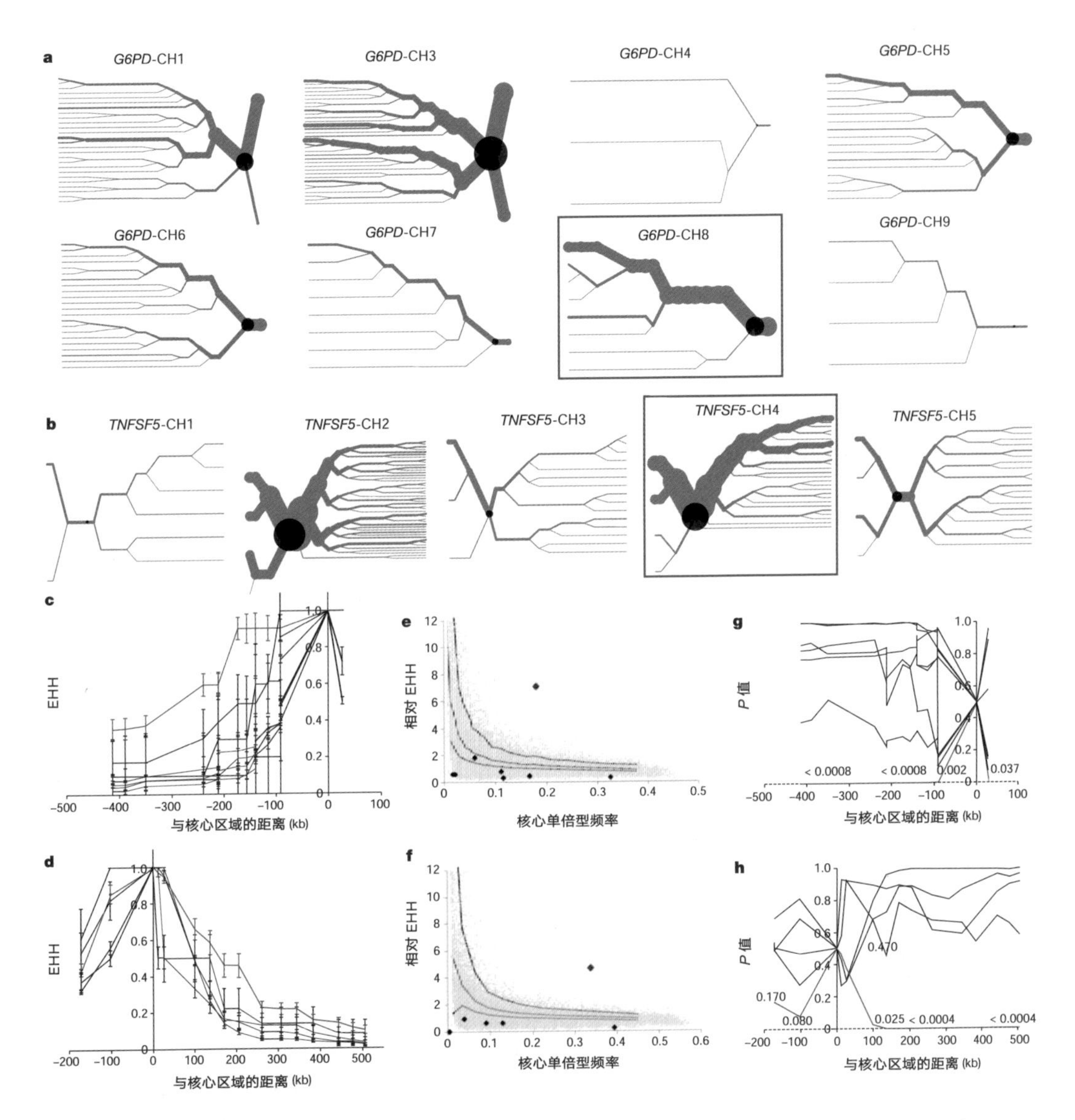


图 2. *G6PD* 和 *TNFSF5* 的核心单倍型频率和相对 EHH。**a**、**b**，在总体非洲人群中，*G6PD*(**a**) 和 *TNFSF5*(**b**) 的每个核心单倍型的分支图（见方法）表明，*G6PD*-CH8 和 *TNFSF5*-CH4（红色框）具有长范围的纯和性，这在其频率下是不寻常的。**c**、**d**，*G6PD*(**c**) 和 *TNFSF5*(**d**) 的每个核心单倍型中核心区域不同距离的 EHH 显示了 *G6PD*-CH8 和 *TNFSF5*-CH4 具有持续高水平的 EHH 值。**e**、**f**，*G6PD*(**e**) 和 *TNFSF5*(**f**) 核心区域最远处的 SNP，其相对 EHH 与核心单倍型的频率进行作图，并与模拟的核心单倍型分布图进行对比（根据 5,000 个数据点的模拟；以灰点表示，并给出第 95、75 和 50 个百分位数）。我们的数据中观察到的非选择的核心单倍型用黑色的菱形表示。**g**、**h**，我们观察到了在离核心单倍型每一距离处模拟分布的数据，并计算了这些数据偏差的显著性差异。*G6PD*-CH8(**g**) 和 *TNFSF5*-CH4(**h**) 显示出在距离核心区域更远的两个方向上，偏离中性漂移模型的程度更大。

We next applied our approach to the CD40 ligand gene (Fig. 1). We defined a core region of 10 kb and genotyped 5 SNPs. The SNPs defined seven core haplotypes (Table 1b). The *TNFSF5*-726C allele, which has been associated with protection from malaria, was present on *TNFSF5*-CH4, which is common in Africa (34%), but is absent outside of Africa. *TNFSF5*-CH4 demonstrates high LD as seen in the haplotype bifurcation diagrams (Fig. 2b) and has high EHH at long distances (Fig. 2d). *TNFSF5*-CH4 is a clear outlier when compared with other haplotypes (Fig. 2f), and its frequency and LD properties are incompatible with neutral evolution under multiple demographic models (*P*-values at 506 kb distal are: constant-sized population, $P < 0.0012$; expansion, $P < 0.0008$; bottleneck, $P < 0.0012$; population structure, $P < 0.0008$; see Methods and Supplementary Table 2 for details). Again, the *P*-value is increasingly significant at further distances from the core region both proximally and distally (Fig. 2h). When each African population was analysed separately, a signal of selection was significant (Yoruba, $P < 0.0008$; Beni, $P < 0.0023$; Shona, $P < 0.0242$). These results thus provide independent evidence supporting the proposed role of CD40 ligand in malaria resistance[2].

We tested our conclusion of positive selection by performing a similar analysis on 17 randomly chosen control regions across the human genome in the same African populations. We only used data from each control if it was closely matched to our data in terms of the number of chromosomes studied and the homozygosity at the core haplotype and at long distances from the core (Fig. 3). *G6PD*-CH8 and *TNFSF5*-CH4 clearly stand out from the other loci, showing that the *P*-values determined by simulation are also supported by direct, empirical comparison. In measuring *P*-values for the controls where there is no prior hypothesis of selection, a Bonferonni correction for multiple-hypothesis testing was applied. Notably, one core haplotype, from the monocyte chemotactic protein 1 region, shows frequency and LD properties similar to *G6PD* and *TNFSF5*, although this nominally significant result may be simply a false-positive owing to the large number of hypotheses examined.

We used a linkage-disequilibrium-based technique[12] to estimate dates of origin of the two resistance variants. The estimates were about 2,500 years for *G6PD* and about 6,500 years for *TNFSF5* (see Supplementary Information for details). The date for *G6PD* is consistent with a recent independent age estimate for *G6PD*-202A based primarily on microsatellite data[13].

Finally, we explored whether positive selection could have been detected with traditional tests (Supplementary Table 3)[14]. We performed Tajima's *D*-test[15], Fu and Li's *D*-test[16], Fay and Wu's *H*-test[17], the Ka/Ks test[18], the McDonald and Kreitman test[19], and the Hudson–Kreitman–Aguadè (HKA) test[20]. None showed significant deviation from neutral evolution for either *G6PD* or *TNFSF5*, consistent with their low power to detect recent selection.

接下来我们将方法应用到CD40配体基因中（图1）。我们定义了一个10 kb的核心区域，对5个SNP进行基因分型。这些SNP确定了7个核心单倍型（表1b）。与疟疾抗性相关的*TNFSF5*-726C等位基因存在于*TNFSF5*-CH4上，后者在非洲很常见（34%），但是在非洲以外地区不存在。*TNFSF5*-CH4在单倍型分支图中显示高LD（图2b），在长距离时显示高EHH（图2d）。*TNFSF5*-CH4与其他单倍型相比是一个明确的离群值（图2f），其频率和LD特性在多种人口统计学模型下都与中性进化相矛盾（506 kb远端的$P$值是：数量稳定的人群$P < 0.0012$；扩张的人群$P < 0.0008$；瓶颈效应$P < 0.0012$；群体结构$P < 0.0008$；细节见方法及附表2）。同样，$P$值显著性随着距核心区域的距离增加而增加，无论是近端还是远端（图2h）。当分别分析非洲群体时，选择的信号都有显著性意义（约鲁巴人$P < 0.0008$；贝尼人$P < 0.0023$；绍纳人$P < 0.0242$）。因此这些结果为支持CD40配体在疟疾抗性中的作用提供了独立证据[2]。

我们通过在相同的非洲人群中对人类基因组中的17个随机选取的对照区域进行类似的分析，验证了我们关于正选择的结论。只有当对照区域在染色体数目、核心单倍型所在位置以及距离核心区较远处的纯合性方面与我们的数据密切匹配时，我们才使用其数据（图3）。*G6PD*-CH8和*TNFSF5*-CH4与其他基因座明显不同，表明通过模拟得出的$P$值也得到了直接的、实证比较分析的支持。在没有关于选择的优先假设的情况下，我们测量了对照$P$值，并采用邦费罗尼校正方法进行多假设检验。显然，来自单核细胞趋化蛋白1区域的一个核心单倍型显示了类似于*G6P*D和*TNFSF5*的频率和LD特征，然而这一表面上显著的结果可能仅仅是由于大量的假设被检验，而呈现的假阳性。

我们使用一种基于连锁不平衡的技术[12]估计这两个抗性变异体的起源日期。*G6PD*估计大约2,500年前，*TNFSF5*大约6,500年前（详情见补充信息）。*G6PD*的起源日期与最近主要基于微卫星数据预估的*G6PD*-202A的独立年龄相一致[13]。

最后，我们探讨了正选择是否能够用传统的方法检测到（补充表3）[14]。我们进行了Tajima的$D$检验[15]，Fu和Li的$D$检验[16]，Fay和Wu的$H$检验[17]，Ka/Ks检验[18]，McDonald和Kreitman检验[19]以及Hudson-Kreitman-Aguadè（HKA）检验[20]。无论是*G6PD*还是*TNFSF5*，均未表现出明显的中性进化偏差，这与它们不能检测到最近的选择相一致。

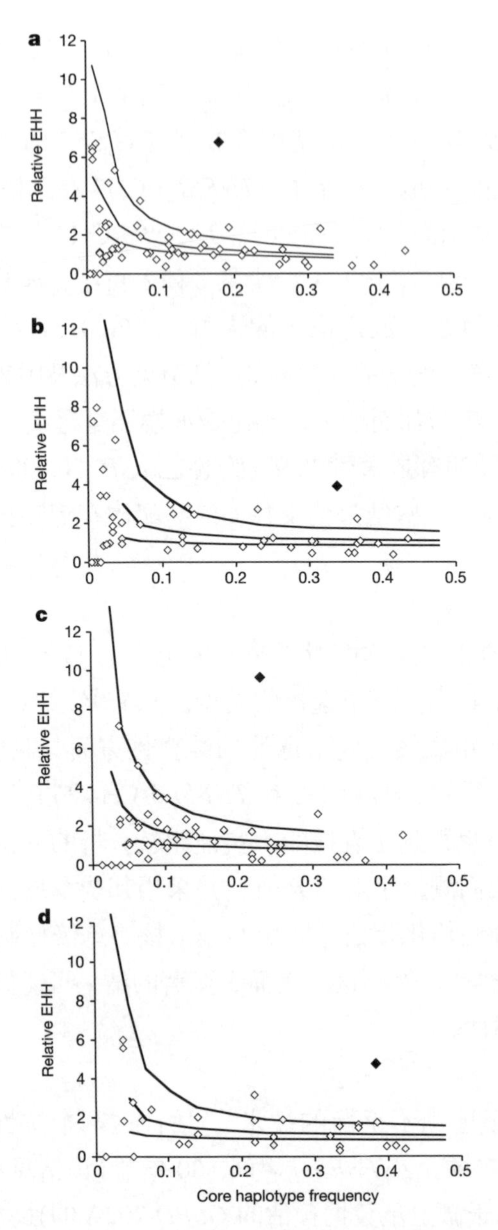


Fig. 3. Control regions: core haplotype frequency against relative EHH. To provide an empirical, non-simulation-based evaluation of the signal of selection, we compared the frequency and relative EHHs for *G6PD* (**a**) and *TNFSF5* (**b**) with patterns observed in randomly chosen genes in the genome (see Methods). **c**, **d**, We performed the entire analysis again on *G6PD* (**c**) and *TNFSF5* (**d**) for a subset of 78 Yoruban haplotypes (using family trios where phase could be determined). We were able to match 30 to 87 core haplotypes (indicated by outlined diamonds) from the control regions to our data. The 95th, 75th and 50th percentiles for simulated data are also shown. *G6PD*-CH8 and *TNFSF5*-CH4 (indicated by black diamonds) clearly stand out from the pattern seen at other loci in the genome, suggesting a true signal of selection.

Our approach, which we refer to as the long-range haplotype (LRH) test, provides a way to detect recent positive selection by analysing haplotype structure in random individuals from a population. How far back in human history can one detect positive selection?

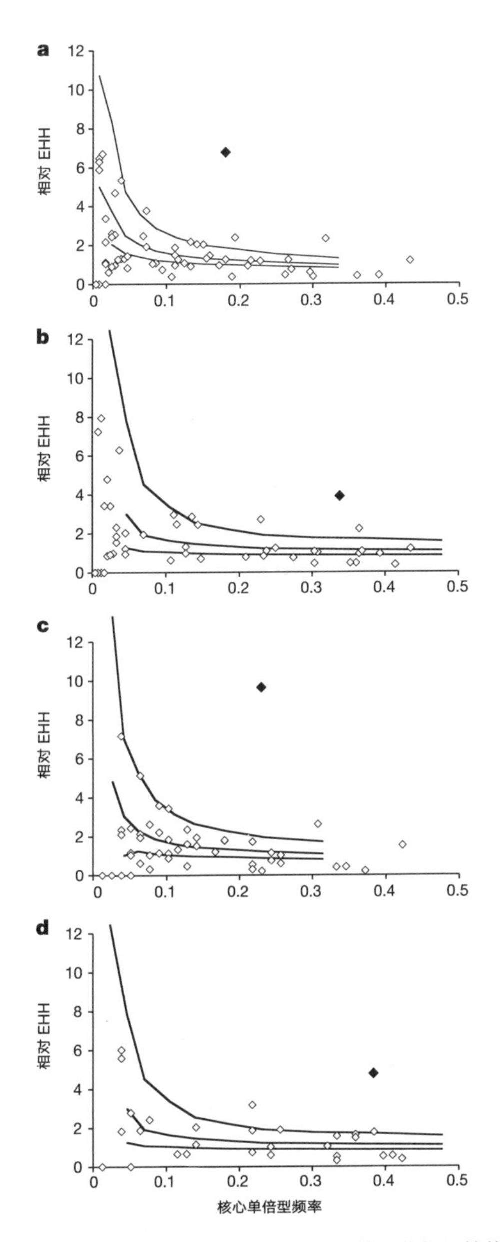


图 3. 对照区域：核心单倍型频率与相对 EHH。为了提供实验性、非模拟性的选择信号评估，我们将 *G6PD* (**a**) 和 *TNFSF5* (**b**) 的频率和相对 EHH 与基因组中随机选择的基因中观察到的模式进行了比较（见方法）。**c**、**d**，我们再次对 78 个约鲁巴人的单倍型子集的 *G6PD*(**c**) 和 *TNFSF5*(**d**) 进行了完整的分析（使用能够确定阶段的三体家系）。我们能够将对照区域的 30 到 87 个核心单倍型（以空心的菱形表示）与我们的数据匹配起来。图中也显示了模拟数据的第 95、75 和 50 个百分位数。*G6PD*-CH8 和 *TNFSF5*-CH4（以黑色菱形表示）与基因组中其他基因座所表现的形式完全不同，表明存在真正的选择信号。

我们称为长范围单倍型（LRH）检测的方法，提供了一种通过分析群体中随机个体的单倍型结构来检测近期正选择的方法。在人类历史上向前追溯多久才能检测到正选择呢？发生在不到 400 代之前的选择性事件（假设 25 年一代的话就是 10,000

Selective events occurring less than 400 generations ago (10,000 years assuming 25 years per generation) should leave a clear imprint at distances of over 0.25 centiMorgans (cM). The signal of such long-range LD should be distinguishable from the background extent of LD for common haplotypes in the genome[21], which are typically tens of thousands of generations old[4] and hence extend 0.02 cM or less. Over many tens of thousands of years, the signal of selection will become lost as recombination whittles the long-range haplotypes to the typical size of haplotype blocks in the human genome[21].

The LRH test can be used to search for evidence of positive selection by testing each common haplotype in a gene, without prior knowledge of a specific variant or selective advantage. Once the signature of selection is found, one must then decipher its cause. The LRH test could be applied to scan the entire human genome for evidence of recent positive selection simply by applying it to each haplotype block in reference data sets from human populations, as will be collected by the Human Haplotype Map project[21]. In this fashion, it should be possible to shed light on how the human genome was shaped by recent changes in culture and environment. The LRH test should also be useful for studying selection in other organisms, including domestic animals and parasites such as the malaria parasite *P. falciparum*[22].

## Methods

### Human subjects

DNA samples from 252 males from Africa were used in the study: 92 Yoruba and 73 Beni from Nigeria, and 87 Shona from Zimbabwe. Additional DNA samples from 29 Yoruban trios (father–mother–child clusters) were genotyped at the 17 control regions. The Yoruba and Shona males were healthy individuals obtained as part of the International Collaborative Study of Hypertension in Blacks. The Beni samples were from civil servants in Benin City. Samples from four non-African populations and four primates were also used (Supplementary Information).

### SNP genotyping

We genotyped 49 SNPs distributed around *G6PD* and 37 SNPs distributed around *TNFSF5* using mass spectrometry (Sequenom)[23]. The SNPs were identified by our own resequencing and through previous discovery efforts[2,24,25]. For *G6PD*, we focused on genotyping SNPs proximal to the gene, as SNPs in the repetitive subtelomeric sequence distal to *G6PD* could not be genotyped. A total of 25 SNPs around *G6PD* and 21 SNPs around *TNFSF5* were successfully genotyped and used in analysis (see Supplementary Information for details).

年）应该在超过 0.25 厘摩（cM）的距离上留下明显的印记。这种长范围的 LD 信号应该能够和基因组中普通单倍型的背景长度的 LD 区别开来 [21]，因为后者通常有几万代的历史 [4]，因此延伸的长度不超过 0.02 cM。经过数万年的时间，随着重组的发生，长范围的单倍型就会变成人类基因组中典型大小的单倍型区域 [21]，选择的信号会逐渐丢失。

LRH 检测可以通过分析基因中的每个常见单倍型来寻找正选择的证据，而无须提前知道特定的变异型或者选择优势。一旦发现了正选择的特征，就必须进一步研究其原因。LRH 检测可用于扫描整个人类基因组，仅仅通过将其用于人类群体中参考数据集内的每个单倍型区域，来寻找近期正选择的证据。这些数据信息将收集在人类单倍型图谱计划中 [21]。通过这种方式，我们有可能弄清楚人类基因组是如何被近期的文化和环境变化塑造的。LRH 检测对研究其他生物的选择也有效，包括家畜和寄生虫（比如恶性疟原虫 [22]）。

## 方　法

### 人群对象

研究中使用了 252 名非洲男性的 DNA 样本：来自尼日利亚的 92 名约鲁巴人、73 名贝尼人，来自津巴布韦的 87 名绍纳人。在 17 个对照区对来源于 29 个约鲁巴三体家系（父亲–母亲–孩子集群）的额外 DNA 样本进行了基因分型。约鲁巴人和绍纳人是健康的男性，是黑人高血压国际合作研究项目的对象。贝尼人样本来自贝宁城的公务员。还使用了 4 个非非洲人群和 4 个灵长类动物的样本（补充资料）。

### SNP 基因分型

我们使用质谱法（Sequenom）[23] 对 *G6PD* 附近的 49 个 SNP 以及 *TNFSF5* 附近的 37 个 SNP 进行了基因分型。这些 SNP 是通过我们自己的重新测序以及先前的发现相结合鉴定出来的 [2,24,25]。对于 *G6PD*，我们主要着眼于对基因近端的 SNP 进行基因分型，因为其远端的重复性亚端粒序列中的 SNP 无法基因分型。总共有 *G6PD* 附近的 25 个 SNP 和 *TNFSF5* 附近的 21 个 SNP 成功基因分型并用于分析（详细内容见补充资料）。

## Haplotype bifurcation diagrams

To visualize the breakdown of LD on core haplotypes, we created bifurcation diagrams using MATLAB. The root of each diagram is a core haplotype, identified by a black circle. The diagram is bi-directional, portraying both proximal and distal LD. Moving in one direction, each marker is an opportunity for a node; the diagram either divides or not based on whether both or only one allele is present. Thus the breakdown of LD on the core haplotype background is portrayed at progressively longer distances. The thickness of the lines corresponds to the number of samples with the indicated long-distance haplotype.

## Extended haplotype homozygosity and relative EHH

EHH at a distance $x$ from the core region is defined as the probability that two randomly chosen chromosomes carrying a tested core haplotype are homozygous at all SNPs[8] for the entire interval from the core region to the distance $x$. EHH is on a scale of 0 (no homozygosity, all extended haplotypes are different) to 1 (complete homozygosity, all extended haplotypes are the same). Relative EHH is the ratio of the EHH on the tested core haplotype compared with the EHH of the grouped set of core haplotypes at the region not including the core haplotype tested. Relative EHH is therefore on a scale of 0 to infinity.

## Coalescent simulations

We used a computer program by Hudson that simulates gene history with recombination[11]. The program was modified to generate data such as we collected. We simulated a long region of DNA (1.3 cM), with one end defined as the "core". We progressively added SNPs at the core until they matched our data (for the *G6PD* or *TNFSF5* core) to within $\pm$ 12.5% in terms of the homozygosity. To mimic the SNP selection strategy used by The SNP consortium[25], which was the source of most of the SNPs in our study, we only included simulated SNPs in our analysis if different alleles were observed at two randomly chosen chromosomes from the sample. At longer distances, we added additional SNPs, only choosing SNPs for analysis that matched our data in terms of frequency (within a $\pm$ 12.5% window) and also broke down EHH to the same extent as was observed in our data (within $\pm$ 12.5%).

We repeated the simulations for 5,000 data sets (each producing typically 6–8 core haplotypes) to generate many data points with which to compare our data. *P*-values were obtained by first binning the simulated data by core haplotype frequency into 30 bins of equal size, each containing about 1,000 data points. We then ranked an observed core haplotype's relative EHH compared with that of all simulated data points within the bin containing haplotypes of the same frequency—the rank determines the *P*-value. For simulations of additional demographic histories, we considered two models of expansion, an extreme bottleneck, and a highly structured population. For expansions, we simulated a population that was constant at size 10,000 until 200 or 5,000 generations ago, when it

## 单倍型分支图

为了可视化核心单倍型中 LD 的分解，我们使用 MATLAB 创建了分支图。每个图标的根部是核心单倍型，以黑圈表示。图表是双向的，表示近端和远端的 LD。向一个方向移动的话，每个标记都是一个形成节点的机会；图标分支与否取决于存在的等位基因是一个还是两个。因此核心单倍型背景中的 LD 的分解就表示为进行性变长的距离。线条的厚度对应于含有表示的长距离单倍型的样本的数量。

## 扩展单倍型纯合性和相对 EHH

距离核心区域 $x$ 的 EHH 被定义为从核心区域到距离 $x$ 的整个区间内，两个随机选择的携带被测的核心单倍型的染色体在所有的 SNP 中是纯合的可能性 [8]。EHH 的范围在 0(没有纯合性，所有的扩展单倍型都是不同的）到 1(完全纯合性，所有扩展单倍型都是一样的）之间。相对 EHH 是被测的核心单倍型的 EHH 与不含有被测单倍型的区域内部的核心单倍型集群的 EHH 的比值，因此相对 EHH 的范围是从 0 到无穷大。

## 联合模拟

我们使用赫德森的计算机程序，通过重组模拟基因历史 [11]。该程序经过修改以生成数据，比如我们之前收集的数据。我们模拟了一个长 DNA 区域（1.3 cM)，其一端定义为“核心”。我们逐渐增加核心部位的 SNP，直到它们匹配我们的数据（*G6PD* 或者 *TNFSF5* 核心）使得纯合性偏差在±12.5% 之内。为了模拟 SNP 联盟使用的 SNP 选择策略 [25]（我们研究中大部分 SNP 来源于此)，我们只在样本中随机选择的两个染色体上观察到不同的等位基因时，才将模拟的 SNP 用于分析。在更远的距离处，我们增加额外的 SNP，仅仅选择那些在频率上与我们的数据匹配（±12.5% 之内)，而且 EHH 的分解程度与我们数据中观察到的一致（±12.5% 之内）的 SNP 用于分析。

我们对 5,000 个数据集合进行重复模拟（每个通常能够产生 6～8 个核心单倍型）以产生多个数据点，用于和我们的数据进行对比。首先将模拟的数据按照核心单倍型频率分类成 30 个大小相等的集合，每个集合约含有 1,000 个数据点，计算 $P$ 值。然后我们对观察到的核心单倍型的相对 EHH 与含有相同频率单倍型的集合内的所有模拟数据点的 EHH 进行排序，这个排列就确定了 $P$ 值。为了模拟更多的人口统计学历史，我们考虑两种扩张模型，一个极端的瓶颈效应和一个高度结构化的人群。对于扩张，我们模拟了一个人群，在 200 或者 5,000

expanded suddenly by a factor of 1,000. For an extreme bottleneck, we simulated a population that was constant at size 10,000 except for a brief bottleneck (inbreeding coefficient 0.18) that occurred 800 generations ago. (An inbreeding coefficient[26] of 0.18 is generated by dropping the population size to 800 chromosomes for 160 generations.) For a structured population, we simulated two equal-sized populations of size $N/2$ that exchanged migrants throughout history with a probability of $1/8N$ per generation per chromosome.

Results of the simulations remained qualitatively similar when we explored additional demographies and when we varied the stringency of matching to SNP allele frequencies and to haplotype homozygosities. A comprehensive, simulation-based exploration of the LRH test will be presented elsewhere, along with explorations of the statistical power of the test and computer code for implementing the LRH test on other data sets, including those with missing or unphased data (D.E.R., manuscript in preparation).

## Control regions

To obtain control data for comparison to the *G6PD* and *TNFSF5* haplotypes, we genotyped the same population samples in 17 randomly chosen autosomal genes (*ACVR2B*, *TGFB1*, *DDR1*, *GTF2H4*, *COL11A2*, *LAMB1*, *WASL*, *SLC6A12*, *KCNA1*, *ARGHDIB*, *PCI*, *PRKCB1*, *NF1*, *SCYA2*, *PAI2*, *IL17R* and *HCF2*) selected previously as part of a genome-wide survey of linkage disequilibrium[26]. For our analyses we randomly picked chromosomes to match the numbers sampled for *G6PD* and *TNFSF5*, and we only included those control regions that we could match to our data in terms of homozygosity at the core and homozygosity at long distance (±25% stringency of matching). After the filtering process, we evaluated *G6PD* at 240.3 kb proximal to the gene and *TNFSF5* at 343.9 kb distal to the gene, because we could not make enough comparisons to control regions at the further distances. Seven genes matched to *G6PD* and seven genes matched to *TNFSF5*. We repeated the analysis using 78 Yoruban chromosomes for which phase information was known experimentally (because of genotyping in trios) and for which phase for the most part did not have to be inferred computationally. Six genes matched to *G6PD* and six genes matched to *TNFSF5* (see Supplementary Information for details).

(**419**, 832-837; 2002)

**Pardis C. Sabeti[*†#], David E. Reich[*], John M. Higgins[*] Haninah Z. P. Levine[*], Daniel J. Richter[*], Stephen F. Schaffner[*], Stacey B. Gabriel[*], Jill V. Platko[*], Nick J. Patterson[*], Gavin J. McDonald[*], Hans C. Ackerman[‡], Sarah J. Campbell[‡], David Altshuler[*§], Richard Cooper[‖], Dominic Kwiatkowski[‡], Ryk Ward[†] & Eric S. Lander[*¶]**

[*] Whitehead Institute/MIT Center for Genome Research, Nine Cambridge Center, Cambridge, Massachusetts 02142, USA

[†] Institute of Biological Anthropology, University of Oxford, Oxford, OX2 6QS, UK

[‡] Wellcome Trust Centre for Human Genetics, Roosevelt Drive, Oxford, OX3 7BN, UK

[§] Departments of Genetics and Medicine, Harvard Medical School, Department of Molecular Biology and Diabetes Unit, Massachusetts General Hospital, Boston, Massachusetts 02114, USA

[‖] Department of Preventive Medicine and Epidemiology, Loyola University Medical School, Maywood, Illinois 60143, USA

[¶] Department of Biology, MIT, Cambridge, Massachusetts 02139, USA

[#] Harvard Medical School, Boston, Massachusetts 02115, USA

代之前规模一直稳定在10,000，然后它突然扩张了1,000倍。对于极端的瓶颈效应，我们模拟了一个大小稳定在10,000的人群，除了在800代前出现了一次短暂的瓶颈效应（近交系数0.18）。（近交系数[26]0.18的产生是在160代内将人群的大小减少至800条染色体。）对于结构化的人群，我们模拟了两个大小均是*N*/2的人群，其在整个历史中以每代每个染色体1/8*N*的概率交换迁移者。

当我们探索其他人口统计学时，当我们改变与SNP等位基因频率和单倍型纯合性匹配的严格程度时，模拟的结果在定性上保持相似。我们还将对LRH测试进行全面的、基于模拟的探索，并探讨该测试的统计能力和在其他数据集（包括那些丢失或者非阶段数据的数据集）上实现LRH测试的计算机代码（赖希，稿件准备中）。

## 对照区域

为了获得对照数据，用来和*G6PD*以及*TNFSF5*单倍型进行比较，我们对同一个人群样本中的17个随机选择的常染色体基因进行基因分型（*ACVR2B*，*TGFB1*，*DDR1*，*GTF2H4*，*COL11A2*，*LAMB1*，*WASL*，*SLC6A12*，*KCNA1*，*ARGHDIB*，*PCI*，*PRKCB1*，*NF1*，*SCYA2*，*PAI2*，*IL17R*和*HCF2*），之前选择它们作为基因组水平连锁不平衡研究的一部分[26]。对于我们的分析，我们随机选取了匹配*G6PD*以及*TNFSF5*样本数目的染色体，而且我们仅仅使用那些能够在核心的纯合性和长距离的纯合性与我们的数据匹配的对照区域（匹配的严格度是±25%）。经过筛选，我们对*G6PD*近端240.3 kb处和*TNFSF5*远端343.9 kb处进行了评估，因为我们无法对更远距离的对照区域进行比较。发现分别有7个基因对应于*G6PD*和*TNFSF5*。我们用78名约鲁巴人的染色体进行了重复分析，它们的阶段信息已经通过实验得到（因为在三体家系中进行了基因分型），而且大部分的阶段信息不需要通过计算机推断。各有6个基因分别对应于*G6PD*和*TNFSF5*（详细内容见补充资料）。

（毛晨晖 翻译；胡松年 审稿）

Received 7 June; accepted 19 September 2002; doi:10.1038/nature01140.
Published online 9 October 2002.

---

References:

1. Ruwende, C. & Hill, A. Glucose-6-phosphate dehydrogenase deficiency and malaria. *J. Mol. Med.* **76**, 581-588 (1998).
2. Sabeti, P. *et al.* CD40L association with protection from severe malaria. *Genes Immun.* **3**, 286-291 (2002).
3. Cavalli-Sforza, L. L., Menozzi, P. & Piazza, A. *The History and Geography of Human Genes* (Princeton Univ. Press, Princeton, 1994).
4. Kimura, M. *The Neutral Theory of Molecular Evolution* (Cambridge Univ. Press, Cambridge/New York, 1983).
5. Stephens, J. C. *et al.* Dating the origin of the CCR5-Delta32 AIDS-resistance allele by the coalescence of haplotypes. *Am. J. Hum. Genet.* **62**, 1507-1515 (1998).
6. Hudson, R. R. & Kaplan, N. L. Statistical properties of the number of recombination events in the history of a sample of DNA sequences. *Genetics* **111**, 147-164 (1985).
7. Lewontin, R. The interaction of selection and linkage. I. General considerations; heterotic models. *Genetics* **49**, 49-67 (1964).
8. Nei, M. *Molecular Evolutionary Genetics* Eqn. 8.4 (Columbia Univ. Press, New York, 1987).
9. Luzatto, L., Mehta, A. & Vulliamy, T. *The Metabolic & Molecular Bases of Inherited Disease* 4517-4553 (McGraw-Hill, New York, 2001).
10. Raymond, M. & Rousset, F. An exact test for population differentiation. *Evolution* **49**, 1280-1283 (1995).
11. Hudson, R. R. Properties of a neutral allele model with intragenic recombination. *Theor. Popul. Biol.* **23**, 183-201 (1983).
12. Reich, D. E. & Goldstein, D. B. *Microsatellites: Evolution and Applications* 128-138 (Oxford Univ. Press, Oxford/New York, 1999).
13. Tishkoff, S. A. *et al.* Haplotype diversity and linkage disequilibrium at human G6PD: recent origin of alleles that confer malarial resistance. *Science* **293**, 455-462 (2001).
14. Rozas, J. & Rozas, R. DnaSP version 3: an integrated program for molecular population genetics and molecular evolution analysis. *Bioinformatics* **15**, 174-175 (1999).
15. Tajima, F. Statistical method for testing the neutral mutation hypothesis by DNA polymorphism. *Genetics* **123**, 585-595 (1989).
16. Fu, Y. X. & Li, W. H. Statistical tests of neutrality of mutations. *Genetics* **133**, 693-709 (1993).
17. Fay, J. C. & Wu, C. I. Hitchhiking under positive Darwinian selection. *Genetics* **155**, 1405-1413 (2000).
18. Hughes, A. L. & Nei, M. Pattern of nucleotide substitution at major histocompatibility complex class I loci reveals overdominant selection. *Nature* **335**, 167-170 (1988).
19. McDonald, J. H. & Kreitman, M. Adaptive protein evolution at the Adh locus in *Drosophila*. *Nature* **351**, 652-654 (1991).
20. Hudson, R. R., Kreitman, M. & Aguade, M. A test of neutral molecular evolution based on nucleotide data. *Genetics* **116**, 153-159 (1987).
21. Gabriel, S. B. *et al.* The structure of haplotype blocks in the human genome. *Science* **23**, 2225-2229 (2002).
22. Wootton, J. C. *et al.* Genetic diversity and chloroquine selective sweeps in *Plasmodium falciparum*. *Nature* **418**, 320-323 (2002).
23. Tang, K. *et al.* Chip-based genotyping by mass spectrometry. *Proc. Natl Acad. Sci. USA* **96**, 10016-10020 (1999).
24. Vulliamy, T. J. *et al.* Linkage disequilibrium of polymorphic sites in the G6PD gene in African populations and the origin of G6PD A. *Gene Geogr.* **5**, 13-21 (1991).
25. Sachidanandam, R. *et al.* A map of human genome sequence variation containing 1.42 million single nucleotide polymorphisms. *Nature* **409**, 928-933 (2001).
26. Reich, D. E. *et al.* Linkage disequilibrium in the human genome. *Nature* **411**, 199-204 (2001).

**Supplementary Information accompanies the paper on** *Nature*'s website (http://www.nature.com/nature).

**Acknowledgements**. We thank B. Blumenstiel, M. DeFelice, A. Lochner, J. Moore, H. Nguyen and J. Roy for assistance in genotyping the 17 control regions. We also thank L. Gaffney, S. Radhakrishna, T. DiCesare and T. Lavery for graphics and technical support, B. Ferrell for the Beni samples, and A. Adeyemo and C. Rotimi for helping to collect the Yoruba and Shona samples. Finally, we thank M. Daly, E. Cosman, B. Gray, V. Koduri, T. Herrington and L. Peterson for comments on the manuscript. P.C.S. was supported by grants from the Rhodes Trust, the Harvard Office of Enrichment, and by a Soros Fellowship. This work was supported by grants from the National Institute of Health.

**Competing interests statement.** The authors declare that they have no competing financial interests.

**Correspondence** and requests for materials should be addressed to E.S.L. (e-mail: lander@genome.wi.mit.edu).

# Initial Sequencing and Comparative Analysis of the Mouse Genome*

Mouse Genome Sequencing Consortium

## Editor's Note

**Here, a high-quality draft sequence of the mouse genome is made freely available for the first time, thanks to the work of an international research consortium. At the time, the data provided an invaluable reference point for geneticists struggling to understand the draft human genome which had been published a year earlier. But it continues to shed light on human biology and disease today, as the mouse remains an invaluable and stalwart model in biomedical research. Comparisons here between the DNA of mouse and man reveal immense similarities, confirming the credentials of this tiny rodent to help fathom the workings of the human genome. But the study also highlights some 300 mouse-specific genes, with the biggest disparities linked to smell, immunity and detoxification.**

The sequence of the mouse genome is a key informational tool for understanding the contents of the human genome and a key experimental tool for biomedical research. Here, we report the results of an international collaboration to produce a high-quality draft sequence of the mouse genome. We also present an initial comparative analysis of the mouse and human genomes, describing some of the insights that can be gleaned from the two sequences. We discuss topics including the analysis of the evolutionary forces shaping the size, structure and sequence of the genomes; the conservation of large-scale synteny across most of the genomes; the much lower extent of sequence orthology covering less than half of the genomes; the proportions of the genomes under selection; the number of protein-coding genes; the expansion of gene families related to reproduction and immunity; the evolution of proteins; and the identification of intraspecies polymorphism.

WITH the complete sequence of the human genome nearly in hand[1,2], the next challenge is to extract the extraordinary trove of information encoded within its roughly 3 billion nucleotides. This information includes the blueprints for all RNAs and proteins, the regulatory elements that ensure proper expression of all genes, the structural elements that govern chromosome function, and the records of our evolutionary history.

* This is a shortened version of the original paper that appeared in *Nature*. In particular, it omits much of the detailed quantitative data in the original figures and tables, and the full list of authors and affiliations has been removed. The original text is freely available in the *Nature* online archive. Citations are numbered herein as they are in the original paper; only those references that are cited in this version are listed in the "References" section.

# 小鼠基因组的初步测序和对比分析*

小鼠基因组测序协作组

## 编者按

感谢国际协作组的工作，本文首次免费提供一个高质量的小鼠基因组序列草图。当时，这些数据为遗传学家努力理解一年前发表的人类基因组草图提供了宝贵的参考点。但是现在，它仍然继续为揭示人类生物学和疾病提供帮助。因为在生物医学研究中，小鼠始终是一个宝贵而有力的模型。小鼠和人类DNA之间的比较揭示了两者巨大的相似性，证实了这种小型啮齿动物能帮助了解人类基因组的运行。但是，该研究也强调了大约300个小鼠特异性基因，其中最大的不同与嗅觉、免疫和解毒作用有关。

---

小鼠基因组序列是理解人类基因组内容的关键信息工具，也是生物医学研究的关键实验工具。在此，我们报告国际合作的成果——一个高质量的小鼠基因组序列草图。我们还对小鼠和人类基因组进行了初步对比分析，描述了从这两个序列中获得的一些启示。我们讨论的主题包括对影响基因组大小、结构和序列的进化力的分析；存在于大部分基因组中的同源染色单体(同线性)的大规模保留；比之低得多只覆盖不到一半基因组的序列直系同源；受选择基因组的比例；蛋白质编码基因的数量；与生殖和免疫相关的基因家族的扩增；蛋白质的进化；物种内多态性的确定。

---

随着即将掌握人类基因组的完整序列[1,2]，下一个挑战是提取大约30亿个核苷酸中编码的非常巨大的信息宝藏。这些信息包括所有RNA和蛋白质的蓝图、确保所有基因正确表达的调控元件、控制染色体功能的结构元件以及我们进化史的记录。这些特征当中，一些可以很容易在人类序列中识别，但很多特征是微妙而且难辨别

* 这是《自然》杂志中原文的缩略版。此版本在于省略了原始图表中的大部分详细量化数据并去除了所有作者及单位名称。原文可在《自然》杂志在线免费获取。参考文献的编码与原文一致，文末“References”仅保留了此版本中有所引用的文献。

Some of these features can be recognized easily in the human sequence, but many are subtle and difficult to discern. One of the most powerful general approaches for unlocking the secrets of the human genome is comparative genomics, and one of the most powerful starting points for comparison is the laboratory mouse, *Mus musculus*.

Metaphorically, comparative genomics allows one to read evolution's laboratory notebook. In the roughly 75 million years since the divergence of the human and mouse lineages, the process of evolution has altered their genome sequences and caused them to diverge by nearly one substitution for every two nucleotides (see below) as well as by deletion and insertion. The divergence rate is low enough that one can still align orthologous sequences, but high enough so that one can recognize many functionally important elements by their greater degree of conservation. Studies of small genomic regions have demonstrated the power of such cross-species conservation to identify putative genes or regulatory elements[3-12]. Genome-wide analysis of sequence conservation holds the prospect of systematically revealing such information for all genes. Genome-wide comparisons among organisms can also highlight key differences in the forces shaping their genomes, including differences in mutational and selective pressures[13,14].

Literally, comparative genomics allows one to link laboratory notebooks of clinical and basic researchers. With knowledge of both genomes, biomedical studies of human genes can be complemented by experimental manipulations of corresponding mouse genes to accelerate functional understanding. In this respect, the mouse is unsurpassed as a model system for probing mammalian biology and human disease[15,16]. Its unique advantages include a century of genetic studies, scores of inbred strains, hundreds of spontaneous mutations, practical techniques for random mutagenesis, and, importantly, directed engineering of the genome through transgenic, knockout and knockin techniques[17-22].

For these and other reasons, the Human Genome Project (HGP) recognized from its outset that the sequencing of the human genome needed to be followed as rapidly as possible by the sequencing of the mouse genome. In early 2001, the International Human Genome Sequencing Consortium reported a draft sequence covering about 90% of the euchromatic human genome, with about 35% in finished form[1]. Since then, progress towards a complete human sequence has proceeded swiftly, with approximately 98% of the genome now available in draft form and about 95% in finished form.

Here, we report the results of an international collaboration involving centres in the United States and the United Kingdom to produce a high-quality draft sequence of the mouse genome and a broad scientific network to analyse the data. The draft sequence was generated by assembling about sevenfold sequence coverage from female mice of the C57BL/6J strain (referred to below as B6). The assembly contains about 96% of the sequence of the euchromatic genome (excluding chromosome Y) in sequence contigs linked together into large units, usually larger than 50 megabases (Mb).

With the availability of a draft sequence of the mouse genome, we have undertaken an

的。解开人类基因组秘密最有力的通用方法之一是比较基因组学，而进行比较最有力的出发点就是实验用小鼠，拉丁文名称为 *Mus musculus*。

打个比方，比较基因组学允许人们阅读进化的实验笔记。在人类和小鼠的谱系分开后的大约 7,500 万年中，进化历程改变了它们的基因组序列，通过几乎每两个核苷酸就替换一个核苷酸(见下文)以及核苷酸的缺失和插入产生分化。分歧率足够低的部分可以比对直系同源序列，但分歧率较高的部分可以通过它们更高的保守性来识别许多功能上重要的元件。对基因组小块区域的研究已经证实了这种跨物种的保守性在识别推测基因或调控元件方面的能力 [3-12]。全基因组范围的序列保守性分析有望系统地揭示所有基因的这类信息。物种之间的全基因组范围比较还可以使得包括突变和选择压力差异在内的形成基因组的关键差异作用更加突出地显示出来 [13,14]。

从字面上看，比较基因组学可以让人们将临床和基础研究人员的实验笔记联系起来。有了这两个基因组的知识，人类基因的生物医学研究可以通过实验操作相应的小鼠基因来补充，加速对于基因功能的理解。在这方面，小鼠是探索哺乳动物生物学和人类疾病无与伦比的模型系统 [15,16]。它的独特优势包括百年的遗传学研究、大量的近交系、成百个自发突变、随机诱变的实用实验技术，以及尤其重要的通过转基因、基因敲除和敲入技术进行的基因组定向操作 [17-22]。

基于这些和其他原因，人类基因组计划(HGP)从一开始就意识到人类基因组的测序需要尽可能快地接着进行小鼠基因组的测序。2001 年初，国际人类基因组测序协作组报告了一份覆盖了 90% 人类常染色质的基因组序列草图，其中 35% 是已经完成了的形式 [1]。从那时起，向着完整人类序列的发展进展迅速，目前大约 98% 的基因组以草图形式提供，约 95% 的基因组以完成图形式提供。

在此，我们报告由美国和英国的多个中心参加的国际合作所产生的一个高质量小鼠基因组序列草图和一个数据分析的广泛科学网络。这个序列草图是由 C57BL/6J 雌性小鼠(以下称为 B6)，大约 7 倍的序列覆盖度组装而成。组装结果包括 96% 的常染色质基因组序列(不包括 Y 染色体)，由测序序列片段重叠拼接在一起形成大的单元，通常大于 50 兆碱基(Mb)。

利用小鼠基因组的序列草图，我们进行了初步的对比分析，以检测人类和小鼠

initial comparative analysis to examine the similarities and differences between the human and mouse genomes. Some of the important points are listed below.

•The mouse genome is about 14% smaller than the human genome (2.5 Gb compared with 2.9 Gb). The difference probably reflects a higher rate of deletion in the mouse lineage.

•Over 90% of the mouse and human genomes can be partitioned into corresponding regions of conserved synteny, reflecting segments in which the gene order in the most recent common ancestor has been conserved in both species.

•At the nucleotide level, approximately 40% of the human genome can be aligned to the mouse genome. These sequences seem to represent most of the orthologous sequences that remain in both lineages from the common ancestor, with the rest likely to have been deleted in one or both genomes.

•The neutral substitution rate has been roughly half a nucleotide substitution per site since the divergence of the species, with about twice as many of these substitutions having occurred in the mouse compared with the human lineage.

•By comparing the extent of genome-wide sequence conservation to the neutral rate, the proportion of small (50–100 bp) segments in the mammalian genome that is under (purifying) selection can be estimated to be about 5%. This proportion is much higher than can be explained by protein-coding sequences alone, implying that the genome contains many additional features (such as untranslated regions, regulatory elements, non-protein-coding genes, and chromosomal structural elements) under selection for biological function.

•The mammalian genome is evolving in a non-uniform manner, with various measures of divergence showing substantial variation across the genome.

•The mouse and human genomes each seem to contain about 30,000 protein-coding genes. These refined estimates have been derived from both new evidence-based analyses that produce larger and more complete sets of gene predictions, and new *de novo* gene predictions that do not rely on previous evidence of transcription or homology. The proportion of mouse genes with a single identifiable orthologue in the human genome seems to be approximately 80%. The proportion of mouse genes without any homologue currently detectable in the human genome (and vice versa) seems to be less than 1%.

•Dozens of local gene family expansions have occurred in the mouse lineage. Most of these seem to involve genes related to reproduction, immunity and olfaction, suggesting that these physiological systems have been the focus of extensive lineage-specific innovation in rodents.

•Mouse–human sequence comparisons allow an estimate of the rate of protein evolution

基因组之间的相似性和差异。下面列出了一些要点。

- 小鼠基因组大概比人类基因组小约 14%(2.5 Gb 比 2.9 Gb)。这种差异可能反映了小鼠谱系中较高的缺失率。

- 超过 90% 的小鼠和人类基因组可以被划分成相应的保守同线性区域，反映了两个物种所保存的最近共同祖先区段上的基因顺序。

- 在核苷酸水平上，大约 40% 的人类基因组可以比对到小鼠基因组上。这些序列似乎代表了来自于共同祖先的两个谱系中的大部分直系同源序列，余下的序列可能在一个或两个基因组中被删除了。

- 自物种分化以来，每个位点的中性替代率大约是核苷酸替代率的一半，与人类谱系相比较，在小鼠中这些替代出现的次数大概是人类的两倍。

- 通过对全基因组序列的保守程度和中性率进行比较，估计哺乳动物基因组中处于(纯化)选择的小片段(50 ~ 100 bp)的比例大约为 5%。这个比例远远高于仅用蛋白质编码序列所能解释的，这意味着基因组很多额外性质(如非翻译区、调节元件、非蛋白质编码基因和染色体结构元件)也受到了生物学功能的选择。

- 哺乳动物基因组正在以不均匀的方式进化，对于分化的不同度量结果显示了基因组中的巨大变化。

- 小鼠和人类的基因组似乎都包含大约 30,000 个蛋白质编码基因。这些精准的估计分别来自基于新证据获得的更大、更完整的基因预测集和不依赖于先前转录或同源性证据的基因重新预测。具有单一可识别直系同源基因的小鼠基因在人类基因组中的比例大约为 80%。没有在人类基因组中检测到任何同源性的小鼠基因的比例似乎小于 1%，反之亦然。

- 小鼠谱系中已经发生了数十个局部基因家族的扩张。大部分看来与生殖、免疫和嗅觉相关，提示这些生理系统是啮齿动物的谱系特异性更新的焦点。

- 小鼠–人类序列的比较可以预测哺乳动物蛋白质的进化速率。某些与生殖、宿

in mammals. Certain classes of secreted proteins implicated in reproduction, host defence and immune response seem to be under positive selection, which drives rapid evolution.

- Despite marked differences in the activity of transposable elements between mouse and human, similar types of repeat sequences have accumulated in the corresponding genomic regions in both species. The correlation is stronger than can be explained simply by local (G+C) content and points to additional factors influencing how the genome is moulded by transposons.

- By additional sequencing in other mouse strains, we have identified about 80,000 single nucleotide polymorphisms (SNPs). The distribution of SNPs reveals that genetic variation among mouse strains occurs in large blocks, mostly reflecting contributions of the two subspecies *Mus musculus domesticus* and *Mus musculus musculus* to current laboratory strains.

The mouse genome sequence is freely available in public databases (GenBank accession number CAAA01000000) and is accessible through various genome browsers (http://www.ensembl.org/Mus_musculus/, http://genome.ucsc.edu/ and http://www.ncbi.nlm.nih.gov/genome/guide/mouse/).

In this paper, we begin with information about the generation, assembly and evaluation of the draft genome sequence, the conservation of synteny between the mouse and human genomes, and the landscape of the mouse genome. We then explore the repeat sequences, genes and proteome of the mouse, emphasizing comparisons with the human. This is followed by evolutionary analysis of selection and mutation in the mouse and human lineages, as well as polymorphism among current mouse strains. A full and detailed description of the methods underlying these studies is provided as Supplementary Information. In many respects, the current paper is a companion to the recent paper on the human genome sequence[1]. Extensive background information about many of the topics discussed below is provided there.

## Background to the Mouse Genome Sequencing Project

### Origins of the mouse

The precise origin of the mouse and human lineages has been the subject of recent debate. Palaeontological evidence has long indicated a great radiation of placental (eutherian) mammals about 65 million years ago (Myr) that filled the ecological space left by the extinction of the dinosaurs, and that gave rise to most of the eutherian orders[23]. Molecular phylogenetic analyses indicate earlier divergence times of many of the mammalian clades. Some of these studies have suggested a very early date for the divergence of mouse from other mammals (100–130 Myr[23-25]) but these estimates partially originate from the fast molecular clock in rodents (see below).Recent molecular studies that are less sensitive to the differences in evolutionary rates have suggested that the eutherian mammalian radiation

主防御和免疫应答相关的分泌蛋白质似乎是处于正选择，这些推动了蛋白质快速进化。

- 尽管小鼠和人类的转座元件的活性有明显差异，但是两个物种的相应基因组区域中积累了相似类型的重复序列。这种相关性要比简单通过局部(G+C)含量来解释更强，并且指向影响基因组如何被转座子塑造的其他因素。

- 通过对其他品系小鼠的额外测序，我们已经鉴定出了大概 80,000 个单核苷酸多态性(SNP)。SNP 的分布表明小鼠品系间的遗传变异发生在大的区块内，主要反映了两个小鼠亚种：*Mus musculus domesticus* 和 *Mus musculus musculus* 对当前实验室品系的贡献。

小鼠基因组序列可在公共数据库(GenBank 编号 CAAA01000000)中免费获得，并可通过多种基因组浏览器(http://www.ensembl.org/Mus_musculus/，http://genome.ucsc.edu/ 和 http://www.ncbi.nlm.nih.gov/genome/guide/mouse/)访问。

在本文中，我们从基因组序列草图的产生、组装和评估，小鼠和人类基因组之间的同线性保守性，以及小鼠基因组的概貌信息入手。然后，我们探索小鼠的重复序列、基因和蛋白质组，尤其是与人类的比较。接着是对小鼠和人类谱系的选择和突变以及目前小鼠品系中的多态性的进化分析。补充材料对这些研究的方法进行了完整而详细的描述。在很多方面，这篇论文与最近一篇关于人类基因组序列的文章[1]的相配套。那里提供了关于下面讨论的许多主题的充分的背景信息。

## 小鼠基因组测序项目的背景

### 小鼠的起源

小鼠和人类谱系的确切起源是最近的争论话题。古生物学的证据长期表明，大约 6,500 万年前胎盘哺乳类动物(真兽亚纲动物)的大量扩张，填充了恐龙灭绝留下的生态空间，真兽亚纲大多数目的动物由此产生[23]。分子系统发育分析表明，许多哺乳动物分支的分化时间较早。其中一些研究表明，小鼠从其他哺乳动物中分化出来的时间非常早(1 亿 ~ 3 亿年前[23-25])，但是这些估计部分来自于啮齿动物的快速分子钟(见下文)。最近对进化率差异不那么敏感的分子研究表明真兽亚纲动物的扩

took place throughout the Late Cretaceous period (65–100 Myr), but that rodents and primates actually represent relatively late-branching lineages[26,27]. In the analyses below, we use a divergence time for the human and mouse lineages of 75 Myr for the purpose of calculating evolutionary rates, although it is possible that the actual time may be as recent as 65 Myr.

## Origins of the mouse genetics

With the rediscovery of Mendel's laws of inheritance in 1900, pioneers of the new science of genetics (such as Cuenot, Castle and Little) were quick to recognize that the discontinuous variation of bred ("fancy") mice was analogous to that of Mendel's peas, and they set out to test the new theories of inheritance in mice. Mating programmes were soon established to create inbred strains, resulting in many of the modern, well-known strains (including C57BL/6J)[30].

Genetic mapping in the mouse began with Haldane's report[31] in 1915 of linkage between the pink-eye dilution and albino loci on the linkage group that was eventually assigned to mouse chromosome 7, just 2 years after the first report of genetic linkage in *Drosophila*. The genetic map grew slowly over the next 50 years as new loci and linkage groups were added—chromosome 7 grew to three loci by 1935 and eight by 1954. The accumulation of serological and enzyme polymorphisms from the 1960s to the early 1980s began to fill out the genome, with the map of chromosome 7 harbouring 45 loci by 1982 (refs 29, 31).

The real explosion, however, came with the development of recombinant DNA technology and the advent of DNA-sequence-based polymorphisms. Initially, this involved the detection of restriction-fragment length polymorphisms (RFLPs)[32]; later, the emphasis shifted to the use of simple sequence length polymorphisms (SSLPs; also called microsatellites), which could be assayed easily by polymerase chain reaction (PCR)[33-36] and readily revealed polymorphisms between inbred laboratory strains.

## Origins of mouse genomics

When the Human Genome Project (HGP) was launched in 1990, it included the mouse as one of its five central model organisms, and targeted the creation of genetic, physical and eventually sequence maps of the mouse genome.

By 1996, a dense genetic map with nearly 6,600 highly polymorphic SSLP markers ordered in a common cross had been developed[34], providing the standard tool for mouse genetics. Subsequent efforts filled out the map to over 12,000 polymorphic markers, although not all of these loci have been positioned precisely relative to one another. With these and other loci, Haldane's original two-marker linkage group on chromosome 7 had now swelled to about 2,250 loci.

张发生在整个白垩纪晚期(6,500万年～1亿年前)，但啮齿类动物和灵长类动物实际代表了相对较晚的分支谱系[26,27]。在下面的分析中，我们用7,500万年的人类和小鼠的谱系分歧时间来计算进化速率，尽管实际时间可能近至6,500万年前。

## 小鼠遗传学的起源

1900年，随着孟德尔遗传定律的重新发现，遗传的新科学先驱们(如屈埃诺、卡斯尔和利特尔)很快意识到培育出来的精致小鼠的不连续变异类似于孟德尔的豌豆，于是他们开始测试小鼠的新遗传理论。很快建立了交配方案培育近交系，从而产生了很多现代众所周知的品系(包括C57BL/6J)[30]。

小鼠的遗传作图始于1915年霍尔丹[31]关于连锁群上粉色眼稀释和白化基因座之间连锁的报告，该连锁群最终被定位于小鼠7号染色体，距首次报道果蝇的遗传连锁仅仅两年。随着新的基因座和连锁群的增加，遗传图谱在随后的50年中慢慢发展起来——7号染色体到1935年增长到3个基因座，到1954年增长到8个基因座。从20世纪60年代到80年代早期，血清学和酶学多态性的积累开始充实基因组，到1982年，7号染色体上包含了45个基因座(见参考文献29、31)。

然而，真正的爆发来自于重组DNA技术的发展和基于DNA序列的多态性的出现。最初，这只涉及限制性内切酶片段长度的多态性(RFLP)检测[32]；后来，重点转移到了简单序列长度多态性(SSLP，也称作微卫星)的应用上，这种多态性可以很容易地通过聚合酶链反应(PCR)进行检测[33-36]，并且容易发现实验室近交系之间的多态性。

## 小鼠基因组学的起源

1990年人类基因组计划(HGP)启动时就把小鼠作为五个中心模式生物之一，将构建遗传、物理和最终小鼠基因组序列图谱列为目标。

到1996年，已绘制出一个密度包括了常见杂交中顺序排列的近6,600个高多态性SSLP标记的遗传图谱[34]，为小鼠遗传学提供了标准工具。随后的工作将该图谱上补充为超过12,000个多态性标记，尽管并不是所有的基因座都精确地一一相对定位。利用这些和其他基因座，霍尔丹当初在7号染色体上的双标记连锁群现在已经扩大到含有约2,250个标记。

Physical maps of the mouse genome also proceeded apace, using sequence-tagged sites (STS) together with radiation-hybrid panels[37,38] and yeast artificial chromosome (YAC) libraries to construct dense landmark maps[39]. Together, the genetic and physical maps provide thousands of anchor points that can be used to tie clones or DNA sequences to specific locations in the mouse genome.

With these resources, it became straightforward (but not always easy) to perform positional cloning of classic single-gene mutations for visible, behavioural, immunological and other phenotypes. Many of these mutations provide important models of human disease, sometimes recapitulating human phenotypes with uncanny accuracy. It also became possible for the first time to begin dissecting polygenic traits by genetic mapping of quantitative trait loci (QTL) for such traits.

Continuing advances fuelled a growing desire for a complete sequence of the mouse genome. The development of improved random mutagenesis protocols led to the establishment of large-scale screens to identify interesting new mutants, increasing the need for more rapid positional cloning strategies. QTL mapping experiments succeeded in localizing more than 1,000 loci affecting physiological traits, creating demand for efficient techniques capable of trawling through large genomic regions to find the underlying genes. Furthermore, the ability to perform directed mutagenesis of the mouse germ line through homologous recombination made it possible to manipulate any gene given its DNA sequence, placing an increasing premium on sequence information. In all of these cases, it was clear that genome sequence information could markedly accelerate progress.

## Origin of the Mouse Genome Sequencing Consortium

With the sequencing of the human genome well underway by 1999, a concerted effort to sequence the entire mouse genome was organized by a Mouse Genome Sequencing Consortium (MGSC). The MGSC originally consisted of three large sequencing centres—the Whitehead/Massachusetts Institute of Technology (MIT) Center for Genome Research, the Washington University Genome Sequencing Center, and the Wellcome Trust Sanger Institute—together with an international database, Ensembl, a joint project between the European Bioinformatics Institute and the Sanger Institute.

In addition to the genome-wide efforts of the MGSC, other publicly funded groups have been contributing to the sequencing of the mouse genome in specific regions of biological interest. Together, the MGSC and these programmes have so far yielded clone-based draft sequence consisting of 1,859 Mb (74%, although there is redundancy) and finished sequence of 477 Mb (19%) of the mouse genome. Furthermore, Mural and colleagues[45] recently reported a draft sequence of mouse chromosome 16 containing 87 Mb (3.5%).

To analyse the data reported here, the MGSC was expanded to include the other publicly funded sequencing groups and a Mouse Genome Analysis Group consisting of scientists

通过联合使用序列标签位点(STS)、辐射杂交组合[37,38]和酵母人工染色体(YAC)文库来建立高密度标记物图谱[39]，小鼠基因组的物理图谱进展也很快。遗传和物理图谱提供了数千个锚定点，可以将克隆或DNA序列与小鼠基因组中的特定位置联系起来。

有了这些资源，对于可见的、行为的、免疫的以及其他表型的经典单基因突变进行定位克隆变得很直接(但并非总是容易的)。这些突变中的很多突变为人类疾病提供了重要模型，有时以不可思议的精准性描述人类表型。而且，通过进行数量性状基因座(QTL)的遗传作图也使首次开始解析多基因性状成为可能。

持续的进步激发了人们对小鼠基因组完整序列不断增长的渴望。改良的随机诱变方案的发展促成对于感兴趣的新突变的大规模筛选，增加了对更快的定位克隆策略的需求。数量性状基因座绘图实验成功地定位了超过1,000个影响生理性状的基因座，产生了对搜寻大片基因组区域定位基因的高效性技术的需求。而且通过同源重组技术对小鼠胚系进行定向诱变使得人们可以操控给定DNA序列的任何基因，从而增加了对序列信息的重视。所有这些例子中显然基因组序列信息最能显著加速进展。

## 小鼠基因组测序协作组的起源

1999年，随着人类基因组测序工作的顺利进行，小鼠基因组测序协作组(MGSC)组织协调了对小鼠整个基因组的测序。MGSC最初由三个大型测序中心——怀特黑德研究所/麻省理工学院(MIT)基因组研究中心、华盛顿大学基因组测序中心和威康信托桑格研究所，以及由欧洲生物信息研究所和桑格研究所合作建立的国际数据库Ensembl组成。

除了MGSC在全基因组范围的努力，其他公共资助的团体也一直在生物学感兴趣的特定区域为小鼠基因组测序做贡献。目前为止，MGSC和这些项目一起产生了基于克隆的小鼠基因组的1,859 Mb序列草图(74%，尽管存在冗余)和477 Mb(19%)的完成序列。此外，穆拉及其同事们[45]最近报道了一个包含87 Mb(3.5%)的小鼠16号染色体的序列草图。

为了分析这里报告的数据，MGSC扩大到包括其他公共项目资助的测序小组和

from 27 institutions in 6 countries.

## Generating the Draft Genome Sequence

### Sequencing strategy

Sanger and co-workers developed the strategy of random shotgun sequencing in the early 1980s, and it has remained the mainstay of genome sequencing over the ensuing two decades. The approach involves producing random sequence "reads", generating a preliminary assembly on the basis of sequence overlaps, and then performing directed sequencing to obtain a "finished" sequence with gaps closed and ambiguities resolved[46]. Ansorge and colleagues[47] extended the technique by the use of "paired-end sequencing", in which sequencing is performed from both ends of a cloned insert to obtain linking information, which is then used in sequence assembly. More recently, Myers and co-workers[48], and others, have developed efficient algorithms for exploiting such linking information.

The ultimate aim of the MGSC is to produce a finished, richly annotated sequence of the mouse genome to serve as a permanent reference for mammalian biology. In addition, we wished to produce a draft sequence as rapidly as possible to aid in the interpretation of the human genome sequence and to provide a useful intermediate resource to the research community. Accordingly, we adopted a hybrid strategy for sequencing the mouse genome. The strategy has four components: (1) production of a BAC-based physical map of the mouse genome by fingerprinting and sequencing the ends of clones of a BAC library[44]; (2) whole-genome shotgun (WGS) sequencing to approximately sevenfold coverage and assembly to generate an initial draft genome sequence; (3) hierarchical shotgun sequencing[46] of BAC clones covering the mouse genome combined with the WGS data to create a hybrid WGS-BAC assembly; and (4) production of a finished sequence by using the BAC clones as a template for directed finishing. This mixed strategy was designed to exploit the simpler organizational aspects of WGS assemblies in the initial phase, while still culminating in the complete high-quality sequence afforded by clone-based maps.

We chose to sequence DNA from a single mouse strain, rather than from a mixture of strains[45], to generate a solid reference foundation, reasoning that polymorphic variation in other strains could be added subsequently (see below). After extensive consultation with the scientific community[52], the B6 strain was selected because of its principal role in mouse genetics, including its well-characterized phenotype and role as the background strain on which many important mutations arose. We elected to sequence a female mouse to obtain equal coverage of chromosome X and autosomes. Chromosome Y was thus omitted, but this chromosome is highly repetitive (the human chromosome Y has multiple duplicated regions exceeding 100 kb in size with 99.9% sequence identity[53]) and seemed an unwise target for the WGS approach. Instead, mouse chromosome Y is being sequenced by a purely clone-based (hierarchical shotgun) approach.

由6个国家27个科研机构的科学家组成的小鼠基因组分析小组。

## 生成基因组草图

### 测序策略

20世纪80年代早期，桑格和同事们研发了随机的鸟枪法测序策略，在随后的20年里，它依然是基因组测序的主流手段。该方法包括产生随机序列“读取片段”，基于序列重叠产生初步组装，然后进行定向测序，以获得填补了空缺和解决了歧义的“完成”序列[46]。安索奇及其同事们[47]通过进行“双末端测序”扩展了该技术，从克隆插入片段的两端进行测序，获得连接信息，然后用于序列组装。最近，迈尔斯和同事们[48]以及其他人已经开发了利用这种连接信息的有效算法。

MGSC的最终目标是产生一个完整的、注释丰富的小鼠基因组序列，作为哺乳动物生物学的永久参考。此外，我们希望尽可能快地产生一个序列草图，以帮助解释人类基因组序列，并为学界提供一个有用的中间资源。因此，我们采用混合策略对小鼠基因组进行测序。这个策略包括四个部分：(1)通过指纹法和对BAC文库克隆末端的测序产生基于BAC的小鼠基因组物理图谱[44]；(2)全基因组鸟枪(WGS)测序到大约7重覆盖范围，组装产生最初的基因组序列草图；(3)覆盖小鼠基因组的BAC克隆的分级鸟枪测序[46]与WGS数据结合，以构建WGS-BAC组装；以及(4)利用BAC克隆作为模板进行定向整理，生产完成序列。这种混合策略的设计主要是在初始阶段先解析较为简单的WGS组装的组织构架，最终通过克隆图谱获得完整的高质量序列。

我们选择一个单一的小鼠品系进行DNA测序，而不是通过品系的混合[45]产生坚实的参考基础，理由是随后可以不断添加其他品系的多态性变化(见下文)。经过科学界的广泛磋商[52]，B6品系因其在小鼠遗传中的重要作用而被选中。B6品系有良好的特征表型，是一个背景品系，许多重要的突变都发生在该品系上。我们选择对雌性小鼠进行测序，以获得X染色体和常染色体的相等覆盖率。因此染色体Y被省略了，但是这条染色体具有高度重复性(人类Y染色体具有多个重复区域，这些区域超过100 kb，具有99.9%的序列一致性[53])，似乎并不是WGS方法的明智目标。取而代之的是，将通过纯克隆的方法(分级鸟枪)对小鼠Y染色体进行测序。

## Sequencing and assembly

The genome assembly was based on a total of 41.4 million sequence reads derived from both ends of inserts (paired-end reads) of various clone types prepared from B6 female DNA. The inserts ranged in size from 2 to 200 kb. The three large MGSC sequencing centres generated 40.4 million reads, and 0.6 million reads were generated at the University of Utah. In addition, we used 0.4 million reads from both ends of BAC inserts reported by The Institute for Genome Research[54].

A total of 33.6 million reads passed extensive checks for quality and source, of which 29.7 million were paired; that is, derived from opposite ends of the same clone. The assembled reads represent approximately 7.7-fold sequence coverage of the euchromatic mouse genome (6.5-fold coverage in bases with a Phred quality score of > 20)[55]. Together, the clone inserts provide roughly 47-fold physical coverage of the genome.

The sequence reads, together with the pairing information, were used as input for two recently developed sequence-assembly programs, Arachne[56,57] and Phusion[58]. No mapping information and no clone-based sequences were used in the WGS assembly, with the exception of a few reads (< 0.1% of the total) derived from a handful of BACs, which were used as internal controls. The assembly programs were tested and compared on intermediate data sets over the course of the project and were thereby refined. The programs produced comparable outputs in the final assembly. The assembly generated by Arachne was chosen as the draft sequence described here because it yielded greater short-range and long-range continuity with comparable accuracy.

The assembly contains 224,713 sequence contigs, which are connected by at least two read-pair links into supercontigs (or scaffolds). There are a total of 7,418 supercontigs at least 2 kb in length, plus a further 37,125 smaller supercontigs representing < 1% of the assembly. The contigs have an N50 length of 24.8 kb, whereas the supercontigs have an N50 length that is approximately 700-fold larger at 16.9 Mb (N50 length is the size $x$ such that 50% of the assembly is in units of length at least $x$). In fact, most of the genome lies in supercontigs that are extremely large: the 200 largest supercontigs span more than 98% of the assembled sequence, of which 3% is within sequence gaps.

## Anchoring to chromosomes

We assigned as many supercontigs as possible to chromosomal locations in the proper order and orientation. Supercontigs were localized largely by sequence alignments with the extensively validated mouse genetic map[34], with some additional localization provided by the mouse radiation-hybrid map[37] and the BAC map[44]. We found no evidence of incorrect global joins within the supercontigs (that is, multiple markers supporting two discordant locations within the genome), and thus were able to place them directly. Altogether, we

## 测序和组装

基因组组装基于共计 4,140 万个序列读取片段而成，均来自 B6 雌鼠 DNA 制备的各种克隆类型的双末端插入片段(成对末端序列读取片段)。插入片段大小为 2 ~ 200 kb。三个大型 MGSC 测序中心产生了 4,040 万个序列读取片段，犹他大学产生了 60 万个序列读取片段。此外，我们使用了基因组研究所报告的 BAC 插入片段双末端的 40 万个序列读取片段 [54]。

总计 3,360 万个序列读取片段通过了质量和来源的充分检查，其中 2,970 万个是配对的；也就是说，来自同一个克隆的两端。组装了的序列读取片段代表了小鼠常染色体基因组约 7.7 重序列覆盖率(Phred 质量分数大于 20 的碱基覆盖率为 6.5 重)[55]。克隆插入片段总共提供了大约 47 倍的基因组物理覆盖。

序列读取片段以及配对信息被最近开发的两个序列组装程序(Arachne[56,57] 和 Phusion[58])用来作为输入信息进行组装。除了从少量作为内参的 BAC 中得到一些序列读取片段(小于总数的 0.1%)，WGS 组装中不使用图谱信息和基于克隆的序列。在项目过程中，对组装程序进行了中间数据集的测试和比较，从而进行了改进。这些程序在最终的组装中产生可比较的序列输出。Arachne 生成的组装被选为本文所描述的序列草图，因为它产生了更好的短区段和长区段的连续性，具有相当的精确度。

这个组装包括 224,713 个序列重叠群，它们通过至少两个配对序列读取片段连接成超级重叠群(或者支架)。总共有 7,418 个超级重叠群至少 2 kb 长，另外还有 37,125 个代表了小于 1% 组装的较小超级重叠群。重叠群的 N50 长度为 24.8 kb，而超级重叠群的 N50 长度大概是其 700 倍达到 16.9 Mb(如果 N50 长度是 $x$，则 50% 组装件的长度至少为 $x$)。事实上，大部分基因组都包括在非常大的超级重叠群中：200 个最大的超级重叠群跨越了 98% 以上的组装序列，其中还有 3% 处于测序空缺区域内。

## 染色体锚定

我们将尽可能多的超级重叠群按照适当的顺序和方向配到染色体位置上。超级重叠群主要通过与充分验证了的小鼠遗传图谱 [34] 进行比对而定位，一些额外的定位由小鼠辐射杂交图谱 [37] 和 BAC 图谱 [44] 提供。我们没有发现超级重叠群中有不正确的总体连接的证据(即两个不一致位置受到基因组中多个标记的支持)，因此可以直接将它们放入染色体中。我们总共放置了 377 个超级重叠群，包括所有长度超过

placed 377 supercontigs, including all supercontigs > 500 kb in length.

Once much of the sequence was anchored, it was possible to exploit additional read-pair and physical mapping information to obtain greater continuity. For example, some adjacent supercontigs were connected by BAC-end (or other) links, satisfying appropriate length and orientation constraints, including single links. Furthermore, some adjacent extended supercontigs were connected by means of fingerprint contigs in the BAC-based physical map. These additional links were used to join sequence into ultracontigs. In the end, a total of 88 ultracontigs with an N50 length of 50.6 Mb (exclusive of gaps) contained 95.7% of the assembled sequence (Fig. 1).

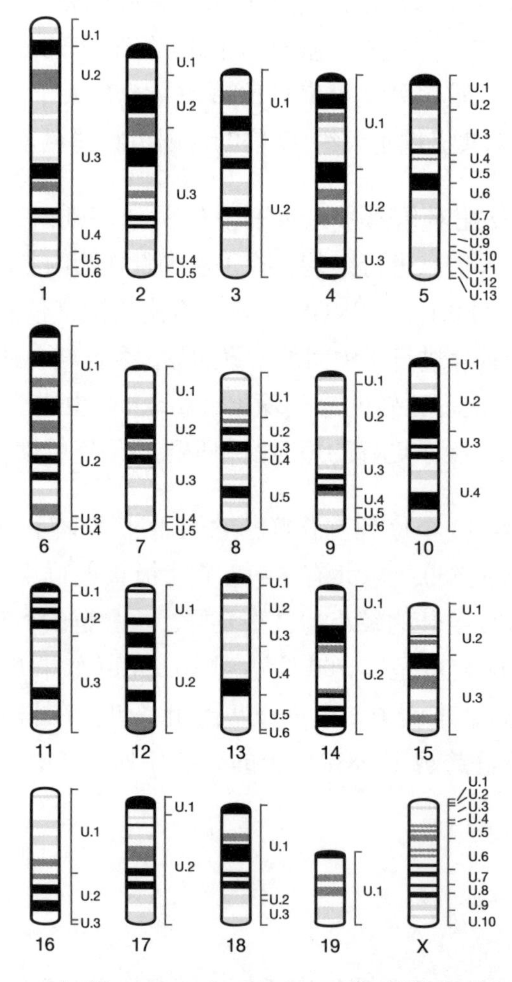


Fig. 1. The mouse genome in 88 sequence-based ultracontigs. The position and extent of the 88 ultracontigs of the MGSCv3 assembly are shown adjacent to ideograms of the mouse chromosomes. All mouse chromosomes are acrocentric, with the centromeric end at the top of each chromosome. The supercontigs of the sequence assembly were anchored to the mouse chromosomes using the MIT genetic map. Neighbouring supercontigs were linked together into ultracontigs using information from single BAC links and the fingerprint and radiation-hybrid maps, resulting in 88 ultracontigs containing 95% of the bases in the euchromatic genome.

500 kb 的超级重叠群。

一旦锚定了大部分序列，就可以利用额外的配对序列读取片段和物理图谱信息来获得更大的连接。例如，一些相邻的超级重叠群通过 BAC 的末端（或者其他）连接相接，如果这些连接包括单连接满足长度和方向约束。此外，在以 BAC 为基础的物理图谱中，一些连接是通过指纹图谱重叠群连接一些相邻延伸的超级重叠群。这些额外的连接用于将序列连成特大超级重叠群。最终，共有 88 个特大超级重叠群，N50 长度为 50.6 Mb（不包括空缺），包括了 95.7% 的组装序列（图 1）。

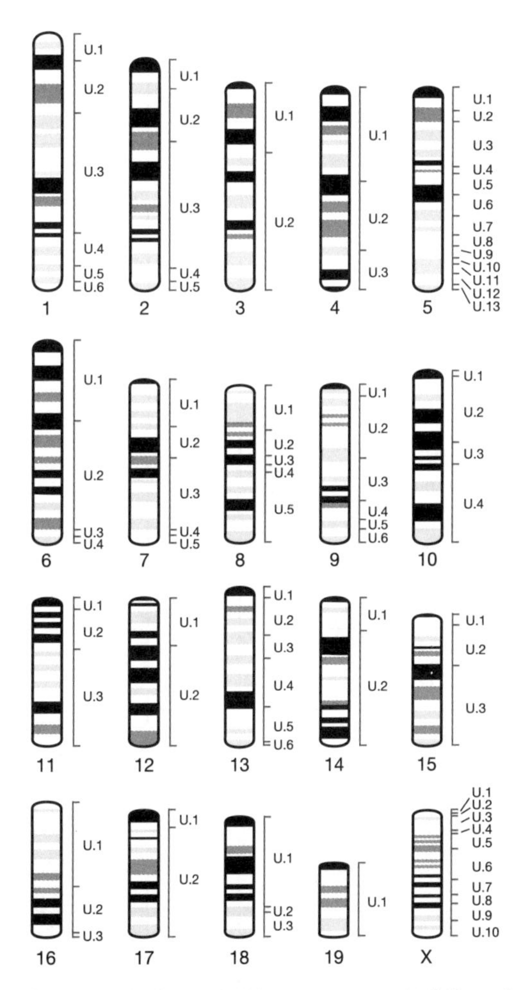


图 1. 88 个基于序列的特大超级重叠群中的小鼠基因组。MGSCv3 组装的 88 个特大超级重叠群的位置和范围显示在小鼠染色体核型模式图相邻的位置。小鼠所有染色体都是近端着丝粒的，着丝粒位于每条染色体的顶端。利用 MIT 基因图谱，将序列组装的超级重叠群锚定在小鼠染色体上。利用来自单个 BAC 连接、指纹图谱和辐射杂交图谱的信息，将相邻的超级重叠群连在一起形成特大超级重叠群，产生 88 个包括常染色体基因组中 95% 碱基的特大超级重叠群。

Of the 187 Mb of finished mouse sequence, 96% was contained in the anchored assembly. This finished sequence, however, is not a completely random cross-section of the genome (it has been cloned as BACs, finished, and in some cases selected on the basis of its gene content). Of 11,452 cDNA sequences from the curated RefSeq collection, 99.3% of the cDNAs could be aligned to the genome sequence (see Supplementary Information). These alignments contained 96.4% of the cDNA bases. Together, this indicates that the draft genome sequence includes approximately 96% of the euchromatic portion of the mouse genome, with about 95% anchored.

Table 3. Mouse chromosome size estimates

| Chromosome | Actual bases in sequence (Mb) | Ultracontigs (Mb) | | Gaps within supercontigs | | Gaps between supercontigs | | | | | Total estimated size (Mb) ‡ |
|---|---|---|---|---|---|---|---|---|---|---|---|
| | | | | | | Captured by additional read pairs | | Captured by fingerprint contigs* | | Uncaptured† | |
| | | Number | N50 size | Number | Mb | Number | Mb | Number | Mb | Number | |
| All | 2,372 | 88 | 52.7 | 176,094 | 104.5 | 252 | 14.0 | 37 | 2.30 | 68 | 2,493 |
| 1 | 183 | 6 | 52.7 | 13,178 | 7.8 | 16 | 1.1 | 1 | 0.32 | 5 | 192 |
| 2 | 169 | 5 | 111.1 | 12,141 | 6.5 | 4 | 0.1 | 1 | 0.20 | 4 | 176 |
| 3 | 149 | 2 | 108.9 | 10,630 | 6.8 | 17 | 0.7 | 3 | 0.16 | 1 | 157 |
| 4 | 140 | 3 | 83.1 | 10,745 | 6.3 | 14 | 0.4 | 3 | 0.26 | 2 | 147 |
| 5 | 137 | 13 | 17.8 | 11,288 | 6.7 | 11 | 0.5 | 3 | 0.11 | 12 | 144 |
| 6 | 138 | 4 | 91.4 | 10,021 | 6.6 | 19 | 1.1 | 2 | 0.26 | 3 | 146 |
| 7 | 122 | 5 | 45.1 | 9,484 | 5.7 | 55 | 3.4 | 4 | 0.12 | 4 | 131 |
| 8 | 119 | 5 | 35.0 | 9,186 | 6.1 | 7 | 0.2 | 2 | 0.12 | 4 | 125 |
| 9 | 116 | 6 | 26.8 | 8,479 | 4.5 | 6 | 0.6 | 1 | 0.06 | 5 | 121 |
| 10 | 121 | 4 | 50.4 | 9,490 | 5.4 | 9 | 0.6 | 0 | 0 | 3 | 127 |
| 11 | 115 | 3 | 80.4 | 8,681 | 4.3 | 2 | 0.0 | 1 | 0.05 | 2 | 119 |
| 12 | 105 | 2 | 77.4 | 7,577 | 4.0 | 27 | 1.2 | 2 | 0.00 | 1 | 110 |
| 13 | 107 | 6 | 28.0 | 7,910 | 4.7 | 13 | 0.8 | 4 | 0.19 | 5 | 113 |
| 14 | 107 | 2 | 93.6 | 7,605 | 4.0 | 10 | 0.5 | 2 | 0.12 | 1 | 112 |
| 15 | 96 | 3 | 65.3 | 7,025 | 4.3 | 2 | 0.1 | 0 | 0 | 2 | 100 |
| 16 | 91 | 3 | 62.3 | 6,695 | 4.4 | 1 | 0.0 | 0 | 0 | 2 | 95 |
| 17 | 85 | 2 | 80.8 | 6,584 | 3.7 | 17 | 1.2 | 4 | 0.19 | 1 | 90 |
| 18 | 84 | 3 | 73.5 | 6,192 | 3.2 | 2 | 0.0 | 0 | 0 | 2 | 87 |
| 19 | 55 | 1 | 57.7 | 3,934 | 2.4 | 7 | 0.6 | 2 | 0.12 | 0 | 58 |
| X | 134 | 10 | 19.9 | 9,249 | 7.0 | 13 | 0.8 | 2 | 0.00 | 9 | 142 |

*These gaps had fingerprint contigs spanning them. The size for 18 out of 37 were estimated using conserved synteny to determine the size of the region in the human genome. The remaining gaps were arbitrarily given the average size of the assessed gaps (59 kb), adjusted to reflect the 16% difference in genome size.

† Uncaptured gaps were estimated by mouse–human synteny to have a total size of 5 Mb. However, because some of these gaps are due to repetitive expansions in mouse (absent in human), the actual total for the uncaptured gaps

在 187 Mb 已完成的小鼠序列中，96% 包含在锚定到染色体的组装中。然而，这个已完成的序列不是一个完全随机的基因组横截面(它已经被克隆成 BAC，完成了测序，并且在一些情况下是根据其基因含量被选择的)。在 RefSeq 数据库收集的 11,452 个 cDNA 序列中，99.3% 的 cDNA 可以和基因组序列比对(见补充信息)。这些比对包含 96.4% 的 cDNA 碱基。这表明基因组序列草图包括小鼠基因组大约 96% 的常染色体部分，其中约 95% 被锚定在染色体上。

表 3. 小鼠染色体长度估算

| 染色体 | 序列中实际碱基(Mb) | 特大超级重叠群(Mb) | | 超级重叠群内的测序空缺 | | 超级重叠群之间的测序空缺 | | | | | 总估计长度(Mb)‡ |
|---|---|---|---|---|---|---|---|---|---|---|---|
| | | 数量 | N50 大小 | 数量 | Mb | 由其他读取片段对捕获 | | 由指纹图谱重叠群捕获 * | | 未捕获 † | |
| | | | | | | 数量 | Mb | 数量 | Mb | 数量 | |
| 合计 | 2,372 | 88 | 52.7 | 176,094 | 104.5 | 252 | 14.0 | 37 | 2.30 | 68 | 2,493 |
| 1 | 183 | 6 | 52.7 | 13,178 | 7.8 | 16 | 1.1 | 1 | 0.32 | 5 | 192 |
| 2 | 169 | 5 | 111.1 | 12,141 | 6.5 | 4 | 0.1 | 1 | 0.20 | 4 | 176 |
| 3 | 149 | 2 | 108.9 | 10,630 | 6.8 | 17 | 0.7 | 3 | 0.16 | 1 | 157 |
| 4 | 140 | 3 | 83.1 | 10,745 | 6.3 | 14 | 0.4 | 3 | 0.26 | 2 | 147 |
| 5 | 137 | 13 | 17.8 | 11,288 | 6.7 | 11 | 0.5 | 3 | 0.11 | 12 | 144 |
| 6 | 138 | 4 | 91.4 | 10,021 | 6.6 | 19 | 1.1 | 2 | 0.26 | 3 | 146 |
| 7 | 122 | 5 | 45.1 | 9,484 | 5.7 | 55 | 3.4 | 4 | 0.12 | 4 | 131 |
| 8 | 119 | 5 | 35.0 | 9,186 | 6.1 | 7 | 0.2 | 2 | 0.12 | 4 | 125 |
| 9 | 116 | 6 | 26.8 | 8,479 | 4.5 | 6 | 0.6 | 1 | 0.06 | 5 | 121 |
| 10 | 121 | 4 | 50.4 | 9,490 | 5.4 | 9 | 0.6 | 0 | 0 | 3 | 127 |
| 11 | 115 | 3 | 80.4 | 8,681 | 4.3 | 2 | 0.0 | 1 | 0.05 | 2 | 119 |
| 12 | 105 | 2 | 77.4 | 7,577 | 4.0 | 27 | 1.2 | 2 | 0.00 | 1 | 110 |
| 13 | 107 | 6 | 28.0 | 7,910 | 4.7 | 13 | 0.8 | 4 | 0.19 | 5 | 113 |
| 14 | 107 | 2 | 93.6 | 7,605 | 4.0 | 10 | 0.5 | 2 | 0.12 | 1 | 112 |
| 15 | 96 | 3 | 65.3 | 7,025 | 4.3 | 2 | 0.1 | 0 | 0 | 2 | 100 |
| 16 | 91 | 3 | 62.3 | 6,695 | 4.4 | 1 | 0.0 | 0 | 0 | 2 | 95 |
| 17 | 85 | 2 | 80.8 | 6,584 | 3.7 | 17 | 1.2 | 4 | 0.19 | 1 | 90 |
| 18 | 84 | 3 | 73.5 | 6,192 | 3.2 | 2 | 0.0 | 0 | 0 | 2 | 87 |
| 19 | 55 | 1 | 57.7 | 3,934 | 2.4 | 7 | 0.6 | 2 | 0.12 | 0 | 58 |
| X | 134 | 10 | 19.9 | 9,249 | 7.0 | 13 | 0.8 | 2 | 0.00 | 9 | 142 |

* 这些测序空缺含有横跨的指纹图谱重叠群。利用人类基因组中该区域长度进行保守同线性估计，得到了 37 个中的 18 个的大小。剩余的空缺且用被估算空缺的平均值(59 kb)给出，由此调节反映了基因组的 16% 差异。

† 通过小鼠–人类同线性估计未捕获的测序空缺共有 5 Mb。然而，因为这些空缺中的一部分是由于小鼠中的重复扩张(人类中缺失)，所以未捕获空缺的实际总数可能会高得多。例如，1 号染色体(Sp-100rs 区域)上一个大的未捕获空缺就大约有 6 Mb(见正文)。

is probably substantially higher. For example, one large uncaptured gap on chromosome 1 (the Sp-100rs region) is roughly 6 Mb (see text).

‡ Omitting centromeres and telomeres. These would add, on average, approximately 8 Mb per chromosome, or about 160 Mb to the genome. Also omitting uncaptured gaps between supercontigs.

On the basis of the estimated sizes of the ultracontigs and gaps between them, the total length of the euchromatic mouse genome was estimated to be about 2.5 Gb (see Supplementary Information), or about 14% smaller than that of the euchromatic human genome (about 2.9 Gb) (Table 3).

## Quality assessment at intermediate scale

Although no evidence of large-scale misassembly was found when anchoring the assembly onto the mouse chromosomes, we examined the assembly for smaller errors.

To assess the accuracy at an intermediate scale, we compared the positions of well-studied markers on the mouse genetic map and in the genome assembly (see Supplementary Information). Out of 2,605 genetic markers that were unambiguously mapped to the sequence assembly (BLAST match using $10^{-100}$ or better as an *E*-value to a single location) we found 1.8% in which the chromosomal assignment in the genetic map conflicted with that in the sequence. This is well within the known range of erroneous assignments within the genetic map[34].

We also found 19 instances (0.7%) of conflicts in local marker order between the genetic map and sequence assembly. A conflict was defined as any instance that would require changing more than a single genotype in the data underlying the genetic map to resolve. We studied ten cases by re-mapping the genetic markers, and eight were found to be due to errors in the genetic map. On the basis of this analysis, we estimate that chromosomal misassignment and local misordering affects $< 0.3\%$ of the assembled sequence.

At the single nucleotide level in the assembly, the observed discrepancy rates varied in a manner consistent with the quality scores assigned to the bases in the WGS assembly (see Supplementary Information). Overall, 96% of nucleotides in the assembly have Arachne quality scores $\geqslant 40$, corresponding to a predicted error rate of 1 per 10,000 bases. Such bases had an observed discrepancy rate against finished sequence of 0.005%, or 5 errors per 100,000 bases.

## Collapse of duplicated regions

The human genome contains many large duplicated regions, estimated to comprise roughly 5% of the genome[59], with nearly identical sequence. If such regions are also common in the mouse genome, they might collapse into a single copy in the WGS assembly. Such artefactual collapse could be detected as regions with unusually high read coverage, compared with the average depth of 7.4-fold in long assembled contigs. We searched for

‡ 去除着丝粒和端粒。这些将使每个染色体平均增加大约 8 Mb，或者基因组增加大约 160 Mb。也除去了超级重叠群之间的未捕获空缺。

以特大超级重叠群和它们之间的空缺的预估尺寸为基础，小鼠常染色体基因组的总长度估计约为 2.5 Gb(见补充信息)，或者比人类常染色体基因组(约 2.9 Gb)小了约 14%(表 3)。

## 中等规模的质量评估

虽然在将组装锚定到小鼠染色体上时没有大规模错误组装的证据，我们还是检查了组装以发现较小的错误。

为了在中等规模上评估准确性，我们比较了小鼠遗传图谱和基因组组装中研究透彻的标记位置(见补充信息)。在明确定位到序列组装的 2,605 个遗传标记中(用 $10^{-100}$ 或更好值作为 $E$ 值与单个位置进行 BLAST 匹配)，我们发现遗传图谱中有 1.8% 的染色体分配和序列不符。这完全在遗传图谱已知的错误分配范围内[34]。

我们还发现局部标记的顺序在遗传图谱和序列组装之间有 19 例(0.7%)不符。矛盾的定义是需要在遗传图谱基础数据中进行多于一个基因型的更改才能解决不符的任何情况。我们通过重新放置映射遗传标记研究了 10 个例子，发现其中 8 个是遗传图谱的错误。在此分析基础上，我们估计染色体分配错误和局部顺序错误影响小于 0.3% 的组装序列。

在组装的单核苷酸水平上，观测到的差异率的变化一定程度上与 WGS 组装中的匹配碱基的质量分数相一致(参见补充信息)。总体来说，该组装中 96% 的核苷酸 Arachne 质量分数 ≥40，相当于预测误差率为每 10,000 个碱基有 1 个错误。这类错误碱基在完成序列上所观测到的差异率为 0.005%，或者每 100,000 个碱基有 5 个错误。

## 复制区域的分解

人类基因组包含很多大规模的复制区域，估计约占基因组的 5%[59]，它们具有几乎相同的序列。如果这些区域在小鼠的基因组中也很常见，在 WGS 组装中它们可能会分解成单个拷贝。与平均深度 7.4 倍的长组装重叠群相比，这种人工折叠分解会被检测为具有异常高的读取片段覆盖度的区域。我们搜索了长度大于 20 kb 且其

contigs that were > 20 kb in size and contained > 10 kb of sequence in which the read coverage was at least twofold higher than the average. Such regions comprised only a tiny fraction ( < 0.0001) of the total assembly, of which only half had been anchored to a chromosome. None of these windows had coverage exceeding the average by more than threefold. This may indicate that the mouse genome contains fewer large regions of near-exact duplication than the human. Alternatively, regions of near-exact duplication may have been systematically excluded by the WGS assembly programme. This issue is better addressed through hierarchical shotgun than WGS sequencing and will be examined more carefully in the course of producing a finished mouse genome sequence.

### Unplaced reads and large tandem repeats

We expected that highly repetitive regions of the genome would not be assembled or would not be anchored on the chromosomes. Indeed, 5.9 million of the 33.6 million passing reads were not part of anchored sequence, with 88% of these not assembled into sequence contigs and 12% assembled into small contigs but not chromosomally localized.

### Evaluation of WGS assembly strategy

The WGS assembly described here involved only random reads, without any additional map-based information. By many criteria, the assembly is of very high quality. The N50 supercontig size of 16.9 Mb far exceeds that achieved by any previous WGS assembly, and the agreement with genome-wide maps is excellent. The assembly quality may be due to several factors, including the use of high-quality libraries, the variety of insert lengths in multiple libraries, the improved assembly algorithms, and the inbred nature of the mouse strain (in contrast to the polymorphisms in the human genome sequences). Another contributing factor may be that the mouse differs from the human in having less recent segmental duplication to confound assembly.

Notwithstanding the high quality of the draft genome sequence, we are mindful that it contains many gaps, small misassemblies and nucleotide errors. It is likely that these could not all be resolved by further WGS sequencing, therefore directed sequencing will be needed to produce a finished sequence. The results also suggest that WGS sequencing may suffice for large genomes for which only draft sequence is required, provided that they contain minimal amounts of sequence associated with recent segmental duplications or large, recent interspersed repeat elements.

### Adding finished sequence

As a final step, we enhanced the WGS sequence assembly by substituting available finished BAC-derived sequence from the B6 strain. In total, we replaced 3,528 draft sequence

中大于 10 kb 序列的读取覆盖度至少是平均值两倍的重叠群。这些区域只包括总组装的小部分（小于 0.0001），其中只有一半锚定在染色体上。这些窗口的覆盖度都没有超过平均值三倍以上。这可能表明小鼠基因组含有比人类更少的近精准复制的大区域。另一种可能是，WGS 组装程序可能已经系统地排除了近精确复制的区域。与 WGS 测序相比，分级鸟枪法能更好解决这个问题，因此将在构建小鼠基因组完成图序列的过程中更仔细地检查这个问题。

### 未入选的序列读取片段和大串联重复

我们预计基因组中高度重复区域不会被组装或者不会被锚定到染色体上。确实，3,360 万个合格的序列读取中有 590 万个不是被锚定序列的一部分，其中的 88% 不会被组装成序列重叠群，12% 组装成小的重叠群，但是没定位在染色体上。

### WGS 组装策略的评估

这里描述的 WGS 组装只涉及随机的序列读取片段，没有任何附加的基于图谱本身的信息。按照很多标准，这种组装是高质量的。超级重叠群的 N50 长度为 16.9 Mb，远远超过之前任何 WGS 组装的结果，并且与全基因组图谱有极好的一致性。组装质量可能归因于几个因素，包括高质量文库的使用、多个文库中插入片段长度的多样性、改进的组装算法以及小鼠品系的近交特性（与人类基因组序列的多态性相比较）。另外一个因素可能是小鼠与人类的不同之处，它具有较少的混淆组装的近期片段复制。

尽管基因组序列草图的质量很高，我们注意到了它包含很多测序空缺、小的组装错误和核苷酸错误。这些问题很可能无法通过进一步的 WGS 测序来解决，因此需要定向测序来产生一个完成图序列。这些结果也提示，WGS 测序对于只需要序列草图的大基因组可能就足够了，前提是近期片段复制或者大量近期分散开的重复元件相关序列最少。

### 添加完成序列

作为最后一步，我们通过替换来自 B6 品系可用的 BAC 完成序列来增强 WGS 序列组装。总共用组装时能够提供的 210 个完成 BAC 的 48.2 Mb 完成序列替换了

contigs with 48.2 Mb of finished sequence from 210 finished BACs available at the time of the assembly. The resulting draft genome sequence, MGSCv3, was submitted to the public databases and is freely available in electronic form through various sources (see below).

As the MGSC produces additional BAC assemblies and finished sequence, we plan to continue to revise and release enhanced versions of the genome sequence *en route* to a completely finished sequence[66], thereby providing a permanent foundation for biomedical research in the twenty-first century.

## Conservation of Synteny between Mouse and Human Genomes

With the draft sequence in hand, we began our analysis by investigating the strong conservation of synteny between the mouse and human genomes. Beyond providing insight into evolutionary events that have moulded the chromosomes, this analysis facilitates further comparisons between the genomes.

Starting from a common ancestral genome approximately 75 Myr, the mouse and human genomes have each been shuffled by chromosomal rearrangements. The rate of these changes, however, is low enough that local gene order remains largely intact. It is thus possible to recognize syntenic (literally "same thread") regions in the two species that have descended relatively intact from the common ancestor.

The earliest indication that genes reside in similar relative positions in different mammalian species traces to the observation that the albino and pink-eye dilution mutants are genetically closely linked in both mouse and rat[67,68]. Significant experimental evidence came from genetic studies of somatic cells[69]. A recent gene-based synteny map[37] used more than 3,600 orthologous loci to define about 200 regions of conserved synteny. However, it is recognized that such maps might still miss regions owing to insufficient marker density.

With a robust draft sequence of the mouse genome and > 90% finished sequence of the human genome in hand, it is possible to undertake a more comprehensive analysis of conserved synteny. Rather than simply relying on known human–mouse gene pairs, we identified a much larger set of orthologous landmarks as follows. We performed sequence comparisons of the entire mouse and human genome sequences using the PatternHunter program[71] to identify regions having a similarity score exceeding a high threshold ( > 40, corresponding to a minimum of a 40-base perfect match, with penalties for mismatches and gaps), with the additional property that each sequence is the other's unique match above this threshold. Such regions probably reflect orthologous sequence pairs, derived from the same ancestral sequence.

About 558,000 orthologous landmarks were identified; in the mouse assembly, these sequences have a mean spacing of about 4.4 kb and an N50 length of about 500 bp. The landmarks had a total length of roughly 188 Mb, comprising about 7.5% of the mouse

3,528 个序列草图重叠群。最终的基因组序列草图 MGSCv3 被提交给各公共数据库，并且通过各种资源以电子形式免费提供（见下文）。

随着 MGSC 产生更多的 BAC 组装和完成序列，我们计划继续修订和发布基因组序列的改进版本，使其成为完整的完成序列[66]，从而为二十一世纪的生物医学研究提供永久的基础。

## 小鼠和人类基因组同线性的保守性

随着序列草图在手，我们开始分析小鼠和人类基因组之间同线性的强保守性。这项研究不仅可以深入了解染色体构造中的进化事件，还可以促进不同物种基因组之间的进一步比较。

从大约 7,500 万年前的共同的祖先基因组开始，小鼠和人类的基因组都被染色体重排彻底改变了。然而，这个改变率相当低，以至于局部基因序列仍然基本保持完整。因此我们才有可能识别两个物种从共同祖先那里相对完整地传下来的同线性区域（字面意思是“同一条线”）。

最早的迹象表明，在不同哺乳动物物种中，基因位于相似的相对位置，这可追溯到对小鼠和大鼠中白化和粉色眼稀释突变都在遗传学上紧密连锁的观察[67,68]。重要的实验证据来自体细胞遗传学研究[69]。最近发布的一项基于基因的同线性图谱[37]使用超过 3,600 个直系同源基因座，定义了约 200 个保守的同线性区域。然而，大家还是意识到由于标记密度不足，这些图谱仍然可能遗漏一些同线性区域。

利用稳定的小鼠基因组序列草图和人类基因组已完成的 90% 的序列，可以进行更全面的保守同线性分析。这项工作并不是简单地依赖于已知的人类–小鼠基因对，而是确定了如下所示的更多数量的直系同源标记物。我们使用 PatternHunter 软件[71]对整个小鼠和人类基因组序列进行对比，识别出相似分值较高（大于 40 分，即至少有 40 碱基完全匹配，错配和空缺会罚分）的序列区域，此外每个序列必须是另一序列高于阈值的唯一匹配。这些区域可能反映了来自同一个祖先序列的直系同源序列对。

我们识别出约 558,000 个直系同源标记；在小鼠基因组组装中，这些序列的平均间隔约为 4.4 kb，N50 长度约为 500 bp。这些标记物总长度大约为 188 Mb，约占小鼠基因组的 7.5%。应当强调的是，这些标记物仅代表了小鼠和人类基因组中可以

genome. It should be emphasized that the landmarks represent only a small subset of the sequences, consisting of those that can be aligned with the highest similarity between the mouse and human genomes. (Indeed, below we show that about 40% of the human genome can be aligned confidently with the mouse genome.)

The locations of the landmarks in the two genomes were then compared to identify regions of conserved synteny. We define a syntenic segment to be a maximal region in which a series of landmarks occur in the same order on a single chromosome in both species. A syntenic block in turn is one or more syntenic segments that are all adjacent on the same chromosome in human and on the same chromosome in mouse, but which may otherwise be shuffled with respect to order and orientation. To avoid small artefactual syntenic segments owing to imperfections in the two draft genome sequences, we only considered regions above 300 kb and ignored occasional isolated interruptions in conserved order (see Supplementary Information). Thus, some small syntenic segments have probably been omitted—this issue will be addressed best when finished sequences of the two genomes are completed.

Marked conservation of landmark order was found across most of the two genomes (Fig. 2). Each genome could be parsed into a total of 342 conserved syntenic segments. On average, each landmark resides in a segment containing 1,600 other landmarks. The segments vary greatly in length, from 303 kb to 64.9 Mb, with a mean of 6.9 Mb and an N50 length of 16.1 Mb. In total, about 90.2% of the human genome and 93.3% of the mouse genome unambiguously reside within conserved syntenic segments. The segments can be aggregated into a total of 217 conserved syntenic blocks, with an N50 length of 23.2 Mb.

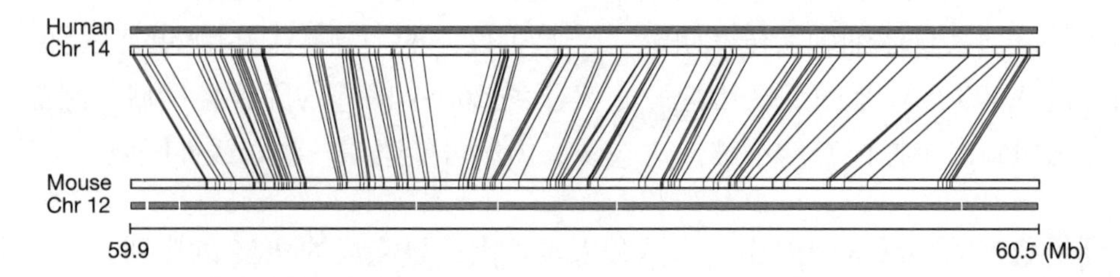


Fig. 2. Conservation of synteny between human and mouse. We detected 558,000 highly conserved, reciprocally unique landmarks within the mouse and human genomes, which can be joined into conserved syntenic segments and blocks (defined in text). A typical 510-kb segment of mouse chromosome 12 that shares common ancestry with a 600-kb section of human chromosome 14 is shown. Blue lines connect the reciprocal unique matches in the two genomes. The cyan bars represent sequence coverage in each of the two genomes for the regions. In general, the landmarks in the mouse genome are more closely spaced, reflecting the 14% smaller overall genome size.

The nature and extent of conservation of synteny differs substantially among chromosomes (Fig. 3). In accordance with expectation, the X chromosomes are represented as single, reciprocal syntenic blocks[72]. Human chromosome 20 corresponds entirely to a portion of mouse chromosome 2, with nearly perfect conservation of order along almost the entire length, disrupted only by a small central segment (Fig. 4a, d). Human chromosome 17

用于比对的高度一致序列的一个小子集。(的确，下文中我们表明大约 40% 人类基因组肯定能够与小鼠的基因组比对。)

随后在两个基因组中比较直系同源标记物的位置以识别保守的同线性区域。我们将同线性片段定义为在两个基因组中的单条染色体按相同顺序出现的一系列直系同源标记的最长区域。一个同线性区域首先是一个或多个在人类基因组和小鼠基因组中一条染色体上的毗邻同线性区段，但是在两个基因组上的排列方向和顺序可以打乱而不同。为了避免由两个基因组序列草图不完善而导致的人为的小同线性片段，我们只考虑大于 300 kb 的区域，并忽略保守排列顺序中偶然出现的孤立的中断部分(见补充信息)。因此，一些小的同线性片段可能会被遗漏了，但当两个基因组完整序列的测序完成时，这个问题会得到彻底解决。

我们在两个基因组的大部分序列中发现了标记顺序的显著保守性(图 2)。每个基因组可以被分解为共计 342 个保守的同线性片段。平均每个标记位于包含其他 1,600 个标记的片段中。这些片段长度差异很大，从 303 kb 到 64.9 Mb，平均 6.9 Mb，N50 长度为 16.1 Mb。总体而言，大约 90.2% 的人类基因组和 93.3% 的小鼠基因组明确置于保守的同线性片段中。这些片段可以组成共计 217 个保守的同线区，N50 长度为 23.2 Mb。

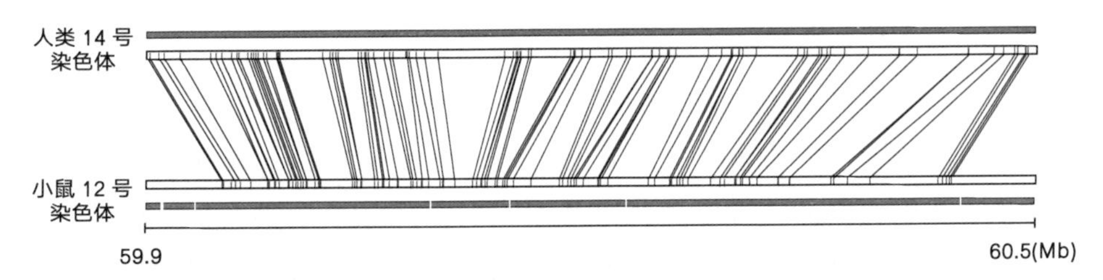


图 2. 人类和小鼠之间保守的同线性。我们在小鼠和人类基因组中检测到 558,000 个高度保守且相互独立的标记，这些标记可以连成保守的同线性片段和区块(定义见正文)。这里显示小鼠 12 号染色体一个 510 kb 与人类 14 号染色体的一个 600 kb 片段具有共同祖先的典型片段。深蓝线连接了两个基因组中相互独立匹配的片段。浅蓝色条带代表该区域在每个基因组的序列覆盖度。一般而言，小鼠基因组中的标记间隔更紧密，反映出其总基因组比人类减少了 14%。

不同染色体之间同线性的保守性质和程度有很大差异(图 3)。与预期一致，X 染色体表现为单一、相互同线性区块 [72]。整条人类 20 号染色体和小鼠 2 号染色体的一部分完全对应，保守性的顺序在整个长度上近乎完美，仅被一个小的中心片段打断(图 4a、4d)。整条人类 17 号染色体与小鼠 11 号染色体的一部分完全对应，但

corresponds entirely to a portion of mouse chromosome 11, but extensive rearrangements have divided it into at least 16 segments (Fig. 4b, e). Other chromosomes, however, show evidence of much more extensive interchromosomal rearrangement than these cases (Fig. 4c, f).

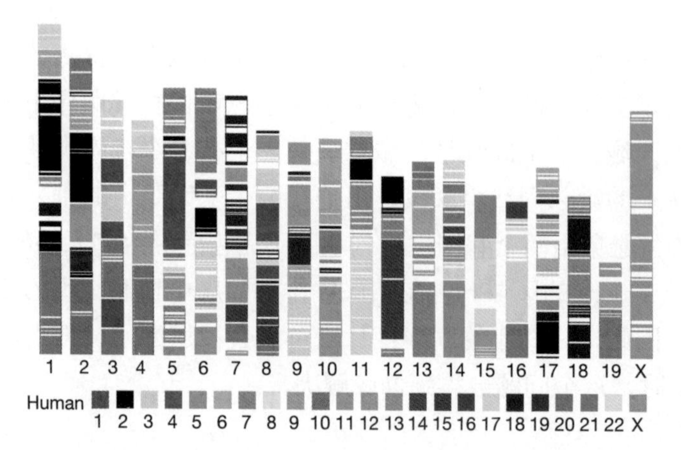


Fig. 3. Segments and blocks > 300 kb in size with conserved synteny in human are superimposed on the mouse genome. Each colour corresponds to a particular human chromosome. The 342 segments are separated from each other by thin, white lines within the 217 blocks of consistent colour.

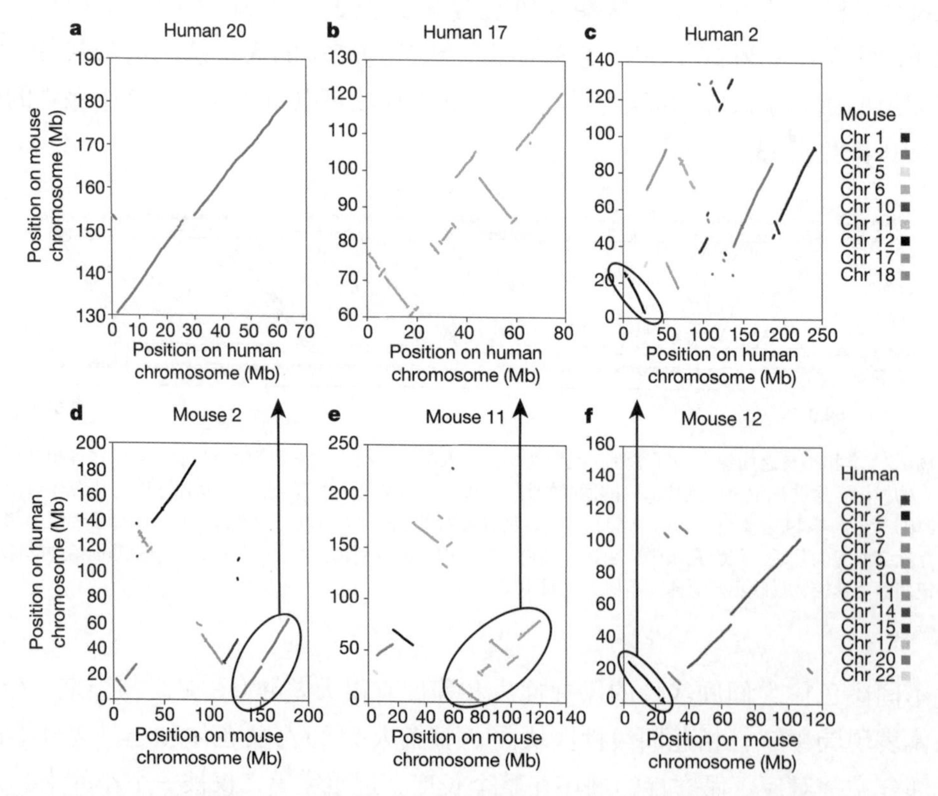


Fig. 4. Dot plots of conserved syntenic segments in three human and three mouse chromosomes. For each of three human (**a–c**) and mouse (**d–f**) chromosomes, the positions of orthologous landmarks are plotted along the *x* axis and the corresponding position of the landmark on chromosomes in the other genome

是大量的重组将其分成了至少 16 个片段（图 4b、4e）。对其他染色体来说，证据则显示比这些例子更为剧烈的染色体之间的重组（图 4c、4f）。

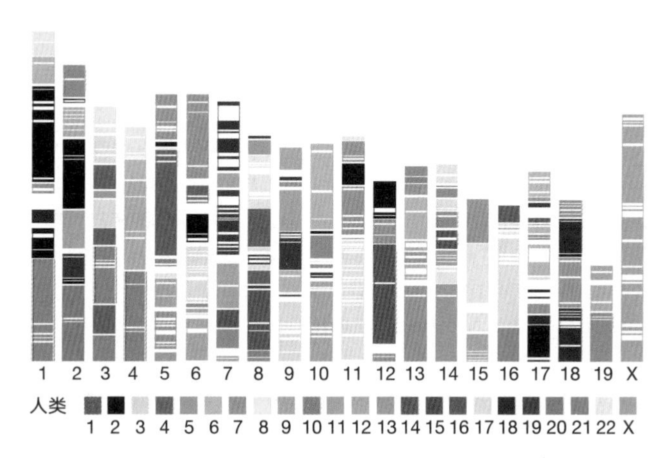


图 3. 人类基因组中大于 300 kb 的保守同线性片段和区块叠加在小鼠的基因组上。每种颜色对应一个特定的人类染色体。在 217 个具有同一颜色的同线性区块中，共有 342 个同线性片段通过细白线彼此分开。

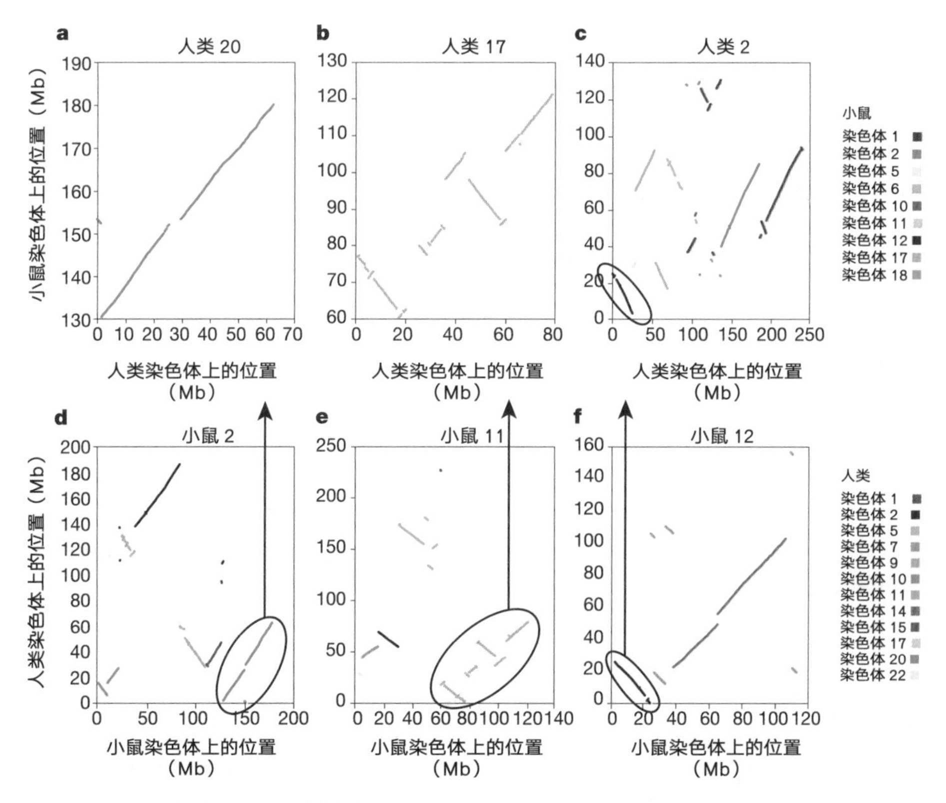


图 4. 人类三个染色体和小鼠三个染色体中保守同线性片段的点图。$x$ 轴为每条人类（**a**～**c**）和小鼠（**d**～**f**）染色体上直系同源标记的位置，$y$ 轴为该标记在另一个基因组染色体上的对应位置。不同颜色区分位于同一条染色体上的多个同源标记对应的另一个物种的不同染色体。保守同线性的一个典型例子是

is plotted on the *y* axis. Different chromosomes in the corresponding genome are differentiated with distinct colours. In a remarkable example of conserved synteny, human chromosome 20 (**a**) consists of just three segments from mouse chromosome 2 (**d**), with only one small segment altered in order. Human chromosome 17 (**b**) also shares segments with only one mouse chromosome (11) (**e**), but the 16 segments are extensively rearranged. However, most of the mouse and human chromosomes consist of multiple segments from multiple chromosomes, as shown for human chromosome 2 (**c**) and mouse chromosome 12 (**f**). Circled areas and arrows denote matching segments in mouse and human.

We compared the new sequence-based map of conserved synteny with the most recent previous map based on 3,600 loci[30]. The new map reveals many more conserved syntenic segments (342 compared with 202) but only slightly more conserved syntenic blocks (217 compared with 170). Most of the conserved syntenic blocks had previously been recognized and are consistent with the new map, but many rearrangements of segments within blocks had been missed (notably on the X chromosome).

The occurrence of many local rearrangements is not surprising. Compared with interchromosomal rearrangements (for example, translocations), paracentric inversions (that is, those within a single chromosome and not including the centromere) carry a lower selective disadvantage in terms of the frequency of aneuploidy among offspring. These are also seen at a higher frequency in genera such as *Drosophila*, in which extensive cytogenetic comparisons have been carried out[73,74].

The block and segment sizes are broadly consistent with the random breakage model of genome evolution[75]. At this gross level, there is no evidence of extensive selection for gene order across the genome. Selection in specific regions, however, is by no means excluded, and indeed seems probable (for example, for the major histocompatibility complex). Moreover, the analysis does not exclude the possibility that chromosomal breaks may tend to occur with higher frequency in some locations.

With a map of conserved syntenic segments between the human and mouse genomes, it is possible to calculate the minimal number of rearrangements needed to "transform" one genome into the other[70,76,77]. When applied to the 342 syntenic segments above, the most parsimonious path has 295 rearrangements. The analysis suggests that chromosomal breaks may have a tendency to reoccur in certain regions. With only two species, however, it is not yet possible to recover the ancestral chromosomal order or reconstruct the precise pathway of rearrangements. As more mammalian species are sequenced, it should be possible to draw such inferences and study the nature of chromosome rearrangement.

## Genome Landscape

We next sought to analyse the contents of the mouse genome, both in its own right and in comparison with corresponding regions of the human genome.

All of the mouse genome information is accessible in electronic form through various

人类 20 号染色体（**a**）仅由来自小鼠 2 号染色体（**d**）上的三个同源片段组成，其中只有一个很小的片段的顺序不一致。人类 17 号染色体（**b**）也仅与小鼠一条染色体（11）（**e**）同源片段共享，不过 16 个片段都被彻底重组了。然而，小鼠和人类大多数染色体由来自多个染色体的多个同源片段组成，如人类 2 号染色体（**c**）和小鼠 12 号染色体（**f**）所示。圆圈圈住的区域和箭头表示小鼠和人类相匹配的片段。

我们比较了全新的基于序列保守同线性图谱和最近发布的基于 3,600 个基因座的前期图谱 [30]。新图谱展示的保守同线性片段数量增加了很多（342 个比 202 个），但保守同线性区域数量只有少量增加（217 个比 170 个）。这表明大多数保守同线性区先前已经被识别了，并且与新图谱一致，但是位于区域内的很多重组片段被错过了（特别是 X 染色体）。

发生许多局部重组并不令人意外。就子代间非整倍体频率而言，臂内倒位（即单个染色体内且不包括着丝粒的倒位）相比于染色体间重组（如易位）选择劣势更低。这种局部重组在果蝇等广泛用于细胞遗传学比较的物种中的频率也很高 [73,74]。

同源区域和片段的长度与基因进化的随机断裂模型大体一致 [75]。总体而言，没有证据表明在基因组中存在广泛的对基因排布顺序的选择。然而，在特殊区域绝不排除这种选择，而且看起来确实可能存在（例如，对主要组织相容性复合体的选择）。此外，分析并不排除染色体断裂可能倾向于在某些位置以更高频率发生的可能性。

利用人类和小鼠基因组之间的保守同线性片段图谱，可以计算将一个基因组“转化”为另一个基因组所需要的最小重组数目 [70,76,77]。应用于上述的 342 个同线性片段，最简捷的途径有 295 个重组。分析表明，染色体断裂可能倾向于在某个区域重复出现。然而仅凭两个物种不太可能复原祖先染色体上的排列顺序或重建精确的染色体重组过程。随着更多哺乳动物物种的基因组测序完成，理论上可以得出这样的推论并研究染色体重组的特性。

## 基因组概貌

下一步，我们试图分析小鼠基因组的内容，包括其本身以及与人类基因组的相应区域的比较。

所有小鼠基因组信息可以通过各种浏览器以电子形式访问：Ensembl(http://

browsers: Ensembl (http://www.ensembl.org), the University of California at Santa Cruz (http://genome.ucsc.edu) and the National Center for Biotechnology Information (http://www.ncbi.nlm.nih.gov). These browsers allow users to scroll along the chromosomes and zoom in or out to any scale, as well as to display information at any desired level of detail. The mouse genome information has also been integrated into existing human genome browsers at these same organizations. In this section, we compare general properties of the mouse and human genomes.

## Genome expansion and contraction

The projected total length of the euchromatic portion of the mouse genome (2.5 Gb) is about 14% smaller than that of the human genome (2.9 Gb). To investigate the source of this difference, we examined the relative size of intervals between consecutive orthologous landmarks in the human and mouse genomes. The mouse/human ratio has a mean at 0.91 for autosomes, but varies widely, with the mouse interval being larger than the human in 38% of cases. Chromosome X, by contrast, shows no net relative expansion or contraction, with a mouse/human ratio of 1.03. What accounts for the smaller size of the mouse genome? We address this question below in the sections on repeat sequences and on genome evolution.

## (G+C) content

The overall distribution of local (G+C) content is significantly different between the mouse and human genomes. Such differences have been noted in biochemical studies[78-81] and in comparative analyses of fourfold degenerate sites in codons of mouse and human genes[82-85], but the availability of nearly complete genome sequences provides the first detailed picture of the phenomenon.

The mouse has a slightly higher overall (G+C) content than the human (42% compared with 41%), but the distribution is tighter. When local (G+C) content is measured in 20-kb windows across the genome, the human genome has about 1.4% of the windows with (G+C) content > 56% and 1.3% with (G+C) content < 33%. Such extreme deviations are virtually absent in the mouse genome. The contrast is even seen at the level of entire chromosomes. The human has extreme outliers with respect to (G+C) content (the most extreme being chromosome 19), whereas the mouse chromosomes tend to be far more uniform.

There is a strong positive correlation in local (G+C) content between orthologous regions in the mouse and human genomes, but with the mouse regions showing a clear tendency to be less extreme in (G+C) content than the human regions. This tendency is not uniform, with the most extreme differences seen at the tails of the distribution.

In mammalian genomes, there is a positive correlation between gene density and (G+C)

www.ensembl.org)、加州大学圣克鲁斯分校(http://genome.ucsc.edu)和美国国家生物技术信息中心(http://www.ncbi.nlm.nih.gov)。这些浏览器允许用户沿着染色体滚动查看，放大或缩小到任何比例，并且以任何期望的细节水平来显示信息。这些组织机构在现有的人类基因组浏览器中也整合了小鼠基因组。在本节中，我们比较了小鼠和人类基因组的一般特性。

## 基因组扩增和收缩

小鼠基因组的常染色体部分(2.5 Gb)的预计总长度比人类基因组(2.9 Gb)小约14%。为了研究这种差异的来源，我们检测了人类和小鼠基因组中连续直系同源标记之间的间隔的相对大小。常染色体的小鼠/人的比率平均值为0.91，但是变化范围很大，38%的情况是小鼠的间隔大于人类。相比之下，X染色体没有表现出对应的净扩张或收缩，小鼠/人的比率为1.03。小鼠基因组较小的原因是什么？我们在下面关于重复序列和基因组进化部分中讨论了这个问题。

## (G+C)含量

小鼠和人类基因组之间的局部(G+C)含量的总体分布有显著差异。在以往生化研究[78-81]以及小鼠与人类基因组密码子的四倍简并位点的比较分析[82-85]中已经注意到了这些差异，但是近乎完整的可用基因组序列为这种现象提供了第一个详细的图像。

小鼠总体(G+C)含量略高于人类(42%比41%)，但是分布更紧密。在基因组中以20 kb窗口测量局部(G+C)含量时，人类基因组具有(G+C)含量大于56%的窗口约为1.4%，(G+C)含量小于33%的约1.3%。这种极端偏差在小鼠基因组中几乎不存在。这种差别甚至在染色体水平上都能看到。人类有极端的(G+C)含量的离群值(最极端的是19号染色体)，而小鼠的染色体则趋向于更加均匀。

在小鼠和人类基因组的直系同源区域之间，局部(G+C)含量有很强的正相关性，但小鼠区域(G+C)含量的趋势明显不如人类区域那样极端。这种趋势并不均匀，最大差异分布在尾部。

在哺乳动物基因组中，基因密度和(G+C)含量呈正相关[81,86-89]。鉴于人类和小

content[81,86-89]. Given the differences in (G+C) content between human and mouse, we compared the distribution of genes—using the sets of orthologous mouse and human genes described below—with respect to (G+C) content for both genomes. The density of genes differed markedly when expressed in terms of absolute (G+C) content, but was nearly identical when expressed in terms of percentiles of (G+C) content. For example, both species have 75–80% of genes residing in the (G+C)-richest half of their genome. Mouse and human thus show similar degrees of homogeneity in the distribution of genes, despite the overall differences in (G+C) content. Notably, the mouse shows similar extremes of gene density despite being less extreme in (G+C) content.

What accounts for the differences in (G+C) content between mouse and human? Does it reflect altered selection for (G+C) content[90,91], altered mutational or repair processes[92-94], or possibly both? Data from additional species will probably be needed to address these issues. Any explanation will need to account for various mysterious phenomena. For example, although overall (G+C) content in mouse is slightly higher than in human (42% compared with 41%), the (G+C) content of chromosome X is slightly lower (39.0% compared with 39.4%). The effect is even more pronounced if one excludes lineage-specific repeats (see below), thereby focusing primarily on shared DNA. In that case, mouse autosomes have an overall (G+C) content that is 1.5% higher than human autosomes (41.2% compared with 39.7%) whereas mouse chromosome X has a (G+C) content that is 1% lower than human chromosome X (37.8% compared with 36.8%).

## CpG islands

In mammalian genomes, the palindromic dinucleotide CpG is usually methylated on the cytosine residue. Methyl-CpG is mutated by deamination to TpG, leading to approximately fivefold under-representation of CpG across the human[1,95] and mouse genomes. In some regions of the genome that have been implicated in gene regulation, CpG dinucleotides are not methylated and thus are not subject to deamination and mutation. Such regions, termed CpG islands, are usually a few hundred nucleotides in length, have high (G+C) content and above average representation of CpG dinucleotides.

We applied a computer program that attempts to recognize CpG islands on the basis of (G+C) and CpG content of arbitrary lengths of sequence[96,97] to the non-repetitive portions of human and mouse genome sequences. The mouse genome contains fewer CpG islands than the human genome (about 15,500 compared with 27,000), which is qualitatively consistent with previous reports[98]. The absolute number of islands identified depends on the precise definition of a CpG island used, but the ratio between the two species remains fairly constant.

The reason for the smaller number of predicted CpG islands in mouse may relate simply to the smaller fraction of the genome with extremely high (G+C) content[99] and its effect on the computer algorithm. Approximately 10,000 of the predicted CpG islands in each species show significant sequence conservation with CpG islands in the orthologous intervals in

鼠之间（G+C）含量的差异，我们用下面描述的小鼠和人类的同源基因集合来对比两种基因组的（G+C）含量来比较基因的分布。两物种间基因密度就绝对（G+C）含量而言，差异显著，但是就（G+C）含量百分数而言，几乎相同。例如，两个物种都有75%～80%的基因位于它们基因组中（G+C）含量最高的一半的区域。因此，尽管在（G+C）含量上存在总体差异，但小鼠和人类在基因分布上显示出相似程度的同质性。值得注意的是，尽管在（G+C）含量上不那么极端，但在基因密度上，小鼠表现出了相似的极端性。

小鼠和人类（G+C）含量差异的原因是什么？它是否反映了对（G+C）含量的不同选择[90,91]，不同的突变或修复过程[92-94]，或者可能两者都有？可能需要来自于其他物种的数据来解决这些问题。各种神秘现象都需要解释。例如，尽管小鼠总（G+C）含量略高于人类（42%比41%），但是X染色体的（G+C）含量略低于人类（39.0%比39.4%）。如果排除了特定谱系的重复序列（见下文），这一现象更加明显，因此主要关注共享DNA。在这种情况下，小鼠常染色体总（G+C）含量比人类常染色体高1.5%（41.2%比39.7%），而小鼠X染色体的总（G+C）含量比人类X染色体低1%（37.8%比36.8%）。

## CpG岛

在哺乳动物的基因组中，回文二核苷酸CpG通常在胞嘧啶残基上甲基化。甲基化的CpG通过脱氨基作用突变为TpG，导致人类和小鼠基因组中CpG的含量大概降低为原来的1/5[1,95]。在一些与基因调控有关的基因组区域，CpG二核苷酸不甲基化，因此不受脱氨基和突变的影响。这些区域被称为CpG岛，通常是几百个核苷酸的长度，高（G+C）含量和高于平均数量的CpG二核苷酸。

我们用一个计算机程序根据（G+C）含量和任意长度序列CpG的含量[96,97]识别人类和小鼠基因组序列中非重复部分的CpG岛。小鼠基因组包含的CpG岛比人类基因组略少（大约15,500比27,000），这与之前的报道[98]在数量级上是一致的。确定的CpG岛的绝对数量依赖于所使用的CpG岛的精确界定，但是两个物种之间的比例保持相当一致。

小鼠中预测的CpG岛数量较少的原因可能与基因组中仅有较少区域具有极高（G+C）含量[99]及这一现象对计算机算法的影响有关。每个物种中约有10,000个预测的CpG岛显示出与其他物种的直系同源区间中的CpG岛存在显著的序列保守性，

the other species, falling within the orthologous landmarks described above. Perhaps these represent functional CpG islands, a proposition that can now be tested experimentally[84].

## Repeats

The single most prevalent feature of mammalian genomes is their repetitive sequences, most of which are interspersed repeats representing "fossils" of transposable elements. Transposable elements are a principal force in reshaping the genome, and their fossils thus provide powerful reporters for measuring evolutionary forces acting on the genome. A recent paper on the human genome sequence[1] provided extensive background on mammalian transposons, describing their biology and illustrating many applications to evolutionary studies. Here, we will focus primarily on comparisons between the repeat content of the mouse and human genomes.

### Mouse has accumulated more new repetitive sequence than human

Approximately 46% of the human genome can be recognized currently as interspersed repeats resulting from insertions of transposable elements that were active in the last 150–200 million years. The total fraction of the human genome derived from transposons may be considerably larger, but it is not possible to recognize fossils older than a certain age because of the high degree of sequence divergence. Because only 37.5% of the mouse genome is recognized as transposon-derived, it is tempting to conclude that the smaller size of the mouse genome is due to lower transposon activity since the divergence of the human and mouse lineages. Closer analysis, however, shows that this is not the case. As we discuss below, transposition has been more active in the mouse lineage. The apparent deficit of transposon-derived sequence in the mouse genome is mostly due to a higher nucleotide substitution rate, which makes it difficult to recognize ancient repeat sequences.

### Lineage-specific versus ancestral repeats

Interspersed repeats can be divided into lineage-specific repeats (defined as those introduced by transposition after the divergence of mouse and human) and ancestral repeats (defined as those already present in a common ancestor). Such a division highlights the fact that transposable elements have been more active in the mouse lineage than in the human lineage. Approximately 32.4% of the mouse genome (about 818 Mb) but only 24.4% of the human genome (about 695 Mb) consists of lineage-specific repeats. Contrary to initial appearances, transposon insertions have added at least 120 Mb more transposon-derived sequence to the mouse genome than to the human genome since their divergence. This observation is consistent with the previous report that the rate of transposition in the human genome has fallen markedly over the past 40 million years[1,100].

属于之前描述的直系同源性标记。或许这些区域代表了功能性的CpG岛，这个假设可以通过实验进行验证[84]。

## 重复序列

哺乳动物基因组中最普遍的特征是它们的重复序列，其中大多数是离散的重复序列，代表着转座元件的“化石”。转座元件是重塑基因组的主要力量，因此它们的化石为测量作用于基因组的进化力提供了有力的报告。最近一篇关于人类基因组序列的论文[1]给出了哺乳动物转座子的大量背景，描述了它们的生物学特性并说明了在进化研究中的许多应用。这里，我们将主要关注于小鼠和人类基因组中重复内容之间的比较。

### 小鼠比人类积累了更多的新的重复序列

目前，大约46%的人类基因组是由过去1.5亿至2亿年中活跃的转座子插入引起的离散重复序列。来自于转座子的人类基因组的总比例可能相当大，但是由于序列的高度分化，不可能识别比某个时代更久远的化石。因为只有37.5%的小鼠基因组被认为来自于转座子，很容易得出结论，小鼠基因组更小是由于人类和小鼠谱系分化后，转座子活性降低。然而，更仔细的分析显示情况并非如此。正如我们下面讨论的，转座现象在小鼠谱系中更为活跃。小鼠基因组中转座衍生序列的明显缺失主要是由于更高的核苷酸替换率，以至于很难识别古老的重复序列。

### 谱系特异性与祖源重复

离散的重复可以分为谱系特异性重复（定义为小鼠和人类分化后通过转座引入的那些重复）和祖源重复（定义为存在于共同祖先中的重复）。这样的划分凸显了一个事实，即转座元件在小鼠谱系中比在人类谱系中更为活跃。约32.4%的小鼠基因（大约818 Mb）但只有24.4%的人类基因（大约695 Mb）由谱系特异性重复组成。与最初的表现相反，自从小鼠和人类的祖先分开之后，在小鼠基因组的转座子插入中，其转座子衍生序列比人类基因组增加了至少120 Mb。这个观察与之前报道的人类基因组的转座率在过去的4,000万年中显著下降一致[1,100]。

The overall lower interspersed repeat density in mouse is the result of an apparent lack of ancestral repeats: they comprise only 5% of the mouse genome compared with 22% of the human genome. The ancestral repeats recognizable in mouse tend to be those of more recent origin, that is, those that originated closest to the mouse–human divergence. This difference may be due partly to a higher deletion rate of non-functional DNA in the mouse lineage, so that more of the older interspersed repeats have been lost. However, the deficit largely reflects a much higher neutral substitution rate in the mouse lineage than in the human lineage, rendering many older ancestral repeats undetectable with available computer programs.

## Higher substitution rate in mouse lineage

The hypothesis that the neutral substitution rate is higher in mouse than in human was suggested as early as 1969 (refs 101–103). The idea has continued to be challenged on the basis that the apparent differences may be due to inaccuracies in mammalian phylogenies[104,105]. The explanation, however, remains unclear, with some attributing it to generation time[101,106] and others pointing to a closer correlation with body size[107,108].

Ancestral repeats provide a powerful measure of neutral substitution rates, on the basis of comparing thousands of current copies to the inferred consensus sequence of the ancestral element. The large copy number and ubiquitous distribution of ancestral repeats overcome issues of local variation in substitution rates (see below). Most notably, differences in divergence levels are not affected by phylogenetic assumptions, as the time spent by an ancestral repeat family in either lineage is necessarily identical.

The median divergence levels of 18 subfamilies of interspersed repeats that were active shortly before the human–rodent speciation indicates an approximately twofold higher average substitution rate in the mouse lineage than in the human lineage, corresponding closely to an early estimate by Wu and Li[109]. In human, the least-diverged ancestral repeats have about 16% mismatch to their consensus sequences, which corresponds to approximately 0.17 substitutions per site. In contrast, mouse repeats have diverged by at least 26–27% or about 0.34 substitutions per site, which is about twofold higher than in the human lineage. The total number of substitutions in the two lineages can be estimated at 0.51. Below, we obtain an estimate of a combined rate of 0.46–0.47 substitutions per site, on the basis of an analysis that counts only substitutions since the divergence of the species.

Assuming a speciation time of 75 Myr, the average substitution rates would have been $2.2 \times 10^{-9}$ and $4.5 \times 10^{-9}$ in the human and mouse lineages, respectively. This is in accord with previous estimates of neutral substitution rates in these organisms. (Reports of highly similar substitution rates in human and mouse lineages relied on a much earlier divergence time of rodents from other mammals[104].)

Comparison of ancestral repeats to their consensus sequence also allows an estimate of the

小鼠总的离散重复密度较低归因于祖源重复序列的明显缺乏：它们仅占小鼠基因组的 5%，而在人类基因组中占了 22%。在小鼠基因组中可识别的祖源重复序列倾向于那些更为近期起源的，也就是起源于最接近小鼠和人类分化的那些。这种差异存在的部分可能原因是小鼠谱系中非功能性 DNA 的删除率较高，因此丢失了更多的古老的离散重复序列。但是，这种缺陷很大程度上反映了小鼠谱系中的中性替代率明显高于人类谱系，使得许多更古老的祖源重复无法用可用的计算机程序检测到。

## 小鼠谱系中的高替代率

早在 1969 年就提出了小鼠的中性替代率高于人类的假说（参考文献 101 ~ 103）。这一观点持续受到挑战，因为这种明显的差异可能是由哺乳动物系统发育不准确造成的[104,105]。然而，这一解释还是不够清楚，一些人将其归因于世代时间[101,106]，而另一些人则指出与身体尺寸关系更紧密[107,108]。

祖源重复序列为中性替代率提供了一个强有力的度量方法，该方法基于将数千份当前拷贝与所推断的祖源元件的共有序列进行比较。祖源重复的大量拷贝数和广泛分布克服了替代率局部差异的问题（见下文）。更值得注意的是，分化水平的差异不受系统发育假设的影响，因为祖源重复家族在任何一个谱系中花费的时间是必然相同的。

在人类–啮齿动物物种形成之前不久活跃的 18 个离散重复序列亚家族的中位分化水平表明，小鼠谱系中的平均替代率大约是人类谱系的两倍，这与 Wu 和 Li[109] 的早期估计非常接近。在人类中，分化最小的祖源重复序列与其共有序列有大约 16% 的错配，相当于每个位点大约 0.17 个替换。相比之下，小鼠的重复序列有至少 26% ~ 27% 的差异或是每个位点大约 0.34 个替换，大约是人类谱系的两倍。这两个谱系的替换总数估计为 0.51。下面，我们根据仅计算物种分化以来产生的替换的分析，得出了每个位点 0.46 ~ 0.47 个替换的综合估计值。

假设物种形成时间为 7,500 万年，人类和小鼠谱系的平均替换率分别为 $2.2 \times 10^{-9}$ 和 $4.5 \times 10^{-9}$。这与之前对这些生物体的中性替代率的估计是一致的。（关于人类和小鼠谱系中高度相似的替代率的报道取决于啮齿动物与其他哺乳动物更早的分化时间[104]。）

将祖源重复序列和它们的一致性序列进行比较，还可以估计小片段插入（小于

rate of occurrence of small ( < 50 bp) insertions and deletions (indels). Both species show a net loss of nucleotides (with deleted bases outnumbering inserted bases by at least 2–3-fold), but the overall loss owing to small indels in ancestral repeats is at least twofold higher in mouse than in human. This may contribute a small amount (1–2%) to the difference in genome size noted above.

It should be noted that the roughly twofold higher substitution rate in mouse represents an average rate since the time of divergence, including an initial period when the two lineages had comparable rates. Comparison with more recent relatives (mouse–rat and human–gibbon, each about 20–25 Myr) indicate that the current substitution rate per year in mouse is probably much higher, perhaps about fivefold higher (see Supplementary Information). Also, note that these estimates refer to substitution rate per year, rather than per generation. Because the human generation time is much longer than that of the mouse (by at least 20-fold), the substitution rate is greater in human than mouse when measured per generation.

## Higher substitution rate obscures old repeats

We measured the impact of the higher substitution rate in mouse on the ability to detect ancestral repeats in the mouse genome. By computer simulation, the ability of the RepeatMasker[100] program to detect repeats was found to fall off rapidly for divergence levels above about 37%. If we simulate the events in the mouse lineage by adjusting the ancestral repeats in the human genome for the higher substitution levels that would have occurred in the mouse genome, the proportion of the genome that would still be recognizable as ancestral repeats falls to only 6%. This is in close agreement with the proportion actually observed for the mouse. Thus, the current analysis of repeated sequences allows us to see further back into human history (roughly 150–200 Myr) than into mouse history (roughly 100–120 Myr).

A higher rate of interspersed repeat insertion does not explain the larger size of the human genome. Below, we suggest that the explanation lies in a higher rate of large deletions in the mouse lineage.

## Comparison of mouse and human repeats

All mammals have essentially the same four classes of transposable elements: (1) the autonomous long interspersed nucleotide element (LINE)-like elements; (2) the LINE-dependent, short RNA-derived short interspersed nucleotide elements (SINEs); (3) retrovirus-like elements with long terminal repeats (LTRs); and (4) DNA transposons. The first three classes procreate by reverse transcription of an RNA intermediate (retroposition), whereas DNA transposons move by a cut-and-paste mechanism of DNA sequence (see refs 1, 100 for further information about these classes).

50 bp）和缺失的发生率。两个物种都有核苷酸的净损失（缺失的碱基数量至少是插入碱基的 2 ~ 3 倍），但是祖源重复中小片段插入缺失导致的整体损失在小鼠中至少是在人类中的两倍。这可能是导致上述基因组大小差异的一小部分原因（1% ~ 2%）。

应该注意的是，小鼠中大约 2 倍高的替代率代表了自分化时间以来的平均比率，包括两个谱系的替代率相当的初始时期。与最近的近源物种（小鼠–大鼠及人类–猩猩，每个大约 2,000 ~ 2,500 万年）的比较表明，小鼠每年的替代率可能要高得多，大约是 5 倍（见补充信息）。此外，注意这些估计指的是每年的替代率，而不是每代的替代率。因为人类的世代时间要比小鼠长很多（至少 20 倍），所以当测量每个世代时，人类的替代率比小鼠高很多。

### 较高的替代率掩盖了旧的重复

我们测量了小鼠高替代率对检测小鼠基因组祖源重复序列能力的影响。通过计算机模拟，发现当差异水平超过 37% 时，RepeatMasker[100] 程序检测重复的能力迅速下降。如果我们模拟小鼠谱系中的事件时利用人类基因组中的祖源重复来校正小鼠基因组中的高替代水平，那么仍然可识别的祖源重复在基因组中的比例下降到仅有 6%。这与实际观察到的小鼠比例非常一致。因此，对重复序列进行的当前分析，可以让我们追溯人类历史（大约 1.5 亿 ~ 2 亿年）到比小鼠历史更远的年代（大约 1 亿 ~ 1.2 亿年）。

比较高的散布的重复插入率并不能解释为何人类的基因组比较大。通过下面的分析，我们认为原因在于小鼠谱系中较高的缺失率。

### 小鼠和人类重复序列的比较

所有哺乳动物基本上都有四类转座元件：（1）自主的长散在核苷酸元件（LINE）一类元件；（2）依赖于 LINE、由短 RNA 衍生的短散在核苷酸元件（SINE）；（3）具有长末端重复（LTR）的逆转录病毒样元件；（4）DNA 转座子。前三类通过 RNA 中间体介导的逆转录（逆转录转座）产生，而 DNA 转座通过 DNA 的剪切–粘贴机制进行（更多信息见参考文献 1、100）。

A comparison of these repeat classes in the mouse and human genomes can be enlightening. On the one hand, differences between the two species reveal the dynamic nature of transposable elements; on the other hand, similarities in the location of lineage-specific elements point to common biological factors that govern insertion and retention of interspersed repeats.

## Differences between mouse and human

The most notable difference is in the changing rate of transposition over time: the rate has remained fairly constant in mouse, but markedly increased to a peak at about 40 Myr in human, and then plummeted. This phenomenon was noted in our initial analysis of the human genome; the availability of the mouse genome sequence now confirms and sharpens the observation. Beyond this overall tendency, there are specific differences in each of the four repeat classes.

The first class that we discuss is LINEs. Copies of LINE1 (L1) form the single largest fraction of interspersed repeat sequence in both human and mouse. No other LINE seems to have been active in either lineage. The extant L1 elements in both species derive from a common ancestor by means of a series of subfamilies defined primarily by the rapidly evolving 3′ non-coding sequences[110]. The L1 5′-untranslated regions (UTRs) in both lineages have been even more variable, occasionally through acquisition of entirely new sequences[111]. Indeed, the three active subfamilies in mouse, which are otherwise > 97% identical, have unrelated or highly diverged 5′ ends[112-114]. L1 seems to have remained highly active in mouse, whereas it has declined in the human lineage. Goodier and co-workers[113] estimated that the mouse genome contains at least 3,000 potentially active elements (full-length with two intact open reading frames (ORFs)). The current draft sequence of the mouse genome contains only 400 young, full-length elements; of these only 12 have two intact ORFs. This is probably a reflection of the WGS shotgun approach used to assemble the genome. Indeed, most of the young elements in the draft genome sequence are incomplete owing to internal sequence gaps, reflecting the difficulty that WGS assembly has with highly similar repeat sequences. This is a notable limitation of the draft sequence.

The second repeat class is SINEs. Whereas only a single SINE (Alu) was active in the human lineage, the mouse lineage has been exposed to four distinct SINEs (B1, B2, ID, B4). Each is thought to rely on L1 for retroposition, although none share sequence similarity, as is the rule for other LINE–SINE pairs[115,116]. The mouse B1 and human Alu SINEs are unique among known SINEs in being derived from 7SL RNA; they probably have a common origin[117]. The mouse B2 is typical among SINEs in having a transfer RNA-derived promoter region. Recent ID elements seem to be derived from a neuronally expressed RNA gene called *BC1*, which may itself have been recruited from an earlier SINE. This subfamily is minor in mouse, with 2–4,000 copies, but has expanded rapidly in rat where it has produced more than 130,000 copies since the mouse–rat speciation[118]. Both B2 and ID closely resemble Ala-tRNA, but seem to have independent origins. The B4 family resembles

对小鼠和人类基因组中这些重复类型进行比较可以获得启发。一方面，两个物种之间的差异揭示了转座元件的动态性质；另一方面，谱系特殊元件位置上的相似性表明控制离散重复的插入和保留具有共同的生物学因素。

## 小鼠和人类之间的差异

最显著的区别在于转座率随着时间发生的变化：小鼠的转座率保持相当稳定，但人类在大约 4,000 万年的时候，转座率明显增加并达到峰值，然后下降。我们在对人类基因组进行初步分析时注意到了这个现象，小鼠基因组序列的获得则证实并印证了这一观察。除了这个整体趋势外，四个重复类型中的每一个都有特定的差异。

我们首先讨论第一类 LINE。在人类和小鼠中，LINE1(L1)的拷贝构成了离散重复序列中的最大部分。在这两个谱系中似乎都没有其他具有活性的 LINE。两个物种中存在的 L1 元件是从一个共同祖先，通过 3′非编码序列的快速进化形成的一系列亚家族衍生而来[110]。两个谱系中 L1 的 5′非翻译区变异性很大，偶尔会获得全新的序列[111]。事实上，小鼠中三个活跃的亚家族在其他方面的相似性大于 97%，但却具有不相关或者高度分化的 5′端[112-114]。L1 似乎在小鼠中保持高度活性，而在人类谱系中却已经下降了。古迪尔和同事们[113]估计，小鼠基因组中至少包含 3,000 个潜在活性元件(全长可达两个完整的开放阅读框)。目前小鼠基因组序列草图只含有 400 个年轻的全长元件，其中只有 12 个含有两个完整的开放阅读框。这可能是用 WGS 鸟枪法组装基因组的一种现象。事实上，由于内部序列空缺，基因组序列草图中大多数年轻元件是不完整的，这反映了 WGS 在组装高度相似重复序列时的难度。这是该序列草图的一个显著不足。

第二类重复是 SINE。虽然人类谱系中只有一个 SINE(Alu)有活性，但是在小鼠谱系中已经显露了四个不同的 SINE(B1、B2、ID、B4)。尽管它们没有共享序列相似性，每个都被认为依赖于 L1 进行逆转录转座，这也是其他 LINE-SINE 配对的规则[115,116]。小鼠 B1 和人类 Alu 的 SINE 在已知的 SINE 中比较特殊，均衍生于 7SL RNA，所以它们可能有共同起源[117]。小鼠的 B2 则是 SINE 中比较典型的，具有转运 RNA 衍生的启动子区域。近期出现的 ID 元件似乎来自于神经组织表达的称为 *BC1* 的 RNA 基因，它本身可能是被一个早期 SINE 募集的。这个亚家族在小鼠体内是次要的，有 2～4,000 个拷贝数，但是在大鼠中迅速扩张，自小鼠–大鼠物种形成以来已经产生超过 130,000 个拷贝数[118]。B2 和 ID 都与 Ala-tRNA(丙氨酸–转

a fusion between B1 and ID[119,120]. We found that 25% of the 75,000 identified ID elements were located within 50 bp of a B1 element of similar orientation, suggesting that perhaps most older ID elements are mislabelled or truncated B4 SINEs.

The third repeat class is LTR elements. All interspersed LTR-containing elements in mammals are derivatives of the vertebrate-specific retrovirus clade of retrotransposons. The earliest infectious retroviruses probably originated from endogenous retroviral-like (ERV) elements that acquired mechanisms for horizontal transmission[121], whereas many current endogenous retroviral elements have probably arisen from infection by retroviruses.

Endogenous retroviruses fall into three classes (I–III), which show a markedly dissimilar evolutionary history in human and mouse. Notably, ERVs are nearly extinct in human whereas all three classes have active members in mouse.

Class III accounts for 80% of recognized LTR element copies predating the human–mouse speciation. This class includes the non-autonomous MaLRs: with 388,000 recognizable copies in mouse, it is the single most successful LTR element. It is still active in mouse (represented by MERVL and the MT and ORR1 MaLRs), but died out some 50 Myr in human[122].

Copies of class II elements are tenfold denser in mouse than in human. In contrast, class I element copies are fourfold more common in the human than the mouse genome (although it is possible that some have not yet been recognized in mouse). In mouse, this class includes active ERVs, such as the murine leukaemia virus, MuRRS, MuRVY and VL30 (several of which have caused insertional mutations in mouse)—no similar activity is known to exist in human.

The fourth repeat class is the DNA transposons. Although most transposable elements have been more active in mouse than human, DNA transposons show the reverse pattern. Only four lineage-specific DNA transposon families could be identified in mouse (the mariner element MMAR1, and the hAT elements URR1, RMER30 and RChar1), compared with 14 in the primate lineage.

For evolutionary survival, DNA transposons are thought to depend on frequent horizontal transfer to new host genomes by means of vectors such as viruses and other intracellular parasites[116,125]. The mammalian immune system probably forms a large obstacle to the successful invasion of DNA transposons. Perhaps the rodent germ line has been harder to infiltrate by horizontal transfer than the primate genome. Alternatively, it is possible that highly diverged families active in early rodent evolution have not been detected yet. Notably, most copies in the human genome were deposited early in primate evolution.

Some of the above differences in the nature of interspersed repeats in human and mouse could reflect systematic factors in mouse and human biology, whereas others may represent

运 RNA）相似，但似乎是独立起源。B4 家族类似于 B1 和 ID 之间的融合[119,120]。我们发现，在 75,000 个已经识别的 ID 元件中，25% 定位于具有相似方向的 B1 元件的 50 bp 中，这表明可能大多古老的 ID 元件是错误标记或者截短的 B4 SINE。

第三类重复是 LTR 元件。哺乳动物中所有离散的含 LTR 元件都是脊椎动物特有的逆转录病毒分支的逆转录转座子衍生物。最早的感染性逆转录病毒可能起源于获得了水平传播机制的内源性逆转录病毒（ERV）样元件[121]，而现在很多内源性逆转录病毒元件可能来自于逆转录病毒的感染。

内源性逆转录病毒分为三类（I ~ III），显示了人类和小鼠进化史明显不同。值得注意的是，ERV 在人类中几乎灭绝，而这三类在小鼠中都有活跃的成员。

在人类–小鼠物种形成之前，第 III 类元件占可识别的 LTR 元件拷贝数的 80%。这类包括非自主的 MaLR：小鼠中包括 388,000 个可识别的拷贝数，它是唯一最成功的 LTR 元件。仍然活跃在小鼠身上（以 MERVL、MT 和 ORR1 MaLR 为代表），但在人类中已经死亡了约 5,000 万年[122]。

第 II 类元件的拷贝数在小鼠中的密度是人类中的十倍。相比之下，第 I 类元件拷贝数在人类基因组中是在小鼠基因组中的四倍多（尽管在小鼠中可能还有一些没有被识别）。在小鼠中，这类包括活跃的 ERV（如小鼠白血病病毒）、MuRRS、MuRVY 和 VL30（其中一些已经导致小鼠插入突变）——在人类中已经没有类似的活性。

第四类重复是 DNA 转座子。尽管大多数转座元件在小鼠体内比在人类中更活跃，但 DNA 转座子显示出相反的模式。在小鼠中，只鉴定出四个谱系特异性 DNA 转座家族（mariner 元件 MMAR1，hAT 元件 URR1、RMER30 和 RChar1），而灵长类谱系中有 14 个。

为了进化生存，DNA 转座子被认为依赖于通过病毒和其他细胞内的寄生虫等载体频繁水平转移到新的宿主基因组[116,125]。哺乳动物的免疫系统可能对 DNA 转座子成功入侵形成巨大的屏障。或许啮齿动物的生殖种系比灵长类基因组更难被水平转移渗透。也可能在啮齿动物进化中活跃的高分化家族可能还没有被检测到。值得注意的是，人类基因组中 DNA 转座子的大多拷贝在灵长类进化早期就被保存下来了。

上述人类和小鼠离散重复的性质差异可以反映小鼠和人类生物学中的系统因素，

random fluctuations. Deeper understanding of the biology of transposable elements and detailed knowledge of interspersed repeat populations in other mammals should clarify these issues.

## Similar repeats accumulate in orthologous locations

One of the most notable features about repeat elements is the contrast in the genomic distribution of LINEs and SINEs. Whereas LINEs are strongly biased towards (A+T)-rich regions, SINEs are strongly biased towards (G+C)-rich regions. The contrast is all the more notable because both elements are inserted into the genome through the action of the same endonuclease[126,127].

Such preferences were studied in detail in the initial analysis of the human genome[1], and essentially equivalent preferences are seen in the mouse genome. With the availability of two mammalian genomes, however, it is possible to extend this analysis to explore whether (A+T) and (G+C) content are truly causative factors or merely reflections of an underlying biological process.

Towards that end, we studied the insertion of lineage-specific repeat elements in orthologous segments in the human and mouse genomes. Each insertion represents a new, independent event occurring in one lineage, and thus any correlation between the two species reflects underlying proclivity to insert or retain repeats in particular regions. Visual inspection reveals a strong correlation in the sites of lineage-specific repeats of the various classes. Lineage-specific repeats also correlate with other genomic features, as discussed in the section on genome evolution.

The correlation of local lineage-specific SINE density is extremely strong. Moreover, local SINE density in one species is better predicted by SINE density in the other species than it is by local (G+C) content. The local density of each distinct rodent-specific type of SINE is a strong predictor of Alu density at the orthologous locus in human, although the Alu equivalent B1 SINEs show the strongest correlation ($r^2 = 0.784$).

We interpret these results to mean that SINE density is influenced by genomic features that are correlated with (G+C) content but that are distinct from (G+C) content *per se*. The fact that (G+C) content alone does not determine SINE density is consistent with the observation that some (G+C)-rich regions of the human genome are not Alu rich[128,129].

## Simple sequence repeats

Mammalian genomes are scattered with simple sequence repeats (SSRs), consisting of short perfect or near-perfect tandem repeats that presumably arise through slippage during DNA replication. SSRs have had a particularly important role as genetic markers

而其他可能代表了随机波动。对转座元件生物机制更深入的了解和对其他哺乳动物的离散重复序列的详细了解应该可以澄清这些问题。

### 类似重复在直系同源位置的积累

重复元件的一个最显著的特征是 LINE 和 SINE 的基因组分布差异。LINE 强烈倾向于(A+T)富集区，而 SINE 强烈倾向于(G+C)富集区。这种差异如此显著是因为两种元件都是通过相同的内切核酸酶的作用插入到基因组中[126,127]。

在人类基因组的最初分析中就对这种倾向性进行了详细研究[1]，在小鼠基因组中也发现了类似的倾向。但是，随着两个哺乳动物基因组的获得，就可以扩展这一分析以探索(A+T)和(G+C)含量是真正的影响因素或者仅仅是对于潜在生物学过程的一种反映。

为此，我们研究了在人类和小鼠基因组的直系同源片段中插入的谱系特异性重复元件。每个插入代表了一个新的、独立的事件发生在一个谱系中，因此两个物种之间的任何相关性反映了在特定区域插入或保留重复序列的潜在倾向。目测显示，不同类型的谱系特异的重复位点有很强的相关性。谱系特异的重复也与其他基因组特征相关，如在基因组进化一节中讨论所述。

局部谱系特异的 SINE 密度的相关性非常强。此外，与根据局部(G+C)含量进行预测相比，一个物种的局部 SINE 密度可以通过另一个物种的 SINE 密度进行更好的预测。尽管 Alu 与 B1 SINE 表现出最强的关联性($r^2 = 0.784$)，每种啮齿动物特定类型的 SINE 局部密度都是人类直系同源性基因座上的 Alu 密度的强力预测因子。

我们将这些结果解释为 SINE 密度受到与(G+C)含量相关但与(G+C)含量本身不同的基因组特征的影响。(G+C)含量不单独决定 SINE 密度，这一事实与人类基因组中某些(G+C)富集的区域 Alu 并不富集的观察结果一致[128,129]。

### 简单序列重复

哺乳动物基因组散在分布着简单重复序列(SSR)，可能由 DNA 复制过程中的滑脱引起的短的完全或近似完全的串联重复序列组成。SSR 在小鼠和人类的连锁研

in linkage studies in both mouse and human, because their lengths tend to be polymorphic in populations and can be readily assayed by PCR. It is possible that such SSRs, arising as they do through replication errors, would be largely equivalent between mouse and human; however, there are impressive differences between the two species[135].

Overall, mouse has 2.25–3.25-fold more short SSRs (1–5 bp unit) than human; the precise ratio depends on the percentage identity required in defining a tandem repeat. The mouse seems to represent an exception among mammals on the basis of comparison with the small amount of genomic sequence available from dog (4 Mb) and pig (5 Mb), both of which show proportions closer to human[136] (E. Green, unpublished data).

The analysis can be refined, however, by excluding transposable elements that contain SSRs at their 3′ ends. When these sources are eliminated, the contrast between mouse and human grows to roughly fourfold.

The reason for the greater density of SSRs in mouse is unknown. Analysis of the distribution of SSRs across chromosomes also reveals an interesting feature common to both organisms. In both human and mouse, there is a nearly twofold increase in density of SSRs near the distal ends of chromosome arms. Because mouse chromosomes are acrocentric, they show the effect only at one end. The increased density of SSRs in telomeric regions may reflect the tendency towards higher recombination rates in subtelomeric regions[1].

## Mouse Genes

Genes comprise only a small portion of the mammalian genome, but they are understandably the focus of greatest interest. One of the most notable findings of the initial sequencing and analysis of the human genome[1] was that the number of protein-coding genes was only in the range of 30,000–40,000, far less than the widely cited textbook figure of 100,000, but in accord with more recent, rigorous estimates[55,139-141]. The lower gene count was based on the observed and predicted gene counts, statistically adjusted for systematic under- and overcounting.

Our goal here is to produce an improved catalogue of mammalian protein-coding genes and to revisit the gene count. Genome analysis has been enhanced by a number of recent developments. These include burgeoning mammalian EST and cDNA collections, knowledge of the genomes and proteomes of a growing number of organisms, increasingly complete coverage of the mouse and human genomes in high-quality sequence assemblies, and the ability to use *de novo* gene prediction methodologies that exploit information from two mammalian genomes to avoid potential biases inherent in using known transcripts or homology to known genes.

We focus here on protein-coding genes, because the ability to recognize new RNA genes remains rudimentary. As used below, the terms "gene catalogue" and "gene count" refer to

究中作为遗传标记物具有特别重要的作用，因为它们的长度在群体中倾向于多态形式存在，可以轻易地通过PCR进行检测。这类SSR可能在人类和小鼠之间大致相同，都是由于复制错误而产生的，尽管其序列在两个物种之间存在着显著的差异[135]。

总体而言，小鼠的短SSR(1～5 bp单位)是人类的2.25～3.25倍；精确比率取决于定义一个串联重复所需要的百分比的一致性。根据与狗(4 Mb)和猪(5 Mb)这两个比例都接近于人类的物种[136]的可获得的少量基因组序列的比较，小鼠似乎代表了哺乳动物中的例外。(格林，未发表数据)

然而，通过排除3′端包含SSR的转座元件可以优化分析。当这些来源被消除后，小鼠和人类之间的差异将增加为大约四倍。

小鼠体内的SSR密度较大的原因未知。对SSR跨染色体分布的分析也揭示了两种生物体共有的一个有趣特征。在人类和小鼠中，靠近染色体臂末端的SSR密度几乎增加了两倍。因为小鼠的染色体是近端着丝点的，所以它们仅在一端显示出效果。端粒区的SSR密度增加可能反映了亚端粒区高重组率的趋势[1]。

## 小鼠基因

虽然基因只占哺乳动物基因组的一小部分，但它们是最受关注的焦点。人类基因组初期测序分析[1]最显著的发现之一是，编码蛋白质的基因数量只在30,000～40,000范围内，远低于被广泛引用的教科书上的数字100,000，但是符合近期更为严谨的估计[55,139-141]。较低的基因计数是基于对于观察的和预测的基因计数进行系统的不足和过度计数的统计校正。

我们的目标是建立一个改进的哺乳动物蛋白质编码基因目录和重新审视基因计数。最近的很多进展增强了基因组分析能力。这包括迅速发展的哺乳动物EST和cDNA收集，对于越来越多的生物的基因组和蛋白质组的知识，对小鼠和人类基因组的覆盖度越来越完整的高质量序列组装，以及使用新型基因预测方法从两个哺乳动物基因组获取信息，从而避免因使用已知转录本或已知同源基因时固有的潜在偏差的能力。

我们在此关注蛋白质编码基因，因为识别新的RNA基因的能力仍然处于初级阶段。如下文所用，术语“基因目录”和“基因计数”仅涉及蛋白质编码基因。我们在

protein-coding genes only. We briefly discuss RNA genes at the end of the section.

## Evidence-based gene prediction

We constructed catalogues of human and mouse gene predictions on the basis of available experimental evidence. The main computational tool was the Ensembl gene prediction pipeline[142] augmented with the Genie gene prediction pipeline[143]. Briefly, the Ensembl system uses three tiers of input. First, known protein-coding cDNAs are mapped onto the genome. Second, additional protein-coding genes are predicted on the basis of similarity to proteins in any organism using the GeneWise program[144]. Third, *de novo* gene predictions from the GENSCAN program[145] that are supported by experimental evidence (such as ESTs) are considered. These three strands of evidence are reconciled into a single gene catalogue by using heuristics to merge overlapping predictions, detect pseudogenes and discard misassemblies. These results are then augmented by using conservative predictions from the Genie system, which predicts gene structures in the genomic regions delimited by paired 5′ and 3′ ESTs on the basis of cDNA and EST information from the region.

The computational pipeline produces predicted transcripts, which may represent fragmentary products or alternative products of a gene. They may also represent pseudogenes, which can be difficult in some cases to distinguish from real genes. The predicted transcripts are then aggregated into predicted genes on the basis of sequence overlaps. The computational pipeline remains imperfect and the predictions are tentative.

## Initial and current human gene catalogue

The initial human gene catalogue[1] contained about 45,000 predicted transcripts, which were aggregated into about 32,000 predicted genes containing a total of approximately 170,000 distinct exons (Table 10). Many of the predicted transcripts clearly represented only gene fragments, because the overall set contained considerably fewer exons per gene (mean 4.3, median 3) than known full-length human genes (mean 10.2, median 8).

This initial gene catalogue was used to estimate the number of human protein-coding genes, on the basis of estimates of the fragmentation rate, false positive rate and false negative rate for true human genes. Such corrections were particularly important, because a typical human gene was represented in the predictions by about half of its coding sequence or was significantly fragmented. The analysis suggested that the roughly 32,000 predicted genes represented about 24,500 actual human genes (on the basis of fragmentation and false positive rates) out of the best-estimate total of approximately 31,000 human protein-coding genes on the basis of estimated false negatives[1]. We suggested a range of 30,000–40,000 to allow for additional genes.

本节末尾简要讨论了 RNA 基因。

### 基于证据的基因预测

我们在现有实验证据的基础上构建了人类和小鼠基因预测目录。主要的计算工具 Ensembl 基因预测流程 [142]，并用 Genie 基因预测流程 [143] 加强预测结果。简而言之，Ensembl 系统使用了三层输入。首先，已知编码蛋白质的 cDNA 标记到基因组上。第二，利用 GeneWise 程序 [144]，根据在任何生物体中蛋白质相似性预测剩下的蛋白质编码基因。第三，基于 GENSCAN 程序 [145] 中的由实验证据（如 EST）支持的新基因预测。通过使用试探法合并重叠预测，检测假基因和去掉错误组装，这三组证据被整合成单个基因目录。然后，这些结果通过 Genie 系统的保守预测得到进一步证实。Genie 是基于 cDNA 和 EST 信息，由配对的 5′和 3′ EST 界定基因组区域的基因结构预测系统。

计算流程产生预测的转录产物，这些可能代表了一个基因的片段产物或可变产物。它们也可能代表假基因，在某些情况下很难将其与真正的基因区分开。然后预测的转录产物根据序列重叠汇集到预测的基因中。目前计算流程还不完善，因此预测的产物也是试验性的。

### 初始和目前的人类基因目录

最初的人类基因目录 [1] 包含大概 45,000 个预测的转录本，被比对到大约 32,000 个预测基因上，其中包括总计约 170,000 个不同的外显子（表 10）。大多数预测的转录本明显只代表基因片段，因为整个序列中每个基因的外显子数目（平均值为 4.3，中位数为 3）比在已知的人类全长基因组上的数目（平均值为 10.2，中位数为 8）少很多。

基于对于碎片化比率、假阳性率和假阴性率的评估，这个初始基因目录被用来估算人类蛋白质编码基因的数量。这种校正非常重要，因为一个典型的人类基因在预测中只代表了其编码序列的一半，或者呈现明显的片段化。分析表明，约 32,000 个预测基因代表了基于假阴性估计的大约 31,000 个人类蛋白编码基因的最佳估计总数 [1] 中的 24,500 个实际的人类基因（基于碎片化和假阳性率）。我们建议使用 30,000 ~ 40,000 的范围以允许增加其他基因。

Table 10. Gene count in human and mouse genomes

| Genome feature | Human | | Mouse | |
|---|---|---|---|---|
| | Initial | Current | Initial* | Extended† |
| | (Feb. 2001) | (Sept. 2002) | (this paper) | (this paper) |
| Predicted transcripts | 44,860 | 27,048 | 28,097 | 29,201 |
| Predicted genes | 31,778 | 22,808 | 22,444 | 22,011 |
| Known cDNAs | 14,882 | 17,152 | 13,591 | 12,226 |
| New predictions | 16,896 | 5,656 | 8,853 | 9,785 |
| Mean exons/transcript‡ | 4.2 (3) | 8.7 (6) | 8.2 (6) | 8.4 (6) |
| Total predicted exons | 170,211 | 198,889 | 191,290 | 213,562 |

* Without RIKEN cDNA set.
† With RIKEN cDNA set.
‡ Median values are in parentheses.

Since the initial paper[1], the human gene catalogue has been refined as sequence becomes more complete and methods are revised. The current catalogue (Ensembl build 29) contains 27,049 predicted transcripts aggregated into 22,808 predicted genes containing about 199,000 distinct exons (Table 10). The predicted transcripts are larger, with the mean number of exons roughly doubling (to 8.7), and the catalogue has increased in completeness, with the total number of exons increasing by nearly 20%. We return below to the issue of estimating the mammalian gene count.

## Mouse gene catalogue

We sought to create a mouse gene catalogue using the same methodology as that used for the human gene catalogue (Table 10). An initial catalogue was created by using the same evidence set as for the human analysis, including cDNAs and proteins from various organisms. This set included a previously published collection of mouse cDNAs produced at the RIKEN Genome Center[41].

We also created an extended mouse gene catalogue by including a much larger set of about 32,000 mouse cDNAs with significant ORFs that were sequenced by RIKEN (see ref. 150). These additional mouse cDNAs improved the catalogue by increasing the average transcript length through the addition of exons (raising the total from about 191,000 to about 213,000, including many from untranslated regions) and by joining fragmented transcripts. The set contributed roughly 1,200 new predicted genes. The total number of predicted genes did not change significantly, however, because the increase was offset by a decrease due to mergers of predicted genes. These mouse cDNAs have not yet been used to extend the human gene catalogue. Accordingly, comparisons of the mouse and human gene catalogues below use the initial mouse gene catalogue.

表 10. 人类和小鼠基因组中的基因计数

| 基因组特性 | 人类 | | 小鼠 | |
|---|---|---|---|---|
| | 初始 | 现在 | 初始 * | 扩展 † |
| | (2001 年 2 月) | (2002 年 9 月) | (本文) | (本文) |
| 预测的转录本 | 44,860 | 27,048 | 28,097 | 29,201 |
| 预测的基因 | 31,778 | 22,808 | 22,444 | 22,011 |
| 已知 cDNA | 14,882 | 17,152 | 13,591 | 12,226 |
| 新预测 | 16,896 | 5,656 | 8,853 | 9,785 |
| 平均外显子数值 / 转录本 ‡ | 4.2(3) | 8.7(6) | 8.2(6) | 8.4(6) |
| 预测的外显子合计 | 170,211 | 198,889 | 191,290 | 213,562 |

* 不包括 RIKEN 的 cDNA 集。
† 包括 RIKEN 的 cDNA 集。
‡ 括号中为中位数。

从最初的文献 [1] 开始，人类基因目录随着序列的完整和方法的修订而不断完善。目前的目录（Ensembl 版本 29）包含 27,049 个预测转录本，比对到 22,808 个预测基因上，其中包含大约 199,000 个不同的外显子（表 10）。预测的转录本更大，外显子的平均数大约翻了一番（至 8.7），目录的完整性增加，外显子的总数增加了近 20%。下面我们重新回到估计哺乳动物基因计数问题。

## 小鼠基因目录

我们试图利用与人类基因目录相同的方式创建一个小鼠基因目录（表 10）。最初的目录是使用与人类分析相同的证据集合创建的，包括不同生物体的 cDNA 和蛋白质。这一集合包括 RIKEN（日本理化学研究所）基因组中心之前发表的小鼠 cDNA 数据集 [41]。

我们还创建了一个扩展的小鼠基因目录，包括一个由 RIKEN 测序且具有重要开放阅读框的更大的约 32,000 个小鼠 cDNA 的集合（见参考文献 150）。这些额外的小鼠 cDNA 通过增加平均转录本长度，即通过增加外显子（总数从约 191,000 增加到约 213,000，包括许多来自于非翻译区）和通过连接片段化的转录产物来改进目录。该集合贡献了大约 1,200 个新的预测基因。然而，预测基因的总数没有明显变化，因为增加量被预测基因的合并造成的减少相抵消了。这些小鼠 cDNA 尚未用来扩展人类基因目录。因此，下面的小鼠和人类基因目录的比较使用的是初始小鼠基因目录。

The extended mouse gene catalogue contains 29,201 predicted transcripts, corresponding to 22,011 predicted genes that contain about 213,500 distinct exons. These include 12,226 transcripts corresponding to cDNAs in the public databases, with 7,481 of these in the well-curated RefSeq collection[151]. There are 9,785 predicted transcripts that do not correspond to known cDNAs, but these are built on the basis of similarity to known proteins.

The new mouse and human gene catalogues contain many new genes not previously identified in either genome. It should be emphasized that the human and mouse gene catalogues, although increasingly complete, remain imperfect. Both genome sequences are still incomplete. Some authentic genes are missing, fragmented or otherwise incorrectly described, and some predicted genes are pseudogenes or are otherwise spurious. We describe below further analysis of these challenges.

## Pseudogenes

An important issue in annotating mammalian genomes is distinguishing real genes from pseudogenes, that is, inactive gene copies. Processed pseudogenes arise through retrotransposition of spliced or partially spliced mRNA into the genome; they are often recognized by the loss of some or all introns relative to other copies of the gene. Unprocessed pseudogenes arise from duplication of genomic regions or from the degeneration of an extant gene that has been released from selection. They sometimes contain all exons, but often have suffered deletions and rearrangements that may make it difficult to recognize their precise parentage. Over time, pseudogenes of either class tend to accumulate mutations that clearly reveal them to be inactive, such as multiple frameshifts or stop codons. More generally, they acquire a larger ratio of non-synonymous to synonymous substitutions than functional genes. These features can sometimes be used to recognize pseudogenes, although relatively recent pseudogenes may escape such filters.

To assess the impact of pseudogenes on gene prediction, we focused on two classes of gene predictions: (1) those that lack a corresponding gene prediction in the region of conserved synteny in the human genome (2,705); and (2) those that are members of apparent local gene clusters and that lack a reciprocal best match in the human genome (5,143). A random sample of 100 such predicted genes was selected, and the predictions were manually reviewed. We estimate that about 76% of the first class and about 30% of the second class correspond to pseudogenes. Overall, this would correspond to roughly 4,000 of the predicted genes in mouse.

## Comparison of mouse and human gene sets

We then sought to assess the extent of correspondence between the mouse and human gene sets. Approximately 99% of mouse genes have a homologue in the human genome. For 96% the homologue lies within a similar conserved syntenic interval in the human genome.

扩展的小鼠基因目录包括 29,201 个预测转录本，对应的 22,011 个预测基因包含约 213,500 个不同的外显子。这些包括对应于 cDNA 公共数据库中的 12,226 个转录本，其中 7,481 个在 RefSeq 集合 [151] 中有很好的注释。有 9,785 个预测转录本与已知的 cDNA 不对应，但这些转录本是建立在已知蛋白质相似性的基础上。

新的小鼠和人类基因目录包括很多新基因，这些新基因以前没有在任何一个基因组中被识别出来。应该强调的是，人类和小鼠基因目录虽然日趋完整，但仍然不完善。两个基因组序列仍然不完整。一些真实的基因缺失，碎片化或以其他方式被错误描述，一些预测基因是假基因或其他形式的假象。下面我们将进一步叙述对这些挑战的分析。

## 假基因

哺乳动物基因组注释的一个重要问题是区分真基因和假基因，即非活性基因拷贝。经过加工的假基因是通过剪接或部分剪接 mRNA 再逆转录到基因组中而产生的；相比于基因的其他拷贝，假基因经常通过部分或全部内含子的丢失而被识别。未经过加工的假基因产生于基因组区域的重复或是自然选择所释放的某个现存基因的退化。它们有时包含所有的外显子，但是经常遭受缺失和重排，这可能使得很难辨别它们的精确亲缘关系。随着时间的推移，任何一类的假基因都倾向于积累突变，明确显示它们是非活性的，例如存在多个移码或者终止密码子。一般而言，它们比功能基因获得更大比例的非同义或同义替代。这些特征有的时候会被用来识别假基因，尽管相对较新的假基因可能会逃过这种筛选。

为了评估假基因对基因预测的影响，我们关注两类基因预测：(1) 与人类基因组的保守同线性区域内缺乏相应基因预测的 (2,705)；(2) 在人类基因组中缺乏相互最佳匹配但有明显局部基因簇的成员 (5,143)。随机抽取 100 个这样的预测基因样本，并对预测进行人工检查。我们估计大约 76% 的第一类和大约 30% 的第二类与假基因相对应。总体而言，这相当于小鼠中大约有 4,000 个预测基因。

## 小鼠和人类基因集合的比较

然后，我们试图评估小鼠和人类基因集合之间的对应程度。大约 99% 的小鼠基因在人类基因组中具有同源性。96% 的同源基因位于人类基因组中相似的保守同线性间隔内。对于 80% 的小鼠基因，人类基因组中的最佳匹配反过来在保守同线性间

For 80% of mouse genes, the best match in the human genome in turn has its best match against that same mouse gene in the conserved syntenic interval. These latter cases probably represent genes that have descended from the same common ancestral gene, termed here 1:1 orthologues.

Comprehensive identification of all orthologous gene relationships, however, is challenging. If a single ancestral gene gives rise to a gene family subsequent to the divergence of the species, the family members in each species are all orthologous to the corresponding gene or genes in the other species. Accordingly, orthology need not be a 1:1 relationship and can sometimes be difficult to discern from paralogy (see protein section below concerning lineage-specific gene family expansion).

There was no homologous predicted gene in human for less than 1% (118) of the predicted genes in mouse. Genes that seem to be mouse-specific may correspond to human genes that are still missing owing to the incompleteness of the available human genome sequence. Alternatively, there may be true human homologues present in the available sequence, but the genes could be evolving rapidly in one or both lineages and thus be difficult to recognize.

## Mammalian gene count

To re-estimate the number of mammalian protein-coding genes, we studied the extent to which exons in the new set of mouse cDNAs sequenced by RIKEN[132] were already represented in the set of exons contained in our initial mouse gene catalogue, which did not use this set as evidence in gene prediction.

Our estimates suggest that the mammalian gene count may fall at the lower end of (or perhaps below) our previous prediction of 30,000–40,000 based on the human draft sequence[1]. Although small, single-exon genes may add further to the count, the total seems unlikely to greatly exceed 30,000. This lower estimate for the mammalian gene number is consistent with other recent extrapolations[141].

## RNA genes

The genome also encodes many RNAs that do not encode proteins, including abundant RNAs involved in mRNA processing and translation (such as ribosomal RNAs and tRNAs), and more recently discovered RNAs involved in the regulation of gene expression and other functions (such as micro RNAs)[165,166]. There are probably many new RNAs not yet discovered, but their computational identification has been difficult because they contain few hallmarks. Genomic comparisons have the potential to significantly increase the power of such predictions by using conservation to reveal relatively weak signals, such as those arising from RNA secondary structure[167]. We illustrate this by showing how comparative

隔中与相同的小鼠基因也具有最佳匹配。后面这些例子可能代表来自同一共同祖先的基因，这里称为 1∶1 直系同源性基因。

然而，全面鉴定所有直系同源基因关系具有挑战性。如果单个祖先基因在物种分化后产生了基因家族，则每个物种的家族成员都与相应基因或其他物种中的基因是直系同源的。因此，直系同源不一定是 1∶1 关系，有时很难从种内同源中辨别出来（参考下文关于谱系特异性基因家族扩增的蛋白质章节）。

小鼠中，少于 1%(118) 的预测的基因与人类中预测的基因没有同源性。看起来是小鼠特有的基因如果在人类基因中还是缺失状态可能缘于现有的人类基因组序列仍然不完整。也有可能是在现有的序列中存在真正的人源基因，但是基因可能在一个或两个谱系中快速进化，因此很难识别。

## 哺乳动物基因计数

为了重新估计哺乳动物蛋白编码基因，我们研究了 RIKEN 测序 [132] 的小鼠 cDNA 新集合中的外显子在我们最初的小鼠基因目录中的外显子集合中的富集程度，而且没有使用这一组数据作为基因预测的证据。

我们的估计表明，哺乳动物基因计数可能会处于（或可能低于）我们先前基于人类序列草图 [1] 预测的 30,000 ~ 40,000 的下限值。虽然小的单外显子基因可能进一步增加计数，但总数似乎不可能大大超过 30,000。对于哺乳动物基因数的这种较低估计与最近的其他推断相符 [141]。

## RNA 基因

基因组还编码许多不编码蛋白质的 RNA，包括参与 mRNA 加工和翻译的大量 RNA（如核糖体 RNA 和 tRNA），以及最近发现的参与基因表达调控和其他功能的 RNA（例如微 RNA）[165,166]。可能还有很多新 RNA 没有被发现，但由于其特征较少导致它们的计算识别很困难。通过基因组之间的比较，使用保守性来揭示相对较弱的信号（例如由 RNA 二级结构产生的信号 [167]）能显著增加此类预测的潜在能力。我们通过展示比较基因组学如何提高一个即便是非常了解的基因家族（tRNA 基因）的识

genomics can improve the recognition of even an extremely well understood gene family, the tRNA genes.

In our initial analysis of the human genome[1], the program tRNAscan-SE[168] predicted 518 tRNA genes and 118 pseudogenes. A small number (about 25 of the total) were filtered out by the RepeatMasker program as being fossils of the MIR transposon, a long-dead SINE element that was derived from a tRNA[169,170].

To improve discrimination of functional tRNA genes, we exploited comparative genomic analysis of mouse and human. True functional tRNA genes would be expected to be highly conserved. Indeed, the 498 putative mouse tRNA genes differ on average by less than 5% (four differences in about 75 bp) from their nearest human match, and nearly half are identical. In contrast, non-genic tRNA-related sequences (those labelled as pseudogenes by tRNAscan-SE or as SINEs by RepeatMasker) differ by an average of 38% and none is within 5% divergence. Notably, the 19 suspect predictions that violate the wobble rules show an average of 26% divergence from their nearest human homologue, and none is within 5% divergence.

On the basis of these observations, we identified the set of tRNA genes having cross-species homologues with < 5% sequence divergence. The set contained 335 tRNA genes in mouse and 345 in human. In both cases, the set represents all 46 expected anti-codons and exactly satisfies the expected wobble rules. The sets probably more closely represent the true complement of functional tRNA genes.

Although the excluded putative genes (163 in mouse and 167 in human) may include some true genes, it seems likely that our earlier estimate of approximately 500 tRNA genes in human is an overestimate. The actual count in mouse and human is probably closer to 350.

## Mouse Proteome

Eukaryotic protein invention appears to have occurred largely through two important mechanisms. The first is the combination of protein domains into new architectures. (Domains are compact structures serving as evolutionarily conserved functional building blocks that are often assembled in various arrangements (architectures) in different proteins[174].) The second is lineage-specific expansions of gene families that often accompany the emergence of lineage-specific functions and physiologies[175] (for example, expansions of the vertebrate immunoglobulin superfamily reflecting the invention of the immune system[1], receptor-like kinases in *A. thaliana* associated with plant-specific self-incompatibility and disease-resistance functions[49], and the trypsin-like serine protease homologues in *D. melanogaster* associated with dorsal–ventral patterning and innate immune response[176,177]).

The availability of the human and mouse genome sequences provides an opportunity to

别度来说明这一点。

在我们对人类基因组最初的分析中[1]，tRNAscan-SE[168]程序预测了518个tRNA基因和118个假基因。少量（一共25个）被RepeatMasker程序识别为MIR转座子化石（来源于tRNA的早已死亡的SINE元件[169,170]）而被过滤。

为了提高对功能性tRNA基因的识别，我们进行了小鼠和人类的比较基因组分析。真正的功能性tRNA基因预期是高度保守的。事实上，498个假定的小鼠tRNA基因与最接近的人类匹配基因的平均差异度小于5%（75 bp中大约有4个差异），近一半是完全相同的。相比之下，非tRNA基因相关序列（那些被tRNAscan-SE标记为假基因或者被RepeatMasker标记为SINE的序列）平均差异度为38%，没有一个在5%的差异范围内。值得注意的是，违反了摆动法则的19个可疑预测显示，与它们最接近的人类同系物的差异度平均为26%，没有一个在5%的差异范围内。

在此观察基础上，我们确定了具有小于5%序列差异的跨物种同系物的tRNA基因集合。这个集合包括335个小鼠tRNA基因和345个人类tRNA基因。这两个集合代表了所有46个预期的反密码子，并且完全满足预期的摆动法则。这些集合可能更接近功能性tRNA基因的真实补充。

尽管被排除掉的假定基因（小鼠163个，人类167个）中可能包括一些真正的基因，但我们早期以为人类大约有500个tRNA基因似乎是高估了。小鼠和人类的实际计数可能更接近350个。

## 小鼠蛋白质学

真核蛋白的出现看来主要通过两个重要机制。第一种是将蛋白质结构域重组到新结构里。（结构域是一种紧凑的结构，作为进化上保守的功能构建块，经常以不同的排列（结构）组装到不同的蛋白质中[174]。）第二种是伴随着谱系特异功能和生理现象出现的基因家族的谱系特异性扩张[175]（例如，脊椎动物免疫球蛋白超家族的扩增反映了免疫系统的出现[1]，拟南芥的受体样激酶与植物特异性自交不亲和性和抗病性功能相关[49]，黑腹果蝇中的胰蛋白酶样丝氨酸蛋白酶同系物与背腹图案和先天免疫反应相关[176,177]）。

人类和小鼠这两个密切相关的基因组序列为探索蛋白质进化课题提供了可能，

explore issues of protein evolution that are best addressed through the study of more closely related genomes. The great similarity of the two proteomes allows extensive comparison of orthologous proteins (those that descended by speciation from a single gene in the common ancestor rather than by intragenome duplication), permitting an assessment of the evolutionary pressures exerted on different classes of proteins. The differences between the mouse and human proteomes, primarily in gene family expansions, might reveal how physiological, anatomical and behavioural differences are reflected at the genome level.

## Overall proteome comparison

We compared the largest transcript for each gene in the mouse gene catalogue to the National Center for Biotechnology Information (NCBI) database ("nr" set; ftp://ftp.ncbi.nih.gov/blast/db/nr.z) using the BLASTP program[178]. Mouse proteins predicted to be homologues ($E < 10^{-4}$) of other proteins were classified into one of six taxonomic groupings: (1) rodent-specific; (2) mammalian-specific; (3) chordate-specific; (4) metazoan-specific; (5) eukaryote-specific; and (6) other (Fig. 17). The results were similar to those from an analysis of human proteins[1].

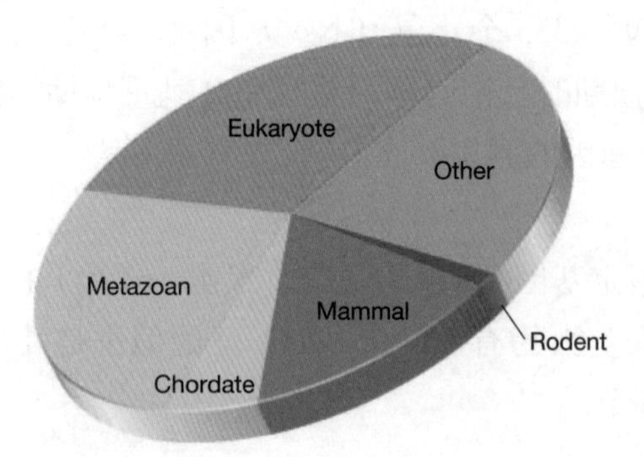


Fig. 17. Taxonomic breakdown of homologues of mouse proteins according to taxonomic range. Note that only a small fraction of genes are possibly rodent-specific ( < 1%) as compared with those shared with other mammals (14%, not rodent-specific); shared with chordates (6%, not mammalian-specific); shared with metazoans (27%, not chordate-specific); shared with eukaryotes (29%, not metazoan-specific); and shared with prokaryotes and other organisms (23%, not eukaryotic-specific).

As expected, most of the protein or domain families have similar sizes in human and mouse. However, 12 of the 50 most populous InterPro[179] families in mouse show significant differences in numbers between the two proteomes, most notably high mobility group HMG1/2 box and ubiquitin domains. On close analysis, the differences for six of these families can be accounted for by differential expansion of endogenous retroviral sequences in the genomes. We return below to the issue of expansion of gene families.

这些问题通过研究更密切相关的基因组得到了最好解决。这两个蛋白质组的巨大相似性允许同源蛋白质(那些由共同祖先的单个基因形成而不是基因组内重复形成的蛋白质)进行广泛比较，从而可以评估施加在不同类型蛋白质上的进化压力。小鼠和人类蛋白质组之间的差异，主要是基因家族扩展，可能揭示了生理、解剖和行为的差异如何在基因组水平上反映出来。

## 总蛋白质组比较

我们使用 BLASTP 程序[178]将小鼠基因目录中每个基因的最大转录本与美国生物技术信息中心(NCBI)数据库(“nr”集合；ftp://ftp.ncbi.nih.gov/blast/db/nr.z)进行了比较。根据与其他物种蛋白质的同源性($E < 10^{-4}$)比较，小鼠蛋白质的同系性被分类为六个类群中的一个：(1)啮齿动物特异性；(2)哺乳动物特异性；(3)脊索动物特异性；(4)后生动物特异性；(5)真核生物特异性；(6)其他(图 17)。这些结果与人类蛋白质分析结果相似[1]。

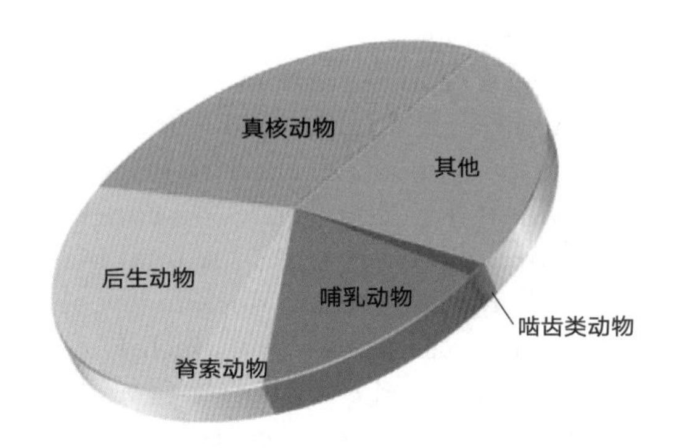


图 17. 按照范围对小鼠蛋白质进行同系物分类。注意，与其他哺乳动物(14%，非啮齿动物特有)共享的基因相比，只有小部分基因可能是啮齿动物特有的(小于 1%)；与脊索动物共享(6%，非哺乳动物特有)；与后生动物共享(27%，非脊索动物特有)；与真核生物共享(29%，非后生动物特有)；与原核生物和其他生物体共享(23%，非真核生物特有)。

正如所料，大多数蛋白质或是其结构域家族在人类和小鼠中的大小相似。然而，小鼠中最常见的 InterPro 家族[179]的 50 个蛋白质中有 12 个在两个蛋白质组之间的数量上存在显著差异，最显著的是高迁移率一组中 HMG1/2 盒和泛素化结构域。仔细分析，这些家族中的六个的差异可以通过基因组中内源性逆转录病毒序列的差异扩张来解释。下面我们回到基因家族的扩张这一问题。

## Evolution of gene families in mouse

As noted above, 80% of mouse proteins seem to have strict 1:1 orthologues in the human genome. Many of the remainder belong to gene families that have undergone differential expansion in at least one of the two genomes, resulting in the lack of a strict 1:1 relationship. Such gene family changes represent an insight into aspects of physiology that have emerged since the last common ancestor.

A well-documented example of family expansion is the olfactory receptor gene family, which represents a branch of the larger G-protein-coupled receptor superfamily tree[193,194]. Duplication of olfactory receptor genes seems to have occurred frequently in both rodent and primate lineages, and differences in number and sequence have been seen as distinguishing the degrees and repertoires of odorant detection between mice and humans. Moreover, an estimated 20% of the mouse olfactory receptor homologues[194] and a higher percentage of human homologues[195,196] are pseudogenes, indicating that there is a dynamic interplay between gene birth and gene death in the recent evolution of this family. The importance of these genes in reproductive behaviour is evident from defects in pheromone responses that result from deletion of the VR1 vomeronasal olfactory receptor gene cluster[197].

To explore systematically recent evolution of the mouse proteome, we searched for mouse-specific gene clusters. We identified genomic regions containing four or more homologous mouse genes that descended from a single gene in the human–mouse common ancestor; these represent local expansions in the mouse lineage. To detect such clusters, we compared all transcripts of each gene with those of five genes on either side (using the BLAST-2-Sequences program with a threshold of $E < 10^{-4}$). A total of 4,563 mouse genes were found to have at least one such homologue within this window. A total of 147 such clusters containing at least four homologues was identified, of which 47 contained multiple olfactory receptor genes, which have been studied elsewhere[193,199] and are not discussed further here. For the remaining 100 clusters, we then constructed dendrograms to examine the evolutionary relationship among the mouse proteins and their human homologues. This allowed us to identify those clusters containing mouse genes that are descendants of a single ancestral gene or for which multiple gene deletions had occurred in the human lineage.

In total, 25 such mouse-specific clusters were identified. In most cases (16), the mouse-specific cluster corresponds to only a single gene in the human genome. Among these 25 clusters, two major functional themes emerge: 14 contain genes involved in rodent reproduction and 5 contain genes involved in host defence and immunity. Each of the 14 "reproduction" clusters contains at least one gene whose expression is modulated by androgens, is involved in the biosynthesis or metabolism of hormones, has an established role in the placenta, gonads or spermatozoa, or has documented roles in mate selection, including pheromone olfaction. The fact that so many of the 25 clusters are related to reproduction is unlikely to be coincidental. Many of the most pronounced physiological differences between rodents and primates relate to reproduction, including substantial

### 小鼠基因家族的进化

如上所述，80% 的小鼠蛋白质在人类基因组中似乎有严格的 1:1 的直系同源物。其余的大多属于基因家族，它们在两个基因组中的至少一个发生了差异性扩增，导致缺乏严格的 1:1 关系。这种基因家族变化代表了自上一个共同祖先生理学方面出现变化的现象。

一个有文献记载的例子是嗅觉受体基因家族的扩张，它隶属于 G 蛋白偶联受体超家族树的一个分支 [193,194]。嗅觉受体基因的重复似乎在啮齿动物和灵长类动物谱系中经常出现，并且其数量和序列的差异被视为小鼠和人类对于气味的检测程度和组成的区分。此外，估计 20% 的小鼠嗅觉受体同源物 [194] 和更高比例的人类同源物 [195,196] 是假基因，这表明在该家族最近的进化过程中，基因出现和基因死亡之间存在着动态的相互作用。这些基因在生殖行为中的重要性从 VR1 鼻骨的嗅觉受体基因簇的缺失所导致的信息素反应缺陷中显而易见 [197]。

为了系统地探索小鼠蛋白质组的最新进化，我们寻找了小鼠特异性基因簇。我们确定了包含四个或者更多同源小鼠基因的基因组区域，这些同源小鼠基因是人类–小鼠共同祖先中一个单基因的后代；这些基因代表了小鼠谱系中的局部扩增。为了检测这些基因簇，我们将每个基因的所有转录产物与人类和小鼠任意一边的五个基因的转录本进行了比较(使用 BLAST2 序列程序，阈值 $E < 10^{-4}$)。发现总共有 4,563 个小鼠基因在这个窗口中至少有一个这样的同源基因。共计鉴定出 147 个这样的簇，包含至少 4 个同源性基因，其中 47 个包括多种嗅觉受体基因，这些基因已经在别处进行了研究 [193,199]，在此不再作进一步讨论。对于剩下的 100 个簇，我们构建了树状图来研究小鼠蛋白质和它们的人类同源物之间的进化关系。这使我们能够辨别那些小鼠基因簇，它们在小鼠中是单个祖先基因的后代，在人类谱系中则发生了多个基因的删除。

总共鉴定出 25 个这样的小鼠特异性基因簇。在大多数情况下(16)，小鼠特异性簇只对应人类基因组中的单个基因。在这 25 个簇中，出现了两个主要的功能主题：14 个包括啮齿动物生殖相关基因，5 个包括宿主防御和免疫相关基因。14 个“生殖”簇中的每一个都包含至少一个其表达受雄激素调节的基因，参与荷尔蒙的生物合成或代谢，在胎盘、性腺或者精子中具有确定的作用，或在配偶选择中有确实的作用，包括信息素嗅觉。25 个簇中这么多与生殖相关的事实不太可能是巧合。啮齿动物和灵长类动物之间许多显著的生理差异与生殖相关，包括胎盘结构、产仔数、

variations in placental structures, litter sizes, oestrous cycles and gestation periods. It seems likely that reproductive traits have been responsible for some of the most powerful evolutionary pressures on the mouse genome, and that the demand for innovation has been met through gene family expansions. Examination of the human genome in this way may similarly reveal gene clusters that reflect particular aspects of human reproduction.

The five mouse clusters that encode genes involved in immunity suggest that another major evolutionary force is acting on host defence genes. The five clusters include the major histocompatibility complex (MHC) class Ib genes, two clusters of antimicrobial β-defensins, a cluster of WAP domain antimicrobial proteins and a cluster of type A ribonucleases. Ribonuclease A genes appear to have been under strong positive selection, possibly due to their significant role in host-defence mechanisms[224].

The two major themes—reproduction and immunity—may not be entirely unrelated; that is, the MHC class Ib genes have roles in both pregnancy and immunity. MHC genotype is also known from ethological studies to influence mate selection, although the molecular mechanisms underlying this effect remain unknown. Within the MHC complex, the class I genes are the most divergent, having arisen after the rodent–human divergence[227].

## Genome Evolution: Selection

Investigation of the two principal forces that shape the evolution of the mouse and human genomes—mutation and selection—requires looking beyond coarse-scale identification of regions of conserved synteny and purely codon-based analysis of orthologues, to fine-scale alignment of the two genomes at the nucleotide level.

The substantial sequence divergence between the mouse and human genomes is still low enough that orthologous sequences undergoing neutral drift remain conserved enough for them to be aligned reliably. The challenge then is to use such alignments to tease apart the effects of neutral drift, which can teach us about underlying mutational processes, and selection, which can inform us about functionally important elements. It should be emphasized that sequence similarity alone does not imply functional constraint.

In this section, we use whole-genome alignments to explore the extent of sequence conservation in neutral sites (such as ancestral repeat sequences), known functional elements (such as coding regions) and the genome as a whole. By comparing these, we are able to estimate the proportion of regions of the mammalian genome under evolutionary selection (about 5%), which far exceeds the amount attributable to protein-coding sequences. In the next section, we then use the neutral sites to study how mutational forces vary across the genome.

发情周期和妊娠期的实质性变化。生殖相关性状似乎对应了小鼠基因组上受到的一些最强大的进化压力，并且通过基因家族扩张满足了创新的需求。以这种方式对人类基因组进行检查，可以类似地揭示反映人类生殖的特定方面的基因簇。

编码免疫基因的五个小鼠簇显示作用于宿主防御基因的另一个主要的进化力量。这五个簇包括主要组织相容性复合物（MHC）Ib 类基因，两组抗微生物的 β-防御素，一组 WAP 结构域抗微生物蛋白和一组 A 型核糖核酸酶。核糖核酸酶 A 基因似乎一直处于强正选择中，可能是由于它们在宿主防御机制中的重要作用[224]。

生殖和免疫这两个主题可能并非完全没有关系；也就是说，MHC 中 Ib 类基因在妊娠和免疫中都有作用。从行为学研究中也知道 MHC 基因型会影响配偶选择，尽管这种影响的分子机制仍然不清楚。在 MHC 复合体中，I 类基因是最为分化的，出现在啮齿动物–人类分化之后[227]。

## 基因组进化：选择

研究塑造小鼠和人类基因组进化的两个主要力量——突变和选择——需要超越粗略识别保守同线性区域和完全基于密码子的直系同源性分析，达到在核苷酸水平对两个基因组进行精准比对。

小鼠和人类基因组之间的实质性序列差异仍然很小，使得经过中性漂变的同源序列仍足够保守到能够可靠地比对到一起。接下来的挑战就是使用这种比对梳理中性漂变和选择的影响，中性漂变可以告诉我们潜在的突变过程，选择可以告诉我们功能性重要元件。应该强调的是序列相似性本身并不意味着功能约束。

在本节中，我们使用全基因组比对来探索中性位点（例如祖源重复序列）、已知功能元件（例如编码区）和整个基因组的序列保守程度。通过比较这些，我们能够估计哺乳动物基因组受到进化选择的区域的比例（大约 5%），远远超过蛋白质编码序列的数量。在下一节，我们用中性位点来研究突变力如何在基因组中变化。

## Fine-scale alignment of genomes

We began by creating a catalogue of sequence alignments between the mouse and human genomes. The alignments were produced by the BLASTZ[328] program by comparing all non-repeat sequences across the genome to identify all high-scoring matches, then, using these as seeds, we extended the alignments into the surrounding regions, including into repeat sequences. Regions that could be aligned clearly at the nucleotide level totalled about 1.1 Gb, corresponding to roughly 40% of the human genome.

## Proportion of genome under selection

To investigate the fraction of a mammalian genome under evolutionary selection for biological function, we estimated the proportion of the genome that is better conserved than would be expected given the underlying neutral rate of substitution. We compared the overall distribution $S_{genome}$ of conservation scores for the genome to the neutral distribution $S_{neutral}$ of conservation scores for ancestral repeats using a genome-wide set of 14.3 million non-overlapping 50-bp (human) windows, each containing at least 45 bp (mean 48.67 bp) of aligned sequence. The genome-wide score distribution for these windows has a prominent tail extending to the right, reflecting a substantial excess of windows with high conservation scores relative to the neutral rate. The excess can be estimated by decomposing the genome-wide distribution $S_{genome}$ as a mixture of two components: $S_{neutral}$ and $S_{selected}$ (reflecting windows under selection).

The mixture coefficients indicate that at least 20.8% of the windows are under selection, with the remainder consistent with neutral substitution. Because about 25.2% of all human bases are contained in the windows, this suggests that at least 5.25% (25.2% of 20.8%) of the 50-base windows in the human genome is under selection. Repeating the analysis on more stringently filtered alignments (with non-syntenic and non-reciprocal best matches removed) requiring different numbers of aligned bases per window and with 100-bp windows, yields similar estimates, ranging mostly from 4.8% to about 6.1% of windows under selection (D. Haussler, unpublished data), as does using an alternative score function that considers flanking base context effects and uses a gap penalty[330].

The analysis thus suggests that about 5% of small segments (50 bp) in the human genome are under evolutionary selection for biological functions common to human and mouse. This corresponds to regions totalling about 140 Mb of human genomic DNA, although not all of the nucleotides in these windows are under selection. In addition, some bases outside these windows are likely to be under selection. In a loose sense, these regions might be regarded as containing the "functional" conserved subset of the mammalian genome. Of course, it should be noted that nonconserved sequence may have important roles, for example, as a passive spacer or providing a function specific to one lineage. Notably, protein-coding regions of genes can account for only a fraction of the genome under selection. From

### 基因组的精细比对

我们首先创建了小鼠和人类基因组之间的序列比对目录。通过 BLASTZ[328] 程序完成比对，该程序通过比较基因组中所有非重复序列来确定所有高得分匹配，然后用这些序列作为种子，我们扩展这些对比到周围区域，包括重复序列在内。在核苷酸水平上可以清楚比对的区域大约为 1.1 Gb，对应于人类基因组的大约 40%。

### 处于选择的基因组比例

为了研究生物功能进化选择下哺乳动物基因组的比例，我们估计这部分基因组比预期的潜在中性替代更保守。我们使用全基因组的 1,430 万个非重叠的 50 bp(人类)窗口，将基因组保守分数 $S_{genome}$ 的总分布和祖源重复序列的保守分数 $S_{neutral}$ 的中性分布进行了比较，每个窗口包含至少 45 bp(平均 48.67 bp)的比对序列。这些窗口的全基因组得分分布具有向右延伸的突出尾部，反映出相对于中性率而言，具有高保守分数的窗口明显超量。超量值可通过将全基因组分布 $S_{genome}$ 分解为两个成分的混合物——$S_{neutral}$ 和 $S_{selected}$(反映经历选择的窗口)进行估计。

混合系数表明，至少 20.8% 的窗口处于选择中，其余的窗口与中性替换一致。因为人类所有碱基的大约 25.2% 都包含在窗口中，这表明人类基因组中 50 个碱基窗口中至少有 5.25%(20.8% 中的 25.2%)处于选择中。以更严格筛选比对条件重复上述分析(去除了非同线性和单向的最佳匹配)，每个窗口需要数量不同的比对碱基和 100 bp 的窗口，得到了类似的估计值，经历选择作用的窗口多数在 4.8% 到 6.1% 范围内(豪斯勒，未发表数据)，如同使用一个考虑碱基两端关系效应和使用空缺惩罚的替代评分函数 [330]。

因此，分析表明人类基因组中大约 5% 的小片段(50 bp)正处于人类和小鼠共同具有的生物学功能的选择性进化中。这相当于人类基因组 DNA 总量中的大约 140 Mb 的区域，尽管并非所有处于此窗口的核苷酸都处在选择中。除此之外，这些窗口之外的一些碱基可能也在经历选择。非严格意义上讲，这些区域可能被认为包含哺乳动物基因组的“功能性”保守亚群。当然，应该注意的是，非保守序列可能具有重要作用，例如作为被动间隔子或者在某一特定谱系发挥功能。值得注意的是，基因的蛋白编码区域只占被选择基因组的小部分。从我们对基因的数量和性质的

our analysis of the number and properties of genes, coding regions comprise only about 1.5% of the human genome and account for less than half of the segments under selection.

What accounts for the remainder of the genome under selection? About 1% of the genome is contained in untranslated regions of protein-coding genes, and some of this sequence is under some functional constraint. Another main class of interest are those sequences that control gene expression; if a typical gene contains a few such regulatory sequences, there may be tens to hundreds of thousands of such elements. In addition, conserved sequences probably encode non-protein-coding RNAs (which remain difficult to discern) and chromosomal structural elements. Furthermore, some of the conserved fraction may correspond to sequences that were under selection for some period of time but are no longer functional; these could include recent pseudogenes. Characterization of the conserved sequences should be a high priority for genomics in the years ahead.

The analysis above allows us to infer the proportion of the genome under selection by decomposing the curve $S_{genome}$ into curves $S_{neutral}$ and $S_{selected}$. Importantly, it does not definitively assign an individual conserved sequence as being neutral or selected. One can calculate, for a sequence with conservation score $S$, the probability $P_{selected}(S)$ that the window of sequence belongs to the selected subset. The probability exceeds 83% for sequences with $S > 3$ and 93% for $S > 4$, but is only 52% for $S = 2$. In other words, some functionally important sequence cannot be separated cleanly from the tail of the distribution of neutral conservation.

## Genome Evolution: Mutation

Genome-wide alignments also allow us to investigate how the patterns of neutral substitution, deletion and insertion vary across the genome, providing an insight on the underlying mutational processes.

### Substitution rate varies across the genome

Significant variation in the level of sequence conservation has been reported in several small-scale studies of human and mouse genomic regions[10,248-254] and in several larger-scale studies of coding sequences[255-260]. It has not been clear in all cases whether the variation reflects differences in neutral substitution rates or in selection. The human–mouse genome alignments allow us to address the variation more comprehensively and to test for co-variation with the rates of other processes, such as insertions of transposable elements[255] and meiotic recombination[258].

We used the collection of aligned ancestral repeats and aligned fourfold degenerate sites to calculate the apparent neutral substitution rate for about 2,500 overlapping 5-Mb windows across the human genome. To accurately follow fluctuations while accounting for

分析来看，编码区域仅大约占人类基因组的 1.5%，可解释少于一半的处于选择中的片段。

是什么原因导致基因组的剩余部分处于选择中？大约 1% 的基因组是在蛋白质编码基因的非翻译区，其中一些序列受到某些功能限制。另一类引起关注的主要类别是那些控制基因表达的序列；如果一个典型的基因包括一些这样的调节序列，则可能会有数以万计的这种元件的存在。此外，保守序列可能编码非蛋白质的 RNA（这些仍然难以识别）和染色体结构元件。进而，部分这些保守区域可能与某段时间内被选择但不再具有功能的序列相符；这些可能包括最近出现的假基因。未来几年中，解析保守序列应该是基因组学研究的高度优先事项。

上述分析允许我们通过将 $S_{genome}$ 曲线分解为 $S_{neutral}$ 曲线和 $S_{selected}$ 曲线来推断处于选择的基因组的比例。重要的是，它没有明确地将单个保守序列指定为中性或被选择。对于具有保守分数 $S$ 的序列，可以计算出属于选择作用子集的序列窗口的概率 $P_{selected}(S)$。$S > 3$ 的序列概率超过 83%，$S > 4$ 的序列概率超过 93%，但 $S = 2$ 的序列只占 52%。换句话说，一些重要功能序列不能从中性保守分布的尾部被清楚地分离出来。

## 基因组进化：突变

全基因组的比对还使得我们能够研究中性替代、缺失和插入在基因组中如何发生和变化，从而提供了对潜在突变过程的解析。

### 替代率在基因组中的变化

在人类和小鼠基因组区域的几个小规模研究[10,248-254]和编码序列的几个大规模研究[255-260]中，已经报道了序列保守性上的显著变化。尚不清楚在所有情况中，这个变化是否反映了中性替代率或选择上的差异。人类–小鼠基因组比对使我们更全面地研究变异，检测与其他过程比率的共突变情况，例如转座元件的插入[255]和减数分裂的重组[258]。

我们通过收集比对的祖源重复序列和比对的四倍简并位点来计算人类基因组中大约 2,500 个重叠 5 Mb 窗口的中性替代率。为了在考虑碱基组成的区域变化的同时能准确地跟踪波动，通过仅使用窗口中的祖源重复位点（平均大约 280,000 个 / 窗

regional changes in base composition, the regional nucleotide substitution rate in ancestral repeat sites, $t_{AR}$, was calculated separately for each 5-Mb window by maximum likelihood estimation of the parameters of the REV model using only the ancestral repeat sites in the window (average of about 280,000 sites per window). The regional nucleotide substitution rate in fourfold degenerate sites, $t_{4D}$, was calculated similarly from an average of about 3,700 fourfold degenerate sites per window. Windows with fewer than 800 ancestral repeats or fourfold degenerate sites were discarded.

The mean and standard deviations across the windows were $t_{AR} = 0.467 \pm 0.022$ and $t_{4D} = 0.447 \pm 0.067$ substitutions per site. The standard deviation is much larger (over tenfold and threefold, respectively) than would be expected from sampling variance. These data clearly indicate substantial regional fluctuation.

What properties of chromosomal DNA could account for the variation in substitution rate? One possible explanation is local (G+C) content, but previous studies disagree on whether it correlates strongly with divergence[92,255,262,263]. We find that $t_{AR}$ and $t_{4D}$ vary with local (G+C) content, although the dependence is nonlinear[262,264] and is better fitted by regression with a quadratic curve[263]. In other words, the substitution rate seems to be higher in regions of extremely high or low (G+C) content, with the sign of the correlation differing in regions with high versus low (G+C) content. This pattern persists if CpG substitutions are removed from the analysis (data not shown).

All three forces that alter the genome (nucleotide substitution, deletion and insertion) thus vary substantially across the genome. Moreover, they are significantly correlated and tend to co-vary along chromosomes. Notably, these three measures of interspecies divergence are also correlated with recent substitutions in the human genome, as measured by the density of SNPs identified by the SNP Consortium[265].

## Possible explanations for variation

What explains the correlation among these many measures of genome divergence? It seems unlikely that direct selection would account for variation and co-variation at such large scales (about 5 Mb) and involving abundant neutral sites taken from ancestral transposon relics. Selection against deleterious mutations can remove linked polymorphisms[270,271], but it is not clear that such effects or related effects[272] could extend to such large scales or to interspecies divergence over such large time periods[273].

It seems more probable that these features reflect local variation in underlying mutation rate, caused by differences in DNA metabolism or chromosome physiology. The causative factors may include recombination-associated mutagenesis[258,266], transcription-associated mutagenesis[274], transposon-associated deletion and genomic rearrangement[275-278], and replication timing[279,280]. Nuclear location may also be involved, including proximity to matrix attachment sites, heterochromatin, nuclear membrane, and origins of replication.

口）对 REV 模型的参数进行最大似然估计，分别计算每 5 Mb 窗口中祖源重复位点 $t_{AR}$ 的区域核苷酸替代率。四倍简并位点 $t_{4D}$ 的区域核苷酸替代率类似地由每个窗口大约 3,700 个四倍简并位点的平均值计算得到。祖源重复或者四倍简并位点少于 800 个的窗口被放弃。

窗口的平均值和标准偏差为每个位点 $t_{AR} = 0.467 \pm 0.022$，$t_{4D} = 0.447 \pm 0.067$ 个碱基替换。标准偏差比抽样方差的估计值大很多（分别超过 10 倍和 3 倍）。这些数据明确显示出巨大的区域波动。

染色体 DNA 的哪些特性可以解释替代率的变化呢？一种可能的解释是局部（G+C）含量，但之前的研究在它是否与这种变化有很强的相关性上结果并不一致 [92,255,262,263]。我们发现，$t_{AR}$ 和 $t_{4D}$ 随着局部（G+C）含量的变化而变化，尽管这种依赖性是非线性的 [262,264]，用二次曲线回归更合适 [263]。换句话说，在（G+C）含量极高或极低的区域，替代率似乎更高，在（G+C）含量高和低的区域，具有不同的相关性。如果从分析中删除 CpG 位点上的碱基替换，这种模式仍然存在（数据未显示）。

因此，这三种改变基因组的力量（核苷酸替换、删除和插入）在基因组中都有很大的差异。此外，它们之间明显相关而且倾向于沿着染色体共同变异。值得注意的是，这三种变异的种间差异的测量也与人类基因组中的最近替换具有相关性，这是由 SNP 协作组进行的 SNP 密度测量所确定的 [265]。

### 变化的可能解释

用什么解释基因组分化的各种测量指标之间的关联性？直接选择似乎不太能解释如此大规模（大约 5 Mb）的变异和共变异，而且这些变异还涉及大量从祖源转座子遗物中获取的大量中性位点。对有害突变的选择可以消除连锁多态性 [270,271]，但是尚不清楚这种效应或者相关效应 [272] 能否扩展到如此大的规模或在如此长的时间内 [273] 扩展到种间分化。

这些特征似乎更可能反映了 DNA 代谢或染色体生理学上的差异所引起的潜在突变率的局部变化。诱发因素可能包括重组相关的突变 [258,266]，转录相关的突变 [274]，转座相关缺失和基因组重排 [275-278] 和复制时间 [279,280]。也可能涉及核定位，包括接近基质附着点、异染色质、核膜和复制起始点。

It is clear that the mammalian genome is evolving under the influence of non-uniform local forces. It remains an important challenge to unravel the mechanistic basis and evolutionary consequences of such variation.

## Genetic Variation among Strains

### Implications for the laboratory mouse

The promise of genomics is the ability to connect phenotypes with genotypes for a wide variety of traits and to use the resulting molecular insights to develop new approaches for the cure and prevention of disease. The laboratory mouse occupies a central place in this vision, both as a prototype for all mammalian biology and as a well-characterized organism for modelling human disease states[15,16,123]. In this section, we briefly discuss ways in which the mouse genome sequence will accelerate biomedical progress in the future. Because the sequence has been made available in public databases in advance of publication, examples for many of the predictions can already be cited.

### Positional cloning of genes for mendelian phenotypes

More than 1,000 spontaneously arising and radiation-induced mouse mutants causing heritable mendelian phenotypes are catalogued in the Mouse Genome Informatics (MGI) database (http://www.informatics.jax.org). Largely through positional cloning, the molecular defect is now known for about 200 of these mutants. The availability of an annotated mouse genome sequence now provides the most efficient tool yet in the gene hunter's toolkit. One can move directly from genetic mapping to identification of candidate genes, and the experimental process is reduced to PCR amplification and sequencing of exons and other conserved elements in the candidate interval. With this streamlined protocol, it is anticipated that many decades-old mouse mutants will be understood precisely at the DNA level in the near future. An example of how the draft genome sequence has already been successfully used is the recent identification of the mouse mutation "chocolate" in the melanosome protein Rab38 (ref. 284).

### Identification of quantitative trait loci

The availability of more than 50 commonly used laboratory inbred strains of mice, each with its own phenotype for multiple continuously variable traits, has provided an important opportunity to map QTLs that underlie heritable phenotypic variation. A systematic initiative is currently underway[285] to define parameters such as body weight, behavioural patterns, and disease susceptibility among a standard set of inbred lines, and to make these data freely available to the scientific community in the Mouse Phenome Database (www.jax.org/phenome). Appropriate crosses between such lines, followed by genotyping, will enable

很明显，哺乳动物基因组是在非均匀局部力量的影响下进化的。揭示这种变异的机制基础和进化后果仍然是一个很重要的挑战。

## 品系间遗传变异

### 对实验小鼠的影响

基因组学的前景是将各种性状的表型和基因型联系起来，并利用由此产生的分子发现来开发治疗和预防疾病的新方案。实验小鼠在这里占据了中心位置，既是作为所有哺乳动物生物学的原型，也是模拟人类疾病状态的具有良好特征的生物体[15,16,123]。在本节中，我们简单讨论小鼠基因组序列在未来将如何加速生物医学进程。因为该序列已经在出版前公布在公共数据库中，所以可以引用许多预测的示例。

### 孟德尔表型基因的定位克隆

在小鼠基因组信息(MGI)数据库(http://www.informatics.jax.org)中收录了超过1,000个自发产生和辐射诱导产生的小鼠突变体，这些突变体引发可遗传的孟德尔表型。主要通过定位克隆，目前已知大约200个这些突变体的分子缺陷。小鼠基因组序列的注释信息提供了基因猎人工具箱中最有效的工具。研究人员可以直接从基因定位转移到候选基因的鉴定，实验过程简化为对候选区间外显子和其他保守元件进行PCR扩增和测序。通过这种简化的方案，预计在不久的将来，将在DNA水平上精确地解释许多几十年前的小鼠突变体。成功使用基因组序列草图的一个例子就是最近在黑素体蛋白Rab38中鉴定出了小鼠突变体“chocolate”(参考文献284)。

### 数量性状基因座的鉴定

常用的实验室近交系小鼠超过50种，每种小鼠具有多种连续变化的性状作为自身表型，这为定位可遗传表型的变异的数量性状基因座(QTL)提供了重要机会。目前正在进行一项系统性举措[285]，以确定标准近交系中例如体重、行为模式和疾病易感性等参数，并在小鼠表观数据库(www.jax.org/phenome)中将这些数据免费提供给科学界。这些品系之间适当杂交，然后进行基因分型，能够进行QTL的定位，然

the mapping of QTLs, which can then be subjected to positional cloning. The degree of difficulty is substantially greater for a QTL cloning project than for a mendelian disorder, however, as the responsible intervals are usually much larger, the boundaries more difficult to delineate precisely, and the causative variant often much more subtle[286]. For these reasons, only a handful of the approximately 1,000 mapped QTLs have been identified at the molecular level. The availability of the mouse sequence should greatly improve the chances for future success.

## Creation of knockout and knockin mice

The wide application of homologous recombination in embryonic stem cells has provided a remarkable abundance of "custom" mice with specifically engineered loss- or gain-of-function mutations in specific genes of biological or medical interest. Yet this remains a time-consuming process. The design of recombinant DNA constructs for injection has often been delayed by incomplete knowledge of gene structure, requiring tedious restriction mapping or sequencing, and occasionally giving rise to unsatisfying outcomes due to incorrect information. The availability of the mouse genome sequence will both speed the design of such constructs and reduce the likelihood of unfortunate choices. Furthermore, the long-range continuity of the sequence should facilitate the generation of models of contiguous gene-deletion syndromes.

## Creation of transgenic animals

For many transgenic experiments, it is important to maintain copy-dependent, tissue-specific expression of the transgene. This is most readily accomplished through BAC transgenesis. The availability of a deep, end-sequenced BAC library from the B6 strain mapped to the genome sequence now makes it straightforward to obtain a desired gene in a BAC for such experiments; end-sequenced BAC libraries from other strains should be available in the future. BACs also provide the ability to make mutant alleles with relative ease, by taking advantage of powerful genetic engineering techniques for custom mutagenesis in the *Escherichia coli* host.

## Applications to cancer

The mouse genome sequence also has powerful applications to the molecular characterization of the somatic mutations that result in neoplasia. High-density SNP mapping to identify loss of heterozygosity[288,289], combined with comparative genomic hybridization using cDNA or BAC arrays[290,291], can be used to identify chromosomal segments showing loss or gain of copy number in particular tumour types. The combination of such approaches with expression arrays that include all mouse genes should further enhance the ability to pinpoint the molecular lesions that result in carcinogenesis. Full sequencing of all the exons and

后进行定位克隆。然而，QTL 克隆项目的难度远远大于孟德尔疾病，因为相关的区间通常要大得多，边界更难精确描绘，而致病突变通常也更加微妙[286]。由于这些原因，在大约 1,000 个定位的 QTL 中，只有少数在分子水平上得到鉴定。而小鼠序列的产生在未来应该大大提高成功的机会。

### 构建敲除和敲入小鼠

同源重组技术在胚胎干细胞中的广泛应用提供了大量“定制”小鼠，它们携带了专门设计的生物学或医学感兴趣的特定基因功能丧失或获得的突变。然而，这仍然是一个耗时的过程。用于注射的重组 DNA 构建体的设计常常因为对基因结构的不完全了解而延迟，需要繁琐的限制性作图或测序，并且偶尔由于不正确的信息而导致令人不满意的结果。小鼠基因组序列的出现将加速这种结构设计并降低产生不幸选择的可能性。此外，序列的长程连续性有助于构建连续基因缺失综合征的模型。

### 构建转基因动物

对于很多转基因实验来说，保持转基因的拷贝依赖性、组织特异性表达是很重要的。这是最容易通过 BAC 转基因实现的。来自于 B6 品系基因组序列的深度末端测序的 BAC 库的应用，使得现在在此类实验中可以直接获得 BAC 中所需的基因；其他品系的末端测序 BAC 库在将来也将可用。通过利用强大的基因工程技术在大肠杆菌宿主中进行定制诱变，BAC 还能够相对容易地产生突变等位基因。

### 应用于癌症

小鼠基因组序列在研究导致瘤形成的体细胞突变的分子特征上也具有强大的应用。用于鉴定杂合性缺失的高密度 SNP[288,289] 比对、结合使用 cDNA 或 BAC 阵列的基因组杂交进行比较[290,291] 都可以用于鉴定在特定肿瘤类型中拷贝数缺失或增加的染色体区段。这些方法与包含所有小鼠基因的表达阵列的结合使用，将进一步精确定位导致癌症产生的分子损伤。目前可以对已知肿瘤抑制基因、癌基因和其他候

regulatory regions of known tumour suppressors, oncogenes, and other candidate genes can now be contemplated, as has been initiated in a few centres for human tumours[292].

### Making better mouse models

Not all mouse models replicate the human phenotype in the expected way. The availability of the full human and mouse sequences provides an opportunity to anticipate these differences, and perhaps to compensate for them. In some instances, it may turn out that the murine mutation did not reside in the true orthologue of the human disease gene. Alternatively, in a circumstance where the human genome contains only a single gene family member, but the mouse genome contains a paralogue as well as the orthologue, one can anticipate that knockout of the orthologue alone may give a much milder phenotype (or none at all). Such was the case, for instance, with the occulocerebrorenal syndrome described by Lowe and colleagues[296]. Creating double knockout mice may then provide a closer match to the human disease phenotype.

### Understanding gene regulation

Of the approximately 5% of windows of the mammalian genome that are under selection, most do not appear to code for protein. Much of this sequence is probably involved in the regulation of gene expression. It should be possible to pinpoint these regulatory elements more precisely with the availability of additional related genomes. However, mouse is likely to provide the most powerful experimental platform for generating and testing hypotheses about their function. An example is the recent demonstration, based on mouse–human sequence alignment followed by knockout manipulation, of several long-range locus control regions that affect expression of the Il4/Il13/Il5 cluster[4].

## Conclusion

The mouse provides a unique lens through which we can view ourselves. As the leading mammalian system for genetic research over the past century, it has provided a model for human physiology and disease, leading to major discoveries in such fields as immunology and metabolism. With the availability of the mouse genome sequence, it now provides a model and informs the study of our genome as well.

Comparative genome analysis is perhaps the most powerful tool for understanding biological function. Its power lies in the fact that evolution's crucible is a far more sensitive instrument than any other available to modern experimental science: a functional alteration that diminishes a mammal's fitness by one part in $10^4$ is undetectable at the laboratory bench, but is lethal from the standpoint of evolution.

选基因的所有外显子和调控区域进行测序，类似工作在几个人类肿瘤中心已经开始进行[292]。

### 制作更好的小鼠模型

并非所有的小鼠模型都以预期的方式复制人类表型。完整的人类和小鼠序列提供了一个机会去预测这些差异，并且进行弥补。在某些状况下，可能会发现，小鼠的突变不在人类疾病基因的真正的直系同源物中。或者，在一种情况下，人类基因组仅包含单个基因家族，但是小鼠基因组包含了旁系同源物和直系同源物，人们可以预测单独的直系同源物的敲除可能会产生更温和的表型(或者没有表型)。例如，洛及其同事们[296]描述的闭塞性肾综合征就是这种情况。通过建立双敲除小鼠可以获得与人类疾病表型更接近的模型。

### 了解基因调控

哺乳动物基因组中大约5%的窗口序列处于选择，大多数的区域似乎不编码蛋白质。大部分这样的序列可能参与基因表达的调控。通过更多相关基因组可以更精准地确定这些调控元件。然而，小鼠很可能提供最强大的实验平台用于产生和测试关于这些元件功能的假设。最近的一个例子是，基于小鼠–人类序列比对后进行的基因敲除证明了影响Il4/Il13/Il5基因簇表达的几个长程调控区域[4]。

## 结　论

小鼠提供了一个独特的镜头，通过它我们可以看到自己。在过去一个世纪里，作为主要的哺乳动物遗传研究系统，它为人类生理和疾病提供了一个模型，从而在免疫学和新陈代谢领域有了重大发现。随着小鼠基因组序列的获得，它正在为我们的基因组研究提供模型和信息。

比较基因组分析可能是了解生物学功能最有力的工具。它的作用在于，进化的熔炉是一种比现代实验科学中任何其他仪器都敏感的仪器：一个会降低哺乳动物身体健康度万分之一的功能性改变是在实验台上检测不到，但是从进化角度来看却是致命的。

Comparative analysis of genomes should thus make it possible to discern, by virtue of evolutionary conservation, biological features that would otherwise escape our notice. In this way, it will play a crucial role in our understanding of the human genome and thereby help lay the foundation for biomedicine in the twenty-first century.

The initial sequence of the mouse genome reported here is merely a first step in this intellectual programme. The sequencing of many additional mammalian and other vertebrate genomes will be needed to extract the full information hidden within our chromosomes. Moreover, as we begin to understand the common elements shared among species, it may also become possible to approach the even harder challenge of identifying and understanding the functional differences that make each species unique.

## Methods

### Production of sequence reads

Paired-end reads from libraries with different insert sizes were produced as previously described[1] using 384-well trays to ensure linkages.

### Availability of sequence and assembly data

Unprocessed sequence reads are available from the NCBI trace archive (ftp://ftp.ncbi.nih.gov/pub/TraceDB/mus_musculus/). Raw assembly data (before removal of contaminants, anchoring to chromosomes, and addition of finished sequence) are available from the Whitehead Institute for Biomedical Research (WIBR) (ftp://wolfram.wi.mit.edu/pub/mouse_contigs/Mar10_02/). The released assembly MGSCv3 is available from Ensembl (http://www.ensembl.org/Mus_musculus/), NCBI (ftp://ftp.ncbi.nih.gov/genomes/M_musculus/MGSCv3_Release1/), UCSC (http://genome.ucsc.edu/downloads.html) and WIBR (ftp://wolfram.wi.mit.edu/pub/mouse_contigs/MGSC_V3/). (See Supplementary Information for detailed Methods.)

(**420**, 520-562; 2002)

Received 18 September; accepted 31 October 2002.

References:

1. International Human Genome Sequencing Consortium Initial sequencing and analysis of the human genome. *Nature* **409**, 860-921 (2001).

2. Venter, J. C. *et al.* The sequence of the human genome. *Science* **291**, 1304-1351 (2001).

3. O'Brien, S. J. *et al.* The promise of comparative genomics in mammals. *Science* **286**, 458-462, 479-481 (1999).

4. Loots, G. G. *et al.* Identification of a coordinate regulator of interleukins 4, 13, and 5 by cross-species sequence comparisons. *Science* **288**, 136-140 (2000).

5. Pennacchio, L. A. & Rubin, E. M. Genomic strategies to identify mammalian regulatory sequences. *Nature Rev. Genet.* **2**, 100-109 (2001).

6. Oeltjen, J. C. *et al.* Large-scale comparative sequence analysis of the human and murine Bruton's tyrosine kinase loci reveals conserved regulatory domains. *Genome Res.* **7**, 315-329 (1997).

7. Ellsworth, R. E. *et al.* Comparative genomic sequence analysis of the human and mouse cystic fibrosis transmembrane conductance regulator genes. *Proc. Natl*

因此，基因组的比较分析应该能够通过进化保守性分辨出我们不会注意到的生物学特征。通过这种方式，它将对我们理解人类基因组发挥重要作用，从而为二十一世纪的生物医学奠定基础。

这里报道的小鼠基因组的初始序列仅仅是这个充满智慧的项目的第一步。我们还需要对许多其他哺乳动物和其他脊椎动物进行测序，以提取隐藏在我们染色体中的全部信息。此外，当我们开始了解物种之间共享的相同元件时，也可能接近识别和理解每个物种所拥有的独特功能及其彼此差异。

## 方 法

### 序列读数的产生

如前所述 [1]，利用 384 孔板获得不同片段长度的文库并生成配对末端读取片段，以确保序列的连锁性。

### 序列和组装数据的可用性

未处理的序列读取片段可以从 NCBI 存档中获得（ftp://ftp.ncbi.nih.gov/pub/TraceDB/mus_musculus/）。原始组装数据（在去掉污染物、锚定染色体和添加完成序列之前）可以从怀特黑德生物医学研究所（WIBR）获得（ftp://wolfram.wi.mit.edu/pub/mouse_contigs/Mar10_02/）。程序集 MGSCv3 可以从 Ensembl（http://www.ensembl.org/Mus_musculus/）、NCBI（ftp://ftp.ncbi.nih.gov/genomes/M_musculus/MGSCv3_Release1/）、UCSC（http://genome.ucsc.edu/downloads.html）和 WIBR（ftp://wolfram.wi.mit.edu/pub/mouse_contigs/MGSC_V3/）获得。（具体方法见补充信息。）

（张瑶楠 翻译；曾长青 审稿）

*Acad. Sci. USA* **97**, 1172-1177 (2000).

8. Mallon, A. M. *et al.* Comparative genome sequence analysis of the Bpa/Str region in mouse and man. *Genome Res.* **10**, 758-775 (2000).
9. Dehal, P. *et al.* Human chromosome 19 and related regions in mouse: conservative and lineage-specific evolution. *Science* **293**, 104-111 (2001).
10. DeSilva, U. *et al.* Generation and comparative analysis of approximately 3.3 Mb of mouse genomic sequence orthologous to the region of human chromosome 7q11.23 implicated in Williams syndrome. *Genome Res.* **12**, 3-15 (2002).
11. Toyoda, A. *et al.* Comparative genomic sequence analysis of the human chromosome 21 down syndrome critical region. *Genome Res.* **12**, 1323-1332 (2002).
12. Ansari-Lari, M. A. *et al.* Comparative sequence analysis of a gene-rich cluster at human chromosome 12p13 and its syntenic region in mouse chromosome 6. *Genome Res.* **8**, 29-40 (1998).
13. Lercher, M. J., Williams, E. J. & Hurst, L. D. Local similarity in evolutionary rates extends over whole chromosomes in human-rodent and mouse-rat comparisons: implications for understanding the mechanistic basis of the male mutation bias. *Mol. Biol. Evol.* **18**, 2032-2039 (2001).
14. Makalowski, W. & Boguski, M. S. Evolutionary parameters of the transcribed mammalian genome: an analysis of 2,820 orthologous rodent and human sequences. *Proc. Natl Acad. Sci. USA* **95**, 9407-9412 (1998).
15. Rossant, J. & McKerlie, C. Mouse-based phenogenomics for modelling human disease. *Trends Mol. Med.* **7**, 502-507 (2001).
16. Paigen, K. A miracle enough: the power of mice. *Nature Med.* **1**, 215-220 (1995).
17. Hogan, B., Beddington, R., Costantini, F. & Lacy, E. *Manipulating the Mouse Embryo: A Laboratory Manual* (Cold Spring Harbor Laboratory Press, Woodbury, New York, 1994).
18. Joyner, A. L. *Gene Targeting: A Practical Approach* (Oxford Univ. Press, New York, 1999).
19. Copeland, N. G., Jenkins, N. A. & Court, D. L. Recombineering: a powerful new tool for mouse functional genomics. *Nature Rev. Genet.* **2**, 769-779 (2001).
20. Yu, Y. & Bradley, A. Engineering chromosomal rearrangements in mice. *Nature Rev. Genet.* **2**, 780-790 (2001).
21. Bucan, M. & Abel, T. The mouse: genetics meets behaviour. *Nature Rev. Genet.* **3**, 114-123 (2002).
22. Silver, L. M. *Mouse Genetics: Concepts and Practice* (Oxford Univ. Press, New York, 1995).
23. Bromham, L., Phillips, M. J. & Penny, D. Growing up with dinosaurs: molecular dates and the mammalian radiation. *Trends Ecol. Evol.* **14**, 113-118 (1999).
24. Nei, M., Xu, P. & Glazko, G. Estimation of divergence times from multiprotein sequences for a few mammalian species and several distantly related organisms. *Proc. Natl Acad. Sci. USA* **98**, 2497-2502 (2001).
25. Kumar, S. & Hedges, S. B. A molecular timescale for vertebrate evolution. *Nature* **392**, 917-920 (1998).
26. Madsen, O. *et al.* Parallel adaptive radiations in two major clades of placental mammals. *Nature* **409**, 610-614 (2001).
27. Murphy, W. J. *et al.* Molecular phylogenetics and the origins of placental mammals. *Nature* **409**, 614-618 (2001).
29. Morse, H. *The Mouse in Biomedical Research* (eds Foster, H. L., Small, J. D. & Fox, J. G.) 1-16 (Academic, New York, 1981).
30. Morse, H. C. *Origins of Inbred Mice* (ed. Morse, H. C.) 1-21 (Academic, New York, 1978).
31. Haldane, J. B. S., Sprunt, A. D. & Haldane, N. M. Reduplication in mice. *J. Genet.* **5**, 133-135 (1915).
32. Botstein, D., White, R. L., Skolnick, M. & Davis, R. W. Construction of a genetic linkage map in man using restriction fragment length polymorphisms. *Am. J. Hum. Genet.* **32**, 314-331 (1980).
33. Dietrich, W. *et al. Genetic Maps* (ed. O'Brien, S.) 4.110-4.142, (1992).
34. Dietrich, W. F. *et al.* A comprehensive genetic map of the mouse genome. *Nature* **380**, 149-152 (1996).
35. Love, J. M., Knight, A. M., McAleer, M. A. & Todd, J. A. Towards construction of a high resolution map of the mouse genome using PCR-analysed microsatellites. *Nucleic Acids Res.* **18**, 4123-4130 (1990).
36. Weber, J. L. & May, P. E. Abundant class of human DNA polymorphisms which can be typed using the polymerase chain reaction. *Am. J. Hum. Genet.* **44**, 388-396 (1989).
37. Hudson, T. J. *et al.* A radiation hybrid map of mouse genes. *Nature Genet.* **29**, 201-205 (2001).
38. Van Etten, W. J. *et al.* Radiation hybrid map of the mouse genome. *Nature Genet.* **22**, 384-387 (1999).
39. Nusbaum, C. *et al.* A YAC-based physical map of the mouse genome. *Nature Genet.* **22**, 388-393 (1999).
41. Kawai, J. *et al.* Functional annotation of a full-length mouse cDNA collection. *Nature* **409**, 685-690 (2001).
44. Gregory, S. G. *et al.* A physical map of the mouse genome. *Nature* **418**, 743-750 (2002).
45. Mural, R. J. *et al.* A comparison of whole-genome shotgun-derived mouse chromosome 16 and the human genome. *Science* **296**, 1661-1671 (2002).
46. Green, E. D. Strategies for the systematic sequencing of complex genomes. *Nature Rev. Genet.* **2**, 573-583 (2001).
47. Edwards, A. *et al.* Automated DNA sequencing of the human HPRT locus. *Genomics* **6**, 593-608 (1990).
48. Huson, D. H. *et al.* Design of a compartmentalized shotgun assembler for the human genome. *Bioinformatics* **17**, S132-S139 (2001).
49. Analysis of the genome sequence of the flowering plant *Arabidopsis thaliana*. *Nature* **408**, 796-815 (2000).
52. Battey, J., Jordan, E., Cox, D. & Dove, W. An action plan for mouse genomics. *Nature Genet.* **21**, 73-75 (1999).
53. Kuroda-Kawaguchi, T. *et al.* The AZFc region of the Y chromosome features massive palindromes and uniform recurrent deletions in infertile men. *Nature Genet.* **29**, 279-286 (2001).
54. Zhao, S. *et al.* Mouse BAC ends quality assessment and sequence analyses. *Genome Res.* **11**, 1736-1745 (2001).
55. Ewing, B. & Green, P. Analysis of expressed sequence tags indicates 35,000 human genes. *Nature Genet.* **25**, 232-234 (2000).
56. Batzoglou, S. *et al.* ARACHNE: a whole-genome shotgun assembler. *Genome Res.* **12**, 177-189 (2002).

57. Jaffe, D. B. *et al.* Whole-genome sequence assembly for mammalian genomes: Arachne **2**. *Genome Res.* (in the press).
58. Mullikin, J. & Ning, Z. The Phusion Assembler. *Genome Res.* (in the press).
59. Bailey, J. A. *et al.* Recent segmental duplications in the human genome. *Science* **297**, 1003-1007 (2002).
66. Mouse Genome Sequencing Consortium Progress in sequencing the mouse genome. *Genesis* **31**, 137-141 (2001).
67. Clark, F. H. Inheritance and linkage relations of mutant characteristics in the deermouse. *Contrib. Lab. Vert. Biol.* 7, 1-11 (1938).
68. Castle, W. W. Observations of the occurrence of linkage in rats and mice. *Car. Inst. Wash. Pub.* **288**, 29-36 (1919).
69. Lalley, P. A., Minna, J. D. & Francke, U. Conservation of autosomal gene synteny groups in mouse and man. *Nature* **274**, 160-163 (1978).
70. Nadeau, J. H. & Taylor, B. A. Lengths of chromosomal segments conserved since divergence of man and mouse. *Proc. Natl Acad. Sci. USA* **81**, 814-818 (1984).
71. Ma, B., Tromp, J. & Li, M. PatternHunter: faster and more sensitive homology search. *Bioinformatics* **18**, 440-445 (2002).
72. Ohno, S. *Sex Chromosomes and Sex-Linked Genes* (Springer, Berlin, 1996).
73. Sturtevant, A. H. & Beadle, G. W. The relations of inversions in the X chromosome of *Drosophila melanogaster* to crossing over and disjunction. *Genetics* **21**, 554-604 (1936).
74. Ranz, J. M., Casals, F. & Ruiz, A. How malleable is the eukaryotic genome? Extreme rate of chromosomal rearrangement in the genus *Drosophila*. *Genome Res.* **11**, 230-239 (2001).
75. Nadeau, J. H. & Sankoff, D. The lengths of undiscovered conserved segments in comparative maps. *Mamm. Genome* **9**, 491-495 (1998).
76. Ferretti, V., Nadeau, J. H. & Sankoff, D. *Combinatorial Pattern Matching, 7th Annual Symposium* (eds Hirschberg, D. & Myers, G.) 159-167 (Springer, Berlin, 1996).
77. Bourque, G. & Pevzner, P. A. Genome-scale evolution: reconstructing gene orders in the ancestral species. *Genome Res.* **12**, 26-36 (2002).
78. Thiery, J. P., Macaya, G. & Bernardi, G. An analysis of eukaryotic genomes by density gradient centrifugation. *J. Mol. Biol.* **108**, 219-235 (1976).
79. Salinas, J., Zerial, M., Filipski, J. & Bernardi, G. Gene distribution and nucleotide sequence organization in the mouse genome. *Eur. J. Biochem.* **160**, 469-478 (1986).
80. Sabeur, G., Macaya, G., Kadi, F. & Bernardi, G. The isochore patterns of mammalian genomes and their phylogenetic implications. *J. Mol. Evol.* **37**, 93-108 (1993).
81. Zerial, M., Salinas, J., Filipski, J. & Bernardi, G. Gene distribution and nucleotide sequence organization in the human genome. *Eur. J. Biochem.* **160**, 479-485 (1986).
82. Mouchiroud, D., Fichant, G. & Bernardi, G. Compositional compartmentalization and gene composition in the genome of vertebrates. *J. Mol. Evol.* **26**, 198-204 (1987).
83. Mouchiroud, D., Gautier, C. & Bernardi, G. The compositional distribution of coding sequences and DNA molecules in humans and murids. *J. Mol. Evol.* **27**, 311-320 (1988).
84. Mouchiroud, D. & Gautier, C. Codon usage changes and sequence dissimilarity between human and rat. *J. Mol. Evol.* **31**, 81-91 (1990).
85. Robinson, M., Gautier, C. & Mouchiroud, D. Evolution of isochores in rodents. *Mol. Biol. Evol.* **14**, 823-828 (1997).
86. Bernardi, G. *et al.* The mosaic genome of warm-blooded vertebrates. *Science* **228**, 953-958 (1985).
87. Mouchiroud, D. *et al.* The distribution of genes in the human genome. *Gene* **100**, 181-187 (1991).
88. Zoubak, S., Clay, O. & Bernardi, G. The gene distribution of the human genome. *Gene* **174**, 95-102 (1996).
89. Saccone, S., Pavlicek, A., Federico, C., Paces, J. & Bernard, G. Genes, isochores and bands in human chromosomes 21 and 22. *Chromosome Res.* **9**, 533-539 (2001).
90. Bernardi, G. Compositional constraints and genome evolution. *J. Mol. Evol.* **24**, 1-11 (1986).
91. Bernardi, G., Mouchiroud, D. & Gautier, C. Compositional patterns in vertebrate genomes: conservation and change in evolution. *J. Mol. Evol.* **28**, 7-18 (1988).
92. Wolfe, K. H., Sharp, P. M. & Li, W. H. Mutation rates differ among regions of the mammalian genome. *Nature* **337**, 283-285 (1989).
93. Sueoka, N. Directional mutation pressure and neutral molecular evolution. *Proc. Natl Acad. Sci. USA* **85**, 2653-2657 (1988).
94. Sueoka, N. On the genetic basis of variation and heterogeneity of DNA base composition. *Proc. Natl Acad. Sci. USA* **48**, 582-592 (1962).
95. Bird, A. P. DNA methylation and the frequency of CpG in animal DNA. *Nucleic Acids Res.* **8**, 1499-1504 (1980).
96. Larsen, F., Gundersen, G., Lopez, R. & Prydz, H. CpG islands as gene markers in the human genome. *Genomics* **13**, 1095-1107 (1992).
97. Gardiner-Garden, M. & Frommer, M. CpG islands in vertebrate genomes. *J. Mol. Biol.* **196**, 261-282 (1987).
98. Antequera, F. & Bird, A. Number of CpG islands and genes in human and mouse. *Proc. Natl Acad. Sci. USA* **90**, 11995-11999 (1993).
99. Adams, R. L. & Eason, R. Increased G+C content of DNA stabilizes methyl CpG dinucleotides. *Nucleic Acids Res.* **12**, 5869-5877 (1984).
100. Smit, A. F. Interspersed repeats and other mementos of transposable elements in mammalian genomes. *Curr. Opin. Genet. Dev.* **9**, 657-663 (1999).
101. Laird, C. D., McConaughy, B. L. & McCarthy, B. J. Rate of fixation of nucleotide substitutions in evolution. *Nature* **224**, 149-154 (1969).
102. Kohne, D. E. Evolution of higher-organism DNA. *Q. Rev. Biophys.* **3**, 327-375 (1970).
103. Goodman, M., Barnabas, J., Matsuda, G. & Moore, G. W. Molecular evolution in the descent of man. *Nature* **233**, 604-613 (1971).
104. Kumar, S. & Subramanian, S. Mutation rates in mammalian genomes. *Proc. Natl Acad. Sci. USA* **99**, 803-808 (2002).
105. Easteal, S., Collet, C. & Betty, D. *The Mammalian Molecular Clock* (Landes, Austin, Texas, 1995).
106. Li, W. H., Ellsworth, D. L., Krushkal, J., Chang, B. H. & Hewett-Emmett, D. Rates of nucleotide substitution in primates and rodents and the generation-time effect hypothesis. *Mol. Phylogenet. Evol.* **5**, 182-187 (1996).
107. Martin, A. P. & Palumbi, S. R. Body size, metabolic rate, generation time, and the molecular clock. *Proc. Natl Acad. Sci. USA* **90**, 4087-4091 (1993).
108. Bromham, L. Molecular clocks in reptiles: life history influences rate of molecular evolution. *Mol. Biol. Evol.* **19**, 302-309 (2002).
109. Wu, C. I. & Li, W. H. Evidence for higher rates of nucleotide substitution in rodents than in man. *Proc. Natl Acad. Sci. USA* **82**, 1741-1745 (1985).
110. Smit, A. F., Toth, G., Riggs, A. D. & Jurka, J. Ancestral, mammalian-wide subfamilies of LINE-1 repetitive sequences. *J. Mol. Biol.* **246**, 401-417 (1995).

111. Adey, N. B. *et al.* Rodent L1 evolution has been driven by a single dominant lineage that has repeatedly acquired new transcriptional regulatory sequences. *Mol. Biol. Evol.* **11**, 778-789 (1994).

112. Mears, M. L. & Hutchison, C. A. III The evolution of modern lineages of mouse L1 elements. *J. Mol. Evol.* **52**, 51-62 (2001).

113. Goodier, J. L., Ostertag, E. M., Du, K. & Kazazian, H. H. Jr A novel active L1 retrotransposon subfamily in the mouse. *Genome Res.* **11**, 1677-1685 (2001).

114. Hardies, S. C. *et al.* LINE-1 (L1) lineages in the mouse. *Mol. Biol. Evol.* **17**, 616-628 (2000).

115. Ohshima, K., Hamada, M., Terai, Y. & Okada, N. The 3′ ends of tRNA-derived short interspersed repetitive elements are derived from the 3′ ends of long interspersed repetitive elements. *Mol. Cell Biol.* **16**, 3756-3764 (1996).

116. Smit, A. F. The origin of interspersed repeats in the human genome. *Curr. Opin. Genet. Dev.* **6**, 743-748 (1996).

117. Quentin, Y. A master sequence related to a free left Alu monomer (FLAM) at the origin of the B1 family in rodent genomes. *Nucleic Acids Res.* **22**, 2222-2227 (1994).

118. Kim, J. & Deininger, P. L. Recent amplification of rat ID sequences. *J. Mol. Biol.* **261**, 322-327 (1996).

119. Lee, I. Y. *et al.* Complete genomic sequence and analysis of the prion protein gene region from three mammalian species. *Genome Res.* **8**, 1022-1037 (1998).

120. Serdobova, I. M. & Kramerov, D. A. Short retroposons of the B2 superfamily: evolution and application for the study of rodent phylogeny. *J. Mol. Evol.* **46**, 202-214 (1998).

121. Coffin, J. M., Hughes, S. H. & Varmus, H. E. (eds) *Retroviruses* (Cold Spring Harbor Laboratory Press, Cold Spring Harbor, New York, 1997).

122. Smit, A. F. Identification of a new, abundant superfamily of mammalian LTR-transposons. *Nucleic Acids Res.* **21**, 1863-1872 (1993).

123. Hamilton, B. A. & Frankel, W. N. Of mice and genome sequence. *Cell* **107**, 13-16 (2001).

125. Kidwell, M. G. Horizontal transfer. *Curr. Opin. Genet. Dev.* **2**, 868-873 (1992).

126. Feng, Q., Moran, J. V., Kazazian, H. H. Jr & Boeke, J. D. Human L1 retrotransposon encodes a conserved endonuclease required for retrotransposition. *Cell* **87**, 905-916 (1996).

127. Jurka, J. Sequence patterns indicate an enzymatic involvement in integration of mammalian retroposons. *Proc. Natl Acad. Sci. USA* **94**, 1872-1877 (1997).

128. Bernardi, G. The isochore organization of the human genome. *Annu. Rev. Genet.* **23**, 637-661 (1989).

129. Holmquist, G. P. Chromosome bands, their chromatin flavors, and their functional features. *Am. J. Hum. Genet.* **51**, 17-37 (1992).

132. Lyon, M. F. X-chromosome inactivation: a repeat hypothesis. *Cytogenet. Cell Genet.* **80**, 133-137 (1998).

135. Beckman, J. S. & Weber, J. L. Survey of human and rat microsatellites. *Genomics* **12**, 627-631 (1992).

136. Toth, G., Gaspari, Z. & Jurka, J. Microsatellites in different eukaryotic genomes: survey and analysis. *Genome Res.* **10**, 967-981 (2000).

139. Dunham, I. *et al.* The DNA sequence of human chromosome 22. *Nature* **402**, 489-495 (1999).

140. Hattori, M. *et al.* The DNA sequence of human chromosome 21. *Nature* **405**, 311-319 (2000).

141. Roest Crollius, H. *et al.* Estimate of human gene number provided by genome-wide analysis using *Tetraodon nigroviridis* DNA sequence. *Nature Genet.* **25**, 235-238 (2000).

142. Hubbard, T. *et al.* The Ensembl genome database project. *Nucleic Acids Res.* **30**, 38-41 (2002).

143. Kulp, D., Haussler, D., Reese, M. G. & Eeckman, F. H. Integrating database homology in a probabilistic gene structure model. *Pac. Symp. Biocomput.* 232-244 (1997).

144. Birney, E. & Durbin, R. Using GeneWise in the *Drosophila* annotation experiment. *Genome Res.* **10**, 547-548 (2000).

145. Burge, C. & Karlin, S. Prediction of complete gene structures in human genomic DNA. *J. Mol. Biol.* **268**, 78-94 (1997).

150. The FANTOM Consortium and the RIKEN Genome Exploration Research Group Phase I & II Team. Analysis of the mouse transcriptome based on functional annotation of 60,770 full-length cDNAs. *Nature* **420**, 563-573 (2002).

151. Pruitt, K. D. & Maglott, D. R. RefSeq and LocusLink: NCBI gene-centered resources. *Nucleic Acids Res.* **29**, 137-140 (2001).

165. Eddy, S. R. Non-coding RNA genes and the modern RNA world. *Nature Rev. Genet.* **2**, 919-929 (2001).

166. Storz, G. An expanding universe of noncoding RNAs. *Science* **296**, 1260-1263 (2002).

167. Eddy, S. R. Computational genomics of noncoding RNA genes. *Cell* **109**, 137-140 (2002).

168. Lowe, T. M. & Eddy, S. R. tRNAscan-SE: a program for improved detection of transfer RNA genes in genomic sequence. *Nucleic Acids Res.* **25**, 955-964 (1997).

169. Daniels, G. R. & Deininger, P. L. Repeat sequence families derived from mammalian tRNA genes. *Nature* **317**, 819-822 (1985).

170. Lawrence, C., McDonnell, D. & Ramsey, W. Analysis of repetitive sequence elements containing tRNA-like sequences. *Nucleic Acids Res.* **13**, 4239-4252 (1985).

174. Ponting, C. P. & Russell, R. R. The natural history of protein domains. *Annu. Rev. Biophys. Biomol. Struct.* **31**, 45-71 (2002).

175. Lespinet, O., Wolf, Y. I., Koonin, E. V. & Aravind, L. The role of lineage-specific gene family expansion in the evolution of eukaryotes. *Genome Res.* **12**, 1048-1059 (2002).

176. Ponting, C. P., Mott, R., Bork, P. & Copley, R. R. Novel protein domains and repeats in *Drosophila melanogaster*: insights into structure, function, and evolution. *Genome Res.* **11**, 1996-2008 (2001).

177. Rubin, G. M. *et al.* Comparative genomics of the eukaryotes. *Science* **287**, 2204-2215 (2000).

178. Altschul, S. F. *et al.* Gapped BLAST and PSI-BLAST: a new generation of protein database search programs. *Nucleic Acids Res.* **25**, 3389-3402 (1997).

179. Zdobnov, E. M. & Apweiler, R. InterProScan—an integration platform for the signature-recognition methods in InterPro. *Bioinformatics* **17**, 847-848 (2001).

193. Young, J. M. *et al.* Different evolutionary processes shaped the mouse and human olfactory receptor gene families. *Hum. Mol. Genet.* **11**, 535-546 (2002).

194. Zhang, X. & Firestein, S. The olfactory receptor gene superfamily of the mouse. *Nature Neurosci.* **5**, 124-133 (2002).

195. Glusman, G., Yanai, I., Rubin, I. & Lancet, D. The complete human olfactory subgenome. *Genome Res.* **11**, 685-702 (2001).

196. Rouquier, S. *et al.* Distribution of olfactory receptor genes in the human genome. *Nature Genet.* **18**, 243-250 (1998).

197. Del Punta, K. *et al.* Deficient pheromone responses in mice lacking a cluster of vomeronasal receptor genes. *Nature* **419**, 70-74 (2002).

199. Lane, R. P. *et al.* Genomic analysis of orthologous mouse and human olfactory receptor loci. *Proc. Natl Acad. Sci. USA* **98**, 7390-7395 (2001).

224. Zhang, J., Dyer, K. D. & Rosenberg, H. F. Evolution of the rodent eosinophil-associated RNase gene family by rapid gene sorting and positive selection. *Proc. Natl Acad. Sci. USA* **97**, 4701-4706 (2000).

227. Yeager, M. & Hughes, A. L. Evolution of the mammalian MHC: natural selection, recombination, and convergent evolution. *Immunol. Rev.* **167**, 45-58 (1999).

248. Koop, B. F. Human and rodent DNA sequence comparisons: a mosaic model of genomic evolution. *Trends Genet.* **11**, 367-371 (1995).

249. DeBry, R. W. & Seldin, M. F. Human/mouse homology relationships. *Genomics* **33**, 337-351 (1996).

250. Gottgens, B. *et al.* Long-range comparison of human and mouse SCL loci: localized regions of sensitivity to restriction endonucleases correspond precisely with peaks of conserved noncoding sequences. *Genome Res.* **11**, 87-97 (2001).

251. Shiraishi, T. *et al.* Sequence conservation at human and mouse orthologous common fragile regions, FRA3B/FHIT and Fra14A2/Fhit. *Proc. Natl Acad. Sci. USA* **98**, 5722-5727 (2001).

252. Wilson, M. D. *et al.* Comparative analysis of the gene-dense ACHE/TFR2 region on human chromosome 7q22 with the orthologous region on mouse chromosome 5. *Nucleic Acids Res.* **29**, 1352-1365 (2001).

253. Hardison, R. C. Conserved noncoding sequences are reliable guides to regulatory elements. *Trends Genet.* **16**, 369-372 (2000).

254. Chiaromonte, F. *et al.* Association between divergence and interspersed repeats in mammalian noncoding genomic DNA. *Proc. Natl Acad. Sci. USA* **98**, 14503-14508 (2001).

255. Matassi, G., Sharp, P. M. & Gautier, C. Chromosomal location effects on gene sequence evolution in mammals. *Curr. Biol.* **9**, 786-791 (1999).

256. Williams, E. J. & Hurst, L. D. The proteins of linked genes evolve at similar rates. *Nature* **407**, 900-903 (2000).

257. Chen, F. C., Vallender, E. J., Wang, H., Tzeng, C. S. & Li, W. H. Genomic divergence between human and chimpanzee estimated from large-scale alignments of genomic sequences. *J. Hered.* **92**, 481-489 (2001).

258. Lercher, M. J. & Hurst, L. D. Human SNP variability and mutation rate are higher in regions of high recombination. *Trends Genet.* **18**, 337-340 (2002).

259. Castresana, J. Genes on human chromosome 19 show extreme divergence from the mouse orthologs and a high GC content. *Nucleic Acids Res.* **30**, 1751-1756 (2002).

260. Smith, N. G. C., Webster, M. & Ellegren, H. Deterministic mutation rate variation in the human genome. *Genome Res.* **12**, 1350-1356 (2002).

262. Bernardi, G. The human genome: organization and evolutionary history. *Ann. Rev. Genet.* **23**, 637-661 (1995).

263. Hurst, L. D. & Willliams, E. J. B. Covarication of GC content and the silent site substitution rate in rodents: implications for methodology and for the evolution of isochores. *Gene* **261**, 107-114 (2000).

264. Bernardi, G. Misunderstandings about isochores. Part 1. *Gene* **276**, 3-13 (2001).

265. The SNP Consortium An SNP map of the human genome generated by reduced representation shotgun sequencing. *Nature* **407**, 513-516 (2000).

266. Perry, J. & Ashworth, A. Evolutionary rate of a gene affected by chromosomal position. *Curr. Biol.* **9**, 987-989 (1999).

270. Charlesworth, B. The effect of background selection against deleterious mutations on weakly selected, linked variants. *Genet. Res.* **63**, 213-227 (1994).

271. Hudson, R. R. & Kaplan, N. L. Deleterious background selection with recombination. *Genetics* **141**, 1605-1617 (1995).

272. Maynard Smith, J. & Haigh, J. The hitch-hiking effect of a favourable gene. *Genet. Res.* **23**, 23-35 (1974).

273. Birky, C. W. & Walsh, J. B. Effects of linkage on rates of molecular evolution. *Proc. Natl Acad. Sci. USA* **85**, 6414-6418 (1988).

274. Francino, M. P. & Ochman, H. Strand asymmetries in DNA evolution. *Trends Genet.* **13**, 240-245 (1997).

275. Gilbert, N., Lutz-Prigge, S. & Moran, J. Genomic deletions created upon LINE-1 retrotransposition. *Cell* **110**, 315-325 (2002).

276. Symer, D. *et al.* Human l1 retrotransposition is associated with genetic instability *in vivo*. *Cell* **110**, 327-338 (2002).

277. Moran, J. *et al.* High frequency retrotransposition in cultured mammalian cells. *Cell* **87**, 917-927 (1996).

278. Hughes, J. F. & Coffin, J. M. Evidence for genomic rearrangements mediated by human endogenous retroviruses during primate evolution. *Nature Genet.* **29**, 487-489 (2001).

279. Wolfe, K. H. Mammalian DNA replication: mutation biases and the mutation rate. *J. Theor. Biol.* **149**, 441-451 (1991).

280. Gu, X. & Li, W. H. A model for the correlation of mutation rate with GC content and the origin of GC-rich isochores. *J. Mol. Evol.* **38**, 468-475 (1994).

284. Loftus, S. K. *et al.* Mutation of melanosome protein RAB38 in chocolate mice. *Proc. Natl Acad. Sci. USA* **99**, 4471-4476 (2002).

285. Paigen, K. & Eppig, J. T. A mouse phenome project. *Mamm. Genome* **11**, 715-717 (2000).

286. Doerge, R. W. Mapping and analysis of quantitative trait loci in experimental populations. *Nature Rev. Genet.* **3**, 43-52 (2002).

288. Mei, R. *et al.* Genome-wide detection of allelic imbalance using human SNPs and high-density DNA arrays. *Genome Res.* **10**, 1126-1137 (2000).

289. Lindblad-Toh, K. *et al.* Loss-of-heterozygosity analysis of small-cell lung carcinomas using single-nucleotide polymorphism arrays. *Nature Biotechnol.* **18**, 1001-1005 (2000).

290. Heiskanen, M. *et al.* CGH, cDNA and tissue microarray analyses implicate FGFR2 amplification in a small subset of breast tumors. *Anal. Cell Pathol.* **22**, 229-234 (2001).

291. Cai, W. W. *et al.* Genome-wide detection of chromosomal imbalances in tumors using BAC microarrays. *Nature Biotechnol.* **20**, 393-396 (2002).

292. Davies, H. *et al.* Mutations of the *BRAF* gene in human cancer. *Nature* **417**, 949-954 (2002).

296. Janne, P. A. *et al.* Functional overlap between murine Inpp5b and Ocrl1 may explain why deficiency of the murine ortholog for OCRL1 does not cause Lowe syndrome in mice. *J. Clin. Invest.* **101**, 2042-2053 (1998).

328. Schwartz, S. *et al.* Human-mouse alignments with Blastz. *Genome Res.* (in the press).

330. Roskin, K. M. Score Functions for Assessing Conservation in Locally Aligned Regions of DNA from Two Species. UCSC Tech Report UCSC-CRL-02-30, School of Engineering, Univ. California (2002).

**Supplementary Information** accompanies the paper on *Nature*'s website (http://www. nature. com/nature).

**Acknowledgements.** We thank J. Takahashi and M. Johnston for comments on the manuscript; the Mouse Liaison Group for strategic advice; L. Gaffney, D. Leja and K.-S. Toh for graphical help; B. Graham and G. Roberts for administrative work on sequencing of individual mouse BACs; and P. Kassos and M. McMurtry for secretarial assistance. We thank D. Hill and L. Corbani of the Mouse Genome Informatics Group for their contributions to the GO analysis for mouse and human, and the members of the Bork group at EMBL for discussions. Funding was provided by the National Institutes of Health (National Human Genome Research Institute, National Cancer Institute, National Institute of Dental and Craniofacial Research, National Institute of Diabetes and Digestive and Kidney Diseases, National Institute of General Medical Sciences, National Eye Institute, National Institute of Environmental Health Sciences, National Institute of Aging, National Institute of Arthritis and Musculoskeletal and Skin Diseases, National Institute on Deafness and Other Communication Disorders, National Institute of Mental Health, National Institute on Drug Abuse, National Center for Research Resources, the National Heart Lung and Blood Institute and The Fogarty International Center); the Wellcome Trust; the Howard Hughes Medical Institute; the United States Department of Energy; the National Science Foundation; the Medical Research Council; NSERC; BMBF (German Ministry for Research and Education); the European Molecular Biology Laboratory; Plan Nacional de I+D and Instituto Carlos III; Swiss National Science Foundation, NCCR Frontiers in Genetics, the Swiss Cancer League and the "Childcare" and "J. Lejeune" Foundations; and the Ministry of Education, Culture, Sports, Science and Technology of Japan. The initial threefold sequence coverage was partly supported by the Mouse Sequencing Consortium (GlaxoSmithKline, Merck and Affymetrix) through the Foundation for the National Institutes of Health. We acknowledge A. Holden for coordinating the Mouse Sequencing Consortium. We thank the Sanger Institute systems group for maintenance and provision of the computer resource. The MGSC also used Hewlett-Packard Company's BioCluster, a configuration of 27 HP AlphaServer ES40 systems with 100 CPUs and 1 terabyte of storage. The BioCluster is housed in Hewlett-Packard's IQ Solutions Center, and was accessed remotely. The computing resource greatly accelerated the analysis.

**Competing interests statement.** The authors declare that they have no competing financial interests.

**Correspondence** and requests for materials should be addressed to R.H.W. (e-mail: waterston@gs.washington.edu), K.L.T. (e-mail: kersli@genome.wi.mit.edu) or E.S.L. (e-mail: lander@genome.wi.mit.edu).

**Authors' contributions.** The following authors contributed to project leadership: R. H. Waterston, K. Lindblad-Toh, E. Birney, J. Rogers, M. R. Brent, F. S. Collins, R. Guigó, R. C. Hardison, D. Haussler, D. B. Jaffe, W. J. Kent, W. Miller, C. P. Ponting, A. Smit, M. C. Zody and E. S. Lander.

# Detection of Polarization in the Cosmic Microwave Background Using DASI

J. M. Kovac *et al.*

## Editor's Note

**The characterization of the ripples in the cosmic microwave background radiation, the "afterglow" of the Big Bang, led to quite accurate determination of various cosmological parameters, such as the density of matter and energy, and the speed of expansion. The very success of the standard cosmological model that emerged prompted researchers to look for independent ways of testing it. The model makes very specific predictions about how the microwave background with be polarized (that is, how the oscillating electromagnetic fields of the microwave photons will be aligned). Here John Kovac and coworkers used a telescope at the South Pole to observe this polarization, and confirm that it is just as predicted.**

---

The past several years have seen the emergence of a standard cosmological model, in which small temperature differences in the cosmic microwave background (CMB) radiation on angular scales of the order of a degree are understood to arise from acoustic oscillations in the hot plasma of the early Universe, arising from primordial density fluctuations. Within the context of this model, recent measurements of the temperature fluctuations have led to profound conclusions about the origin, evolution and composition of the Universe. Using the measured temperature fluctuations, the theoretical framework predicts the level of polarization of the CMB with essentially no free parameters. Therefore, a measurement of the polarization is a critical test of the theory and thus of the validity of the cosmological parameters derived from the CMB measurements. Here we report the detection of polarization of the CMB with the Degree Angular Scale Interferometer (DASI). The polarization is deteced with high confidence, and its level and spatial distribution are in excellent agreement with the predictions of the standard theory.

---

THE CMB radiation provides a pristine probe of the Universe roughly 14 billion years ago, when the seeds of the complex structures that characterize the Universe today existed only as small density fluctuations in the primordial plasma. As the physics of such a plasma is well understood, detailed measurements of the CMB can provide critical tests of cosmological models and can determine the values of cosmological parameters with high precision. The CMB has accordingly been the focus of intense experimental and theoretical investigations since its discovery nearly 40 years ago[1]. The frequency spectrum of the CMB was well determined by the COBE FIRAS instrument[2,3]. The initial detection of temperature anisotropy was made on large angular scales by the COBE DMR

# 利用 DASI 探测宇宙微波背景的偏振

科瓦奇等

## 编者按

宇宙微波背景辐射中的涟漪特征，又被称为宇宙大爆炸的"余辉"。我们可以非常准确地通过它来测定各种宇宙学参数，例如物质和能量的密度，以及膨胀速度等。标准宇宙学模型的成功进一步促使研究人员去寻找独立检测它的方法。该模型对微波背景如何被偏振（即微波光子产生的振荡电磁场如何排列）进行了非常具体的预测。约翰·科瓦奇及其同事在南极使用望远镜观测了这种偏振，并且确认观测结果与预测一致。

---


在过去几年已经出现了的标准宇宙学模型中，宇宙微波背景（CMB）辐射在一定的度角尺度上的小温差被理解为是由早期宇宙中热等离子体的声学振荡引起的，而后者又来源于原初密度扰动。最近在该模型下对温度扰动的测量导出了关于宇宙起源、演化和组成的深刻结论。仅仅利用温度扰动的测量，理论框架对预测 CMB 的偏振水平基本上没有自由参数可言。因此，偏振的测量是理论以及从 CMB 测量得到的宇宙学参数的有效性的重要检验。本文中，我们将报道用度角尺度干涉仪（DASI）检测 CMB 所得的探测结果。偏振的检测置信度很高，其水平和空间分布与标准理论的预测非常一致。


---

CMB 辐射可以作为约 140 亿年前宇宙状态的原始探测仪。现今宇宙复杂结构的种子在当时仅存在于原初等离子体的微小密度扰动中。由于人们已经很好地理解了这种等离子体的物理特性，所以对 CMB 更细节性地测量可以提供宇宙学模型的关键检测，并且可以高精度地确定宇宙学参数的值。自发现至今的将近 40 年时间里，CMB 一直是大量实验和理论研究关注的焦点[1]。CMB 的频谱由宇宙背景探测器（COBE）远红外绝对分光光度计（FIRAS）很好地确定[2,3]。通过 COBE 较差微波辐射计（DMR）[4]，研究人员在大角度尺度上进行了温度各向异性的初始检测，而最近

instrument[4] and recently there has been considerable progress in measuring the anisotropy on finer angular scales[5-10].

In the now standard cosmological model (see, for example, ref. 11), the shape of the CMB angular power spectrum directly traces acoustic oscillations of the photon-baryon fluid in the early Universe. As the Universe expanded, it cooled; after roughly 400,000 years, the intensity of the radiation field was no longer sufficient to keep the Universe ionized, and the CMB photons decoupled from the baryons as the first atoms formed. Acoustic oscillations passing through extrema at this epoch are observed in the CMB angular power spectrum as a harmonic series of peaks. Polarization is also a generic feature of these oscillations[12-18], and thus provides a model-independent test of the theoretical framework[19-21]. In addition, detection of polarization can in principle triple the number of observed CMB quantities, enhancing our ability to constrain cosmological parameters.

CMB polarization arises from Thompson scattering by electrons of a radiation field with a local quadrupole moment[22]. In the primordial plasma, the local quadrupole moment is suppressed until decoupling, when the photon mean free path begins to grow. At this time, the largest contribution to the local quadrupole is due to Doppler shifts induced by the velocity field of the plasma[23]. CMB polarization thus directly probes the dynamics at the epoch of decoupling. For a spatial Fourier mode of the acoustic oscillations, the velocities are perpendicular to the wavefronts, leading only to perpendicular or parallel alignment of the resulting polarization, which we define as positive and negative respectively. We refer to these polarization modes as scalar $E$-modes in analogy with electric fields; they have no curl component. Because the level of the polarization depends on velocity, we expect that the peaks in the scalar $E$-mode power spectrum correspond to density modes that are at their highest velocities at decoupling and are thus at minimum amplitude. The location of the harmonic peaks in the scalar $E$-mode power spectrum is therefore expected to be out of phase with the peaks in the temperature ($T$) spectrum[15-17].

In light of the above discussion, it is clear that the CMB temperature and polarization anisotropy should be correlated at some level[24]. For a given multipole, the sign of the $TE$ correlation should depend on whether the amplitude of the density mode was increasing or decreasing at the time of decoupling. We therefore expect a change in the sign of the $TE$ correlation at maxima in the $T$ and the $E$ power spectra, which correspond to modes at maximum and minimum amplitude, respectively. The $TE$ correlation therefore offers a unique and powerful test of the underlying theoretical framework.

Primordial gravity waves will lead to polarization in the CMB[14,25] with an $E$-mode pattern as for the scalar density perturbations, but will also lead to a curl component, referred to as $B$-mode polarization[26-28]. The $B$-mode component is due to the intrinsic polarization of the gravity waves. In inflationary models, the level of the $B$-mode polarization from gravity waves is set by the fourth power of the inflationary energy scale. While the detection of $B$-mode polarization would provide a critical test of inflation, the signal is likely to be very weak and may have an amplitude that is effectively unobservable[29]. Furthermore, distortions

在更精细的角度尺度上测量各向异性的工作也已经取得了相当大的进展[5-10]。

在现代标准的宇宙学模型中(如参考文献11)，CMB角功率谱的形状可以直接追踪到早期宇宙中光子-重子流体的声学振荡。随着宇宙的扩张，它冷却了下来；大约在此40万年后，辐射场的强度已经不足以使宇宙保持电离状态，从而在第一个原子形成的同时，CMB光子与重子退耦。声学振荡在这个时期达到了极值，并在CMB角功率谱中的一系列简谐波峰被观察到。偏振也是这些振荡[12-18]的普遍特征，因而为理论框架[19-21]提供了与模型无关的检测。此外，偏振的探测原则上可以使观察到的CMB数量增加到原来的三倍，从而也增强了我们约束宇宙学参数的能力。

CMB偏振是由具有局部四极矩的辐射场中电子的汤普森散射产生的[22]。在原始等离子体中，当光子平均自由程开始增加时，局部四极矩被抑制直到退耦。此时，对局部四极矩贡献最大的是由等离子体的速度场引起的多普勒频移[23]。因此，CMB偏振将直接探测到退耦时期的动力学过程。对于声学振荡的空间傅里叶模式，速度垂直于波阵面，导致所得偏振按垂直或平行方向排列，我们将其分别定义为正向和负向。这些偏振模式称为标量$E$模式，类似于电场；它们是无旋的。因为偏振水平取决于速度，我们期望标量$E$模式功率谱中的峰值对应于退耦时处于其最高速度的密度模式，也就是处于最小的幅度之时。因此可以预期标量$E$模式功率谱中的谐波峰值的位置与温度($T$)频谱中的峰值异相[15-17]。

鉴于上述讨论，很明显可以看出CMB温度和偏振的各向异性应该在某种程度上相关联[24]。对于给定的多极矩，$TE$关联的迹象应当取决于退耦时的密度幅度是增加还是减小。所以我们期望$TE$关联中$T$和$E$功率谱中最大值处的变化，分别对应于最大和最小幅度的模式。因此，$TE$相关性为基础理论框架提供了独特而强大的检测手段。

原初引力波将导致CMB中具有与标量密度扰动有关的$E$模式的偏振[14,25]，但也将产生一个旋度分量，称为$B$模式偏振[26-28]。$B$模式分量源于引力波的内禀偏振。在暴胀模型中，来自引力波的$B$模式偏振的水平由暴胀能量标度的四次幂决定。虽然$B$模式偏振的探测将为暴胀提供关键的检测，但是其信号可能非常微弱，而且事实上其幅度可能无法有效观测[29]。此外，宇宙大尺度结构引力透镜的干预对标量$E$

to the scalar $E$-mode polarization by the gravitational lensing of the intervening large scale structure in the Universe will lead to a contaminating $B$-mode polarization signal which will severely complicate the extraction of the polarization signature from gravity waves[30-33]. The possibility, however, of directly probing the Universe at energy scales of order $10^{16}$ GeV by measuring the gravity-wave induced polarization (see, for example, ref. 34) is a compelling goal for CMB polarization observations.

Prior to the results presented in this paper, only upper limits have been placed on the level of CMB polarization. This is due to the low level of the expected signal, demanding sensitive instruments and careful attention to sources of systematic uncertainty (see ref. 35 for a review of CMB polarization measurements).

The first limit to the degree of polarization of the CMB was set in 1965 by Penzias and Wilson, who stated that the new radiation that they had discovered was isotropic and unpolarized within the limits of their observations[1]. Over the next 20 years, dedicated polarimeters were used to set much more stringent upper limits on angular scales of order several degrees and larger[36-41]. The current best upper limits for the $E$-mode and $B$-mode polarizations on large angular scales are 10 μK at 95% confidence for the multipole range $2 \leqslant l \leqslant 20$, set by the POLAR experiment[42].

On angular scales of the order of one degree, an analysis of data from the Saskatoon experiment[43] set the first upper limit (25 μK at 95% confidence for $l \approx 75$); this limit is also noteworthy in that it was the first limit that was lower than the level of the CMB temperature anisotropy. The current best limit on similar angular scales was set by the PIQUE experiment[44]—a 95% confidence upper limit of 8.4 μK to the $E$-mode signal, assuming no $B$-mode polarization. A preliminary analysis of cosmic background imager (CBI) data[45] indicates an upper limit similar to the PIQUE result, but on somewhat smaller scales. An attempt was also made to search for the $TE$ correlation using the PIQUE polarization and Saskatoon temperature data[46].

Polarization measurements have also been pursued on much finer angular scales (of the order of an arcminute), resulting in several upper limits (for example, refs 47 and 48). However, at these angular scales, corresponding to multipoles of about 5,000, the level of the primary CMB anisotropy is strongly damped and secondary effects due to interactions with large-scale structure in the Universe are expected to dominate[11].

In this paper, we report the detection of polarized anisotropy in the CMB radiation with the Degree Angular Scale Interferometer (DASI) located at the National Science Foundation (NSF) Amundsen–Scott South Pole research station. The polarization data were obtained during the 2001 and 2002 austral winter seasons. DASI was previously used to measure the temperature anisotropy from $140 < l < 900$, during the 2000 season. We presented details of the instrument, the measured power spectrum and the resulting cosmological constraints in a series of three papers: refs 49, 6 and 50. Prior to the start of the 2001 season, DASI was modified to allow polarization measurements in all four Stokes parameters over the same

模式偏振的扭曲将导致 $B$ 模式偏振信号的污染，这将严重地使引力波[30-33]的偏振特征的提取复杂化。然而，探索能否通过测量引力波引起的偏振（如参考文献 34），进而在 $10^{16}$ GeV 的能量尺度上直接探测宇宙，仍然是 CMB 偏振观测的一个引人入胜的目标。

在本文得出的结果之前，只有关于 CMB 偏振水平上限的报道。这是由于预期信号的水平较低，需要敏锐的仪器并仔细地留意系统不确定性的来源（参考文献 35 给出了有关 CMB 偏振测量的文献综述）。

CMB 偏振程度的第一个极限是由彭齐亚斯和威耳孙在 1965 年确定的，他们阐明其发现的新辐射在观察极限内是各向同性和非偏振的[1]。在接下来的 20 年中，人们借助专用的起偏器以确定更严格的上限，其角度的量级为几度或者更大[36-41]。对于多极矩而言，在 $2 \leqslant l \leqslant 20$ 的范围中，当前大角度范围内 $E$ 模式和 $B$ 模式偏振的最佳上限为 10 μK，由 POLAR 实验[42]给出。

在 1 度量级的角度尺度上，来自萨斯卡通实验的数据分析[43]确定了第一个上限（25 μK，置信度为 95%，$l \approx 75$）；这个极限也是值得注意的，因为它是第一个低于 CMB 温度各向异性水平的限制。假设没有 $B$ 模式偏振，当前最佳的情况是由设置了相类似角度标度的 PIQUE 实验[44]给出的——$E$ 模式信号上限为 8.4 μK，置信度为 95%。宇宙背景成像仪（CBI）数据[45]的初步分析显示了一个类似于 PIQUE 结果的上限，但是在稍微小的尺度上给出的。另外一个尝试是使用 PIQUE 偏振和萨斯卡通温度数据[46]研究 $TE$ 相关性。

偏振测量也在更精细的角度范围（1 角分量级）内进行，研究人员已经得到了几个上限（如参考文献 47 和 48）。然而，在这些角度尺度上，对应于大约 5,000 的多极矩，主要的 CMB 各向异性效果被强烈阻尼掉，并且预计与宇宙中的大尺度结构相互作用而产生的次级效应会占主导地位[11]。

在本文中，我们报告了使用位于美国国家科学基金会（NSF）阿蒙森–斯科特南极研究站的度角尺度干涉仪（DASI）检测到的 CMB 辐射中的偏振各向异性。偏振数据是在 2001 年和 2002 年南半球冬季获得的。DASI 前期用于测量 2000 年观测季中 $140 < l < 900$ 的温度各向异性。我们在三篇系列文章（参考文献 49、6 和 50）中提供了仪器的详细信息、测量的功率谱和由此产生的宇宙学约束。在 2001 观测季开始之前，DASI 被调整为可以在所有四个斯托克斯参数下进行偏振观测，其中 $l$ 的取值范

$l$ range as the previous measurements. The modifications to the instrument, observational strategy, data calibration and data reduction are discussed in detail in a companion paper in this issue[51].

The measurements reported here were obtained within two 3.4° full-width at half-maximum (FWHM) fields separated by 1 h in right ascension. The fields were selected from the subset of fields observed with DASI in 2000 in which no point sources were detected, and are located in regions of low Galactic synchrotron and dust emission. The temperature angular power spectrum is found to be consistent with previous measurements and its measured frequency spectral index is −0.01 (−0.16 to 0.14 at 68% confidence), where 0 corresponds to a 2.73 K Planck spectrum. Polarization of the CMB is detected at high confidence ($\geqslant 4.9\sigma$) and its power spectrum is consistent with theoretical predictions, based on the interpretation of CMB anisotropy as arising from primordial scalar adiabatic fluctuations. Specifically, assuming a shape for the power spectrum consistent with previous temperature measurements, the level found for the $E$-mode polarization is 0.80 (0.56 to 1.10 at 68% confidence interval), where the predicted level given previous temperature data is 0.9 to 1.1. At 95% confidence, an upper limit of 0.59 is set to the level of $B$-mode polarization parameterized with the same shape and normalization as the $E$-mode spectrum. The $TE$ correlation of the temperature and $E$-mode polarization is detected at 95% confidence and is also found to be consistent with predictions.

With these results contemporary cosmology has passed a long anticipated and crucial test. If the test had not been passed, the underpinnings of much of what we think we know about the origin and early history of the Universe would have been cast into doubt.

## Measuring Polarization with DASI

DASI is a compact interferometric array optimized for the measurement of CMB temperature and polarization anisotropy[49,51]. Because they directly sample Fourier components of the sky, interferometers are well suited to measurements of the CMB angular power spectrum. In addition, an interferometer gathers instantaneous two-dimensional information while inherently rejecting large-scale gradients in atmospheric emission. For observations of CMB polarization, interferometers offer several additional features. They can be constructed with small and stable instrumental polarization. Furthermore, linear combinations of the data can be used to construct quantities with essentially pure $E$- and $B$-mode polarization response patterns on a variety of scales. This property of the data greatly facilitates the analysis and interpretation of the observed polarization in the context of cosmological models.

DASI is designed to exploit these advantages in the course of extremely long integrations on selected fields of sky. The 13 horn/lens antennas that comprise the DASI array are compact, axially symmetric, and sealed from the environment, yielding small and extremely stable instrumental polarization. Additional systematic control comes from multiple levels of phase

围与之前的研究相同。另一篇与本文一并发表的文章详细讨论了我们在仪器、观测策略、数据校准和数据归算方面的修改[51]。

本文报告的测量结果是在两个视场获得的，二者具有 3.4° 半高宽（FWHM），赤经上间隔 1 小时。这些视场是从 2000 年利用 DASI 观测到的视场的子集中选出的。在这之中没有检测到点源，且位于银河系同步辐射和尘埃辐射较低的区域。温度角功率谱与先前的测量值一致，并且其测量的频率谱指数为 –0.01（置信度为 68% 的参数区间为 –0.16 至 0.14），其中 0 对应 2.73 K 的普朗克谱。基于对原始标量绝热扰动引起的 CMB 各向异性的解释，探测到的 CMB 的偏振具有很高的可信度（$\geqslant 4.9\sigma$），并且其功率谱与理论预测一致。具体来讲，假设功率谱的形状与先前的温度测量一致，针对 $E$ 模式的偏振水平为 0.80（置信度为 68% 的参数区间为 0.56 至 1.10），若温度数据为先前给定的数据，预测水平将为 0.9 至 1.1。在 95% 置信度下，与 $E$ 模式谱有着相同的形状和归一化的 $B$ 模式偏振参数化水平的上限被限定为 0.59。温度与 $E$ 模式偏振的 $TE$ 相关性以 95% 置信度被探测到，并且发现与预测一致。

包含这些结果在内，当代宇宙学已经通过了长期的预测和关键的检验。如果没有通过这些检验，那么我们对于宇宙的起源及其早期历史的认知基础恐怕就要动摇了。

## 用 DASI 测量偏振

DASI 是一款紧凑型干涉测量阵列，是专为测量 CMB 温度和偏振各向异性而优化的[49,51]。因为它们直接采样天空的傅里叶分量，所以干涉仪非常适合测量 CMB 角功率谱。而且干涉仪会收集瞬时二维信息，同时本身还具有排除大气辐射中大尺度梯度的特性。对于 CMB 偏振的观察，干涉仪提供了几个附加功能。观测数据可以用微小且稳定的仪器偏振构造出来。此外，数据的线性组合可用于在各种尺度上构建具有基本上纯的 $E$ 和 $B$ 模式的偏振响应模式。数据的这种性质极大地方便了在宇宙学模型下分析和解释观察到的偏振。

DASI 旨在利用这些优势，在选定的天空视场进行极长的累积观测。构成 DASI 阵列的 13 个喇叭/透镜天线结构紧凑、轴对称、与环境隔离且密封，产生的仪器偏振非常小而极其稳定。额外的系统化的控制来自多级相位切换和视场作差，目的是

switching and field differencing designed to remove instrumental offsets. The DASI mount is fully steerable in elevation and azimuth with the additional ability to rotate the entire horn array about the faceplate axis. The flexibility of this mount allows us to track any field visible from the South Pole continuously at constant elevation angle, and to observe it in redundant faceplate orientations which allow sensitive tests for residual systematic effects.

## Instrumental response and calibration

Each of DASI's 13 receivers may be set to admit either left or right circular polarization. An interferometer measures the correlations between the signals from pairs of receivers, called visibilities; as indicated by equation (4) in the "Theory covariance matrix" subsection recovery of the full complement of Stokes parameters requires the correlation of all four pairwise combinations of left and right circular polarization states (RR, LL, RL and LR), which we refer to as Stokes states. The co-polar states (RR, LL) are sensitive to the total intensity, while the cross-polar states (RL, LR) measure linear combinations of the Stokes parameters $Q$ and $U$.

Each of DASI's analogue correlator channels can accommodate only a single Stokes state, so measurement of the four combinations is achieved via time-multiplexing. The polarizer for each receiver is switched on a 1-h Walsh cycle, with the result that over the full period of the cycle, every pair of receivers spends an equal amount of time in all four Stokes states.

In ref. 51, we detail the calibration of the polarized visibilities for an interferometer. In order to produce the calibrated visibilities as defined in equation (4) below, a complex gain factor which depends on the Stokes state must be applied to each raw visibility. Although the cross-polar gain factors could easily be determined with observations of a bright polarized source, no suitable sources are available, and we therefore derive the full calibration through observations of an unpolarized source. The gains for a given pair of receivers (a baseline) can be decomposed into antenna-based factors, allowing us to construct the cross-polar gains from the antenna-based gain factors derived from the co-polar visibilities. DASI's calibration is based on daily observations of the bright $H_{II}$ region RCW38, which we described at length in ref. 49. We can determine the individual baseline gains for all Stokes states with statistical uncertainties $< 2\%$ for each daily observation. Systematic gain uncertainties for the complete data set are discussed in the "Systematic uncertainties" section.

The procedure for deriving the baseline gains from antenna-based terms leaves the cross-polar visibilities multiplied by an undetermined overall phase offset (independent of baseline). This phase offset effectively mixes $Q$ and $U$, and must be measured to obtain a clean separation of CMB power into $E$- and $B$-modes. Calibration of the phase offset requires a source whose polarization angle is known, and we create one by observing RCW38 through polarizing wire grids attached to DASI's 13 receivers. From the wire-grid observations, we can derive the phase offset in each frequency band with an uncertainty of $\lesssim 0.4°$.

消除仪器错位。DASI承载底座在高度和方位角上具有完全可操纵性，并且能够围绕面板对整个喇叭阵列进行轴旋转。这种承载架的灵活性使我们能够以恒定的仰角连续追踪来自南极的任何可见的视场，并以其他冗余的面板方向观察它，从而可以对剩余的系统效应进行灵敏性检测。

### 仪器响应和校准

DASI的13个接收器中的每一个都允许被设置为左或右圆偏振。干涉仪测量的是来自成对的接收器信号之间的相关，称为可视性；如本文“理论协方差矩阵”部分中的等式(4)所示，斯托克斯参数的完整获得要求左右圆偏振状态(RR、LL、RL和LR)的所有四个成对组合相关，我们称之为斯托克斯状态。共极态(RR、LL)对总强度敏感，而交叉极态(RL、LR)可以测量斯托克斯参数$Q$和$U$的线性组合。

每个DASI的模拟相关器通道只能容纳单个斯托克斯状态，因此通过时分多路复用实现四种状态组合的测量。每个接收器的起偏器接通1小时沃尔什周期，最终使得在整个周期内，每对接收器在所有的四个斯托克斯状态下花费相等的时间。

在参考文献51中，我们详细说明了干涉仪偏振可视性的校准。为了产生如下文的等式(4)中定义的校准可视性，必须将依赖于斯托克斯状态的复增益因子应用在每个初始可视性上。尽管通过观察明亮的偏振源可以很容易地确定交叉偏振增益因子，但是没有合适的源可用，因此我们通过观察非偏振源得到完全的校准。一对给定的接收器(即一条基线)的增益可以分解为多个天线增益因子，这允许我们从来自共极可视性的天线增益因子来导出交叉偏振增益。DASI的校准基于每天对RCW38的$H_{II}$区域的观测，我们在参考文献[49]中对其进行了详细的描述。我们可以确定所有斯托克斯状态的单个基线增益，每日观察的系统不确定性$<2\%$。“系统不确定性”部分讨论了完整数据集的系统增益不确定性。

从基于天线的项获得基线增益的过程中剩下了交叉偏振可视性乘以未确定的总相位偏移(与基线无关)。该相位偏移有效地混合$Q$和$U$，并且必须进行测量才能获得从CMB功率到$E$模式和$B$模式的无污染的分离。相位偏移的校准需要一个偏振角度已知的光源，我们通过观测DASI的13个接收器附属的偏振丝栅RCW38来创建一个这样的光源。根据丝栅观测，我们可以推导出每个频带的相位偏移，不确定度$\lesssim 0.4^\circ$。

As an independent check of this phase offset calibration, the Moon was observed at several epochs during 2001–02. Although the expected amplitude of the polarized signal from the Moon is not well known at these frequencies, the polarization pattern is expected to be radial to high accuracy, and this can be used to determine the cross-polar phase offset independently of the wire grid observations. We show in ref. 51 that these two determinations of the phase offset are in excellent agreement.

## On-axis leakage

For ideal polarizers, the cross-polar visibilities are strictly proportional to linear combinations of the Stokes parameters $Q$ and $U$. For realistic polarizers, however, imperfect rejection of the unwanted polarization state leads to additional terms in the cross-polar visibilities proportional to the total intensity $I$. These leakage terms are the sum of the complex leakage of the two antennas which form each baseline. During the 2000–01 austral summer, DASI's 13 receivers were retrofitted with multi-element broadband circular polarizers designed to reject the unwanted polarization state to high precision across DASI's 26–36 GHz frequency band. Before installation on the telescope, the polarizers were tuned to minimize these leakages.

At several epochs during 2001–02, the antenna-based leakages were determined with a fractional error of 0.3% from deep observations of the calibrator source RCW38. We show in ref. 51 that antenna-based leakages are $\lesssim 1\%$ (of $I$) at all frequency bands except the highest, for which they approach 2%; this performance is close to the theoretical minimum for our polarizer design. Comparison of the measurements from three epochs separated by many months indicates that the leakages are stable with time.

Given the low level of DASI's leakages, the mixing of power from temperature into polarization in the uncorrected visibilities is expected to be a minor effect at most (see the "Systematic uncertainties" section). Nonetheless, in the analysis presented in this paper, the cross-polar data have in all cases been corrected to remove this effect using the leakages determined from RCW38.

## Off-axis leakage

In addition to on-axis leakage from the polarizers, the feeds will contribute an instrumental polarization that varies across the primary beam. Offset measurements of RCW38 and the Moon indicate that the off-axis instrumental polarization pattern is radial, rising from zero at the beam centre to a maximum of about 0.7% near 3°, and then tapering to zero (see also ref. 51).

With the on-axis polarizer leakage subtracted to $\lesssim 0.3\%$ (see above), this residual leakage, while still quite small compared to the expected level of polarized CMB signal (again,

作为这种相位偏移校准的独立检查，在2001年～2002年期间的几个时期将月球作为观测对象。虽然在这些频率下，来自月球的偏振信号的预期幅度并不十分清楚，但在高精度下预计的偏振模式是发散状的，这可以用来独立地确定丝栅观测的交叉偏振相位偏移。我们在参考文献51中表明这两个相位偏移的测定非常一致。

### 共轴渗漏

对于理想的起偏器，交叉偏振可视性严格正比于斯托克斯参数 $Q$ 和 $U$ 的线性组合。然而，对于真实的起偏器，不需要的偏振态的不完全抑制导致交叉偏振可视性中的附加项与总强度 $I$ 成正比。这些渗漏项是从每个基线形成的两个天线的复合渗漏的总和。在2000年～2001年南半球夏季期间，研究人员采用多元宽带圆形起偏器更新改造了DASI的13个接收器，这是为了使通过DASI的26 GHz～36 GHz频带在高精度时抑制不必要的偏振态。在安装到望远镜上之前，起偏器被调整到最大限度地减少这些渗漏的状态。

在2001年～2002年间的几个时期内，通过对校准器源RCW38的深度观察，天线基线的渗漏被确定会导致0.3%的相对误差。参考文献51显示天线基线的渗漏在除最高频段之外的所有频段都 $\lesssim 1\%(I)$，而最高频段接近2%；这种性能接近我们起偏器设计的理论最小值。通过三个时期的测量值(时期之间相隔数个月)的比较，表明渗漏随时间推移是稳定的。

鉴于DASI渗漏水平较低，预计未校正的可视性中，从温度到极化的功率混合最多也只是产生次要的影响(见“系统不确定性”部分)。尽管如此，在本文提出的分析方案中，所有情况下都使用RCW38确定的渗漏来纠正交叉极化数据，从而消除上述影响。

### 离轴渗漏

除了来自起偏器的共轴渗漏之外，反馈将导致仪器偏振，这个偏振在横穿初级光束的过程中不断变化。RCW38和月球的偏移测量表明，离轴仪器偏振模式是径向的，光束中心为0，在接近3°时上升到最大值0.7%，然后逐渐减小到0(参考文献51)。

随着共轴起偏器的渗漏减去 $\lesssim 0.3\%$(见上文)的值，这种残余的渗漏虽然与偏振CMB信号的预期水平相比仍然非常小(见“系统不确定性”部分)，但却是仪器贡献

see the "Systematic uncertainties" section), is the dominant instrumental contribution. Although the visibilities cannot be individually corrected to remove this effect (as for the on-axis leakage), it may be incorporated in the analysis of the CMB data. Using fits to the offset data (see ref. 51 for details), we account for this effect by modelling the contribution of the off-axis leakage to the signal covariance matrix as described in the "Theory covariance matrix" subsection.

## CMB Observations and Data Reduction

### Observations

For the observations presented here, two fields separated by 1 h of right ascension, at RA = 23 h 30 min and RA = 00 h 30 min with declination −55°, were tracked continuously. The telescope alternated between the fields every hour, tracking them over precisely the same azimuth range so that any terrestrial signal can be removed by differencing. Each 24-h period included 20 h of CMB observations and 2.3 h of bracketing calibrator observations, with the remaining time spent on skydips and miscellaneous calibration tasks.

The fields were selected from the 32 fields previously observed by DASI for the absence of any detectable point sources (see ref. 49). The locations of the 32 fields were originally selected to lie at high elevation angle and to coincide with low emission in the IRAS 100 μm and 408 MHz maps of the sky[52].

The data presented in this paper were acquired from 10 April to 27 October 2001, and again from 14 February to 11 July 2002. In all, we obtained 162 days of data in 2001, and 109 days in 2002, for a total of 271 days before the cuts described in the next section.

### Data cuts

Observations are excluded from the analysis, or cut, if they are considered suspect owing to hardware problems, inadequate calibration, contamination from Moon or Sun signal, poor weather or similar effects. In the "Data consistency and $\chi^2$ tests" section, we describe consistency statistics that are much more sensitive to these effects than are the actual results of our likelihood analysis, allowing us to be certain that the final results are unaffected by contamination. Here we briefly summarize the principal categories of data cuts; we describe each cut in detail in ref. 51.

In the first category of cuts, we reject visibilities for which monitoring data from the telescope indicate obvious hardware malfunction, or simply non-ideal conditions. These include cryogenics failure, loss of tuning for a receiver, large offsets between real/imaginary multipliers in the correlators, and mechanical glitches in the polarizer stepper motors. All data are rejected for a correlator when it shows evidence for large offsets, or excessive noise.

的主要成分。虽然可视性不能通过独立地校正以消除这种影响(对于共轴渗漏)，但它却可能包含在 CMB 数据的分析中。使用拟合偏移数据(参考文献 51)，我们通过模拟离轴渗漏对信号协方差矩阵的贡献来解释这种影响，如“理论协方差矩阵”部分所述。

## CMB 观测和数据归算

### 观测

这里呈现的观测是在赤经为 23 h 30 min 和 00 h 30 min，赤纬为 −55° 的位置上，对赤经相隔一个小时的两个视场连续追踪所获得的。望远镜每小时在各个视场之间交替观测，在精确相同的方位角上追踪它们，以便通过作差去除所有地面信号。每个 24 小时周期包括 20 小时的 CMB 观察和 2.3 小时的托架校准器观察，剩余的时间用于天空倾角校准和其他校准任务。

这些视场是从 DASI 先前观察到的 32 个视场中选择的，因为缺乏任何可检测的点源(参考文献 49)。这 32 个视场的位置最初被选择为处于高仰角并且与 IRAS 100 μm 和 408 MHz 天图中的低辐射一致的地方[52]。

本文提供的数据是在 2001 年 4 月 10 日至 10 月 27 日以及 2002 年 2 月 14 日至 7 月 11 日分两次获得的。我们在 2001 年总共获得了 162 天的数据，在 2002 年获得了 109 天的数据，在下一节描述的数据舍弃之前，共计 271 天。

### 数据舍弃

如果观测结果被认为具有硬件问题、校准不充分、月亮或太阳信号污染、恶劣天气或其他类似的可疑影响，将被排除在分析之外，或者说被舍弃。在“数据一致性和 $\chi^2$ 检验”部分中我们指出，比较而言，一致性统计对于这些影响的敏感度要远高于获得这一统计量的似然分析的实际结果，这要求我们要能够确定最终结果不受污染的影响。在这里，我们简要总结一下数据舍弃的主要类别；在参考文献 51 中我们详细描述了每种数据的舍弃。

第一类舍弃的可视性是在具有明显的硬件故障的或仅仅是非理想条件下的望远镜的监测数据。其中包括低温失效、接收机失调、相关器中实/虚乘法器之间的大偏移，以及起偏器步进电机中的机械假信号。当显示有过大偏移或过大噪声的证据时，

An additional cut, and the only one based on the individual data values, is a $> 30\sigma$ outlier cut to reject rare ($\ll 0.1\%$ of the data) hardware glitches. Collectively, these cuts reject about 26% of the data.

In the next category, data are cut on the phase and amplitude stability of the calibrator observations. Naturally, we reject data for which bracketing calibrator observations have been lost due to previous cuts. These cuts reject about 5% of the data.

Cuts are also based on the elevation of the Sun and Moon. Co-polar data are cut whenever the Sun was above the horizon, and cross-polar data whenever the solar elevation exceeded $5°$. These cuts reject 8% of the data.

An additional cut, which is demonstrably sensitive to poor weather, is based on the significance of data correlations as discussed in the "Noise model" subsection. An entire day is cut if the maximum off-diagonal correlation coefficient in the data correlation matrix exceeds $8\sigma$ significance, referred to gaussian uncorrelated noise. A total of 22 days are cut by this test in addition to those rejected by the solar and lunar cuts.

### Reduction

Data reduction consists of a series of steps to calibrate and reduce the data set to a manageable size for the likelihood analysis. Phase and amplitude calibrations are applied to the data on the basis of the bracketing observations of our primary celestial calibrator, RCW38. The raw 8.4-s integrations are combined over each 1-h observation for each of 6,240 visibilities (78 complex baselines × 10 frequency bands × 4 Stokes states). In all cases, on-axis leakage corrections are applied to the data, and sequential 1-h observations of the two fields in the same $15°$ azimuth range are differenced to remove any common ground signal, using a normalization $(\text{field1}-\text{field2})/\sqrt{2}$ that preserves the variance of the sky signal. Except in the case where the data set is split for use in the $\chi^2$ consistency tests in the "$\chi^2$ tests" subsection, observations from different faceplate rotation angles, epochs and azimuth ranges are all combined, as well as the two co-polar Stokes states, LL and RR. The resulting data set has $N \leqslant 4{,}680$ elements ($6{,}240 \times 3/4 = 4{,}680$, where the 3/4 results from the combination of LL and RR). We call this the uncompressed data set, and it contains all of the information in our observations of the differenced fields for Stokes parameters $I$, $Q$ and $U$.

## Data Consistency and $\chi^2$ Tests

We begin our analysis by arranging the data into a vector, considered to be the sum of actual sky signal and instrumental noise: $\mathbf{\Delta} = \mathbf{s} + \mathbf{n}$. The noise vector $\mathbf{n}$ is hypothesized to be gaussian and random, with zero mean, so that the noise model is completely specified by a known covariance matrix $C_N \equiv \langle \mathbf{nn}^t \rangle$. Any significant excess variance observed in the

该相关器所有数据都被拒绝。额外的数据舍弃，以及唯一一种基于数据值个体本身的舍弃是$>30\sigma$的异常值，以拒绝罕见的（$<<0.1\%$的数据）硬件假信号。总的来说，这些舍弃拒绝了大约26%的数据。

下一类数据舍弃的依据是校准器观察的相位和幅度稳定性。于是我们拒绝因先前数据舍弃导致托架校准器观察丢失的数据。这些削减拒绝了大约5%的数据。

数据也会因为太阳和月亮的升高带来的影响而被舍弃。每当太阳高于地平线时，同极数据都会被舍弃，而当太阳升高超过5°时，就会出现交叉偏振数据。这些舍弃拒绝了8%的数据。

另一个明显对恶劣天气敏感的舍弃，依据的是"噪声模型"部分对数据相关的显著性的讨论。如果数据相关矩阵中的最大非对角线相关系数的显著性超过$8\sigma$，则视为高斯不相关噪声，全天的数据都将被舍弃。除了被太阳和月球影响而拒绝那些数据之外，该检测总共减少了22天的数据量。

## 归算

数据归算包括一系列步骤，用于校准数据集并将数据集压缩到可管理的大小以进行似然分析。根据我们的初级天体校准器RCW38的基础观测，相位和幅度校准将应用在数据上。原始的8.4 s集成结合了每小时观测中所有6,240种可视性（78个复基线 ×10个频带 ×4个斯托克斯状态）。共轴渗漏校正被正应用于所有情况下的数据，两个视场在相同的15° 方位角范围内的连续1 h观测值相减以消除任何常见的地面信号，并且使用（视场1 − 视场2）$/\sqrt{2}$来归一化，从而保留天空信号的方差。除了在"$\chi^2$检验"部分对数据集进行分划以实现$\chi^2$一致性检验的情况之外，来自不同的面盘旋转角度、时期和方位角范围的观测结果都被组合起来，同时被组合起来的还包括两个共极斯托克斯状态，LL和RR。得到的数据集具有$N\leqslant 4{,}680$个元素（$6{,}240\times 3/4=4{,}680$，其中3/4由LL和RR的组合产生）。我们将其称为未压缩数据集，它包含经过作差处理之后的视场斯托克斯参数$I$、$Q$和$U$的所有观测信息。

## 数据一致性和$\chi^2$检验

我们通过将数据排列成一个矢量来开始分析，该矢量$\Delta$被认为是实际天空信号和仪器噪声的总和：$\Delta=\mathbf{s}+\mathbf{n}$。噪声矢量$\mathbf{n}$被假设为高斯的且随机的，具有零均值，因此噪声模型完全由已知的协方差矩阵$C_N\equiv\langle\mathbf{nn}^t\rangle$确定。在数据矢量$\Delta$中观察到的

data vector $\mathbf{\Delta}$ will be interpreted as signal. In the likelihood analysis of the next section, we characterize the total covariance of the data set $C = C_T(\kappa) + C_N$ in terms of parameters $\kappa$ that specify the covariance $C_T$ of this sky signal. This is the conventional approach to CMB data analysis, and it is clear that for it to succeed, the assumptions about the noise model and the accuracy of the noise covariance matrix must be thoroughly tested. This is especially true for our data set, for which long integrations have been used to reach unprecedented levels of sensitivity in an attempt to measure the very small signal covariances expected from the polarization of the CMB.

## Noise model

The DASI instrument and observing strategy are designed to remove systematic errors through multiple levels of differencing. Slow and fast phase switching as well as field differencing is used to minimize potentially variable systematic offsets that could otherwise contribute a non-thermal component to the noise. The observing strategy also includes Walsh sequencing of the Stokes states, observations over multiple azimuth ranges and faceplate rotation angles, and repeated observations of the same visibilities on the sky throughout the observing run to allow checks for systematic offsets and verification that the sky signal is repeatable. We hypothesize that after the cuts described in the previous section, the noise in the remaining data is gaussian and white, with no noise correlations between different baselines, frequency bands, real/imaginary pairs, or Stokes states. We have carefully tested the noise properties of the data to validate the use of this model.

Noise variance in the combined data vector is estimated by calculating the variance in the 8.4-s integrations over the period of 1 h, before field differencing. To test that this noise estimate is accurate, we compare three different short-timescale noise estimates: calculated from the 8.4-s integrations over the 1-h observations before and after field differencing and from sequential pairs of 8.4-s integrations. We find that all three agree within 0.06% for co-polar data and 0.03% for cross-polar data, averaged over all visibilities after data cuts.

We also compare the noise estimates based on the short-timescale noise to the variance of the 1-h binned visibilities over the entire data set (up to 2,700 1-hour observations, over a period spanning 457 days). The ratio of long-timescale to short-timescale noise variance, averaged over all combined visibilities after data cuts, is 1.003 for the co-polar data and 1.005 for the cross-polar data, remarkably close to unity. Together with the results of the $\chi^2$ consistency tests described in the "$\chi^2$ tests" subsection, these results demonstrate that the noise is white and integrates down from timescales of a few seconds to thousands of hours. We find that scaling the diagonal noise by 1% makes a negligible difference in the reported likelihood results (see the "Systematic uncertainties" section).

To test for potential off-diagonal correlations in the noise, we calculate a $6{,}240 \times 6{,}240$ correlation coefficient matrix from the 8.4-s integrations for each day of observations. To increase our sensitivity to correlated noise, we use only data obtained simultaneously for

任何显著的方差超出将被解释为信号。在下一部分的似然分析中，我们依据确定该天空信号协方差 $C_T$ 的参数 $\kappa$ 来表征数据集的总协方差 $C = C_T(\kappa) + C_N$。这是 CMB 数据分析的传统方法，为了能成功实现，很显然必须彻底检测噪声模型的假设和噪声协方差矩阵的准确性。这对于我们的数据集尤甚，因而为了测量由 CMB 偏振产生的非常小的信号的协方差，我们使用长期积分来达到前所未有的灵敏度水平。

## 噪声模型

DASI 仪器和观测策略被设计为通过多级作差消除系统误差。慢速和快速相位切换以及视场作差被用于最小化潜在可变的系统偏移，否则可能增加噪声的非热分量。观测策略还包括斯托克斯状态的沃尔什测序，多个方位角范围和面盘旋转角度的观测，以及通过允许检查系统与对偏移与天空信号的检查是可重复的观测，进而实现在整个观测过程中对天空的相同可视性进行重复观测。我们假设在如前一节所描述的数据舍弃之后，剩余数据中的噪声是高斯的白噪声，不同基线、频带、实/虚对或斯托克斯状态之间没有噪声相关。我们仔细检测了数据的噪声属性，以验证该模型的实用性。

通过在视场作差之前 1 小时的时段内计算 8.4 s 积分的方差，组合而成的数据矢量中的噪声方差可以被估算出来。为了检测这种噪声估计是否准确，我们比较了三段不同的短时间尺度内的噪声估计：计算视场差前后 1 h 观测值的 8.4 s 积分以及 8.4 s 积分的连续对。我们发现三者在共极数据中均为 0.06%，在交叉偏振数据中为 0.03%，在数据舍弃后平均了所有可视性。

我们还将基于短时间尺度的噪声估计值与整个数据集中 1 小时分格可视性的方差进行比较（在时间跨度为 457 天的情况下，最多可得到 2,700 个 1 小时观测值）。数据舍弃后，所有组合可视性的长时间尺度与短时间尺度噪声方差的比率可以给出，其中共极数据为 1.003，而交叉偏振数据为 1.005，非常接近于 1。这与“$\chi^2$ 检验”部分描述的 $\chi^2$ 一致性检验的结果一起表明噪声是白色的，并且随着时间尺度从几秒合并为几千小时，噪声也同时下降。我们发现将对角线噪声缩放 1% 使似然结果产生的差异可以忽略不计（参见“系统不确定性”部分）。

为了检测噪声中潜在的非对角线相关性，我们计算了每天观测的 8.4 s 积分的 $6{,}240 \times 6{,}240$ 相关系数矩阵。为了提高数据对相关噪声的灵敏度，我们仅使用特定的一对数据矢量元素同时获得的数据。由于 8.4 s 积分的变量数 $M$ 曾用于计算每个

a given pair of data vector elements. Owing to the variable number of 8.4-s integrations $M$ used to calculate each off-diagonal element, we assess the significance of the correlation coefficient in units of $\sigma = 1/\sqrt{M-1}$. Our weather cut statistic is the daily maximum off-diagonal correlation coefficient significance (see "Data cuts" subsection).

We use the mean data correlation coefficient matrix over all days, after weather cuts, to test for significant correlations over the entire data set. We find that 1,864 (0.016%) of the off-diagonal elements exceed a significance of $5.5\sigma$, when about one such event is expected for uncorrelated gaussian noise. The outliers are dominated by correlations between real/imaginary pairs of the same baseline, frequency band, and Stokes state, and between different frequency bands of the same baseline and Stokes state. For the real/imaginary pairs, the maximum correlation coefficient amplitude is 0.14, with an estimated mean amplitude of 0.02; for interband correlations the maximum amplitude and estimated mean are 0.04 and 0.003, respectively. We have tested the inclusion of these correlations in the likelihood analysis and find that they have a negligible impact on the results, see the "Systematic uncertainties" section.

## $\chi^2$ tests

As a simple and sensitive test of data consistency, we construct a $\chi^2$ statistic from various splits and subsets of the visibility data. Splitting the data into two sets of observations that should measure the same sky signal, we form the statistic for both the sum and difference data vectors, $\chi^2 = \mathbf{\Delta}^t C_N^{-1} \mathbf{\Delta}$, where $\mathbf{\Delta} = (\mathbf{\Delta}_1 \pm \mathbf{\Delta}_2)/2$ is the sum or difference data vector, and $C_N = (C_{N1} + C_{N2})/4$ is the corresponding noise covariance matrix. We use the difference data vector, with the common sky signal component subtracted, to test for systematic offsets and mis-estimates of the noise. The sum data vector is used to test for the presence of a sky signal in a straightforward way that is independent of the likelihood analyses that will be used to parameterize and constrain that signal.

We split the data for the difference and sum data vectors in five different ways:

(1) Year—2001 data versus 2002 data;

(2) Epoch—the first half of observations of a given visibility versus the second half;

(3) Azimuth range—east five versus west five observation azimuth ranges;

(4) Faceplate position—observations at a faceplate rotation angle of 0° versus a rotation angle of 60°; and

(5) Stokes state—co-polar observations in which both polarizers are observing left circularly polarized light (LL Stokes state) versus those in which both are observing right circularly polarized light (RR Stokes state).

非对角线元素，故我们以 $\sigma = 1/\sqrt{M-1}$ 为单位评估相关系数的显著性。我们基于天气因素的数据舍弃统计量是每日非对角线相关系数的显著性的最大值（参见“数据舍弃”部分）。

在天气因素的舍弃数据后，所有观测日均使用平均数据相关系数矩阵来检测整个数据集的显著相关性。我们发现 1,864 个（0.016%）非对角线元素具有超过 $5.5\sigma$ 的显著性，而原本这个相关系数矩阵应该只存在不相关的高斯噪声。异常值主要受相同基线、频带和斯托克斯状态的实/虚对之间的相关性，以及相同基线和斯托克斯状态的不同频带之间的相关性影响。对于实/虚对，最大相关系数幅度为 0.14，估计平均幅度为 0.02；对于带间相关性，最大幅度和估计平均值分别为 0.04 和 0.003。我们已经在似然分析中检测了这些相关性，发现它们对结果的影响可以忽略不计，请参阅“系统不确定性”部分。

## $\chi^2$ 检验

作为对数据一致性的简单且灵敏的检验，我们从可视性数据的各种划分和子集中构建 $\chi^2$ 统计量。将数据分成两组观测值，它们应该测量相同的天空信号，我们构造了和与差数据矢量的统计量，$\chi^2 = \Delta^t C_N^{-1} \Delta$，其中 $\Delta = (\Delta_1 \pm \Delta_2)/2$，是和或差数据矢量，而 $C_N = (C_{N1} + C_{N2})/4$ 是相应的噪声协方差矩阵。我们使用差数据矢量，减去公共天空信号分量，来检验系统偏移和噪声的误差估计。数据矢量之和是一种直接用于检验天空信号存在的方式，该方式独立于即将应用于参数化和信号约束的似然分析。

我们以五种不同的方式划分数据矢量的差与和：

（1）年份——2001 年的数据与 2002 年的数据；

（2）时期——对于给定可视性的前半次观测和后半次观测；

（3）方位角范围——东方的和西方的五个观测方位角；

（4）面盘位置——观察面板旋转角度为 0° 和旋转角度为 60°；

（5）斯托克斯状态——共极观测，其中两个起偏器都观察到左旋圆偏振光（LL 斯托克斯状态），而另两个起偏器都观察到右旋圆偏振光（RR 斯托克斯状态）。

These splits were done on the combined 2001–02 data set and (except for the first split type) on 2001 and 2002 data sets separately, to test for persistent trends or obvious differences between the years. The faceplate position split is particularly powerful, since the six-fold symmetry of the $(u, v)$ plane coverage allows us to measure a sky signal for a given baseline with a different pair of receivers, different backend hardware, and at a different position on the faceplate with respect to the ground shields, and is therefore sensitive to calibration and other offsets that may depend on these factors. The co-polar split tests the amplitude and phase calibration between polarizer states, and tests for the presence of circularly polarized signal.

For each of these splits, different subsets can be examined: co-polar data only, cross-polar data only (for all except the Stokes state split), various $l$-ranges (as determined by baseline length in units of wavelength), and subsets formed from any of these which isolate modes with the highest expected signal to noise ($s/n$). In constructing the high $s/n$ subsets, we must assume a particular theoretical signal template in order to define the $s/n$ eigenmode basis[53] appropriate for that subset. For this we use the concordance model defined in the "Likelihood parameters" subsection, although we find the results are not strongly dependent on the choice of model. Note that the definitions of which modes are included in the high $s/n$ subsets are made in terms of average theoretical signal, without any reference to the actual data. In Table 1, we present the difference and sum $\chi^2$ values for a representative selection of splits and subsets. In each case we give the degrees of freedom, $\chi^2$ value, and probability to exceed (PTE) this value in the $\chi^2$ cumulative distribution function. For the 296 different split/subset combinations that were examined, the $\chi^2$ values for the difference data appear consistent with noise; among these 296 difference data $\chi^2$ values, there are two with a PTE $< 0.01$ (the lowest is 0.003), one with a PTE $> 0.99$, and the rest appear uniformly distributed between this range. There are no apparent trends or outliers among the various subsets or splits.

Table 1. $\chi^2$ Consistency tests for a selection of data splits and subsets

| Temperature data | | | | | | |
|---|---|---|---|---|---|---|
| Split type | Subset | Difference | | | Sum | |
| | | d.f. | $\chi^2$ | PTE | $\chi^2$ | PTE |
| Year | Full | 1,448 | 1,474.2 | 0.31 | 23,188.7 | $< 1 \times 10^{-16}$ |
| | $s/n > 1$ | 320 | 337.1 | 0.24 | 21,932.2 | $< 1 \times 10^{-16}$ |
| | $l$ range 0–245 | 184 | 202.6 | 0.17 | 10,566.3 | $< 1 \times 10^{-16}$ |
| | $l$ range 0–245 high $s/n$ | 36 | 38.2 | 0.37 | 10,355.1 | $< 1 \times 10^{-16}$ |
| | $l$ range 245–420 | 398 | 389.7 | 0.61 | 7,676.0 | $< 1 \times 10^{-16}$ |
| | $l$ range 245–420 high $s/n$ | 79 | 88.9 | 0.21 | 7,294.4 | $< 1 \times 10^{-16}$ |
| | $l$ range 420–596 | 422 | 410.5 | 0.65 | 3,122.5 | $< 1 \times 10^{-16}$ |
| | $l$ range 420–596 high $s/n$ | 84 | 73.5 | 0.79 | 2,727.8 | $< 1 \times 10^{-16}$ |
| | $l$ range 596–772 | 336 | 367.8 | 0.11 | 1,379.5 | $< 1 \times 10^{-16}$ |

这些划分整合了2001年～2002年的数据集(除第一种划分类型之外)，是2001年和2002年分开进行的，以检测这两年持续的趋势或者明显的差异。面盘位置的划分方法特别有效，因为$(u, v)$平面覆盖的六重对称允许我们在给定基线上使用不同的接收器对、不同的后端硬件和相对于接地屏蔽的面盘的不同位置来测量天空信号，因此校准和其他偏移可能敏感地取决于这些因素。共极划分检测了起偏器状态之间的幅度和相位校准，并检测圆偏振信号是否存在。

对于上述每一个划分方法，可以检查不同的子集：单一共极数据、单一交叉偏振数据(除了斯托克斯状态划分之外的所有数据)、各种多级矩$l$范围(由波长为单位的基线长度确定)，以及由其中具有任意最高的预估信噪比$(s/n)$的孤立模式形成的子集。在构造高$s/n$子集时，我们必须假设特定的理论信号拟合模板，以便定义适合于该子集的$s/n$基本本征模[53]。为此，我们使用“似然参数”部分定义的一致性模型，尽管发现结果并不强烈依赖于模型的选择。需要注意的是，高$s/n$子集中包括哪些模式的定义是根据平均理论信号进行的，而不参考实际数据。在表1中，我们给出了划分和子集的代表选择的差值与和值的$\chi^2$。在每种情况下，我们都给出自由度、$\chi^2$值以及这个值在$\chi^2$累积分布函数的超出概率(PTE)。对于测试的296种不同的划分/子集组合，差值数据的$\chi^2$值似乎与噪声一致；在这296个差值数据的$\chi^2$值中，有两个PTE < 0.01(最低为0.003)，一个PTE > 0.99，其余均匀分布在该范围之间。各种子集或划分中没有明显的趋势或异常值。

表1. 用于选择数据划分与数据子集的$\chi^2$一致性检验

| 温度数据 | | | | | | |
|---|---|---|---|---|---|---|
| 划分方式 | 子集 | 差值 | | | 和值 | |
| | | d.f. | $\chi^2$ | PTE | $\chi^2$ | PTE |
| 年份 | 完整 | 1,448 | 1,474.2 | 0.31 | 23,188.7 | $< 1 \times 10^{-16}$ |
| | $s/n > 1$ | 320 | 337.1 | 0.24 | 21,932.2 | $< 1 \times 10^{-16}$ |
| | $l$范围0～245 | 184 | 202.6 | 0.17 | 10,566.3 | $< 1 \times 10^{-16}$ |
| | $l$范围0～245高$s/n$ | 36 | 38.2 | 0.37 | 10,355.1 | $< 1 \times 10^{-16}$ |
| | $l$范围245～420 | 398 | 389.7 | 0.61 | 7,676.0 | $< 1 \times 10^{-16}$ |
| | $l$范围245～420高$s/n$ | 79 | 88.9 | 0.21 | 7,294.4 | $< 1 \times 10^{-16}$ |
| | $l$范围420～596 | 422 | 410.5 | 0.65 | 3,122.5 | $< 1 \times 10^{-16}$ |
| | $l$范围420～596高$s/n$ | 84 | 73.5 | 0.79 | 2,727.8 | $< 1 \times 10^{-16}$ |
| | $l$范围596～772 | 336 | 367.8 | 0.11 | 1,379.5 | $< 1 \times 10^{-16}$ |

*Continued*

| Temperature data | | | | | | |
|---|---|---|---|---|---|---|
| Split type | Subset | Difference | | | Sum | |
| | | d.f. | $\chi^2$ | PTE | $\chi^2$ | PTE |
| Year | $l$ range 596–772 high $s/n$ | 67 | 82.3 | 0.10 | 991.8 | $<1\times10^{-16}$ |
| | $l$ range 772–1,100 | 108 | 103.7 | 0.60 | 444.4 | $<1\times10^{-16}$ |
| | $l$ range 772–1,100 high $s/n$ | 21 | 22.2 | 0.39 | 307.7 | $<1\times10^{-16}$ |
| Epoch | Full | 1,520 | 1,546.3 | 0.31 | 32,767.2 | $<1\times10^{-16}$ |
| | $s/n>1$ | 348 | 366.5 | 0.24 | 31,430.0 | $<1\times10^{-16}$ |
| Azimuth range | Full | 1,520 | 1,542.6 | 0.34 | 32,763.8 | $<1\times10^{-16}$ |
| | $s/n>1$ | 348 | 355.2 | 0.38 | 31,426.9 | $<1\times10^{-16}$ |
| Faceplate position | Full | 1,318 | 1,415.2 | 0.03 | 27,446.5 | $<1\times10^{-16}$ |
| | $s/n>1$ | 331 | 365.3 | 0.09 | 26,270.1 | $<1\times10^{-16}$ |
| Stokes state | Full | 1,524 | 1,556.6 | 0.27 | 33,050.6 | $<1\times10^{-16}$ |
| | $s/n>1$ | 350 | 358.2 | 0.37 | 31,722.5 | $<1\times10^{-16}$ |
| Polarization data | | | | | | |
| Split type | Subset | Difference | | | Sum | |
| | | d.f. | $\chi^2$ | PTE | $\chi^2$ | PTE |
| Year | Full | 2,896 | 2,949.4 | 0.24 | 2,925.2 | 0.35 |
| | $s/n>1$ | 30 | 34.4 | 0.27 | 82.4 | $8.7\times10^{-7}$ |
| | $l$ range 0–245 | 368 | 385.9 | 0.25 | 315.0 | 0.98 |
| | $l$ range 0–245 high $s/n$ | 73 | 61.0 | 0.84 | 64.5 | 0.75 |
| | $l$ range 245–420 | 796 | 862.2 | 0.05 | 829.4 | 0.20 |
| | $l$ range 245–420 high $s/n$ | 159 | 176.0 | 0.17 | 223.8 | $5.4\times10^{-7}$ |
| | $l$ range 420–596 | 844 | 861.0 | 0.33 | 837.3 | 0.56 |
| | $l$ range 420–596 high $s/n$ | 168 | 181.3 | 0.23 | 189.7 | 0.12 |
| | $l$ range 596–772 | 672 | 648.1 | 0.74 | 704.4 | 0.19 |
| | $l$ range 596–772 high $s/n$ | 134 | 139.5 | 0.35 | 160.0 | 0.06 |
| | $l$ range 772–1,100 | 216 | 192.3 | 0.88 | 239.1 | 0.13 |
| | $l$ range 772–1,100 high $s/n$ | 43 | 32.3 | 0.88 | 47.6 | 0.29 |
| Epoch | Full | 3,040 | 2,907.1 | 0.96 | 3,112.2 | 0.18 |
| | $s/n>1$ | 34 | 29.2 | 0.70 | 98.6 | $3.3\times10^{-8}$ |
| Azimuth range | Full | 3,040 | 3,071.1 | 0.34 | 3,112.9 | 0.17 |
| | $s/n>1$ | 34 | 38.7 | 0.27 | 98.7 | $3.3\times10^{-8}$ |
| Faceplate position | Full | 2,636 | 2,710.4 | 0.15 | 2,722.2 | 0.12 |
| | $s/n>1$ | 32 | 43.6 | 0.08 | 97.5 | $1.6\times10^{-8}$ |

Results of $\chi^2$ consistency tests for a representative selection of splits and subsets of the combined 2001–02 data set. Visibility data containing the same sky signal is split to form two data vectors; using the instrument noise model, the $\chi^2$ statistic is then calculated on both the difference and sum data vectors. Also tabulated are the number of degrees

续表

| 温度数据 | | | | | | |
|---|---|---|---|---|---|---|
| 划分方式 | 子集 | 差值 | | | 和值 | |
| | | d.f. | $\chi^2$ | PTE | $\chi^2$ | PTE |
| 年份 | $l$ 范围 596～772 高 $s/n$ | 67 | 82.3 | 0.10 | 991.8 | $<1\times10^{-16}$ |
| | $l$ 范围 772～1,100 | 108 | 103.7 | 0.60 | 444.4 | $<1\times10^{-16}$ |
| | $l$ 范围 772～1,100 高 $s/n$ | 21 | 22.2 | 0.39 | 307.7 | $<1\times10^{-16}$ |
| 时期 | 完整 | 1,520 | 1,546.3 | 0.31 | 32,767.2 | $<1\times10^{-16}$ |
| | $s/n>1$ | 348 | 366.5 | 0.24 | 31,430.0 | $<1\times10^{-16}$ |
| 方位角范围 | 完整 | 1,520 | 1,542.6 | 0.34 | 32,763.8 | $<1\times10^{-16}$ |
| | $s/n>1$ | 348 | 355.2 | 0.38 | 31,426.9 | $<1\times10^{-16}$ |
| 面盘位置 | 完整 | 1,318 | 1,415.2 | 0.03 | 27,446.5 | $<1\times10^{-16}$ |
| | $s/n>1$ | 331 | 365.3 | 0.09 | 26,270.1 | $<1\times10^{-16}$ |
| 斯托克斯状态 | 完整 | 1,524 | 1,556.6 | 0.27 | 33,050.6 | $<1\times10^{-16}$ |
| | $s/n>1$ | 350 | 358.2 | 0.37 | 31,722.5 | $<1\times10^{-16}$ |
| 偏振数据 | | | | | | |
| 划分方式 | 子集 | 差值 | | | 和值 | |
| | | d.f. | $\chi^2$ | PTE | $\chi^2$ | PTE |
| 年份 | 完整 | 2,896 | 2,949.4 | 0.24 | 2,925.2 | 0.35 |
| | $s/n>1$ | 30 | 34.4 | 0.27 | 82.4 | $8.7\times10^{-7}$ |
| | $l$ 范围 0～245 | 368 | 385.9 | 0.25 | 315.0 | 0.98 |
| | $l$ 范围 0～245 高 $s/n$ | 73 | 61.0 | 0.84 | 64.5 | 0.75 |
| | $l$ 范围 245～420 | 796 | 862.2 | 0.05 | 829.4 | 0.20 |
| | $l$ 范围 245～420 高 $s/n$ | 159 | 176.0 | 0.17 | 223.8 | $5.4\times10^{-7}$ |
| | $l$ 范围 420～596 | 844 | 861.0 | 0.33 | 837.3 | 0.56 |
| | $l$ 范围 420～596 高 $s/n$ | 168 | 181.3 | 0.23 | 189.7 | 0.12 |
| | $l$ 范围 596～772 | 672 | 648.1 | 0.74 | 704.4 | 0.19 |
| | $l$ 范围 596～772 高 $s/n$ | 134 | 139.5 | 0.35 | 160.0 | 0.06 |
| | $l$ 范围 772～1,100 | 216 | 192.3 | 0.88 | 239.1 | 0.13 |
| | $l$ 范围 772～1,100 高 $s/n$ | 43 | 32.3 | 0.88 | 47.6 | 0.29 |
| 时期 | 完整 | 3,040 | 2,907.1 | 0.96 | 3,112.2 | 0.18 |
| | $s/n>1$ | 34 | 29.2 | 0.70 | 98.6 | $3.3\times10^{-8}$ |
| 方位角范围 | 完整 | 3,040 | 3,071.1 | 0.34 | 3,112.9 | 0.17 |
| | $s/n>1$ | 34 | 38.7 | 0.27 | 98.7 | $3.3\times10^{-8}$ |
| 面盘位置 | 完整 | 2,636 | 2,710.4 | 0.15 | 2,722.2 | 0.12 |
| | $s/n>1$ | 32 | 43.6 | 0.08 | 97.5 | $1.6\times10^{-8}$ |

结合了2001年～2002年数据集划分和子集的代表性选择的 $\chi^2$ 一致性检验结果。包含相同天空信号的可视性数据被划分成两个数据矢量；利用仪器噪声模型，我们计算了数据矢量差值与和值的 $\chi^2$ 统计量。这里还列出了自由度(d.f.)的大小与 $\chi^2$ 累积分布函数中的超出概率(PTE)，来显示结果的显著性(在我们计算的 $\chi^2$ 累积分布函数的精度

of freedom (d.f.), and probability to exceed (PTE) the value in the $\chi^2$ cumulative distribution function, to show the significance of the result (PTE values indicated as $< 1 \times 10^{-16}$ are zero to the precision with which we calculate the $\chi^2$ cumulative distribution function). Difference data $\chi^2$ values test for systematic effects in the data, while comparisons with sum data values test for the presence of a repeatable sky signal. Temperature (co-polar) data are visibility data in which the polarizers from both receivers are in the left (LL Stokes state) or right (RR Stokes state) circularly polarized state; polarization (cross-polar) data are those in which the polarizers are in opposite states (LR or RL Stokes state). The "$s/n > 1$" subset is the subset of $s/n$ eigenmodes > 1 and the $l$ range high $s/n$ subsets are the 20% highest $s/n$ modes within the given $l$ range. See "$\chi^2$ tests" section for further description of the data split types and subsets. We have calculated 296 $\chi^2$ values for various split types and subsets, with no obvious trends that would indicate systematic contamination of the data.

The high $s/n$ mode subsets are more sensitive to certain classes of systematic effects in the difference data vector and more sensitive to the expected sky signal in the sum data vector, that otherwise may be masked by noise. Also, the number of modes with expected $s/n > 1$ gives an indication of the power of the experiment to constrain the sky signal. The co-polar data, which are sensitive to the temperature signal, have many more high $s/n$ modes than the cross-polar data, which measure polarized radiation. Within the context of the concordance model used to generate the $s/n$ eigenmode basis, we have sensitivity with an expected $s/n > 1$ to ~340 temperature (co-polar) modes versus ~34 polarization (cross-polar) modes.

## Detection of signal

Given that the data show remarkable consistency in $\chi^2$ tests of the difference data vectors, the $\chi^2$ values of the sum data vectors can be used to test for the presence of sky signal, independently of the likelihood analysis methods described below. In the co-polar data, all splits and subsets show highly significant $\chi^2$ values (PTE $< 1 \times 10^{-16}$, the precision to which we calculate the cumulative distribution function).

For the cross-polar data, the sum data vector $\chi^2$ values for the high $s/n$ subsets show high significance, with the PTE $< 1 \times 10^{-6}$ for all $s/n > 1$ subsets in Table 1. This simple and powerful test indicates that we have detected, with high significance, the presence of a polarized signal in the data, and that this signal is repeatable in all of the data splits. The polarization map shown in Fig. 1 gives a visual representation of this repeatable polarization signal. Shown are the epoch split sum and difference polarization maps, constructed using the same using the same 34 modes with $s/n > 1$ of the polarization data in the concordance model $s/n$ eigenmode basis that appear in Table 1. The sum map shows a repeatable polarized signal, while the difference map is consistent with instrument noise.

下，PTE 的值 $<1\times10^{-16}$ 即为零）。差值数据 $\chi^2$ 值检验的是数据中的系统效应，同时与数据值的和值对照检验是否存在可重复的天空信号。温度（共极）数据是可视性数据，来自两个接收器的起偏器处于左（LL 斯托克斯状态）或右（RR 斯托克斯状态）圆偏振状态；偏振（交叉偏振）数据是起偏器处于相反状态（LR 或 RL 斯托克斯状态）的数据。"$s/n>1$" 子集是 $s/n$ 本征模 $>1$ 的子集，而 $l$ 范围内的高 $s/n$ 子集是给定 $l$ 范围内最高的 20%的 $s/n$ 模式。有关数据划分类型和子集的进一步说明，请参阅"$\chi^2$ 检验"部分。我们计算了各种划分类型和子集的 296 个 $\chi^2$ 值，没有明显的趋势表明数据存在系统污染。

高 $s/n$ 模式的子集对差值数据矢量中的某些类别的系统效应更敏感，并且对和值数据矢量中预期的天空信号更敏感，否则它就有可能被噪声掩盖。此外，$s/n>1$ 预期的模式数量表示实验约束天空信号的能力。对温度信号敏感的共极数据比测量偏振辐射的交叉偏振数据的 $s/n$ 模式更高。根据生成 $s/n$ 基础本征模的一致性模型，我们预期约 340 种温度（共极）模式与约 34 种偏振（交叉偏振）模式具有 $s/n>1$ 的灵敏度。

## 信号检测

假设数据在差值数据矢量的 $\chi^2$ 检验中出现了显著的一致性，则可以使用和值数据矢量的 $\chi^2$ 值来检测天空信号的存在，而与下面描述的似然分析方法无关。在共极数据中，所有划分和子集显示高度显著的 $\chi^2$ 值（PTE $<1\times10^{-16}$，这个精度是我们计算累积分布函数的精度）。

对于交叉偏振数据，高 $s/n$ 子集的和值数据向量 $\chi^2$ 值显示出高显著性，表 1 中所有 $s/n>1$ 子集的 PTE $<1\times10^{-6}$。这个简单而强大的检测表示我们已经高度显著地检测到数据中存在偏振信号，并且该信号在所有数据拆分方法中都是可重复的。图 1 中所示的偏振图给出了该可重复偏振信号的视觉呈现，展示出了根据时期划分的和值与差值偏振图，使用相同的 34 种模式构建，其中 $s/n>1$ 的偏振数据在一致性模型 $s/n$ 基础本征模中，如表 1 所示。和值图显示了可重复的偏振信号，而差值图与仪器噪声一致。

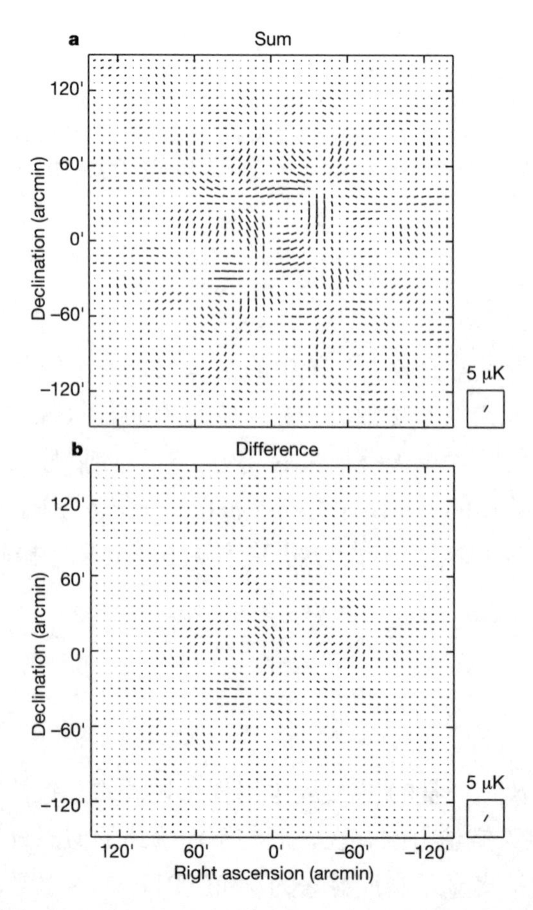


Fig. 1. Polarization maps formed from high signal/noise eigenmodes. Shown are maps constructed from polarized data sets that have been split by epoch, and formed into sum (**a**) or difference (**b**) data vectors, as reported in section "$\chi^2$ tests". In order to isolate the most significant signal in our data, we have used only the subset of 34 eigenmodes which, under the concordance model, are expected to have average signal/noise $(s/n) > 1$. Because the maps have only 34 independent modes, they exhibit a limited range of morphologies, and unlike conventional interferometer maps, these $s/n$ selected eigenmodes reflect the taper of the primary beam, even when no signal is present. This is visually apparent in the difference map (**b**), which is statistically consistent with noise. Comparison of the difference map to the sum map (**a**) illustrates a result also given numerically for this split/subset in Table 1: that these individual modes in the polarized data set show a significant signal.

It is possible to calculate a similar $\chi^2$ statistic for the data vector formed from the complete, unsplit data set. Combining all the data without the requirement of forming two equally weighted subsets should yield minimal noise, albeit without an exactly corresponding null (that is, difference) test. Recalculating the $s/n$ eigenmodes for this complete cross-polar data vector gives 36 modes with expected $s/n > 1$, for which $\chi^2 = 98.0$ with a PTE $= 1.2 \times 10^{-7}$. This significance is similar to those from the sum data vectors under the various splits, which actually divide the data fairly equally and so are nearly optimal. It should be noted that our focus so far has been to test the instrumental noise model, and we have not dealt with the small cross-polar signal expected as a result of the off-axis leakage. As noted in the "Off-axis leakage" subsection, it is not possible to correct the data elements directly for this effect,

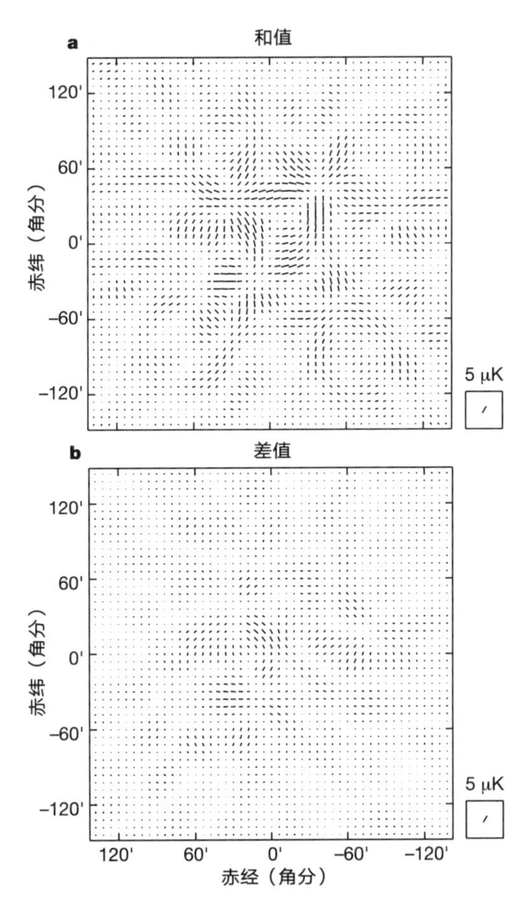


图1. 根据高信号/噪声本征模绘制的偏振图。图中显示了由已经按时期划分的偏振数据集所构建的映射，并且生成了和值(**a**)或差值(**b**)数据矢量，如"$\chi^2$检验"部分中所报告的。为了分离数据中最重要的信号，我们只使用了34个本征模的子集，在协调模型下，预计平均信号/噪声($s/n$) > 1。因为这个图只有34个独立模式，它们表现的形态范围是有限的，并且与传统的干涉仪图不同，即使没有信号存在，这些$s/n$选择的本征模也反映了初级光束的锥度。这在差值图(**b**)中表现得很明显，其在统计上与噪声一致。差值图与和值图(**a**)的比较阐释了表1对于该划分/子集数值给出的结果：偏振数据集中的这些独立模式表现为显著的信号。

从一个完整的未划分数据集形成的数据矢量计算出相似的$\chi^2$统计量是可能的。没有形成两个同等加权的子集的情况下，组合所有数据应该产生最小的噪声，尽管没有完全对应的零(即差值)检测。重新计算该完整交叉偏振数据矢量的$s/n$本征模，得到36个预期$s/n>1$的模式，其中$\chi^2=98.0$，PTE = $1.2\times10^{-7}$。这种显著性类似于那些来自各种划分下的和值数据矢量，其实际上相当均等地划分了数据，因此这几乎是最佳的方式。应该指出的是，到目前为止我们的重点是已经检测了的仪器噪声模型，而我们还没有处理能预期到的由于离轴渗漏而产生的小的交叉偏振信号。如"离轴渗漏"部分中所述，不可能直接针对此效应校正数据元素，但我们可以通过计

but we can account for it in calculating these $\chi^2$ results by including the expected covariance of this leakage signal (see "Theory covariance matrix" subsection) in the noise matrix $C_N$. Again recalculating the $s/n$ eigenmodes, we find 34 cross-polar modes with $s/n > 1$ which give a $\chi^2 = 97.0$ and a PTE $= 5.7 \times 10^{-8}$, a significance similar to before. The off-axis leakage is also included in the likelihood analyses, where it is again found to have an insignificant impact on the results.

The likelihood analysis described in the following sections makes use of all of the information in our data set. Such an analysis, in principle, may yield statistically significant evidence of a signal even in cases of data sets for which it is not possible to isolate any individual modes which have average $s/n > 1$. However, the existence of such modes in our data set, which has resulted from our strategy of integrating deeply on a limited patch of sky, allows us to determine the presence of the signal with the very simple analysis described above. It also reduces sensitivity to the noise model estimation in the likelihood results that we report next. Finally, it gives our data set greater power to exclude the possibility of no signal than it might have had if we had observed more modes but with less $s/n$ in each.

## Likelihood Analysis Formalism

The preceding section gives strong evidence for the presence of a signal in our polarization data. We now examine that signal using the standard tool of likelihood analysis. In such an analysis, the covariance of the signal, $C_T(\kappa)$, is modelled in terms of parameters $\kappa$ appropriate for describing the temperature and polarization anisotropy of the CMB. The covariance of the data vector is modelled $C(\kappa) \equiv C_T(\kappa) + C_N$; where $C_N$ is the noise covariance matrix. Given our data vector $\mathbf{\Delta}$, the likelihood of the model specified by the parameter vector $\kappa$ is the probability of our data vector given that model:

$$L(\kappa) = P(\mathbf{\Delta}|\kappa) \propto \det(C(\kappa))^{-1/2} \exp(-\frac{1}{2} \mathbf{\Delta}^t C(\kappa)^{-1} \mathbf{\Delta}) \quad (1)$$

Although the full likelihood function itself is the most basic result of the likelihood analysis, it is useful to identify and report the values of the parameters that maximize the likelihood (so-called maximum likelihood (ML) estimators). Uncertainties in the parameter values can be estimated by characterizing the shape of the likelihood surface, as discussed further in the "Reporting of likelihood results" subsection.

### The CMB power spectra

The temperature and polarization anisotropy of the CMB can be characterized statistically by six angular power spectra: three that give the amplitudes of temperature, $E$-mode and $B$-mode polarization anisotropy as a function of angular scale, and three that describe correlations between them. These spectra are written $C_l^X$, with $X = \{T, E, B, TE, TB, EB\}$. In our likelihood analyses, we choose various parameterizations of these spectra to constrain.

算这些噪声矩阵 $C_N$ 中渗漏信号的预期协方差来计算 $\chi^2$ 结果（见“理论协方差矩阵”部分）。接着重新计算 $s/n$ 本征模，我们发现 34 个交叉偏振模式，其中 $s/n > 1$，其给出 $\chi^2 = 97.0$ 及 PTE $= 5.7 \times 10^{-8}$，具有与之前类似的显著性。离轴渗漏也包括在似然分析中，而且再次发现它对结果的影响不大。

以下部分中描述的似然分析使用了我们数据集中的所有信息。即使对于整个数据可能无法分离出任何平均信噪比 $s/n > 1$ 的独立模式，但是原则上这样的分析还是可能得到一个信号具有统计学显著性的证据。但是，在我们数据集当中这种模式的存在，是我们在天空的有限区域中做深度的积分策略造成的，这使我们能够通过上述非常简单的分析来确定信号的存在。它还降低了我们接下来报告的结果中对噪声模型估计敏感的可能性。最后，它使我们的数据集可以更好地排除那些没有信号的可能性，而不是排除那些如果我们观测更多的模且在每个模中具有更小的 $s/n$ 就可以观测到的情况。

### 似然分析形式化

上一节为我们的偏振数据中存在信号提供了有力证据。我们现在使用似然分析这一标准方法来检查信号。在这样的分析中，信号的协方差 $C_T(\kappa)$ 根据适合于描述 CMB 的温度和各向异性偏振的参数 $\kappa$ 来建模。建模数据矢量的协方差为 $C(\kappa) \equiv C_T(\kappa) + C_N$；其中 $C_N$ 是噪声协方差矩阵。考虑到我们的数据矢量为 $\Delta$，由参数矢量 $\kappa$ 给出的模型的似然函数是该模型的数据矢量的概率：

$$L(\kappa) = P(\Delta|\kappa) \propto \det(C(\kappa))^{-1/2}\exp(-\frac{1}{2}\Delta^t C(\kappa)^{-1}\Delta) \tag{1}$$

虽然完整的似然函数本身是似然分析的最基本结果，但确定和报告最大似然函数的参数值（所谓的最大似然（ML）估计量）是很有用的。参数值的不确定性可以通过似然函数的边缘概率函数来估计，“似然函数结果报告”部分中会有进一步的讨论。

## CMB 功率谱

CMB 的温度和偏振各向异性可以通过六个角功率谱来统计表征：其中三个是温度幅度、$E$ 模式和 $B$ 模式各向异性偏振作为角度尺度的函数，另外三个是它们之间的相关性的描述。这些谱写成 $C_l^X$, $X = \{T, E, B, TE, TB, EB\}$。在似然分析中，我们选择这些功率谱的不同参数化来进行约束。

For a given cosmological model, these spectra can be readily calculated using efficient, publicly available Boltzmann codes[54]. Details of how to define these spectra in terms of all-sky multipole expansions of the temperature and linear polarization of the CMB radiation field are available in the literature (see refs 15 and 16). For DASI's 3.4° field of view, a flat sky approximation is appropriate[55], so that the spectra may be defined somewhat more simply[26]. In this approximation the temperature angular power spectrum is defined:

$$C_l^T \simeq C^T(u) \equiv \langle \frac{\widetilde{T}^*(\mathbf{u})\widetilde{T}(\mathbf{u})}{T_{\rm CMB}^2} \rangle \tag{2}$$

where $\widetilde{T}(\mathbf{u})$ is the Fourier transform of $T(\mathbf{x})$, $T_{\rm CMB}$ is the mean temperature of the CMB, and $l/2\pi = u$ gives the correspondence between multipole $l$ and Fourier radius $u = |\mathbf{u}|$. The order spectra in the flat sky approximation are similarly defined, for example, $C^{TE}(u) \equiv \langle \widetilde{T}^*(\mathbf{u})\widetilde{E}(\mathbf{u})/T_{\rm CNB}^2 \rangle$. The relationship between $\widetilde{E}$, $\widetilde{B}$ and the linear polarization Stokes parameters $Q$ and $U$ is:

$$\begin{aligned} \widetilde{Q}(\mathbf{u}) &= \cos(2\chi)\,\widetilde{E}(\mathbf{u}) - \sin(2\chi)\,\widetilde{B}(\mathbf{u}) \\ \widetilde{U}(\mathbf{u}) &= \sin(2\chi)\,\widetilde{E}(\mathbf{u}) - \cos(2\chi)\,\widetilde{B}(\mathbf{u}) \end{aligned} \tag{3}$$

where $\chi = \arg(\mathbf{u})$ and the polarization orientation angle defining $Q$ and $U$ are both measured on the sky from north through east.

## Theory covariance matrix

The theory covariance matrix is the expected covariance of the signal component of the data vector, $\mathbf{C}_T \equiv \langle \mathbf{s}\mathbf{s}^t \rangle$. The signals measured by the visibilities in our data vector for a given baseline $\mathbf{u}_i$ (after calibration and leakage correction) are:

$$\begin{aligned} V^{\rm RR}(\mathbf{u}_i) &= \alpha_i \int d\mathbf{x}\, A(\mathbf{x}, \nu_i)[T(\mathbf{x}) + V(\mathbf{x})]e^{-2\pi i \mathbf{u}_i \cdot \mathbf{x}} \\ V^{\rm LL}(\mathbf{u}_i) &= \alpha_i \int d\mathbf{x}\, A(\mathbf{x}, \nu_i)[T(\mathbf{x}) - V(\mathbf{x})]e^{-2\pi i \mathbf{u}_i \cdot \mathbf{x}} \\ V^{\rm RL}(\mathbf{u}_i) &= \alpha_i \int d\mathbf{x}\, A(\mathbf{x}, \nu_i)[Q(\mathbf{x}) + iU(\mathbf{x})]e^{-2\pi i \mathbf{u}_i \cdot \mathbf{x}} \\ V^{\rm LR}(\mathbf{u}_i) &= \alpha_i \int d\mathbf{x}\, A(\mathbf{x}, \nu_i)[Q(\mathbf{x}) - iU(\mathbf{x})]e^{-2\pi i \mathbf{u}_i \cdot \mathbf{x}} \end{aligned} \tag{4}$$

where $A(\mathbf{x}, \nu_i)$ specifies the beam power pattern at frequency $\nu_i$, $T(\mathbf{x})$, $Q(\mathbf{x})$, $U(\mathbf{x})$,and $V(\mathbf{x})$ are the four Stokes parameters in units of CMB temperature (μK), and $\alpha_i = \partial B_{\rm Plank}(\nu_i, T_{\rm CMB})/\partial T$ is the appropriate factor for converting from these units to flux density (Jy). The co-polar visibilities $V^{\rm RR}$ and $V^{\rm LL}$ are sensitive to the Fourier transform of the temperature signal $T(\mathbf{x})$ and circular polarization component $V(\mathbf{x})$ (expected to be zero). The cross-polar visibilities $V^{\rm RL}$ and $V^{\rm LR}$ are sensitive to the Fourier transform of the linear polarization components

对于给定的宇宙学模型，使用有效的公开可用的玻尔兹曼代码[54]可以很容易地计算这些功率谱。有关如何根据CMB辐射场的温度和线性偏振的全天多极展开来定义这些功率谱的细节可在文献中获得(参考文献15和16)。对于DASI的3.4°视场，平坦天空近似是合适的[55]，因此可以更简单地定义功率谱[26]。在该近似中，温度角功率谱定义为：

$$C_l^T \simeq C^T(u) \equiv \langle \frac{\widetilde{T}^*(\mathbf{u})\,\widetilde{T}(\mathbf{u})}{T_{\mathrm{CMB}}^2} \rangle \tag{2}$$

其中 $\widetilde{T}(\mathbf{u})$ 是 $T(\mathbf{x})$ 的傅里叶变换，$T_{\mathrm{CMB}}$ 是CMB的平均温度，$l/\pi = u$ 给出了多极矩 $l$ 和傅里叶半径 $u = |\mathbf{u}|$ 之间的对应关系。平坦天空的有序谱定义也是类似的，例如，$C^{TE}(u) \equiv \langle \widetilde{T}^*(\mathbf{u})\widetilde{E}(\mathbf{u})/T_{\mathrm{CNB}}^2 \rangle$。$\widetilde{E}$、$\widetilde{B}$，以及线性偏振斯托克斯参数 $Q$ 和 $U$ 之间的关系是：

$$\begin{aligned} \widetilde{Q}(\mathbf{u}) &= \cos(2\chi)\,\widetilde{E}(\mathbf{u}) - \sin(2\chi)\,\widetilde{B}(\mathbf{u}) \\ \widetilde{U}(\mathbf{u}) &= \sin(2\chi)\,\widetilde{E}(\mathbf{u}) - \cos(2\chi)\,\widetilde{B}(\mathbf{u}) \end{aligned} \tag{3}$$

其中用来定义 $Q$ 和 $U$ 的 $\chi = \arg(\mathbf{u})$ 与偏振方向角都是从天空上由北到东测量的。

### 理论协方差矩阵

理论协方差矩阵是数据矢量的信号分量的预期协方差，$\mathbf{C}_T \equiv \langle \mathbf{s}\mathbf{s}^t \rangle$。对于给定的基线 $\mathbf{u}_i$，我们的数据矢量中可视性测量的信号(在校准和渗漏校正之后)是：

$$\begin{aligned} V^{\mathrm{RR}}(\mathbf{u}_i) &= \alpha_i \int \mathrm{d}\mathbf{x}\, A(\mathbf{x}, \nu_i)[T(\mathbf{x}) + V(\mathbf{x})]\mathrm{e}^{-2\pi i \mathbf{u}_i \cdot \mathbf{x}} \\ V^{\mathrm{LL}}(\mathbf{u}_i) &= \alpha_i \int \mathrm{d}\mathbf{x}\, A(\mathbf{x}, \nu_i)[T(\mathbf{x}) - V(\mathbf{x})]\mathrm{e}^{-2\pi i \mathbf{u}_i \cdot \mathbf{x}} \\ V^{\mathrm{RL}}(\mathbf{u}_i) &= \alpha_i \int \mathrm{d}\mathbf{x}\, A(\mathbf{x}, \nu_i)[Q(\mathbf{x}) + iU(\mathbf{x})]\mathrm{e}^{-2\pi i \mathbf{u}_i \cdot \mathbf{x}} \\ V^{\mathrm{LR}}(\mathbf{u}_i) &= \alpha_i \int \mathrm{d}\mathbf{x}\, A(\mathbf{x}, \nu_i)[Q(\mathbf{x}) - iU(\mathbf{x})]\mathrm{e}^{-2\pi i \mathbf{u}_i \cdot \mathbf{x}} \end{aligned} \tag{4}$$

其中 $A(\mathbf{x}, \nu_i)$ 是指定频率 $\nu_i$ 下的光束功率模式，$T(\mathbf{x})$、$Q(\mathbf{x})$、$U(\mathbf{x})$ 和 $V(\mathbf{x})$ 是以CMB温度(μK)为单位的四个斯托克斯参数，且 $\alpha_i = \partial B_{\mathrm{Plank}}(\nu_i, T_{\mathrm{CMB}})/\partial T$ 是从这些单位转换为通量密度(Jy)的适当因子。共极性 $V^{\mathrm{RR}}$ 和 $V^{\mathrm{LL}}$ 对温度信号 $T(\mathbf{x})$ 和圆偏振分量 $V(\mathbf{x})$(预期为零)的傅里叶变换敏感。交叉极性可视性 $V^{\mathrm{RL}}$ 和 $V^{\mathrm{LR}}$ 对线性偏振分量 $Q$ 和 $U$ 的傅里叶变换敏感。由等式(3)可以看出，可视性的成对组合是天空上几乎纯

$Q$ and $U$. Using equation (3), it can be seen that pairwise combinations of the visibilities are direct measures of nearly pure $T$, $E$ and $B$ Fourier modes on the sky, so that the data set easily lends itself to placing independent constraints on these power spectra.

We construct the theory covariance matrix as the sum of components for each parameter in the analysis:

$$C_{\mathrm{T}}(\kappa) = \sum_p \kappa_p B_{\mathrm{T}}^p \tag{5}$$

From equations (2)–(4), it is possible to derive a general expression for the matrix elements of a theory matrix component:

$$B_{\mathrm{T}ij}^p = \frac{1}{2}\alpha_i \alpha_j T_{\mathrm{CMB}}^2 \int \mathrm{d}\mathbf{u} C^X(u) \widetilde{A}(\mathbf{u}_i - \mathbf{u}, \nu_i) \times [\zeta_1 \widetilde{A}(\mathbf{u}_j - \mathbf{u}, \nu_j) + \zeta_2 \widetilde{A}(\mathbf{u}_j + \mathbf{u}, \nu_j)] \tag{6}$$

The coefficients $\zeta_1$ and $\zeta_2$ can take values $\{0, \pm 1, \pm 2\} \times \{\cos\{2\chi, 4\chi\}, \sin\{2\chi, 4\chi\}\}$ depending on the Stokes states (RR, LL, RL, LR) of each of the two baselines $i$ and $j$ and on which of the six spectra ($T$, $E$, $B$, $TE$, $TB$, $EB$) is specified by $X$. The integration may be limited to annular regions which correspond to $l$-ranges over which the power spectrum $C^X$ is hypothesized to be relatively flat, or else some shape of the spectrum may be postulated.

Potentially contaminated modes in the data vector may be effectively projected out using a constraint matrix formalism[53]. This formalism can be used to remove the effect of point sources of known position without knowledge of their flux densities, as we described in ref. 6. This procedure can be generalized to include the case of polarized point sources. Although we have tested for the presence of point sources in the polarization power spectra using this method, in the final analysis we use constraint matrices to project point sources out of the temperature data only, and not the polarization data (see "Point sources" subsection).

The off-axis leakage, discussed in the "Off-axis leakage" subsection and in detail in ref. 51, has the effect of mixing some power from the temperature signal $T$ into the cross-polar visibilities. Our moolel of the off-axis leakage allow us to write an expression for it analogous to equation (4), and to construct a corresponding theory covariance matrix component to account for it. In practice, this is a small effect, as discussed in the "Systematic uncertainties" section.

## Likelihood parameters

In the "Likelihood results" section we present the results from nine separate likelihood analyses involving the polarization data, the temperature data, or both. Our choice of parameters with which to characterize the six CMB power spectra is a compromise between maximizing sensitivity to the signal and constraining the shape of the power spectra. In

的 $T$、$E$ 和 $B$ 的傅里叶模的直接测量，这使得数据集很容易对这些功率谱设置独立的约束。

在分析中，我们将理论协方差矩阵构造为每个参数部分的总和：

$$C_{\mathrm{T}}(\kappa)=\sum_{p}\kappa_{p}\,B_{\mathrm{T}}^{P} \tag{5}$$

从等式(2)~(4)，可以推导出理论矩阵分量的矩阵元素的一般表达式：

$$B_{\mathrm{T}ij}^{p}=\frac{1}{2}\alpha_{i}\alpha_{j}T_{\mathrm{CMB}}^{2}\int \mathrm{d}\mathbf{u}C^{X}(u)\,\widetilde{A}\,(\mathbf{u}_{i}-\mathbf{u},\nu_{i})\times[\zeta_{1}\,\widetilde{A}\,(\mathbf{u}_{j}-\mathbf{u},\nu_{j})+\zeta_{2}\,\widetilde{A}\,(\mathbf{u}_{j}+\mathbf{u},\nu_{j})] \tag{6}$$

系数 $\zeta_1$ 和 $\zeta_2$ 可取 $\{0,\pm1,\pm2\}\times\{\cos\{2\chi,4\chi\},\sin\{2\chi,4\chi\}\}$，取决于两个基线 $i$ 和 $j$ 各自的斯托克斯状态(RR，LL，RL，LR)，同时取决于 $X$ 的谱($T, E, B, TE, TB, EB$)。积分可以限于环形区域，其对应于范围为 $l$ 的功率谱 $C^X$；$C^X$ 被假设为相对平坦的，或者可以假设为某种形状的功率谱。

可以使用约束矩阵形式[53]有效地投射出数据矢量中的潜在污染模式。这种形式可以用来消除已知位置，但不知道它们的通量密度的点源的影响，正如我们在参考文献 6 中所描述的那样。这个程序可以运用推广到包括偏振点源的情况中。虽然我们已经使用这种方法检测了偏振功率谱中点源的存在，但在最后的分析中，我们仅从温度数据中用约束矩阵投射点源，而不是用偏振数据(参见“点源”部分)。

正如在“离轴渗漏”部分中讨论过的，详见参考文献 51，离轴渗漏具有来自温度信号 $T$ 的一些功率混合到交叉偏振可视性中的效果。我们的离轴渗漏的模式允许我们为它写出类似于等式(4)的表达式，并构造相应的理论协方差矩阵分量来解释它。在实践中，这是一个很小的影响因素，如“系统不确定性”部分所述。

### 似然参数

在“似然结果”部分中，我们呈现了九组包括对偏振数据、温度数据或两者联合数据的不同的似然分析结果。对用于描述六张 CMB 功率谱的参数的选择，我们针对信号灵敏度最大化和约束功率谱形状这两方面进行折中。针对不同的分析，我们

the different analyses we either characterize various power spectra with a single amplitude parameter covering all angular scales, or split the $l$-range into five bands over which spectra are approximated as piecewise-flat, in units of $l(l+1)C_l/(2\pi)$. Five bands were chosen as a compromise between too many for the data to bear and too few to capture the shape of the underlying power spectra. The $l$-ranges of these five bands are based on those of the nine-band analysis we used in ref. 6; we have simply combined the first four pairs of these bands, and kept the ninth as before. In some analyses we also constrain the frequency spectral indices of the temperature and polarization power spectra as a test for foreground contamination.

The $l$-range to which DASI has non-zero sensitivity is $28 < l < 1{,}047$. That range includes the first three peaks of the temperature power spectrum, and within it the amplitude of that spectrum, which we express in units $l(l+1)C_l/(2\pi)$, varies by a factor of about 4. Over this same range, the $E$-mode polarization spectrum is predicted to have four peaks while rising roughly as $l^2$ (in the same units), varying in amplitude by nearly two orders of magnitude[17]. The $TE$ correlation is predicted to exhibit a complex spectrum that in fact crosses zero five times in this range.

For the single bandpower analyses which maximize our sensitivity to a potential signal, the shape of the model power spectrum assumed will have an effect on the sensitivity of the result. In particular, if the assumed shape is a poor fit to the true spectrum preferred by the data, the results will be both less powerful and difficult to interpret. For temperature spectrum measurements, the most common choice in recent years has been the so-called flat bandpower, $l(l+1)C_l \propto$ constant, which matches the gross large-angle "scale-invariant" power-law shape of that spectrum. Because of extreme variations predicted in the $E$ and $TE$ spectra over DASI's $l$-range, we do not expect a single flat bandpower parameterization to offer a good description of the entire data set (although in the "$E/B$ analysis" subsection we describe results of applying such an analysis to limited $l$-range subsets of data). A more appropriate definition of "flat bandpower" for polarization measurements sensitive to large ranges of $l < 1{,}000$ might be $C_l \propto$ constant (or $l(l+1)C_l \propto l^2$). Other shapes have been tried, notably the gaussian autocorrelation function (by the PIQUE group[56]) which reduces to $C_l \propto$ constant at large scales and perhaps offers a better fit to the gross shape of the predicted $E$ spectrum.

In our single band analyses, we have chosen a shape for our single bandpower parameters based on the predicted spectra for a cosmological model currently favoured by observations. The specific model that we choose—which we will call the concordance model—is a ΛCDM model with flat spatial curvature, 5% baryonic matter, 35% dark matter, 60% dark energy, and a Hubble constant of 65 km $s^{-1}Mpc^{-1}$, ($\Omega_b = 0.05$, $\Omega_{cdm} = 0.35$, $\Omega_\Lambda = 0.60$, $h = 0.65$) and the exact normalization $C_{10} = 700\ \mu K^2$. This concordance model was defined in ref. 50 as a good fit to the previous DASI temperature power spectrum and other observations. The concordance model spectra for $T$, $E$, and $TE$ are shown in Fig. 4. The five flat bandpower likelihood results shown in Fig. 4, and discussed in the next section, suggest that the concordance shaped spectra do indeed characterize the data better than any

或假设功率谱在全部的角尺度范围上具有相同的幅度参数，或将功率谱的多极矩 $l$ 分为 5 个频段并假设每一段的功率谱（以 $l(l+1)C_l/(2\pi)$ 为单位）基本平坦。之所以选择 5 个频段，是考虑到若分段太多，则需加载的数据太多；若分段太少，则无法捕捉到功率谱的基本形状。这五个频段的 $l$ 的范围是基于我们在参考文献 6 用过的九频段分析。这里，我们只是将这九个频段的前八个频段按四对合并在一起，并保持第九个频段不变。在某些分析中，我们会约束温度和偏振功率谱的谱指数以检验前景污染。

DASI 实验的多极矩探测范围为 $28 < l < 1,047$。这个范围包含了温度功率谱的前三个震荡峰，其中谱的幅度（以 $l(l+1)C_l/(2\pi)$ 为单位）约为四。在同样的多极矩范围内，$E$ 模式偏振功率谱预计会包含四个峰，大致随 $l^2$（和温度谱相同的单位）上升而上升，其幅度的变化程度几乎达到两个量级[17]。$TE$ 相关功率谱预计会比较复杂，实际上，它在这段 $l$ 的范围内会有五次等于零。

对于单频段功率分析实现对一个潜在信号的灵敏度的最大化，其分析结果的灵敏度会受到假定的模型功率谱形状的影响，尤其当假设的形状和数据支持的真实功率谱的形状相差甚远时，结果就不太可靠且难以被解释。近几年来，人们普遍选择所谓的“平坦频段功率”方法，即 $l(l+1)\ C_l \propto$ 常数，来测量温度谱。“平坦频段功率”和温度谱的大角度“标度不变”幂律分布的形状相匹配。根据预测，在 DASI 的 $l$ 范围内 $E$ 和 $TE$ 功率谱变化剧烈，因此我们并不指望一个单平坦频段功率参数化就能很好地描述整个数据集（尽管在“$E/B$ 分析”部分中，我们所描述的结果就是将这种分析方法用于有限 $l$ 范围的数据子集给出的）。对偏振测量敏感的范围（$l < 1,000$），更准确的“平坦频段功率”的定义可以是 $C_l \propto$ 常数（或者 $l(l+1)\ C_l \propto l^2$）。其他已经被测试过的形状，特别是高斯自相关函数（被 PIQUE[56] 组使用过），在大尺度上近似为 $C_l \propto$ 常数，或许会更符合 $E$ 谱的总体形状。

作单频段分析的时候，我们会根据目前观测所支持的宇宙学模型预测的功率谱，为单频带功率参数选定一个形状。被选择的特定模型——我们称之为协调模型——是一个 ΛCDM 模型，该模型具有平坦的空间曲率、5% 重子物质、35% 暗物质、60% 暗能量、65 km · s$^{-1}$ · Mpc$^{-1}$ 的哈勃常数（即 $\Omega_b = 0.05$，$\Omega_{cdm} = 0.35$，$\Omega_\Lambda = 0.60$，$h = 0.65$），以及确定的归一化 $C_{10} = 700\ \mu K^2$。这个协调模型定义在参考文献 50 中，与先前 DASI 温度功率谱和其他观测匹配得很好。图 4 展示了协调模型的 $T$、$E$ 和 $TE$ 功率谱。五个平坦频段功率似然结果（参见图 4）表明协调模型给出的形状的确比任何幂律分布近似都更符合数据，我们会在下一节对该结果进行讨论。在“$E/B$ 分

power-law approximation. In the "$E/B$ analysis" subsection, we explicitly test the likelihood of the concordance model parameterization against that of the two power laws mentioned above, and find that the concordance model shape is strongly preferred by the data.

It should be noted that the likelihood analysis is always model dependent, regardless of whether a flat or shaped model is chosen for parameterization. To evaluate the expectation value of the results for a hypothesized theoretical power spectrum, we must use window functions appropriate for the parameters of the particular analysis. The calculation of such parameter window functions has previously been described, both generally[57,58], and with specific reference to polarization spectra[59]. In general, the parameter window function has a non-trivial shape (even for a flat bandpower analysis) which is dependent on the shape of the true spectra as well as the intrinsic sensitivity of the instrument as a function of angular scale. Parameter window functions for the $E/B$ and $E5/B5$ polarization analysis are shown in Fig. 2, and will also be made available on our website (http://astro.uchicago.edu/dasi).

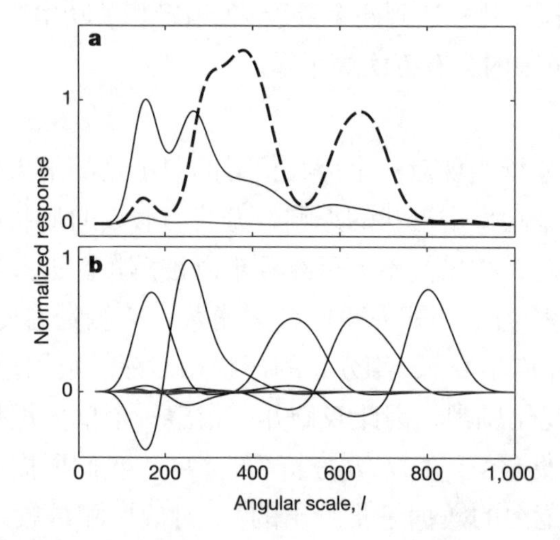


Fig. 2. Parameter window functions, which indicate the angular scales over which the parameters in our analyses constrain the power spectra. **a**, The functions for the $E$ parameter of our $E/B$ analysis, with the solid blue curve indicating response to the $E$ power spectrum and the solid red (much lower) curve indicating response of the same $E$ parameter to the $B$ spectrum. The blue dashed curve shows the result of multiplying the $E$ window function by the concordance $E$ spectrum, illustrating that for this CMB spectrum, most of the response of our experiment's $E$ parameter comes from the region of the second peak ($300 \lesssim l \lesssim 450$), with a substantial contribution also from the third peak and a smaller contribution from the first. **b**, The $E1$ to $E5$ parameter window functions for the $E$ power spectrum (blue) and $B$ power spectrum (red, again much lower) from our $E5/B5$ analysis. All of these window functions are calculated with respect to the concordance model discussed in the text. DASI's response to $E$ and $B$ is very symmetric, so that $E$ and $B$ parameter window functions which are calculated with respect to a model for which $E = B$ are nearly identical, with the $E$ and $B$ spectral response reversed.

## Likelihood evaluation

Prior to likelihood analysis, the data vector and the covariance matrices can be compressed

析”部分，我们会有明确地检测以比较协调模型参数化和前面提到过的两种幂律参数化的似然。结果表明，数据强烈支持协调模型给出的形状。

需要注意的是，不管采用平坦模型或者给定成形模型来参数化，似然分析总是依赖于模型的。对于一个假设的理论功率谱，为了估计其分析结果的期望值，我们必须采用适用于这个特定分析的参数的窗口函数。这种参数窗口函数的计算方法曾经被概括地描述过[57,58]，也有特别针对偏振功率谱的文献[59]。一般而言，参数窗口函数具有非平凡的形状(即使对平坦频段功率分析也是如此)，其形状依赖于真实功率谱的形状和仪器的内秉灵敏度，而后者是一个关于角度的函数。$E/B$、$E5/B5$ 偏振分析的参数窗口函数如图 2 所示，也可以在我们的网站(http://astro.uchicago.edu/dasi)上找到。

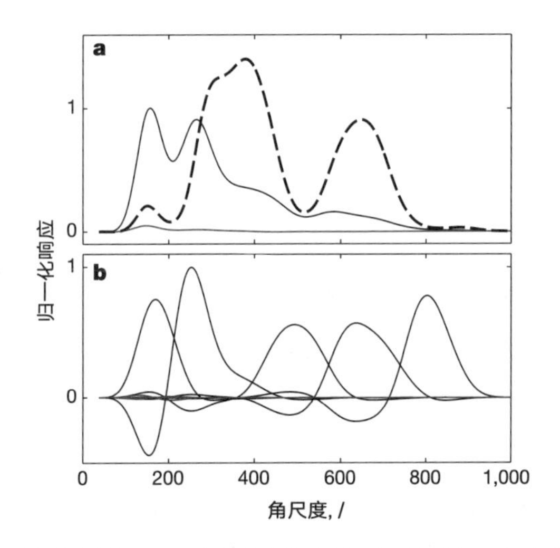


图 2. 参数窗口函数，它标示我们分析中的参数约束的是哪一段角度范围上的功率谱。**a**，$E/B$ 分析的 $E$ 参数窗口函数。其中蓝色实线表示对 $E$ 功率谱的响应，红色实线(位置低得多)表示相同的 $E$ 参数对 $B$ 功率谱的响应。蓝色虚线表示的是将 $E$ 窗口函数和协调模型 $E$ 谱相乘的结果，这表明对于此 CMB 功率谱，我们实验的 $E$ 参数的响应大部分来自于第二峰的对应的 $l$ 区域($300 \leqslant l \leqslant 450$)，也有相当一部分来自第三峰，第一峰的贡献更小些。**b**，$E$ 功率谱(蓝色)、$B$ 功率谱(红色，同样位置但低很多)的 $E1$ 到 $E5$ 参数窗口函数。所有这些窗口函数都是根据文中提到过的协调模型计算的。DASI 对 $E$ 和 $B$ 模式的响应非常对称，因此根据 $E=B$ 的模型计算给出的 $E$ 和 $B$ 参数窗口函数基本相同，而 $E$ 和 $B$ 谱响应是相反的。

## 似然评估

做似然分析之前，通过合并 $(u, v)$ 平面上邻近的、信号高度相关的数据点，我

by combining visibility data from nearby points in the $(u, v)$ plane, where the signal is highly correlated. This reduces the computational time required for the analyses without a significant loss of information about the signal. All analyses were run on standard desktop computers.

For each analysis, we use an iterated quadratic estimator technique to find the ML values of our parameters[53]. To further characterize the shape of the likelihood function, in ref. 6 we used an offset log-normal approximation. Here, for improved accuracy in calculating confidence intervals and likelihood ratios, we explicitly map out the likelihood function by evaluating equation (1) over a uniform parameter grid large enough to enclose all regions of substantial likelihood. A single likelihood evaluation typically takes several seconds, so this explicit grid evaluation is impractical for the analyses which include five or more parameters. For each analysis we also implement a Markov chain evaluation of the likelihood function[60]. We find this to be a useful and efficient tool for mapping the likelihoods of these high-dimensional parameter spaces in the region of substantial likelihood. We have compared the Markov technique to the grid evaluation for the lower-dimensional analyses and found the results to be in excellent agreement. In all cases, the peak of the full likelihood evaluated with either technique is confirmed to coincide with the ML values returned by the iterated quadratic estimator.

## Simulations and parameter recovery tests

The likelihood analysis software was extensively tested through analysis of simulated data. The analysis software and data simulation software were independently authored, as a check for potential coding errors.

Simulated sky maps were generated from realizations of a variety of smooth CMB power spectra, including both the concordance spectrum and various non-concordance models, both with and without $E$ and $B$ polarization and $TE$ correlations. Independent realizations of the sky were "observed" to construct simulated visibilities with Fourier-plane sampling identical to the real data. The simulations were designed to replicate the actual data as realistically as possible and include artefacts of the instrumental polarization response and calibration, such as the on-axis and off-axis leakages and the cross-polar phase offset, described in the "Measuring polarization with DASI" section, allowing us to test the calibration and treatment of these effects implemented in the analysis software.

Each of the analyses described in the "Likelihood results" section was performed on hundreds of these simulated data sets with independent realizations of sky and instrument noise, both with noise variances that matched the real data, and with noise a factor of ten lower. In all cases, we found that the means of the ML estimators recovered the expectation values $\langle \kappa_p \rangle$ of each parameter without evidence of bias, and that the variance of the ML estimators was found to be consistent with the estimated uncertainty given by $F^{-1}$ evaluated at $\langle \kappa \rangle$, where $F$ is the Fisher matrix.

们可以压缩数据矢量和协方差矩阵，实现在不显著丢失信号信息的同时，减少分析所需要的计算时间。所有分析计算均在标准台式计算机上运行。

对每一次的分析，我们使用一个迭代二次估计量寻找参数的最大似然(ML)值[53]。为进一步表征似然函数的形状，在参考文献6中，我们采用了一个偏移对数正态近似。这里，为提高置信区间和似然比的计算精度，我们在足够大以至覆盖所有实际可能区域的均匀参数网格上求解方程(1)以明确地画出似然函数。一次似然估计一般需要花费几秒钟，因此这种显式网格评估对于包含五个或更多参数的分析是不切实际的。对每次分析我们也用马尔可夫链求似然函数[60]，我们发现它是将这些高维度参数空间的似然函数在实际可能的区域内映射出来的一个高效且实用的工具。对于低维度分析，我们发现马尔可夫法和格点估计法的结果完全吻合。在所有情况下，这两种方法给出的似然函数的峰值和迭代二次估计量给出的最大似然值都相符。

**模拟和参数恢复检测**

我们用模拟数据对似然分析软件进行了大量检测，分析软件和数据模拟软件被独立编写，以检测潜在的编程错误。

模拟的天图是通过各种光滑的CMB功率谱的实现生成的，功率谱包括协调模型的，各种非协调模型的，包括或者不包括$E$、$B$模式偏振和$TE$相关性的。我们通过和真实数据同样的傅里叶平面采样“观测”独立的天图的实现来构造模拟的可视度。模拟旨在尽可能真实地重复实际数据，并加入仪器偏振响应和校准的人造信号，比如“用DASI测量偏振”部分中讲到过的共轴渗漏、离轴渗漏和交叉偏振相位偏移效应，这样做使我们可以检测分析软件执行过程中对这些效应的校准和处理。

“似然结果”部分中描述的每个分析都是在数百个模拟数据集上进行的，模拟数据都由独立的天图实现和仪器误差给出，两者的噪声方差都和实际数据相匹配，其中噪声低于信号的十分之一。我们发现，在任何情况下，最大似然估计量给出的均值和每个参数的均值$\langle\kappa_p\rangle$相比均无明显偏差，并且最大似然估计量的方差和用$F^{-1}$方法在$\langle\kappa\rangle$处估算给出的不确定度一致，这里，$F$是费希尔矩阵。

## Reporting of likelihood results

Maximum likelihood (ML) parameter estimates reported in this paper are the global maxima of the multidimensional likelihood function. Confidence intervals for each parameter are determined by integrating (marginalizing) the likelihood function over the other parameters; the reported intervals are the equal-likelihood bounds which enclose 68% of this marginal likelihood distribution. This prescription corresponds to what is generally referred to as the highest posterior density (HPD) interval. When calculating these intervals we consider all parameter values, including non-physical ones, because our aim is simply to summarize the shape of the likelihood function. Values are quoted in the text using the convention "ML (HPD-low to HPD-high)" to make clear that the confidence range is not directly related to the maximum likelihood value.

In the tabulated results, we also report marginalized uncertainties obtained by evaluating the Fisher matrix at the maximum likelihood model, that is, $(F^{-1})_{ii}^{1/2}$ for parameter $i$. Although in most cases, the two confidence intervals are quite similar, we regard the HPD interval as the primary result.

For parameters which are intrinsically positive we consider placing (physical) upper limits by marginalizing the likelihood distribution as before, but excluding the unphysical negative values. We then test whether the 95% integral point has a likelihood smaller than that at zero; if it does we report an upper limit rather than a confidence interval.

We also report the parameter correlation matrices for our various likelihood analyses to allow the reader to gauge the degree to which each parameter has been determined independently. The covariance matrix is the inverse of the Fisher matrix and the correlation matrix, $R$, is defined as the covariance matrix normalized to unity on the diagonal, that is, $C = F^{-1}$ and $R_{ij} = C_{ij}/\sqrt{C_{ii}C_{jj}}$.

## Goodness-of-fit tests

Using the likelihood function, we wish to determine if our results are consistent with a given model. For example, we would like to examine the significance of any detections by testing for the level of consistency with zero signal models, and we would like to determine if the polarization data are consistent with predictions of the standard cosmological model. We define as a goodness-of-fit statistic the logarithmic ratio of the maximum of the likelihood to its value for some model $\mathscr{H}_0$ described by parameters $\kappa_0$:

$$\Lambda(\mathscr{H}_0) \equiv -\log\left(\frac{\mathrm{L}(\kappa_{\mathrm{ML}})}{\mathrm{L}(\kappa_0)}\right)$$

The statistic $\Lambda$ simply indicates how much the likelihood has fallen from its peak value down

### 似然结果汇报

本文中报告的最大似然（ML）参数估计是多维似然函数的全局最大值。每个参数的置信区间是将似然函数对其他参数求积分（边缘化）确定的。本文报告的区间包含 68%边缘似然分布的等概率边界，这个区间一般被认为是最高后验概率密度（HPD）区间。我们的目标仅仅是给出似然函数的形状，因此在计算这个区间的时候，我们会考虑所有的参数值，包括非物理的参数。本文约定用“ML(HPD 低到 HPD 高)”的格式引述参数值，以此来说明置信区间和最大似然值之间并没有直接关系。

通过计算最大似然模型处的费希尔矩阵，即对参数 $i$ 求 $(F^{-1})_{ii}^{1/2}$，我们给出边缘化后的不确定度，也将其报告在结果列表中。虽然在大多数情况下，这两个置信区间非常相近，但我们采用 HPD 区间作为基本结果。

对于那些固有取正的参数，我们考虑通过如前所述的似然函数边缘化的方法来设置（物理）上界，消去非物理的负值。然后我们检查 95% 积分点的似然是否比参数值为零处的小。如果是，则报告一个上限而不是置信区间。

我们还报告了各个似然分析的参数相关矩阵，方便读者能够评估每个参数被独立测定的程度。协方差矩阵是费希尔矩阵的逆，把它的对角元素归一化后的矩阵被定义为相关矩阵 $R$，即 $C=F^{-1}$，$R_{ij}=C_{ij}/\sqrt{C_{ii}C_{jj}}$。

### 拟合优度检验

我们想使用似然函数来确定我们的结果是否支持一个给定的模型。比如说，我们想检查所有探测信号的显著性，则需要检测与零信号模型一致的程度。此外，我们想确定偏振数据是否和标准宇宙学模型预言的一致。我们定义了一个拟合优度统计量，它是似然函数的最大值和某些模型 $\mathscr{H}_0$ 的似然的对数比，记模型 $\mathscr{H}_0$ 的参数为 $\kappa_0$，则有：

$$\Lambda(\mathscr{H}_0) \equiv -\log\left(\frac{\mathrm{L}(\kappa_{\mathrm{ML}})}{\mathrm{L}(\kappa_0)}\right)$$

简单来说，统计量 $\Lambda$ 表明当参数从最大似然处的值变为 $\kappa_0$ 时，似然函数下降的程度。

to its value at $\kappa_0$. Large values indicate inconsistency of the likelihood result with the model $\mathcal{H}_0$. To assess significance, we perform Monte Carlo (MC) simulations of this statistic under the hypothesis that $\mathcal{H}_0$ is true. From this, we can determine the probability, given $\mathcal{H}_0$ true, to obtain a value of $\Lambda$ that exceeds the observed value, which we hereafter refer to as PTE.

When considering models which the data indicate to be very unlikely, sufficient sampling of the likelihood statistic becomes computationally prohibitive; our typical MC simulations are limited to only 1,000 realizations. In the limit that the parameter errors are normally distributed, our chosen statistic reduces to $\Lambda = \Delta\chi^2/2$. The integral over the $\chi^2$ distribution is described by an incomplete gamma function;

$$\mathrm{PTE} = \frac{1}{\Gamma(N/2)} \int_{\Lambda}^{\infty} \mathrm{e}^{-x} x^{\frac{N}{2}-1} \mathrm{d}x$$

where $\Gamma(x)$ is the complete gamma function, and $N$ is the number of parameters. Neither the likelihood function nor the distribution of the ML estimators is, in general, normally distributed, and therefore this approximation must be tested. In all cases where we can compute a meaningful PTE with MC simulations, we have done so and found the results to be in excellent agreement with the analytic approximation. Therefore, we are confident that adopting this approximation is justified. All results for PTE in this paper are calculated using this analytic expression unless otherwise stated.

## Likelihood Results

We have performed nine separate likelihood analyses to constrain various aspects of the signal in our polarization data, in our temperature data, or in both analysed together. The choice of parameters for these analyses and the conventions used for reporting likelihood results have been discussed in "Likelihood parameters" and "Reporting of likelihood results" subsections. Numerical results for the analyses described in this section are given in Tables 2, 3 and 4.

### Polarization data analyses and *E* and *B* results

***E*/*B* analysis.** The $E/B$ analysis uses two single-bandpower parameters to characterize the amplitudes of the $E$ and $B$ polarization spectra. As discussed in the "Likelihood parameters" subsection, this analysis requires a choice of shape for the spectra to be parameterized. DASI has instrumental sensitivity to $E$ and $B$ that is symmetrical and nearly independent. Although the $B$ spectrum is not expected to have the same shape as the $E$ spectrum, we choose the same shape for both spectra in order to make the analysis also symmetrical.

We first compute the likelihood using a $l(l+1)C_l/2\pi$ = constant bandpower (commonly referred to as "flat" bandpower) including data only from a limited $l$ range in which DASI

$\Lambda$ 值大，则暗示似然结果不支持模型 $\mathscr{H}_0$。为了评估显著性程度，我们在 $\mathscr{H}_0$ 为真的假设下对该统计量进行蒙特卡罗（MC）模拟。这样，我们可以求出当 $\mathscr{H}_0$ 为真的情况下得到一个比观测到的 $\Lambda$ 值更大的值的概率大小。我们后面称这个概率为 PTE。

考虑在数据十分不支持的模型的情况下，似然统计的充分抽样在计算上会变得难以承受，我们标准的 MC 模拟次数限制在仅 1,000 次实现。在参数误差是正态分布的情况下，我们选择的统计量 $\Lambda$ 简化为 $\Lambda = \Delta\chi^2/2$。对 $\chi^2$ 分布的积分可以描述为一个不完全伽马函数：

$$\mathrm{PTE} = \frac{1}{\Gamma(N/2)}\int_{\Lambda}^{\infty} \mathrm{e}^{-x} x^{\frac{N}{2}-1}\mathrm{d}x$$

其中 $\Gamma(x)$ 是完全伽马函数，$N$ 是参数的数目。由于似然函数和最大似然估计量的分布一般都不是正态分布，我们还需要验算这个近似是否成立。任何情况下，我们都可以用 MC 模拟计算出一个有意义的 PTE，这样处理后，我们发现 MC 给出的结果和解析近似的结果非常一致。因此我们认为这个近似是可信的。若无特殊说明，本文中 PTE 的结果都是用这个解析表达式计算得到的。

## 似然结果

我们已经进行了九次不同的似然分析，对偏振数据、温度数据或两者的联合数据的各个方面作出约束。我们已经在“似然参数”和“似然结果报告”部分中对这些分析的参数的选择和报告似然结果所使用的约定进行了讨论。本节讨论似然分析的数值结果并在表 2、3 和 4 中给出。

### 偏振数据分析和 $E$、$B$ 结果

**$E/B$ 分析** $E/B$ 分析用到两个单频带功率参数描述 $E$ 和 $B$ 偏振谱的幅度。如同“似然参数”部分所讨论过的，该分析需选择要被参数化的功率谱的形状。DASI 对 $E$ 和 $B$ 的仪器灵敏度是对称的且几乎独立的。虽然 $B$ 功率谱的形状预计和 $E$ 功率谱的不同，但为使该分析是对称的，我们还是为这两种功率谱选择了相同的形状。

我们首先用 $l(l+1)C_l/2\pi$ = 常数频段功率（一般称为“平坦频段功率”）来计算似然函数，数据仅包括来自 DASI 且具有高灵敏度的有限的 $l$ 范围上的数据，见图 2。利

has high sensitivity; see Fig. 2. Using the range $300 < l < 450$ which includes 24% of the complete data set, we find the ML at flat bandpower values $E = 26.5\ \mu K^2$ and $B = 0.8\ \mu K^2$. The likelihood falls dramatically for the zero polarization "nopol" model $E = 0$, $B = 0$. Marginalizing over $B$, we find $\Lambda(E = 0) = 16.9$ which, assuming the uncertainties are normally distributed, corresponds to a PTE of $5.9 \times 10^{-9}$ or a significance of $E$ detection of $5.8\sigma$. As expected, changing the $l$ range affects the maximum likelihood values and the confidence of detection; for example, shifting the centre of the above $l$ range by $\pm 25$ reduces the confidence of detection to $5.6\sigma$ and $4.8\sigma$, respectively. Clearly it is desirable to perform the analysis over the entire $l$ range sampled by DASI using a well motivated bandpower shape for the parameterization.

We considered three *a priori* shapes to check which is most appropriate for our data: the concordance $E$ spectrum shape (as defined in "Likelihood parameters"), and two power law alternatives, $l(l+1)C_l \propto$ constant (flat) and $l(l+1)C_l \propto l^2$. For each of these three cases, the point at $E = 0$, $B = 0$ corresponds to the same zero-polarization "nopol" model, so that the likelihood ratios $\Lambda$(nopol) may be compared directly to assess the relative likelihoods of the best-fit models. For the $l(l+1)C_l \propto$ constant case, the ML flat bandpower values are $E = 6.8\ \mu K^2$ and $B = -0.4\ \mu K^2$, with $\Lambda(\text{nopol}) = 4.34$. For the $l(l+1)C_l \propto l^2$ case, the ML values are $E = 5.1\ \mu K^2$ and $B = 1.2\ \mu K^2$ (for $l(l+1)C_l/2\pi$) at $l = 300$, with $\Lambda(\text{nopol}) = 8.48$. For the concordance shape, the ML values are $E = 0.80$ and $B = 0.21$ in units of the concordance $E$ spectrum amplitude, with $\Lambda(\text{nopol}) = 13.76$. The likelihood of the best-fit model in the concordance case is a factor of 200 and 12,000 higher than those of the $1(l+1)C_l \propto l^2$ and $l(l+1)C_l \propto$ constant cases, respectively, and so compared to the concordance shape either of these is a very poor model for the data. The data clearly prefer the concordance shape, which we therefore use for the $E/B$ and other single bandpower analyses presented in our results tables.

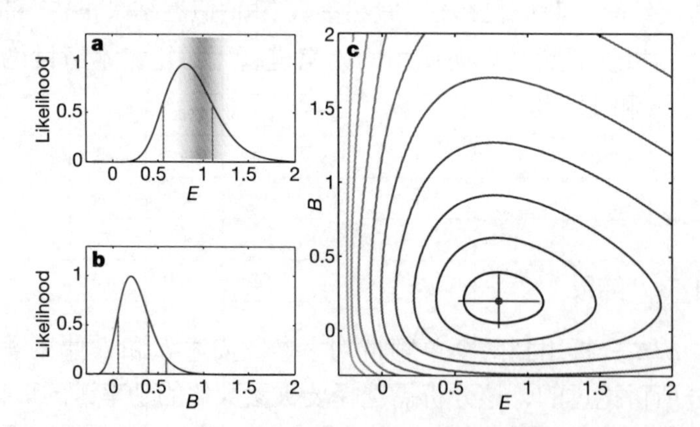


Fig. 3. Results from the two-parameter shaped bandpower $E/B$ polarization analysis. An $E$-mode power spectrum shape as predicted for the concordance model is assumed, and the units of amplitude are relative to that model. The same shape is assumed for the $B$-mode spectrum. **c**, The point shows the maximum likelihood value with the cross indicating Fisher matrix errors. Likelihood contours are placed at levels $\exp(-n^2/2)$, $n = 1,2...$, relative to the maximum, that is, for a normal distribution, the extrema of these contours along either dimension would give the marginalized $n$-sigma interval. **a**, **b**, The corresponding single parameter likelihood distributions marginalized over the other parameter. Note

用 $300 < l < 450$（含完整数据集的24%）的数据，我们发现平坦频段功率的最大似然值为 $E = 26.5\ \mu K^2$，$B = 0.8\ \mu K^2$。零偏振（$E = 0$，$B = 0$）的"无偏振"模型的似然值急剧下降。把 $B$ 边缘化后，我们得到 $\Lambda(E = 0) = 16.9$。若假设误差呈正态分布，则该值对应的PTE为 $5.9 \times 10^{-9}$，或者说探测到 $E$ 的显著程度为 $5.8\sigma$。正如我们预计的，改变 $l$ 范围会改变最大似然值和探测的显著水平。举个例子，把上面的 $l$ 范围中心值移动 $\pm25$，则探测的显著水平分别下降至 $5.6\sigma$ 和 $4.8\sigma$。采用一个合理的频段功率形状进行参数化，并在DASI采样的整个 $l$ 范围内做这样的分析，这显然是我们所希望实现的。

我们考虑了三个先验形状来检查哪个最适合我们的数据，它们分别是协调模型的 $E$ 谱（定义在"似然参数"部分中）、$l(l+1)C_l \propto$ 常数（平坦）和 $l(l+1)C_l \propto l^2$。后两者为幂律分布。对这三种情况，$E = 0$，$B = 0$ 的点都对应偏振为零的"无偏振"模型，因此直接比较似然比 $\Lambda$（无偏振）就可以估计最佳拟合模型的相对似然值。对 $l(l+1)C_l \propto$ 常数的情况，其在 $l = 300$ 处的最大似然值为 $E = 6.8\ \mu K^2$，$B = -0.4\ \mu K^2$，$\Lambda$（无偏振）$= 4.34$；对 $l(l+1)C_l \propto l^2$ 模型，同样在 $l = 300$ 处的最大似然值是 $E = 5.1\ \mu K^2$，$B = 1.2\ \mu K^2$，$\Lambda$（无偏振）$= 8.48$。对于协调模型形状，取协调模型 $E$ 谱幅度为单位幅度，ML值为 $E = 0.8$，$B = 0.21$，$\Lambda$（无偏振）$= 13.76$。协调模型最佳拟合下的似然比模型 $(l+1)C_l \propto l^2$ 和 $l(l+1)C_l \propto$ 常数分别增大为原来的200倍和12,000倍，因此比起协调形状，其他两个幂律模型都很糟糕。数据明显支持协调模型形状，故我们将此模型用于那些展示在结果列表中的 $E/B$ 及其他单频段功率的分析。

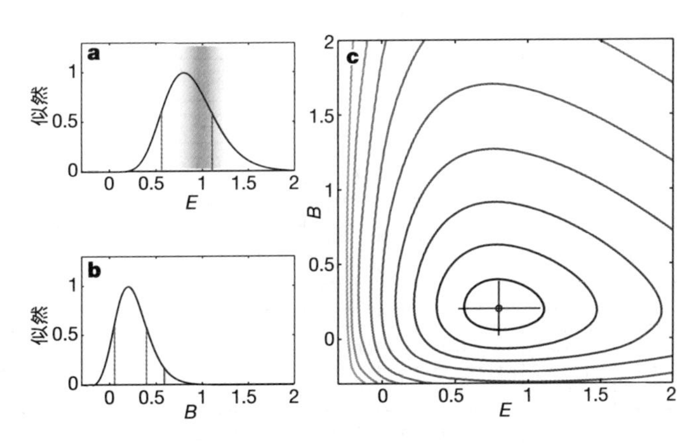


图3. 双参数成形频段功率 $E/B$ 偏振分析结果。假设 $E$ 模式功率谱形状由协调模型给出，其幅度的单位是相对于该模型的幅度，假设 $B$ 模式功率谱也具有相同的形状。**c**，图中的点表示最大似然值，十字表示费希尔矩阵误差。似然等高线设置为最大似然值的 $\exp(-n^2/2)$ 水平上，其中 $n = 1$，$2\cdots$。对于一个正态分布，沿着任何一个方向，这些等高线的极值就是边缘化的 $n$-$\sigma$ 区间。**a** 与 **b**，单参数似然分布（将其余参数边缘化）图。注意似然函数朝着低功率值快速下降，这种似然函数形状（和 $\chi^2$ 分布类似）是那

the steep fall in likelihood toward low power values; this likelihood shape (similar to a $\chi^2$ distribution) is typical for positive-definite parameters for which a limited number of high $s/n$ modes are relevant. The grey lines enclose 68% of the total likelihood. The red line indicates the 95% confidence upper limit on $B$-mode power. The green band shows the distribution of $E$ expectation values for a large grid of cosmological models weighted by the likelihood of those models given our previous temperature result (see ref. 50).

Figure 3 illustrates the result of the $E/B$ concordance shape polarization analysis. The maximum likelihood value of $E$ is 0.80 (0.56 to 1.10 at 68% confidence). For $B$, the result should clearly be regarded as an upper limit; 95% of the $B > 0$ likelihood (marginalized over $E$) lies below 0.59.

Figure 2a shows the parameter window functions relevant for this analysis. Note that the $E$ parameter has very little sensitivity to $B$ and vice versa—the purity with which DASI can separate these is remarkable. This is also demonstrated by the low correlation (−0.046) between the $E$ and $B$ parameters (see Table 2).

Table 2. Results of likelihood analyses from polarization data

| Analysis | Parameter | $l_{low}$–$l_{high}$ | ML est. | 68% interval | | | | |
|---|---|---|---|---|---|---|---|---|
| | | | | $(F^{-1})_{ii}^{1/2}$ error | Lower | Upper | UL(95%) | Units |
| $E/B$ | $E$ | – | 0.80 | ±0.28 | 0.56 | 1.10 | – | Fraction of concordance $E$ |
| | $B$ | – | 0.21 | ±0.18 | 0.05 | 0.40 | 0.59 | Fraction of concordance $E$ |
| $E5/B5$ | $E1$ | 28–245 | -0.50 | ±0.8 | -1.20 | 1.45 | 2.38 | μK² |
| | $E2$ | 246–420 | 17.1 | ±6.3 | 11.3 | 31.2 | – | μK² |
| | $E3$ | 421–596 | -2.7 | ±5.2 | -10.0 | 4.3 | 24.9 | μK² |
| | $E4$ | 597–772 | 17.5 | ±16.0 | 3.8 | 40.3 | 47.2 | μK² |
| | $E5$ | 773–1,050 | 11.4 | ±49.0 | -32.5 | 92.3 | 213.2 | μK² |
| | $B1$ | 28–245 | -0.65 | ±0.65 | -1.35 | 0.52 | 1.63 | μK² |
| | $B2$ | 246–420 | 1.3 | ±2.4 | -0.7 | 5.0 | 10.0 | μK² |
| | $B3$ | 421–596 | 4.8 | ±6.5 | -0.6 | 13.5 | 17.2 | μK² |
| $E5/B5$ | $B4$ | 597–772 | 13.0 | ±14.9 | 1.6 | 31.0 | 49.1 | μK² |
| | $B5$ | 773–1,050 | -54.0 | ±28.9 | -77.7 | -4.4 | 147.4 | μK² |
| $E/\beta_E$ | $E$ | – | 0.84 | ±0.28 | 0.55 | 1.08 | – | Fraction of concordance $E$ |
| | $\beta_E$ | – | 0.17 | ±1.96 | -1.63 | 1.92 | – | Temperature spectral index |
| Scalar/Tensor | $S$ | – | 0.87 | ±0.29 | 0.62 | 1.18 | – | Fraction of concordance $S$ |
| | $T$ | – | -14.3 | ±7.5 | -20.4 | -3.9 | 25.4 | $T/(S=1)$ |

些只和有限个高信噪比模式相关的、定义为正的参数所特有的。灰色线是 68% 置信区间边界，红色线表示 $B$ 模式功率的 95% 置信上限。绿色的带是大量的宇宙学模型给出的 $E$ 期望值的分布，其权重是用以前的温度结果计算得到的这些模型的似然值(参考文献 50)。

图 3 示意了 $E/B$ 协调形状偏振分析的结果。$E$ 的最大似然值为 0.8(置信度为 68%的参数区间为 0.56 至 1.10)。对于 $B$，结果显然应该视为上限；$B>0$ 的概率为 95%(将 $E$ 边缘化)时，$B$ 小于 0.59。

图 2a 展示了和本次分析相关的参数窗口函数。注意 $E$ 参数几乎和 $B$ 参数无关，反之亦然——DASI 区分两者的能力是非凡的，这一点也反映在 $E$ 和 $B$ 参数之间相关关系(相关系数为 0.046)弱上(见表 2)。

表 2. 偏振数据似然分析结果

| 分析 | 参数 | $l_{low} \sim l_{high}$ | ML est. | 68% 区间 | | | | |
|---|---|---|---|---|---|---|---|---|
| | | | | $(F^{-1})_{ii}^{1/2}$ 误差 | 下界 | 上界 | UL(95%) | 单位 |
| $E/B$ | $E$ | – | 0.80 | ±0.28 | 0.56 | 1.10 | – | 一致的 $E$ 的比例 |
| | $B$ | – | 0.21 | ±0.18 | 0.05 | 0.40 | 0.59 | 一致的 $E$ 的比例 |
| $E5/B5$ | $E1$ | 28 ~ 245 | −0.50 | ±0.8 | −1.20 | 1.45 | 2.38 | $\mu K^2$ |
| | $E2$ | 246 ~ 420 | 17.1 | ±6.3 | 11.3 | 31.2 | – | $\mu K^2$ |
| | $E3$ | 421 ~ 596 | −2.7 | ±5.2 | −10.0 | 4.3 | 24.9 | $\mu K^2$ |
| | $E4$ | 597 ~ 772 | 17.5 | ±16.0 | 3.8 | 40.3 | 47.2 | $\mu K^2$ |
| | $E5$ | 773 ~ 1,050 | 11.4 | ±49.0 | −32.5 | 92.3 | 213.2 | $\mu K^2$ |
| | $B1$ | 28 ~ 245 | −0.65 | ±0.65 | −1.35 | 0.52 | 1.63 | $\mu K^2$ |
| | $B2$ | 246 ~ 420 | 1.3 | ±2.4 | −0.7 | 5.0 | 10.0 | $\mu K^2$ |
| | $B3$ | 421 ~ 596 | 4.8 | ±6.5 | −0.6 | 13.5 | 17.2 | $\mu K^2$ |
| $E5/B5$ | $B4$ | 597 ~ 772 | 13.0 | ±14.9 | 1.6 | 31.0 | 49.1 | $\mu K^2$ |
| | $B5$ | 773 ~ 1,050 | −54.0 | ±28.9 | −77.7 | −4.4 | 147.4 | $\mu K^2$ |
| $E/\beta_E$ | $E$ | – | 0.84 | ±0.28 | 0.55 | 1.08 | – | 一致的 $E$ 的比例 |
| | $\beta_E$ | – | 0.17 | ±1.96 | −1.63 | 1.92 | – | 温度谱指数 |
| 标量/张量 | $S$ | – | 0.87 | ±0.29 | 0.62 | 1.18 | – | 一致的 $S$ 的比例 |
| | $T$ | – | −14.3 | ±7.5 | −20.4 | −3.9 | 25.4 | $T/(S=1)$ |

*Continued*

| The four corresponding parameter correlation matrices | | | | | | | | | | | |
|---|---|---|---|---|---|---|---|---|---|---|---|
| *E*1 | *E*2 | *E*3 | *E*4 | *E*5 | *B*1 | *B*2 | *B*3 | *B*4 | *B*5 | *E* | *B* |
| 1 | −0.137 | 0.016 | −0.002 | 0.000 | −0.255 | 0.047 | −0.004 | 0.000 | 0.000 | 1 | −0.046 |
| | 1 | −0.117 | 0.014 | −0.002 | 0.024 | −0.078 | 0.004 | 0.000 | 0.000 | | 1 |
| | | 1 | −0.122 | 0.015 | −0.003 | 0.010 | −0.027 | 0.003 | −0.001 | | |
| | | | 1 | −0.119 | 0.000 | −0.001 | 0.002 | −0.016 | 0.003 | *E* | $\beta_E$ |
| | | | | 1 | 0.000 | 0.000 | 0.000 | 0.002 | −0.014 | 1 | −0.046 |
| | | | | | 1 | −0.226 | 0.022 | −0.002 | 0.000 | | 1 |
| | | | | | | 1 | −0.097 | 0.011 | −0.002 | | |
| | | | | | | | 1 | −0.111 | 0.018 | *S* | *T* |
| | | | | | | | | 1 | −0.164 | 1 | −0.339 |
| | | | | | | | | | 1 | | 1 |

ML est., maximum likelihood estimate. $(F^{-1})_{ii}^{1/2}$, Fisher matrix uncertainty for parameter $i$ is evaluated at ML. UL, upper limit.

Assuming that the uncertainties in $E$ and $B$ are normally distributed ("Goodness-of-fit tests" section), the likelihood ratio $\Lambda(\text{nopol}) = 13.76$ implies a probability that our data are consistent with the zero-polarization hypothesis of PTE $= 1.05 \times 10^{-6}$. Our data are highly incompatible with the no-polarization hypothesis. Marginalizing over $B$, we find $\Lambda(E = 0) = 12.1$ corresponding to detection of $E$-mode polarization at a PTE of $8.46 \times 10^{-7}$ (or a significance of $4.92\sigma$).

The likelihood ratio for the concordance model gives $\Lambda(E = 1, B = 0) = 1.23$, for which the Monte Carlo and analytic PTE are both 0.28. We conclude that our data are consistent with the concordance model.

However, given the precision to which the temperature power spectrum of the CMB is currently known, even within the ~7-parameter class of cosmological models often considered, the shape and amplitude of the predicted $E$-mode spectrum are still somewhat uncertain. To quantify this, we have taken the model grid generated for ref. 50 and calculated the expectation value of the shaped band $E$ parameter for each model using the window function shown in Fig. 2. We then take the distribution of these predicted $E$ amplitudes, weighted by the likelihood of the corresponding model given our previous temperature results (using a common calibration uncertainty for the DASI temperature and polarization measurements). This yields a 68% credible interval for the predicted value of the $E$ parameter of 0.90 to 1.11. As illustrated in Fig. 3a, our data are compatible with the expectation for $E$ on the basis of existing knowledge of the temperature spectrum.

***E5*/*B5*.** Figure 4a and b show the results of a ten-parameter analysis characterizing the $E$ and $B$-mode spectra using five flat bandpowers for each. Figure 2b shows the corresponding parameter window functions. Note the extremely small uncertainty in the measurements of the first bands $E1$ and $B1$.

续表

| 四个对应参数的相关矩阵 | | | | | | | | | | | |
|---|---|---|---|---|---|---|---|---|---|---|---|
| *E*1 | *E*2 | *E*3 | *E*4 | *E*5 | *B*1 | *B*2 | *B*3 | *B*4 | *B*5 | *E* | *B* |
| 1 | −0.137 | 0.016 | −0.002 | 0.000 | −0.255 | 0.047 | −0.004 | 0.000 | 0.000 | 1 | −0.046 |
| | 1 | −0.117 | 0.014 | −0.002 | 0.024 | −0.078 | 0.004 | 0.000 | 0.000 | | 1 |
| | | 1 | −0.122 | 0.015 | −0.003 | 0.010 | −0.027 | 0.003 | −0.001 | | |
| | | | 1 | −0.119 | 0.000 | −0.001 | 0.002 | −0.016 | 0.003 | *E* | $\beta_E$ |
| | | | | 1 | 0.000 | 0.000 | 0.000 | 0.002 | −0.014 | 1 | −0.046 |
| | | | | | 1 | −0.226 | 0.022 | −0.002 | 0.000 | | 1 |
| | | | | | | 1 | −0.097 | 0.011 | −0.002 | | |
| | | | | | | | 1 | −0.111 | 0.018 | *S* | *T* |
| | | | | | | | | 1 | −0.164 | 1 | −0.339 |
| | | | | | | | | | 1 | | 1 |

ML est.，最大似然估计。$(F^{-1})_{ii}^{1/2}$，在 ML 处参数 $i$ 的费希尔矩阵不确定度。UL，上限。

假定 $E$ 和 $B$ 的不确定度呈正态分布（见“拟合优度检验”部分），似然比 $\Lambda$(无偏振) = 13.76 意味着我们的数据和零偏振假设一致的概率 PTE = $1.05 \times 10^{-6}$，我们的数据和无偏振的假设非常不相容。将 $B$ 边缘化，我们发现 $\Lambda(E=0) = 12.1$，意味着探测到 $E$ 模式偏振的 PTE 为 $8.46 \times 10^{-7}$（或者说显著程度为 $4.92\sigma$）。

协调模型给出的似然比为 $\Lambda(E=1, B=0) = 1.23$，蒙特卡洛和解析近似给出的 PTE 都是 0.28，由此推断我们的数据支持协调模型。

然而，考虑到目前 CMB 的温度功率谱的精度，即使是经常被考虑的具有 ~ 7 个参数的这类宇宙学模型，预计给出 $E$ 模式谱的形状和幅度仍然有些不确定。为了量化这种不确定性，我们采用了为参考文献 50 构造的模型网格，并使用图 2 所示的窗口函数计算每个模型的成形的频带 $E$ 参数的期望值。然后我们取这些 $E$ 模式幅度的预测值，并根据原先给出温度结果的对应模型下的似然值作加权（对 DASI 的温度和偏振测量使用相同的校准不确定度）。这样做给出一个 $E$ 参数的 68% 置信区间为 0.90 到 1.11。如图 3a 所示，基于目前我们对温度谱的知识给出的 $E$ 的期望值和我们的数据是相匹配的。

***E*5/*B*5**　图 4a 和 b 是 $E$ 和 $B$ 模式在分别使用五个平坦频段功率时的十参数分析结果。表 2b 是对应的参数窗口函数。注意第一频带 $E1$ 和 $B1$ 的测量不确定度非常小。

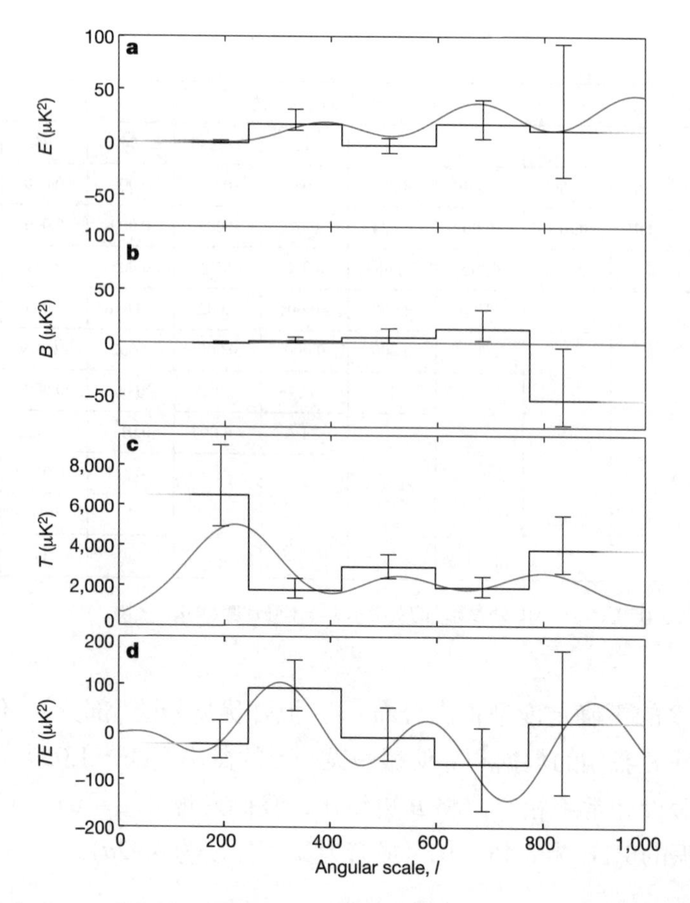


Fig. 4. Results from several likelihood analyses. The ten-parameter $E5/B5$ polarization analysis is shown in **a** and **b**. The $T5$ temperature analysis is shown in **c** and the five $TE$ bands from the $T/E/TE5$ joint analysis are shown in **d**. All the results are shown in flat bandpower units of $l(l+1)C_l/(2\pi)$. The blue line shows the piecewise flat bandpower model for the maximum likelihood parameter values, with the error bars indicating the 68% central region of the likelihood of each parameter, marginalizing over the other parameter values (analogous to the grey lines in Fig. 3a and b). In each case the green line is the concordance model.

Table 3. Results of likelihood analyses from temperature data

| Analysis | Parameter | $l_{low}$–$l_{high}$ | ML est. | 68% interval | | | |
|---|---|---|---|---|---|---|---|
| | | | | $(F^{-1})_{ii}^{1/2}$ error | Lower | Upper | Units |
| $T/\beta_T$ | $T$ | – | 1.19 | ±0.11 | 1.09 | 1.30 | Fraction of concordance $T$ |
| | $\beta_T$ | – | -0.01 | ±0.12 | -0.16 | 0.14 | Temperature spectral index |
| $T5$ | $T1$ | 28–245 | 6,510 | ±1,610 | 5,440 | 9,630 | μK² |
| | $T2$ | 246–420 | 1,780 | ±420 | 1,480 | 2,490 | μK² |
| | $T3$ | 421–596 | 2,950 | ±540 | 2,500 | 3,730 | μK² |
| | $T4$ | 597–772 | 1,910 | ±450 | 1,530 | 2,590 | μK² |
| | $T5$ | 773–1,050 | 3,810 | ±1,210 | 3,020 | 6,070 | μK² |

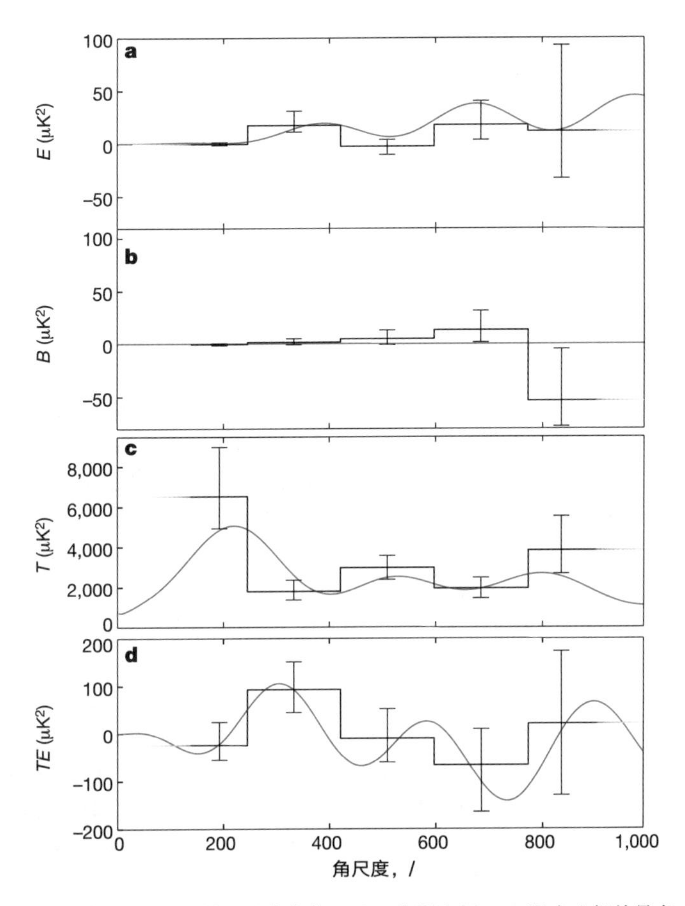


图 4. 数个似然分析的结果。**a** 和 **b** 展示了十参数 *E*5/*B*5 偏振分析。*T*5 温度分析结果在 **c** 中，*T*/*E*/*TE*5 联合分析的五个 *TE* 频段结果在 **d** 中。图示所有结果都以 $l(l+1)C_l/2\pi$ 为单位。蓝色线表示最大似然参数值给出的分段平坦模型下的形状，误差棒是各个参数 68% 置信区间范围的区域大小（将其他参数边缘化，和图 3a 和 b 中的灰色线类似）。每个图中的绿色线，都对应协调模型给出的形状。

表 3. 温度数据似然分析结果

| 分析 | 参数 | $l_{low} \sim l_{high}$ | ML est. | 68% 区间 | | | |
|---|---|---|---|---|---|---|---|
| | | | | $(F^{-1})_{ii}^{1/2}$ 误差 | 下界 | 上界 | 单位 |
| $T/\beta_T$ | $T$ | – | 1.19 | ±0.11 | 1.09 | 1.30 | 一致的 $T$ 的比例 |
| | $\beta_T$ | – | −0.01 | ±0.12 | −0.16 | 0.14 | 温度谱指数 |
| $T5$ | $T1$ | 28 ~ 245 | 6,510 | ±1,610 | 5,440 | 9,630 | μK² |
| | $T2$ | 246 ~ 420 | 1,780 | ±420 | 1,480 | 2,490 | μK² |
| | $T3$ | 421 ~ 596 | 2,950 | ±540 | 2,500 | 3,730 | μK² |
| | $T4$ | 597 ~ 772 | 1,910 | ±450 | 1,530 | 2,590 | μK² |
| | $T5$ | 773 ~ 1,050 | 3,810 | ±1,210 | 3,020 | 6,070 | μK² |

*Continued*

| The two corresponding parameter correlation matrices | | | | | | |
|---|---|---|---|---|---|---|
| $T1$ | $T2$ | $T3$ | $T4$ | $T5$ | $T$ | $\beta_T$ |
| 1 | −0.101 | 0.004 | −0.004 | −0.001 | 1 | 0.023 |
| | 1 | −0.092 | −0.013 | −0.011 | | 1 |
| | | 1 | −0.115 | −0.010 | | |
| | | | 1 | −0.147 | | |
| | | | | 1 | | |

To test whether these results are consistent with the concordance model, we calculate the expectation value for the nominal concordance model in each of the five bands, yielding $E = (0.8, 14, 13, 37, 16)$ and $B = (0, 0, 0, 0, 0)$ $\mu K^2$. The likelihood ratio comparing this point in the ten-dimensional parameter space to the maximum gives $\Lambda = 5.1$, which for ten degrees of freedom results in a PTE of 0.42, indicating that our data are consistent with the expected polarization parameterized in this way. The $E5/B5$ results are highly inconsistent with the zero-polarization "nopol" hypothesis, for which $\Lambda = 15.2$ with a PTE = 0.00073. This statistic is considerably weaker than the equivalent one obtained for the single band analysis in the "$E/B$ analysis" section, as expected from the higher number of degrees of freedom in this analysis. In this ten-dimensional space, all possible random deviations from the "nopol" expectation values $E = (0, 0, 0, 0, 0)$, $B = (0, 0, 0, 0, 0)$ are treated equally in constructing the PTE for our $\Lambda$ statistic. Imagining the "nopol" hypothesis to be true, it would be far less likely to obtain a result in this large parameter space that is both inconsistent with "nopol" at this level and at the same time is consistent with the concordance model, than it would be to obtain a result that is merely inconsistent with "nopol" in some way at this level. It is the latter probability that is measured by the PTE for our $\Lambda$(nopol), explaining why this approach to goodness-of-fit weakens upon considering increasing numbers of parameters.

***$E/\beta_E$.*** We have performed a two-parameter analysis to determine the amplitude of the $E$-mode polarization signal as above and the frequency spectral index $\beta_E$ of this signal relative to CMB (Fig. 5). As expected, the results for the $E$-mode amplitude are very similar to those for the $E/B$ analysis described in the previous section. The spectral index constraint is not strong; the maximum likelihood value is $\beta_E = 0.17$ (−1.63 to 1.92). The result is nevertheless interesting in the context of ruling out possible foregrounds (see the "Diffuse foregrounds" subsection below).

续表

| 两个对应参数的相关矩阵 | | | | | | |
|---|---|---|---|---|---|---|
| $T1$ | $T2$ | $T3$ | $T4$ | $T5$ | $T$ | $\beta_T$ |
| 1 | −0.101 | 0.004 | −0.004 | −0.001 | 1 | 0.023 |
| | 1 | −0.092 | −0.013 | −0.011 | | 1 |
| | | 1 | −0.115 | −0.010 | | |
| | | | 1 | −0.147 | | |
| | | | | 1 | | |

为了检测这些结果是否与协调模型一致，我们计算了五个频段中每个频段的(名义上的)协调模型的期望值，得到 $E = (0.8, 14, 13, 37, 16)$，$B = (0, 0, 0, 0, 0)$ $\mu K^2$。该点(在十维参数空间上)和最大处相比得到似然比 $\Lambda = 5.1$，对于十个自由度下对应的 PTE 为 0.42，表明我们的数据和这种参数化方式给出的预测是相符的。$E5/B5$ 结果和零偏振的“无偏振”模型高度不一致，其中 $\Lambda = 15.2$，PTE = 0.00073。该统计量远低于“$E/B$ 分析”部分中单频段分析给出的对应的值，正如从该分析中较高的自由度数目所能预期的那样。在这个十维空间中，为我们的 $\Lambda$ 统计量构造 PTE 时候，所有来自“无偏振”期望值 $E = (0, 0, 0, 0, 0)$，$B = (0, 0, 0, 0, 0)$ 的可能的随机偏离都有相同的贡献。假设“无偏振”是正确的，在这样一个大参数空间中获得一个既在上述的水平上与“无偏振”模型相悖，同时又和协调模型匹配的结果，比按照同样的方法得到一个仅仅和“无偏振”模型相悖的结果的可能性要小得多，我们的 $\Lambda$(无偏振)对应的 PTE 是用后者的概率给出的，这解释了为什么增加参数数量情况下该拟合优度方法的结论会减弱。

$E/\beta_E$　我们已经进行了双参数分析以确定如上所述的 $E$ 模式极化信号的幅度以及该信号相对于 CMB 的谱指数 $\beta_E$(图 5)。正如我们预测的，$E$ 模式幅度的结果和上一部分“$E/B$ 分析”给出的差不多，谱指数的约束不是很强，最大似然估计值为 $\beta_E = 0.17$(置信度为 68% 的参数区间为 −1.63 到 1.92)。不管怎样，若不考虑前景(参见“弥散前景”部分)的可能性，这个结果还是值得关注的。

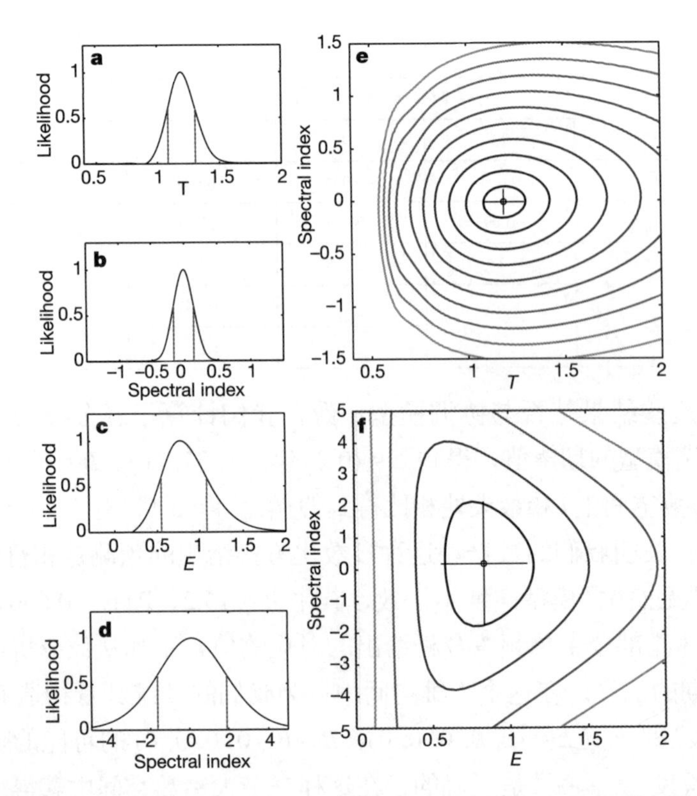


Fig. 5. Results of shaped bandpower amplitude/spectral-index analyses. **a**, **b**, **e**, The $T/\beta_T$ temperature analysis assuming the $T$ power spectrum shape as predicted for the concordance model, and in units relative to that model. The layout of the plot is analogous to Fig. 3. Spectral index is relative to the CMB blackbody—in these units, synchrotron emission would be expected to have an index of approximately −3. **c**, **d**, **f**, Results of the similar $E/\beta_E$ analysis performed on the polarization data.

**Scalar/tensor.** Predictions exist for the shape of the $E$ and $B$-mode spectra which would result from primordial gravity waves, also known as tensor perturbations, although their amplitude is not well constrained by theory. In a concordance-type model such tensor polarization spectra are expected to peak at $l \approx 100$. Assuming reasonable priors, current measurements of the temperature spectrum (in which tensor and scalar contributions will be mixed) suggest $T/S < 0.2$ (ref. 61), where this amplitude ratio is defined in terms of the tensor and scalar contributions to the temperature quadrupole $C_2^T$. We use the distinct polarization angular power spectra for the scalars (our usual concordance $E$ shape, with $B = 0$) and the tensors ($E_T$ and $B_T$) as two components of a likelihood analysis to constrain the amplitude parameters of these components. In principle, because the scalar $B$-mode spectrum is zero, this approach avoids the fundamental sample variance limitations arising from using the temperature spectrum alone. However, the $E5/B5$ analysis (see subsection "$E5/B5$") indicates that we have only upper limits to the $E$- and $B$-mode polarization at the angular scales most relevant ($l \lesssim 200$) for the tensor spectra. It is therefore not surprising that our limits on $T/S$ derived from the polarization spectra as reported in Table 2 are quite weak.

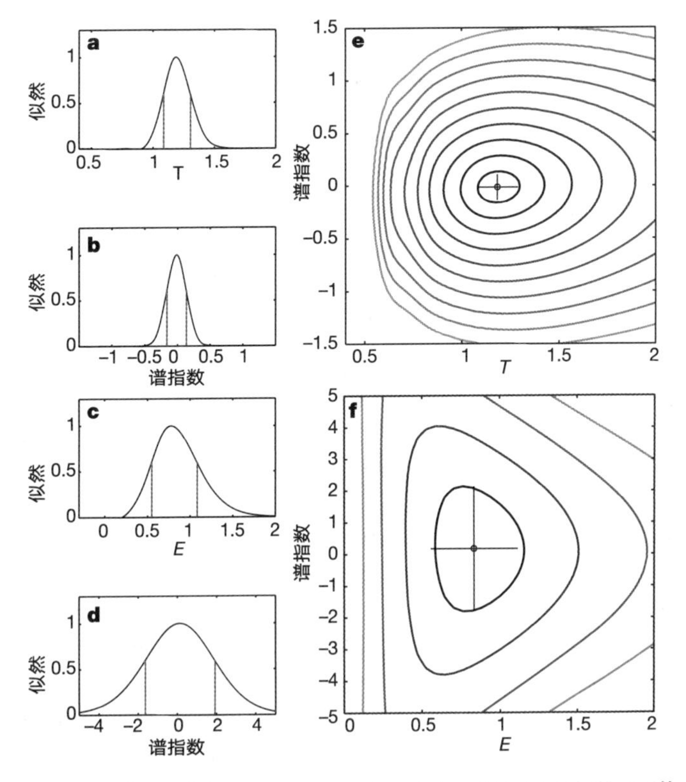


图5. 成形带功率幅度/谱指数分析结果。图**a**、**b**、**e**，假设温度功率谱满足协调模型下的 $T/\beta_T$ 温度分析结果，与模型相对单位一致。此图的布局和图3相似，谱指数对应CMB黑体谱。在这些单位下，预计同步加速辐射的谱指数约为−3。图**c**、**d**、**f**，对偏振数据做类似 $E/\beta_E$ 分析的结果。

**标量/张量** 对于 $E$ 和 $B$ 模式谱的形状存在多个预测，预测认为这些谱可能由原初引力波(即张量扰动)产生，尽管理论并不能很好地就它们的幅度给出约束。在协调型模型中，这种张量偏振谱预计在 $l=100$ 处有峰值。假定一个合理的先验，目前的温度谱(张量和标量的贡献混合在一起)测量暗示 $T/S<0.2$(参考文献61)，这个幅度比的定义是依据张量和标量对温度四极矩 $C_2^T$ 的贡献给出的：我们把标量(我们惯用的协调模型 $E$ 的形状，$B=0$)和张量($E_T$ 和 $B_T$)各自的偏振角功率谱作为似然分析的两个部分来约束这两个组分的幅度参数。原则上，由于标量 $B$ 模式功率谱是零，这个方法能够避免单独使用温度谱引入基本样本方差的局限性。但是 $E5/B5$ 分析(参见“$E5/B5$”部分)暗示我们在张量功率谱最相关的 $l\leqslant200$ 的角尺度上，只能得到 $E$ 和 $B$ 模式的上限。这样，在表2中报告中，我们从偏振谱得到的 $T/S$ 的限制相当弱也就不令人费解了。

## Temperature data analyses and results for *T* spectrum

***T*/$\beta_T$.** Results are shown in Fig. 5 for a two-parameter analysis to determine the amplitude and frequency spectral index of the temperature signal. The bandpower shape used is that of the concordance $T$ spectrum, and the amplitude parameter is expressed in units relative to that spectrum. The spectral index is relative to the CMB, so that 0 corresponds to a 2.73-K Planck spectrum. The amplitude of $T$ has a maximum likelihood value of 1.19 (1.09 to 1.30), and the spectral index $\beta_T = -0.01$ ($-0.16$ to 0.14). Although the uncertainty in the temperature amplitude is dominated by sample variance, the spectral index is limited only by the sensitivity and fractional bandwidth of DASI. Owing to the extremely high $s/n$ of the temperature data, the constraints on the spectral index are superior to those from previous DASI observations (ref. 6).

***T5*.** Fig. 4c shows the results of an analysis using five flat bands to characterize the temperature spectrum. These results are completely dominated by the sample variance in the differenced field. They are consistent with, although less precise than our previous temperature power spectra described in ref. 6; we include them here primarily to emphasize that DASI makes measurements simultaneously in all four Stokes parameters and is therefore able to measure temperature as well as polarization anisotropy. Note that these results and those for $T/\beta_T$ have not been corrected for residual point sources.

## Joint analyses and cross spectra results

***T/E/TE*.** Figure 6 shows the results of a three-parameter single bandpower analysis of the amplitudes of the $T$ and $E$ spectra, and the $TE$ cross-correlation spectrum. As before, bandpower shapes based on the concordance model are used. The $T$ and $E$ constraints are, as expected, very similar to those from the $E/B$, $E/\beta_E$ and $T/\beta_T$ analyses described above. The new result here is $TE$ which has a maximum likelihood value of 0.91 (0.45 to 1.37). Note that in contrast to the two-dimensional likelihoods shown in other figures, here the contours show apparent evidence of correlation between the two parameters; the parameter correlation coefficients from Table 4 are 0.28 for $E/TE$ and 0.21 for $T/TE$.

Table 4. Results of likelihood analyses from joint temperature-polarization data set

| Analysis | Parameter | $l_{low}$–$l_{high}$ | ML est. | 68% interval | | | |
|---|---|---|---|---|---|---|---|
| | | | | $(F^{-1})_{ii}^{1/2}$ error | Lower | Upper | Units |
| *T/E/TE* | $T$ | – | 1.13 | ±0.10 | 1.05 | 1.29 | Fraction of concordance $T$ |
| | $E$ | – | 0.77 | ±0.27 | 0.57 | 1.10 | Fraction of concordance $E$ |
| | $TE$ | – | 0.91 | ±0.38 | 0.45 | 1.37 | Fraction of concordance $TE$ |

### 温度数据分析和 *T* 谱结果

*T*/$\beta_T$　限制温度信号的幅度和谱指数的双参数分析，结果如图5所示，选用协调模型 *T* 谱的频带功率形状，幅度参数以该谱相对应的单位来表示。该谱的谱指数是相对于CMB的，因此0对应2.73 K的普朗克谱。*T* 幅度的最大似然估计值为1.19（置信度为68%的参数区间为1.09到1.30），谱指数 $\beta_T = -0.01$（置信度为68%的参数区间为−0.16到0.14）。尽管温度幅度的误差主要来源于样本误差，但是谱指数的误差仅受限于DASI的灵敏度和带宽比。得益于温度数据极高的信噪比，温度谱指数的限制优于从先前DASI观测给出的（参考文献6）。

*T5*　图4c展示了一个用五平坦频带描述温度谱的分析结果。这些结果完全由离散区域内的样本方差主导。尽管不如我们先前在参考文献6中所描述的温度功率谱那么精确，但它们是吻合的。我们把它们列在此处，主要是为了强调DASI同时测量四个斯托克斯参数，因此能够在测量温度的同时测量偏振的各向异性。注意，这些结果以及 *T*/$\beta_T$ 的结果还未做残留点光源校准。

### 联合分析和交叉谱结果

*T*/*E*/*TE*　图6为功率 *T* 谱、*E* 谱和 *TE* 相关谱的幅度的三参数单频带分析结果。和前面一样，频带功率选用协调模型的形状。和预想的一样，*T* 和 *E* 的约束与前面描述过的 *E*/*B*、*E*/$\beta_E$ 和 *T*/$\beta_T$ 分析非常相似。这部分的新结果是 *TE* 的最大似然估计值，为0.91（置信度为68%的参数区间为0.45到1.37）。注意和其他图中的二维似然分布相反，这里两个参数的二维（似然）等高线图呈现明显的相关关系，从表4可知，参数的相关系数为0.28（*E*/*TE*）和0.21（*T*/*TE*）。

表4. 温度–偏振数据集联合似然分析结果

| 分析 | 参数 | $l_{low} \sim l_{high}$ | ML est. | 68% 区间 | | | |
|---|---|---|---|---|---|---|---|
| | | | | $(F^{-1})_{ii}^{1/2}$ 误差 | 下界 | 上界 | 单位 |
| *T*/*E*/*TE* | *T* | – | 1.13 | ±0.10 | 1.05 | 1.29 | 一致的 *T* 的比例 |
| | *E* | – | 0.77 | ±0.27 | 0.57 | 1.10 | 一致的 *E* 的比例 |
| | *TE* | – | 0.91 | ±0.38 | 0.45 | 1.37 | 一致的 *TE* 的比例 |

*Continued*

| Analysis | Parameter | $l_{low}$–$l_{high}$ | ML est. | 68% interval | | | |
|---|---|---|---|---|---|---|---|
| | | | | $(F^{-1})_{ii}^{1/2}$ error | Lower | Upper | Units |
| *T*/*E*/*TE*5 | *T* | – | 1.12 | ±0.10 | 1.09 | 1.31 | Fraction of concordance *T* |
| | *E* | – | 0.81 | ±0.28 | 0.71 | 1.36 | Fraction of concordance *E* |
| | *TE*1 | 28–245 | -24.8 | ±32.2 | -55.3 | 24.7 | μK² |
| | *TE*2 | 246–420 | 92.3 | ±38.4 | 44.9 | 151.1 | μK² |
| | *TE*3 | 421–596 | −10.5 | ±48.2 | -60.1 | 52.0 | μK² |
| | *TE*4 | 597–772 | -66.7 | ±74.3 | -164.6 | 9.5 | μK² |
| | *TE*5 | 773–1,050 | 20.0 | ±167.9 | -130.3 | 172.3 | μK² |
| *T*/*E*/*B*/ *TE*/*TB*/*EB* | *T* | – | 1.13 | ±0.10 | 1.03 | 1.27 | Fraction of concordance *T* |
| | *E* | – | 0.75 | ±0.26 | 0.59 | 1.19 | Fraction of concordance *E* |
| | *B* | – | 0.20 | ±0.18 | 0.11 | 0.52 | Fraction of concordance *E* |
| | *TE* | – | 1.02 | ±0.37 | 0.53 | 1.49 | Fraction of concordance *TE* |
| | *TB* | – | 0.53 | ±0.32 | 0.08 | 0.82 | Fraction of concordance *TE* |
| | *EB* | – | −0.16 | ±0.16 | −0.38 | 0.01 | Fraction of concordance *E* |

The three parameter correlation matrices

| *T* | *E* | *TE*1 | *TE*2 | *TE*3 | *TE*4 | *TE*5 |
|---|---|---|---|---|---|---|
| 1 | 0.026 | −0.071 | 0.202 | −0.018 | −0.075 | 0.008 |
| | 1 | −0.067 | 0.339 | −0.023 | −0.090 | 0.008 |
| | | 1 | −0.076 | 0.006 | 0.011 | −0.001 |
| | | | 1 | −0.078 | −0.039 | 0.004 |
| | | | | 1 | −0.056 | 0.004 |
| | | | | | 1 | −0.066 |
| | | | | | | 1 |

| *T* | *E* | *B* | *TE* | *TB* | *EB* |
|---|---|---|---|---|---|
| 1 | 0.026 | 0.004 | 0.230 | 0.136 | 0.033 |
| | 1 | −0.027 | 0.320 | −0.040 | −0.182 |
| | | 1 | −0.027 | 0.219 | −0.190 |
| | | | 1 | −0.150 | 0.109 |
| | | | | 1 | 0.213 |
| | | | | | 1 |

| *T* | *E* | *TE* |
|---|---|---|
| 1 | 0.017 | 0.207 |
| | 1 | 0.282 |
| | | 1 |

Marginalizing over $T$ and $E$, we find that the likelihood of $TE$ peaks very near 1, so that $\Lambda(TE=1)=0.02$ with a PTE of 0.857. For the "no cross-correlation" hypothesis, $\Lambda(TE=0)=1.85$ with an analytic PTE of 0.054 (the PTE calculated from Monte Carlo simulations is 0.047). This result represents a detection of the expected $TE$ correlation at 95% confidence and is particularly interesting in that it suggests a common origin for the observed temperature and polarization anisotropy.

It has been suggested that an estimator of $TE$ cross-correlation constructed using a $TE=0$ prior may offer greater immunity to systematic errors[59]. We have confirmed that applying

续表

| 分析 | 参数 | $l_{low} \sim l_{high}$ | ML est. | 68% 区间 | | | |
|---|---|---|---|---|---|---|---|
| | | | | $(F^{-1})_{ij}^{1/2}$ 误差 | 下界 | 上界 | 单位 |
| *T*/*E*/*TE*5 | *T* | – | 1.12 | ±0.10 | 1.09 | 1.31 | 一致的 *T* 的比例 |
| | *E* | – | 0.81 | ±0.28 | 0.71 | 1.36 | 一致的 *E* 的比例 |
| | *TE*1 | 28 ~ 245 | −24.8 | ±32.2 | −55.3 | 24.7 | μK² |
| | *TE*2 | 246 ~ 420 | 92.3 | ±38.4 | 44.9 | 151.1 | μK² |
| | *TE*3 | 421 ~ 596 | −10.5 | ±48.2 | −60.1 | 52.0 | μK² |
| | *TE*4 | 597 ~ 772 | −66.7 | ±74.3 | −164.6 | 9.5 | μK² |
| | *TE*5 | 773 ~ 1,050 | 20.0 | ±167.9 | −130.3 | 172.3 | μK² |
| *T*/*E*/*B*/*TE*/*TB*/*EB* | *T* | – | 1.13 | ±0.10 | 1.03 | 1.27 | 一致的 *T* 的比例 |
| | *E* | – | 0.75 | ±0.26 | 0.59 | 1.19 | 一致的 *E* 的比例 |
| | *B* | – | 0.20 | ±0.18 | 0.11 | 0.52 | 一致的 *E* 的比例 |
| | *TE* | – | 1.02 | ±0.37 | 0.53 | 1.49 | 一致的 *TE* 的比例 |
| | *TB* | – | 0.53 | ±0.32 | 0.08 | 0.82 | 一致的 *TE* 的比例 |
| | *EB* | – | −0.16 | ±0.16 | −0.38 | 0.01 | 一致的 *E* 的比例 |

| 三参数的相关矩阵 | | | | | | | | | | | | |
|---|---|---|---|---|---|---|---|---|---|---|---|---|
| *T* | *E* | *TE*1 | *TE*2 | *TE*3 | *TE*4 | *TE*5 | *T* | *E* | *B* | *TE* | *TB* | *EB* |
| 1 | 0.026 | −0.071 | 0.202 | −0.018 | −0.075 | 0.008 | 1 | 0.026 | 0.004 | 0.230 | 0.136 | 0.033 |
| | 1 | −0.067 | 0.339 | −0.023 | −0.090 | 0.008 | | 1 | −0.027 | 0.320 | −0.040 | −0.182 |
| | | 1 | −0.076 | 0.006 | 0.011 | −0.001 | | | 1 | −0.027 | 0.219 | −0.190 |
| | | | 1 | −0.078 | −0.039 | 0.004 | | | | 1 | −0.150 | 0.109 |
| | | | | 1 | −0.056 | 0.004 | | | | | 1 | 0.213 |
| | | | | | 1 | −0.066 | | | | | | 1 |
| | | | | | | 1 | | | | | | |

| *T* | *E* | *TE* |
|---|---|---|
| 1 | 0.017 | 0.207 |
| | 1 | 0.282 |
| | | 1 |

将 $T$ 和 $E$ 边缘化，我们发现 $TE$ 峰处的似然非常接近 1，因而 $\Lambda(TE = 1) = 0.02$，PTE 为 0.857。在 $T$ 和 $E$“无交叉相关关系”的假设下，$\Lambda(TE = 1) = 1.85$，其解析 PTE 为 0.054(用蒙特卡洛模拟计算出的 PTE 为 0.047)。这个结果表明在 95% 置信度下，$TE$ 具有相关性。更为有趣的是，这暗示着观测到的温度和偏振各向异性具有相同的起源。

也有人提出，使用 $TE = 0$ 的先验构造的一个 $TE$ 交叉相关估计量或许对系统误差更不敏感[59]。我们已经证明对我们的数据使用这样的技巧会产生和上面的似然分

such a technique to our data yields similar results to the above likelihood analysis, with errors slightly increased as expected.

***T*/*E*/*TE*5.** We have performed a seven-parameter analysis using single shaped band powers for $T$ and $E$, and five flat bandpowers for the $TE$ cross-correlation; the $TE$ results from this are shown in Fig. 4d. In this analysis the $B$-mode polarization has been explicitly set to zero. Again, the $T$ and $E$ constraints are similar to the values for the other analyses where these parameters appear. The $TE$ bandpowers are consistent with the predictions of the concordance model.

***T*/*E*/*B*/*TE*/*TB*/*EB*.** Finally, we describe the results of a six shaped bandpower analysis for the three individual spectra $T$, $E$ and $B$, together with the three possible cross-correlation spectra $TE$, $TB$ and $EB$. We include the $B$ cross-spectra for completeness, though there is little evidence for any $B$-mode signal. Because there are no predictions for the shapes of the $TB$ or $EB$ spectra (they are expected to be zero), we preserve the symmetry of the analysis between $E$ and $B$ by simply parameterizing them in terms of the $TE$ and $E$ spectral shapes. The results for $T$, $E$, $B$ and $TE$ are similar to those obtained before, with no detection of $EB$ or $TB$.

## Systematic Uncertainties

### Noise, calibration, offsets and pointing

To assess the effect of systematic uncertainties on the likelihood results, we have repeated each of the nine analyses with alternative assumptions about the various effects that we have identified which reflect the range of uncertainty on each.

Much of the effort of the data analysis presented in this paper has gone into investigating the consistency of the data with the noise model as discussed in the "Noise model" subsection. We find no discrepancies between complementary noise estimates on different timescales, to a level $\ll 1\%$. As discussed in the "$\chi^2$ tests" subsection, numerous consistency tests on subsets of the co-polar and cross-polar visibility data show no evidence for an error in the noise scaling to a similar level. When we re-evaluate each of the analyses described in the "Likelihood results" section with the noise scaled by 1%, the shift in the maximum likelihood values for all parameters is entirely negligible.

In the "Noise model" subsection, we reported evidence of some detectable noise correlations between real/imaginary visibilities and between visibilities from different bands of the same baseline. When either or both of these noise correlations are added to the covariance matrix at the measured level, the effects are again negligible: the most significant shift is in the highest-$l$ band of the $E$ spectrum from the $E5/B5$ analysis (see the "$E5/B5$" subsection), where the power shifts by about 2 $\mu K^2$.

析类似的结果，不过和预测的一样，误差会稍稍增大。

*T*/*E*/*TE*5　我们也进行了一个七参数分析，其中对 *T* 和 *E* 采用了单形状频带功率，对 *TE* 交叉相关谱采用了五平坦频段功率的形状，*TE* 结果见图 4d。此分析中，*B* 模式偏振被明确地设置为零。同样，对 *T* 和 *E* 的约束与其他出现这两个参数的分析类似，*TE* 频段功率和调和模型给出的理论预言一致。

*T*/*E*/*B*/*TE*/*TB*/*EB*　最后，我们描述了对 *T*、*E* 和 *B* 三个功率谱以及三个可能的交叉相关谱 *TE*, *TB* 和 *EB* 的六个成型频带功率分析。虽然几乎没有迹象表明存在 *B* 模式信号，但是为了完备性，我们还是把 *B* 交叉谱列入了分析。由于没有对 *TB* 谱和 *EB* 谱的形状的预测(它们被认为是零)，我们简单地根据 *TE* 和 *E* 谱形状对它们进行参数化，以保持 *E* 和 *B* 之间分析的对称性。*T*、*E*、*B* 和 *TE* 的结果与之前得到结果的类似，*EB* 和 *TB* 没有被探测到。

## 系统不确定度

### 噪声、校准、补偿和指向

为了评估系统不确定度对似然结果的影响，我们重复了这九组的每一个分析，但是对那些我们已经知道的反映各自不同的不确定范围的效应作了替代的假设。

本文展示的大部分数据分析都尝试研究数据和噪声模型(在“噪声模型”部分讨论)的一致性。我们发现不同时间尺度的互补噪声估计量之间并无差异，均为 $\ll 1\%$ 的水平。正如“$\chi^2$ 检验”部分讨论过的，我们对自相关偏振和交叉相关偏振的可视性数据的子集进行了大量的一致性检验，没有迹象表明有一个噪声误差能够达到类似的水平。我们将噪声按比例缩放 1%，重复“似然结果”部分的每一个分析，发现所有参数的最大似然估计值的偏移完全可以忽略不计。

在“噪声模型”部分，我们报告了存在于实部/虚部可视度之间，以及相同基线上不同频带之间可视性的某些可探测噪声相关性的证据。把这些噪声相关的其中一个或两个全部添加到协方差矩阵(在实际测量到的水平上)，它们的影响再次可以忽略不计：最显著的偏移出现在 *E*5/*B*5 分析中 *E* 谱的最高 $l$ 频段上(参见“*E*5/*B*5”部分)，其中功率偏移了约 2 μK$^2$。

Errors in the determination of the absolute cross-polar phase offsets will mix power between $E$ and $B$; these phase offsets have been independently determined from wire-grid calibrations and observations of the Moon, and found to agree to within the measurement uncertainties of about 0.4° (ref. 51). Reanalysis of the data with the measured phase offsets shifted by 2° demonstrates that the likelihood results are immune to errors of this magnitude: the most significant effect occurs in the highest-$l$ band of the $TE$ spectrum from the $T$, $E$, $TE5$ analysis (see the "$T/E/TE5$" subsection), where the power shifts by about 30 $\mu K^2$.

The on-axis leakages described in the "On-axis leakage" subsection will mix power from $T$ into $E$ and $B$, and the data are corrected for this effect in the course of reduction, before input to any analyses. When the likelihood analyses are performed without the leakage correction, the largest effects appear in the shaped $TE$ amplitude analysis (see "$T/E/TE$" subsection), and the lowest-$l$ band of $TE5$ from the $T$, $E$, $TE5$ analysis (see the "$T/E/TE5$" subsection); all shifts are tiny compared to the 68% confidence intervals. As the leakage correction itself has little impact on the results, the uncertainties in the correction, which are at the < 1% level, will have no noticeable effect.

As described in the "Off-axis leakage" subsection, the off-axis leakage from the feeds is a more significant effect, and is accounted for in the likelihood analysis by modelling its contribution to the covariance matrix. When this correction is not applied, the $E$, $B$ results (see "$E/B$ analysis" subsection) shift by about 4% and 2%, respectively, as expected from simulations of this effect. Although this bias is already small, the simulations show that the correction removes it completely to the degree to which we understand the off-axis leakage. Uncertainties in the leakage profiles of the order of the fit residuals (see ref. 51) lead to shifts of less than 1%.

The pointing accuracy of the telescope is measured to be better than 2 arcmin and the root-mean-square tracking errors are < 20 arcsec; as we discussed in refs 49 and 6, this is more than sufficient for the characterization of CMB anisotropy at the much larger angular scales measured by DASI.

Absolute calibration of the telescope was achieved through measurements of external thermal loads, transferred to the calibrator RCW38. The dominant uncertainty in the calibration is due to temperature and coupling of the thermal loads. As reported in ref. 6, we estimate an overall calibration uncertainty of 8% ($1\sigma$), expressed as a fractional uncertainty on the $C_l$ bandpowers (4% in $\Delta T/T$). This applies equally to the temperature and polarization data presented here.

## Foregrounds

**Point sources.** The highest-sensitivity point-source catalogue in our observing region is the 5-GHz PMN survey[62]. For our first-season temperature analysis described in refs 49

确定绝对交叉–偏振相位补偿时的误差会混合在 $E$ 和 $B$ 的功率之间，这些相位补偿是通过导线网格校准和对月球的观测独立确定的，且发现其在约 0.4° 的测量误差值（参考文献 51）之内。用测量到的偏移量为 2° 的相位补偿对数据进行重新分析表明，似然结果不受这个程度的误差影响：最显著的影响发生在 $T$、$E$、$TE5$ 分析（参见“$T/E/TE$”部分）中 $TE$ 谱的最高 $l$ 频段上，其中功率偏移了约 30 $\mu K^2$。

“共轴渗漏”部分描述过的共轴渗漏会将 $T$ 功率混入到 $E$ 和 $B$ 功率中，在作任何分析之前，数据会在还原过程中校正掉该效应。若似然分析之前并未做渗漏校准，受影响最显著的是在成形的 $TE$ 幅度分析（参见“$T/E/TE$”部分）和 $T$、$E$、$TE5$ 分析（参见“$T/E/TE5$”部分）中 $TE5$ 的最低 $l$ 频段上。相比较于 68% 置信区间，所有的偏移都是微小的。由于渗漏校准本身对结果几乎没有影响，校准时候的误差——小于 1% 的程度——将不会有明显的影响。

如同在“离轴渗漏”部分中所描述的，来自馈源的离轴渗漏是一个更加显著的效应。我们根据其在协方差矩阵的贡献进行建模以解释该效应对似然分析的影响。若不使用该校准，如同模拟该效应时所预料的那样，$E$、$B$ 结果（参见“$E/B$ 分析”部分）分别偏移约 4%和 2%。虽然这种偏差已经很小，但模拟表明，就我们所理解的离轴渗漏的程度，校正能将其完全消除。达到拟合残差（参考文献 51）量级的渗漏带来的不确定性会导致的偏移小于 1%。

我们测得望远镜的指向精度优于 2 角分，均方根跟踪误差 $<20$ 角秒。正如我们在参考文献 49 和 6 中所讨论的那样，这个精度对刻画在 DASI 测量的更大角尺度上的 CMB 各向异性已经足够了。

望远镜的绝对校准是通过测量外部热负荷给出的，并传输到 RCW38 定标器上。校准的最重要的不确定性来源于温度和热负荷的耦合。如参考文献 6 中报告的，我们估计存在 8%($1\sigma$) 的总体校准不确定度，$C_l$ 频段功率的不确定度表示为分数（在 $\Delta T/T$ 中占 4%）。这同样适用于温度和偏振数据。

## 前景

**点源** 我们观测区域中灵敏度最高的点源星表是 5 GHz PMN 巡天[62]。对于参考文献 49 和 6 所描述的第一季度温度分析，我们就是利用此目录来剔除已知点源

and 6 we projected out known sources using this catalogue. We have kept this procedure for the temperature data presented here, projecting the same set of sources as before.

Unfortunately the PMN survey is not polarization sensitive. We note that the distribution of point-source polarization fractions is approximately exponential (see below). Total intensity is thus a poor indicator of polarized intensity and it is therefore not sensible to project out the PMN sources in our polarization analysis.

Our polarization fields were selected for the absence of any significant point-source detections in the first-season data. No significant detections are found in the 2001–02 data, either in the temperature data, which are dominated by CMB anisotropy, or in the polarization data.

To calculate the expected contribution of undetected point sources to our polarization results we would like to know the distribution of polarized flux densities, but unfortunately no such information exists in our frequency range. However, to make an estimate, we use the distribution of total intensities, and then assume a distribution of polarization fractions. We know the former distribution quite well from our own first-season 32-field data where we detect 31 point sources and determine that $dN/dS_{31} = (32 \pm 7)S^{(-2.15\pm0.20)}$ $Jy^{-1}sr^{-1}$ in the range 0.1 to 10 Jy. This is consistent, when extrapolated to lower flux densities, with a result from the CBI experiment valid in the range 5–50 mJy (ref. 63). The distribution of point source polarization fractions at 5 GHz can be characterized by an exponential with a mean of 3.8% (ref. 64); data of somewhat lower quality at 15 GHz are consistent with the same distribution[65]. Qualitatively, we expect the polarization fraction of synchrotron-dominated sources initially to rise with frequency, and then reach a plateau or fall, with the break point at frequencies $\ll$5 GHz (see ref. 66 for an example). In the absence of better data we have conservatively assumed that the exponential distribution mentioned above continues to hold at 30 GHz.

We proceed to estimate the effect of point sources by Monte Carlo simulation, generating realizations using the total intensity and polarization fraction distributions mentioned above. For each realization, we generate simulated DASI data by adding a realization of CMB anisotropy and appropriate instrument noise. The simulated data are tested for evidence of point sources and those realizations that show statistics similar to the real data are kept. The effect of off-axis leakage, which we describe and quantify in ref. 51, is included in these calculations.

When the simulated data are passed through the *E*/*B* analysis described in the "*E*/*B* analysis" subsection, the mean bias of the *E* parameter is 0.04 with a standard deviation of 0.05; in 95% of cases the shift distance in the *E*/*B* plane is less than 0.13. We conclude that the presence of point sources consistent with our observed data has a relatively small effect on our polarization results.

**Diffuse foregrounds.** In ref. 51, we find no evidence for contamination of the

的。我们对这里所呈现的温度数据保留了同样的处理，和以前一样剔除掉了相同的点源集。

很遗憾的是，PMN巡天对偏振并不敏感，我们注意到，点源偏振部分的分布大致呈指数分布（见下文），故而总强度不能代表偏振强度，因此做偏振分析时，剔除PMN源是不明智的。

在第一季度数据中没有发现任何显著的点源信号，据此我们选择偏振视场。在2001年~2002年数据中，无论是由CMB各向异性主导的温度数据还是偏振数据，我们未发现显著的信号。

为了预先计算未被检测到的点源对我们的偏振结果的贡献，我们希望知道被偏振的通量密度的分布，但遗憾的是在我们的频率范围内并不存在这样的信息。但是，为了对其进行估算，我们使用总强度的分布，然后假设偏振部分的分布函数。从第一季度32视场数据中，我们对前者的分布有了很好的了解，在那里我们探测到31个点源并确定在0.1到10 Jy范围内有 $dN/dS_{31} = (32 \pm 7)S^{(-2.15 \pm 0.20)}\mathrm{Jy}^{-1} \cdot \mathrm{sr}^{-1}$。当外推到较低的通量密度时，它的结果和从CBI实验（参考文献63）在5~50 mJy范围的结果是一致的。5 GHz处的点源偏振部分的分布可以用平均值为3.8%（参考文献64）的指数分布表征，15 GHz处的那些质量稍低的数据具有相同的分布[65]。定性地说，我们预计同步加速机制主导的源的偏振部分首先随着频率上升而增大，然后达到平稳水平或达到频率 $<<5$ GHz的转折点后下降（例子请参见参考文献66）。在没有更好的数据的情况下，我们保守地假设上面提到的指数分布直到30 GHz仍适用。

我们继续使用蒙特卡罗模拟估计点源的影响，使用上面提到的总强度和偏振部分的分布生成实现。对于每个实现，我们通过加入一个CMB各向异性信号和适当的仪器噪声来生成模拟的DASI数据。我们就点源的痕迹对模拟数据进行检测，然后保留那些具有同实际数据相近的统计量的实现。离轴渗漏——我们在文献51中对其进行描述和量化——的影响已经包含在这些计算中了。

当模拟数据通过“$E/B$分析”部分描述的方法进行$E/B$分析时，$E$参数的平均偏差为0.04，其中标准差为0.05；在占95%的案例中，$E/B$平面中的偏移距离小于0.13。我们得出结论，若点源与我们观察到的数据一致，则它们的存在对我们的偏振结果的影响相对较小。

**弥散前景**　正如“$T/\beta_T$”部分对温度谱指数的限制所证实的，在参考文献51中，

temperature data by synchrotron, dust or free–free emission, as confirmed by the limits on the temperature spectral index presented in the "$T/\beta_T$" subsection. The expected fractional polarization of the CMB is of order 10%, while the corresponding number for free–free emission is less than 1%. Diffuse thermal dust emission may be polarized by several per cent (see for example ref. 67), although we note that polarization of the admixture of dust and free–free emission observed with DASI in NGC 6334 is $\ll 1\%$ (see ref. 51). Likewise, emission from spinning dust is not expected to be polarized at detectable levels[68]. Therefore if free–free and dust emission did not contribute significantly to our temperature anisotropy results they are not expected to contribute to the polarization. Synchrotron emission on the other hand can in principle be up to 70% polarized, and is by far the greatest concern; what was a negligible contribution in the temperature case could be a significant one in polarization.

There are no published polarization maps in the region of our fields. Previous attempts to estimate the angular power spectrum of polarized synchrotron emission have been guided by surveys of the Galactic plane at frequencies of 1–3 GHz (refs 69 and 70). These maps show much more small-scale structure in polarization than in temperature, but this is mostly induced by Faraday rotation[71], an effect which is negligible at 30 GHz. In addition, because synchrotron emission is highly concentrated in the disk of the Galaxy it is not valid to assume that the angular power spectrum at low Galactic latitudes has much to tell us about that at high latitudes[72].

Our fields lie at Galactic latitude −58.4° and −61.9°. The brightness of the IRAS 100 μm and Haslam 408 MHz (ref. 52) maps within our fields lie at the 6% and 25% points, respectively, of the integral distributions taken over the whole sky. There are several strong pieces of evidence from the DASI data set itself that the polarization results described in this paper are free of significant synchrotron contamination. The significant $TE$ correlation shown in Fig. 6 indicates that the temperature and $E$-mode signal have a common origin. The tight constraints on the temperature anisotropy spectral index require that this common origin has a spectrum consistent with CMB. Galactic synchrotron emission is known to have a temperature spectral index of −2.8 (ref. 73), with evidence for steepening to −3.0 at frequencies above 1–2 GHz (ref. 74). At frequencies where Faraday depolarization is negligible (> 10 GHz), the same index will also apply for polarization. The dramatically tight constraint on the temperature spectral index of $\beta_T = 0.01$ (−0.16 to 0.14) indicates that any component of the temperature signal coming from synchrotron emission is negligibly small in comparison to the CMB. More directly, the constraint on the $E$-mode spectral index $\beta_E = 0.17$ (−1.63 to 1.92) disfavours synchrotron polarization at nearly $2\sigma$. A third, albeit weaker, line of argument is that a complex synchrotron emitting structure is not expected to produce a projected brightness distribution which prefers $E$-mode polarization over $B$-mode[75]. Therefore, the result in Fig. 3 could be taken as further evidence that the signal we are seeing is not due to synchrotron emission.

我们没有发现同步加速、尘埃或者自由–自由辐射污染温度数据的迹象。CMB的偏振部分预计达到10%的程度，而自由–自由辐射对应的偏振部分小于1%。尽管我们注意到用DASI观察到的NGC 6334尘埃和自由–自由辐射的混合偏振 <<1%(参考文献51)，但是弥散的热尘埃辐射可能会被偏振几个百分点(如参考文献67)。同样地，自旋尘埃辐射的偏振程度预计不会达到可探测水平[68]。因此，如果自由–自由和尘埃热辐射对我们的温度各向异性结果没有显著贡献，则可以预料它们不会对偏振产生影响。另一方面，同步加速辐射的偏振率原则上可以高达70%，这也是目前最被关注的问题。它对温度的贡献可以忽略不计，但可能对偏振有显著贡献。

我们视场区域不存在已经公布过的偏振天图。之前尝试估计计划的同步辐射角功率谱是根据频率为1–3 GHz的银道面巡天(参考文献69和70)来做的。这些天图显示的偏振结构尺度比温度的更小，但这主要是由法拉第旋转[71]引起的，该效应在30 GHz处可以忽略不计。此外，由于同步辐射高度集中在银河星系盘中，因此通过假设的低银河系纬度的角功率谱是无法知道高纬度地区的角功率谱的[72]。

我们的视场位于银河纬度 −58.4° 和 −61.9° 之间。在我们的视场内，IRAS 100 μm和哈斯拉姆408 MHz(参考文献52)天图的亮度分别占整个天空6%和25%点的分布。几个来自DASI自身数据集的有力证据表明本文所述的偏振结果没有受到显著的同步辐射污染。图6所示的*TE*强相关性表明温度和*E*模信号具有共同的起源。对温度各向异性谱指数的严格约束要求这个共同起源具有与CMB相一致的频谱。已知银河同步辐射的温度谱指数为 −2.8(参考文献73)，有证据表明它在频率高于1 ~ 2 GHz时变陡峭，谱指数为 −3.0(参考文献74)。在对法拉第消偏振可忽略不计(> 10 GHz)的频率处，相同的谱指数同样适用于偏振。在温度谱指数 $\beta_T = 0.01$(置信度为68%的参数区间为 −0.16 至 0.14)的严格约束下，与CMB相比，任何来自同步辐射的温度信号都可忽略不计。更直接地，*E*模式谱指数 $\beta_E = 0.17$(置信度为68%的参数区间为 −1.63 至 1.92)的限制在近 $2\sigma$ 水平下和同步辐射偏振不符。第三个(虽然较弱的)论证思路是，复杂的同步辐射结构预计不会产生一个*E*模式优于*B*模式的投影亮度分布[75]。因此，图3中的结果可以进一步证明我们看到的信号不是来自同步辐射。

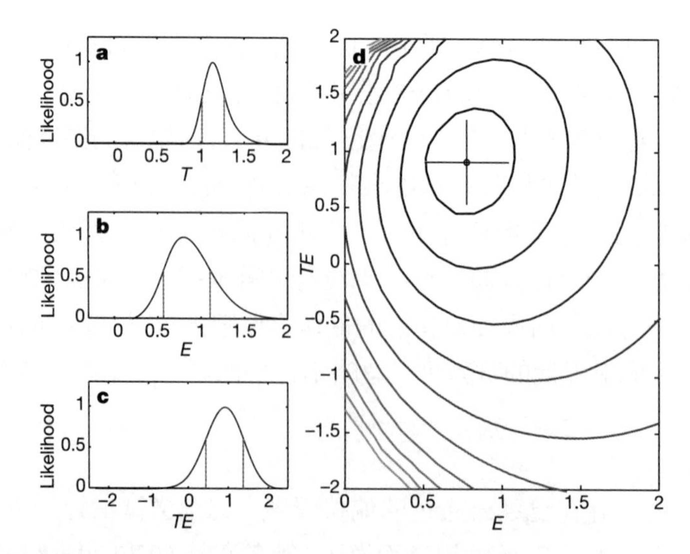


Fig. 6. Results from the three-parameter shaped bandpower $T/E/TE$ joint analysis. Spectral shapes as predicted for the concordance model are assumed (**a**–**c**) and the units are relative to that model. The layout of the plot is analogous to Fig. 3. The two-dimensional distribution in **d** is marginalized over the $T$ dimension.

## Discussion

This paper presents several measures of the confidence with which CMB polarization has been detected with DASI. Which measure is preferred depends on the desired level of statistical rigour and independence from *a priori* models of the polarization. The $\chi^2$ analyses in the "$\chi^2$ tests" subsection offer the most model-independent results, although the linear combinations of the data used to form the $s/n$ eigenmodes are selected by consideration of theory and the noise model. For the high $s/n$ eigenmodes of the polarization data, the probability to exceed (PTE) the measured $\chi^2$ for the sum of the various data splits ranges from $1.6 \times 10^{-8}$ to $8.7 \times 10^{-7}$, while the $\chi^2$ for the differences are found to be consistent with noise. The PTE for the $\chi^2$ found for the total (that is, not split) high $s/n$ polarization eigenmodes, corrected for the beam offset leakage, is $5.7 \times 10^{-8}$.

Likelihood analyses are in principle more model dependent, and the analyses reported make different assumptions for the shape of the polarization power spectrum. Using theory to select the angular scales on which DASI should be most sensitive, we calculate the likelihood for a flat bandpower in $E$ and $B$ over the multipole range $300 < l < 450$ and find that data are consistent with $B = 0$ over this range, but that $E = 0$ can be rejected with a PTE of $5.9 \times 10^{-9}$.

The choice of the model bandpower is more important when the full $l$ range of the DASI data is analysed. In this case, the likelihood analyses indicate that a $l(l+1)C_l \propto l^2$ model for the $E$-mode spectrum is 60 times more likely than a $l(l+1)C_l =$ constant model. Further, a bandpass shape based on the concordance model is found to be 12,000 times more

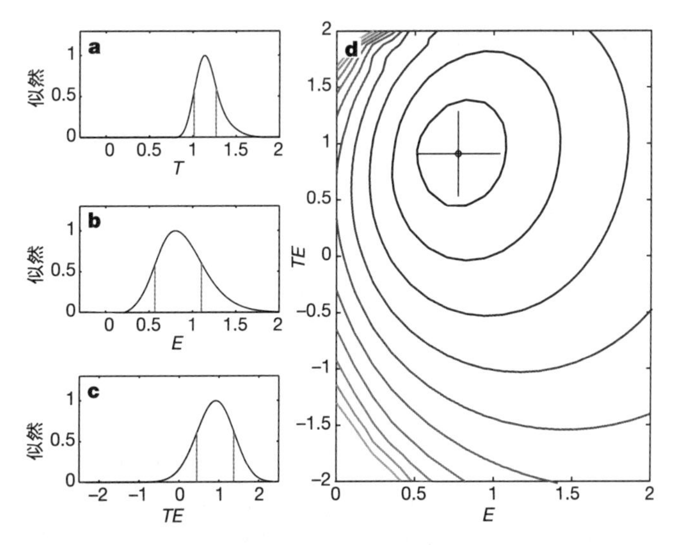


图 6. 三参数型频带功率 $T/E/TE$ 联合分析的结果。假设谱形状满足协调模型（**a**～**c**），而单位则是相对于该模型的。该图的布局类似于图 3。**d** 中的二维分布是边缘化 $T$ 维度得到的。

## 讨　论

本文提出了用 DASI 检测 CMB 偏振的几种置信度测量方法。首选哪种测量方法取决于所需的统计精度和是否独立于偏振的先验模型。尽管用于构成信噪比 $(s/n)$ 本征模的数据的线性组合是基于理论和噪声模型来选择的，"$\chi^2$ 检验"部分中的 $\chi^2$ 分析还是给出了一个最不依赖于模型的结果。对于偏振数据的高信噪比本征模，用不同数据分割的求和计算给出的超过测量得到的 $\chi^2$ 的概率（PTE）范围从 $1.6\times10^{-8}$ 到 $8.7\times10^{-7}$ 不等。另外，对分割数据的差值计算给出的 $\chi^2$ 被发现和噪声是一致的。将光束偏移渗漏校准后，全部的（即不是分割的）高信噪比偏振本征模给出的 $\chi^2$ 的 PTE 是 $5.7\times10^{-8}$。

似然分析原则上更加依赖于模型，且就本文报告的分析，我们对偏振功率谱的形状做出了不同的假设。根据理论来选择 DASI 理应最为敏感的角度范围，我们计算了多极矩范围为 $300<l<450$ 的一个平坦频段模型的 $E$ 和 $B$ 的似然函数。我们发现，在此范围内数据与 $B=0$ 一致，但是 $E=0$ 的 PTE 为 $5.9\times10^{-9}$，故该假设可以被拒绝。

分析完整 $l$ 范围的 DASI 数据时，模型频段功率的选择更为重要。在这种情况下，似然分析表明对于 $E$ 模式频谱，$l(l+1)C_l\propto l^2$ 的模型是 $l(l+1)C_l=$ 常数的模型的可能性的 60 倍。此外我们发现，基于协调模型的带通形状是平坦频段功率的可能性

likely than the flat bandpower. The likelihood ratio test leads to a PTE of $8.5 \times 10^{-7}$ for the concordance shaped bandpower, corresponding to a confidence of detection of $4.9\sigma$. In all three of these tests, $B$ is found to be consistent with zero, as expected in the concordance model.

The concordance model is also supported by the results of the five-band piecewise-flat analyses, $E5/B5$. The upper limit for the first $E$ band at $28 \leqslant l \leqslant 245$ is only 2.38 $\mu K^2$, a factor of 30 lower in power than the previous upper limit. The next band at $246 \leqslant l \leqslant 420$ is detected with a maximum likelihood value of 17.1 $\mu K^2$. Such a sharp rise in power with increasing $l$ is expected in the $E$-mode spectrum (see Fig. 4) owing to the length scale introduced by the mean free path to photon scattering. The polarization of the larger modes is suppressed because the velocity differences are not as large on the scale of the mean free path. In fact, the $E$ spectrum is expected to increase as $l(l+1)C_l \propto l^2$ at small $l$. This dependence is not expected to continue to the higher $l$ values to which DASI is sensitive, owing to diffusion damping, which suppresses power on scales smaller than the mean free path. The maximum likelihood values of the higher $l$ bands of the DASI five-band $E$ analysis are consistent with the damped concordance model, but lie below a simple extrapolation of the $l^2$ power law. Again, the five-band $B$-mode spectrum is consistent with the concordance prediction of zero.

The $TE$ analysis provides further confidence in our detection of CMB polarization and for the concordance model. From the $T/E/TE$ likelihood analysis, the $TE = 0$ hypothesis is rejected with a PTE of 0.054. Note that $TE$ could be negative as well as positive and therefore the $TE \leqslant 0$ hypothesis is rejected with higher confidence.

Lastly, the measured $T$ frequency spectral index, 0.01 (−0.16 to 0.14), is remarkably well constrained to be thermal and is inconsistent with any known foregrounds. The $E$ frequency spectral index, 0.17 (−1.63 to 1.92), is also consistent with the CMB, and although less well constrained than the $T$ index, is inconsistent with diffuse foreground synchrotron emission at nearly $2\sigma$.

In summary, the analyses reported in this paper all indicate a robust detection of $E$-mode CMB polarization with a confidence level $\geqslant 4.9\sigma$. The measured properties of the polarization are in good agreement with predictions of the concordance model and, as discussed in the "Foregrounds" subsection, are inconsistent with expectations from known sources of foreground emission. These results therefore provide strong support for the underlying theoretical framework for the generation of CMB anisotropy. They lend confidence to the values of the cosmological parameters and to the extraordinary picture of the origin and structure of the Universe derived from CMB measurements. The prospect of further refining our understanding of the Universe using precision polarization measurements is the goal of many ongoing and planned CMB experiments. The detection of polarization at the predicted level reported in this paper points to a promising future for the field.

(**420**, 772-787; 2002)

的 12,000 倍。似然比检验给出一致形状频段功率的 PTE 为 $8.5 \times 10^{-7}$，对应探测置信度为 $4.9\sigma$。在所有这三个检验中，我们发现，正如在协调模型中所预期的那样，$B$ 与零一致。

五频段分段–平坦分析（$E5/B5$）的结果同样支持协调模型。在 $28 \leqslant l \leqslant 245$ 处的第一个 $E$ 频段的上限仅为 2.38 $\mu K^2$，功率是前一个上限的三十分之一。在 $246 \leqslant l \leqslant 420$ 处的第二个频段的最大似然值为 17.1 $\mu K^2$。由于光子散射的平均自由程引入的长度标度，出现在 $E$ 模式功率谱（参见图 4）的这种随着 $l$ 的增加功率急剧上升的现象是在预料之中的。由于较大模式上的速度差异没有平均自由程尺度上的那么大，因此这些模式的偏振会被抑制。实际上，$l$ 较小时，$E$ 谱预计会以 $l(l+1)C_l \propto l^2$ 形式上升，但由于扩散阻尼抑制了小于平均自由程尺度上的功率，故这种依赖性不会保持到 DASI 敏感的较高的 $l$ 值。在 DASI 五频段 $E$ 分析中，较高 $l$ 频段的最大似然值的和带阻尼的协调模型一致，但位于 $l^2$ 幂律的简单外推模型的下方。同样，正如协调模型所预测的，五频段 $B$ 模式功率谱与零一致。

$TE$ 分析进一步使我们增强了对于 CMB 偏振信号的存在和协调模型的信心。根据 $T/E/TE$ 似然分析，$TE$ 为零的假设以 0.054 的 PTE 被拒绝。注意，$TE$ 可正可负，因此 $TE \leqslant 0$ 假设以更高的置信度被拒绝。

最后，$T$ 的谱指数的测量值为 0.01（置信度为 68% 的参数区间为 −0.16 至 0.14），极好地被约束在热的且不属于任何已知的前景。$E$ 的谱指数为 0.17（置信度为 68% 的参数区间为 −1.63 至 1.92）也与 CMB 一致，虽然比起 $T$、$E$ 的谱指数约束得不那么好，但与弥散前景同步辐射存在近 $2\sigma$ 的不一致性。

总之，本文报道的分析均表明存在一个显著的 CMB 的 $E$ 偏振模式信号，其中置信水平 $\geqslant 4.9\sigma$。测量到的偏振的特性和调和模型给出的预测非常一致，并且如“前景”部分所讨论的那样，与目前已知的前景辐射都不匹配。因此，这些结果为 CMB 各向异性的起源的基础理论框架提供了强而有力的支持，从而使得从 CMB 测量推导得出的宇宙学参数的值以及宇宙起源和结构的庞大图景更为可信。使用精确偏振测量进一步完善我们对宇宙的理解是许多正在进行的和计划中的 CMB 实验的目标。本文报道的探测到预计水平的偏振信号为该领域的光明未来指明了方向。

（谭秀慧 吕孟珍 翻译；夏俊卿 审稿）

J. M. Kovac[*†‡], E. M. Leitch[§†‡], C. Pryke[§†‡‖], J. E. Carlstrom[§*†‡‖], N. W. Halverson[¶†] & W. L. Holzapfel[¶†]
[*] Department of Physics; [†] Center for Astrophysical Research in Antarctica; [‡] Center for Cosmological Physics; [§] Department of Astronomy & Astrophysics and [‖]Enrico Fermi Institute, University of Chicago, 5640 South Ellis Avenue, Chicago, Illinois 60637, USA
[¶] Department of Physics, University of California at Berkeley, Le Conte Hall, California 94720, USA

Received 7 October; accepted 5 November 2002; doi:10.1038/nature 01269.

---

References:

1. Penzias, A. A. & Wilson, R. W. A measurement of excess antenna temperature at 4080 Mc/s. *Astrophys. J.* **142**, 419-421 (1965).
2. Mather, J. C. *et al.* Measurement of the cosmic microwave background spectrum by the COBE FIRAS instrument. *Astrophys. J.* **420**, 439-444 (1994).
3. Fixsen, D. J. *et al.* The cosmic microwave background spectrum from the full COBE FIRAS data set. *Astrophys. J.* **473**, 576-587 (1996).
4. Smoot, G. F. *et al.* Structure in the COBE differential microwave radiometer first-year maps. *Astrophys. J.* **396**, L1-L5 (1992).
5. Miller, A. D. *et al.* A measurement of the angular power spectrum of the cosmic microwave background form l = 100 to 400. *Astrophys. J.* **524**, L1-L4 (1999).
6. Halverson, N. W. *et al.* Degree angular scale interferometer first results: A measurement of the cosmic microwave background angular power spectrum. *Astrophys. J.* **568**, 38-45 (2002).
7. Netterfield, C. B. *et al.* A measurement by BOOMERANG of multiple peaks in the angular power spectrum of the cosmic microwave background. *Astrophys. J.* **571**, 604-614 (2002).
8. Lee, A. T. *et al.* A high spatial resolution analysis of the MAXIMA-1 cosmic microwave background anisotropy data. *Astrophys. J.* **561**, L1-L5 (2001).
9. Pearson, T. J. *et al.* The anisotropy of the microwave background to l = 3500: Mosaic observations with the cosmic background imager. *Astrophys. J.* (submitted); preprint astro-ph/0205388 at ⟨http://xxx.lanl.gov⟩ (2002).
10. Scott, P. F. *et al.* First results from the Very Small Array - III. The CMB power spectrum. *Mon. Not. R. Astron. Soc.* (submitted); preprint astro-ph/0205380 at ⟨http://xxx.lanl.gov⟩ (2002).
11. Hu, W. & Dodelson, S. Cosmic microwave background anisotropies. *Annu. Rev. Astron. Astrophys.* **40**, 171-216 (2002).
12. Kaiser, N. Small-angle anisotropy of the microwave background radiation in the adiabatic theory. *Mon. Not. R. Astron. Soc.* **202**, 1169-1180 (1983).
13. Bond, J. R. & Efstathiou, G. Cosmic background radiation anisotropies in universes dominated by nonbaryonic dark matter. *Astrophys. J.* **285**, L45-L48 (1984).
14. Polnarev, A. G. Polarization and anisotropy induced in the microwave background by cosmological gravitational waves. *Sov. Astron.* **29**, 607-613 (1985).
15. Kamionkowski, M., Kosowsky, A. & Stebbins, A. Statistics of cosmic microwave background polarization. *Phys. Rev. D.* **55**, 7368-7388 (1997).
16. Zaldarriaga, M. & Seljak, U. All-sky analysis of polarization in the microwave background. *Phys. Rev.D.* **55**, 1830-1840 (1997).
17. Hu, W. & White, M. A CMB polarization primer. *New Astron.* **2**, 323-344 (1997).
18. Kosowsky, A. Introduction to microwave background polarization. *New Astron. Rev.* **43**, 157-168 (1999).
19. Hu, W., Spergel, D. N. & White, M. Distinguishing causal seeds from inflation. *Phys. Rev. D* **55**, 3288-3302 (1997).
20. Kinney, W. H. How to fool cosmic microwave background parameter estimation. *Phys. Rev. D* **63**, 43001 (2001).
21. Bucher, M., Moodley, K. & Turok, N. Constraining isocurvature perturbations with cosmic microwave background polarization. *Phys. Rev. Lett.* **87**, 191301 (2001).
22. Rees, M. J. Polarization and spectrum of the primeval radiation in an anisotropic universe. *Astrophys. J.* **153**, L1-L5 (1968).
23. Zaldarriaga, M. & Harari, D. D. Analytic approach to the polarization of the cosmic microwave background in flat and open universes. *Phys. Rev. D* **52**, 3276-3287 (1995).
24. Coulson, D., Crittenden, R. G. & Turok, N. G. Polarization and anisotropy of the microwave sky. *Phys. Rev. Lett.* **73**, 2390-2393 (1994).
25. Crittenden, R., Davis, R. L. & Steinhardt, P. J. Polarization of the microwave background due to primordial gravitational waves. *Astrophys. J.* **417**, L13-L16 (1993).
26. Seljak, U. Measuring polarization in the cosmic microwave background. *Astrophys. J.* **482**, 6-16 (1997).
27. Kamionkowski, M., Kosowsky, A. & Stebbins, A. A probe of primordial gravity waves and vorticity. *Phys. Rev. Lett.* **78**, 2058-2061 (1997).
28. Seljak, U. & Zaldarriaga, M. Signature of gravity waves in the polarization of the microwave background. *Phys. Rev. Lett.* **78**, 2054-2057 (1997).
29. Lyth, D. H. What would we learn by detecting a gravitational wave signal in the Cosmic Microwave Background anisotropy? *Phys. Rev. Lett.* **78**, 1861-1863 (1997).
30. Zaldarriaga, M. & Seljak, U. Gravitational lensing effect on cosmic microwave background polarization. *Phys. Rev. D* **58**, 23003 (1998).
31. Hu, W. & Okamoto, T. Mass reconstruction with cosmic microwave background polarization. *Astrophys. J.* **574**, 566-574 (2002).
32. Knox, L. & Song, Y. A limit on the detectability of the energy scale of inflation. *Phys. Rev. Lett.* **89**, 011303 (2002).
33. Kesden, M., Cooray, A. & Kamionkowski, M. Separation of gravitational-wave and cosmic-shear contributions to cosmic microwave background polarization. *Phys. Rev. Lett.* **89**, 011304 (2002).
34. Kamionkowski, M. & Kosowsky, A. The cosmic microwave background and particle physics. *Annu. Rev. Nucl. Part. Sci.* **49**, 77-123 (1999).
35. Staggs, S. T., Gunderson, J. O. & Church, S. E. *ASP Conf. Ser. 181, Microwave Foregrounds* (eds de Oliveira-Costa, A. & Tegmark, M.) 299-309 (Astronomical Society of the Pacific, San Francisco, 1999).
36. Caderni, N., Fabbri, R., Melchiorri, B., Melchiorri, F. & Natale, V. Polarization of the microwave background radiation. II. An infrared survey of the sky. *Phys. Rev. D* **17**, 1908-1918 (1978).

37. Nanos, G. P. Polarization of the blackbody radiation at 3.2 centimeters. *Astrophys. J.* **232**, 341-347 (1979).
38. Lubin, P. M. & Smoot, G. F. Search for linear polarization of the cosmic background radiation. *Phys. Rev. Lett.* **42**, 129-132 (1979).
39. Lubin, P. M. & Smoot, G. F. Polarization of the cosmic background radiation. *Astrophys. J.* **245**, 1-17 (1981).
40. Lubin, P., Melese, P. & Smoot, G. Linear and circular polarization of the cosmic background radiation. *Astrophys. J.* **273**, L51-L54 (1983).
41. Sironi, G. *et al.* A 33 GHZ polarimeter for observations of the cosmic microwave background. *New Astron.* **3**, 1-13 (1997).
42. Keating, B. G. *et al.* A limit on the large angular scale polarization of the cosmic microwave background. *Astrophys. J.* **560**, L1-L4 (2001).
43. Wollack, E. J., Jarosik, N. C., Netterfield, C. B., Page, L. A. & Wilkinson, D. A measurement of the anisotropy in the cosmic microwave background radiation at degree angular scales. *Astrophys. J.* **419**, L49-L52 (1993).
44. Hedman, M. M. *et al.* New limits on the polarized anisotropy of the cosmic microwave background at subdegree angular scales. *Astrophys. J.* **573**, L73-L76 (2002).
45. Cartwright, J. K. *et al.* Polarization observations with the cosmic background imager. in *Moriond Workshop 37, The Cosmological Model* (in the press).
46. de Oliveira-Costa, A. *et al.* First attempt at measuring the CMB cross-polarization. *Phys. Rev. D* (submitted); preprint astro-ph/0204021 at 〈http://xxx.lanl.gov〉 (2002).
47. Partridge, R. B., Richards, E. A., Fomalont, E. B., Kellermann, K. I. & Windhorst, R. A. Small-scale cosmic microwave background observations at 8.4 GHz. *Astrophys. J.* **483**, 38-50 (1997).
48. Subrahmanyan, R., Kesteven, M. J., Ekers, R. D., Sinclair, M. & Silk, J. An Australia telescope survey for CMB anisotropies. *Mon. Not. R. Astron. Soc.* **315**, 808-822 (2000).
49. Leitch, E. M. *et al.* Experiment design and first season observations with the degree angular scale interferometer. *Astrophys. J.* **568**, 28-37 (2002).
50. Pryke, C. *et al.* Cosmological parameter extraction from the first season of observations with the degree angular scale interferometer. *Astrophys. J.* **568**, 46-51 (2002).
51. Leitch, E. M. *et al.* Measurement of polarization with the Degree Angular Scale Interferometer. *Nature* **420**, 763-771 (2002).
52. Haslam, C. G. T. *et al.* A 408 MHz all-sky continuum survey. I—Observations at southern declinations and for the north polar region. *Astron. Astrophys.* **100**, 209-219 (1981).
53. Bond, J. R., Jaffe, A. H. & Knox, L. Estimating the power spectrum of the cosmic microwave background. *Phys. Rev. D* **57**, 2117-2137 (1998).
54. Seljak, U. & Zaldarriaga, M. A line-of-sight integration approach to cosmic microwave background anisotropies. *Astrophys. J.* **469**, 437-444 (1996).
55. White, M., Carlstrom, J. E., Dragovan, M. & Holzapfel, W. H. Interferometric observation of cosmic microwave background anisotropies. *Astrophys. J.* **514**, 12-24 (1999).
56. Hedman, M. M., Barkats, D., Gundersen, J. O., Staggs, S. T. & Winstein, B. A limit on the polarized anisotropy of the cosmic microwave background at subdegree angular scales. *Astrophys. J.* **548**, L111-L114 (2001).
57. Knox, L. Cosmic microwave background anisotropy window functions revisited. *Phys. Rev. D* **60**, 103516 (1999).
58. Halverson, N. W. *A Measurement of the Cosmic Microwave Background Angular Power Spectrum with DASI.* PhD thesis, Caltech (2002).
59. Tegmark, M. & de Oliveira-Costa, A. How to measure CMB polarization power spectra without losing information. *Phys. Rev. D* **64**, 063001 (2001).
60. Christensen, N., Meyer, R., Knox, L. & Luey, B. Bayesian methods for cosmological parameter estimation from cosmic microwave background measurements. *Class. Quant. Gravity* **18**, 2677-2688 (2001).
61. Wang, X., Tegmark, M. & Zaldarriaga, M. Is cosmology consistent? *Phys. Rev. D* **65**, 123001 (2002).
62. Wright, A. E., Griffith, M. R., Burke, B. F. & Ekers, R. D. The Parkes-MIT-NRAO (PMN) surveys. 2: Source catalog for the southern survey ($-87.5° < \delta < -37°$). *Astrophys. J. Suppl.* **91**, 111-308 (1994).
63. Mason, B. S., Pearson, T. J., Readhead, A. C. S., Shepherd, M. C. & Sievers, J. L. The anisotropy of the microwave background to 1 = 3500: Deep field observations with the cosmic background imager. *Astrophys. J.* (submitted); preprint astro-ph/0205384 at 〈http://xxx.lanl.gov〉 (2002).
64. Zukowski, E. L. H., Kronberg, P. P., Forkert, T. & Wielebinski, R. Linear polarization measurements of extragalactic radio sources at λ6.3 cm. *Astron. Astrophys. Suppl.* **135**, 571-577 (1999).
65. Simard-Normandin, M., Kronberg, P. P. & Neidhoefer, J. Linear polarization observations of extragalactic radio sources at 2 cm and at 17-19 cm. *Astron. Astrophys. Suppl.* **43**, 19-22 (1981).
66. Simard-Normandin, M., Kronberg, P. P. & Button, S. The Faraday rotation measures of extragalactic radio sources. *Astrophys. J. Suppl.* **45**, 97-111 (1981).
67. Hildebrand, R. H. *et al.* A primer on far-infrared polarimetry. *Publ. Astron. Soc. Pacif.* **112**, 1215-1235 (2000).
68. Lazarian, A. & Prunet, S. *AIP Conf. Proc. 609, Astrophysical Polarized Backgrounds* (eds Cecchini, S., Cortiglioni, S., Sault, R. & Sbarra, C.) 32-43 (AIP, Melville, New York, 2002).
69. Tegmark, M., Eisenstein, D. J., Hu, W. & de Oliveria-Costa, A. Foregrounds and forecasts for the cosmic microwave background. *Astrophys. J.* **530**, 133-165 (2000).
70. Giardino, G. *et al.* Towards a model of full-sky Galactic synchrotron intensity and linear polarisation: A re-analysis of the Parkes data. *Astron. Astrophys.* **387**, 82-97 (2002).
71. Gaensler, B. M. *et al.* Radio polarization from the inner galaxy at arcminute resolution. *Astrophys. J.* **549**, 959-978 (2001).
72. Gray, A. D. *et al.* Radio polarimetric imaging of the interstellar medium: Magnetic field and diffuse ionized gas structure near the W3/W4/W5/HB 3 complex. *Astrophys. J.* **514**, 221-231 (1999).
73. Platania, P. *et al.* A determination of the spectral index of galactic synchrotron emission in the 1–10 GHz range. *Astrophys. J.* **505**, 473-483 (1998).
74. Banday, A. J. & Wolfendale, A. W. Fluctuations in the galactic synchrotron radiation—I. Implications for searches for fluctuations of cosmological origin. *Mon. Not. R. Astron. Soc.* **248**, 705-714 (1991).

75. Zaldarriaga, M. The nature of the E-B decomposition of CMB polarization. *Phys. Rev. D* (submitted); preprint astro-ph/0106174 at ⟨http://xxx.lanl.gov⟩ (2001).

**Acknowledgements**. We are grateful for the efforts of B. Reddall and E. Sandberg, who wintered over at the National Science Foundation (NSF) Amundsen–Scott South Pole research station to keep DASI running smoothly. We are indebted to M. Dragovan for his role in making DASI a reality, and to the Caltech CBI team led by T. Readhead, in particular, to S. Padin, J. Cartwright, M. Shepherd and J. Yamasaki for the development of key hardware and software. We are indebted to the Center for Astrophysical Research in Antarctica (CARA), in particular to the CARA polar operations staff. We are grateful for contributions from K. Coble, A. Day, G. Drag, J. Kooi, E. LaRue, M. Loh, R. Lowenstein, S. Meyer, N. Odalen, R. Pernic, D. Pernic and E. Pernic, R. Spotz and M. Whitehead. We thank Raytheon Polar Services and the US Antarctic Program for their support of the DASI project. We have benefitted from many interactions with the Center for Cosmological Physics members and visitors. In particular, we gratefully acknowledge many conversations with W. Hu on CMB polarization and suggestions from S. Meyer, L. Page, M. Turner and B. Winstein on the presentation of these results. We thank L. Knox and A. Kosowsky for bringing the Markov technique to our attention. We thank the observatory staff of the Australia Telescope Compact Array, in particular B. Sault and R. Subrahmanyan, for providing point source observations of the DASI fields. This research was initially supported by the NSF under a cooperative agreement with CARA, a NSF Science and Technology Center. It is currently supported by an NSF-OPP grant. J.E.C. gratefully acknowledges support from the James S. McDonnell Foundation and the David and Lucile Packard Foundation. J.E.C. and C.P. gratefully acknowledge support from the Center for Cosmological Physics.

**Competing interests statement**. The authors declare that they have no competing financial interests.

Correspondence and requests for materials should be addressed to J.M.K. (e-mail: john@hyde.uchicago.edu).

# Upper Limits to Submillimetre-range Forces from Extra Space-time Dimensions

J. C. Long *et al.*

## Editor's Note

**Gravity is the weakest of the four fundamental forces, being 40 orders of magnitude weaker than the electromagnetic force. In string theory—an attempt to unify the description of particles and forces—this weakness arises because the force is dissipated in six extra spatial dimensions that are "curled up" and therefore invisible to us. It has been suggested that the effect of these curled-up dimensions might become evident at very small length scales, whereby objects would feel more gravitation force at those scales. Here physicist Joshua Long and colleagues describe an experiment that looked for a deviation of the gravitational force from its familiar inverse-square law on length scales of 100 micrometres. They found no such deviation.**

---

String theory is the most promising approach to the long-sought unified description of the four forces of nature and the elementary particles[1], but direct evidence supporting it is lacking. The theory requires six extra spatial dimensions beyond the three that we observe; it is usually supposed that these extra dimensions are curled up into small spaces. This "compactification" induces "moduli" fields, which describe the size and shape of the compact dimensions at each point in space-time. These moduli fields generate forces with strengths comparable to gravity, which according to some recent predictions[2-7] might be detected on length scales of about 100 μm. Here we report a search for gravitational-strength forces using planar oscillators separated by a gap of 108 μm. No new forces are observed, ruling out a substantial portion of the previously allowed parameter space[4] for the strange and gluon moduli forces, and setting a new upper limit on the range of the string dilaton[2,3] and radion[5-7] forces.

---

THE combined potential energy $V$ due to a modulus force and newtonian gravity may be written:

$$V = -\int d\mathbf{r}_1 \int d\mathbf{r}_2 \, \frac{G\rho_1(\mathbf{r}_1)\rho_2(\mathbf{r}_2)}{r_{12}} \, [1+\alpha \exp(-r_{12}/\lambda)] \tag{1}$$

The first term is Newton's universal gravitation law, with $G$ the gravitational constant, $r_{12}$ the distance between two points $\mathbf{r}_1$ and $\mathbf{r}_2$ in the test masses, and $\rho_1$, $\rho_2$ the mass densities of the two bodies. The second term is a Yukawa potential, with $\alpha$ the strength of the new force relative to gravity, and $\lambda$ the range. As for any force-mediating field, the range of the modulus force is related to its mass $m$ by $\lambda = \hbar/mc$. Previous tests of Newtonian gravity

# 来自额外空间维度在亚毫米尺度上力程的上限

朗等

编者按

引力是四种基本力中最弱的，比电磁力要小四十个数量级。在弦论——一个试图统一描述粒子和力的尝试——中，引力的弱性是由于引力向着六个额外的卷曲不可见维度泄漏导致的。已经有一些建议说这些卷曲维度的效应可能在很小的尺度下会变得很明显，在这些尺度上物体将会感受到更大的引力。本文中，物理学家乔舒亚·朗和合作者们介绍了一个实验，这个实验在 100 μm 的尺度上寻找引力偏离通常熟知的平方反比律的迹象。他们发现没有这样的偏离。

---

弦理论是人们一直在寻找的、统一自然界四种力和基本粒子的最有希望的方法[1]，但是现在仍然缺乏直接支持它的证据。除了我们观察到的三个维度之外，弦理论还要求存在六个额外空间维度。通常假定这些额外的维度卷曲成了很小的空间。这个“紧致化”诱导出了“模”场，用来描述时空中每一点紧致维度的大小和形状。根据最近的预测[2-7]，这些模场产生了和引力大小可比的力，可能能在 100 μm 尺度上被探测到。本文中，我们报告了以下工作：用相隔 108 μm 的两个平面振子来搜寻这一引力强度量级的力。结果未观察到新的力，从而排除了一大块之前对于奇异模力和胶子模力来说是允许的参数空间[4]，对弦伸缩子力[2,3]和径向子力[5-7]的力程给出了一个新的上限。

---

由模力和牛顿引力结合得到的势能 $V$ 可以写作

$$V=-\int d\mathbf{r}_1\int d\mathbf{r}_2\,\frac{G\rho_1(\mathbf{r}_1)\rho_2(\mathbf{r}_2)}{r_{12}}\,[1+\alpha\exp(-r_{12}/\lambda)] \tag{1}$$

第一项是牛顿万有引力定律，$G$ 是引力常数，$r_{12}$ 是检验质量位置 $\mathbf{r}_1$ 和 $\mathbf{r}_2$ 两点间的距离，$\rho_1$，$\rho_2$ 是两个物体的质量密度。第二项是汤川势，$\alpha$ 是新的力相对于引力的强度，$\lambda$ 是力程。对于任何传递相互作用的场，模力的力程和它的质量 $m$ 存在如下关系 $\lambda=\hbar/mc$。之前关于牛顿引力的检验和对于新的宏观力的搜寻涵盖了从光年到纳

and searches for new macroscopic forces have covered length scales from light-years to nanometres[8-10], and it has been found that new forces of gravitational strength can be excluded for ranges $\lambda$ from 200 μm to nearly a light-year[8,11,12], but limits on new forces become poor very rapidly below 200 μm (refs 9, 13).

The strange and gluon moduli arise in a scenario where supersymmetry (a hypothesized symmetry incorporated in string theory that relates bosons to fermions) is broken at 10–100 TeV, and all compactification occurs near the ultrahigh Planck scale of $10^{19}$ GeV where gravity is unified (in these scenarios) with other forces. In this case a range of approximately a millimetre is specifically predicted. The dilaton is a scalar field required for the consistency of string theory whose vacuum value fixes the strength of the interaction between strings. It may be interpreted as the modulus describing the size of the tenth space dimension that is compactified in M-theory to yield string theory in nine space dimensions. (M-theory is the structure that unifies the various types of string theory into a single framework[1].) The strength $\alpha$ associated with the dilaton can be computed, but the range $\lambda$ at present can only be constrained by experiment; little is known about the mechanism which gives the dilaton mass. The radion results from a very different scenario in which one or more dimensions compactify at TeV energy scales, but here a range $\lambda$ accessible to our experiment is again predicted.

The special significance of millimetre scales derives from a mass formula of the form $m \approx M^2/M_P$, where $M_P$ is the mass associated with the Planck scale, $10^{19}$ GeV$/c^2$, and $M$ is a mass of 1–100 TeV$/c^2$, leading to $\lambda$ in the centimetre to micrometre range. In the strange and gluon moduli case, $M$ is the scale where supersymmetry breaking occurs, whereas in the radion modulus case it corresponds to the size of one or more compact dimensions. This formula also applies to the ADD theory[14] (named for its authors) in which two compact extra dimensions of millimetre size modify gravity itself. In the ADD theory $M$ is the fundamental length scale where all physics is unified while $m$ is a mass corresponding to the size of the large compact extra dimensions. Many scenarios of compactification, symmetry breaking and mass generation are still viable, so although the possibility of observing new forces at millimetre scales is exciting, such experiments cannot currently falsify string theory.

The planar geometry of our source and detector masses (Fig. 1) is chosen to concentrate as much mass density as possible at the length scale of interest. It is approximately null with respect to the $1/r^2$ newtonian background, a helpful feature in the context of a new force search. A cantilever mode (similar to the motion of a diving board) of the tungsten source mass is driven to a tip amplitude of 19 μm at the resonant frequency of the detector mass. The tungsten detector mass is a double torsional oscillator[15]; in the resonant mode of interest the two rectangular sections of the detector counter-rotate about the torsion axis, with most of the amplitude confined to the smaller rectangle under the source mass and shield. Torsional motion of the detector will be driven if a mass coupled force is present between the source and detector. The motions are detected with a capacitive transducer, followed by a preamplifier, filters and a lock-in amplifier. To suppress background forces due to electrostatics and residual gas, a stiff conducting shield is fixed between the test masses. With

米的尺度[8-10]，已有结果表明，在 200 μm 到接近一个光年的尺度范围内不存在引力强度量级的新力[8,11,12]，但是在 200 μm 之下的尺度，对出现新力的限制会快速变弱(参考文献 9 和 13)。

奇异模空间和胶子模空间出现在 10 ~ 100 TeV 下超对称(一个弦论中猜想的将玻色子和费米子联系起来的对称性)破缺的时候，而所有的紧致化过程则是发生在临近超高普朗克尺度($10^{19}$ GeV)上的(这些情况下的引力和其他力是统一的)。在这种情况下，可以预测出一个大约是毫米级的力程。伸缩子是弦理论自洽性所要求的一个标量场，它的真空值决定了弦之间相互作用的强度。它或许可以被理解成是 M 理论中描述第十个空间维度的尺度的模参数，这个空间维度在 M 理论中被做了紧致化来得到具有九个空间维度的弦论(M 理论是将所有不同的弦论整合成一个框架的理论结构[1])。和伸缩子相关的强度 $\alpha$ 可以通过计算得到，但是力程 $\lambda$ 目前只能通过实验来限制，并且我们几乎不知道给出伸缩子质量的机制是什么。径向子是由一个完全不同的机制得到的，这一机制下一个或者多个额外维度的紧致化发生在 TeV 能标尺度上，但这里实验可以得到的力程 $\lambda$ 也是预测出来的。

毫米尺度特殊的重要性可以通过形如 $m \approx M^2/M_{\mathrm{P}}$ 的质量公式导出，这里 $M_{\mathrm{P}}$ 是和普朗克尺度相关的质量 $10^{19}$ GeV/$c^2$，$M$ 是 1 ~ 100 TeV/$c^2$ 的质量，导致 $\lambda$ 的尺度在厘米到微米范围之间。在奇异模空间和胶子模空间中，$M$ 是超对称破缺发生的尺度，而在径向子模空间的情况下，它对应于一个或者多个紧致维度的尺度。这个公式适用于 ADD 理论[14](根据其提出者们的姓名命名)，这一理论中两个毫米尺度的紧致化额外维会修改引力。在 ADD 理论中，$M$ 是一个基本尺度，在这个尺度上所有的物理理论都是统一起来的，而 $m$ 是对应于大的紧致化额外维尺度的质量。在许多紧致化图像中，对称性破缺和质量产生依然是可靠的，因此，尽管在毫米尺度上观察到新力的可能性是令人激动的，但这些实验目前还不能证伪弦理论。

我们选择的源和探测器质量呈现为图 1 所示的平面几何，是为了使其尽可能地在感兴趣的尺度上集中更高的质量密度。它几乎不受 $1/r^2$ 牛顿背景的影响，这在寻找一个新力的时候是一个有帮助的特征。钨制的源的悬臂模式(和跳水板的运动类似)在探测器共振频率的驱动下出现了 19 μm 的尖端振幅。钨探测器是双扭转振子[15]，在我们感兴趣的共振模式下，探测器的两个矩形部分绕着扭转轴反转，绝大部分振幅都被限制在防护屏和源下面的小矩形中。如果源和探测器之间存在一个质量耦合力的话将会驱动探测器的扭转运动。运动用电容传感器探测，跟着是一个前置放大器、过滤器和锁定放大器。为了压低由于静电和剩余气体产生的背景力，一个刚性的导电防护屏固定在两个检验质量之间。对于静止的源，源和探测器的距离

the source at rest, the gap between source and detector is adjusted to 108 μm, and the entire apparatus is placed in a vacuum enclosure and maintained at pressures below $2 \times 10^{-7}$ torr.

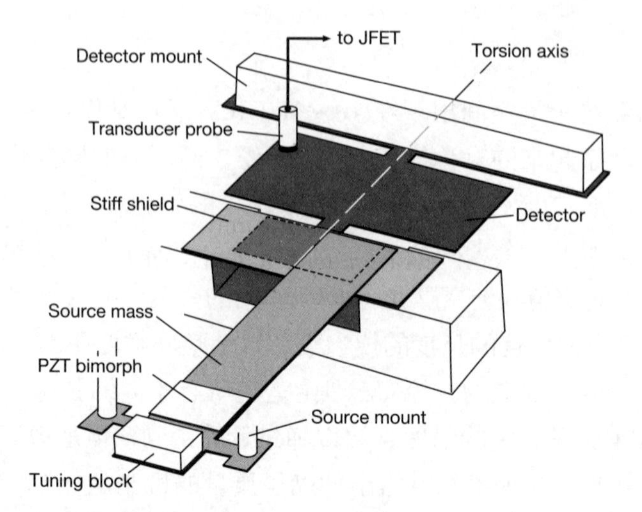


Fig. 1. Major components of the apparatus. The smaller rectangle of the tungsten detector (under the shield) is 11.455 mm wide, 5.080 mm long and 195 μm thick. The detector is annealed at 1,300 °C to increase its mechanical $Q$ to 25,000. In operation the 1,173.085-Hz resonant frequency of the detector is stabilized by actively controlling the detector temperature to 305 K. The tungsten source mass is 35 mm long, 7 mm wide and 305 μm thick. The source mass resonant frequency is tuned to the detector and driven by the PZT (lead zirconate titanate) piezoelectric bimorph. The shield is a 60-μm-thick sapphire plate coated with 100 nm of gold on both sides. The test masses and the shield are supported by three separate five-stage passive vibration isolation stacks[23], each providing approximately 200-dB attenuation at 1 kHz. Mechanical probes are used to directly measure the relative orientation and position of each component, and to measure the source mass amplitude. Detector motions are sensed by a cylindrical capacitive probe supported 100 μm above a rear corner of the large rectangle of the detector mass. The probe is biased at 200 V through a 100 GΩ resistor, and connected to an SK 152 junction field-effect transistor (JFET) through a blocking capacitor. The JFET noise temperature of 100 mK is more than sufficient to detect 305 K thermal motions of the detector mass. The JFET preamplifier is followed by a second preamplifier, filters, and finally a two-phase lock-in amplifier. The total voltage gain from the capacitive probe to the lock-in input is approximately 1,600. A crystal-controlled oscillator provides a reference signal for the lock-in amplifier and drives the source mass PZT through a 1:10 step-up transformer.

Figure 2 shows histograms of the raw data, collected with the source mass drive tuned both on and off the detector resonant frequency. Each plot contains data from one channel of the lock-in amplifier, corresponding to one of two orthogonal phases of the detector motion at the drive frequency. The widths of the off-resonance distributions are due to preamplifier noise, whereas the on-resonance distributions are due to the sum of preamplifier noise and detector thermal motions. The on- and off-resonance means shown in Fig. 3 agree within their standard deviations, indicating the absence of a significant resonant force signal.

调整到 108 μm，整个仪器放在真空罩中，然后将压强控制在 $2 \times 10^{-7}$ torr 以下。

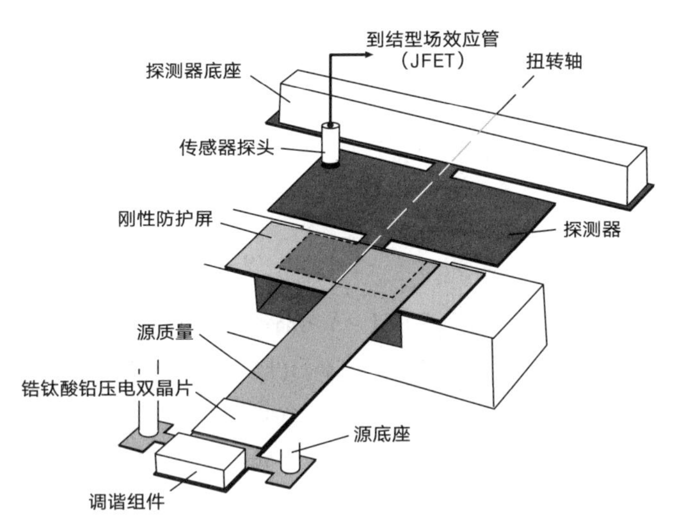


图 1. 仪器的主要构成。钨探测器的小矩形（在防护屏下面）有 11.455 mm 宽，5.080 mm 长，195 μm 厚。探测器在 1,300 ℃下退火以便增加它的机械品质因子 $Q$ 到 25,000。在操作上，探测器 1,173.085 Hz 的共振频率是通过主动地将探测器的温度控制在 305 K 来稳定下来。钨制的源有 35 mm 长，7 mm 宽，305 μm 厚。源的共振频率调制到探测器的共振频率，并由锆钛酸铅（PZT）压电双晶片驱动。防护屏是一个 60 μm 厚的蓝宝石板，两侧镀上了 100 nm 厚的金。检验质量和防护屏由三个分立的五步被动隔振层支撑[23]，每一个在 1 kHz 时提供了大约 200 dB 的噪音减弱。机械探头用来直接探测每个组分的相对取向和位置，还用来测量源的振幅。探测器的运动通过在探测器大矩形后面的角的上方 100 μm 处的一个圆柱形电容式探头来感知。探头通过一个 100 GΩ 的电阻在 200 V 电压上偏置，并且通过一个隔直流电容器连着一个 SK 152 的结型场效应晶体管（JFET）。JFET 100 mK 的噪声温度足够用来探测 305 K 的探测器的热运动了。JFET 前置放大器接着第二个前置放大器、过滤器，最后是一个两相锁定放大器。从电容式探头到锁定放大器输入的总电压放大了约 1,600 倍。一个晶体调控的振子给锁定放大器提供了一个参考信号，并且通过一个 1∶10 的升压变压器驱动源质量的 PZT。

图 2 展示了原始数据的直方图，当源的驱动频率在探测器的共振频率时和不在探测器的共振频率时都收集了数据。每个图都包含来自锁定放大器的一个通道的信号，对应于在驱动频率下探测器运动的两个正交模式中的一个。非共振分布的展宽是因为前置放大器的噪声，而共振分布的展宽来自前置放大器的噪声和探测器的热运动之和。图 3 中这个共振和非共振的平均值在标准偏差之内是吻合的，表明不存在一个显著的共振力的信号。

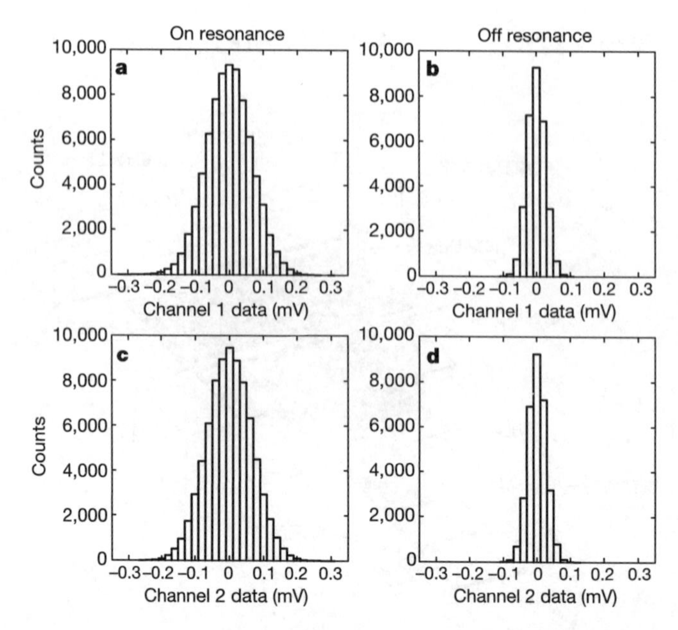


Fig. 2. Distributions of data samples. Data were recorded at 1 Hz with a lock-in bandwidth chosen to include the noise power of the detector thermal oscillations, which was used for calibration. Each data cycle began with five 120-sample diagnostic runs with a direct-current bias of 5–10 V applied to the shield to induce a large test force, transmitted from source to detector via deflections of the shield. The biased runs were recorded at five drive frequencies separated by 15 mHz to cover the detector resonance. The shield was then grounded and the cycle continued with 720 samples with the drive tuned on-resonance (**a, c**) and 288 samples with the drive tuned 2 Hz below the detector resonance (**b, d**). The off-resonance run provided a continuous zero check. A total of 108 such cycles were acquired over five days yielding 77,760 on-resonance samples. The biased diagnostic data show that the source mass amplitude, the detector $Q$ and resonant frequency, and the electronic gain were all stable throughout the data set.

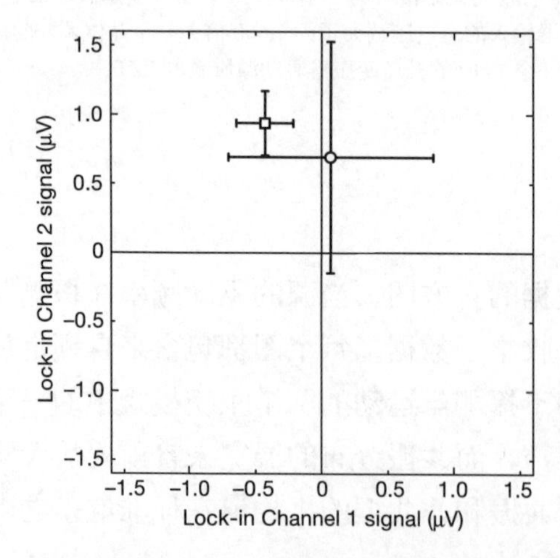


Fig. 3. Means of the off- and on-resonance data samples. The circular point with the larger standard deviations is the on-resonance mean. Correlations between nearby samples have been accounted for in computing the standard deviations shown by error bars. The small offset from the origin is due to leakage of the reference signal internal to the lock-in. Measurements with the shield removed and a bias voltage between the source and detector show that the phase for an attractive force is 189°. The on-resonance mean minus the off-resonance mean at 189° is $-0.44 \pm 0.82$ μV. Based on calibration via the equipartition

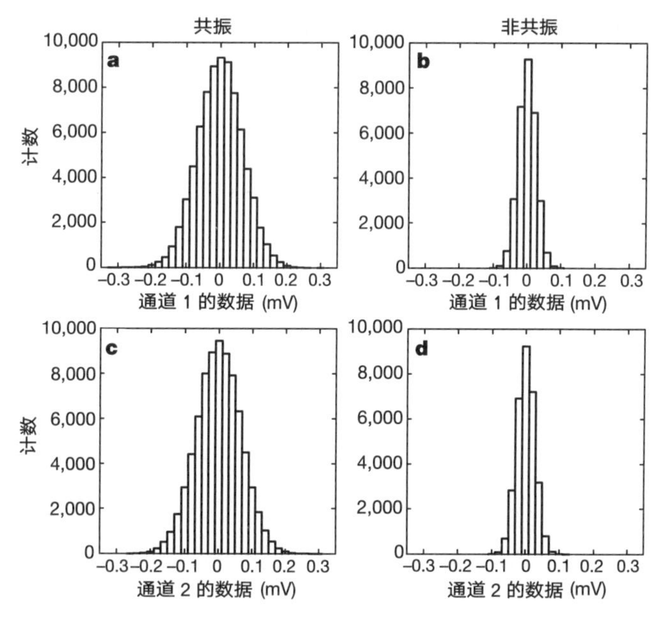


图 2. 数据样本的分布。数据在 1 Hz 处收集，选取的锁定带宽包含了探测器热振动的噪声功率以用于校正。每个数据周期从 120 个样本的 5 次诊断运行开始，在防护屏上应用 5 ~ 10 V 的直流偏压，用来产生一个很大的检测力，然后通过防护屏的偏转在源到探测器之间传递。偏置运行在五个驱动频率下记录，这五个驱动频率之间间隔 15 mHz 从而覆盖探测器共振频率。防护屏之后接地，然后这个周期继续在驱动频率调到共振频率的 720 个样本(**a** 和 **c**)和驱动频率调到探测器共振频率以下 2 Hz 处的 288 个的样本(**b** 和 **d**)中进行。非共振运行提供了一个连续的零检验。五天时间内获得了共 108 个这样的循环，产生了 77,760 个共振样本。偏置诊断的数据显示源质量的振幅、探测器的品质因子 $Q$ 和共振频率以及电子增益在整个数据集上都是稳定的。

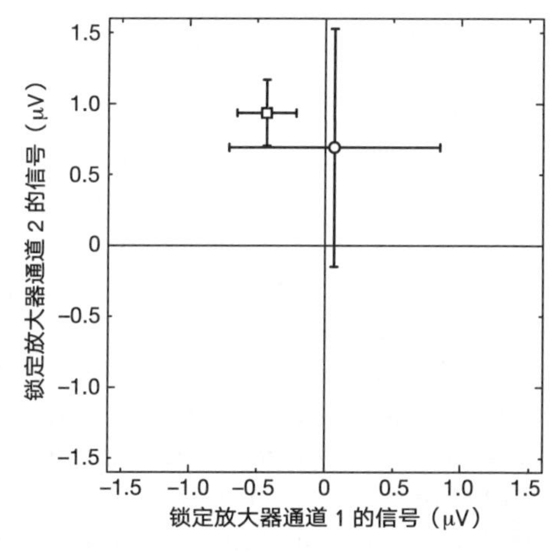


图 3. 非共振和共振数据样本的平均值。具有更大标准偏差的圆点是共振的平均值。邻近样本之间的关联在计算标准偏差时已经考虑进去了，标准偏差用误差棒来体现。原点处的微小偏移来源于锁定放大器内部的参考信号的溢出。去掉防护屏，并在源和探测器之间加上偏置电压进行了探测，结果显示吸引力的相位是 189°。在 189°，共振的平均值减去非共振的平均值是 −0.44 ± 0.82 μV。基于通过能均分定理

theorem this corresponds to a lumped force amplitude at the edge of the detector (smaller rectangle in Fig. 1) of $-1.2 \pm 2.2$ fN, where the negative sign indicates repulsion.

Additional cycles of data were acquired with the source mass on the opposite side of the detector, with a larger, 1-mm gap, and with reduced overlap between the source and detector. No resonant signal was observed in any of these sessions, making it unlikely that the observed null result is due to a fortuitous cancellation of surface potential, magnetic, and/or acoustic effects. Several on-resonance runs were acquired with different transducer probe bias voltage settings. The observed linear dependence on bias voltage of the root-mean-square (r.m.s.) fluctuations in these data is consistent with detector motion due only to thermal noise and rules out additional motion from transducer back-action noise. This check is important because the magnitude of the detector thermal motion is used for calibration.

Data from diagnostic runs with a bias voltage applied to the shield can be used to estimate the minimum size of the residual potential difference between the shield and the (grounded) test masses needed to produce a resonant signal. We find that at least 1.5 V would be needed to generate an effect above detector thermal noise, about an order of magnitude larger than the residual potential difference actually measured between the shield and test masses. The most important magnetic background effect involves eddy currents generated when the source mass moves in an external magnetic field. Fields produced by the source currents create eddy currents in the detector, which then interact with the external field. Studies of this effect with large applied magnetic fields show that the ambient field actually present cannot generate a signal greater than one-fifth of the thermal-noise-limited sensitivity.

The instrument can be calibrated in several ways, but the most accurate method is to use the r.m.s. thermal motion of the detector, which dominates the on-resonance distributions in Fig. 2. According to equipartition, the average kinetic energy in each normal mode of the detector is equal to $\frac{1}{2}kT$; where $k$ is Boltzmann's constant and $T$ is the temperature. The normal mode amplitude corresponding to the thermal energy can be calculated, and by comparing this with the observed voltage fluctuations a calibration can be established relating mode amplitude to voltage. A further calculation must be done to find the mode amplitude resulting from any hypothesized force.

For given values of $\alpha$ and $\lambda$ the driving force due to equation (1) is computed at 30 values of the source mass phase using Monte Carlo integration, and then the Fourier amplitude of the driving force at the resonant frequency is computed. Using the observed statistics of the data we construct a likelihood function for $\alpha$ for each value of $\lambda$ and compute 95% confidence-level upper limits on $\alpha$, assuming a uniform prior probability density function (PDF) for $\alpha$. This analysis is complicated by the presence of uncertainties in the geometrical and mechanical parameters needed to compute the driven displacement. To include these effects we actually construct a likelihood function of $\alpha$ and the uncertain parameters, and then integrate out the uncertain parameters using prior PDFs based on their experimental uncertainties. The most important of these parameters is the 108-

的校正，这对应于一个在探测器边缘(图 1 中的小矩形)大小为 −1.2±2.2 fN 的集总力振幅，负号表明它是排斥力。

更多的数据周期通过将源放在探测器的另外一侧获得，源和探测器之间有一个比之前大的、大小为 1 mm 的间隙，并且有着更小的交叠。在这些测量中都没有看到任何的共振信号，使得观测到的零结果不太可能源于表面势、磁和(或)声学效应的偶然抵消。设定不同的传感器探头的偏置电压获得了几次共振。在这些数据中观测到的方均根误差的涨落对于偏置电压的线性依赖和只由热噪声产生的探测器运动是自洽的，排除了传感器反作用噪声导致的额外运动。这个检查是重要的，因为探测器热运动的大小是被用来做校正的。

来自诊断运行(在防护屏上施加偏置电压)的数据可以用来估计防护屏和(接地的)检验质量之间的剩余电势差的最小尺度，这个电势差是需要的，它可以产生一个共振信号。我们发现至少需要 1.5 V 的电压来产生一个高于探测器热噪声的效应，1.5 V 大约比测得的在防护屏和检验质量之间的剩余电势差高一个数量级。最重要的磁背景效应是当源在外磁场运动时产生的涡流。源电流产生的场在探测器中产生了涡流，涡流之后会和外部的场相互作用。关于这个大施加磁场带来的效应的研究表明，周围实际存在的场所产生的信号，不能大于热噪声限制给出的灵敏度的五分之一。

这个实验器材可以通过几种方式校正，但是最准确的方法是用探测器的方均根热运动进行校正，它在图 2 的共振分布中占主导。根据能均分定理，每个探测器正则模式的平均动能等于$\frac{1}{2}kT$；$k$ 是玻尔兹曼常数而 $T$ 是温度。对应于热能的正则模式的振幅可以算出来，比较它和观测到的电压涨落，可以通过将模式振幅和电压联系起来建立一个校正。如果要寻找任何假设存在的力所产生的模式振幅，需要进行更进一步的计算。

对于给定的 $\alpha$ 和 $\lambda$ 的值，由方程(1)给出的驱动力在源相位取的 30 个值下用蒙特卡罗积分计算，然后算出驱动力在共振频率下的傅里叶振幅。用观测到的数据统计，我们对每个 $\lambda$ 的值构建了一个 $\alpha$ 的似然函数，然后在假定了一个 $\alpha$ 的均匀先验概率密度分布函数(PDF)的基础上，在 95% 的置信度上计算了 $\alpha$ 的上限。由于计算驱动位移时所需要的几何和动力学参数存在不确定性，这个分析会变得复杂。为了包括这些效应，我们实际上构建了一个 $\alpha$ 和这些不确定参数的似然函数，然后基于实验不确定性，利用先验概率密度分布函数积掉了不确定的参数。这些参数中最重要的是在源和探测器之间具有 6 μm 不确定性的 108 μm 平衡间隙。分析方法的进一

μm equilibrium gap between source and detector which has a 6-μm uncertainty. Further details of our analysis methods are given elsewhere[16].

Our results are shown in Fig. 4, together with other experimental limits and the theoretical predictions. Our limit is the strongest available between 10 and 100 μm. When the gluon and strange moduli forces were first proposed, the areas of their allowed regions in the ($\alpha$, $\lambda$) parameter space were 4.4 and 5.2 square decades respectively. By now they are nearly excluded with only 0.75 and 1.4 square decades still available. Our limit on the dilaton range for $\alpha = 2{,}000$ is $\lambda < 23$ μm (corresponding to $m > 8.6$ meV), a factor of two better than the previous limit. For the radion modulus we set an upper limit on $\lambda$ of 88 μm, close to the value of 40 μm estimated in the theory (for one extra dimension). For the ADD theory with two large extra compact dimensions, we do not quite reach the limit on the size of these dimensions already set in refs 11 and 12.

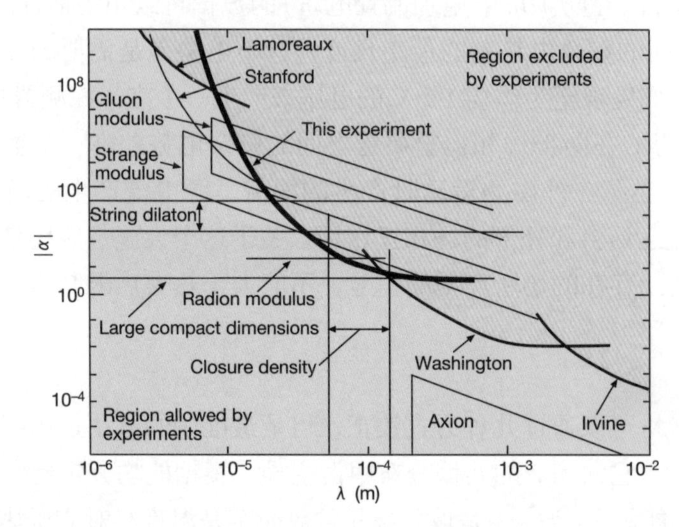


Fig. 4. Current limits on new gravitational strength forces between 1 μm and 1 cm. Our result is a 95% confidence-level upper limit on the Yukawa strength $\alpha$ as a function of range $\lambda$ (solid bold curve). It is shown together with limits from previous experiments (Lamoreaux[24], Washington[12], Irvine[25]) and theoretical predictions newly constrained (gluon modulus[4], strange modulus[4], string dilaton[2], radion modulus[7]). An unpublished limit from the Stanford experiment[26] is also shown; it is derived in the presence of a background force. The dilaton strength is somewhat model-dependent and there is a range of values reported in the literature[2]. We have chosen the region $200 < \alpha < 3{,}000$, which includes most values. Also shown are predictions from the ADD theory with two large compact extra dimensions[14,27,28], axion mediated forces[20,21], and the $\lambda$ region corresponding to a cosmological energy density between 1.0 and 0.1 times the closure density[17]. For the moduli, dilaton, and ADD theories, the upper bounds on $\lambda$ of the regions shown are set at the experimental limits that were known at the time the theories were proposed.

Besides forces from extra dimensions, two other ideas have suggested new weak forces at submillimetre scales. The cosmological energy density needed to close the universe, if converted to a length by taking its inverse fourth root (in natural units where $\hbar = c = 1$), corresponds to about 100 μm. This fact has led to repeated attempts[17-19] to address difficulties connected with the very small observed size of Einstein's cosmological constant

步细节在其他论文中给出[16]。

图 4 展示了我们的结果，其中也包括其他实验的限制和理论的预言。我们的限制在 10 μm 到 100 μm 范围内是已知的方法中最强的。当胶子模力和奇异模力第一次被提出的时候，它们在 $(\alpha, \lambda)$ 参数空间中所允许的区域分别是 4.4 和 5.2 个单位。现在它们除了 0.75 和 1.4 个允许的单位外，其余的都被排除了。关于 $\alpha = 2{,}000$ 的伸缩子的范围，我们的限制是 $\lambda < 23$ μm（对应于 $m > 8.6$ MeV），比之前的极限好上一倍。对于径向子的模空间，我们给出了一个 $\lambda$=88 μm 的上限，接近理论（一个额外维）预言的 40 μm 的值。对于有两个大的额外紧致维度的 ADD 理论，我们并没有达到文献 11 和 12 中给出的关于这些维度的大小的极限。

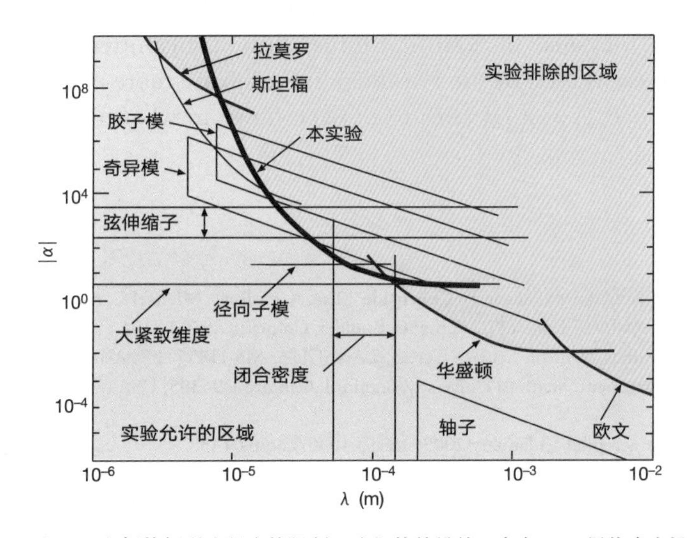


图 4. 目前 1 μm 到 1 cm 之间的新引力强度的限制。我们的结果是一个在 95% 置信度上汤川强度 $\alpha$ 函数的上限，它是关于力程 $\lambda$ 的函数（粗体实线）。之前的实验（拉莫罗[24]，华盛顿[12]，欧文[25]）给出的极限以及理论预言给出的新极限也一并展示在图中（胶子模[4]，奇异模[4]，弦伸缩子[2]，径向子模[7]）。一个来自斯坦福的未发表的实验[26]也展示在图中，它是在存在背景力的前提下得到的。伸缩子的强度某种意义上是依赖于模型，在文献中有一个取值范围[2]。我们选择区域 $200 < \alpha < 3{,}000$，它包含了大部分的值。图中也展示了分别由具有两个大的紧致额外维度的 ADD 理论[14,27,28]，轴子传递的力[20,21]以及宇宙学能量密度在 1.0 到 0.1 倍闭合能量密度之间的 $\lambda$ 区域[17]所给出的预言结果。对于模空间、伸缩子和 ADD 理论，展示的 $\lambda$ 区域的上界设置在理论提出时人们已知的实验所能达到的极限。

除了来自额外维的力，还有其他两个想法可以在亚毫米尺度给出新的弱的相互作用。用来闭合宇宙的宇宙学（临界）能量密度，如果通过做它的负四次方根来得到一个长度（在自然单位制下 $\hbar = c = 1$），结果大概是 100 μm。这个事实使得人们不断尝试[17-19]通过引入一个力程大约 100 μm 的力来解决爱因斯坦的宇宙学常数的观测

by introducing new forces with a range near 100 μm. Our result is the best upper bound on $\alpha$ in this region, but we have not quite reached gravitational sensitivity. Finally, the oldest of these predictions, still out of reach, is the very feeble axion-mediated force[20,21]. The axion is a field intended to explain why the violation of charge-parity symmetry is so small in quantum chromodynamics, the theory of the strong nuclear force.

Experiments of the sort reported here constrain string-inspired scenarios by setting very restrictive limits on predicted submillimetre forces. Of course, the actual observation of any new force would be a major advance. Because several theoretical scenarios point especially to these length scales, it is an important goal for the future to reach gravitational strength at even shorter distances, perhaps down to 10 μm. Experiments attempting to reach such distances will confront rapidly increasing background forces, especially electrostatic forces arising from the spatially non-uniform surface potentials of metals[22]. Electric fields due to surface potentials can be shielded with good conductors, but because of the finite stiffness of any shield they still cause background forces to be transmitted between test masses. Stretched membranes (as used by the Washington group) are more effective than stiff plates at the shortest distances, but it remains to be seen down to what distance the background forces can be effectively suppressed.

(**421**, 922-925; 2003)

**Joshua C. Long*†, Hilton W. Chan*†, Allison B. Churnside*, Eric A. Gulbis*, Michael C. M. Varney* & John C. Price***

* Physics Department, University of Colorado, UCB 390, Boulder, Colorado 80309, USA

† Present addresses: Los Alamos Neutron Science Center, LANSCE-3, MS-H855, Los Alamos, New Mexico 87545, USA (J.C.L.); and Physics Department, Stanford University, Stanford, California 94305, USA (H.W.C.)

Received 21 October 2002; accepted 13 January 2003; doi:10.1038/nature01432.

---

References:

1. Greene, B. *The Elegant Universe: Superstrings, Hidden Dimensions, and the Quest for the Ultimate Theory* (Norton, New York, 1999).
2. Kaplan, D. B. & Wise, M. B. Couplings of a light dilaton and violations of the equivalence principle. *J. High Energy Phys.* **8**, 37 (2000).
3. Taylor, T. R. & Veneziano, G. Dilaton couplings at large distances. *Phys. Lett. B* **213**, 450-454 (1988).
4. Dimopoulos, S. & Giudice, G. Macroscopic forces from supersymmetry. *Phys. Lett. B* **379**, 105-114 (1996).
5. Antoniadis, I. A possible new dimension at a few TeV. *Phys. Lett. B* **246**, 377-384 (1990).
6. Antoniadis, I., Dimopoulos, S. & Dvali, G. Millimeter-range forces in superstring theories with weak- scale compactification. *Nucl. Phys. B* **516**, 70-82 (1998).
7. Chacko, Z. & Perazzi, E. Extra dimensions at the weak scale and deviations from Newtonian gravity. Preprint hep-ph/0210254 available at ⟨arXiv.org⟩ (2002).
8. Fischbach, E. & Talmadge, C. *The Search for Non-Newtonian Gravity* (Springer, New York, 1999).
9. Bordag, M., Mohideen, U. & Mostepanenko, V. M. New Developments in the Casimir effect. *Phys. Rep.* **353**, 1-205 (2001).
10. Long, J. C., Chan, H. W. & Price, J. C. Experimental status of gravitational-strength forces in the sub-centimeter regime. *Nucl. Phys. B* **539**, 23-34 (1999).
11. Hoyle, C. D. *et al.* Sub-millimeter tests of the gravitational inverse-square law: A search for "large" extra dimensions. *Phys. Rev. Lett.* **86**, 1418-1421 (2001).
12. Adelberger, E. G. Sub-mm tests of the gravitational inverse-square law. Preprint hep-ex/0202008 available at ⟨arXiv.org⟩ (2002).
13. Fischbach, E., Krause, D. E., Mostepanenko, V. M. & Novello, M. New constraints on ultrashort-ranged Yukawa interactions from atomic force microscopy. *Phys. Rev. D* **64**, 075010 (2001).
14. Arkani-Hamed, N., Dimopoulos, S. & Dvali, G. The hierarchy problem and new dimensions at a millimeter. *Phys. Lett. B* **429**, 263-272 (1998).
15. Kleiman, R. N., Kaminsky, G. K., Reppy, J. D., Pindak, R. & Bishop, D. J. Single-crystal silicon high-Q torsional oscillators. *Rev. Sci. Instrum.* **56**, 2088-2091 (1985).
16. Long, J. C. *et al.* New experimental limits on macroscopic forces below 100 microns. Preprint hep-ph/0210004 available at ⟨arXiv.org⟩ (2002).
17. Beane, S. R. On the importance of testing gravity at distances less than 1 cm. *Gen. Rel. Grav.* **29**, 945-951 (1997).
18. Sundrum, R. Towards an effective particle-string resolution of the cosmological constant problem. *J High Energy Phys.* **7**, 1 (1999).

值很小所带来的困难。我们的结果是这个区域内能得到的最好的上界，但是我们还没有达到引力的灵敏度。最后，这些预言中最古老的一个，也是目前依然探测不到的，是很微弱的轴子传递的相互作用[20,21]。轴子的引入是为了解释为什么在量子色动力学（强核力的理论）中电荷–宇称对称性的破缺是如此的微小。

本文报道的这类实验对预测在亚毫米尺度上存在额外力的类弦理论给出了一个很强约束性的极限。当然，能够观测到任何新的力都将是一个巨大的进步。因为一些理论特别地指向了这些尺度，未来在更小的距离（可能低至 10 μm 以下）上到达引力的强度将是一个重要的目标。试图到达这个距离的实验将会面临更快速增长的背景力，尤其是由于金属表面空间不均匀的势所产生的静电力[22]。表面势产生的电场可以被良导体屏蔽，但是由于任何防护屏都有有限的刚性系数，它们依然会造成表面力在检验质量之间传递。延展的膜（即华盛顿课题组所使用的材料）在小距离下比刚性的盘更加有效，但是低至多大的距离下背景力可以被有效地压低还不能确定。

（安宇森 翻译；蔡荣根 审稿）

19. Schmidhuber, C. Old puzzles. Preprint hep-th/0207203 available at ⟨arXiv.org⟩ (2002).

20. Moody, J. E. & Wilczek, F. New macroscopic forces? *Phys. Rev. D* **30**, 130-138 (1984).

21. Rosenberg, L. J. & van Bibber, K. A. Searches for invisible axions. *Phys. Rep.* **325**, 1-39 (2000).

22. Price, J. C. in *Proc. Int. Symp. on Experimental Gravitational Physics* (eds Michelson, P., En-ke, H. & Pizzella, G.) 436-439 (World Scientific, Singapore, 1988).

23. Chan, H. W., Long, J. C. & Price, J. C. Taber vibration isolator for vacuum and cryogenic applications. *Rev. Sci. Instrum.* **70**, 2742-2750 (1999).

24. Lamoreaux, S. K. Demonstration of the Casimir force in the 0.6 to 6 μm range. *Phys. Rev. Lett.* **78**, 5-8 (1997).

25. Hoskins, J. K., Newman, R. D., Spero, R. & Shultz, J. Experimental tests of the gravitational inverse-square law for mass separations from 2 to 105 cm. *Phys. Rev. D* **32**, 3084-3095 (1985).

26. Chiaverini, J., Smullin, S. J., Geraci, A. A., Weld, D. M. & Kapitulnik, A. New experimental constraints on non-Newtonian forces below 100 microns. Preprint hep-ph/0209325 available at ⟨arXiv.org⟩ (2002).

27. Floratos, E. G. & Leontaris, G. K. Low scale unification, Newton's law and extra dimensions. *Phys. Lett. B* **465**, 95-100 (1999).

28. Kehagias, A. & Sfetsos, K. Deviations from the $1/r^2$ Newton law due to extra dimensions. *Phys. Lett. B* **472**, 39-44 (2000).

**Acknowledgements.** We thank E. Lagae for work in the laboratory, and C. Briggs, T. Buxkemper, L. Czaia, H. Green, S. Gustafson and H. Rohner of the University of Colorado and JILA instrument shops for technical assistance. We also gratefully acknowledge discussions with S. de Alwis, B. Dobrescu and S. Dimopoulos. This work is supported by grants from the US National Science Foundation.

**Competing interests statement.** The authors declare that they have no competing financial interests.

Correspondence and requests for materials should be addressed to J.C.P. (e-mail: john.price@colorado.edu).

# An Extended Upper Atmosphere Around the Extrasolar Planet HD209458b

A. Vidal-Madjar *et al.*

## Editor's Note

**The extrasolar planet HD 209458b crosses (transits) its parent star as seen from our perspective on Earth, which enables astronomers to determine its mass and radius. Here Alfred Vidal-Madjar and colleagues show that the planet has an extended atmosphere—the first such observation for an extrasolar planet. They obtained a spectrum of the parent star during three transits, and saw that some of the light was absorbed by the planet's atmosphere. They propose that the extended atmosphere probably consists of hydrogen atoms escaping from the planet's gravitational field, though it was later proposed that the atoms were ions from the stellar wind that pick up electrons in the outer atmosphere of the planet. Which explanation is correct is still debated.**

---

The planet in the system HD209458 is the first one for which repeated transits across the stellar disk have been observed[1,2]. Together with radial velocity measurements[3], this has led to a determination of the planet's radius and mass, confirming it to be a gas giant. But despite numerous searches for an atmospheric signature[4-6], only the dense lower atmosphere of HD209458b has been observed, through the detection of neutral sodium absorption[7]. Here we report the detection of atomic hydrogen absorption in the stellar Lyman $\alpha$ line during three transits of HD209458b. An absorption of $15\pm4\%$ ($1\sigma$) is observed. Comparison with models shows that this absorption should take place beyond the Roche limit and therefore can be understood in terms of escaping hydrogen atoms.

---

FAR more abundant than any other species, hydrogen is well-suited for searching weak atmospheric absorptions during the transit of an extrasolar giant planet in front of its parent star, in particular over the strong resonant stellar ultraviolet Lyman $\alpha$ emission line at 1,215.67 Å. Depending upon the characteristics of the planet's upper atmosphere, an H I signature much larger than that for Na I at 0.02% (ref. 7) is foreseeable. Three transits of HD209458b (named A, B and C hereafter) were sampled in 2001 (on 7–8 September, 14–15 September and 20 October, respectively) with the Space Telescope Imaging Spectrograph (STIS) onboard the Hubble Space Telescope (HST); the data set is now public in the HST archive. To partially overcome contamination from the Earth's Lyman $\alpha$ geocoronal emission, we used the G140M grating with the $52'' \times 0.1''$ slit (medium spectral resolution: $\sim$20 km s$^{-1}$). For each transit, three consecutive HST orbits (named 1, 2 and 3 hereafter) were scheduled such that the first orbit (1,780 s exposure) ended before the first contact to serve as a reference, and the two following ones (2,100 s exposures each) were

# 太阳系外行星 HD209458b 扩展的高层大气研究

维达尔–马贾尔等

## 编者按

在地球上观测到太阳系外行星 HD209458b 穿越其母恒星圆面的现象（凌星）可以帮助天文学家确定行星的质量与半径。本文中，阿尔弗雷德·维达尔–马贾尔与他的同事报道了一颗具有扩展大气层的行星，这也是第一次在太阳系外观测到这样的星体。他们从三次凌星过程中获得了母恒星的光谱，发现某些光线被行星的大气层吸收了。他们提出行星扩展大气层中可能包含有从行星重力场中逃逸出来的氢原子。尽管不久之后有研究指出这些原子是源于行星外层大气中携带电子的恒星风的离子，然而两种假说的正确性尚无定论。

---

位于 HD209458 系统中的行星是第一颗被观测到多次横越恒星圆面的行星[1,2]。结合凌星观测与视向速度法[3]，可以确定这颗行星的半径和质量，从而确认该行星是一颗气态巨行星。尽管人们对 HD209458 系统的大气特征进行了大量研究[4-6]，但这些研究仅是通过中性钠原子的吸收探测到 HD209458b 存在稠密的低层大气[7]。本文报道的是在 HD209458b 三次凌星过程中，探测到恒星莱曼 α 线被氢原子吸收，观测到的吸收强度为 15±4% (1$\sigma$)。通过与模型的模拟结果对比，发现这种吸收应该在超过洛希极限时才会发生，因此可以认为是逃逸的氢原子的吸收。

---

在宇宙中，氢的丰度远高于其他元素。因此，在太阳系外巨行星的凌星过程中，氢非常适合于探测较弱的大气吸收，尤其是在强共振恒星紫外莱曼 α 发射线(1,215.67Å) 波段。根据该行星的上层大气特征，可以预测 $H_I$ 的特征谱线比 $Na_I$ 的特征谱线要强 0.02%[7]。通过哈勃空间望远镜（HST）上装载的空间望远镜成像光谱仪（STIS），分别在 2001 年 9 月 7 日 ~ 8 日、9 月 14 日 ~ 15 日和 10 月 20 日对 HD209458b 的三次凌星（分别命名为 A、B、C）进行了采样，这些数据现在已经在 HST 档案库中公开。为了部分地避免地冕莱曼 α 发射线的干扰，我们采用了 G140M 光栅，其缝隙大小为 52″ × 0.1″（光谱分辨率约为 20 km · s$^{-1}$）。在每次凌星过程中，安排哈勃望远镜进行连续三个轨道周期（下文以 1、2、3 编号）的观测，其中第一周期（曝光 1,780 s）在发现凌星前就结束以便提供参考，而随后两个周期的观测（分别

partly or entirely within the transit.

The observed Lyman α spectrum of HD209458 is typical for a solar-type star, with a double-peaked emission originating from the stellar chromosphere (Fig. 1). It also shows a wide central absorption feature due to neutral hydrogen in the interstellar medium. The geocoronal emission filled the aperture of the spectrograph, resulting in an extended emission line perpendicular to the dispersion direction. The extent of this emission along the slit allowed us to remove it at the position of the target star. We evaluated its variation both along the slit and from one exposure to another, and excluded the wavelength domain where the corresponding standard deviation per pixel is larger than 20% of the final spectrum. We concluded that the geocoronal contamination can be removed with high enough accuracy outside the central region 1,215.5 Å < λ < 1,215.8 Å, labelled "Geo" in Figs 2, 4).

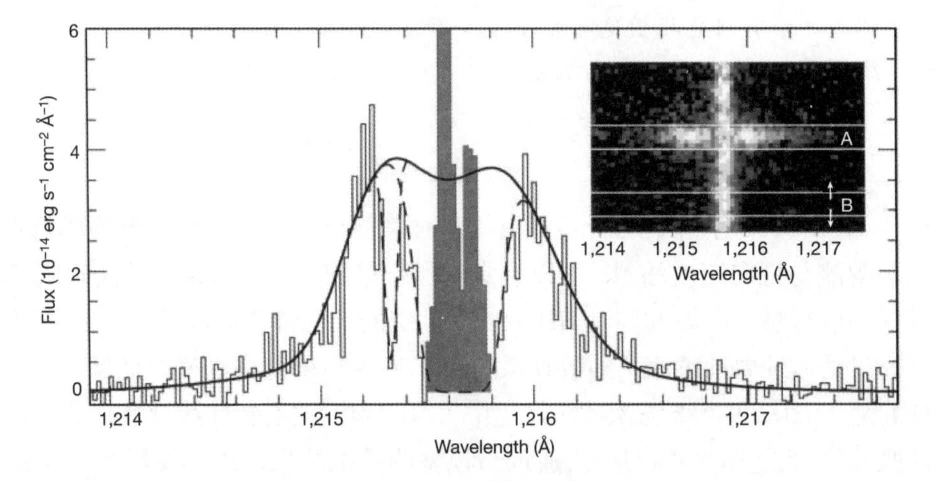


Fig. 1. The HD209458 Lyman α emission line. This high-resolution spectrum (histogram) was obtained with the E140M echelle grating and the 0.2″ × 0.2″ wide slit with a spectral resolution of 5 km s$^{-1}$; it was not used in the present analysis, but it allows the different components of the line profile to be seen. The continuum is a double-peaked emission line originating from the stellar chromosphere: the temperature increase in the lower chromosphere causes an emission line with a central dip due to the high opacity of the abundant hydrogen atoms (solid line). The observed spectrum also has a narrow absorption line (1,215.3 Å, barely seen at lower resolution) and a central wide absorption line (1,215.6 Å) due to the interstellar deuterium and hydrogen, respectively (dashed lines). The grey zone represents the fraction of the spectrum contaminated by the geocoronal emission, which is double-peaked in that case because the plotted spectrum is the average of four exposures obtained at two different epochs. The inset shows a small portion of the 2D image of a G140M first-order spectrum containing the stellar Lyman α profile and a sample of the geocoronal signal. This spectrum is one of the nine spectra used for this analysis. The G140M spectra have lower spectral resolution but higher signal-to-noise ratio. The stellar spectrum is seen as a horizontal line where the two peaks are resolved from the geocoronal emission (vertical line along the slit). The one-dimensional spectra are obtained by vertically adding around ten pixels within the A band. Measurements along the slit direction (~800 pixels available), for example at the position of the B band, allow us to estimate the geocoronal contamination and the background subtraction as well as the corresponding uncertainties.

曝光 2,100 s）则部分或全部在凌星期间内。

如图 1 所示，观测到的 HD209458 莱曼 α 谱线是典型的类太阳恒星光谱，具有恒星色球层产生的特征双峰发射。同时我们发现在中心位置附近还存在一个较宽的吸收带，这是因为在星际介质中存在中性氢。由于地冕发射线进入摄谱仪孔径，垂直色散方向的发射线发生扩展。沿光栅缝隙的发射线的这种扩展使得我们能够将其从观测目标恒星的位置移除。我们计算出发射线沿缝隙以及两次不同曝光之间的变化情况，同时去除波长域中对应单位像素标准差比最终光谱大 20% 的部分。我们从而得出以下结论：对于波长中心区域（1,215.4 Å < λ < 1,215.8 Å，在图 2、4 中用 "Geo" 标出）以外的波段，地冕引起的干扰可以很大程度地被消除掉。

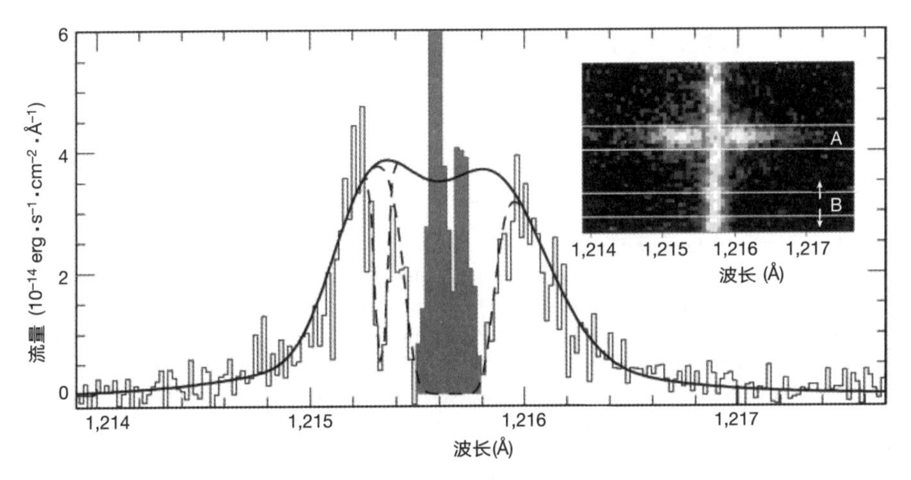


图 1. HD209458 的莱曼 α 发射线。这幅高分辨率光谱图（直方图）由 E140M 阶梯光栅得到的，该光栅的缝隙尺寸为 0.2″ × 0.2″，光谱分辨率为 5 km · s$^{-1}$，可以分离谱线轮廓的不同分量（尽管本次分析中并未使用）。图中连续发射谱具有双峰结构，这是由恒星色球层产生的，即由于大量氢原子的不透明度较高，底层色球层温度升高，这将使得发射线具有中心下陷结构（实线）。同时，观察到光谱还具有一个非常窄的吸收线（位于 1,215.3 Å，较低分辨率下很难观察到）和一个较宽的中心吸收谱线（1,215.6 Å，虚线），二者分别是由于星际间的氘和氢所引起的。图中灰色区域表示混杂着地冕干扰的光谱部分，由于光谱是在两个不同的纪元得到的四次曝光的平均，因此显示出双峰结构。右上角小框图给出了 G140M 一级光谱二维图像的一小部分，包含了恒星莱曼 α 廓线和部分地冕干扰信号。该光谱是我们用于分析的九个光谱之一，尽管 G140M 光谱具有较低的谱分辨率，但是具有较高的信噪比。该恒星光谱可以看作一条水平线，其中可以在地冕发射线中分辨出两个光谱峰（沿缝隙的垂直方向）。通过将 A 条带区域内的大约 10 个像素相加可以得到一维光谱图。通过沿缝隙方向测量（大约 800 个像素），例如在 B 条带位置，可以使我们估算出地冕干扰和背景影响，以及相应的不确定度。

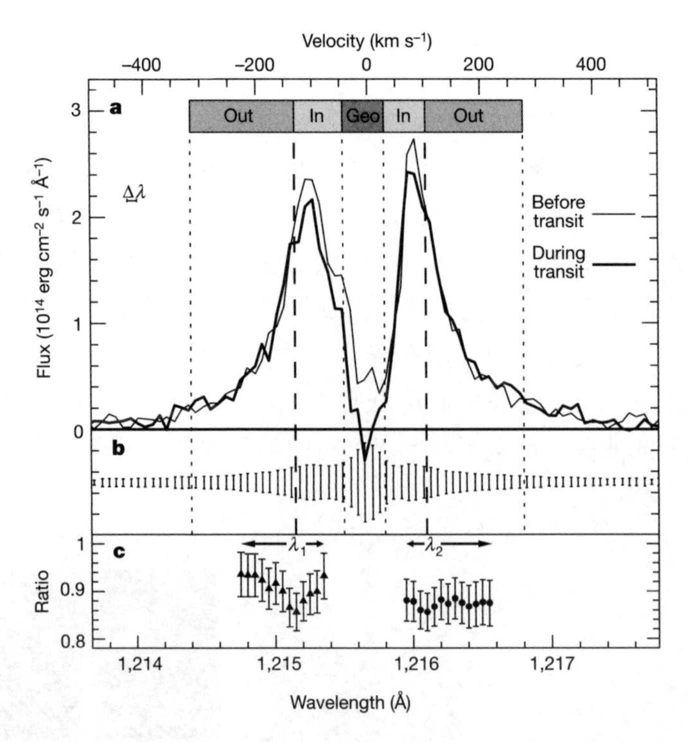


Fig. 2. The HD209458 Lyman α profile observed with the G140M grating. The geocoronal emission has been subtracted; the propagated errors are consequently larger in the central part of the profile, particularly in the Geo domain (see text). Δλ represents the spectral resolution. **a,** The thin line shows the average of the three observations performed before the transits (exposures A1, B1 and C1); the thick line shows the average of the three observations recorded entirely within the transits (exposures A2, B3 and C3). Variations are seen in the In domain as absorption over the blue peak of the line and partially over the red peak (between $-130$ km s$^{-1}$ and 100 km s$^{-1}$). Quoted velocities are in the stellar reference frame, centred on $-13$ km s$^{-1}$ in the heliocentric reference frame. **b,** $\pm 1\sigma$ error bars. **c,** The ratio of the two spectra in the In domain, the spectra being normalized such that the ratio is 1 in the Out domain. This ratio is plotted as a function of $\lambda_1$ using $\lambda_2$ = 1,216.10 Å (triangles), and as a function of $\lambda_2$ using $\lambda_1$ = 1,215.15 Å (circles). The ratio is always significantly below 1, with a minimum at $\lambda_1$ = 1,215.15 Å ($-130$ km s$^{-1}$) and $\lambda_2$ = 1,216.10 Å (100 km s$^{-1}$). In the domain defined by these values, the Lyman α intensity decreases during the transits by 15 ± 4%. The detection does not strongly depend on a particular selection of the domain. While the decrease of the Lyman α intensity is not sensitive to the position of $\lambda_2$, it is more sensitive to the position of $\lambda_1$, showing that most of the absorption occurs in the blue part of the line. Using the whole domain where the absorption is detected, the exoplanetary atmospheric hydrogen is detected at more than 3σ.

From the two-dimensional (2D) images of the far-ultraviolet (FUV) multi-anode microchannel array (MAMA) detector, the STIS standard pipeline extracts one-dimensional spectra in which the dark background has not been removed. The background level was systematically increasing from one exposure to the next within each of the three visits, but still remained below 2% of the peak intensity of the stellar signal. We therefore reprocessed the 2D images by using two independent approaches. The first one uses the 2D images provided by the standard pipeline and interpolates the background (including the Earth's geocoronal emission) with a polynomial fitted per column above and below the spectrum region. The second method starts from the 2D raw images to which is applied a dark

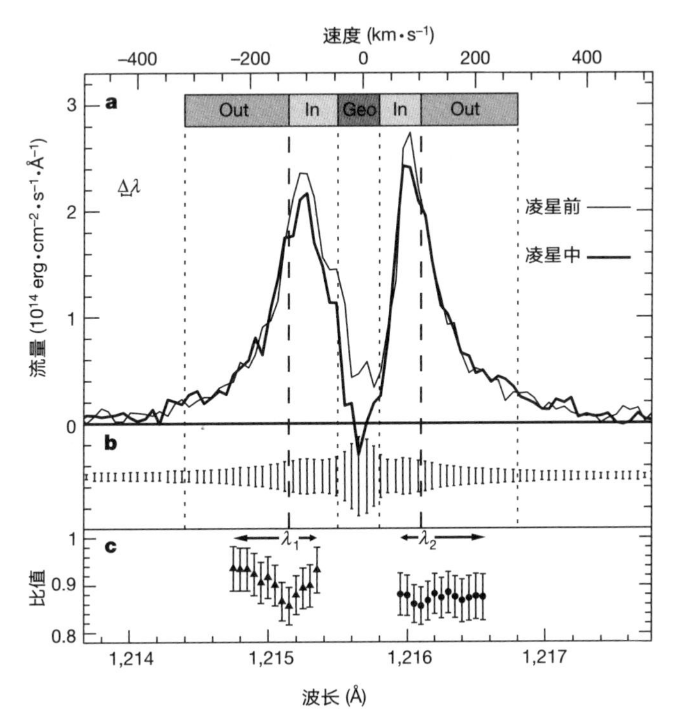


图 2. 利用 G140M 光栅观测到的 HD209458 的莱曼 α 轮廓。其中，地冕发射的干扰已经被剔除，因此传递误差在廓线中心附近变大，特别是在“Geo”区域，Δλ 表示光谱分辨率。**a.** 图中细线给出了凌星前进行的三次观测（即 A1、B1、C1 三次曝光）的平均；粗线表示凌星中的三次观测（即 A2、B3、C3 三次曝光）的平均。很明显，两条曲线在“In”域存在较大变化，特别是在谱线蓝峰以及部分红峰附近（速度在 −130 km · s⁻¹ 到 100 km · s⁻¹ 之间）。括号里的速度是相对恒星参考系的速度，在以太阳为中心的参考系中心速度为 −13 km · s⁻¹。**b.** 该图给出了 ±1σ 的误差棒。**c.** 该图给出了两条光谱曲线在“In”域的比值，归一化满足“Out”域比值为 1。其中，带三角标志的曲线表示 $\lambda_2$ = 1,216.10 Å 条件下比值与 $\lambda_1$ 的关系，而带实心圆标志的曲线给出了 $\lambda_1$ = 1,215.15 Å 条件下比值与 $\lambda_2$ 的关系，两种情况下比值的最小值分别于 $\lambda_1$ = 1,215.15 Å（对应 −130 km · s⁻¹）和 $\lambda_2$ = 1,216.10 Å（对应 100 km · s⁻¹）处获得。在这两个最小值定义域范围内，莱曼 α 谱线强度在凌星过程中减小了 15±4%。值得注意的是，探测结果本身与特定的光谱域并没有很强的关联性。莱曼 α 谱线强度的减小对 $\lambda_2$ 的位置并不敏感，而对于 $\lambda_1$ 的位置则相对敏感，这说明更多地吸收发生于谱线蓝端部分。对整个探测到吸收的区域的分析表明，系外行星大气氢在大于 3σ 的情况下可以被测度。

根据远紫外（FUV）多阳极微通道阵（MAMA）探测器捕获到的二维图像，STIS 标准通道从二维图像中提取出一维光谱，其中的暗背景并未去除。尽管三次凌星的背景水平从某次曝光到下一次曝光会系统性地增加，但是仍然低于星体信号峰值强度的 2%。因此，我们采用了两种独立的方法对二维图像进行了重新处理。第一种方法采用标准通道提供的二维图像，利用多项式拟合方法对背景光谱区上下各区间（包括地冕发射）进行插值。第二种方法是对二维原始图像进行暗背景校正，这种校正参照一种超暗图像（通过周期观测建立），同时减去沿光栅缝隙测量得到的地冕发

background correction from a super-dark image (created for the period of the observations) as well as a subtraction of the geocoronal emission measured along the slit, away from the stellar spectrum. The differences between the results of both approaches are negligible, showing that systematic errors generated through the background corrections are small compared to the statistical errors. Those errors are dominated by photon counting noise to which we added quadratically the error evaluated for both the dark background and geocoronal subtractions.

The Lyman $\alpha$ line profiles observed before and during the transits are plotted in Fig. 2. The three exposures outside the transits (exposures A1, B1 and C1) and the three entirely within the transits (A2, B3 and C3) were co-added to improve the signal-to-noise ratio. An obvious signature in absorption is detected during the transits, mainly over the blue side of the line, and possibly at the top of the red peak.

To characterize this signature better, we have defined two spectral domains: "In" and "Out" of the absorption. The In domain is a wavelength interval limited by two variables $\lambda_1$ and $\lambda_2$ (excluding the Geo geocoronal region). The Out domain is the remaining wavelength coverage within the interval 1,214.4–1,216.8 Å, for which the Lyman $\alpha$ line intensity can be accurately measured at the time of the observation. The corresponding In/Out flux ratio derived for each exposure is shown in Fig. 3, revealing the absorption occurring during the transits. To evaluate whether this detection is sensitive to a particular choice of $\lambda_1$ and $\lambda_2$, we averaged the three ratios before the transits and the three ratios entirely within the transits. We calculated the averaged pre-transits over mid-transits ratio as a function of $\lambda_1$ and $\lambda_2$ and propagated the errors through boot-strap estimations of the ratio calculated with 10,000 randomly generated spectra according to the evaluated errors over each individual pixel (Fig. 2c). The averaged ratio is always significantly below 1, with the minimum at $\lambda_1 = 1{,}215.15$ Å and $\lambda_2 = 1{,}216.1$ Å. In the interval defined by these two wavelengths, the Lyman $\alpha$ line is reduced by $15 \pm 4\%$ ($1\sigma$) during the transit. This is a larger-than-$3\sigma$ detection of an absorption in the hydrogen line profile during the planetary transits.

HD209458 (G0V) is close to solar type, for which time variations are known to occur in the chromospheric Lyman $\alpha$ line[8]. We thus evaluated the In/Out ratio in the solar Lyman $\alpha$ line profile as measured over the whole solar disk by the Solar Ultraviolet Measurements of Emitted Radiation (SUMER) instrument onboard the Solar and Heliospheric Observatory (SOHO)[9] during the last solar cycle from 1996 to 2001, that is, from quiet to active Sun. During this time, the total solar Lyman $\alpha$ flux varies by about a factor of two, while its In/Out ratio varies by less than $\pm 6\%$. Within a few months, a time comparable to our HD209458 observations, the solar In/Out ratio varies by less than $\pm 4\%$. This is an indication that the absorption detected is not of stellar origin but is due to a transient absorption occurring during the planetary transits.

A bright hot spot on the stellar surface hidden during the planetary transit is also excluded. Such a hot spot would have to contribute about 15% of the Lyman $\alpha$ flux over 1.5% of

射线的干扰。以上两种方法所得到的结果差异小到可以忽略不计，表明背景校正产生的系统误差小于统计误差。这些误差受光子计数噪声控制，我们在这些噪声中加入了暗背景和地冕删除两次误差评估。

图 2 给出了凌星前和凌星中观测到的莱曼 α 线廓线。为改善信噪比，我们把三次凌星前的曝光（分别表示为 A1、B1、C1）和三次凌星过程之中的曝光（分别表示为 A2、B3、C3）叠加在了一起。在凌星过程中，一个清晰的吸收信号被探测到，主要位于光谱的蓝端，也可能位于红端峰的顶部。

为了更好地描述该吸收信号特征，我们定义了光谱吸收域“In”和“Out”。其中“In”区域位于波长 $\lambda_1$ 和 $\lambda_2$ 之间（不包括“Geo”区域），“Out”区域位于波长 $\lambda_1$ 和 $\lambda_2$ 外侧，两区域均位于 1,214.4 ~ 1,216.8 Å 之间，该波段的莱曼 α 线强度在观测期间可以精确测量。图 3 给出了每次曝光相对应的 In/Out 通量比，结果表明在凌星过程中发生了吸收现象。为了评估这次探测对于我们选择的 $\lambda_1$、$\lambda_2$ 值是否敏感，我们将凌星前的三个比值和凌星中的三个比值分别进行平均。然后计算出凌星前与凌星中比值的平均值，并将其作为关于 $\lambda_1$ 和 $\lambda_2$ 的函数。根据每个独立像素的估计误差，利用比值的自举估算来传递误差，其中比值是由 10,000 个随机产生的光谱计算产生（图 2c）。平均比值始终显著地小于 1，其中最小值位于位置 $\lambda_1$ = 1,215.15 Å 和 $\lambda_2$ = 1,216.1 Å。在 $\lambda_1$ 和 $\lambda_2$ 之间的波段，凌星过程中莱曼 α 线强度减弱了 15 ± 4%（$1\sigma$）。这是一个在行星凌星中对氢吸收廓线所做的大于 $3\sigma$ 的探测。

HD209458（G0V）是类太阳恒星，其色球莱曼 α 线强度随时间发生变化[8]。在最近一个太阳活动周期（1996 ~ 2001 年），太阳活动从宁静变得逐渐活跃。在这段时间，太阳和日球层探测器（SOHO）上搭载的太阳紫外辐射光谱仪对整个日面进行了测量，从而可以求出太阳莱曼 α 线轮廓中的 In/Out 比率。结果表明，在该活动周期内，太阳莱曼 α 总通量变化因子大约为 2，其 In/Out 比值变化小于 ±6%。在和 HD209458 观测时间差不多的几个月内，太阳的 In/Out 比值起伏变化小于 ±4%。这说明，探测到的吸收光谱并不是由恒星本身产生，而是在凌星过程中产生的短暂吸收。

我们同时也排除了在凌星过程中恒星表面可能存在明亮热斑的可能性。因为

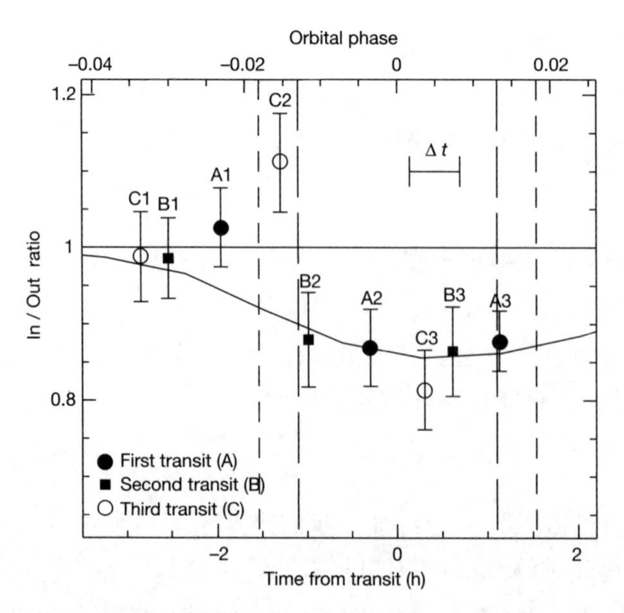


Fig. 3. Relative flux of Lyman $\alpha$ as a function of the HD209458's system phase. The averaged ratio of the flux is measured in the In (1,215.15–1,215.50 Å and 1,215.80–1,216.10 Å) and the Out (1,214.40–1,215.15 Å and 1,216.10–1,216.80 Å) domains in individual exposures of the three observed transits of HD209458b. The central time of each exposure is plotted relative to the transit time. The vertical dashed lines indicate the first and the second contact at the beginning and the end of the transit; the exposures A1, B1 and C1 were performed before the transits, and the exposures A2, B3 and C3 were entirely within the transits. The ratio is normalized to the average value of the three observations completed before the beginning of the transits. The $\pm 1\sigma$ error bars are statistical; they are computed through boot-strap estimations (see text). The In/Out ratio smoothly decreases by around 15% during the transit. The thick line represents the absorption ratio modelled through a particle simulation which includes hydrogen atoms escaping from the planet. In this simulation, hydrogen atoms are sensitive to the radiation pressure above an altitude of 0.5 times the Roche radius, where the density is assumed to be $2 \times 10^5$ cm$^{-3}$; these two parameters correspond to an escape flux of $\sim 10^{10}$ g s$^{-1}$. The stellar radiation pressure is taken to be 0.7 times the stellar gravitation. The mean lifetime of escaping hydrogen atoms is taken to be 4 h. The model yields an atom population in a curved comet-like tail, explaining why the computed absorption lasts well after the end of the transit.

the stellar surface occulted by the planet, in contradiction with Lyman α inhomogeneities observed on the Sun[10]. Furthermore, this spot would have to be perfectly aligned with the planet throughout the transit, at the same latitude as the Earth's direction, and with a peculiar narrow single-peaked profile confined over the In spectral region. It seems unlikely that a stellar spot could satisfy all these conditions.

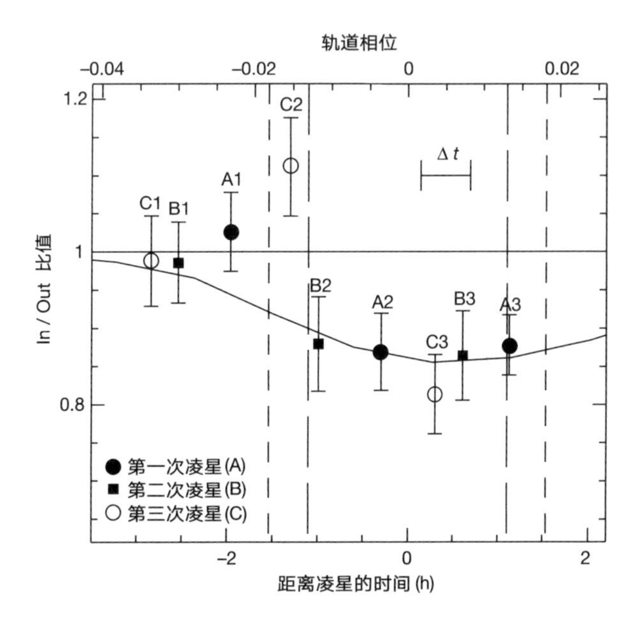


图 3. 莱曼 $\alpha$ 谱线的相对通量与 HD209458 系统相位的函数关系。HD209458b 每次凌星过程中，在 In（1,215.15 ~ 1,215.50 Å 和 1,215.80 ~ 1,216.10 Å）和 Out（1,214.40 ~ 1,215.15 Å 和 1,216.10 ~ 1,216.80 Å）域上测量每次曝光，可以得到通量平均比值。图中每次曝光过程的中心时间点是相对整个凌星的时间值，垂直虚线表示凌星开始和结束时的第一次和第二次接触。图中 A1、B1、C1 三次曝光是在凌星前进行，而 A2、B3、C3 是完全在凌星中进行。其中，图中比值相对凌星前完成的三次观测的平均值进行了归一化。$\pm 1\sigma$ 误差棒所示为统计误差，它们通过自举估算计算得到。可以看出，在整个凌星过程中，In/Out 比值平滑地降低了大约 15%。图中粗线表示通过粒子仿真模型计算得到的吸收比值，其中该模型考虑了从行星逃逸的氢原子。在该仿真中，氢原子对于海拔高度大于 0.5 倍洛希半径处的辐射压较为敏感，其密度假定为 $2\times10^5$ $cm^{-3}$，对应的逃逸通量约为 $10^{10}$ $g\cdot s^{-1}$。恒星辐射压设为 0.7 倍于恒星引力。逃逸氢原子的平均寿命取 4 h。该模型计算表明，大量逃逸氢原子构成了一个类似弯曲彗尾的结构，这就解释了为什么凌星完成后还观测到吸收现象。

如果恒星表面被行星遮挡了 1.5% 的话，那么这样一个热斑将会贡献莱曼 α 通量的 15%，然而这与观测到的太阳的莱曼 α 非均匀性相矛盾 [10]。而且，如果有热斑存在的话，那么在整个凌星过程中，位于地球方向上的相同纬度，热斑都需要与该行星严格对准，此外在 In 光谱区域还会有一个宽度很窄的特征单峰。这些条件过于苛刻，恒星上的热斑不太可能满足所有的这些条件。

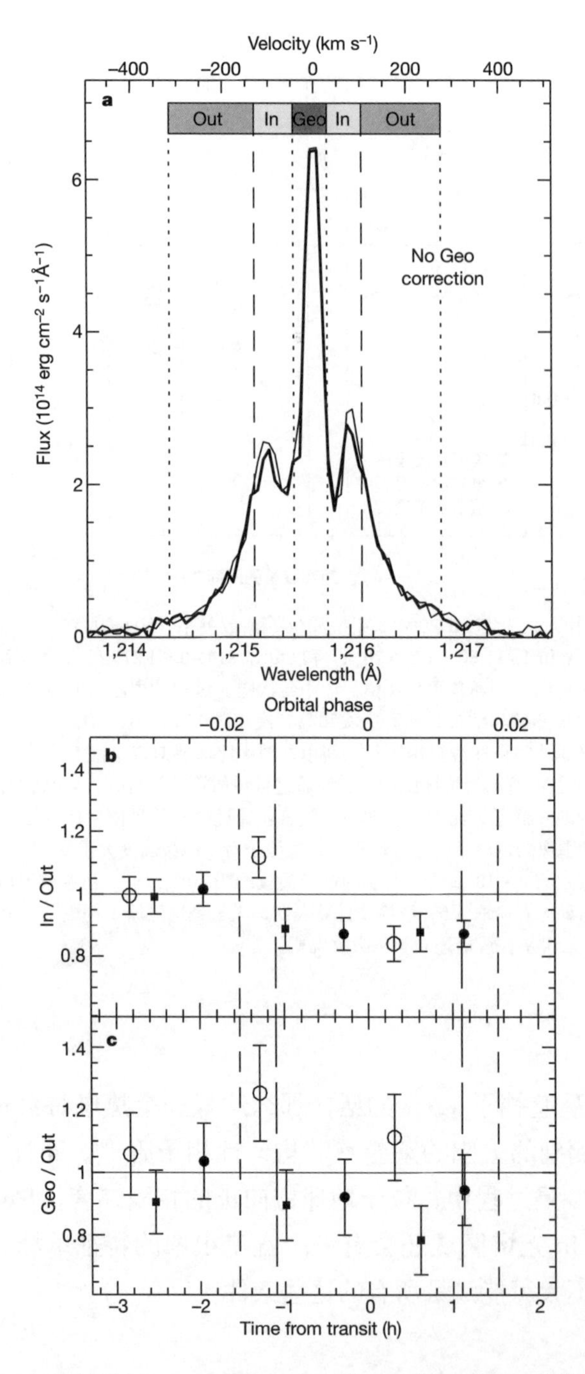


Fig. 4. Spectra and ratios with no geocoronal correction. To investigate the possibility that a bad estimate of the geocoronal correction may cause the detected signal, the spectra and ratios are plotted without this correction. First, note that the geocoronal emission is of the order of the stellar flux, and its peak value is never larger than three times the stellar one. Consequently, the correction is very small outside the Geo domain. **b** shows the same evaluation of the In/Out ratios as made in Fig. 3, with no geocoronal and no background correction. It appears that the impact of the geocoronal correction on these ratios is negligible. To further show that the "wings" of the geocoronal emission do not carry the detected transit signal, **c** shows the Geo/Out ratios, where Geo is the total flux due to the geocorona. If the flux in the In

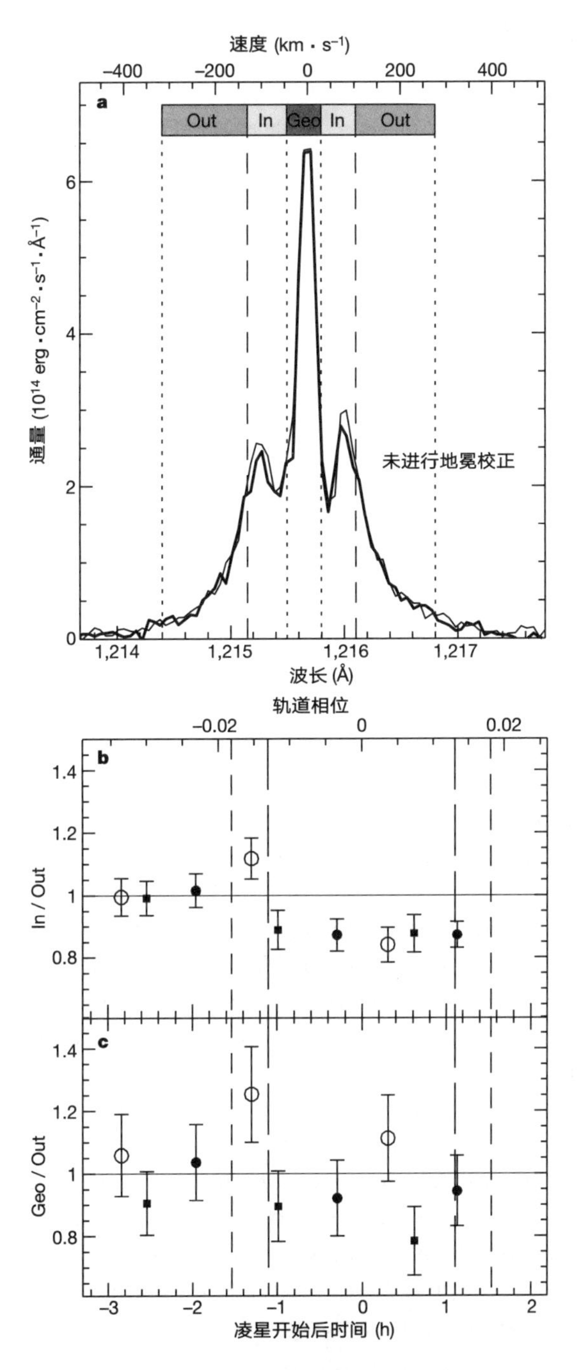


图 4. 未进行地冕校正的光谱和比值。为了研究不良的地冕校正对于探测信号的影响，图中给出了未经校正的光谱和比值。首先，值得注意的是地冕发射通量与恒星通量量级相同，而且其峰值不会高于恒星峰值的三倍。因此，在 Geo 域外的校正很小。图 **b** 给出了 In/Out 比值(与图 3 计算相同)，但没有进行地冕和背景校正。从图中可见，地冕校正对于比值的影响可以忽略。为了进一步说明地冕发射的“两翼”并不具有探测到的凌星信号，图 **c** 给出了 Geo/Out 比值，其中 Geo 代表地冕发射的总通量。如果 In 区域内的通量受地冕发射的严重影响，则会发现 Geo/Out 比值与图 3 中比值相似。从图中可知，Geo/Out 比值在 A1、B1 和 C1 平均值附近随机起伏，并没有随地冕变化出现始终一致的凌星特征。类

domain was significantly affected by the geocorona, the Geo/Out ratios should present a signal similar to the In/Out ratios plotted in Fig. 3. The Geo/Out ratios are found to be random fluctuations around the average of the A1, B1 and C1 values. There is no consistent signature of the transit in the geocoronal variations. As in Fig. 2a, **a** shows the average of the three spectra obtained before and during the transits (thin and thick lines, respectively). Quoted velocities are in the stellar reference frame. Here the spectra are without any geocoronal correction. A simple constant has been subtracted to compensate for the mean background levels in order to have matching spectra in the Out domain. This clearly shows that the absorption signature is present even without the geocoronal correction.

Finally, we confirmed with various tests that there are no correlations between the geocoronal variations and the detected signature in absorption. One method is presented in Fig. 4, showing that a contamination of the In domain by the geocorona is excluded. We thus conclude that the detected profile variation can only be related to an absorption produced by the planetary environment.

The observed 15% intensity drop is larger than expected *a priori* for an atmosphere of a planet occulting only 1.5% of the star. Although the small distance (8.5 stellar radii) between the planet and the star results in an extended Roche lobe[11] with a limit at about 2.7 planetary radii (that is, 3.6 Jupiter radii), the filling up of this lobe gives a maximum absorption of about 10% during the planetary transit. Because a more important absorption is detected, hydrogen atoms must cover a larger area: a drop of 15% corresponds to an occultation by an object of 4.3 Jupiter radii. This is clearly beyond the Roche limit as theoretically predicted[6]. Thus some hydrogen atoms should escape from the planet. The spectral absorption width shows independently that the atoms have large velocities relative to the planet. Thus hydrogen atoms must be escaping the planetary atmosphere.

We have built a particle simulation in which we assumed that hydrogen atoms are sensitive to the stellar radiation pressure inside and outside the Roche lobe. Their motion is evaluated by taking into account both the planetary and stellar gravities. The Lyman $\alpha$ radiation pressure is known to be 0.7 times the stellar gravity, the escape flux and the neutral hydrogen lifetime being free parameters. The lifetime of hydrogen is limited to a few hours owing to stellar extreme ultraviolet ionization. Escaping hydrogen atoms expand in an asymmetric comet-like tail and progressively disappear when moving away from the planet. This simple scenario is consistent with the observations (Fig. 3). In this model, atoms in the evaporating coma and tail cover a large area of the star, and most are blueshifted because of the radiation pressure repelling them away from the star. The detection of most of the absorption in the blue part of the line is consistent with these escaping atoms. On the other hand, more observations are needed to clarify whether an absorption is also present in the red part of the line.

To account for the observed absorption depth, the particle simulation implies a minimum escape flux of around $10^{10}$ g $s^{-1}$. However, owing to saturation effects in the absorption line, a flux larger by several orders of magnitude would produce a similar absorption signature. So, to evaluate the actual escape flux, we need to estimate the vertical distribution of hydrogen atoms up to the Roche limit, in an atmosphere extended by the stellar tidal forces

似图 2a，图 **a** 给出了凌星前和凌星中获得的三个光谱的平均值，分别用细线和粗线画出。图中速度相对恒星参考系。这里的光谱没有进行任何地冕校正。为了在 Out 域获得匹配光谱，我们在光谱中减去了一个常数以补偿背景平均水平。这清楚地表明，即使没有经过地冕校正也存在特征吸收。

最终，我们通过各种检验确认探测到的特征吸收与地冕的变化没有相关性。图 4 给出了其中的一种方法，表明可以消除 In 区域内地冕造成的干扰。因此，我们得出以下结论：探测到的光谱廓线变化只可能与行星环境产生的吸收有关。

鉴于这颗恒星表面被行星大气遮挡住 1.5% 的情况，观测到的光谱强度减小了 15%，这比预期的值要大。尽管行星与恒星之间的短距离（8.5 恒星半径）造成了洛希瓣扩展 [11]，扩展的上限为 2.7 倍行星的半径（即木星半径的 3.6 倍），而对洛希瓣的填充会引起行星凌星过程的最大吸收，约为 10%。探测到的强吸收意味着氢原子一定覆盖着更大的面积，15% 的减小对应着一个 4.3 倍于木星半径的天体被遮挡，这显然超出了洛希极限的理论值 [6]。所以，一部分氢原子应该从这颗行星发生了逃逸。另外，光谱的吸收宽度独立地表明这些原子相对这颗行星具有较大的速度。因此氢原子必定从这颗行星的大气层逃逸了。

为此，我们建立了一个粒子仿真，在该仿真中我们假设氢原子对洛希瓣内外的恒星辐射压较为敏感。通过考虑行星和恒星的引力作用，可以计算出这些氢原子的运动情况。已知莱曼 α 辐射压是恒星引力的 0.7 倍，氢原子的逃逸通量和中性氢原子的寿命都是自由参量。由于恒星极端紫外电离，氢原子的寿命一般为数小时。逃逸的氢原子扩展形成一个非对称的彗状尾，并随着行星的远离逐渐消失。这一图景与图 3 中的观测相符。在这一模型中，位于蒸发中的彗发和彗尾中的逃逸氢原子覆盖了恒星很大的表面积，并且这些原子被辐射压从恒星表面排斥开，因此大多发生蓝移，蓝移可以解释为辐射压将这些原子从恒星上排斥出来。对蓝端部分谱线吸收的检测结果与这些逃逸中原子的检测结果相吻合。另一方面，对于大多数观测需要弄清是否在谱线红端部分也存在吸收现象。

考虑观测到的吸收深度，粒子仿真表明需要存在最小逃逸通量，其值大约为 $10^{10}$ g · $s^{-1}$。由于吸收光谱存在饱和效应，即使比最小通量大若干个数量级的通量也会产生相似的特征吸收。所以，为了计算实际的逃逸通量，我们需要估计氢原子在洛希极限以内的大气层垂直分布。大气层由于受到恒星潮汐力作用而扩展，同时由于许多可能的机制而被加热，这些效应都可能引起更大的逃逸通量。对于通量的详

and heated by many possible mechanisms. These effects may lead to a much larger escape flux. A detailed calculation is beyond the scope of this Letter. This raises the question of the lifetime of evaporating extrasolar planets which may be comparable to the star's lifetime itself. If so, the so-called "hot Jupiters" could evolve faster than their parent star, eventually becoming smaller objects, which could look like "hot hydrogen-poor Neptune-mass" planets. This evaporation process, more efficient for planets close to their star, might explain the very few detections[12,13] of "hot Jupiters" with orbiting periods shorter than three days.

(**422**, 143-146; 2003)

**A. Vidal-Madjar*, A. Lecavelier des Etangs*, J.-M. Désert*, G. E. Ballester†, R. Ferlet*, G. Hébrard* & M. Mayor‡**
* Institut d'Astrophysique de Paris, CNRS/UPMC, 98bis boulevard Arago, F-75014 Paris, France
† Lunar and Planetary Laboratory, University of Arizona, 1040 E. 4th St., Rm 901, Tucson, Arizona 85721-0077, USA
‡ Observatoire de Genève, CH-1290 Sauverny, Switzerland

Received 13 September 2002; accepted 27 January 2003; doi:10.1038/nature01448.

---

References:

1. Henry, G. W., Marcy, G. W., Butler, R. P. & Vogt, S. S. A transiting "51 Peg-like" planet. *Astrophys. J.* **529**, L41-L44 (2000).
2. Charbonneau, D., Brown, T. M., Latham, D. W. & Mayor, M. Detection of planetary transits across a Sun-like star. *Astrophys. J.* **529**, L45-L48 (2000).
3. Mazeh, T. *et al.* The spectroscopic orbit of the planetary companion transiting HD 209458. *Astrophys. J.* **532**, L55-L58 (2000).
4. Bundy, K. A. & Marcy, G. W. A search for transit effects in spectra of 51 Pegasi and HD 209458. *Proc. Astron. Soc. Pacif.* **112**, 1421-1425 (2000).
5. Rauer, H., Bockelée-Morvan, D., Coustenis, A., Guillot, T. & Schneider, J. Search for an exosphere around 51 Pegasi B with ISO. *Astron. Astrophys.* **355**, 573-580 (2000).
6. Moutou, C. *et al.* Search for spectroscopical signatures of transiting HD 209458b's exosphere. *Astron. Astrophys.* **371**, 260-266 (2001).
7. Charbonneau, D., Brown, T. M., Noyes, R. W. & Gilliland, R. L. Detection of an extrasolar planet atmosphere. *Astrophys. J.* **568**, 377-384 (2002).
8. Vidal-Madjar, A. Evolution of the solar lyman alpha flux during four consecutive years. *Sol. Phys.* **40**, 69-86 (1975).
9. Lemaire, P. *et al.* in *Proc. Symp. SOHO 11, From Solar Min to Max: Half a Solar Cycle with SOHO* (ed. Wilson, A.) 219-222 (ESA SP-508, ESA Publications Division, Noordwijk, 2002).
10. Prinz, D. K. The spatial distribution of Lyman alpha on the Sun. *Astrophys. J.* **187**, 369-375 (1974).
11. Paczynski, B. Evolutionary processes in close binary systems. *Annu. Rev. Astron. Astrophys.* **9**, 183-208 (1971).
12. Cumming, A., Marcy, G. W., Butler, R. P. & Vogt, S. S. The statistics of extrasolar planets: Results from the Keck survey. Preprint astro-ph/0209199 available at ⟨http://xxx.lanl.gov⟩ (2002).
13. Konacki, M., Torres, G., Jha, S. & Sasselov, D. D. An extrasolar planet that transits the disk of its parent star. *Nature* **421**, 507-509 (2003).

**Acknowledgements.** This work is based on observations with the NASA/ESA Hubble Space Telescope, obtained at the Space Telescope Science Institute, which is operated by AURA, Inc. We thank M. Lemoine, L. Ben Jaffel, C. Emerich, P. D. Feldman and J. McConnell for comments, J. Herbert and W. Landsman for conversations on STIS data reduction, and J. Valenti for help in preparing the observations.

**Competing interests statement.** The authors declare that they have no competing financial interests.

Correspondence and requests for materials should be addressed to A.V.-M. (e-mail: alfred@iap.fr).

细计算已经超出了本文的研究范围。这也引出了关于太阳系外蒸发中的行星的寿命问题，其寿命可能与母恒星的寿命相当。如果确实如此，这些所谓的“热木星”就会比其母恒星演化更快，最终演变为更小的天体，而这些“热而缺氢”行星和海王星的质量差不多。如果行星距离恒星更近的话，以上蒸发过程将会更快。这也可能解释为什么轨道周期短于三天的“热木星”很少被探测到[12,13]。

（金世超 翻译；胡永云 审稿）

# The Principle of Gating Charge Movement in a Voltage-dependent $K^+$ Channel

Y. Jiang *et al.*

## Editor's Note

**Proteins in cell membranes that act as gates for alkali-metal ions, opening and shutting in response to changes in the electrochemical potential (voltage) across the membrane, play key roles in important biological processes of higher organisms, such as muscle action and nerve signal transmission. But the mechanism by which these "ion channels" operate was not fully known before these papers from biochemist Roderick MacKinnon and colleagues. This understanding demanded a detailed picture of the molecular structure of the proteins, which was hard to obtain because membrane proteins seldom crystallize well for X-ray diffraction studies. Nonetheless, having succeeded in obtaining a detailed crystal structure, reported in a companion paper to this one, the researchers here describe the changes in molecular shape that control potassium-ion movement through the central pore.**

---

The steep dependence of channel opening on membrane voltage allows voltage-dependent $K^+$ channels to turn on almost like a switch. Opening is driven by the movement of gating charges that originate from arginine residues on helical S4 segments of the protein. Each S4 segment forms half of a "voltage-sensor paddle" on the channel's outer perimeter. Here we show that the voltage-sensor paddles are positioned inside the membrane, near the intracellular surface, when the channel is closed, and that the paddles move a large distance across the membrane from inside to outside when the channel opens. KvAP channels were reconstituted into planar lipid membranes and studied using monoclonal Fab fragments, a voltage-sensor toxin, and avidin binding to tethered biotin. Our findings lead us to conclude that the voltage-sensor paddles operate somewhat like hydrophobic cations attached to levers, enabling the membrane electric field to open and close the pore.

---

VOLTAGE-DEPENDENT $K^+$ channel opening follows a very steep function of membrane voltage[1]. To allow channels to switch to the open state, gating charges—charged amino acids on the channel protein—move within the membrane electric field to open the pore[1-3]. The crystal structure of KvAP, a voltage-dependent $K^+$ channel, suggests how these gating charge movements might occur[4]. Four arginine residues are located on a predominantly hydrophobic helix–turn–helix structure called the voltage-sensor paddle. One paddle on each subunit is present at the outer perimeter of the channel. By moving across the membrane near the protein–lipid interface, the paddles could carry the arginine residues through the electric field, coupling pore opening to membrane voltage. We test this

# 电压依赖性 $K^+$ 通道的门控电荷运动的原理

姜有星等

## 编者按

那些位于细胞膜中的、对碱金属离子起着门控作用、随着膜两侧电化学势（电压）的变化而开放和关闭的蛋白质在高等生物的重要生物过程中起着关键作用，比如肌肉动作和神经信号传导。但是，在生物化学家罗德里克·麦金农及其同事们发表相关的文章之前，这些“离子通道”运转的机制并不完全为人们所知。对这种机制的了解需要有对蛋白质的分子结构的详细描述，然而这一点很难做到，因为膜蛋白很难得到足够好的晶体进行 X 射线衍射分析。尽管如此，研究者们还是成功地获得了其详细的晶体结构并在本文的姊妹篇中进行了报道，在本文中他们则描述了这些蛋白分子形状的变化，这些变化控制了钾离子穿过中央小孔的运动。

---

通道开放对膜电压的严格依赖使电压依赖性 $K^+$ 通道像开关一样打开。其开放受门控电荷运动驱动，门控电荷来自这个蛋白的 S4 螺旋片段上的精氨酸残基。每个 S4 片段在通道的外围形成半个“电压感受桨”。在这里，我们证明当通道关闭时，电压感受桨位于膜内，接近细胞内表面，当通道打开时桨从内向外跨膜移动很大的距离。将 KvAP 通道重建到平面脂质膜中，并用单克隆的 Fab(抗原表位)片段、电压传感器毒素和连接了生物素的抗生物素蛋白对 KvAP 通道进行研究。通过研究结果我们可以得出这样的结论：电压感受桨的运转有点像疏水性阳离子依附到杠杆上，使得膜电场可打开和关闭孔道。

---

电压依赖性 $K^+$ 通道开放严格受控于膜电压 [1]。为了让通道切换到开放状态，门控电荷——通道蛋白上的带电氨基酸——在膜电场内运动以打开孔道 [1-3]。KvAP 的晶体结构（一种电压依赖性 $K^+$ 通道），可能表明了这些门控电荷运动是如何发生的 [4]。四个精氨酸残基位于一个显著疏水的被称为电压感受桨的螺旋–转角–螺旋结构上。每个亚基上有一个桨位于通道外围。通过跨膜运动接近蛋白–膜脂质界面，桨可携带精氨酸残基通过电场，将孔道开放与膜电压耦联。我们通过测定通道关闭和打开时膜内电压感受桨的位置来验证该门控电荷运动假说。

hypothesis for the movement of gating charges by estimating the positions of the voltage-sensor paddles inside the membrane when the channel is closed and opened.

## Using Fabs and a Toxin to Detect Paddle Motions

Two monoclonal Fabs, 6E1 and 33H1, were used to crystallize and determine the structures of the full-length KvAP channel and the isolated voltage sensor, respectively[4]. Both Fabs were found attached to the same epitope on the tip of the voltage-sensor paddle between S3b and S4. We used these same Fabs to examine the position of the voltage-sensor paddles when the KvAP channel functions in lipid membranes (Fig. 1a), and to assess whether they change their position when the channel gates open in response to membrane depolarization. Figure 1b shows that both the 6E1 and 33H1 Fabs inhibit channel function when applied to the external solution. By contrast, neither Fab affected channel function from the internal solution.

Inhibition by external Fabs requires membrane depolarization (Fig. 1c). When the 33H1 Fab is added to the external chamber while the channel is held closed at −100 mV for 10 min (Fig. 1c, interval between the black and red data points), no inhibition is observed. The slightly larger current of the first red data point (20-min point) reflects recovery from a small amount of steady-state inactivation of channels occurring during the control pulse period (0–10-min data points)[5]. The important point, however, is that inhibition of current is detectable only on the second pulse following the addition of the Fab, as if the channel has first to open in order for the Fabs to bind. Inhibition progresses as the membrane is repeatedly depolarized. We also observe gradual recovery from inhibition if, once the Fabs are bound, the membrane is held at a negative voltage for a prolonged period (30–40-min interval), implying that negative membrane voltages destabilize the interactions between channels and Fabs, causing the Fabs to dissociate. The same properties of inhibition are observed for Fab 6E1. Based on these results, we conclude that the voltage-sensor paddles must remain inaccessible as long as the channel is held closed by the negative membrane voltage, and that the entire epitope (two helical turns of S3b and one turn of S4) becomes exposed to the external side in response to depolarization.

Why do the Fabs inhibit the channel? If they bind to the voltage-sensor paddles from the external side when the membrane is depolarized, why do the Fabs not simply hold the channel permanently open? Inhibition can be explained by the fact that the KvAP channel, like most voltage-dependent $K^+$ channels, inactivates[5]. That is, its pore stops conducting ions spontaneously during prolonged depolarizations of the membrane, even though the voltage sensors remain in their open conformation. Inactivation could also occur when the Fabs bind and hold the voltage sensors in their open conformation, thus explaining inhibition.

## 使用 Fabs 和毒素检测桨的运动

两个单克隆的 Fabs 片段——6E1 和 33H1，分别用于结晶并确定全长 KvAP 通道和独立的电压传感器的结构[4]。我们发现两个 Fabs 都附在位于 S3b 和 S4 之间的电压感受桨末端的相同抗原表位。当 KvAP 通道在脂质膜内起作用时，我们用相同的 Fabs 检查电压感受桨的位置（图 1a），并评估当膜去极化使通道开放时它们是否改变自己的位置。图 1b 显示，当 Fabs 片段 6E1 和 33H1 被加入外侧溶液中时都能抑制通道功能。相反，当被加入内侧溶液中时两个 Fabs 片段都不影响通道功能。

外侧 Fabs 抑制需要膜去极化（图 1c）。当 33H1 Fab 加到外侧，同时通道在 –100 mV 保持关闭 10 min 时（图 1c，黑色和红色的数据点之间的间隔），没有观察到抑制。第一个红色数据点所显示的略大的电流（20 min 点）反映了在控制脉冲周期内发生的从小量通道稳态失活的恢复（0 ~ 10 min 数据点）[5]。然而，重点是电流抑制只在加入 Fab 后的第二个脉冲检测到，好像通道必须首先打开才能使得 Fabs 进行结合。当膜反复去极化时抑制会继续。若是一旦 Fabs 被结合，且使膜维持更长时间的负电压（30 ~ 40 min 间隔），我们还可以观察到从抑制中逐渐恢复的过程，这意味着负的膜电压使通道和 Fabs 间相互作用不稳定，造成 Fabs 脱离。在 Fab 6E1 的实验中也观察到了相同的抑制特性。基于这些结果，我们得出这样的结论：只要通道通过负膜电压保持关闭，电压感受桨必定是难以接近的，只有去极化时整个抗原表位（S3b 的两个螺旋转角和 S4 的一个转角）才会相应地暴露在外侧。

为什么 Fabs 会抑制通道？当膜去极化时，如果它们从外侧结合到电压感受桨，那为什么 Fabs 不能简单地维持通道永久开放？抑制可以解释为，KvAP 通道会像大多数电压依赖性 $K^+$ 通道一样失活[5]。即当膜去极化时间延长，即使电压传感器保持其开放构象，其孔道也会自发停止导通离子。当 Fabs 结合并保持电压传感器开放的构象中，失活也会发生，从而解释了抑制现象。

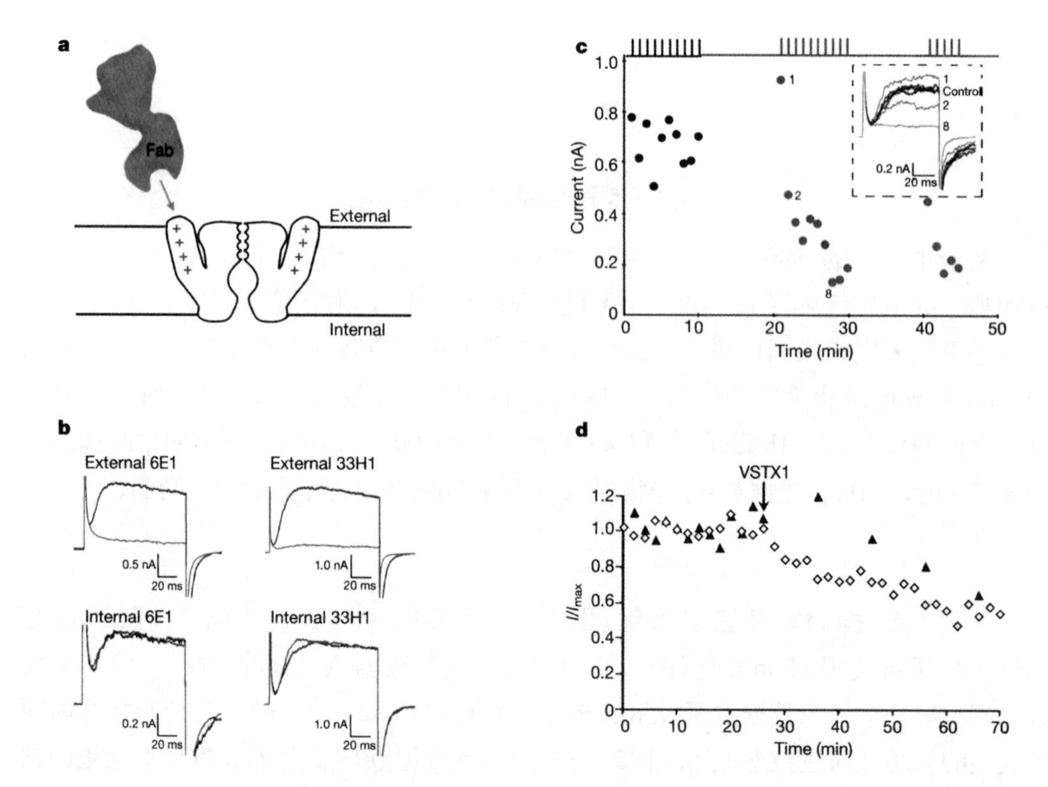


Fig. 1. Inhibition of KvAP channels by Fabs and a tarantula venom toxin that bind to the voltage-sensor paddles. **a**, Experimental strategy: Fab (green) is added to the external or internal side of a planar lipid membrane to determine whether the epitope is exposed. **b**, Fabs used in crystallization (6E1 and 33H1) inhibit from the external (red traces) but not the internal (blue traces) side of the membrane. Currents before (black traces) and after the addition of about 500 nM Fab (red and blue traces) were elicited by membrane depolarization to 100 mV from a holding voltage of −100 mV. **c**, Fab 33H1 binds to the voltage-sensor paddle only when the membrane is depolarized. Current elicited by depolarization from −100 mV to 100 mV at times indicated by the stimulus trace (above) in the absence (black symbols) or presence of 500 nM Fab (red symbols) is presented as a function of time. Selected current traces corresponding to the numbered symbols are shown (inset). **d**, VSTX1 binds from the external side only when the membrane is depolarized. Currents, normalized to the average control value, were elicited by a 100-ms depolarization to 100 mV every 120 s (diamonds) or every 600 s (triangles). VSTX1 (30 nM) was added to the external solution at the point indicated by the arrow.

It is interesting that tarantula venom toxins also bind to residues on S3 and S4 (ref. 6), which we now know to be on the voltage-sensor paddles[4]. On addition of an approximately half-inhibitory concentration of the tarantula venom toxin VSTX1 to the external side of KvAP, we observe that the rate of inhibition is faster if the membrane is depolarized at a higher frequency (Fig. 1d). Therefore, the toxins, like the Fabs, require membrane depolarization.

We conclude from experiments with Fab fragments and a tarantula venom toxin that the voltage-sensor paddles are exposed to the extracellular solution during membrane depolarization (positive inside) but not during hyperpolarization (negative inside). In view of the KvAP crystal structure[4], these results imply that the voltage-sensor paddles can move a large distance across the lipid membrane. How deep inside the membrane do the paddles sit

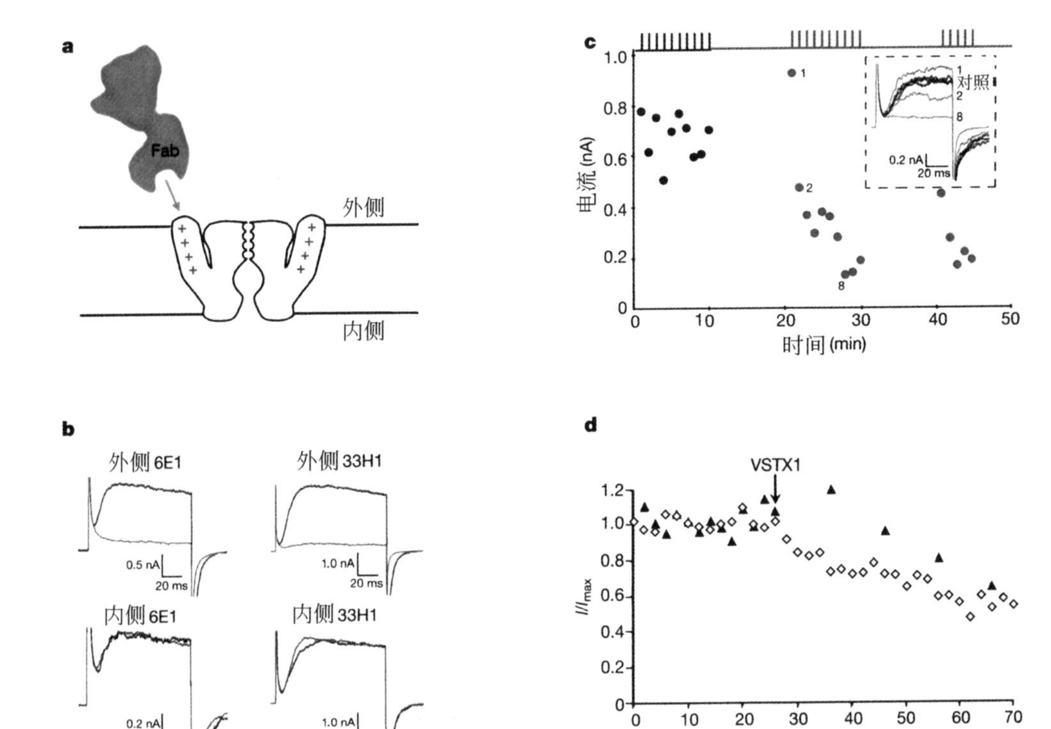


图 1. 结合到电压感受桨的 Fabs 和狼蛛毒液毒素对 KvAP 通道的抑制。**a**，实验策略：将 Fab(绿色)加入平面脂质膜的外侧或内侧，以确定抗原表位是否暴露出来。**b**，结晶所使用的 Fabs(6E1 和 33H1)从膜的外侧(红色轨迹)抑制而非内侧(蓝色轨迹)。加入约 500 nM Fab 之前(黑色轨迹)和之后(红色和蓝色轨迹)的电流是由膜从 −100 mV 钳制电压去极化到 100 mV 时产生的。**c**，只有当膜去极化时，Fab 33H1 才结合到电压感受桨。从 −100 mV 到 100 mV 去极化时产生的电流，在不存在(黑色符号)或存在(红色符号)500 nM Fab 时表示为时间的函数。图中对应于数字符号的电流值，来自选定的电流轨迹(插图中)。**d**，只有当膜去极化时，VSTX1 才从外侧结合。100 ms 去极化至 100 mV，将每 120 s(菱形)或每 600 s(三角形)时间点产生的电流归一化到平均值。在箭头所指示的时间点，将 VSTX1(30 nM)加到外侧溶液。

有趣的是，狼蛛毒液毒素也结合 S3 和 S4 的残基(参考文献 6)，现在我们知道它们位于电压感受桨上[4]。另外加入大约一半抑制浓度的狼蛛毒液毒素 VSTX1 到 KvAP 外侧，我们看到，如果膜的去极化频率越高，则抑制速率也越快(见图 1d)。因此，毒素像 Fabs 一样需要膜去极化。

我们从 Fab 片段和狼蛛毒液毒素实验可得出结论，电压感受桨在膜去极化时(内部为正)，而不是超极化时(内部为负)暴露在细胞外的溶液中。鉴于 KvAP 晶体结构[4]，这些结果意味着，电压感受桨可以跨越脂质膜移动一段较大的距离。当膜

when the membrane is hyperpolarized, and how far do they move when the channel opens?

## Using Biotin and Avidin to Measure Paddle Motions

Finkelstein and co-workers used avidin binding to biotinylated colicin to examine the movements of its components across the lipid bilayer[7-9]. We subjected the KvAP channel to a similar analysis. The idea behind the experiments is outlined in Fig. 2a. We introduce cysteine residues at specific locations in the channel, biotinylate the cysteine, reconstitute channels into planar lipid membranes, and then determine whether avidin binds from the internal or external side, and whether binding depends on membrane depolarization. Wild-type KvAP channels contain a single cysteine on the carboxy-terminus, which was mutated to serine (without affecting function) to work with a channel without background cysteine residues.

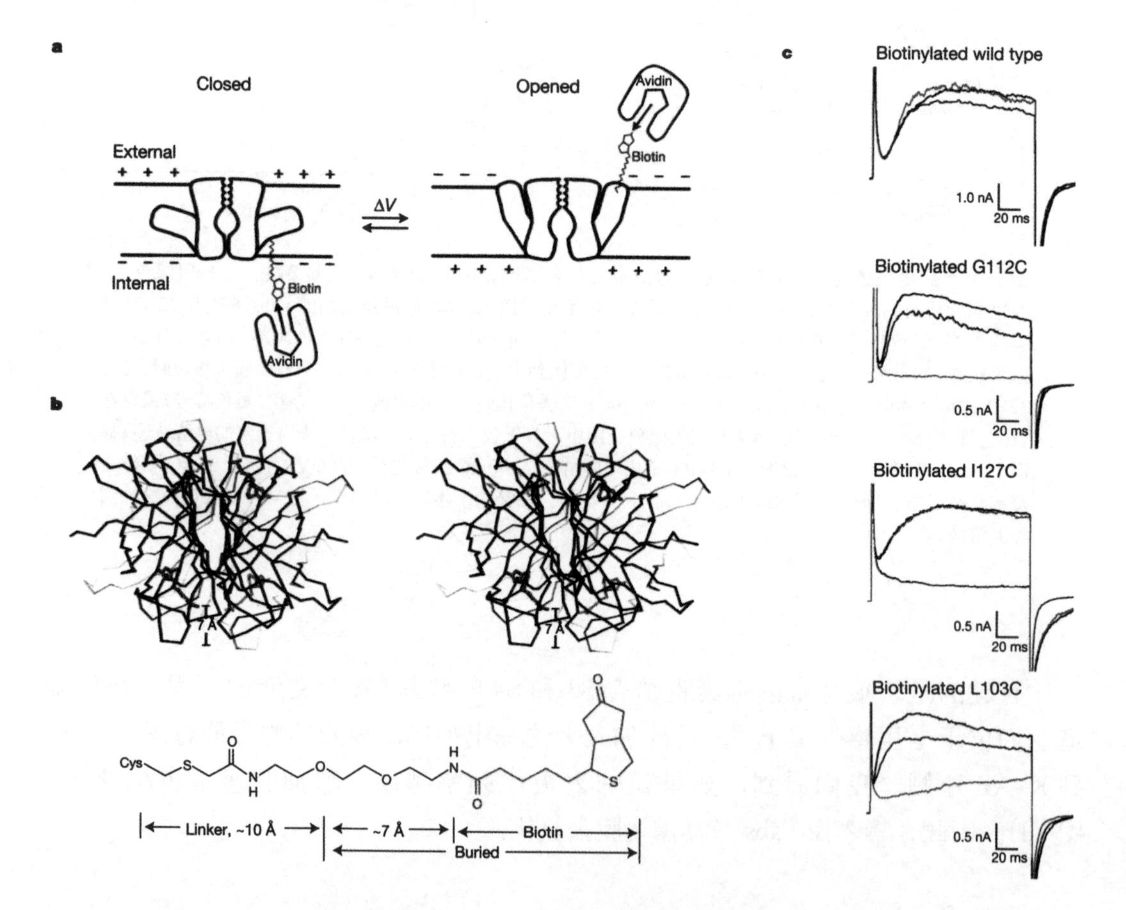


Fig. 2. Using avidin and tethered biotin as a molecular ruler to measure positions of the voltage-sensor paddles. **a**, Experimental strategy: KvAP channels with biotin tethered to a site-directed cysteine can be "grabbed" by avidin in solution to affect channel function. **b**, Stereo view of an avidin tetramer (Cα trace, Protein Data Bank code 1AVD) with biotin (red) in its binding pockets. Chemical structure of biotin and its PEO-iodoacetyl linker with buried (inside avidin) and exposed segments indicated. **c**, Representative traces

超极化时，桨在膜内多深的地方？通道打开时它们会移动多远？

## 利用生物素和抗生物素蛋白测量桨运动

芬克尔斯坦和他的同事们用结合大肠杆菌生物素的抗生物素蛋白来检测其组成部分跨越脂质双层的运动[7-9]。我们对 KvAP 通道做了类似的分析。图 2a 概述了实验构想。我们在通道的特殊位置引入半胱氨酸残基，用生物素标记这些半胱氨酸，将通道重建到平面脂质膜，然后确定抗生物素蛋白是从内侧还是从外侧结合，以及结合是否取决于膜的去极化。野生型 KvAP 通道蛋白在羧基末端包含一个半胱氨酸，将它突变成丝氨酸（不影响功能），与无本底半胱氨酸残基背景的通道一起进行实验。

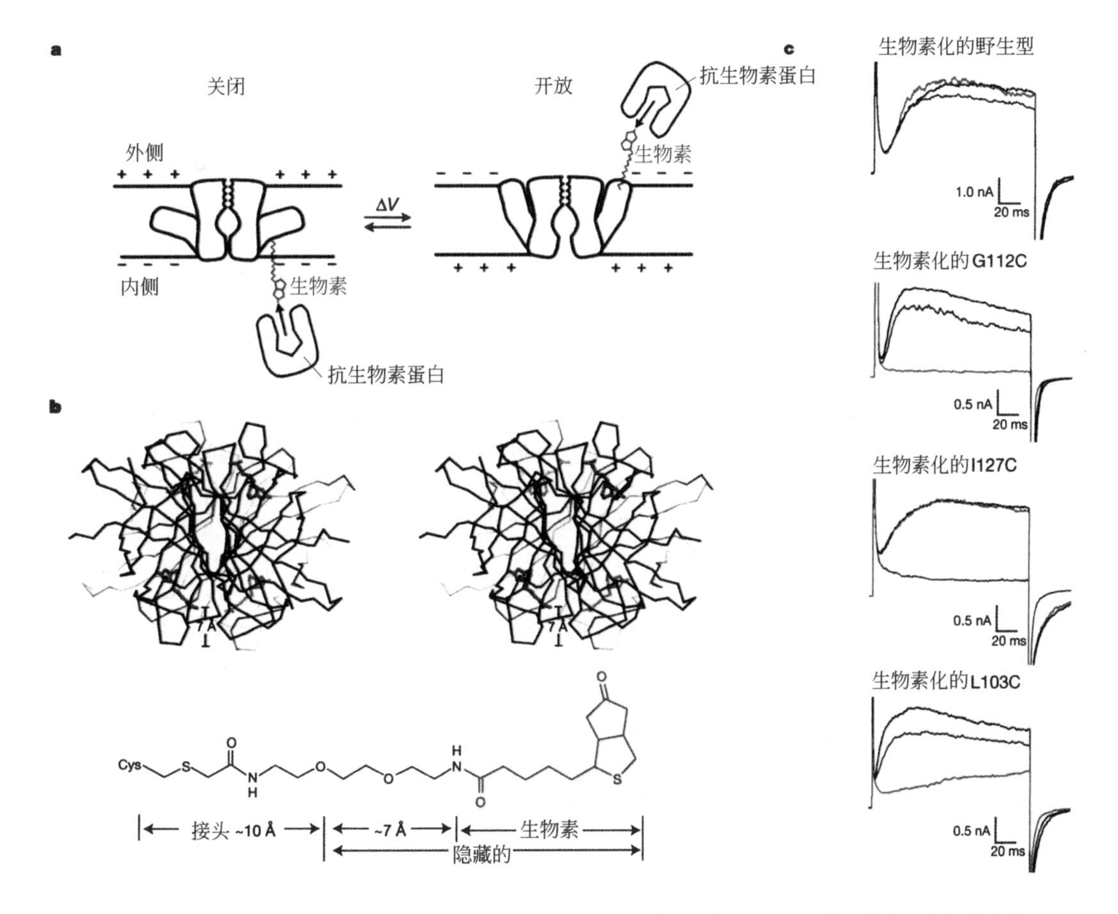


图 2. 用结合了生物素的抗生物素蛋白作为分子尺来测量电压感受桨的位置。**a**，实验策略：生物素定点结合到 KvAP 通道的半胱氨酸上，就可以被溶液内抗生物素蛋白“捕获”进而影响通道功能。**b**，抗生物素蛋白四聚体（α 碳原子示踪，蛋白质数据库索引号 1AVD）与结合在其口袋处的生物素（红色）的立体构象。生物素及其 PEO–碘乙酰基连接接头的化学结构，其中隐藏的（即在抗生物素蛋白内的）和暴

showing the effects of avidin on wild-type and mutant biotinylated KvAP channels. Currents were elicited with depolarizing steps in the absence (black traces) or presence of internal (blue traces) or external (red traces) avidin.

The crystal structure of avidin and the chemical structure of biotin attached to its linker are shown in Fig. 2b. Biotin binds within a deep cleft inside the core of avidin, a rigid protein molecule[10]. The atom on the biotin molecule to which the linker is attached is 7 Å beneath the surface of avidin (Fig. 2b). Therefore, when avidin binds to a biotinylated cysteine on the channel, the distance from the cysteine $\alpha$-carbon to the surface of avidin is 10 Å; this is the pertinent linker length from the $\alpha$-carbon to avidin. Since avidin is large (the tetramer is a 57-kDa protein), it cannot fit into clefts on the channel and therefore cannot penetrate the surface. The important point is that, for avidin to bind to a tethered biotin, the cysteine $\alpha$-carbon has to be within 10 Å of the bulk aqueous solution on either side of the membrane. Applying this restraint, we used linked biotin and avidin as a molecular ruler to measure positions of the voltage-sensor paddles.

Detection of avidin binding to biotinylated channels depends on the demonstration of a functional effect when avidin "grabs" tethered biotin. Control experiments and examples are illustrated in Fig. 2c. Our convention for representing data will be black, red and blue traces corresponding to control (no avidin), external and internal avidin, respectively. Biotinylated wild-type channels are not affected by external avidin and show a small reduction in current when avidin is applied to the internal solution. On the basis of protein gel assays, we conclude that the wild-type channels do not contain detectable biotin on them after the biotinylation procedure (not shown). Three examples of biotinylated cysteine mutant channels are shown. The G112C mutant is inhibited completely by external but not internal avidin; the small reduction by internal avidin is similar to the wild-type control. The I127C mutant is inhibited completely by internal but not by external avidin. The L103C mutant shows an outcome different from complete inhibition: external avidin reduces the current but does not abolish it, and changes the kinetics of current activation. Another outcome that is observed at certain positions is partial inhibition by avidin without changing the gating kinetics (Fig. 3). These cases probably reflect incomplete biotinylation of buried cysteine residues. We excluded the possibility that "no effect" (that is, internal avidin on G112C or external avidin on I127C) represents "silent" avidin binding by adding avidin first to the "no effect" side of the membrane and then to the opposite side (not shown). Because avidin binding to biotin is essentially irreversible, silent binding to one side should protect from binding to the other, but this was never observed. Thus, "no effect" signifies no binding.

露的部分在图中标示出来了。**c**，代表性电流轨迹显示了抗生物素蛋白对野生型和突变体生物素标记的KvAP通道的影响。在没有（黑色轨迹）或存在内侧（蓝色轨迹）或外侧（红色轨迹）抗生物素蛋白的情况下，去极化可以引起电流。

抗生物素蛋白的晶体结构和结合到其接头的生物素的化学结构如图2b所示。生物素结合在一个刚性蛋白分子——抗生物素蛋白——核心处的一个深深的裂缝内[10]。生物素分子上与接头结合的原子在距抗生物素蛋白表面7 Å处（图2b）。因此，当抗生物素蛋白结合到通道上的生物素化的半胱氨酸时，从半胱氨酸α碳到抗生物素蛋白表面的距离就是10 Å；这是从α碳到抗生物素蛋白的合适连接长度。因为抗生物素蛋白很大（该四聚体是一个57 kDa的蛋白），它不能进入通道上的裂缝，因此不能穿越表面。重要的一点是，为使抗生物素蛋白结合到与通道绑定的生物素上，半胱氨酸的α碳必须位于膜任意一侧距大量水溶液10 Å范围内。运用这个尺度，我们将连接的生物素和抗生物素蛋白作为分子尺来测量电压感受桨的位置。

检测结合到生物素标记通道的抗生物素蛋白取决于当抗生物素蛋白“捕获”绑定的生物素时的功能影响的论证。对照组和实验组如图2c所示。我们给出的代表性数据中黑色、红色和蓝色的轨迹分别为对照（无抗生物素）、外侧和内侧加入抗生物素蛋白。生物素标记的野生型通道不受外部抗生物素蛋白影响，并且当抗生物素加入内侧溶液后电流轻微减弱。在蛋白质凝胶电泳分析基础上，我们得出这样的结论：野生型通道在生物素化处理后不含有可检测的生物素（未显示）。这里显示了三个生物素化半胱氨酸突变通道实例。G112C突变体被外侧的抗生物素蛋白完全抑制，而内侧抗生物素蛋白无此作用；内侧抗生物素蛋白引起的轻微减弱类似于野生型对照。I127C突变体被内侧的抗生物素蛋白完全抑制，而外侧抗生物素蛋白无此作用。L103C突变体显示了与完全抑制不同的结果：外侧抗生物素蛋白降低电流，但并不完全消除电流，它还改变了电流激活的动力学。另一项结果是观察到的某些位置被抗生物素蛋白部分抑制而不改变门控动力学（见图3）。这些例子可能反映了包埋的半胱氨酸残基不能被生物素完全标记。我们通过先将抗生物素蛋白加到膜“没有影响”的一侧然后加到另一侧，排除了“没有影响”（内侧抗生物素蛋白对G112C或外侧抗生物素蛋白对I127C）代表“沉默”抗生物素蛋白结合的可能性（未显示）。由于抗生物素蛋白结合生物素基本上是不可逆的，沉默结合到一侧时应不再结合到另一侧，事实上也从未被观察到，因此，“没有影响”意味着没有结合。

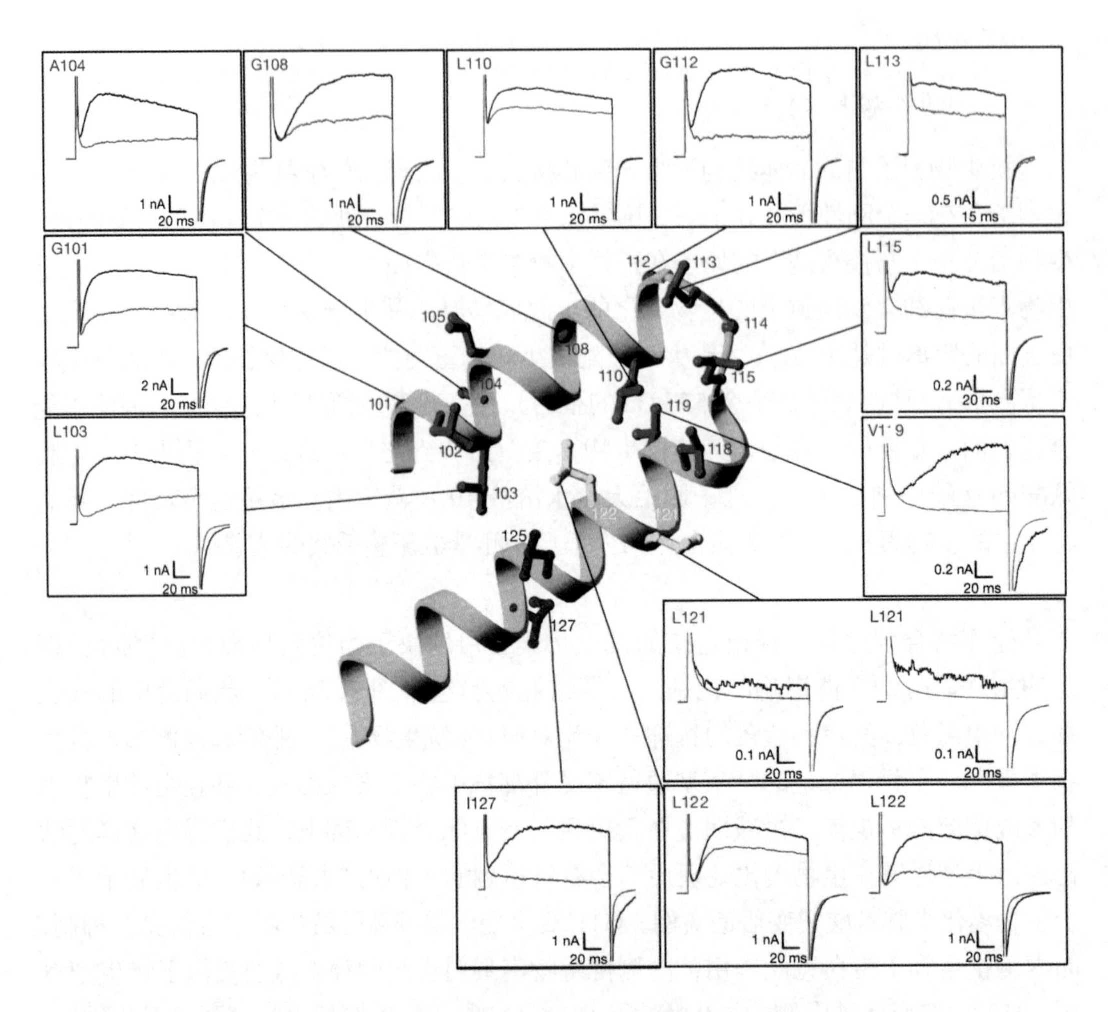


Fig. 3. Accessibility of voltage-sensor paddle residues to the internal and external sides of the membrane. Membrane was depolarized every 120 s in the absence (control) or presence of avidin on the internal and external side of each biotinylated mutant. Red side chains on the voltage-sensor paddle and selected traces indicate inhibition by avidin from the external side only; blue side chains and selected traces indicate inhibition by avidin from the internal side only; yellow side chains and blue and red traces indicate inhibition by avidin from both sides. Black traces show control currents before avidin addition and coloured traces show currents after adding avidin. Each trace is the average of 5–10 measured traces, with the exception of the single traces for position 121.

Data from a scan of the voltage-sensor paddles from amino-acid position 101 on S3b to position 127 on S4 are shown in Fig. 3. Positions were mutated individually to cysteine, biotinylated and studied in planar lipid membranes. Certain biotinylated mutants showed altered gating even before avidin addition (for example, valine 119), or appeared to be less abundant in the membrane (for example, leucine 121). But in all cases, channels could be held closed at negative membrane voltages (typically −100 mV) and opened by membrane depolarizations (typically 100 mV for 200 ms) every 120 s. To compensate for mutations that shifted the voltage-activation curve, we sometimes held the membrane as negative as −140 mV, or used opening depolarizations as positive as 200 mV. Avidin bound to the

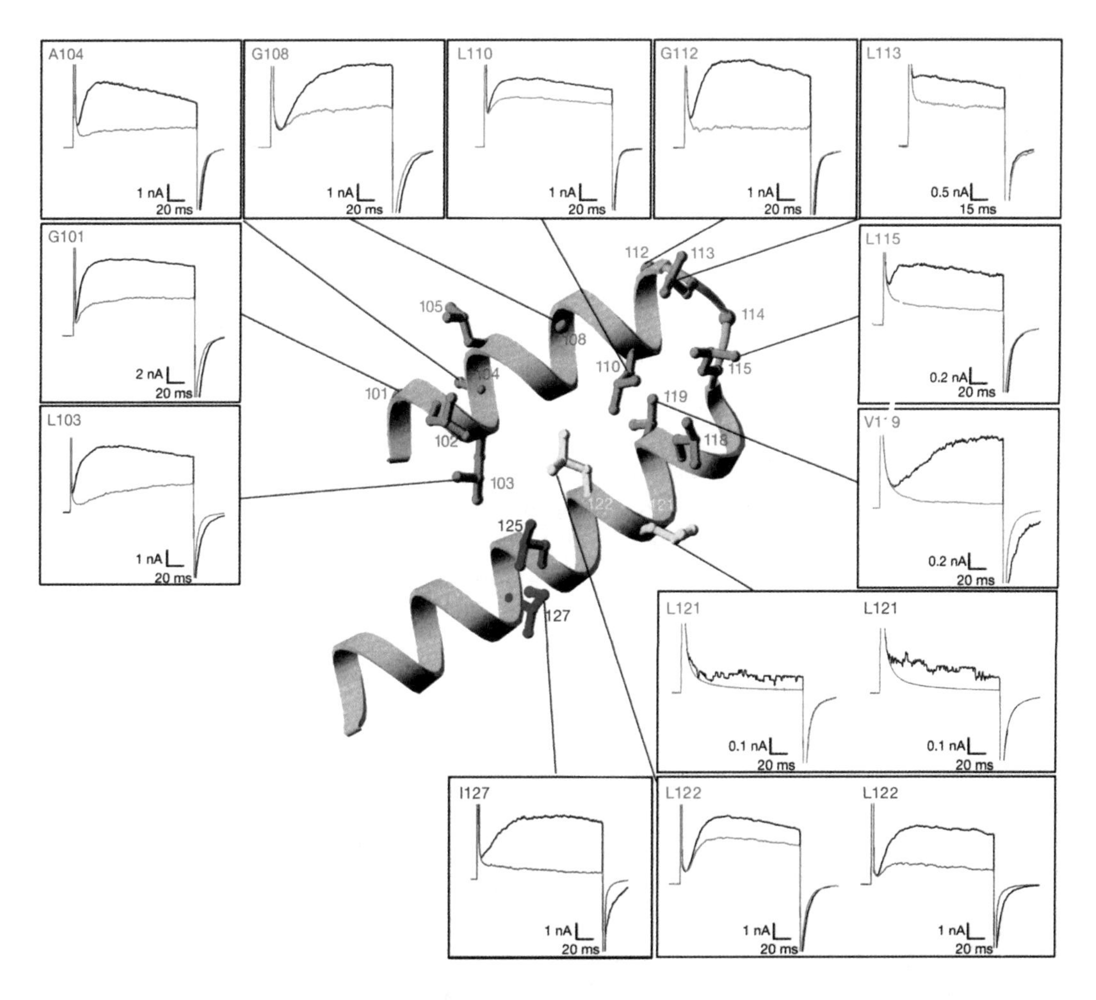


图 3. 电压感受桨残基与膜内侧和外侧的可结合性。在每个生物素标记的突变体内侧和外侧没有(对照)或存在抗生物素蛋白的情况下，每 120 s 膜被去极化。电压感受桨上的红色侧链和对应的电流轨迹表明抗生物素蛋白仅从外侧抑制，蓝色侧链和对应的电流轨迹表明抗生物素蛋白仅从内侧抑制，黄色侧链及蓝色和红色的轨迹表明抗生物素蛋白从两侧抑制。黑色轨迹表明抗生物素蛋白加入前的对照电流，彩色轨迹表明加入抗生物素蛋白后的电流。每个轨迹是 5～10 个测量轨迹的平均值，除了位点 121 是单一的轨迹外。

图 3 显示了对电压感受桨从 S3b 上 101 位点到 S4 上 127 位点的氨基酸进行扫描得到的数据。对突变为半胱氨酸的位点，用生物素标记后在平面脂质膜内对其进行了研究。某些生物素标记的突变体甚至在加入抗生物素蛋白之前就显示了门控的改变(例如，缬氨酸 119)，或在膜中的含量减少(例如，亮氨酸 121)。但在所有情况下，通道可以在负膜电压时维持关闭(通常为 −100 mV)，并通过每 120 s 一次的膜去极化(通常为 100 mV，200 ms)而开放。为了抵消突变体电压激活曲线的移动，我们有时将膜电位维持在 −140 mV，或使用更高的去极化电位(200 mV)使其开放。

tethered biotin and affected channel function at all positions studied. In many cases, avidin caused complete inhibition, and in some cases partial inhibition with altered kinetics. Partial inhibition at certain sites (for example, leucine 110 and leucine 122) can be explained on the basis of incomplete biotinylation because the side chain is buried within the "core" of the voltage-sensor paddle between S3b and S4. This finding is consistent with the proposal that S3b and S4 move together as a voltage-sensor paddle unit.

We are most interested in whether a particular position on the voltage-sensor paddle allows binding to avidin from the external or internal solution. Amino acids on the paddle are colour-coded according to the membrane side from which avidin bound: red, outside; blue, inside; yellow, both. For avidin binding from the external side, we examined whether membrane depolarization is required, by studying the effect of depolarization frequency on the rate of channel inhibition by avidin (Fig. 4). Inhibition from the external solution required depolarization: higher frequencies gave higher rates of inhibition. In other words, the voltage-sensor paddles can be protected from external avidin binding by keeping the membrane at negative voltages, and the paddles are exposed to external avidin at positive voltages. This was the case for all external positions (Fig. 3, red and yellow).

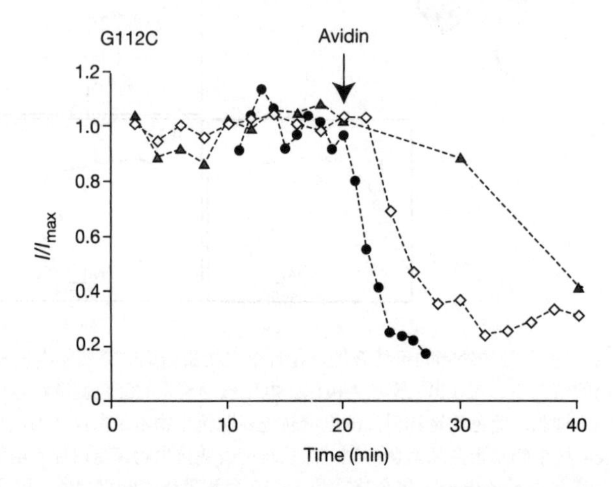


Fig. 4. Exposure of the voltage-sensor paddle to the external solution occurs only when the membrane is depolarized. For the biotinylated G112C mutant, normalized (to the average control) currents elicited by depolarization to 100 mV every 60 s (circles), 120 s (diamonds) and 600 s (triangles) are shown before and after the addition of avidin to the external side.

However, protection from inhibition by external avidin, Fabs and a voltage-sensor toxin, by holding the membrane at negative voltages, is incomplete. In particular, if the wait at negative voltages is long enough, inhibition occurs but at a low rate (Fig. 4). This finding is explained on the basis of thermal fluctuations of the voltage sensors. Probably four voltage-sensor paddles have to move to open the pore[1,11-13]. But individual paddle movements must occasionally occur even at negative voltages. These sensor movements underlying incomplete protection are consistent with gating currents preceding pore opening[2], and longer delays before pore opening when the membrane is depolarized from more negative

抗生物素蛋白结合到绑定的生物素上，在被研究的所有位点上影响通道功能。在许多情况下，抗生物素蛋白引起完全抑制，而在另一些情况下，通过改变分子动力学而引起部分抑制作用。在某些位点上(例如，110 位亮氨酸和 122 位亮氨酸)部分抑制可以在不完全生物素标记的基础上进行解释，因为其侧链被包埋在位于 S3b 和 S4 之间的电压感受桨的“核心”内。这一发现与认为 S3b 和 S4 作为电压感受桨单元一起移动的观点一致。

我们最感兴趣的是抗生物素蛋白是否从外侧或内侧溶液中结合到电压感受桨上某一特定位置。根据抗生物素蛋白从膜的哪一侧结合，对桨上的氨基酸进行了颜色标记：红色，外侧；蓝色，内侧；黄色，两侧。对于抗生物素蛋白从外侧结合，我们通过研究去极化频率对抗生物素蛋白抑制通道的程度的影响，观察膜的去极化是否必需(图 4)。从外侧溶液抑制需要去极化：频率越高抑制率也越高。换言之，通过保持膜的负电压可以保护电压感受桨免受外侧抗生物素蛋白结合，而正电压时桨暴露于外侧的抗生物素蛋白中。这是所有外侧位置的情况(图 3，红色和黄色)。

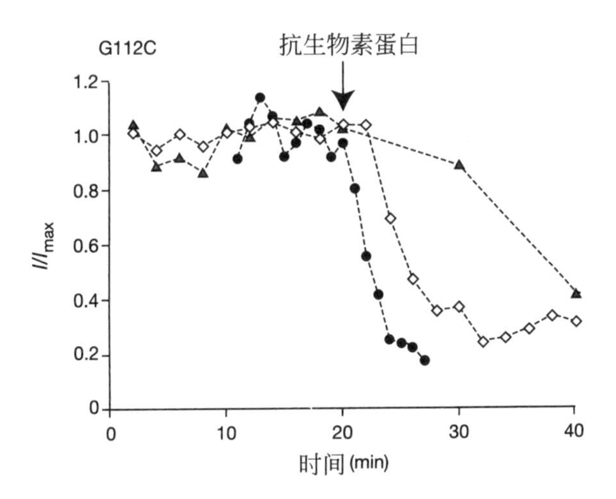


图 4. 只有当膜去极化时，才会发生电压感受桨暴露在外侧溶液中的情况。对于生物素标记的 G112C 突变体，图中显示了抗生物素蛋白加入外侧前后，由每 60 s(圆圈)、120 s(菱形)、600 s(三角形)去极化到 100 mV 所引起的归一化(至平均值)电流。

然而，通过维持膜负电压，并不能完全避免外侧抗生物素、Fabs 和电压传感器毒素造成的抑制。特别是，如果在负电压保留的时间足够长，抑制可发生但发生率低(图 4)。这个发现可用电压传感器的热波动来解释。大概有 4 个电压感受桨必须移动以打开孔道 [1,11-13]。但是，即使在负电压下也必然会偶尔发生个别的桨运动。这些不完全保护下的传感器运动与先于孔道开放的门控电流一致 [2]，且当膜从更负的

holding voltages, a phenomenon known as the Cole–Moore effect[14].

Biotin molecules tethered at two positions were captured by avidin from both sides of the membrane (121 and 122; Fig. 3, yellow). At these positions, inhibition was complete from either side alone, because subsequent addition of avidin to the opposite side caused no further inhibition. Thus, the dual accessibility cannot be ascribed to two structurally distinct populations of channels. We conclude that positions 121 and 122 actually drag biotin and its linker all the way across the lipid membrane from the solution on one side to that on the other when the channel gates. This finding is very important, because it indicates that the voltage-sensor paddles must move a large distance through the membrane, and that they must move through a lipid environment where a bulky chemical structure such as biotin and its linker (Fig. 2b) would be unimpeded. Biotin and its linker could not be dragged through the core of a protein, as would be required by conventional models, which invoke an S4 helix buried within the protein.

## Discussion

Positional constraints on the voltage-sensor paddles are summarized in Fig. 5a, b. Horizontal solid lines show the external and internal surfaces of the cell membrane (~35 Å thick) and dashed lines show the 10 Å limit from the membrane surface set by the linker length (Fig. 2b). If the α-carbon of a cysteine residue comes within 10 Å of the membrane surface, then avidin can bind to the attached biotin, otherwise the biotin is inaccessible. Avidin is too large to enter crevices on the channel, so it cannot penetrate below the membrane surface.

At negative membrane voltages when the channel is closed, no positions are accessible to the external side; all residues on the voltage-sensor paddles in their channel-closed position must lie deeper in the membrane than the 10-Å limit below the external surface (Fig. 5a). At negative voltages, the blue and yellow residues on S4 bind from inside, and therefore must come within 10 Å of the internal solution. The red residues, including all of S3b, the tip of the paddle and the first helical turn into S4, are protected from both sides at negative voltages and must therefore lie further than 10 Å from both surfaces; that is, between the two dashed lines. This pattern of accessibility in the closed (negative voltage) conformation constrains the voltage-sensor paddles to lie near the internal surface of the membrane with S3b above S4, as shown (Fig. 5a), similar to the orientation in the crystal structure[4].

At positive membrane voltage when the channel is opened, the entire S3b helix, the tip of the paddle and the first two-and-a-half helical turns of S4 become accessible to external avidin and must therefore be within 10 Å of the external solution (Fig. 5b). The next helical turn on S4 (positions 125 and 127, blue residues) remains more distant than 10 Å from the external surface. The Fabs inform us that, in the opened conformation, two helical turns of S3b and one turn of S4 must actually protrude clear into the external solution, otherwise the epitope would not be exposed (Fig. 1b). The Fabs and pattern of avidin accessibility in

保持电压去极化时孔道开放前的延迟更长，这一现象被称为科尔–摩尔效应[14]。

抗生物素蛋白从膜两侧捕获标记在两个位置的生物素分子（121 位和 122 位；图 3，黄色）。在这些位置，来自任一侧的抑制都是完全的，因为随后在对侧加入抗生物素蛋白并没有引起进一步的抑制。因此，双重可结合性不能归因于两种在结构上截然不同的通道类型。我们断定当通道处于门控状态时，位于 121 位和 122 位的氨基酸残基实际上拖动生物素和其接头一路跨越脂质膜，从一侧溶液到达另一侧溶液。这一发现很重要，因为它说明电压感受桨必须移动很长一段距离穿过膜，必须移动通过脂质环境，在该脂质环境里大的化学结构如生物素及其接头（图 2b）不受阻碍。生物素及接头无法通过蛋白质的核心，正如传统模型所述，该过程要借助于包埋在蛋白质内的 S4 螺旋。

## 讨　论

图 5a 和 b 总结了电压感受桨上的位置约束。水平实线显示了细胞膜的外表面和内表面（内外表面间的厚度约 35 Å），虚线显示了由接头长度决定的距离膜表面的 10 Å 距离（图 2b）。如果半胱氨酸残基 α 碳到达距膜表面 10 Å 的范围内，那么抗生物素蛋白可以结合到与它连接的生物素上，否则无法接近生物素。抗生物素蛋白太大不能进入通道裂缝，所以不能穿透膜表面。

当膜电压为负，通道关闭时，没有可连接到外侧的位点；电压感受桨上所有残基在通道处于关闭位置时均处于膜外表面下 10 Å 以下的深处（图 5a）。负电压时，S4 上蓝色和黄色残基可从内侧结合，因此，必须到达距细胞内液 10 Å 以内。红色的残基，包括所有 S3b，桨的尖端和 S4 的第一个螺旋，在负电压时它们的两侧都受到保护，因此必须位于离两个表面远于 10 Å 的地方，即两虚线之间。在关闭（负电压）构象中，S3b 在上、S4 在下形成的电压感受桨处于接近膜内表面附近，如图所示（图 5a），类似于晶体结构中的取向[4]。

当膜电压为正，通道开放时，整个 S3b 螺旋、桨的末端和 S4 最初的两个半螺旋转角变得易于接触到外侧抗生物素蛋白，因此必然在距外侧溶液 10 Å 内（图 5b）。S4 上紧接的一个螺旋转角（125 位和 127 位，蓝色残基）与外表面保持着超过 10 Å 的距离。结合位点告诉我们，在开放构象中，S3b 的两个螺旋转角和 S4 的一个转角需从膜内突出进入外侧溶液，否则抗原表位将不会暴露出来（图 1b）。在开放（正电压）构象时，抗原表位和抗生物素蛋白的可结合模式迫使电压感受桨以一个更加垂

the opened (positive voltage) conformation constrain the voltage-sensor paddles to be near the external membrane surface with a more vertical orientation, as shown (Fig. 5b).

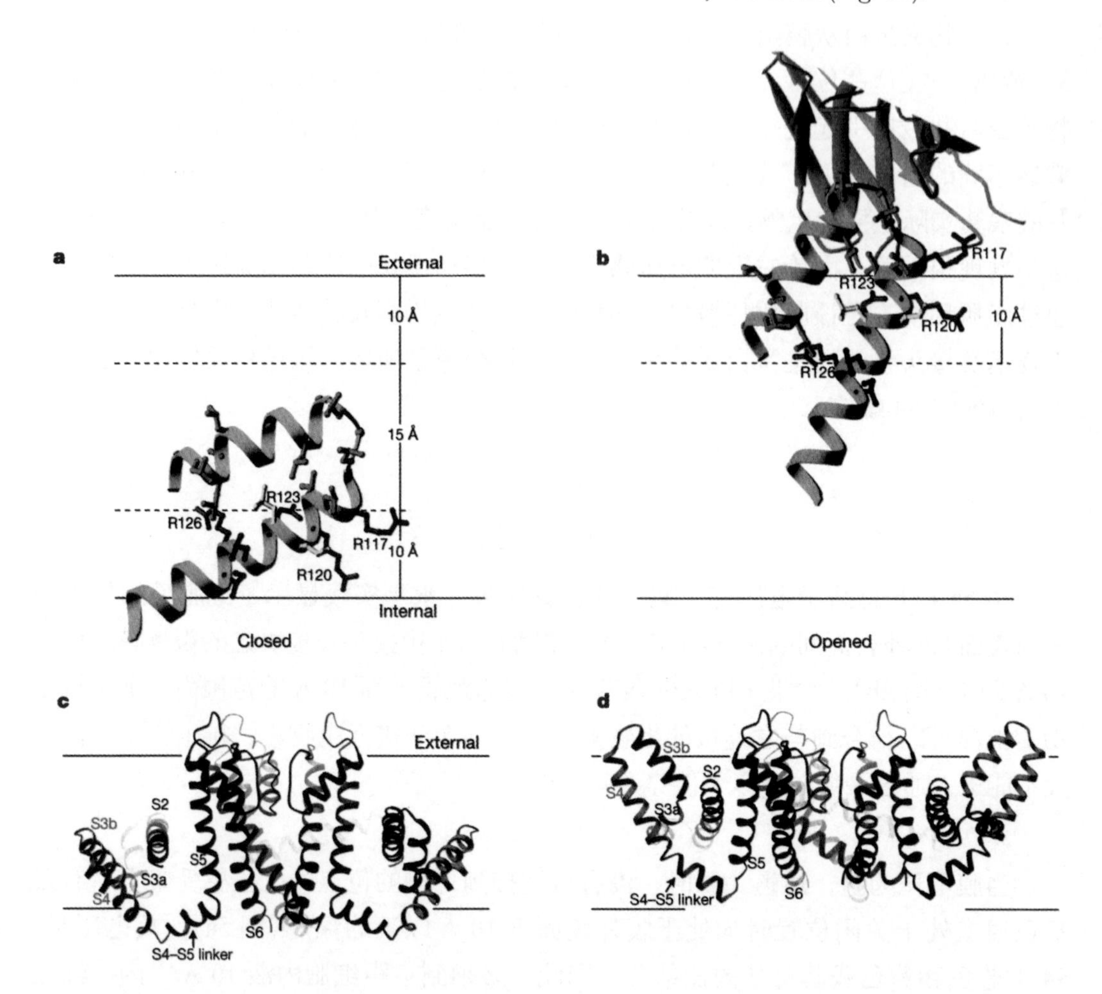


Fig. 5. Positions within the membrane of the voltage-sensor paddles during closed and opened conformations, and a hypothesis for coupling to pore opening. **a**, **b**, Closed (**a**) and opened (**b**) positions of the paddles derived from the tethered biotin–avidin measurements, and structural and functional measurements with Fabs. A voltage-sensor paddle is shown as a cyan ribbon with side chains colour-coded as in Fig. 3. Grey side chains show four arginine residues on the paddle, and the green ribbon (**b**) shows part of a bound Fab from the crystal structures. Solid horizontal lines show the external and internal membrane surfaces, and dashed lines indicate the 10-Å distance from the surfaces set by biotin and its linker. **c**, The closed KvAP structure is based on the paddle depth and orientation in **a** (red), and adjusting the S5 and S6 helices of KvAP to the positions in KcsA, a closed K⁺ channel. **d**, The opened KvAP structure is based on the paddle depth and orientation in **b** (red), and the pore of KvAP.

In moving from their closed to their opened position, the voltage-sensor paddles' centre of mass translates approximately 20 Å (assuming a membrane thickness of 35 Å) through the membrane from inside to out, and the paddles tilt from a somewhat horizontal to a more vertical orientation. Arginines 117, 120, 123 and 126, the first four arginines on S4 (Fig. 1a), are distributed along the paddles. In the closed conformation, arginines 126,

直的取向靠近膜的外表面，如图所示（图 5b）。

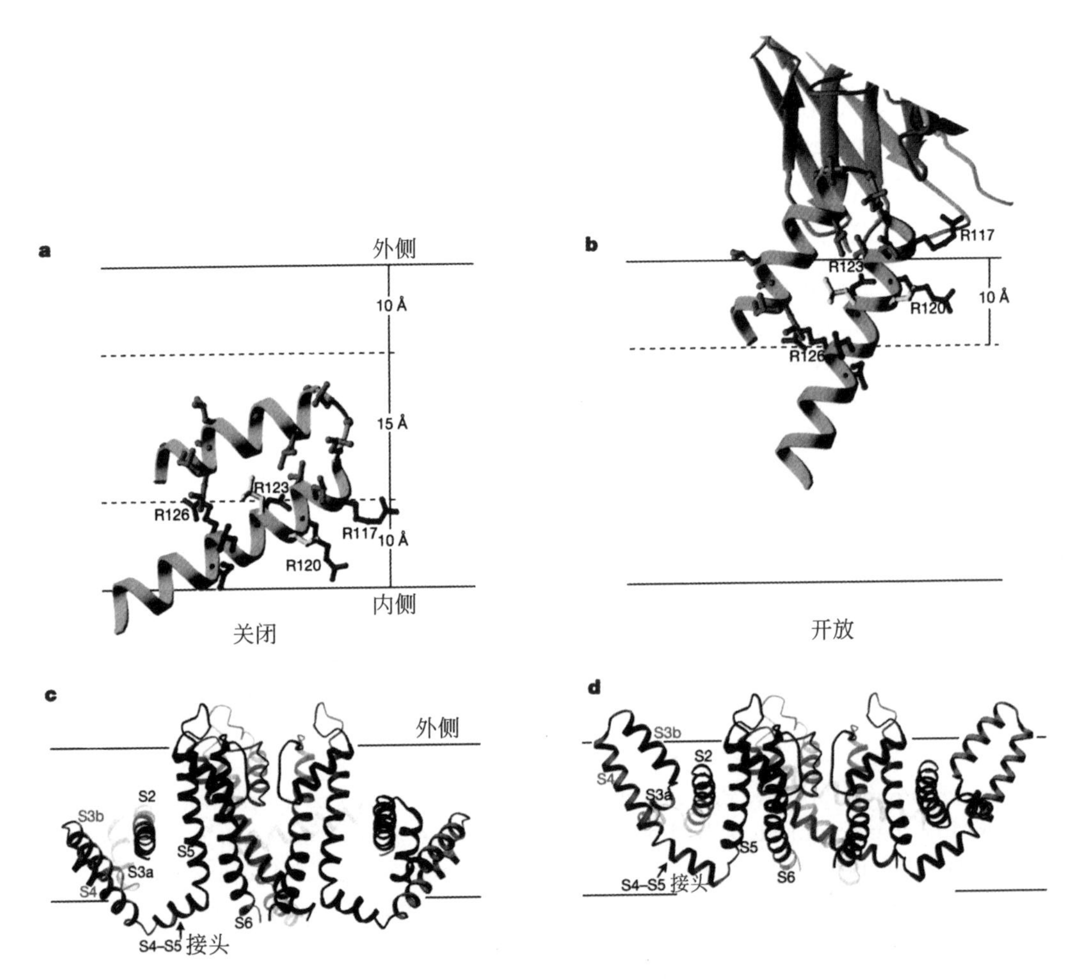


图 5. 关闭和开放构象期间电压感受桨在膜内的位置，以及与孔道开放结合的假设。**a** 和 **b** 分别表示关闭和开放状态时桨的位置，是基于生物素–抗生物素的连接以及借助 Fabs 结合进行的结构和功能测定。如图 3 中显示电压感受桨为一个带有彩色编码侧链的蓝绿色带。灰色侧链显示桨上的 4 个精氨酸残基，(**b**) 中的绿色带为晶体中一个结合的 Fab 的局部结构。水平实线显示外侧和内侧膜表面，虚线表明由生物素及其接头决定的距膜表面 10 Å 的距离。**c**，关闭的 KvAP 结构是基于 **a**(红色) 中桨的深度和位置，并调整 KvAP 的 S5 和 S6 螺旋到 KcsA 中位置，使通道关闭。**d**，开放的 KvAP 结构基于 **b**(红色) 中桨的深度和方向，以及 KvAP 孔道。

在它们从关闭位置移动到开放位置的过程中，电压感受桨的质量中心从内到外穿过膜移动大约 20 Å（假设膜厚度为 35 Å），而桨从相对水平向更加垂直的方向倾斜。S4 上位于 117 位、120 位、123 位和 126 位的前 4 个精氨酸（图 1a）沿桨分布。在关闭的构象中，126 位、123 位，也许还有 120 位的精氨酸可将带正电荷的胍基延

123 and perhaps 120 can probably extend their positively charged guanidinium group to the internal lipid head group layer, whereas arginine 117 is near the internal side but still within the membrane. In the opened conformation, arginines 117, 120 and probably 123 can extend to the external solution or lipid head group layer, whereas 126 is near the external side but still within the membrane. The near-complete transfer of these four arginine residues across the membrane (in each of the four subunits) is compatible with the total gating charge in the *Shaker* $K^+$ channel of 12–14 electrons (3.0–3.5 electrons per subunit)[15-17], and with the demonstration that each of these four arginine residues carries approximately one electron charge unit[16,17].

A positional aspect of the voltage-sensor paddles not constrained by these experiments is whether they lie tangential to the channel's outer surface or whether they point in a radial direction away from it. The flexible S3 loops and S4–S5 linkers probably do not constrain the paddles much. However, we have two reasons for thinking that the paddles are positioned tangentially, which is the way they are positioned in Fig. 5c. First, the paddles are oriented tangentially in the crystal structure of the full channel; and second, the crystal structure of the isolated voltage sensor shows interesting salt-bridge interactions between S4 and S2, and between S3a and S2 (ref. 4). A tangential orientation would favour salt-bridge interactions between arginine residues on the voltage-sensor paddles and acidic residues on the S2 and S3a helices. Studies by Papazian and co-workers on the *Shaker* $K^+$ channel indeed suggest that salt-bridge pairs are important, and that they might exchange as the paddles move between their closed and opened positions on the outer perimeter of the channel[18-20].

To a first approximation, we describe the voltage-sensor paddles as hydrophobic cations that carry gating charges through the lipid bilayer. Ionic interactions between S4 arginines and S2 and S3 acidic residues probably assist the gating charge movement, and the presence of a polar, sometimes acidic loop between S3b and S4 in certain voltage-dependent $K^+$ channels (for example, the *Shaker* $K^+$ channel[21]) raises interesting questions about the structure of the lipid–water interface above the paddles when they are in their closed channel position. However, there is no escaping the basic finding that the paddles are located at the protein–lipid interface, and move while contacting the lipid membrane. Given that the paddles move through a lipid environment, it is interesting to ask why the basic residues on the voltage-sensor paddles are nearly always arginine and not lysine? One reason is that arginine ($pK_a \approx 12.5$ in water) will nearly always move through the membrane with a protonated, charged side chain, whereas lysine ($pK_a \approx 10.5$ in water) will sometimes be unprotonated. Another reason may be that arginine can easily participate in multiple hydrogen bonds (perhaps with acidic side chains) in a spatially directed way. Yet a third possible reason is that it might be energetically less costly to transfer from water to lipid the diffuse positive charge on a guanidinium group (arginine) than the more focused charge on an amino group (lysine). Studies of the transfer of hydrophobic peptides from water to octanol by White and co-workers[22] provide evidence for this idea by showing that the charged lysine side chain is energetically more costly in this assay than the charged arginine by about 1.0 kcal $mol^{-1}$. Given that four voltage-sensor paddles must move through the membrane to open the pore,

伸至内侧的脂质头部基团，而 117 位精氨酸接近内侧但仍然在膜内。在开放的构象中，117 位、120 位，或许还有 123 位的精氨酸可以延伸至外侧溶液或脂质头部基团层，而 126 位精氨酸接近外侧但仍然在膜内。这 4 个精氨酸残基跨膜接近完全的转移（在 4 个亚基中的每一个内）与 12 ~ 14 个电子的 *Shaker* $K^+$ 通道的总门控电荷（每个亚基 3.0 ~ 3.5 个电子）相一致 [15-17]，并证明了每组 4 个精氨酸残基搬运了约 1 个电荷单位 [16,17]。

不受这些实验限制的电压感受桨要么位于通道外表面的切面上，要么指向一个远离通道外表面的径向方向。柔性的 S3 无规卷曲和 S4–S5 接头可能不太限制桨的位置。但是，我们有两个理由认为桨是切向定位的，它们的定位方式见图 5c。首先，在通道的完整晶体结构中桨是切向定位的；其次，独立的电压感受器的晶体结构显示了 S2 与 S4 之间和 S3a 与 S2 之间存在有趣的盐桥相互作用（参考文献 4）。切线方向将有利于电压感受桨上的精氨酸残基和 S2 与 S3a 螺旋上的酸性残基之间的盐桥相互作用。帕帕济安及其同事们对 *Shaker* $K^+$ 通道的研究表明盐桥对确实具有重要意义，当桨在关闭和开放位置间运动时，它们可能在通道外口周围发生交换 [18-20]。

我们已经描述了电压感受器桨作为疏水阳离子可携带门控电荷通过脂质双层。S4 精氨酸与 S2 和 S3 上酸性残基之间的离子相互作用可能协助门控电荷的运动并产生极性，有时候，在某些电压依赖性 $K^+$ 通道（例如，*Shaker* $K^+$ 通道 [21]）中的 S3b 和 S4 之间的酸性环带来当通道在关闭位置时桨上方脂–水界面结构的有趣问题。然而，无法回避的基本结果是，即桨位于蛋白质–脂质界面，且随着脂质膜移动。已知桨是在脂质环境中移动，有趣的是，为什么电压感受桨上的碱性残基几乎总是精氨酸而不是赖氨酸？一个原因是，精氨酸（在水中 $pK_a \approx 12.5$）几乎总是带有一个质子化的、带电的侧链进行跨膜运动，而赖氨酸（在水中 $pK_a \approx 10.5$）有时是非质子化的。另外一个原因可能是精氨酸易于在空间定向上形成多个氢键（也许是与酸性侧链）。第三个可能的原因是，从水向脂质转移在胍基（精氨酸）上分散的正电荷比转移氨基（赖氨酸）上更集中的电荷耗费较少的能量。怀特及其同事 [22] 对疏水多肽从水向辛醇转移的研究为这一想法提供了证据，他们的研究显示带电赖氨酸侧链在这一实验中比带电精氨酸耗费的能量大约多 1.0 kcal · $mol^{-1}$。鉴于 4 个电压感受桨必须通过膜内运动才能打开孔道，我们推测从进化的角度讲更偏向于精氨酸而非赖氨酸。在这方面，有趣的是，最近发现机械敏感性通道 MscS 的结构在其跨膜区段有两个碱性氨基酸，

we might expect there to be a strong evolutionary bias in favour of arginine over lysine. In this regard, it is interesting that the recent structure of MscS, a mechanosensitive channel, has two basic amino acids in its transmembrane segments that are arginine[23]. MscS is not gated by voltage *per se*, but its mechanical force-induced opening can be modulated by voltage[24], and the candidate residues that are proposed to underlie voltage modulation are arginine, not lysine[23].

On the basis of the KvAP crystal structure[4], the deduced positions of its voltage-sensor paddles in the functioning channel (Fig. 5a, b), and previous studies of opened and closed $K^+$ channels[25,26], we propose a model for how membrane voltage gates the pore (Fig. 5c, d). This is a working model to envision how the electromechanical coupling process might occur, and it will need to be revised as more data are obtained. In the closed conformation (Fig. 5c), the positively charged voltage-sensor paddles (red) are near the intracellular membrane surface, held there by the large electric field (mean value > $10^7$ V $m^{-1}$) imposed by the negative resting membrane voltage. In this conformation, the S5 and S6 (outer and inner) helices are arranged as they are in KcsA, a closed $K^+$ channel[25]; favourable packing interactions between the inner helices have been proposed to help to stabilize the closed conformation[27]. In response to depolarization, the voltage-sensor paddles move across the membrane to their external position, which exerts a force on the S4–S5 linker, pulling the S5 helices away from the pore axis (Fig. 5d). The crystal structure of KvAP shows that the S5 helices form a cuff outside the S6 helices, and suggests that when the S5 cuff is expanded, the S6 helices follow, opening the pore[4].

Although previous mutational studies of voltage-dependent gating have been interpreted in the context of the conventional models, many of the data are consistent with the structural model presented here, and indeed help to constrain it. For example, second-site suppressor mutations in the *Shaker* $K^+$ channel suggest that certain salt bridges probably break and reform as the voltage-sensor paddles move[20]. In addition to four S4 arginine residues, a component of the gating charge in *Shaker* was shown to come from an S2 acidic residue (glutamate 293, corresponding to aspartate 72 in KvAP)[16], implying that S2 might change its position in the membrane as the voltage-sensor paddles move; the loose attachment of S2 to the pore in the KvAP crystal structure certainly would allow this to happen. Disulphide cross-linking of S4 to the turret loop between S5 and the pore helix[28], and of two S4 segments within a tetramer to each other[29] (an observation that was very difficult to understand in the context of conventional models), are in agreement with the mobile voltage-sensor paddles in Fig. 5c, d. The accessibility of thiol-reactive compounds to cysteine residues introduced into the *Shaker* $K^+$ channel S4 (refs 30–32), and the ability of histidine residues to shuttle protons across the membrane[33], are in reasonable agreement with our data on voltage-sensor paddle movements. However, our structural and mechanistic interpretation, summarized in Fig. 5, differs fundamentally from past models of voltage-dependent gating. This new picture is based on the elucidation of the voltage-sensor paddle structure, its flexible attachments and disposition relative to the pore, and the paddles' positions in the membrane when the channel is closed and opened.

均是精氨酸[23]。MscS 不受电压本身门控，而它由机械外力诱导的开放可被电压调节[24]，并且认为产生电压调节的候补残基是精氨酸而不是赖氨酸[23]。

基于 KvAP 晶体结构[4]、推导的电压感受桨在功能通道中的位置（图 5a 和 b）和以往对开放和关闭 $K^+$ 通道的研究[25,26]，我们提出了一个膜电压如何门控孔道的模型（图 5c 和 d）。即一个如何产生电–机械耦联的工作模式的假想模型，该模型在获得更多的数据时需要做修正。在关闭构象下（图 5c），带正电荷的电压感受桨（红色）接近细胞内膜表面，在这附近由于负的静息膜电位而存在较大电场（平均值 > $10^7\ V \cdot m^{-1}$）。在此构象中，S5 和 S6（外和内）螺旋的排列与它们在 KcsA（一个关闭的 $K^+$ 通道）中一致[25]；推测内螺旋间的相互作用有利于稳定关闭构象[27]。当发生去极化反应时，电压感受桨跨膜移动到其位于膜外的位置上，也对 S4–S5 连接施加力量，拉动 S5 螺旋远离孔轴（图 5d）。KvAP 晶体结构表明，S5 螺旋在 S6 螺旋外形成一个袖口，并推测当 S5 袖口扩大时，S6 螺旋跟随其后，打开孔道[4]。

虽然已进行的电压依赖性门控的突变研究已经解释了本文介绍的模型，许多数据与这里呈现的结构模型是一致的，并且确实有助于修正它。例如，*Shaker* $K^+$ 通道中第二位点抑制突变表明，当电压感受桨运动时某些盐桥可能断裂或重组[20]。除了 4 个 S4 精氨酸残基以外，*Shaker* 中门控电荷的一个组成被证明来自 S2 上的一个酸性残基（293 位谷氨酸，对应于 KvAP 的 72 位天门冬氨酸）[16]，这意味着当电压感受桨移动时 S2 可能改变其在膜中的位置；在 KvAP 晶体结构中 S2 松散连接到孔上将允许这种情况发生。二硫键将 S4 交联到 S5 和孔区螺旋之间的塔形无规卷曲上[28]，并将四聚体内的 2 个 S4 片段相互交联[29]（这种现象在传统模型背景下很难理解），这种二硫交联与图 5c 和 d 中可移动的电压感受桨一致。巯基化合物与 *Shaker* $K^+$ 通道 S4 的半胱氨酸残基的可结合性（参考文献 30 ~ 32），以及组氨酸残基跨膜穿梭运送质子的能力[33]，与我们关于电压感受桨移动的数据相当一致。然而，我们关于结构和机械运动的解释（归纳在图 5 中）本质上不同于过去的电压依赖性门控模型。这幅新图是以电压感受桨的结构阐明、它与孔道的柔性连接及与孔道功能的关系，以及当通道关闭和开放时桨在膜中的位置这三点为前提的。

## Conclusion

Here and in an accompanying paper[4], we have shown that (1) the gating charges are carried on voltage-sensor paddles, which are helix–turn–helix structures attached to the channel through flexible S3 loops and S4–S5 linkers; (2) the paddles are located at the channel's outer perimeter and move within the lipid membrane; (3) the S2 helices lie beside the pore and contain acidic amino acids that could help to stabilize positive charges on the paddles as they move across the membrane; (4) the total displacement of the voltage-sensor paddles is approximately 20 Å perpendicular to the membrane; and (5) the large displacement of the paddles could open the pore by pulling on the S4–S5 linker. We conclude that the voltage sensor operates by an extraordinarily simple principle based on hydrophobic cations attached to levers, which enables the membrane electric field to perform mechanical work to open and close the ion-conduction pore.

## Methods

### Biotinylation

All biotinylation studies were carried out using a KvAP channel in which the single endogenous cysteine was mutated to serine (C247S). This mutant showed no detectable electrophysiological differences when compared to wild-type KvAP. Single cysteine mutations were then added to the voltage-sensor paddle (positions 101 to 127) using the QuickChange method (Stratagene) and confirmed by sequencing the entire gene. Mutant channels were expressed and purified by the same protocol as wild-type KvAP channels[5], except that before gel filtration, mutant KvAP channels were incubated with 10 mM DTT for 1 h. Immediately after gel filtration, mutant channels (at 0.5–1.0 mg $ml^{-1}$) were incubated with 500 μM PEO-iodoacetylbiotin (Pierce) for 2–3 h at room temperature, and then either reconstituted into lipid vesicles for electrophysiological analysis, or purified away from the excess biotin reagent on a desalting column, complexed with avidin (Pierce) and run on an SDS gel to assess the extent of biotinylation.

### Electrophysiology

Fabs (6E1 and 33H1) and VSTX1 were purified as described in the companion paper[4] and used in electrophysiological assays. Electrophysiological studies of wild-type and biotinylated KvAP channels were carried out as described[5]. To measure inhibition, Fabs (~500 nM), VSTX1 (30 nM) or avidin (40 μg $ml^{-1}$) were added to wild-type and/or cysteine-mutant, biotinylated channels reconstituted into lipid membranes, and studied using various voltage protocols.

(**423**, 42-48; 2003)

## 结　论

在本文及一篇相关论文[4]中，我们已经表明：(1)在电压感受桨上携带有门控电荷，它们通过灵活的S3无规卷曲和S4–S5接头连接到通道上的螺旋–转角–螺旋结构；(2)桨位于通道的外围并在脂质膜内移动；(3)S2螺旋位于孔道旁边并含有酸性氨基酸，当桨跨膜移动时，这些酸性氨基酸可能有助于稳定桨上的正电荷；(4)电压感受桨垂直于膜的总位移约20 Å；(5)桨的较大位移可以通过拉动S4–S5接头打开孔道。我们总结如下：电压传感器的操作是基于一个非常简单的将疏水阳离子连接到杠杆上的原理，使得膜电场转为机械操作来打开和关闭离子导通孔。

## 方　法

### 生物素化

所有的生物素化研究都是在KvAP通道上进行的，在KvAP通道中单个内源性半胱氨酸被突变为丝氨酸(C247S)。与野生型KvAP相比，这个突变体没有可检测到的电生理方面的差异。使用QuickChange法(Stratagene公司)，将电压感受桨上的单个半胱氨酸(101位至127位)突变，并通过全基因测序验证突变。采用与野生型KvAP通道相同的方法表达和纯化了突变通道[5]，与野生型不同的是，在凝胶过滤前，突变的KvAP通道需先与10 mM DTT(二硫苏糖醇)温育1 h。凝胶过滤后，突变的通道(0.5 ~ 1.0 mg · $ml^{-1}$)立即与500 μM PEO–碘代乙酰生物素(皮尔斯公司)在室温下温育2 ~ 3 h，然后将突变通道重建到脂质膜上进行电生理分析，或者用脱盐柱纯化以除去过量的生物素试剂，与抗生物素蛋白(皮尔斯公司)结合并通过SDS凝胶电泳评估生物素化的程度。

### 电生理学

如本文的相关论文中[4]所描述，Fabs(6E1和33H1)以及VSTX1被纯化并用于电生理实验。野生型和生物素化KvAP通道的电生理研究按描述的方法进行[5]。为检测抑制情况，Fabs(约500 nM)，VSTX1(30 nM)或抗生物素蛋白(40 μg · $ml^{-1}$)被加到野生型和(或)半胱氨酸突变体的生物素标记的通道上，通道重建在脂质膜中，并利用各种电压刺激方式对它们进行研究。

(李梅 翻译；王晓良 审稿)

**Youxing Jiang*, Vanessa Ruta, Jiayun Chen, Alice Lee & Roderick MacKinnon**
Howard Hughes Medical Institute, Laboratory of Molecular Neurobiology and Biophysics, Rockefeller University, 1230 York Avenue, New York, New York 10021, USA
* Present address: University of Texas Southwestern Medical Center, Department of Physiology, 5323 Harry Hines Blvd, Dallas, Texas 75390-9040, USA

Received 19 February; accepted 11 March 2003; doi: 10.1038/nature01581.

---

References:

1. Sigworth, F. J. Voltage gating of ion channels. *Q. Rev. Biophys.* **27**, 1-40 (1994).
2. Armstrong, C. M. & Bezanilla, F. Charge movement associated with the opening and closing of the activation gates of the Na+ channels. *J. Gen. Physiol.* **63**, 533-552 (1974).
3. Bezanilla, F. The voltage sensor in voltage-dependent ion channels. *Physiol. Rev.* **80**, 555-592 (2000).
4. Jiang, Y. *et al.* X-ray structure of a voltage-dependent K+ channel. *Nature* **423**, 33-41 (2003).
5. Ruta, V., Jiang, Y., Lee, A., Chen, J. & MacKinnon, R. Functional analysis of an archeabacterial voltage-dependent K+ channel. *Nature* **422**, 180-185; advance online publication, 2 March 2003 (doi:10.1038/nature01473).
6. Swartz, K. J. & MacKinnon, R. Mapping the receptor site for hanatoxin, a gating modifier of voltage-dependent K+ channels. *Neuron* **18**, 675-682 (1997).
7. Slatin, S. L., Qiu, X. Q., Jakes, K. S. & Finkelstein, A. Identification of a translocated protein segment in a voltage-dependent channel. *Nature* **371**, 158-161 (1994).
8. Qiu, X. Q., Jakes, K. S., Finkelstein, A. & Slatin, S. L. Site-specific biotinylation of colicin Ia. A probe for protein conformation in the membrane. *J. Biol. Chem.* **269**, 7483-7488 (1994).
9. Qiu, X. Q., Jakes, K. S., Kienker, P. K., Finkelstein, A. & Slatin, S. L. Major transmembrane movement associated with colicin Ia channel gating. *J. Gen. Physiol.* **107**, 313-328 (1996).
10. Pugliese, L., Coda, A., Malcovati, M. & Bolognesi, M. Three-dimensional structure of the tetragonal crystal form of egg-white avidin in its functional complex with biotin at 2.7 Å resolution. *J. Mol. Biol.* **231**, 698-710 (1993).
11. Zagotta, W. N., Hoshi, T., Dittman, J. & Aldrich, R. W. Shaker potassium channel gating. II. Transitions in the activation pathway. *J. Gen. Physiol.* **103**, 279-319 (1994).
12. Zagotta, W. N., Hoshi, T. & Aldrich, R. W. Shaker potassium channel gating. III. Evaluation of kinetic models for activation. *J. Gen. Physiol.* **103**, 321-362 (1994).
13. Schoppa, N. E. & Sigworth, F. J. Activation of Shaker potassium channels. III. An activation gating model for wild-type and V2 mutant channels. *J. Gen. Physiol.* **111**, 313-342 (1998).
14. Cole, K. S. & Moore, J. W. Potassium ion current in the squid giant axon: dynamic characteristic. *Biophys. J.* **1**, 1-14 (1960).
15. Schoppa, N. E., McCormack, K., Tanouye, M. A. & Sigworth, F. J. The size of gating charge in wild-type and mutant Shaker potassium channels. *Science* **255**, 1712-1715 (1992).
16. Seoh, S. A., Sigg, D., Papazian, D. M. & Bezanilla, F. Voltage-sensing residues in the S2 and S4 segments of the Shaker K+ channel. *Neuron* **16**, 1159-1167 (1996).
17. Aggarwal, S. K. & Mackinnon, R. Contribution of the S4 segment to gating charge in the Shaker K+ channel. *Neuron* **16**, 1169-1177 (1996).
18. Papazian, D. M. *et al.* Electrostatic interactions of S4 voltage sensor in Shaker K+ channel. *Neuron* **14**, 1293-1301 (1995).
19. Tiwari-Woodruff, S. K., Lin, M. A., Schulteis, C. T. & Papazian, D. M. Voltage-dependent structural interactions in the Shaker K+ channel. *J. Gen. Physiol.* **115**, 123-138 (2000).
20. Papazian, D. M., Silverman, W. R., Lin, M. C., Tiwari-Woodruff, S. K. & Tang, C. Y. Structural organization of the voltage sensor in voltage-dependent potassium channels. *Novartis Found. Symp.* **245**, 178-190 (2002).
21. Gonzalez, C., Rosenman, E., Bezanilla, F., Alvarez, O. & Latorre, R. Periodic perturbations in Shaker K+ channel gating kinetics by deletions in the S3-S4 linker. *Proc. Natl Acad. Sci. USA* **98**, 9617-9623 (2001).
22. Wimley, W. C., Creamer, T. P. & White, S. H. Solvation energies of amino acid side chains and backbone in a family of host-guest pentapeptides. *Biochemistry* **35**, 5109-5124 (1996).
23. Bass, R. B., Strop, P., Barclay, M. & Rees, D. C. Crystal structure of *Escherichia coli* MscS, a voltage-modulated and mechanosensitive channel. *Science* **298**, 1582-1587 (2002).
24. Martinac, B., Buechner, M., Delcour, A. H., Adler, J. & Kung, C. Pressure-sensitive ion channel in *Escherichia coli*. *Proc. Natl Acad. Sci. USA* **84**, 2297-2301 (1987).
25. Doyle, D. A. *et al.* The structure of the potassium channel: molecular basis of K+ conduction and selectivity. *Science* **280**, 69-77 (1998).
26. Jiang, Y. *et al.* The open pore conformation of potassium channels. *Nature* **417**, 523-526 (2002).
27. Yifrach, O. & MacKinnon, R. Energetics of pore opening in a voltage-gated K+ channel. *Cell* **111**, 231-239 (2002).
28. Laine, M. *et al.* Structural interactions between voltage sensor and pore in the Shaker K+ channels. *Biophys. J.* **82**, 231a (2002).
29. Aziz, Q. H., Partridge, C. J., Munsey, T. S. & Sivaprasadarao, A. Depolarization induces intersubunit cross-linking in a S4 cysteine mutant of the Shaker potassium channel. *J. Biol. Chem.* **277**, 42719-42725 (2002).
30. Larsson, H. P., Baker, O. S., Dhillon, D. S. & Isacoff, E. Y. Transmembrane movement of the Shaker K+ channel S4. *Neuron* **16**, 387-397 (1996).

31. Yusaf, S. P., Wray, D. & Sivaprasadarao, A. Measurement of the movement of the S4 segment during the activation of a voltage-gated potassium channel. *Pflugers Arch.* **433**, 91-97 (1996).

32. Baker, O. S., Larsson, H. P., Mannuzzu, L. M. & Isacoff, E. Y. Three transmembrane conformations and sequence-dependent displacement of the S4 domain in Shaker $K^+$ channel gating. *Neuron* **20**, 1283-1294 (1998).

33. Starace, D. M., Stefani, E. & Bezanilla, F. Voltage-dependent proton transport by the voltage sensor of the Shaker $K^+$ channel. *Neuron* **19**, 1319-1327 (1997).

**Acknowledgements.** We thank D. Gadsby and O. Andersen for helpful discussions and advice on the manuscript. This work was supported in part by a grant from the National Institutes of Health (NIH) to R.M. V.R. is supported by a National Science Foundation Graduate Student Research Fellowship, and R.M. is an Investigator in the Howard Hughes Medical Institute.

**Competing interests statement.** The authors declare that they have no competing financial interests.

**Correspondence** and requests for materials should be addressed to R.M. (mackinn@rockvax.rockefeller.edu).

# The International HapMap Project

The International HapMap Consortium*

## Editor's Note

**This paper describes the International HapMap Project, a large international collaborative effort to identify and catalog the common patterns of human DNA sequence variation. By making the information freely available, the project aimed to help researchers identify genes related to health, disease and responses to drugs and environmental factors. Here, they describe the strategy and key components of the project, including the genotyping of over a million sequence variants, their frequencies and degrees of association, in DNA samples from populations rooted in parts of Africa, Asia and Europe. The Project, officially launched in 2002, successfully published the completed Haplotype Map in 2005, then a more detailed second-generation map in 2007.**

---

The goal of the International HapMap Project is to determine the common patterns of DNA sequence variation in the human genome and to make this information freely available in the public domain. An international consortium is developing a map of these patterns across the genome by determining the genotypes of one million or more sequence variants, their frequencies and the degree of association between them, in DNA samples from populations with ancestry from parts of Africa, Asia and Europe. The HapMap will allow the discovery of sequence variants that affect common disease, will facilitate development of diagnostic tools, and will enhance our ability to choose targets for therapeutic intervention.

---

COMMON diseases such as cardiovascular disease, cancer, obesity, diabetes, psychiatric illnesses and inflammatory diseases are caused by combinations of multiple genetic and environmental factors[1]. Discovering these genetic factors will provide fundamental new insights into the pathogenesis, diagnosis and treatment of human disease. Searches for causative variants in chromosome regions identified by linkage analysis have been highly successful for many rare single-gene disorders. By contrast, linkage studies have been much less successful in locating genetic variants that affect common complex diseases, as each variant individually contributes only modestly to disease risk[2,3]. A complementary approach to identifying these specific genetic risk factors is to search for an association between a specific variant and a disease, by comparing a group of affected individuals with a group of unaffected controls[4]. In the absence of strong natural selection, there is likely to be a broad spectrum of frequency of such variants, many of which are likely to be common in the population. A number

* A full list of participants and affiliations is given in the original paper.

# 国际人类基因组单体型图计划

国际人类基因组单体型图协作组 *

## 编者按

本文描述了国际人类基因组单体型图计划，这是一个大型的国际合作项目，旨在识别和编目人类 DNA 序列变异的常见模式。该计划通过免费提供这些信息，帮助研究人员识别与健康、疾病以及对药物和环境因素产生响应的相关基因。本文中，他们描述了该计划的策略和关键组成部分，包括从部分非洲、亚洲和欧洲地区提供的 DNA 样本中对 100 多万个序列变异进行基因分型以及它们的频率和关联程度。该计划于 2002 年正式启动，2005 年成功发布了完整的单体型图谱，2007 年又发布了更详细的第二代图谱。

---

国际人类基因组单体型图计划的目标是确定人类基因组中 DNA 序列变异的共同模式，并将这些资料免费向公众开放。一个国际协作组将通过来自部分非洲、亚洲和欧洲族群的 DNA 样本，确定 100 万或更多序列变异的基因型以及它们的频率和它们之间的相关程度，从而绘制整个基因组的模式图谱。HapMap 将支持我们发现影响常见疾病的序列变异，促进疾病诊断工具的发展，并且提高我们选择治疗干预靶标的能力。

---

常见的疾病如心血管疾病、癌症、肥胖、糖尿病，精神疾病和炎症性疾病，是由多种基因和环境因素的组合共同造成的 [1]。发现这些遗传因素将为研究人类疾病的发病机理、诊断和治疗提供重要的新见解。对许多罕见的单基因疾病来说，通过连锁分析的方法鉴定染色体区域中的致病变异已经非常成功了。相比之下，通过连锁分析寻找影响常见复杂疾病的遗传变异却不怎么成功，因为每个单独的变异只会增加轻微的疾病风险 [2,3]。通过比较一组病例和正常对照来寻找具体某个遗传变异和某一种疾病间的关联，是确定这些具体遗传风险因素的一个补充方法 [4]。缺乏强大的自然选择时，这种变异的频率范围可能是广泛的，其中许多变异可能是群体中常见的。经过一些侧重于候选基因、与疾病连锁的区域或更大范围的关联研究，人

---

* 协作组的参与者及所属单位的名单请参见原文。

of association studies, focused on candidate genes, regions of linkage to a disease or more large-scale surveys, have already led to the discovery of genetic risk factors for common diseases. Examples include type 1 diabetes (human leukocyte antigen (HLA[5]), insulin[6] and *CTLA4* (ref. 7)), Alzheimer's disease (*APOE*)[8], deep vein thrombosis (factor V)[9], inflammatory bowel disease (*NOD2* (refs 10, 11) and also 5q31 (ref. 12)), hypertriglyceridaemia (*APOAV*)[13], type 2 diabetes (*PPARG*)[14,15], schizophrenia (neuregulin 1)[16], asthma (*ADAM33*)[17], stroke (*PDE4D*)[18] and myocardial infarction (*LTA*)[19].

One approach to doing association studies involves testing each putative causal variant for correlation with the disease (the "direct" approach)[2]. To search the entire genome for disease associations would entail the substantial expense of whole-genome sequencing of numerous patient samples to identify the candidate variants[3]. At present, this approach is limited to sequencing the functional parts of candidate genes (selected on the basis of a previous functional or genetic hypothesis) for potential disease-associated candidate variants. An alternative approach (the "indirect" approach) has been proposed[20], whereby a set of sequence variants in the genome could serve as genetic markers to detect association between a particular genomic region and the disease, whether or not the markers themselves had functional effects. The search for the causative variants could then be limited to the regions showing association with the disease.

Two insights from human population genetics suggest that the indirect approach is able to capture most human sequence variation, with greater efficiency than the direct approach. First, ~90% of sequence variation among individuals is due to common variants[21]. Second, most of these originally arose from single historical mutation events, and are therefore associated with nearby variants that were present on the ancestral chromosome on which the mutation occurred. These associations make the indirect approach feasible to study variants in candidate genes, chromosome regions or across the whole genome. Prior knowledge of putative functional variants is not required. Instead, the approach uses information from a relatively small set of variants that capture most of the common patterns of variation in the genome, so that any region or gene can be tested for association with a particular disease, with a high likelihood that such an association will be detectable if it exists.

The aim of the International HapMap Project is to determine the common patterns of DNA sequence variation in the human genome, by characterizing sequence variants, their frequencies, and correlations between them, in DNA samples from populations with ancestry from parts of Africa, Asia and Europe. The project will thus provide tools that will allow the indirect association approach to be applied readily to any functional candidate gene in the genome, to any region suggested by family-based linkage analysis, or ultimately to the whole genome for scans for disease risk factors.

Common variants responsible for disease risk will be most readily approached by this strategy, but not all predisposing variants are common. However, it should be noted that even a relatively uncommon disease-associated variant can potentially be discovered using this approach. Reflecting its historical origins, the uncommon variant will be travelling on

们已经发现了一些常见疾病的遗传风险因素。例如1型糖尿病（人类白细胞抗原（HLA[5]），胰岛素[6]和*CTLA4*（参考文献7）），阿尔茨海默病（*APOE*）[8]，深静脉血栓症（V因子）[9]，炎症性肠病（*NOD2*（参考文献10和11）和5q31（参考文献12）），高甘油三酯血症（*APOAV*）[13]，2型糖尿病（*PPARG*）[14,15]，精神分裂症（神经调节蛋白1）[16]，哮喘（*ADAM33*）[17]，中风（*PDE4D*）[18]和心肌梗死（*LTA*）[19]。

进行关联研究的一个方法是测试每个潜在致病变异与疾病的相关性（"直接"方法）[2]。寻找整个基因组中的疾病相关变异，则需要对众多患者样本开展全基因组测序来确定候选变异，这需要投入大量费用[3]。目前，这种做法仅限于对候选基因的功能部分（基于前期的功能或遗传假说进行选择）进行测序以寻找潜在的疾病相关候选变异。人们已提出了另一种方法（"间接"方法）[20]，利用基因组内一些序列变异作为遗传标签，来检测特定基因组区域和疾病之间的关联，无论这些标签本身是否有功能效应。这样一来，就可以限制在与疾病相关的区域寻找致病变异了。

来自人类群体遗传学的两个观点表明，间接方法能捕获绝大多数的人类序列变异，比直接方法的效率更高。首先，个体间约90%序列变异是由常见变异组成的[21]。其次，这些变异中大多数最初产生于单一的历史突变事件，因此，与出现在突变发生的祖先染色体上的邻近变异有关联。这些关联使间接方法易于研究候选基因、染色体区域或整个基因组内的变异。这不需要与假定功能变异相关的先验知识。相反，这个方法用相对较小的一组变异信息来涵盖基因组中绝大多数变异的常见模式，因此可以检测任何基因组区域或基因与特定疾病的关联性，如果存在此种关联，被检测到的可能性很高。

国际人类基因组单体型图计划的目的是通过描述部分非洲、亚洲和欧洲族群的DNA样本序列变异、序列变异的频率和它们之间的相关性等特征，确定人类基因组中DNA序列变异的常见模式。因此该计划将提供工具，使间接关联方法适用于基因组中任何功能候选基因、家族型连锁分析提示的任何区域或最终应用于寻找疾病危险因素的全基因组扫描。

与疾病风险关联的常见变异通过这一策略很容易找到，但并非所有致病变异都是常见的。但是，应该指出的是，即使是相对罕见的疾病，相关变异也可能通过这种方法被发现。罕见的变异将在携带邻近序列变异特有模式的染色体上移动，反映

a chromosome that carries a characteristic pattern of nearby sequence variants. In a group of people affected by a disease, the rare variant will be enriched in frequency compared with its frequency in a group of unaffected controls. This observation, for example, was of considerable assistance in the identification of the genes responsible for cystic fibrosis[22] and diastrophic dysplasia[23], after linkage had pointed to the general chromosomal region.

Below we provide a brief description of human sequence variation, and then describe the strategy and key components of the project. These include the choice of samples and populations for study, the process of community engagement or public consultation, selection of single-nucleotide polymorphisms (SNPs), genotyping, data release and analysis.

## Human DNA Sequence Variation

Any two copies of the human genome differ from one another by approximately 0.1% of nucleotide sites (that is, one variant per 1,000 bases on average)[24-27]. The most common type of variant, a SNP, is a difference between chromosomes in the base present at a particular site in the DNA sequence (Fig. 1a). For example, some chromosomes in a population may have a C at that site (the "C allele"), whereas others have a T (the "T allele"). It has been estimated that, in the world's human population, about 10 million sites (that is, one variant per 300 bases on average) vary such that both alleles are observed at a frequency of $\geqslant 1\%$, and that these 10 million common SNPs constitute 90% of the variation in the population[21,28]. The remaining 10% is due to a vast array of variants that are each rare in the population. The presence of particular SNP alleles in an individual is determined by testing ("genotyping") a genomic DNA sample.

Nearly every variable site results from a single historical mutational event as the mutation rate is very low (of the order of $10^{-8}$ per site per generation) relative to the number of generations since the most recent common ancestor of any two humans (of the order of $10^4$ generations). For this reason, each new allele is initially associated with the other alleles that happened to be present on the particular chromosomal background on which it arose. The specific set of alleles observed on a single chromosome, or part of a chromosome, is called a haplotype (Fig. 1b). New haplotypes are formed by additional mutations, or by recombination when the maternal and paternal chromosomes exchange corresponding segments of DNA, resulting in a chromosome that is a mosaic of the two parental haplotypes[29].

The coinheritance of SNP alleles on these haplotypes leads to associations between these alleles in the population (known as linkage disequilibrium, LD). Because the likelihood of recombination between two SNPs increases with the distance between them, on average such associations between SNPs decline with distance. Many empirical studies have shown highly significant levels of LD, and often strong associations between nearby SNPs, in the human genome[30-34]. These strong associations mean that in many chromosome regions there are only a few haplotypes, and these account for most of the variation among people

其历史起源。相比于不受影响的对照组，罕见变异将会显著富集于受某种疾病影响的人群中。在连锁分析已经显示大致染色体易感区域之后，这个发现对鉴定如引起囊性纤维症[22]和畸形发育不良[23]的基因具有相当大的帮助。

下面我们将简要说明人类序列变异，然后描述这个项目的策略和关键组成部分。其中包括选择研究的样本和群体，社会参与或公共咨询的过程，单核苷酸多态性(SNP)的选择，基因分型，数据的公布和分析。

## 人类 DNA 序列变异

任何两份人类基因组具有大约 0.1%的核苷酸位点差异(即，平均每 1,000 个碱基有一个变异)[24-27]。最常见的变异类型——单核苷酸多态性(SNP)，是染色体间 DNA 序列特定位点的碱基差异(图 1a)。例如，在某一群体中，有些染色体在某个位点可能有一个 C("C 等位基因")，而其他染色体有一个 T("T 等位基因")。据估计，在世界人群中，约 1,000 万位点(即，平均每 300 个碱基有一个变异)在两个等位基因的变异频率 ≥ 1%，并且这 1,000 万常见的 SNP 贡献了种群变异的 90%[21,28]。其余 10%是群体中各种各样的罕见变异。个体中存在特定 SNP 等位基因是通过检测("基因分型")一个基因组 DNA 样本确定的。

相对于任何两个人最近的共同祖先之后的世代数(大约 $10^4$ 代)，几乎每个来自单一历史突变事件的可变异位点的突变率都是非常低的(每代每个位点大约 $10^{-8}$)。出于这个原因，每一个新的等位基因最初都是与它所在的特定染色体背景上的恰好存在的其他等位基因相联系。在单一染色体上或染色体的一部分上观察到的一套具体的等位基因，称为一个单体型(图 1b)。或者额外突变或者母本和父本染色体交换相应 DNA 片段的重组，形成了新的单体型，产生了双方父母单体型嵌合的染色体[29]。

这些单体型上的 SNP 等位基因的共同继承导致群体中这些等位基因间的相互关联(称为连锁不平衡，LD)。因为两个 SNP 间重组的可能性随它们之间距离的增加而增加，因此平均来说，SNP 之间的这种关联会随距离的增加而减少。许多实证研究表明在人类基因组中存在高显著性水平的 LD，而且相邻的 SNP 通常具有显著关联[30-34]。这些紧密关联意味着在许多染色体区域中只有少数单体型，而它们代表了

in those regions[31,35,36].

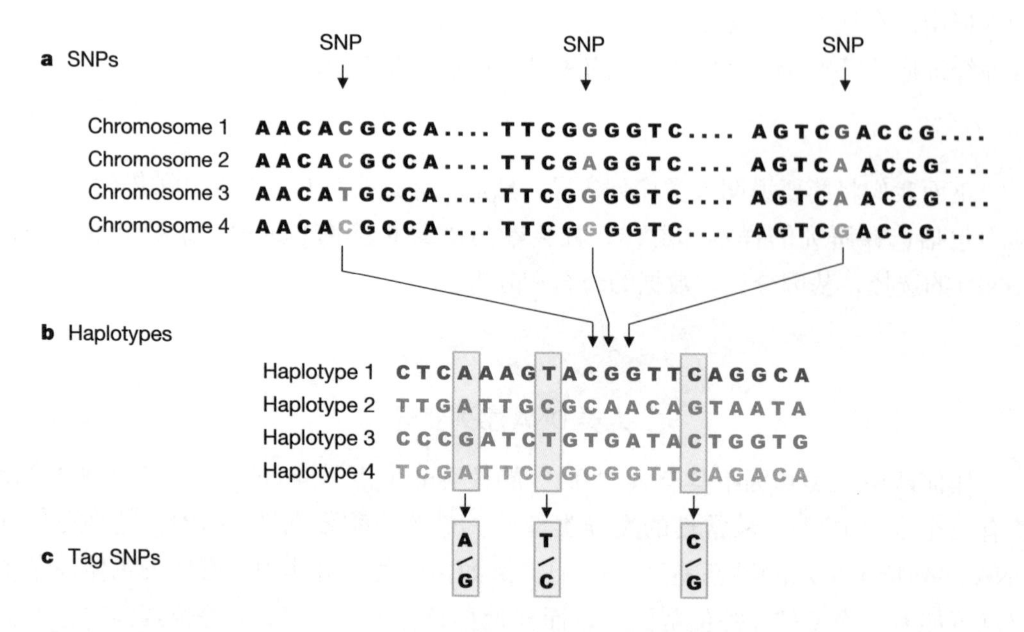


Fig. 1. SNPs, haplotypes and tag SNPs. **a**, SNPs. Shown is a short stretch of DNA from four versions of the same chromosome region in different people. Most of the DNA sequence is identical in these chromosomes, but three bases are shown where variation occurs. Each SNP has two possible alleles; the first SNP in panel **a** has the alleles C and T. **b**, Haplotypes. A haplotype is made up of a particular combination of alleles at nearby SNPs. Shown here are the observed genotypes for 20 SNPs that extend across 6,000 bases of DNA. Only the variable bases are shown, including the three SNPs that are shown in panel **a**. For this region, most of the chromosomes in a population survey turn out to have haplotypes 1–4. **c**, Tag SNPs. Genotyping just the three tag SNPs out of the 20 SNPs is sufficient to identify these four haplotypes uniquely. For instance, if a particular chromosome has the pattern A–T–C at these three tag SNPs, this pattern matches the pattern determined for haplotype 1. Note that many chromosomes carry the common haplotypes in the population.

The strong associations between SNPs in a region have a practical value: genotyping only a few, carefully chosen SNPs in the region will provide enough information to predict much of the information about the remainder of the common SNPs in that region. As a result, only a few of these "tag" SNPs are required to identify each of the common haplotypes in a region[35,37-39] (Fig. 1c).

As the extent of association between nearby markers varies dramatically across the genome[30-32,34,35,40], it is not efficient to use SNPs selected at random or evenly spaced in the genome sequence. Instead, the patterns of association must be empirically determined for efficient selection of tag SNPs. On the basis of empirical studies, it has been estimated that most of the information about genetic variation represented by the 10 million common SNPs in the population could be provided by genotyping 200,000 to 1,000,000 tag SNPs across the genome[31,36,38,39]. Thus, a substantial reduction in the amount of genotyping can be obtained with little loss of information, by using knowledge of the LD present in the genome.

这些区域人群间的大多数变异[31,35,36]。

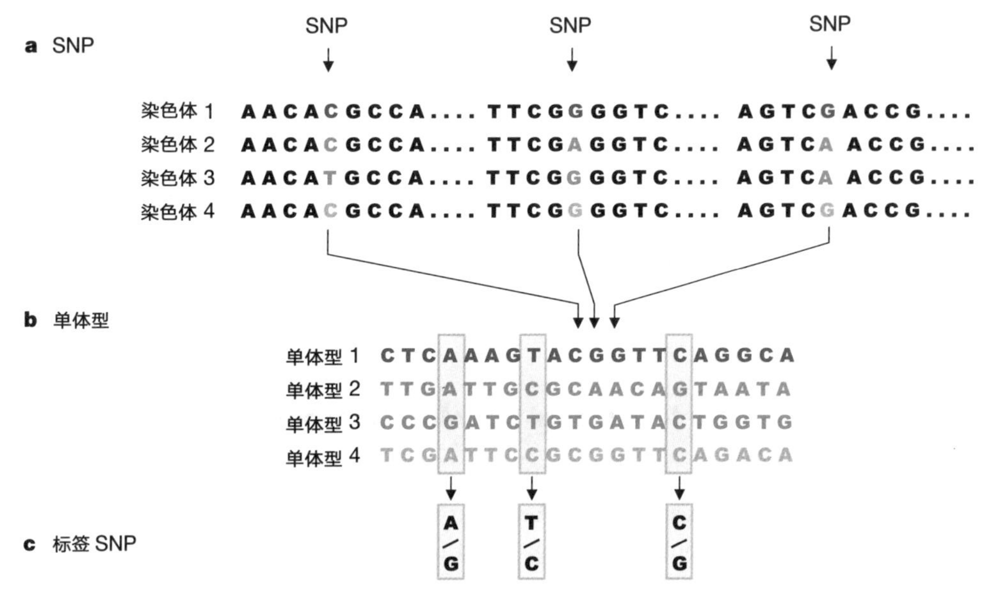


图 1. SNP，单体型和标签 SNP。**a**，SNP。所示的是不同人的同一染色体区域四个版本的一小段 DNA。在这些染色体中大多数 DNA 序列是相同的，但显示了三个发生变异的碱基。每个 SNP 有两个可能的等位基因；第一个 SNP 在 **a** 图中有等位基因 C 和 T。**b**，单体型。单体型是由邻近 SNP 的等位基因的特定组合构成的。此处显示的是观察到的跨越 6,000 个 DNA 碱基的 20 个 SNP 的基因型。只显示变异的碱基，包括 **a** 图中显示的 3 个 SNP。对此区域，调查群体中大多数染色体有单体型 1～4。**c**，标签 SNP。只对 20 个 SNP 中的三个标签 SNP 进行基因分型就足以确定这四个独特的单体型。例如，如果某一特定染色体在这三个标签 SNP 上有 A–T–C 模式，这种模式匹配由单体型 1 确定的模式。请注意，群体中许多染色体携带共同的单体型。

一个区域内 SNP 之间的紧密关联有着实际的意义：只对该区域内少数、精心挑选的 SNP 进行基因分型将提供足够的信息来预测那个区域剩余的常见 SNP 的大量信息。因此，只需要少量这种“标签”SNP，就可以确定一个区域内的每个常见单体型[35,37-39]（图 1c）。

因为基因组内邻近标记间的关联程度变化显著[30-32,34,35,40]，用随机选择或基因组序列间均匀分隔的 SNP 都不是很有效。相反，相互关联的模式必须凭实验确定，以选择更为有效的标签 SNP。在实验研究的基础上，我们估计群体中 1,000 万常见 SNP 所代表的大部分遗传变异信息，可以通过对基因组内的 200,000 至 1,000,000 标签 SNP 进行基因分型来提供[31,36,38,39]。因此，运用基因组中 LD 的知识，基因分型的数量大幅度减少而得到的信息几乎没有损失。

For common SNPs, which tend to be older than rare SNPs, the patterns of LD largely reflect historical recombination and demographic events[41]. Some recombination events occur repeatedly at "hotspots"[30,42]. The result of these processes is that current chromosomes are mosaics of ancestral chromosome regions[29]. This explains the observations that haplotypes and patterns of LD are shared by apparently unrelated chromosomes within a population and generally among populations[43].

These observations are the conceptual and empirical foundation for developing a haplotype map of the human genome, the "HapMap". This map will describe the common patterns of variation, including associations between SNPs, and will include the tag SNPs selected to most efficiently and comprehensively capture this information.

## The International HapMap Consortium

An initial meeting to discuss the scientific and ethical issues associated with developing a human haplotype map was held in Washington DC on 18–19 July 2001 (http://www.genome.gov/10001665). Groups were organized to consider the ethical issues, to develop the scientific plan and to choose the populations to include. The International HapMap Project (http://www.hapmap.org/) was then formally initiated with a meeting in Washington DC on 27–29 October 2002 (http://www.genome.gov/10005336). The participating groups and funding sources are listed in Table 1.

Table 1. Groups participating in the International HapMap Project

| Country | Research Group | Institution | Role | Per cent genome | Chromosomes | Funding agency |
|---|---|---|---|---|---|---|
| Japan | Yusuke Nakamura | RIKEN, Univ. of Tokyo | Genotyping: Third Wave | 25.1% | 5, 11, 14, 15, 16, 17, 19 | Japanese MEXT |
| | Ichiro Matsuda | Health Science Univ. of Hokkaido, Eubios Ethics Inst., Shinshu Univ. | Public consultation, Samples | | | |
| United Kingdom | David Bentley | Sanger Inst. | Genotyping: Illumina | 24.0% | 1, 6, 10, 13, 20 | Wellcome Trust |
| | Peter Donnelly | Univ. of Oxford | Analysis | | | TSC, US NIH |
| | Lon Cardon | Univ. of Oxford | Analysis | | | Wellcome Trust, TSC, US NIH |
| Canada | Thomas Hudson | McGill Univ. and Génome Québec Innovation Centre | Genotyping: Illumina | 10.0% | 2, 4p | Genome Canada, Génome Québec |

常见的 SNP 往往比罕见的 SNP 更古老，其 LD 的模式基本上可以反映历史重组和人口事件[41]。一些重组事件在“热点区”反复发生[30,42]。这些过程的结果是：现在的染色体是祖先染色体区域的嵌合体[29]。这解释了单体型和 LD 的模式被某一群体内显然不相关的染色体或群体之间共享这一观察结果[43]。

这些观察是开发人类基因组单体型图“HapMap”的概念和经验基础。这个图谱将介绍常见的变异模式，包括 SNP 间的关联，未来还将包括用于最有效和全面捕获这些遗传信息的标签 SNP。

## 国际 HapMap 协作组

最初的讨论关于开发人类单体型图的科学和伦理问题的会议是 2001 年 7 月 18 日至 19 日在华盛顿举行的(http://www.genome.gov/10001665)。协作组被组织起来考虑道德问题，制定科学计划，并选择包括的人群。之后国际 HapMap 计划(http://www.hapmap.org/)于 2002 年 10 月 27 日至 29 日在华盛顿会议上正式启动(http://www.genome.gov/10005336)。参加的团体和资金来源列于表 1。

表 1. 参与国际人类基因组单体型图计划的团队

| 国家 | 研究团队 | 机构 | 任务 | 基因组百分数 | 染色体 | 资助机构 |
|---|---|---|---|---|---|---|
| 日本 | 中村佑辅 | 日本理化学研究所，东京大学 | 基因型分型：Third Wave 公司 | 25.1% | 5, 11, 14, 15, 16, 17, 19 | 日本文部科学省 |
| | 松田一郎 | 北海道医疗大学，生物伦理研究所，信州大学 | 公众咨询，样本 | | | |
| 英国 | 戴维 · 本特利 | 桑格尔研究所 | 基因型分型：Illumina 公司 | 24.0% | 1, 6, 10, 13, 20 | 维康基金会 |
| | 彼得 · 唐纳利 | 牛津大学 | 分析 | | | SNP 协作组，美国国立卫生研究院 |
| | 朗 · 卡登 | 牛津大学 | 分析 | | | 维康基金会，SNP 协作组，美国国立卫生研究院 |
| 加拿大 | 托马斯 · 哈德森 | 麦吉尔大学和魁北克基因组创新中心 | 基因型分型：Illumina 公司 | 10.0% | 2, 4p | 加拿大基因组，魁北克基因组 |

*Continued*

<table>
<tr><th>Country</th><th colspan="2">Research Group</th><th>Institution</th><th>Role</th><th colspan="2">Per cent genome</th><th>Chromosomes</th><th>Funding agency</th></tr>
<tr><td rowspan="5">China</td><td rowspan="3">Huanming Yang<br>The Chinese HapMap Consortium</td><td>Changqing Zeng</td><td>Beijing Genomics Inst., Chinese National Human Genome Center at Beijing</td><td>Genotyping: PerkinElmer and Sequenom</td><td>4.8%</td><td rowspan="3">10%</td><td rowspan="3">3, 8p, 21</td><td rowspan="3">Chinese MOST, Chinese Academy of Sciences, Natural Science Foundation of China, Hong Kong Innovation and Technology Commission, University Grants Committee of Hong Kong</td></tr>
<tr><td>Wei Huang</td><td>Chinese National Human Genome Centre at Shanghai, Inst. of Biomedical Sciences (Taiwan)</td><td>Genotyping: Illumina</td><td>3.2%</td></tr>
<tr><td>Lap-Chee Tsui</td><td>Univ. of Hong Kong, Hong Kong Univ. of Sci. & Tech., Chinese Univ. of Hong Kong</td><td>Genotyping: Sequenom</td><td>2.0%</td></tr>
<tr><td colspan="2">Houcan Zhang</td><td>Beijing Normal Univ.</td><td>Community engagement</td><td rowspan="2" colspan="2"></td><td rowspan="2"></td><td rowspan="2">Chinese MOST</td></tr>
<tr><td colspan="2">Changqing Zeng</td><td>Beijing Genomics Inst.</td><td>Samples</td></tr>
<tr><td rowspan="7">United State</td><td colspan="2">Mark Chee</td><td>Illumina</td><td>Genotyping: Illumina</td><td>15.5%</td><td rowspan="5">30.9%</td><td>8q, 9, 18q, 22, X</td><td rowspan="6">US NIH</td></tr>
<tr><td colspan="2" rowspan="2">David Altshuler</td><td rowspan="2">Whitehead Inst.</td><td>Genotyping: Sequenom and Illumina</td><td>9.1%</td><td rowspan="2">4q, 7q, 18p, Y</td></tr>
<tr><td>Analysis</td><td></td></tr>
<tr><td colspan="2">Richard Gibbs</td><td>Baylor College of Medicine, ParAllele</td><td>Genotyping: ParAllele</td><td>4.4%</td><td>12</td></tr>
<tr><td colspan="2">Pui-Yan Kwok</td><td>UCSF, Washington Univ.</td><td>Genotyping: PerkinElmer</td><td>1.9%</td><td>7p</td></tr>
<tr><td colspan="2">Aravinda Chakravarti</td><td>Johns Hopkins Univ.</td><td>Analysis</td><td colspan="2"></td><td></td></tr>
<tr><td colspan="2">Mark Leppert</td><td>Univ. of Utah</td><td>Community engagement Samples</td><td colspan="2"></td><td></td><td>W. M. Keck Foundation, Delores Dore Eccles Foundation, US NIH</td></tr>
<tr><td>Nigeria</td><td colspan="2">Charles Rotimi</td><td>Howard Univ., Univ. of Ibadan</td><td>Community engagement Samples</td><td colspan="2"></td><td></td><td>US NIH</td></tr>
<tr><td></td><td colspan="2">Lincoln Stein</td><td>Cold Spring Harbor Lab., New York</td><td>Data coordination centre</td><td colspan="2"></td><td></td><td>TSC</td></tr>
</table>

## DNA Samples and Populations

Human populations are the products of numerous social, historical and demographic processes. As a result, no populations are typical, special or sharply bounded[44,45]. As

续表

<table>
<tr><th>国家</th><th colspan="2">研究团队</th><th>机构</th><th>任务</th><th colspan="2">基因组百分数</th><th>染色体</th><th>资助机构</th></tr>
<tr><td rowspan="5">中国</td><td rowspan="3">杨焕明<br>中国人类基因组单体型图协作组</td><td>曾长青</td><td>北京基因组研究所，国家人类基因组北方研究中心</td><td>基因型分型：PerkinElmer 公司和 Sequenom 公司</td><td>4.8%</td><td rowspan="3">10%</td><td rowspan="3">3, 8p, 21</td><td rowspan="3">中国科技部，中国科学院，国家自然科学基金，香港创新科技署，香港大学教育资助委员会</td></tr>
<tr><td>黄薇</td><td>国家人类基因组南方研究中心，生物医学科学研究所（台湾）</td><td>基因型分型：Illumina 公司</td><td>3.2%</td></tr>
<tr><td>徐立之</td><td>香港大学，香港科技大学，香港中文大学</td><td>基因型分型：Sequenom 公司</td><td>2.0%</td></tr>
<tr><td colspan="2">张厚粲</td><td>北京师范大学</td><td>社会参与</td><td colspan="2" rowspan="2"></td><td rowspan="2"></td><td rowspan="2">中国科技部</td></tr>
<tr><td colspan="2">曾长青</td><td>北京基因组研究所</td><td>样本</td></tr>
<tr><td rowspan="7">美国</td><td colspan="2">陈马克（音译）</td><td>Illumina 公司</td><td>基因型分型：Illumina 公司</td><td>15.5%</td><td rowspan="5">30.9%</td><td>8q, 9, 18q, 22, X</td><td rowspan="6">美国国立卫生研究院</td></tr>
<tr><td colspan="2" rowspan="2">戴维·阿特舒勒</td><td rowspan="2">怀特黑德研究所</td><td>基因型分型：Sequenom 公司和 Illumina 公司</td><td>9.1%</td><td rowspan="2">4q, 7q, 18p, Y</td></tr>
<tr><td>分析</td><td></td></tr>
<tr><td colspan="2">理查德·吉布斯</td><td>贝勒医学院，ParAllele 公司</td><td>基因型分型：ParAllele 公司</td><td>4.4%</td><td>12</td></tr>
<tr><td colspan="2">郭沛恩</td><td>加州大学旧金山分校，华盛顿大学</td><td>基因型分型：PerkinElmer 公司</td><td>1.9%</td><td>7p</td></tr>
<tr><td colspan="2">阿拉温达·查克拉瓦蒂</td><td>约翰斯·霍普金斯大学</td><td>分析</td><td colspan="2"></td><td></td></tr>
<tr><td colspan="2">马克·莱珀特</td><td>犹他大学</td><td>社会参与，样本</td><td colspan="2"></td><td></td><td>凯克基金会，德洛丽丝·多尔·埃克尔斯基金会，美国国立卫生研究院</td></tr>
<tr><td>尼日利亚</td><td colspan="2">查尔斯·罗蒂米</td><td>霍华德大学，伊巴丹大学</td><td>社会参与，样本</td><td colspan="2"></td><td></td><td>美国国立卫生研究院</td></tr>
<tr><td></td><td colspan="2">林肯·斯坦</td><td>纽约冷泉港实验室</td><td>数据协调中心</td><td colspan="2"></td><td></td><td>SNP 协作组</td></tr>
</table>

## DNA 样本和群体

人类群体是许多社会、历史和人口学过程的产物。因此，没有哪个人群是典型的、特殊的或有明显界限的[44,45]，因为最常见的变异模式可以在任一人群中找到[46]，

most common patterns of variation can be found in any population[46], no one population is essential for inclusion in the HapMap. Nonetheless, we decided to include several populations from different ancestral geographic locations to ensure that the HapMap would include most of the common variation and some of the less common variation in different populations, and to allow examination of various hypotheses about patterns of LD.

Studies of allele frequency distributions suggest that ancestral geography is a reasonable basis for sampling human populations[44,47,48]. Pilot studies using samples from the Yoruba, Japanese, Chinese and individuals with ancestry from Northern and Western Europe have shown substantial similarity in their haplotype patterns, although the frequencies of haplotypes often differ[31,44]. Given these scientific findings, coupled with consideration of ethical, social and cultural issues, these populations were approached for inclusion in the HapMap through a process of community engagement or consultation (see Box 1).

### Box 1. Community engagement, public consultation and individual consent

As no personally identifiable information will be linked to the samples, the risk that an individual will be harmed by a breach of privacy, or by discrimination based on studies that use the HapMap, is minimal. However, because tag SNPs for future disease studies will be chosen on the basis of haplotype frequencies in the populations included in the HapMap, the data will be identified as coming from one of the four populations involved, and it will be possible to make comparisons between the populations. As a result, the use of population identifiers may create risks of discrimination or stigmatization, as might occur if a higher frequency of a disease-associated variant were to be found in a group and this information were then overgeneralized to all or most of its members[64]. It is possible that there are other culturally specific risks that may not be evident to outsiders[65]. To identify and address these group risks, a process of community engagement, or public consultation, was undertaken to confer with members of the populations being approached for sample donation about the implications of their participation in the project[66,67]. The goal was to give people in the localities where donors were recruited the opportunity to have input into the informed consent and sample collection processes, and into such issues as how the populations from which the samples were collected would be named. Community engagement is not a perfect process, but it is an effort to involve potential donors in a more extended consideration of the implications of a research project before being asked to take part in it[68]. Community engagement and individual informed consent were conducted under the auspices of local governments and ethics committees, taking into account local ethical standards and international ethical guidelines. As in any cross-cultural endeavour, the form and outcome of the processes varied from one population to another. A Community Advisory Group is being set up for each community to serve as a continuing liaison with the sample repository, to ensure that future uses of the samples are consistent with the uses described in the informed consent documents. A more detailed article discussing ethical, social and cultural issues relevant to the project, and describing the processes used to engage donor populations in identifying and evaluating these issues, is in preparation.

没有哪个人群是必须列入 HapMap 的。尽管如此，我们依旧决定将几个位于不同祖先地理位置的群体列入其中，以确保 HapMap 可以包括大多数常见变异和不同群体中的一些不太常见的变异，并允许检查各种关于 LD 模式的假说。

对等位基因频率分布的研究表明，祖先地理是选取人群样本的一个合理的依据[44,47,48]。对约鲁巴、日本和中国样本以及对北欧和西欧后裔个体的先导研究表明，虽然单体型的频率常常不同，但其单体型模式有很大相似性[31,44]。鉴于这些科学成果，加上对道德、社会和文化问题的考虑，通过社会参与或咨询后，这些群体被列入了 HapMap 中(见框 1)。

**框 1. 社会参与、公众咨询及个人同意**

由于个人身份信息不会与样本相联系，侵犯个人隐私或基于利用 HapMap 进行研究而受到的歧视的风险非常小。然而，因为未来用于疾病研究的标记 SNP 将在 HapMap 群体单体型频率的基础上选择，这些数据将被认为是来自四个群体中的某一个，而且将有可能在群体间进行比较。因此，使用群体标识可能产生歧视或侮辱的风险，如果在一个组中发现某种疾病有关的变异频率较高，然后泛化过渡到所有或大部分成员中，就会产生这种风险[64]。还有可能存在其他对外不明显的文化特异的风险[65]。为查明并解决这些风险，将通过社会参与或公共咨询来与样本捐助群体的成员协商参与此计划的意义[66,67]。该计划的目标是让居住在招募捐助者地方的人们有机会加入知情同意和样本收集过程，并参与这些问题的讨论，如收集样本的群体将如何被命名。社会参与不是一个完美的过程，但它努力让潜在的捐助者在被要求参加这个计划前更广泛会考虑该计划研究的意义[68]。社会参与和个人知情同意是在地方政府和道德委员会的帮助下实施的，并考虑了当地的道德标准和国际道德准则。正如在任何跨越文化中的努力一样，这个过程的形式和结果在不同群体间是变化的。正在建立的社区咨询团体充当了使每个社区与样本资源库保持持续联系的角色，确保今后样本的用途与知情同意书中所描述的用途一致。目前正在编写一篇更详细的文章，该文章讨论了有关此项目的伦理、社会和文化问题，并说明招募捐助群体参与确定和评价这些问题的过程。

The HapMap developed with samples from these four large populations will include a substantial amount of the genetic variation found in all populations throughout the world. The goal of the HapMap is medical, and the common patterns of variation identified by the project will be useful to identify genes that contribute to disease and drug response in many other populations. Samples from several other populations are being collected for studies that will examine how similar their haplotype patterns are to those in the HapMap. If the patterns found are very different, samples from some of these populations may be genotyped on a large scale to make the HapMap more applicable to them. Further follow-up studies in other populations, small and large, are likely to be undertaken by scientists in many nations for common disease gene discovery.

The project will study a total of 270 DNA samples: 90 samples (see Supplementary Information, part 1) from a US Utah population with Northern and Western European ancestry (samples collected in 1980 by the Centre d'Etude du Polymorphisme Humain (CEPH)[49] and used for other human genetic maps, 30 trios of two parents and an adult child), and new samples collected from 90 Yoruba people in Ibadan, Nigeria (30 trios), 45 unrelated Japanese in Tokyo, Japan, and 45 unrelated Han Chinese in Beijing, China. All donors gave specific consent for their inclusion in the project. Population membership was determined in ways appropriate for each culture: for the Yoruba by asking the donor whether all four grandparents were Yoruba, for the Han Chinese by asking the donor whether at least three of four grandparents were Han Chinese, and for the Japanese by self-identification. The CEPH samples are available from the non-profit Coriell Institute of Medical Research (http://locus.umdnj.edu/nigms/); cell lines and DNA from the new samples will be available from Coriell in early 2004 for future studies with research protocols approved by appropriate ethics committees. It is anticipated that other researchers will genotype additional SNPs in these samples in the future, and that these data will continuously improve the HapMap.

These samples will have population and sex identifiers without information that could link them to individual donors. As the goal of the project is solely to identify patterns of genetic variation, no medical or other phenotypic information will be included. About 50% more samples were collected than will be used, so that inclusion of a sample from any particular donor cannot be known.

Samples of 45 unrelated individuals should be sufficient to find 99% of haplotypes with a frequency of 5% or greater in a population. Studies of LD can use random individual samples, trios or larger pedigrees; each design has advantages (ease of sampling) and disadvantages (decreasing efficiency with increasing numbers of related individuals). Analysis of existing data and computer simulations suggested that unrelated individuals and trios have considerable power for estimating local LD patterns. The trios will provide useful information on the accuracy of the genotyping platforms being used for the project.

用这四大群体的样本开发的 HapMap 将包括世界各地所有人群中发现的大量的遗传变异。HapMap 计划的目标是医疗，通过此计划确定的共同变异模式可用于鉴定在其他许多人群中对疾病和药物响应的基因。其他几个群体的样本正在收集以用于检测它们的单体型模式与 HapMap 中的有多大相似性。如果发现模式有很大的不同，将对它们中的某些群体样本进行更大规模的基因分型，以使 HapMap 更适用于它们。许多国家的科学家为了发现常见的疾病基因，可能会对其他小型和大型群体进行后续研究。

这个项目将总共研究 270 份 DNA 样本：90 份（见补充资料，第 1 部分）来自祖先是北欧和西欧的美国犹他州居民的样本（样本是由人类多态性研究中心（CEPH）于 1980 年采集的[49]，并用于其他人类的遗传图谱，30 个由父母双方和一个成年孩子组成的三体家系），以及从以下来源收集到的新的样本：90 个尼日利亚伊巴丹的约鲁巴人（30 个三体家系）、45 个互不相关的来自日本东京的日本人和 45 个互不相关的来自中国北京的汉族人。所有捐助者都同意加入该计划，通过与各自背景相符的方式确定群体成员资格：对约鲁巴人，询问捐助者他们的四个祖父母是否都是约鲁巴人；对汉族人，询问捐助者他们的四个祖父母是否至少有三个是汉族人；对日本人只是询问祖先是否来自日本。人类多态性研究中心的样本来自非营利机构科里尔医学研究所（http://locus.umdnj.edu/nigms/）。2004 年，研究协议经相应的伦理委员会批准后，科里尔研究所将为进一步的研究提供新样本的细胞系和 DNA。预计，今后其他研究人员将对这些样本中更多的 SNP 进行基因型分析，这些数据将不断改善 HapMap。

这些样本有群体和性别标识而无捐助者个人标识。正如这一计划的目的仅仅是确定遗传变异的模式，不包含医疗或其他表型信息。收集的样本超过 50% 未被采用，因此并不能知道列入的样本来自哪个捐助者。

从 45 个无关的个体样本中应当足以找到群体中 99%的、频率大于 5% 的单体型。LD 研究可以使用随机个体样本、三体家系或较大的家系；每个设计都有优势（方便抽样）和缺点（随着亲缘个体数目的增加而效率降低）。分析现有的数据和计算机模拟表明，无血缘关系的个体和三体家系对估计 LD 模式有相当强的能力。三体家系将对用于此计划的基因分型平台的准确性提供有益的信息。

## Choice of SNPs

A high density of SNPs is needed to describe adequately the genetic variation across the entire genome. When the project started, the average density of markers in the public database dbSNP (http://www.ncbi.nlm.nih.gov/SNP/)[50] was approximately one every kilobase (2.8 million SNPs) but, given their variable distribution, many regions had a lower density of SNPs.

Further SNPs were obtained by random shotgun sequencing from whole-genome and whole-chromosome (flow-sorted) libraries[51], using methods developed for the initial human SNP map[52], and also by collaboration with Perlegen Sciences[36] and through the purchase of sequence traces from Applied Biosystems[53] for SNP detection (see Supplementary Information, part 2). One useful result of this search for more SNPs is the confirmation of SNPs found previously. SNPs for which each allele has been seen independently in two or more samples ("double-hit" SNPs) have a higher average minor allele frequency than do "single-hit" SNPs[28]. This leads to substantial savings in assay development. On 4 November 2003, the number of SNPs (with a unique genomic position) in dbSNP (build 118) was 5.7 million, and the number of double-hit SNPs was over 2 million. By February of 2004, 6.8 million SNPs (with a unique genomic position) are expected to be in dbSNP and available for the project, including 2.7 million double-hit SNPs.

As the extent of LD and haplotypes varies by 100-fold across the genome[30-32,34,35], a hierarchical genotyping strategy has been adopted. In an initial round of genotyping, the project aims to genotype successfully 600,000 SNPs spaced at approximately 5-kilobase intervals and each with a minor allele frequency of at least 5%, in the 270 DNA samples. Priority is being given to previously validated SNPs, double-hit SNPs and SNPs causing amino-acid changes (as these may alter protein function). When these genotyping data are produced (by mid-2004; see below for details of data release), they will be analysed for associations between neighbouring SNPs. Additional SNPs will then be genotyped in the same DNA samples at a higher density only in regions where the associations are weak. Further rounds of analysis and genotyping will be carried out as required. It is expected that more than one million SNPs will be genotyped overall. This hierarchical strategy will permit regions of the genome with the least LD to be characterized at densities of up to one SNP per kilobase, maximizing the characterization of regions with associations only over short distances.

## Genotyping

Each genotyping centre is responsible for genotyping all the samples for all the selected SNPs on the chromosome regions allocated (Table 1). Among the centres, a total of five high-throughput genotyping technologies are being used, which will provide an opportunity to compare their accuracy, success rate, throughput and cost. Access to several platforms is

## SNP 的 选 择

充分描述全基因组的遗传变异需要高密度的 SNP。当这个计划开始时，公共数据库 dbSNP(http://www.ncbi.nlm.nih.gov/SNP/)[50] 中的平均标记密度为大约每千碱基中有一个 SNP(280 万个 SNP)，但鉴于其分布不同，许多区域的 SNP 密度比较低。

使用最初为人类 SNP 图谱开发的方法[52]，通过对全基因组和整个染色体(流式分离)文库[51]随机鸟枪法测序得到更多的 SNP，通过与 Perlegen Sciences[36] 合作并从 Applied Biosystems 购买序列示踪剂[53]来检测 SNP(见补充资料，第 2 部分)也可得到更多的 SNP。通过上述方法搜索更多 SNP 得到的一个有用的结果是对以前发现的 SNP 进行确认。当 SNP 的每个等位基因在两个或多个样本中分别被发现时("两次发现"SNP)，这些 SNP 比"单次发现"的 SNP 具有更高的平均最小等位基因型频率[28]，这在研发分析中节省了大笔开支。2003 年 11 月 4 日，dbSNP(118 版本)中的 SNP(每个具有独特的基因组位置)的数目是 570 万，而"两次发现"SNP 已经超过 200 万。到 2004 年 2 月，预计 dbSNP 有 680 万 SNP(每个具有独特的基因组位置)可用于此计划，其中包括 270 万"两次发现"SNP。

因为 LD 和单体型在基因组内有着 100 倍的变化范围[30-32,34,35]，因此已采取分级基因分型策略。在最初一轮的基因分型中，该计划的目标是在 270 个 DNA 样本中，对间隔为大约 5 kb 的 600,000 个 SNP 成功进行基因分型，并且每一个 SNP 的最小等位基因型频率至少达到 5%。优先考虑先前确定的 SNP、"两次发现"SNP 及造成氨基酸变化的 SNP(因为这些可能会改变蛋白质的功能)。当得到这些基因分型数据后(截至 2004 年中期；发布的详细数据见下文)，将分析相邻 SNP 间的关联性。其余的 SNP 将在相同的 DNA 样本中以高密度方式对相关性很弱的区域进行基因分型，这需要进一步进行几轮分析和基因分型。预计总共进行基因分型的 SNP 超过 100 万。这一分级战略将允许对 LD 密度最小的基因组区域以高达每千碱基一个 SNP 的密度基因分型，最大限度地确定仅有短距离联系的区域。

## 基 因 分 型

每个基因分型中心负责对所有的样本中分配的染色体区域内全部选定的 SNP 进行基因分型(表 1)。在这些中心内，一共使用了 5 个高通量基因分型技术，用来比较其准确性、成功率、通量和成本。可以使用几个平台是这个计划的优势，因为一

an advantage for the project, as a SNP assay that fails on one platform may be developed successfully using another method in order to fill a gap in the HapMap. All platforms will be evaluated using a common set of performance criteria to ensure that the quality of data produced for the project meets a uniformly high standard.

Genotype quality is being assessed in three ways. First, at the beginning of the project, all centres were assigned the same randomly selected set of 1,500 SNPs for assay development and genotyping in the 90 CEPH DNA samples being used for the project. Genotyping centres produced data that were on average more than 99.2% complete and more than 99.5% accurate (as compared to the consensus of at least two other platforms). Second, every genotyping experiment includes samples for internal quality checks, with each 96-well plate containing duplicates of five different samples, and one blank. In addition, the data from trios provide a check for consistent mendelian inheritance of SNP alleles. For all the populations, the data from the unrelated samples provide a check that the SNPs are in Hardy–Weinberg equilibrium (a test of genetic mating patterns). Although a small proportion of SNPs may fail these checks for biological reasons, they more typically fail if a genotyping platform makes consistent errors, such as undercalling heterozygotes. Third, a sample of SNP genotypes deposited by each centre will be selected at random and re-genotyped by other centres. These stringent third-party evaluations of quality will ensure the completeness and reliability of the data produced by the project.

## Data Release

The project is committed to rapid and complete data release, and to ensuring that project data remain freely available in the public domain at no cost to users. The project follows the data-release principles of a "community resource project" (http://www.wellcome.ac.uk/en/1/awtpubrepdat.html).

All data on new SNPs, assay conditions, and allele and genotype frequencies will be released rapidly into the public domain on the internet at the HapMap Data Coordination Center (DCC) (http://www.hapmap.org/) and deposited in dbSNP. Individual genotype and haplotype data initially will be made available at the DCC under a short-term "click-wrap" licence agreement. This strategy has been adopted to ensure that data from the project cannot be incorporated into any restrictive patents, and will thus remain freely available in the long term. The only condition for data access is that users must agree not to restrict use of the data by others and to share the data only with others who have agreed to the same condition. When haplotypes are defined in a region, then the individual genotypes, haplotypes and tag SNPs in that region will be publicly released to dbSNP, where there are no licensing conditions. Project participants have agreed that their own laboratories will access the data through the DCC and under the click-wrap licence, ensuring that all scientists have equal access to the data for research.

The consortium believes that SNP, genotype and haplotype data in the absence of specific

个 SNP 分析在一个平台失败，可以用其他方法成功地开展以填补 HapMap 的缺口。所有平台将使用一套共同的性能标准评价，以确保为此计划产生的数据质量符合一致的高标准。

基因分型质量有三种方法进行评估。首先，在计划开始时，所有的中心分配同一套随机挑选的 1,500 个 SNP 进行实验开发和对用于该计划的 90 个 CEPH DNA 样本进行基因分型。分型中心产生数据的完整性和准确性平均分别超过 99.2%和 99.5%(相对于至少有两个其他平台的一致性)。其次，每个基因分型实验包括样本的内部质量检查，每 96 孔板含有五个不同样本的重复和一个空白对照。此外，三体家系数据检查 SNP 等位基因符合孟德尔遗传。对于所有群体，无血缘关系的样本数据检查 SNP 的哈迪–温伯格平衡(一个遗传交配模式测试)。虽然因为生物学原因，一小部分 SNP 在这些检查中可能会失败，但如果一个基因分型平台产生一致的错误，它们可能会出现更典型的失败，例如复杂杂合子。第三，每个中心存放的 SNP 基因型的样本将随机抽样并由其他中心重新基因分型。这些严格的第三方质量评价将确保这个计划产生的数据的完整性和可靠性。

## 数据公布

该计划致力于迅速和完全发布数据，并确保此计划的数据免费向公众开放。该计划遵循“公众资源计划”的数据发布原则(http://www.wellcome.ac.uk/en/1/awtpubrepdat.html)。

关于新 SNP、实验条件、等位基因和基因型频率的所有数据将迅速公布在位于 HapMap 数据协调中心(DCC)(http://www.hapmap.org/)的公共领域并储存在 dbSNP 中。个体基因型和单体型数据最初将在短期“点击许可”协议下由 DCC 提供。采用这一策略以确保该计划的数据不被纳入任何限制性专利中，因此将继续长期免费提供。数据存取唯一的条件是，用户必须同意不限制其他人使用数据并且仅与其他已同意相同条件的用户共享。当单体型在一个区域被确定后，同一区域的单个基因型、单体型和标签 SNP 将公开发布到 dbSNP 上，没有任何附加条款。该计划的参与者已同意自己的实验室也将在“点击许可”协议下通过 DCC 获得数据，以确保所有科学家有平等机会获得数据进行研究。

协作组认为，没有具体效用的 SNP、基因型和单体型数据，并不适于申请专

utility do not constitute appropriately patentable inventions. Specific utility would involve, for example, finding an association of a SNP or haplotype with a medically important phenotype such as a disease risk or drug response. The project does not include any phenotype association studies. However, the data-release policy does not block users from filing for appropriate intellectual property on such associations, as long as any ensuing patent is not used to prevent others' access to the HapMap data.

## Data Analysis

The project will apply existing and new methods for analysis and display of the data. LD between pairs of markers will be calculated using standard measures such as $D'$ (ref. 54), $r^2$ (refs 55, 56) and others. Various methods are being evaluated to define regions of high LD and haplotypes along chromosomes. Existing methods include "sliding window" LD profiles[57,58], LD unit maps[59], haplotype blocks[31,35] and estimates of meiotic recombination rates along chromosomes[35,60-62]. After analysis of the LD in the first phase of the project, regions in which there is little or no LD will be identified and ranked for further SNP selection and genotyping. Methods to select optimal collections of tag SNPs will be developed and evaluated (see above). The project will thus provide views of the data and tag SNPs that will be useful to the research community. As all data and analysis methods will be made available, other researchers will also be able to analyse the data and improve the analysis methods.

To assist optimization of SNP selection and analysis of LD and haplotypes, a pilot study is underway to produce a dense set of genotypes across large genomic regions. Ten 500-kilobase regions of the genome (see Supplementary Information, part 3) will be sequenced in 48 unrelated HapMap DNA samples (16 CEPH (currently being sequenced), 16 Yoruba, 8 Japanese and 8 Han Chinese). All SNPs identified, as well any additional SNPs in the public databases, will be genotyped in all of the 270 HapMap DNA samples, and the genotype data will be released following the guidelines described above. This study will provide dense genotype data for developing methods for SNP selection and for assessing the completeness of the information extracted, to guide the later stages of genotyping.

When the HapMap is used to examine large genomic regions, the problem of multiple comparisons will arise from testing tens to hundreds of thousands of SNPs and haplotypes for disease associations. This will lead to difficulty in separating true from false-positive results. Thus, new statistical methods, replication studies and functional analyses of variants will be important to confirm the findings and identify the functionally important SNPs.

## Conclusion

The goal of the International HapMap Project is to develop a research tool that will help investigators across the globe to discover the genetic factors that contribute to susceptibility

利。具体的效用应涉及如找到一个 SNP 或单体型与医学重要的表型如疾病风险或药物反应的关联。该计划不包括任何表型关联研究。但是，数据发布政策并不阻止用户对这些关联申请适当的知识产权，只要任何随后的专利不被用于防止他人使用 HapMap 数据。

## 数据分析

该计划将采用现有的和新的方法来分析和显示数据。成对标记间的 LD 将使用如 $D'$(参考文献 54) 和 $r^2$(参考文献 55 和 56) 等标准的措施计算。各种方法正在接受评估以确定高 LD 区域和沿染色体的各个单体型。现有的方法包括“滑动窗口”LD 描绘 [57,58]，LD 单位图 [59]，单体型区块 [31,35] 和沿染色体估计减数分裂重组率 [35,60-62]。经过该计划第一阶段 LD 分析，只有很少或没有 LD 的区域将被确定和分级用于进一步 SNP 选择和基因分型。选择最佳的收集标签 SNP 的方法将被制定和评估(见上文)。因此，该计划将提供对数据和标签 SNP 的看法，这些看法将对研究团体有帮助。所有数据和分析方法将被公开，其他研究人员也可以分析数据，提高分析方法。

为帮助优化 SNP 选择和 LD、单体型分析，我们正在开展一个先导研究对一些大片段的基因组区域进行致密的基因分型。基因组的 10 个 500-kb 区域(见补充资料，第 3 部分)将在 48 个无血缘关系的 HapMap DNA 样本(16 个 CEPH(目前正在测序)、16 个约鲁巴人、8 个日本人和 8 个中国汉族人)中测序。所有确定的 SNP 以及公共数据库中任何更多的 SNP，将在所有的 270 份 HapMap DNA 样本中进行基因分型，并且基因型数据都将按照上文所述的准则发布。这项研究将为开发 SNP 选择方法和评估提取数据的完整性提供密集的基因型数据，以对后期基因分型进行指导。

用 HapMap 来研究大的基因组区域，检测成千上万 SNP 和单体型与疾病的关联时将产生多重比较的问题。这将导致难以从假阳性结果中分离出真正阳性的结果。因此，新的统计方法、验证研究和功能分析对确认发现和鉴定具有重要功能的 SNP 很重要。

## 结　论

国际 HapMap 计划的目标是建立一个研究工具，这将有助于在全球范围内调查发现有助于探索疾病的遗传易感因素、对抗疾病的保护方法和药物反应情况。

to disease, to protection against illness and to drug response. The HapMap will provide an important shortcut to carry out candidate-gene, linkage-based and genome-wide association studies, transforming an unfeasible strategy into a practical one. In its scope and potential consequences, the International HapMap Project has much in common with the Human Genome Project, which sequenced the human genome[63]. Both projects have been scientifically ambitious and technologically demanding, have involved intense international collaboration, have been dedicated to the rapid release of data into the public domain, and promise to have profound implications for our understanding of human biology and human health. Whereas the sequencing project covered the entire genome, including the 99.9% of the genome where we are all the same, the HapMap will characterize the common patterns within the 0.1% where we differ from each other.

For the full potential of the HapMap to be realized, several things must occur. The technology for genotyping must become more cost efficient, and the analysis methods must be improved. Pilot studies with other populations must be completed to confirm that the HapMap is generally applicable, with consideration given to expanding the HapMap if needed so that all major world populations can derive the greatest benefit. To use the tools created by the HapMap, later projects must establish carefully phenotyped sets of affected and unaffected individuals for many common diseases in a way that preserves confidentiality but retains detailed clinical and environmental exposure data. Longitudinal cohort studies of hundreds of thousands of individuals will also be invaluable for assessing the genetic and environmental contributions to disease.

Careful and sustained attention must also be paid to the ethical issues that will be raised by the HapMap and the studies that will use it. By consulting members of donor populations about the consent process and the implications of population-specific findings before sample collection, the project has helped to advance the ethical standard for international population genetics research. Future population genetics projects will continue to refine this approach. It will be an ongoing challenge to avoid misinterpretations or misuses of results from studies that use the HapMap. Researchers using the HapMap should present their findings in ways that avoid stigmatizing groups, conveying an impression of genetic determinism, or attaching incorrect levels of biological significance to largely social constructs such as race.

The HapMap holds much promise as a powerful new tool for discovery—to enhance our understanding of the hereditary factors involved in health and disease. Realizing its full benefits will involve the close partnership of basic science researchers, population geneticists, epidemiologists, clinicians, social scientists, ethicists and the public.

(**426**, 789-796; 2003)

doi:10.1038/nature02168.

HapMap 将提供一个实现候选基因、连锁和全基因组关联研究的捷径，将一个不可行战略变为实际。在适用范围和潜在后果方面，国际 HapMap 计划与为人类基因组测序[63]的人类基因组计划在很多方面有相同之处。这两个项目都有远大的科学雄心并对技术要求很高，涉及广泛的国际合作，一直致力于向公众领域迅速发布数据，承诺将对人类生物学和人类健康的理解产生深刻的影响。测序项目涵盖了整个基因组，包括人类 99.9%的相同基因组，而 HapMap 计划将确定人类互不相同的 0.1%的常见模式。

为充分发挥 HapMap 计划的潜力，有几件事情必须考虑。基因分型技术必须更具成本效益，分析方法必须加以改进。关于其他群体的先导研究必须完成以确认 HapMap 计划是普遍适用的，如果需要的话，HapMap 计划可以考虑扩大以使全球所有主要的群体能够获得最大的利益。如果要使用 HapMap 项目所创造的工具，后来的研究必须仔细建立许多受常见疾病影响和不受影响的个体的表型集合。这些集合除了详细的临床及环境暴露数据，其他都处于机密状态。纵向研究几十万个个体对评估疾病的遗传和环境因素将有重大意义。

由 HapMap 和使用 HapMap 进行的研究引发的道德问题需得到认真持续的关注。通过在样本采集前与捐助群体成员就是否同意这个过程以及种群特异性发现的意义进行商议，该项目帮助推动了国际人口遗传学研究的道德标准，未来群体遗传学项目将继续完善这一做法。这将是一项持续的挑战，以避免误解或误用 HapMap 的研究结果。研究人员利用 HapMap 呈现其研究结果时应避免污蔑某一种群，或传达遗传决定论的思想，或将生物意义与大的社会建构如种族联系在一起。

HapMap 项目很有希望成为强大的新搜索工具，以提高我们对涉及健康和疾病的遗传因素的理解。充分发挥它的作用将涉及基础科学的研究人员、群体遗传学家、流行病学家、临床医生、社会科学家、伦理学家和公众的紧密合作。

（李梅 翻译；常江 审稿）

References:

1. King, R. A., Rotter, J. I. & Motulsky, A. G. *The Genetic Basis of Common Diseases* Vol. 20 (eds Motulsky, A. G., Harper, P. S., Scriver, C. & Bobrow, M.) (Oxford Univ. Press, Oxford, 1992).
2. Risch, N. J. Searching for genetic determinants in the new millennium. *Nature* **405**, 847-856 (2000).
3. Botstein, D. & Risch, N. Discovering genotypes underlying human phenotypes: past successes for mendelian disease, future approaches for complex disease. *Nature Genet.* **33** (Suppl.), 228-237 (2003).
4. Risch, N. & Merikangas, K. The future of genetic studies of complex human diseases. *Science* **273**, 1516-1517 (1996).
5. Dorman, J. S., LaPorte, R. E., Stone, R. A. & Trucco, M. Worldwide differences in the incidence of type I diabetes are associated with amino acid variation at position 57 of the HLA-DQ β chain. *Proc. Natl Acad. Sci. USA* **87**, 7370-7374 (1990).
6. Bell, G. I., Horita, S. & Karam, J. H. A polymorphic locus near the human insulin gene is associated with insulin-dependent diabetes mellitus. *Diabetes* **33**, 176-183 (1984).
7. Nisticò, L. *et al.* The *CTLA-4* gene region of chromosome 2q33 is linked to, and associated with, type 1 diabetes. *Hum. Mol. Genet.* **5**, 1075-1080 (1996).
8. Strittmatter, W. J. & Roses, A. D. Apolipoprotein E and Alzheimer's disease. *Annu. Rev. Neurosci.* 19, 53-77 (1996).
9. Dahlbäck, B. Resistance to activated protein C caused by the factor V $R^{506}Q$ mutation is a common risk factor for venous thrombosis. *Thromb. Haemost.* **78**, 483-488 (1997).
10. Hugot, J. P. *et al.* Association of NOD2 leucine-rich repeat variants with susceptibility to Crohn's disease. *Nature* **411**, 599-603 (2001).
11. Ogura, Y. *et al.* A frameshift mutation in NOD2 associated with susceptibility to Crohn's disease. *Nature* **411**, 603-606 (2001).
12. Rioux, J. D. *et al.* Genetic variation in the 5q31 cytokine gene cluster confers susceptibility to Crohn disease. *Nature Genet.* **29**, 223-228 (2001).
13. Pennacchio, L. A. *et al.* An apolipoprotein influencing triglycerides in humans and mice revealed by comparative sequencing. *Science* **294**, 169-173 (2001).
14. Deeb, S. S. *et al.* A Pro12Ala substitution in PPARγ2 associated with decreased receptor activity, lower body mass index and improved insulin sensitivity. *Nature Genet.* **20**, 284-287 (1998).
15. Altshuler, D. *et al.* The common PPARγ Pro12Ala polymorphism is associated with decreased risk of type 2 diabetes. *Nature Genet.* **26**, 76-80 (2000).
16. Stefansson, H. *et al.* Neuregulin 1 and susceptibility to schizophrenia. *Am. J. Hum. Genet.* **71**, 877-892 (2002).
17. Van Eerdewegh, P. *et al.* Association of the *ADAM33* gene with asthma and bronchial hyperresponsiveness. *Nature* **418**, 426-430 (2002).
18. Gretarsdottir, S. *et al.* The gene encoding phosphodiesterase 4D confers risk of ischemic stroke. *Nature Genet.* **35**, 131-138 (2003).
19. Ozaki, K. *et al.* Functional SNPs in the lymphotoxin-α gene that are associated with susceptibility to myocardial infarction. *Nature Genet.* **32**, 650-654 (2002).
20. Collins, F. S., Guyer, M. S. & Chakravarti, A. Variations on a theme: cataloging human DNA sequence variation. *Science* **278**, 1580-1581 (1997).
21. Kruglyak, L. & Nickerson, D. A. Variation is the spice of life. *Nature Genet.* **27**, 234-236 (2001).
22. Kerem, B. *et al.* Identification of the cystic fibrosis gene: genetic analysis. *Science* **245**, 1073-1080 (1989).
23. Hästbacka, J. *et al.* Linkage disequilibrium mapping in isolated founder populations: diastrophic dysplasia in Finland. *Nature Genet.* **2**, 204-211 (1992).
24. Li, W. H. & Sadler, L. A. Low nucleotide diversity in man. *Genetics* **129**, 513-523 (1991).
25. Wang, D. G. *et al.* Large-scale identification, mapping, and genotyping of single-nucleotide polymorphisms in the human genome. *Science* **280**, 1077-1082 (1998).
26. Cargill, M. *et al.* Characterization of single-nucleotide polymorphisms in coding regions of human genes. *Nature Genet.* **22**, 231-238 (1999).
27. Halushka, M. K. *et al.* Patterns of single-nucleotide polymorphisms in candidate genes for blood-pressure homeostasis. *Nature Genet.* **22**, 239-247 (1999).
28. Reich, D. E., Gabriel, S. B. & Altshuler, D. Quality and completeness of SNP databases. *Nature Genet.* **33**, 457-458 (2003).
29. Pääbo, S. The mosaic that is our genome. *Nature* **421**, 409-412 (2003).
30. Jeffreys, A. J., Kauppi, L. & Neumann, R. Intensely punctate meiotic recombination in the class II region of the major histocompatibility complex. *Nature Genet.* **29**, 217-222 (2001).
31. Gabriel, S. B. *et al.* The structure of haplotype blocks in the human genome. *Science* **296**, 2225-2229 (2002).
32. Reich, D. E. *et al.* Linkage disequilibrium in the human genome. *Nature* **411**, 199-204 (2001).
33. Abecasis, G. R. *et al.* Extent and distribution of linkage disequilibrium in three genomic regions. *Am. J. Hum. Genet.* **68**, 191-197 (2001).
34. Dawson, E. *et al.* A first-generation linkage disequilibrium map of human chromosome 22. *Nature* **418**, 544-548 (2002).
35. Daly, M. J., Rioux, J. D., Schaffner, S. F., Hudson, T. J. & Lander, E. S. High-resolution haplotype structure in the human genome. *Nature Genet.* **29**, 229-232 (2001).
36. Patil, N. *et al.* Blocks of limited haplotype diversity revealed by high-resolution scanning of human chromosome 21. *Science* **294**, 1719-1723 (2001).
37. Johnson, G. C. L. *et al.* Haplotype tagging for the identification of common disease genes. *Nature Genet.* **29**, 233-237 (2001).
38. Carlson, C. S. *et al.* Additional SNPs and linkage-disequilibrium analyses are necessary for whole-genome association studies in humans. *Nature Genet.* **33**, 518-521 (2003).
39. Goldstein, D. B., Ahmadi, K. R., Weale, M. E. & Wood, N. W. Genome scans and candidate gene approaches in the study of common diseases and variable drug responses. *Trends Genet.* **19**, 615-622 (2003).
40. Taillon-Miller, P. *et al.* Juxtaposed regions of extensive and minimal linkage disequilibrium in human Xq25 and Xq28. *Nature Genet.* **25**, 324-328 (2000).
41. Chakravarti, A. Population genetics—making sense out of sequence. *Nature Genet.* **21**, 56-60 (1999).
42. Chakravarti, A. *et al.* Nonuniform recombination within the human β-globin gene cluster. *Am. J. Hum. Genet.* **36**, 1239-1258 (1984).

43. Tishkoff, S. A. *et al.* Global patterns of linkage disequilibrium at the CD4 locus and modern human origins. *Science* **271**, 1380-1387 (1996).

44. Cavalli-Sforza, L. L., Menozzi, P. & Piazza, A. *The History and Geography of Human Genes* (Princeton Univ. Press, Princeton, 1994).

45. Foster, M. W. & Sharp, R. R. Race, ethnicity, and genomics: social classifications as proxies of biological heterogeneity. *Genome Res.* **12**, 844-850 (2002).

46. Barbujani, G., Magagni, A., Minch, E. & Cavalli-Sforza, L. L. An apportionment of human DNA diversity. *Proc. Natl Acad. Sci. USA* **94**, 4516-4519 (1997).

47. Rosenberg, N. A. *et al.* Genetic structure of human populations. *Science* **298**, 2381-2385 (2002).

48. Jorde, L. B. *et al.* Microsatellite diversity and the demographic history of modern humans. *Proc. Natl Acad. Sci. USA* **94**, 3100-3103 (1997).

49. Dausset, J. *et al.* Centre d'Etude du Polymorphisme Humain (CEPH): collaborative genetic mapping of the human genome. *Genomics* **6**, 575-577 (1990).

50. Sherry, S. T. *et al.* dbSNP: the NCBI database of genetic variation. *Nucleic Acids Res.* **29**, 308-311 (2001).

51. Ning, Z., Cox, A. J. & Mullikin, J. C. SSAHA: a fast search method for large DNA databases. *Genome Res.* **11**, 1725-1729 (2001).

52. The International SNP Working Group. A map of human genome sequence variation containing 1.42 million single nucleotide polymorphisms. *Nature* **409**, 928-933 (2001).

53. Venter, J. C. *et al.* The sequence of the human genome. *Science* **291**, 1304-1351 (2001).

54. Lewontin, R. C. The interaction of selection and linkage. I. General considerations: heterotic models. *Genetics* **49**, 49-67 (1964).

55. Hill, W. G. & Robertson, A. Linkage disequilibrium in finite populations. *Theor. Appl. Genet.* **38**, 226-231 (1968).

56. Ohta, T. & Kimura, M. Linkage disequilibrium due to random genetic drift. *Genet. Res.* **13**, 47-55 (1969).

57. Dawson, K. J. The decay of linkage disequilibrium under random union of gametes: how to calculate Bennett's principal components. *Theor. Popul. Biol.* **58**, 1-20 (2000).

58. Langley, C. H. & Crow, J. F. The direction of linkage disequilibrium. *Genetics* **78**, 937-941 (1974).

59. Maniatis, N. *et al.* The first linkage disequilibrium (LD) maps: delineation of hot and cold blocks by diplotype analysis. *Proc. Natl Acad. Sci. USA* **99**, 2228-2233 (2002).

60. Hudson, R. R. Estimating the recombination parameter of a finite population model without selection. *Genet. Res.* **50**, 245-250 (1987).

61. Fearnhead, P. & Donnelly, P. Estimating recombination rates from population genetic data. *Genetics* **159**, 1299-1318 (2001).

62. McVean, G., Awadalla, P. & Fearnhead, P. A coalescent-based method for detecting and estimating recombination from gene sequences. *Genetics* **160**, 1231-1241 (2002).

63. International Human Genome Sequencing Consortium. Initial sequencing and analysis of the human genome. *Nature* **409**, 860-921 (2001).

64. Clayton, E. W. The complex relationship of genetics, groups, and health: what it means for public health. *J. Law Med. Ethics* **30**, 290-297 (2002).

65. Foster, M. W. & Sharp, R. R. Genetic research and culturally specific risks: one size does not fit all. *Trends Genet.* **16**, 93-95 (2000).

66. Sharp, R. R. & Foster, M. W. Involving study populations in the review of genetic research. *J. Law Med. Ethics* **28**, 41-51 (2000).

67. Marshall, P. A. & Rotimi, C. Ethical challenges in community-based research. *Am. J. Med. Sci.* **322**, 241-245 (2001).

68. Juengst, E. T. Commentary: what "community review" can and cannot do. *J. Law Med. Ethics* **28**, 52-54 (2000).

**Supplementary Information** accompanies the paper on www.nature.com/nature.

**Acknowledgements**. We thank many people who contributed to this project: J. Beck, C. Beiswanger, D. Coppock, J. Mintzer and L. Toji at the Coriell Institute for Medical Research for transforming the samples, distributing the DNA and cell lines, and storing the samples for use in future research; J. Greenberg and R. Anderson of the NIH National Institute of General Medical Sciences (NIGMS) for providing funding and support for cell-line transformation and storage in the NIGMS Human Genetic Cell Repository at the Coriell Institute; K. Wakui at Shinshu University for assistance in transforming the Japanese cell lines; N. Carter and D. Willey at the Wellcome Trust Sanger Institute for flow sorting the chromosomes and for library construction, respectively; M. Deschesnes and B. Godard for assistance at the University of Montréal; C. Darmond-Zwaig, J. Olivier and S. Roumy at McGill University and Génome Québec Innovation Centre; C. Allred, B. Gillman, E. Kloss and M. Rieder for help in implementing data flow protocols; S. Olson for work on the website explanations; S. Adeniyi-Jones, D. Burgess, W. Burke, T. Citrin, A. Clark, D. Cowhig, P. Epps, K. Hofman, A. Holt, E. Juengst, B. Keats, J. Levin, R. Myers, A. Obuoforibo, F. Romero, C. Tamura and A. Williamson for providing advice on the project to NIH; A. Peck and J. Witonsky of the National Human Genome Research Institute (NHGRI) for help with project management; E. DeHaut-Combs and S. Saylor of NHGRI for staff support; M. Gray for organizing phone calls and meetings; the people of Tokyo, Japan, the Yoruba people of Ibadan, Nigeria, and the community at Beijing Normal University, who participated in public consultations and community engagements; and the people in these communities who were generous in donating their blood samples. This work was supported in part by Genome Canada, Génome Québec, the Chinese Ministry of Science and Technology, the Chinese Academy of Sciences, the Natural Science Foundation of China, the Hong Kong Innovation and Technology Commission, the University Grants Committee of Hong Kong, the Japanese Ministry of Education, Culture, Sports, Science and Technology, the Wellcome Trust, the SNP Consortium, the US National Institutes of Health (FIC, NCI, NCRR, NEI, NHGRI, NIA, NIAAA, NIAID, NIAMS, NIBIB, NIDA, NIDCD, NIDCR, NIDDK, NIEHS, NIGMS, NIMH, NINDS, OD), the W.M. Keck Foundation and the Delores Dore Eccles Foundation.

**Competing interests statement**. The authors declare that they have no competing financial interests.

**Correspondence** and requests for materials should be addressed to D.B. (drb@sanger.ac.uk) or M.F. (fost1848@msmailhub.oulan.ou.edu).

# A Correlation between the Cosmic Microwave Background and Large-scale Structure in the Universe

S. Boughn and R. Crittenden

## Editor's Note

**The ripples in the cosmic microwave background radiation are relics of the seeds from which the large-scale structure of the Universe grew. If this is so, hot spots in this radiation should correspond to concentrations of galaxies, and cold spots to voids. This paper by Stephen Boughn and Robert Crittenden reports the first observation of such a correlation. X-rays from hot gas inside clusters of galaxies turn out to be correlated with fluctuations in the backgrounds. The distribution of galaxies that are bright radio sources is also correlated with the fluctuations. The authors point out that the correlation supports the idea that dark energy, which drives the acceleration of cosmic expansion, has left an imprint on large-scale structure.**

---

Observations of distant supernovae and the fluctuations in the cosmic microwave background (CMB) indicate that the expansion of the Universe may be accelerating[1] under the action of a "cosmological constant" or some other form of "dark energy". This dark energy now appears to dominate the Universe and not only alters its expansion rate, but also affects the evolution of fluctuations in the density of matter, slowing down the gravitational collapse of material (into, for example, clusters of galaxies) in recent times. Additional fluctuations in the temperature of CMB photons are induced as they pass through large-scale structures[2] and these fluctuations are necessarily correlated with the distribution of relatively nearby matter[3]. Here we report the detection of correlations between recent CMB data[4] and two probes of large-scale structure: the X-ray background[5] and the distribution of radio galaxies[6]. These correlations are consistent with those predicted by dark energy, indicating that we are seeing the imprint of dark energy on the growth of structure in the Universe.

---

IN the standard model of the origin of structure, most of the fluctuations in the CMB were imprinted as the photons last scattered off free electrons, when the universe was just 400,000 years old (at a redshift of $z \approx 1{,}100$). However, once the dark energy (or the spatial curvature[7]) becomes important dynamically ($z \approx 1$), additional fluctuations are induced in the CMB photons by what is known as the integrated Sachs–Wolfe (ISW) effect[2]. The gravitational potentials of large, diffuse concentrations and rarefactions of matter begin to evolve, causing the energy of photons passing through them to change by an amount that depends on the depth of the potentials. The amplitude of these ISW fluctuations tends to

# 宇宙微波背景和大尺度结构的相关

博夫恩，克里滕登

编者按

宇宙微波背景辐射中的波纹是“种子”的遗迹，宇宙的大尺度结构便是从这些“种子”发育而成。如果真如上所言，这种辐射中的热斑应该对应于星系的聚集，而冷斑则对应巨洞。本文作者史蒂夫·博夫恩和罗伯特·克里滕登首次报道了这种相关性。来自星系团内的热气体的X射线与背景中的扰动相关。作为亮射电源的星系的分布也与扰动相关。作者指出，这种相关性支持以下观点：驱动宇宙膨胀加速的暗能量在大尺度结构上留下了印记。

对遥远的超新星和宇宙微波背景辐射(CMB)涨落的观测表明，宇宙很可能在“宇宙学常数”或者某种其他形式的暗能量作用下加速膨胀[1]。暗能量在当前宇宙中占主导地位，它不仅改变了宇宙的膨胀速率，也影响了物质密度扰动的演化，减慢了物质的引力塌缩(例如星系团的生成过程)。宇宙微波背景辐射温度的额外涨落来自于CMB光子穿越大尺度结构时所受到的扰动[2]，因此这些涨落必然与所对应区域邻近物质的分布相关[3]。在这里我们报道最近的CMB数据[4]和两种大尺度结构的观测结果(X射线背景[5]和射电星系分布[6])之间的相关性。这些相关性与暗能量的预言相符，表明我们在宇宙结构演化的过程中观测到了暗能量的印记。

在结构起源的标准模型中，大部分CMB涨落都带有光子被自由电子最后一次散射时的特征，此刻宇宙只有400,000年的历史(红移大约为1,100)。但是一旦暗能量(或者空间曲率[7])逐渐变得重要(红移大约为1)，累计萨克斯–沃尔夫效应(ISW)将会在CMB中引入额外的扰动[2]。由于大量聚集和稀薄的弥散物质的引力势随时间开始演化，穿过引力势阱的光子能量发生变化，这个变化量取决于势阱的深度。除了在很大的尺度上以外，ISW的扰动幅度要比源自最后散射的扰动小得多。然而

be small compared to the fluctuations originating at the epoch of last scattering except on very large scales. However, as ISW fluctuations were created more locally, it is expected that the CMB fluctuations should be partially correlated with galaxies, which serve as tracers of the large-scale matter distribution.

Detecting the relatively weak ISW effect requires correlating the CMB with the distribution of galaxies spread over a large fraction of the observable Universe. This necessitates a survey of galaxies covering much of the sky, out to distances of many billions of light years ($z \approx 1$). Focus thus has been on luminous active galaxies, which are believed to trace the mass distribution on large scales. Although active galaxies emit at a wide range of frequencies, the most useful maps are in the hard X-rays (2–10 keV), where they dominate the X-ray sky, and in the radio, where the number counts are dominated by sources at $z \lesssim 1$. The full sky map of the intensity of the hard X-ray background made by the HEAO-1 satellite[5] and a map of the number density of radio sources provided by the NRAO VLA sky survey (NVSS)[6] are two of the best maps in which to look for this effect. To predict the expected level of the ISW effect in these two surveys, it is essential to know both the inherent clustering of the sources and their distribution in redshift. The former can be determined from previous studies of the X-ray and radio auto-correlation functions[8,9] and the latter can be estimated from deep, pointed surveys[10,11].

Previous attempts at correlating the CMB with both the HEAO-1 and the NVSS maps have yielded only upper limits[8,9,12]. Here we repeat these analyses using the much improved CMB maps recently provided by the WMAP satellite mission[4]. We have compared two different CMB maps generated from this data (the "internal linear combination" map[13], and the "cleaned" map of ref. 14) with the HEAO and NVSS maps. The dominant "noise" in the maps is due to the fluctuations in the CMB itself and is well characterized, while the instrument noise is negligible on the angular scales relevant to the present analysis. To reduce possible contamination by emission from our Milky Way Galaxy as well as from other nearby radio sources, these maps were masked with the most aggressive foreground template provided by the WMAP team[13]. The X-ray map was similarly masked and was corrected for several large-scale systematics including a linear drift of the detectors[9]. Finally, the NVSS catalogue was corrected for a systematic variation of source counts with declination[8]. The sky coverage was 68% for the CMB maps, 56% for the NVSS map, and 33% for the X-ray map. The reduced coverage of the X-ray map was the result of its low resolution (3°) and a larger number of foreground sources. Significantly, however, our results do not depend on the corrections or the level of masking of the maps.

A standard measure of the correspondence of two data sets is the cross-correlation function, CCF($\theta$). It represents the extent to which the two measures of the sky separated by an angle $\theta$ are correlated. Figures 1 and 2 show the CCFs of the WMAP data with the X-ray and radio surveys. The results using the "internal linear combination" and "cleaned" CMB maps were entirely consistent with each other, and we plot the averages of the results for the two maps. Four hundred Monte Carlo simulations of the CMB sky were also cross-correlated with the actual X-ray and radio maps. The r.m.s. values of these trials are

ISW 扰动是在局部空间产生的，CMB 涨落应与星系分布部分相关，故可以以此探测大尺度上的物质分布。

探测相对较弱的 ISW 效应要求研究 CMB 与占可观测宇宙中较大部分的星系分布的相关性。这需要星系巡天具有大比例的天空覆盖，且巡天深度延伸至几十亿光年的距离（红移为 1 左右）。因此巡天目标集中在明亮的活动星系，这被认为可以很好地示踪大尺度的质量分布。虽然活动星系在很宽的波段内都有辐射，但是最有效的成图在硬 X 射线波段（2 ~ 10 keV），它主导了 X 射线天空，同时在射电波段，射电源计数由红移 $\leqslant 1$ 的源所主导。HEAO–1 卫星所作的亮硬 X 射线背景强度全天图 [5] 和 NRAO VLA 巡天（NVSS）所作的射电源数密度巡天图为寻找 ISW 效应提供了最好的图像。为了预测在这两个巡天中 ISW 效应所期望达到的水平，清楚这些源内在的成团性和它们随红移的分布是至关重要的。成团性可由过去已有的对 X 射线和射电源的自相关函数 [8,9] 的研究中得到，分布则可由深度指向巡天 [10,11] 来估计。

过去将 CMB 与 HEAO–1、NVSS 天图做交叉相关的尝试仅得到了相关的上限 [8,9,12]。在这里我们用 WMAP 卫星计划 [4] 最近提供的有显著改进的 CMB 图重复这种相关分析。我们将从 WMAP 得到的两种不同的 CMB 图（“内在线性组合”图 [13] 和“净化”图 [14]）与 HEAO 和 NVSS 的天图相比较。图中“噪声”的主要成分来自 CMB 自身的涨落，其特征能够被很好地描述，而在要分析的角尺度上仪器噪声则可以忽略。为了减少来自银河系和其他附近射电源的辐射污染，这些图均按 WMAP 团队所提供的最苛刻的前景模板进行了掩模 [13]。X 射线图也进行了类似的掩模，同时还修正了几个大尺度上类似探测器线性漂移的系统误差 [9]。最后，NVSS 星表也修正了源计数偏差引起的系统误差 [8]。CMB 图覆盖了整个天空的 68%，NVSS 图覆盖了 56%，X 射线图只覆盖了 33%。X 射线图的覆盖率偏低是因为较低的分辨率（3 度）和大量的前景源。尽管如此，我们的结果与修正或对图像的掩模程度明显地关系不大。

度量两个数据集合相关性的标准方法是求交叉相关函数 CCF($\theta$)。它表征了角间距为 $\theta$ 的两个方向之间的空间相关性。图 1 与图 2 表示 WMAP 分别与 X 射线巡天图和射电巡天图之间的交叉相关函数。用 CMB 的“内在线性组合”图和“洁化”图得到的结果是完全符合的；我们还绘出了这两种图结果的平均。通过四百次蒙特卡洛方法模拟得到的 CMB 图被用于与实际的 X 射线巡天图和射电巡天图做交叉相关。

taken as the errors for each bin and these are highly correlated owing to the large-scale structure in the maps. For comparison, the uncertainties were also estimated from the data themselves by rotating the maps with respect to each other, and the errors determined in this way were very similar to those found using the simulated maps. The significance of the detection of the ISW effect is estimated by fitting the amplitude of the theoretical profile to the data. For the X-ray/CMB CCF, the correlations are detected at a confidence level (CL) of 99.9%, that is, a 3.0σ detection. The radio/CMB CCF is detected at a 99.4% CL, or a 2.5σ detection. These confidence limits are consistent with the number of times the CCF($\theta$) of the Monte Carlo trials exceed the values of the measured CCF($\theta$).

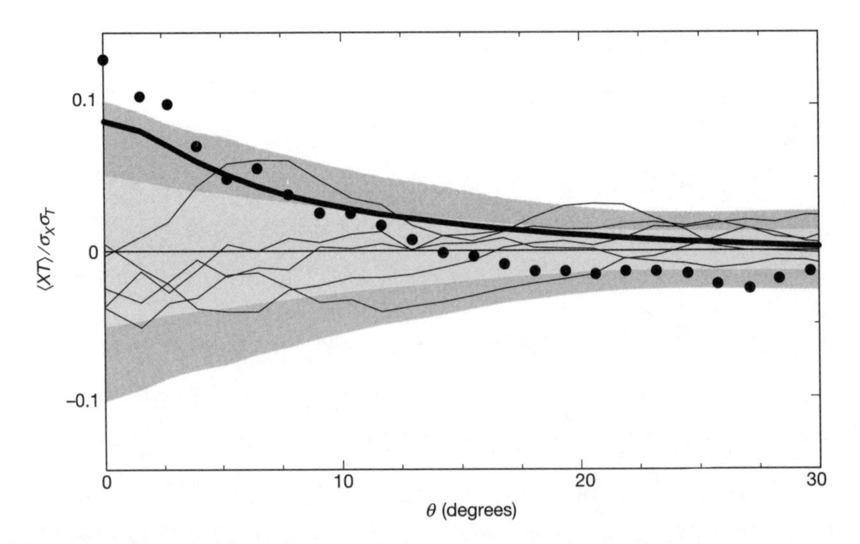


Fig. 1. The X-ray background fluctuations are correlated with the microwave sky at a higher level than would be expected by chance. Here we plot the cross correlation (filled circles) between the X-ray intensity ($X$) measured by HEAO-A1 and the CMB temperature fluctuations ($T$) measured by WMAP. We normalize by the r.m.s. levels of the two maps ($\sigma_x$ and $\sigma_T$, respectively), which makes it independent of any linear biasing of the survey. To give an idea of the level of accidental correlations, the shaded areas show the 1σ and 2σ regions derived from simulated, uncorrelated CMB maps with the same power spectrum as the WMAP data. The thin solid lines show the results for five such Monte Carlo simulations, and we see that the signals in neighbouring bins are highly correlated for a given realization. The signal-to-noise ratio is greatest at smaller angular separations, even though the error is amplified because fewer pairs of pixels contribute to the correlation. For $\theta = 0°$, 1.3° and 2.6°, the Monte Carlo trials exceed the amplitude of the actual X-ray/CMB correlation only 0.3%, 0.8% and 0.3% of the time, respectively. The bold line shows theoretical predictions for the ISW effect in a cosmological-constant-dominated Universe, using the best-fit WMAP model for scale-invariant fluctuations[19] (dark energy fraction $\Omega_\Lambda = 0.73$, Hubble constant $H_0 = 100\ h$ km s$^{-1}$ Mpc$^{-1}$ with $h = 0.72$, matter density $\Omega_m h^2 = 0.14$ and baryon density $\Omega_b h^2 = 0.024$). At larger angular separations, the observed correlations appear to fall faster than predicted by theory, but the low signal-to-noise ratio makes it difficult to say whether this is a real effect.

这些试验的均方根被认为是每一个数据间隔的误差，由于天图存在的大尺度结构，它们是高度相关的。作为比较，不确定性可以通过天图本身的相对旋转由对原始数据的分析获得，它所给出的误差与通过使用模拟图得到的误差非常接近。ISW 效应探测的显著性可以通过理论轮廓的大小和实际数据的拟合来估计。对于 CMB 与 X 射线巡天之间的交叉相关函数，探测的置信水平为 99.9%(即 $3\sigma$ 探测)；对于 CMB 与射电巡天之间的交叉相关函数，探测的置信水平为 99.4%(即 $2.5\sigma$ 探测)。这些置信限与大量蒙特卡洛试验得出的交叉相关函数是一致的。

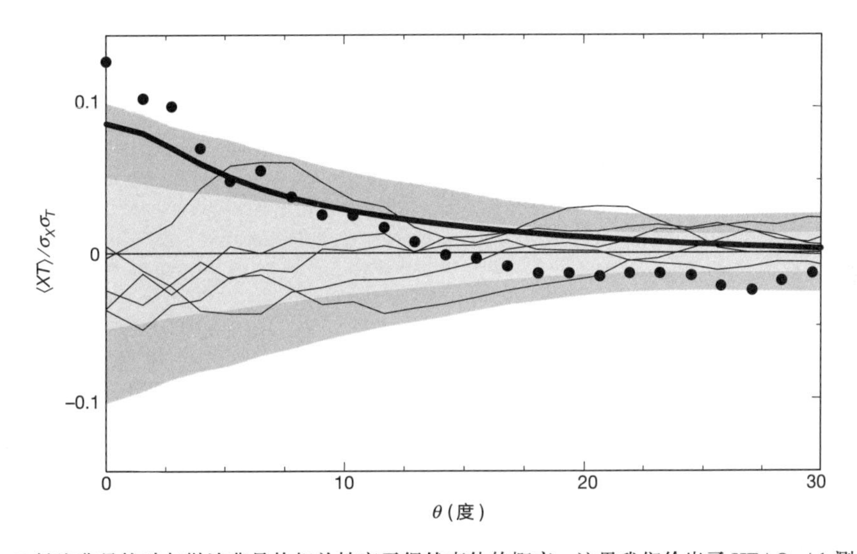


图 1. X 射线背景扰动与微波背景的相关性高于偶然事件的概率，这里我们给出了 HEAO–A1 测得的 X 射线强度 ($X$) 和 WMAP 观测的微波背景辐射温度扰动 ($T$) 的交叉相关（实心圆）。我们用两个图的均方根水平 ($\sigma_X$ 和 $\sigma_T$) 做归一化，使得它与巡天的线性偏袒无关。为了了解偶然事件导致相关的可能性，阴影区域显示与 WMAP 具有相同功率谱且统计无关的 CMB 模拟图给出的 $1\sigma$ 和 $2\sigma$ 区域。细实线显示了 5 个这样的蒙特卡罗模拟，我们注意到，临近数据区间的信号与给定的模拟实现高度相关。尽管贡献相关性的像素较少而使误差被放大，这个信噪比在小角间距下还是最大的。对 $\theta = 1$ 度、1.3 度和 2.6 度，蒙特卡罗的结果只超过真实 X 射线/CMB 相关的 0.3%、0.8% 和 0.3%。图中粗线显示在宇宙学常数主导的宇宙中 ISW 效应的理论预测，它采用了尺度不变扰动的 WMAP 最佳拟合模型 [19](暗物质分数 $\Omega_\Lambda = 0.73$，哈勃常数 $H_0 = 100h\ \mathrm{km \cdot s^{-1} \cdot Mpc^{-1}}$，$h = 0.72$，物质密度 $\Omega_m h^2 = 0.14$，重子密度 $\Omega_b h^2 = 0.024$)。对更大的角间距，观测到的相关性看上去比理论预言的相对性下降得更快，但是信噪比很低，因此很难说这是否是真实的结果。

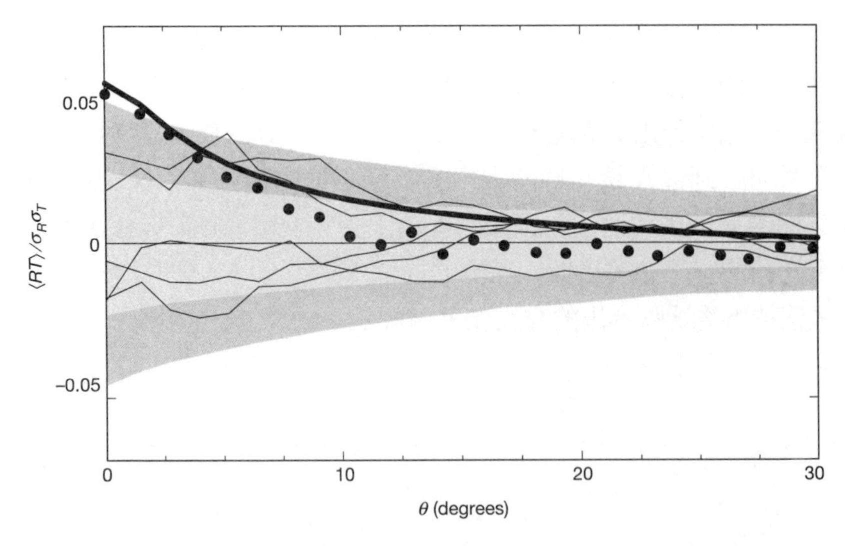


Fig. 2. The distribution of radio galaxies is correlated with the microwave sky. Here we plot the correlation (filled circles) between the NVSS radio galaxy number counts ($R$) and the WMAP temperature maps ($T$). The other curves are as in Fig. 1. The Monte Carlo trials exceed the amplitude of the actual radio/CMB correlation in the lowest three bins 1.2%, 1.9% and 3.4% of the time, respectively. Again, there is good agreement with the theoretical predictions, with the signal again falling off somewhat faster than predicted at larger angles. The lower amplitude compared to the X-ray result is due to the higher resolution of the radio map, which increases the radio r.m.s. ($\sigma_R$) without significantly increasing the cross-correlation. The consistency of the NVSS and X-ray CCFs suggests that the signal is not the result of unknown systematics in either the X-ray or the NVSS map. It is possible that microwave emission from the radio/X-ray sources themselves could result in the positive correlation of the maps. However, extrapolations of the frequency spectra of the radio galaxies indicate that the microwave emission is much smaller than the observed signal[8]. In addition, the clustering of radio sources is on a much smaller angular scale than the apparent signals, so any such contamination would appear only at $\theta \approx 0°$ and not at larger angles.

The theoretical curves for the ISW effect also are shown in Figs 1 and 2, and fall off with angular separation similarly to the observed correlations. For these calculations, we have assumed that the fluctuations in matter density, $\delta\rho$, are traced directly by the fluctuations in the number density of the sources. That is, $\delta N/N = b\delta\rho/\rho$ where the bias factor, $b$, is assumed to be constant and its value is derived from the auto-correlation function of the X-ray or radio sources. The predictions assume a scale-invariant primordial spectrum and the best-fit cosmology determined from WMAP. Although the derived bias factors depend on precisely how the sources are assumed to be distributed in redshift, the expected cross-correlation is only weakly dependent on this or on the possible evolution of the bias factor with redshift. Detailed considerations for calculating the amplitude of the ISW effect can be found elsewhere[3,12].

It is important to consider what else might be contributing to the observed correlation. Unmasked foreground sources are not a major contaminant, because the observed correlations do not change when the Galactic cuts are varied nor when the aggressiveness of the masking is changed. In fact, the amplitudes of the CCFs are essentially unchanged when no masking at all is applied to the CMB maps. Also, because the results from analysing

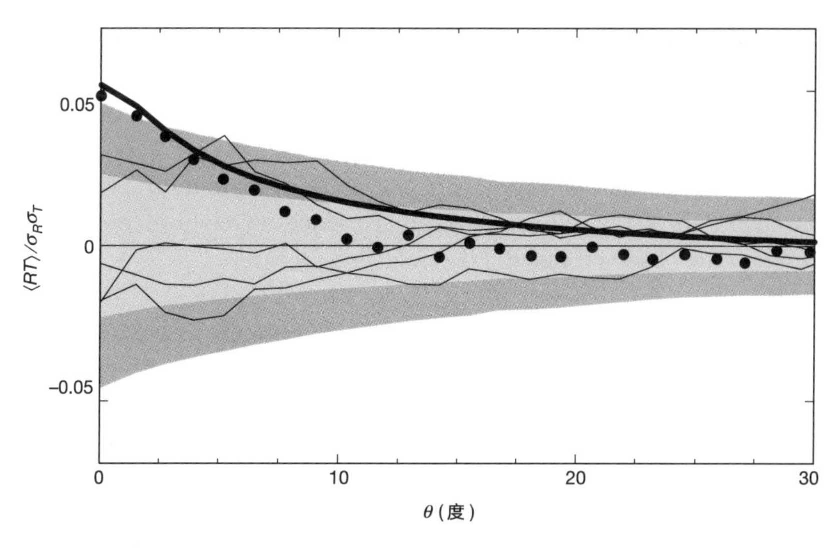


图 2. 射电星系的分布与微波背景的相关。这里我们绘出 NVSS 射电星系计数（$R$）和 WMAP 温度（$T$）的相关图（用实心圆表示）。其他曲线的含义与图 1 相同。蒙特卡罗试验中超过实际射电/CMB 相关的幅度在最低的三个区间分别为 1.2%、1.9%、3.4%。这与理论预言再一次保持一致，而在大角度处信号衰减速度再次快于预言。其幅度比 X 射线低是由于射电图更高的分辨率增加了射电的均方根（$\sigma_R$），而没有显著增加交叉相关。NVSS 和 X 射线交叉相关函数的一致暗示了信号不是来自 X 射线或者 NVSS 图像中未知的系统误差，而来自射电或者 X 射线源的微波辐射可能导致图像的正相关。尽管如此，射电星系的频谱外推指出这种微波辐射远小于观测结果 [8]。而且，射电源的成团性发生在比视信号小得多的角度上，因此所有这种污染应该出现在 $\theta\approx0°$ 而非更大的角度。

ISW 效应的理论曲线同时在图 1 和图 2 中给出，它与观测所得的相关类似，随着角间隔增大而衰减。在这些计算中，我们已经假设物质密度扰动（δρ）可直接由辐射源的数密度扰动所示踪。即 $\delta N/N=b\delta\rho/\rho$，假设偏袒因子 $b$ 为常数且它的值可通过 X 射线或者射电源的自相关函数得出。预测过程中假设了尺度不变的原初功率谱以及由 WMAP 数据的最优拟合得到的宇宙学。虽然由此得到的偏袒因子 $b$ 严格依赖于对源的红移分布的假设，但对期望的交叉相关函数而言，其对红移分布的依赖性比较弱，对偏袒因子 $b$ 可能的红移演化的依赖性也不强。对 ISW 效应幅度计算的详细讨论容易在其他文献中找到 [3,12]。

还有一点非常重要，即要考虑是否还有其他过程对观测到的相关有贡献。未掩模的前景源不是主要的污染，因为当改变银河系扣除方式，或者改变前述的严格掩模处理，观察到的相关性都不会改变。实际上，即使完全不对 CMB 做任何掩模处理，交叉相关函数的幅度都不会有本质的改变。同时，因为对不同半球的巡天图的

the different hemispheres of the maps separately are consistent, it is unlikely that a small number of diffuse foreground sources is responsible for the signal. Fluctuations caused by the Sunyaev–Zeldovich effect—the scattering of CMB photons as they travel through the ionized intergalactic medium of rich clusters of galaxies—are expected to contribute at smaller angles and be anti-correlated with the matter distribution[15]. Finally, microwave emission from the radio/X-ray sources could also produce correlations, but on much smaller angular scales than the observed signal.

The correlated signal evident in the figures is large enough to have been seen in our previous analyses using a COBE satellite combination map[8,9], albeit with a smaller signal-to-noise ratio because of that map's significant instrument noise and low angular resolution. However, we found no evidence for a correlation in the COBE analysis. When the differences in angular resolution are taken into account, the WMAP and COBE maps are consistent given COBE's large instrumental noise (~70 μK per pixel)[4]. The difference between the observed cross-correlations implies that a small part (r.m.s. ≈ 1–2 mK) of the (COBE-WMAP) difference map is anti-correlated with the X-ray and radio maps. This is an unlikely occurrence (1 in 100) if the COBE instrument noise model is correct. There are known systematics in the COBE data[16] above this level, but the large instrument noise makes it difficult to ascertain whether this is the origin of the problem. In any case, the source of the discrepancy almost certainly lies with the COBE data, which contain much larger noise and systematic effects than the WMAP data. Thus, we believe the WMAP correlations presented here are robust. Further discussion of this and other issues are available as Supplementary Information.

The confidence levels that we quote (99.9% and 99.4%) are not primarily limited by experimental error, but by the fact that we can only observe a finite region of the Universe. We find it significant that both signals are consistent with the theoretical predictions with no free parameters. However, it should be emphasized that these two measurements are by no means independent. Both of them use the same CMB maps and, furthermore, the X-ray background is highly correlated with the NVSS radio sources[17]. If the distribution of sources of one of the maps by chance coincided with the fluctuations of the CMB, one would expect the other map also to be correlated with the CMB to some degree. The coincidence of the expected amplitudes of the two signals is encouraging but by no means definitive. On the other hand, the detection of the ISW-type signal in both CCFs gives a strong indication that they are not due to unknown systematic effects in the maps. The radio and X-ray data were gathered by quite distinct methods, and it would be surprising if unknown systematics in the two maps were correlated in any way. We will present more detailed analysis of these issues elsewhere.

We conclude that we have observed the ISW effect. If so, these observations offer the first direct glimpse into the production of CMB fluctuations, and provide important, independent confirmation of the new standard cosmological model: an accelerating universe, dominated by dark energy. It should be pointed out that measurements of the power spectrum of CMB fluctuations do not show evidence of increased power on large

分析得到的结果是一致的，因此很难想象这些相关的信号是由少量的弥散前景源所导致的。苏尼阿耶夫–泽尔多维奇效应（CMB 光子在穿越富星系团中的电离星系际介质时发生的散射）引起的扰动应该贡献在小角度上，并且与物质分布反相关[15]。最后，射电源和 X 射线源发射的微波辐射也可能产生相关，但其发生的角尺度比观测信号要小得多。

图中给出的相关信号的证据已经足够大到即使使用宇宙背景探测器（COBE）卫星组合图也应该在我们前述的分析中被观察到[8,9]，尽管由于图像显著的仪器噪声和低角分辨率，信号的信噪比会比较低。但是我们并没有在对 COBE 的分析中发现存在相关性的证据。当考虑 WMAP 和 COBE 的角分辨率不同，以及在 COBE 具有较大仪器噪声（约 70 μK 每像素）[4] 的情况下，COBE 和 WMAP 的图像是一致的。观测到交叉相关的差别意味着，在 COBE 与 WMAP 的图像差中的一小部分（均方根约为 1 ~ 2 μK）存在与 X 射线图和射电图的反相关。如果 COBE 的仪器噪声的模型是正确的，上述情况不大可能发生（发生概率为 1/100）。虽然已知在 COBE 的数据中系统误差高于这个水平，但仪器噪声过大让我们很难确定这是否是问题的根源。无论如何，这个不一致的来源都被认为来自 COBE 数据自身[16]，因为它包含的噪声和系统性效应比 WMAP 数据要大得多。因此，我们相信 WMAP 的相关性是稳健的。进一步的讨论和更多的问题详见补充材料。

我们报道的置信度（99.9% 和 99.4%）并非主要受限于试验误差，而是由于我们只能观测到宇宙的有限区域。重要的是，理论预言并不需要加入自由参数就能和两种信号很好地吻合。但应该强调，这两种方法并不完全独立。因为两者都使用了相同的 CMB 图，并且 X 射线背景与 NVSS 射电源[17] 高度相关。如果某一种图的源分布偶然地与 CMB 的扰动相符，那么另一种图也会在某种程度上与 CMB 相关。虽然两个信号的期望幅度的一致性令人振奋，但还不是确定性的结果。另一方面，在两个交叉相关函数中都观测到 ISW 型的信号，强烈地表明这并非由图中未知的系统误差所引起的。射电数据和 X 射线数据是用非常不同的方法所采集的，如果两种巡天图中某种未知的系统误差存在关联，那将是令人惊奇的。我们会在其他论文中给出对这一问题的细致分析。

我们的结论是我们已观测到 ISW 效应。如果确实如此，这是第一次直接观测到 CMB 涨落的产生，也为新标准宇宙学模型——一个加速膨胀的、暗能量占主导的宇宙提供了重要的、独立的支持。应该指出的是，CMB 扰动功率谱的测量并没有如 ISW 效应所预言的那样在大的角尺度上（$\theta > 20^\circ$）上升，而是可能存在扰动功

angular scales ($\theta > 20°$) as predicted by the ISW effect, but rather indicate that there may be power missing[4]. Although this deficit is only at the 2σ level, it is intriguing, and may be telling us something about the formation of the very largest structures in the Universe. The consequences of the ISW effect reported here are primarily on intermediate angular scales, and are not in direct conflict with the power deficit on larger angular scales. Finally, we note that the WMAP-NVSS results have recently been independently analysed by the WMAP team, who find a similar level of correlation[18].

(**427**, 45-47; 2004)

**Stephen Boughn[1] & Robert Crittenden[2]**

[1] Department of Astronomy, Haverford College, Haverford, Pennsylvania 19041, USA

[2] Institute of Cosmology and Gravitation, University of Portsmouth, Portsmouth PO1 2EG, UK

Received 15 April; accepted 8 October 2003; doi:10.1038/nature02139.

---

References:

1. Bahcall, N., Ostriker, J. P., Perlmutter, S. & Steinhardt, P. The cosmic triangle: Revealing the state of the universe. *Science* **284**, 1481-1488 (1999).
2. Sachs, R. K. & Wolfe, A. M. Perturbations of a cosmological model and angular variations of the microwave background. *Astrophys. J.* **147**, 73-90 (1967).
3. Crittenden, R. & Turok, N. Looking for a cosmological constant with the Rees-Sciama effect. *Phys. Rev. Lett.* **76**, 575-578 (1996).
4. Bennett, C. L. *et al.* First year *Wilkinson Microwave Anisotropy Probe* observations: Preliminary maps and basic results. *Astrophys. J. Suppl.* **148**, 1-27 (2003).
5. Boldt, E. The cosmic X-ray background. *Phys. Rep.* **146**, 215-257 (1987).
6. Condon, J. *et al.* The NRAO VLA sky survey. *Astron. J.* **115**, 1693-1716 (1998).
7. Kinkhabwala, A. & Kamionkowski, M. New constraint on open cold-dark-matter models. *Phys. Rev. Lett.* **82**, 4172-4175 (1999).
8. Boughn, S. & Crittenden, R. Cross correlation of the CMB with radio sources: Constraints on an accelerating universe. *Phys. Rev. Lett.* **88**, 021302 (2002).
9. Boughn, S., Crittenden, R. & Koehrsen, G. The large scale structure of the X-ray background and its cosmological implications. *Astrophys. J.* **580**, 672-684 (2002).
10. Cowie, L. L., Barger, A. J., Bautz, M. W., Brandt, W. N. & Garmire, G. P. The redshift evolution of the 2-8 keV X-ray luminosity function. *Astrophys. J.* **584**, 57-60 (2003).
11. Dunlop, J. S. & Peacock, J. A. The redshift cut-off in the luminosity function of radio galaxies and quasars. *Mon. Not. R. Astron. Soc.* **247**, 19-42 (1990).
12. Boughn, S., Crittenden, R. & Turok, N. Correlations between the cosmic X-ray and microwave backgrounds: Constraints on a cosmological constant. *New Astron.* **3**, 275-291 (1998).
13. Bennett, C. L. *et al.* First year *Wilkinson Microwave Anisotropy Probe* observations: Foreground emission. *Astrophys. J. Suppl.* **148**, 97-117 (2003).
14. Tegmark, M., de Oliveira-Costa, A. & Hamilton, A. A high resolution foreground cleaned CMB map from WMAP. Preprint at ⟨http://xxx.lanl.gov/astro-ph/0302496⟩ (2003).
15. Peiris, H. V. & Spergel, D. N. Cross-correlating the Sloan Digital Sky Survey with the microwave sky. *Astrophys. J.* **540**, 605-613 (2000).
16. Kogut, A. *et al.* Calibration and systematic error analysis for the COBE DMR 4 year sky maps. *Astrophys. J.* **470**, 653-673 (1996).
17. Boughn, S. Cross-correlation of the 2-10 keV X-ray background with radio sources: Constraining the large-scale structure of the X-ray background. *Astrophys. J.* **499**, 533-541 (1998).
18. Nolta, M.R. *et al.* First year *Wilkinson Microwave Anisotropy Probe* observations: Dark energy induced correlation with radio sources. Preprint at ⟨http://xxx.lanl.gov/astro-ph/0305907⟩ (2003).
19. Spergel, D. N. *et al.* First year *Wilkinson Microwave Anisotropy Probe* observations: Determination of cosmological parameters. *Astrophys. J. Suppl.* **148**, 175-194 (2003).

**Supplementary Information** accompanies the paper on www.nature.com/nature.

**Acknowledgements.** We are grateful to M. Nolta, L. Page and the rest of the WMAP team, as well as to N. Turok, B. Partridge and B. Bassett, for useful conversations. R.C. acknowledges financial support from a PPARC fellowship.

**Competing interests statement.** The authors declare that they have no competing financial interests.

Correspondence and requests for materials should be addressed to R.C. (Robert.Crittenden@port.ac.uk).

率丢失[4]。虽然亏损的置信水平只有 $2\sigma$，但这一现象却十分神秘，这也许告诉了我们一些宇宙中非常大的结构形成的信息。这里对 ISW 效应所下的结论主要在中等角尺度上，并不与更大角尺度上的扰动功率丢失相矛盾。最后，WMAP 团队最近也对 WMAP–NVSS 结果做了独立的分析，他们也发现了类似水平的相关性[18]。

（周杰 翻译；冯珑珑 审稿）

# Mice Cloned from Olfactory Sensory Neurons

K. Eggan *et al.*

## Editor's Note

**By 2004, various mammals including sheep and mice had been cloned by injecting donor DNA into enucleated eggs. But in this paper, American cell biologist Rudolf Jaenisch and colleagues use donor DNA taken from mature olfactory neurons to generate fertile mouse clones by the process of nuclear transfer. The study demonstrates that DNA taken from a specialised cell that has stopped dividing can re-enter the cell cycle and be reprogrammed to an embryonic-like state after nuclear transfer. The cloned mice possessed the full range of organized, odorant receptor genes that were indistinguishable from those of normal mice, suggesting that selection of a particular receptor gene in a mature olfactory neuron does not cause irreversible changes in DNA.**

---

Cloning by nuclear transplantation has been successfully carried out in various mammals, including mice. Until now mice have not been cloned from post-mitotic cells such as neurons. Here, we have generated fertile mouse clones derived by transferring the nuclei of post-mitotic, olfactory sensory neurons into oocytes. These results indicate that the genome of a post-mitotic, terminally differentiated neuron can re-enter the cell cycle and be reprogrammed to a state of totipotency after nuclear transfer. Moreover, the pattern of odorant receptor gene expression and the organization of odorant receptor genes in cloned mice was indistinguishable from wild-type animals, indicating that irreversible changes to the DNA of olfactory neurons do not accompany receptor gene choice.

---

THE chromosome is a dynamic structure that undergoes complex changes that underlie the development and differentiated function of an organism. Alterations in chromosome structure include variation in the complement of regulatory proteins, covalent modifications in chromatin proteins or DNA, and in rare instances, DNA rearrangements[1]. The extent to which such chromosomal changes are reversible can be discerned by cloning experiments involving nuclear transfer. Thus far, cloning experiments using nuclei from post-mitotic cells that have irreversibly exited the cell cycle as part of their programme of differentiation have not generated viable embryos or mice[2,3]. These observations have led to the suggestion that post-mitotic cells might be refractory to epigenetic reprogramming or alternatively might have acquired changes in their DNA that could limit their developmental potential[4,5].

Directed DNA rearrangements are rarely observed as part of a normal differentiation programme and, in vertebrates, they have been described only for the generation of the diverse repertoire of antibodies and T-cell receptors[6-8]. Several observations have suggested

# 从嗅觉神经元克隆小鼠

伊根等

编者按

截至 2004 年，包括绵羊和小鼠在内的多种哺乳动物已通过将供体 DNA 注射到去核卵细胞中被成功克隆。但在本文中，美国细胞生物学家鲁道夫·耶尼施及其同事使用从成熟嗅觉神经元中提取的供体 DNA 通过核移植获得可育的克隆小鼠。该研究表明，从已经停止分裂的特化细胞中提取的 DNA 可以重新进入细胞周期，并在核移植后重编程为胚胎样状态。克隆小鼠具有与正常小鼠无法区分的全系列有组织的气味受体基因，这表明在成熟的嗅觉神经元中选择特定的受体基因不会引起 DNA 的不可逆变化。

---

通过细胞核移植进行的克隆已成功地在包括小鼠在内的多种哺乳动物中进行。到目前为止，还没有报道称可以利用有丝分裂后的细胞（如神经元）克隆小鼠。在这里，我们通过将有丝分裂后的嗅觉神经元的细胞核转移到卵母细胞中，成功获得可育克隆小鼠。这些结果表明，有丝分裂后终末分化的神经元基因组可以重新进入细胞周期，并在核移植后重编程产生全能性。此外，克隆小鼠的气味受体基因表达模式和气味受体基因的组构与野生型动物没有显著区别，这表明嗅觉神经元 DNA 的不可逆变化不随受体基因的选择同时出现。

---

染色体是一种动态结构，经历复杂的变化，是生物发育和分化功能的基础。染色体结构的变化包括调节蛋白的结合，染色质蛋白或 DNA 中的共价修饰，以及在极少数情况下发生的 DNA 重排[1]。染色体变化的可逆程度可以通过涉及核移植的克隆实验确定。有丝分裂后的细胞不可逆地退出细胞周期，这是其分化程序的一部分。迄今为止，使用这种细胞的细胞核进行克隆实验，尚未获得存活的胚胎或小鼠[2,3]。这些观察结果表明，有丝分裂后的细胞可能难以进行表观遗传重编程，或者后天获得的 DNA 变化可能会限制其发育潜力[4,5]。

定向 DNA 重排很少作为正常分化过程的一部分被观察到，而在脊椎动物中，它们被描述成只发生在产生抗体和 T 细胞受体多样性过程中[6-8]。一些观察结果表

that post-mitotic neurons may also use irreversible DNA alterations to generate diversity. First, a population of neurons undergoes apoptosis in mice bearing mutations in the double-strand-break DNA-repair enzymes, which are also required for DNA rearrangements in lymphocytes[9,10]. Second, cortical neurons exhibit a higher incidence of aneuploidy than other cell types, although the functional significance of these changes is unknown[11,12]. These observations and the inability to clone mice from neuronal nuclei have led to models in which the DNA of post-mitotic neurons might undergo rearrangements to supply additional genetic diversity that may enhance neural function[4,5,13-15].

One particularly clear example of neuronal diversity is provided by the olfactory sensory epithelium. In the mouse, each of the 2,000,000 cells in the olfactory epithelium expresses only one of about 1,500 odorant receptor genes, such that the functional identity of a neuron is defined by the nature of the receptor it expresses[16]. Thus, the sensory epithelium consists of at least 1,500 neuronal types. The pattern of receptor expression is apparently random within one of four zones in the epithelium, suggesting that the choice of receptor gene may be stochastic[17,18]. One mechanism to permit the stochastic choice of a single receptor could involve DNA rearrangements[19,20].

Here we report the generation of fertile mouse clones by transferring the nuclei of post-mitotic olfactory neurons into enucleated oocytes. Thus, a post-mitotic nucleus can re-enter the cell cycle and be reprogrammed to totipotency. The DNA of mice derived from sensory neurons reveals no evidence for rearrangements of the expressed olfactory receptor gene. The pattern of receptor expression in these mice was indistinguishable from that of wild-type animals, indicating that irreversible changes in DNA do not accompany olfactory receptor gene choice.

## Genetic Marking of Olfactory Sensory Neurons

Less than 1% of nuclear transfers result in the production of an embryonic stem (ES) cell line or live animal[1]. It is therefore difficult to identify with certainty the origin of the donor nucleus that contributed to the generation of a particular cloned animal[2]. This problem is particularly apparent in the olfactory epithelium, which contains mature sensory neurons intermingled with stem cells, neural progenitors and support cells. Therefore, we generated mice in which the endogenous olfactory marker protein (OMP) promoter drives simultaneous expression of OMP and Cre recombinase by inserting an internal ribosome entry site (IRES)-Cre cassette 3′ of the OMP stop codon. OMP is expressed only in mature olfactory sensory neurons (OSNs). When OMP–IRES-Cre mice are crossed to a reporter mouse strain (Z/EG), Cre expression catalyses the excision of a transcriptional stop sequence and results in green fluorescent protein (GFP) expression solely in mature OSNs (Fig. 1a and Methods). Transfer of marked OSN nuclei into oocytes should generate embryos in which every cell expresses GFP, as the Z/EG reporter uses the ubiquitous actin promoter.

明，有丝分裂后的神经元也可以利用不可逆的DNA变化来产生多样性。首先，一群神经元在带有双链断裂DNA修复酶突变的小鼠中经历细胞凋亡，该酶是淋巴细胞中DNA重排所必需的[9,10]。其次，尽管这些变化的功能意义尚不清楚，但皮质神经元表现出比其他细胞类型更高的非整倍性发生率[11,12]。这些观察结果和无法从神经元核中克隆小鼠的事实，促使新的模式被提出：有丝分裂后的神经元的DNA可经历重排以提供额外的遗传多样性来增强神经功能[4,5,13-15]。

嗅觉上皮提供了一个特别清楚的神经元多样性的例子。在小鼠嗅觉上皮中约有2,000,000个细胞，每一个细胞仅表达大约1,500个气味受体基因中的一种，因而神经元的功能由所表达受体的性质决定[16]。因此，感觉上皮由至少1,500种神经元类型组成。在上皮的四个区域中的某个区域内，受体表达的模式显然是随机的，这表明受体基因的选择可能是随机的[17,18]。一种允许随机选择单一受体的机制可能涉及DNA重排[19,20]。

在这里，我们报道通过将有丝分裂后的嗅觉神经元的细胞核移植到去核卵母细胞中产生的可育克隆小鼠。因此，有丝分裂后的细胞核可以重新进入细胞周期并被重编程和获得全能性。源自感觉神经元的小鼠DNA并未显示出其表达的嗅觉受体基因重排的证据。这些小鼠中受体表达的模式与野生型动物无法区分，表明DNA的不可逆变化不伴随嗅觉受体基因选择。

## 嗅觉神经元的遗传标记

只有不到1%的细胞核移植可以产生胚胎干细胞系或活体动物[1]。因此很难确定克隆动物来源于哪个供体核[2]。这个问题在嗅上皮中特别明显，嗅上皮含有成熟的感觉神经元，它们与干细胞、神经祖细胞和支持细胞混合在一起。因此，我们建立了一种小鼠模型，其中在内源性嗅觉标记蛋白（OMP）终止密码子的3′后端插入了内部核糖体进入位点（IRES）-Cre盒。这样OMP启动子同时驱动OMP和Cre重组酶的表达。OMP仅在成熟的嗅觉神经元（OSN）中表达。当OMP–IRES-Cre小鼠与报告小鼠品系（Z/EG）杂交时，Cre的表达催化了转录终止序列的切除，并导致仅在成熟OSN中表达绿色荧光蛋白（GFP）（图1a和方法）。将标记的OSN核转移到卵母细胞应该可以产生胚胎，并且因为Z/EG报告基因使用普遍存在的肌动蛋白启动子，胚胎中每个细胞都表达GFP。

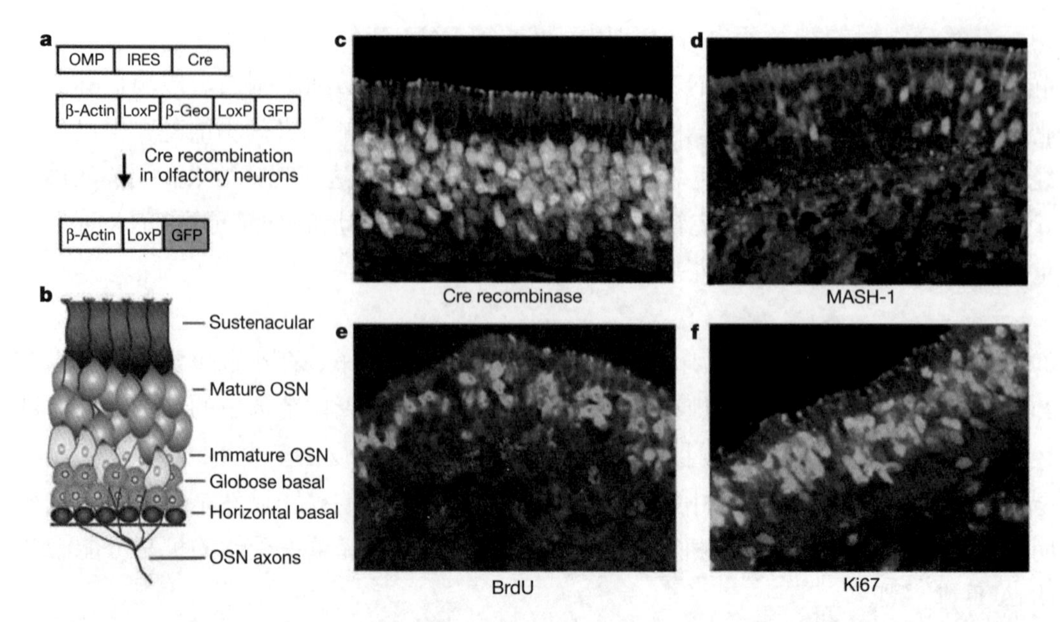


Fig. 1. Genetically marking post-mitotic OSNs. **a**, Donor animals carry one OMP–IRES-Cre allele and one copy of the Z/EG Cre reporter transgene. Cre expression in OSNs catalyses excision of the β-geo/stop cassette, resulting in selective GFP expression and genetic marking of mature OSNs. Thus, $GFP^+$ neurons could be selectively chosen for nuclear transfer. **b**, Schematic diagram of the approximate laminar distribution of cell types in the mouse olfactory epithelium. **c–f**, Three-colour immunofluorescence staining. Blue, nuclear marker TOTO-3; green, GFP protein fluorescence (**c**, **d**) or antibody to GFP protein (**e**, **f**); red, antibodies to Cre recombinase (**c**), progenitor-specific marker MASH-1 (**d**), markers of dividing cells, BrdU (**e**) and Ki67 (**f**).

In mice bearing the OMP–IRES-Cre allele and the Z/EG reporter gene, GFP expression was observed only in the most mature OSNs. Sections through the entire olfactory epithelium of several adult mice revealed GFP expression in the regions that contain mature OSNs (Fig. 1b–f). Cell counts revealed no $GFP^+$ cells in either the basal cell layers, where immature progenitors and stem cells reside, or in the apical support cell layer (0 of 835 $GFP^+$ cells)[21,22]. Double immunostaining for GFP and Cre protein revealed that all $GFP^+$ cells with visible nuclei also expressed Cre recombinase (599 of 599) (Fig. 1c). In addition, none of the MASH-$1^+$ basal cell precursors of OSNs expressed GFP (0 of 350 MASH-$1^+$ cells) (Fig. 1d). These results showed that GFP expression was restricted to mature sensory neurons.

We confirmed that the $GFP^+$ mature OSNs were post-mitotic by injecting mice with 5-bromodeoxyuridine (BrdU) and staining the olfactory epithelium with anti-BrdU and GFP antibodies (Fig. 1e). BrdU incorporation was observed in the more basal GFP-negative layers as well as in rare support cells in the apical layers. Examination of more than 3,000 $GFP^+$ cells never revealed a $GFP^+$ $BrdU^+$ cell. In a separate experiment, labelling with an antibody specific to Ki67, a protein restricted to the nuclei of dividing cells, revealed no staining coincident with GFP[23,24] (Fig. 1f). These data show that $GFP^+$ cells in the olfactory epithelium of donor animals are mature post-mitotic OSNs.

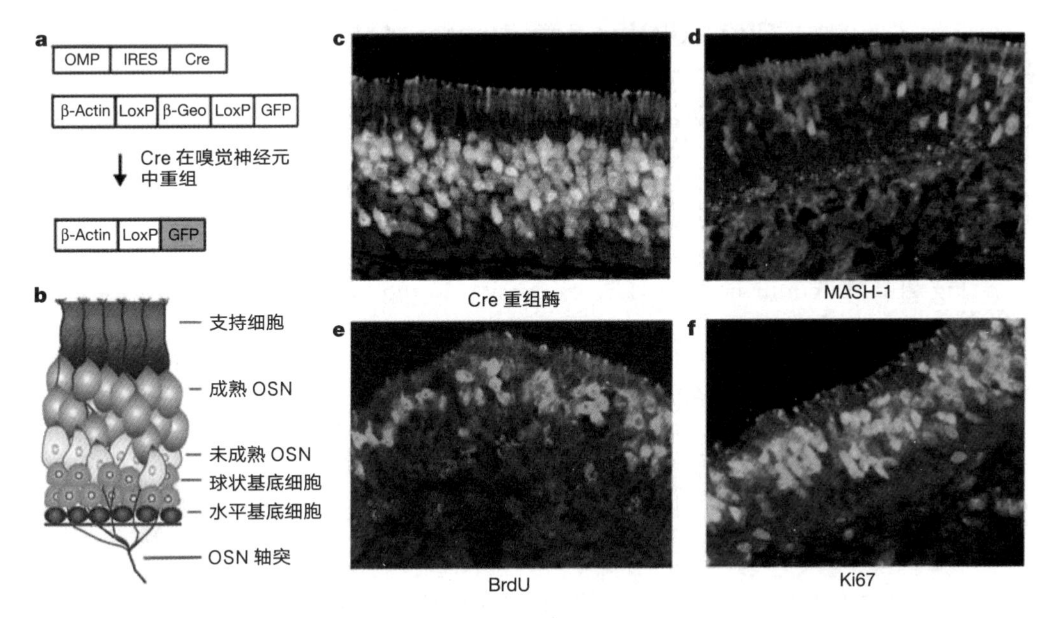


图 1. 有丝分裂后 OSN 的遗传标记。**a**，供体动物携带一个 OMP–IRES-Cre 等位基因和一个 Z/EG Cre 报道转基因拷贝。OSN 中 Cre 的表达催化 β-geo/终止盒的切除，导致选择性 GFP 表达和成熟 OSN 的遗传标记。因此，可以选择性地使用 $GFP^+$ 神经元进行核转移。**b**，小鼠嗅上皮各细胞类型近似层状分布的示意图。**c**～**f**，三色免疫荧光染色。蓝色，核标记物 TOTO-3；绿色，GFP 蛋白荧光（**c**、**d**）或 GFP 蛋白抗体（**e**、**f**）；红色，Cre 重组酶的抗体（**c**），祖细胞特异性标记物 MASH-1（**d**），分裂细胞标记物 BrdU（**e**）和 Ki67（**f**）。

在携带 OMP–IRES-Cre 等位基因和 Z/EG 报告基因的小鼠中，仅在最成熟的 OSN 中观察到 GFP 表达。数只成年小鼠的整个嗅上皮切片显示，GFP 在含有成熟 OSN 的区域中表达（图 1b～1f）。细胞计数显示，无论是存在未成熟祖细胞和干细胞的基底细胞层，抑或顶端支持细胞层中均没有 $GFP^+$ 细胞（835 个细胞中 $GFP^+$ 细胞有 0 个）[21,22]。对 GFP 和 Cre 蛋白的双重免疫染色显示，所有具有可见核的 $GFP^+$ 细胞均表达 Cre 重组酶（599/599）（图 1c）。此外，OSN 的 $MASH\text{-}1^+$ 基底细胞前体均未表达 GFP（350 个 $MASH\text{-}1^+$ 细胞中 $GFP^+$ 细胞有 0 个）（图 1d）。这些结果表明 GFP 表达仅限于成熟的感觉神经元。

我们通过向小鼠注射 5–溴脱氧尿苷（BrdU）并用抗 BrdU 抗体和 GFP 抗体染色嗅上皮，来证实 $GFP^+$ 的成熟 OSN 是经过有丝分裂的（图 1e）。在较多的 GFP 阴性基底细胞和较少的顶层支撑细胞中也观察到 BrdU 整合。检查 3,000 多个 $GFP^+$ 细胞，并未发现一个 $GFP^+$ $BrdU^+$ 细胞。在另一个实验中，用仅使分裂细胞核显色的 Ki67 特异性抗体标记细胞，显示没有 $GFP^+$ $Ki67^+$ 细胞（图 1f）[23,24]。这些数据显示供体动物嗅上皮中的 $GFP^+$ 细胞是成熟的有丝分裂后的 OSN。

## Cloning Mice from Neuronal Nuclei

We asked whether the nucleus of a post-mitotic OSN could re-enter the cell cycle and direct preimplantation development. We dissociated the olfactory epithelium and picked GFP-expressing OSNs for nuclear transfer into enucleated oocytes (Fig. 2a)[25-29]. Of the 352 embryos generated, 48 (14%) developed to the blastocyst stage (Table 1). All blastocysts expressed GFP, demonstrating that they were derived from the mature OSN donor nuclei (Fig. 2b). We confirmed that Cre expression in the OSNs caused constitutive GFP expression after nuclear transfer by generating blastocysts with GFP-negative cells from the donor mice. As expected, none of these blastocysts expressed GFP, demonstrating that Cre is not aberrantly expressed as a result of nuclear transfer or subsequent cloning procedures.

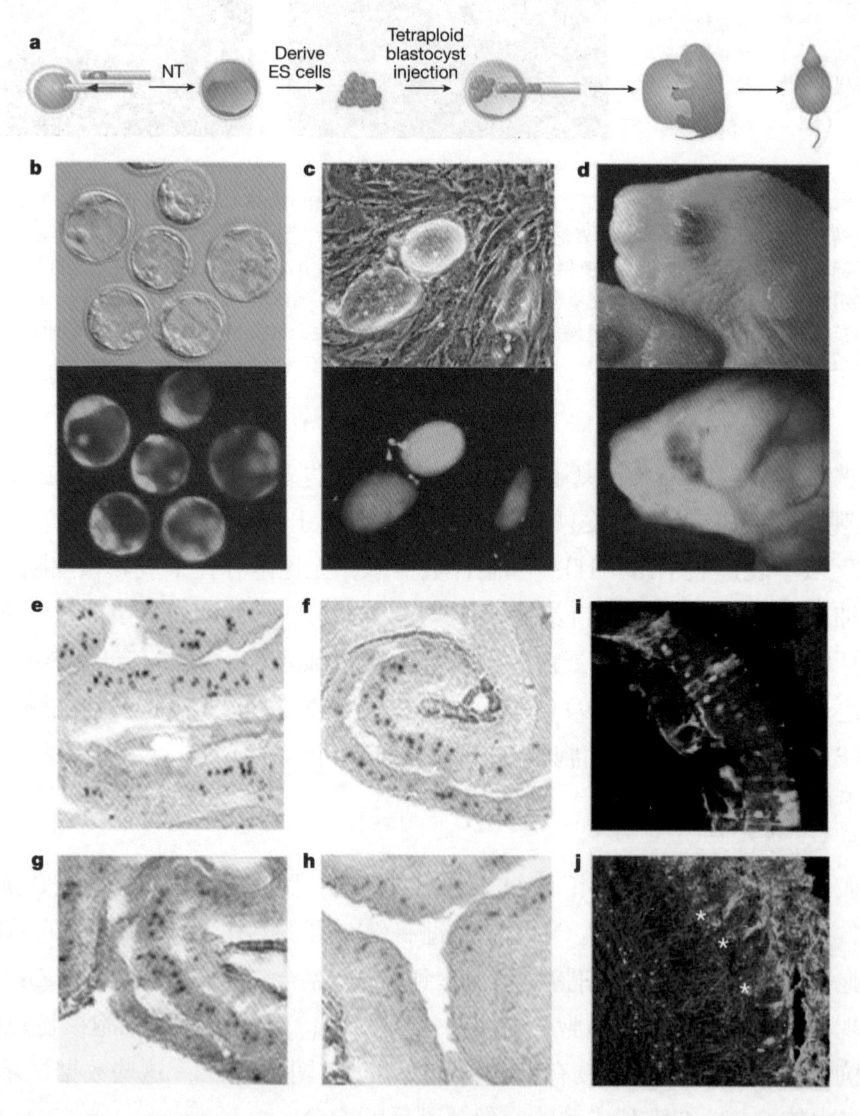


Fig. 2. Mice derived from OSN nuclei. **a**, Strategy to generate cloned ES cell lines and OSN-derived mice by tetraploid complementation with cloned ES cells. **b**, **c**, Bright-field and fluorescence images of nuclear

## 通过神经元细胞的核移植克隆小鼠

我们探究有丝分裂后 OSN 的细胞核是否可以重新进入细胞周期并指导胚胎植入前发育。我们裂解了嗅上皮组织，并选择了表达 GFP 的 OSN，将其核移植到去核卵母细胞中（图 2a）[25-29]。在产生的 352 个胚胎中，48 个（14%）发育到囊胚期（表 1）。所有囊胚均表达 GFP，证明它们来自成熟的 OSN 供体细胞核（图 2b）。我们通过对来自供体的 GFP 阴性细胞进行核移植产生囊胚验证了组成型 GFP 表达是由 OSN 中的 Cre 表达所致。正如预期的那样，这些囊胚中没有一个表达 GFP，这表明 Cre 不会因核移植或随后的克隆程序而异常表达。

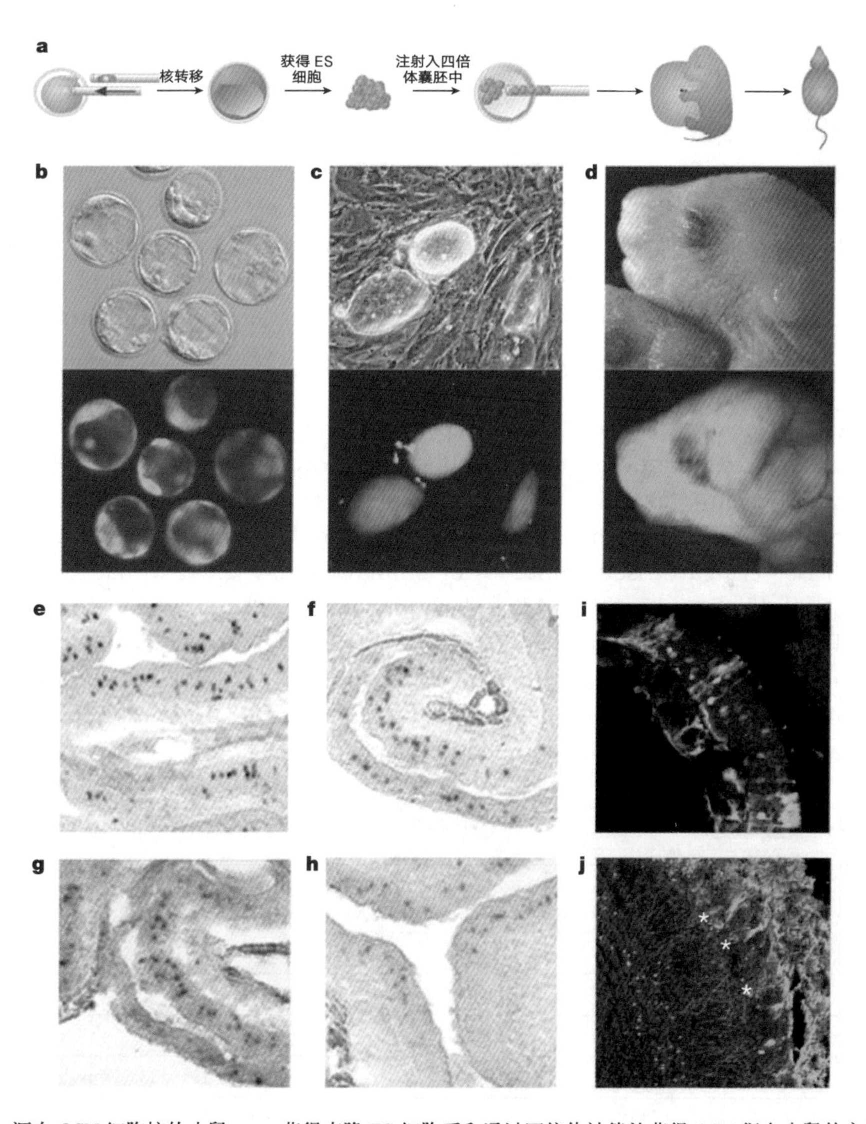


图 2. 源自 OSN 细胞核的小鼠。**a**，获得克隆 ES 细胞系和通过四倍体补偿法获得 OSN 衍生小鼠的方法。**b**、**c**，由 OSN 细胞核产生的核移植囊胚（**b**）和 OSN3 衍生 ES 细胞（**c**）的明场和荧光图像。**d**，通过

transfer blastocysts (**b**) and OSN3 ES cells (**c**) produced from OSN nuclei. **d**, A P0.5 mouse produced by tetraploid complementation with OSN3 ES cells (top) and a P0.5 C57/B6 control (bottom). **e–h**, *In situ* hybridization on the olfactory epithelium of mice wholly derived from OSN3 ES cells with probes for odorant receptors P2 (**e**), M50 (**f**), I7 (**g**) and M71 (**h**). **i, j**, Contribution of GFP$^+$ OSN2 ES cells to the olfactory epithelium (**i**) and olfactory bulb (**j**) of chimaeras generated by injecting OSN2 ES cells into diploid blastocysts. Several representative glomeruli are demarcated with asterisks.

Table 1. Production of ES cell lines and mice by nuclear transfer

| Donor cells | Oocytes surviving (% injected) | Oocytes activated (% surviving) | Two-cell embryos to oviduct (% oocytes activated) | Blastocysts (% oocytes activated) | ES cell lines (% oocytes activated) | Alive at term (% into oviduct) |
|---|---|---|---|---|---|---|
| OSNs | 508 (88) | 352 (69) | — | 48 (14) | 3 (1) | — |
| P2 OSNs | 261 (95) | 141 (54) | — | 18 (13) | 2 (1) | — |
| P2SN1 ES | 50 (95) | 38 (76) | 14 (37) | — | — | 1 (7) |
| P2SN2 ES | 47 (90) | 37 (79) | 18 (49) | — | — | 1 (6) |

The 48 GFP$^+$ blastocysts were used to generate ES cell lines and three gave rise to colonies that resembled ES cells (OSN1–3). All three lines expressed GFP, and Southern blotting confirmed the predicted genomic rearrangement at the Z/EG locus (Fig. 2c and data not shown). When the OSN2 and OSN3 ES cells were injected into diploid blastocysts, they contributed extensively to all tissues of the resulting chimaeras, including the germ line (Fig. 2i, j and data not shown).

We injected these cloned ES cells into tetraploid blastocysts to perform the most stringent test of their developmental potency (Table 2). Tetraploid embryo complementation generates an "ES-fetus" in which all embryonic lineages are derived from the injected ES cells, whereas the extraembryonic lineages develop from the host blastocyst[29,30]. Both OSN2 and OSN3 ES cells gave rise to embryos of embryonic day (E) 19.5 that expressed GFP ubiquitously (Table 2 and Fig. 2d). Mice derived from OSN3 survived to adulthood, were fertile and were overtly normal. Histological examination of serial sections through their brains revealed no obvious abnormalities (data not shown). These results indicate that the genome of a post-mitotic, terminally differentiated neuron can be reprogrammed after nuclear transfer and direct the development of all embryonic lineages.

Table 2. Mice produced from sensory neurons by tetraploid embryo complementation

| ES cell line | Receptor expressed | Tetraploid blastocysts injected | Mice alive at term (% injected) | Breathing normally (% injected) | Cross-fostered | Survival to maturity (% fostered) |
|---|---|---|---|---|---|---|
| OSN1 | ? | 137 | 0 | 0 | 0 | 0 |
| OSN2 | ? | 160 | 1 (1) | 0 | 0 | 0 |
| OSN3 | ? | 273 | 49 (18) | 29 (11) | 21 | 17 (81) |
| P2SN1 | P2 | 89 | 16 (18) | 16 (100) | 13 | 12 (92) |
| P2SN2 | P2 | 151 | 31 (21) | 28 (90) | 28 | 22 (78) |

OSN3 ES 细胞利用四倍体补偿法产生的 P0.5 小鼠（上图）和 P0.5 C57/B6 对照小鼠（下图）。**e～h**，用针对气味受体 P2(**e**)、M50(**f**)、I7(**g**) 和 M71(**h**) 的探针，在完全来自 OSN3 衍生 ES 细胞的小鼠嗅上皮进行原位杂交的结果。**i**、**j**，GFP⁺ OSN2 衍生 ES 细胞对通过将其注射到二倍体囊胚中获得的嵌合体小鼠嗅上皮 (**i**) 和嗅球 (**j**) 的贡献。几个代表性的嗅小球用星号标出。

表 1. 通过核移植获得 ES 细胞系和小鼠的情况

| 供体细胞 | 存活卵母细胞数（% 占总注射数百分比） | 激活卵母细胞数（% 占总存活数百分比） | 进入输卵管的双细胞胚胎数（% 占总激活卵母细胞数百分比） | 囊胚数（% 占总激活卵母细胞数百分比） | ES 细胞系数（% 占总数活卵母细胞数百分比） | 实验过程中存活（% 占总进入输卵管胚胎数百分比） |
|---|---|---|---|---|---|---|
| OSNs | 508 (88) | 352 (69) | — | 48 (14) | 3 (1) | — |
| P2 OSNs | 261 (95) | 141 (54) | — | 18 (13) | 2 (1) | — |
| P2SN1 ES | 50 (95) | 38 (76) | 14 (37) | — | — | 1 (7) |
| P2SN2 ES | 47 (90) | 37 (79) | 18 (49) | — | — | 1 (6) |

48 个 $GFP^+$ 囊胚用于获得 ES 细胞系，其中 3 个产生类似 ES 细胞的集落（OSN1～OSN3）。所有三个品系都表达 GFP，Southern 印迹证实了之前预测的在 Z/EG 基因座处发生的基因组重排（图 2c，未显示数据）。当将 OSN2 和 OSN3 ES 细胞注射到二倍体囊胚中时，它们对所得嵌合体的所有组织（包括生殖细胞系）产生了广泛嵌合（图 2i、2j，未显示数据）。

我们将这些克隆的 ES 细胞注射到四倍体囊胚中，以对其发育潜力进行最严格的测试（表 2）。四倍体胚胎补偿法获得“ES 胎儿”，其中所有胚胎细胞谱系来自注射的 ES 细胞，而胚外细胞谱系来自宿主囊胚 [29,30]。OSN2 和 OSN3 的 ES 细胞均获得可长至 19.5 天的胚胎，其细胞普遍表达 GFP(表 2 和图 2d)。源自 OSN3 的小鼠存活至成年，可育并且明显正常。大脑连续切片的组织学检查显示没有明显的异常（数据未显示）。这些结果表明，有丝分裂后终末分化的神经元的基因组可以在核移植后重编程，并指导所有胚胎细胞谱系的发育。

表 2. 感觉神经元通过四倍胚胎互补法获得小鼠的情况

| ES 细胞系 | 受体表达 | 四倍体囊胚注射数 | 实验过程中小鼠存活数（% 占总注射数百分比） | 正常呼吸的小鼠数（% 占总注射数百分比） | 交叉抚育小鼠数 | 存活至性成熟小鼠数（% 占交叉抚育百分比） |
|---|---|---|---|---|---|---|
| OSN1 | ? | 137 | 0 | 0 | 0 | 0 |
| OSN2 | ? | 160 | 1 (1) | 0 | 0 | 0 |
| OSN3 | ? | 273 | 49 (18) | 29 (11) | 21 | 17 (81) |
| P2SN1 | P2 | 89 | 16 (18) | 16 (100) | 13 | 12 (92) |
| P2SN2 | P2 | 151 | 31 (21) | 28 (90) | 28 | 22 (78) |

## Odorant Receptor Expression in Cloned Mice

We examined the patterns of receptor gene expression in the olfactory epithelium of the cloned mice generated by tetraploid complementation. If irreversible genetic rearrangements are required for receptor choice we might expect an altered profile of receptor expression in animals cloned from an OSN. In the simplest model, a rearrangement involving one receptor gene might persist in all neurons such that all neurons will express the same receptor. This scenario was observed in mice derived from a B-cell nucleus, in which all B cells expressed the same immunoglobulin gene[28]. We therefore asked whether sensory neurons from cloned mice express a single receptor or a repertoire of receptor genes. We performed *in situ* hybridization on the olfactory epithelium of mice derived from OSN3 ES cells using probes specific for seven different odorant receptors (Fig. 2e–h and data not shown). The pattern of expression of all seven receptors was indistinguishable from control mice, excluding a simple model of gene rearrangement that would result in the expression of a single receptor gene in all OSNs.

As a neuron expresses a receptor from only one of the two alleles, it remained possible that a rearranged receptor gene expressed in the donor nucleus would be expressed in only half of the sensory neurons. We therefore analysed the repertoire of receptors expressed by polymerase chain reaction with reverse transcription (RT–PCR) to determine whether a single receptor transcript was enriched in the epithelium of OSN-derived mice. RNA from the olfactory epithelium of cloned animals was used in RT–PCR reactions with degenerate primers that recognize conserved motifs present in the majority of odorant receptors. Forty-four PCR products were cloned and restriction digest analysis indicated that they encoded 38 different receptors. Sequence analysis of 20 of these PCR clones revealed that they encoded 20 different receptors, which were located in seven clusters on six different chromosomes. This suggests that a single odorant receptor did not predominate in the olfactory epithelium of mice derived from OSN nuclei.

One additional assay for the diversity of odorant receptor expression is based on the observation that neurons expressing a given receptor, although randomly distributed within a zone of the epithelium, converge on two spatially invariant loci or glomeruli in the olfactory bulb[18,31]. If half of the OSNs from cloned mice expressed the same receptor, then their axonal projections should preferentially innervate a small set of glomeruli. Analysis of the olfactory bulb of chimaeric mice produced with OSN2 ES cells revealed that the OSN2-derived $GFP^+$ cells innervated most, if not all glomeruli, with no glomerulus receiving predominant input (Fig. 2i, j). These results show that the sensory neurons of mice cloned from an OSN that had expressed a single receptor can express a large repertoire of odorant receptor genes.

## 克隆小鼠中的气味受体表达

我们检查了由四倍体补偿法获得的克隆小鼠的嗅上皮中受体基因表达的模式。如果受体选择需要不可逆的基因重排，我们可能会从 OSN 克隆的动物中发现受体表达的改变。在最简单的模型中，涉及一个受体基因的重排可能在所有神经元中持续存在，使得所有神经元都表达相同的受体。在源自 B 细胞核的小鼠中观察到这种情况；在这些小鼠中，所有 B 细胞表达相同的免疫球蛋白基因[28]。因此，我们想知道克隆小鼠的感觉神经元是表达单一受体还是表达受体基因库。我们使用对七种不同气味受体特异的探针对源自 OSN3 的 ES 细胞的小鼠的嗅上皮进行原位杂交（图 2e ~ 2h，未显示数据）。所有七种受体的表达模式与对照小鼠无法区分，而基因重排的简单模型将导致所有 OSN 中表达单一受体基因，因此排除了这种简单模型。

由于神经元仅从两个等位基因中选择一个表达受体，因此在供体细胞核中表达的重排受体基因仍然可能仅在一半感觉神经元中表达。因此，我们通过逆转录聚合酶链反应（RT–PCR）分析了表达的受体库，以确定单个受体转录物是否富集在 OSN 衍生小鼠的上皮细胞中。来自克隆动物嗅上皮的 RNA 用于 RT–PCR 反应，其中简并引物识别大多数气味受体中存在的保守基序。克隆得到 44 种 PCR 产物，限制性消化分析表明它们编码了 38 种不同的受体。对这些 PCR 克隆中的 20 个进行序列分析，发现它们编码了 20 种不同的受体，这些受体位于 6 个不同染色体上的 7 个簇中。这表明源自 OSN 核的小鼠的嗅上皮，并非只表达单一气味受体。

气味受体表达多样性的另一种测定是基于以下观察：表达给定受体的神经元虽然随机分布在上皮特定区域内，但会富集在嗅球中两个空间位置不变的基因座或嗅小球上[18,31]。如果来自克隆小鼠的 OSN 中有一半表达相同的受体，那么它们的轴突投射应该优先支配一小组嗅小球。对 OSN2 的 ES 细胞衍生的嵌合小鼠的嗅球分析显示，OSN2 衍生的 $GFP^+$ 细胞支配绝大多数嗅小球，没有嗅小球接受（单一受体对应的）优先支配（图 2i、2j）。这些结果表明，由表达单一受体的 OSN 克隆获得的小鼠，其感觉神经元可以表达大量的气味受体基因。

## Mice Cloned from Neurons Expressing the P2 Receptor

The previous experiments suggest that irreversible rearrangements are not required for receptor gene expression. Because these experiments did not allow the prospective identification of the receptor gene expressed by the donor OSN nucleus, we were unable to examine its pattern of expression or its DNA sequence in the cloned mice. Neurons that express the P2 odorant receptor were marked by introducing an IRES directing the translation of GFP into the 3′ untranslated region of the P2 gene (P2–IRES-GFP)[32]. These mice were crossed with strains carrying both the OMP–IRES-Cre alteration and a reporter allele in which the *Rosa26* promoter is separated from a weak GFP gene by a LoxP-flanked transcriptional terminator[33] (Fig. 3a). Mice carrying the P2–IRES-GFP, OMP–IRES-Cre and *Rosa26*–LoxP-Stop-LoxP–weak GFP reporter genes exhibited intense GFP fluorescence in the approximately 0.1% of the OSNs that expressed the P2–IRES-GFP allele, whereas all other neurons appeared dark (Fig. 3b). *Rosa26*-driven GFP was present in mature OSNs, but the GFP signal was too weak to be detected by direct fluorescence. No green cells were observed in animals that contained only the OMP–IRES-Cre and *Rosa26*–LoxP-Stop-LoxP–weak GFP genes (Fig. 3c). Evidence for Cre-mediated excision of the transcription terminator and the resulting expression of weak GFP in mature OSNs was detected by amplifying the signal with GFP antibodies (data not shown).

Fluorescent cells expressing the P2–IRES-GFP allele were picked and their nuclei were injected into enucleated oocytes (Fig. 3d). Eighteen blastocysts developed from 141 reconstructed oocytes generating two ES cell lines: P2SN1 and P2SN2 (Table 1). PCR analysis revealed Cre-mediated excision at the *Rosa26* locus in both cell lines, indicating their origin from cloned nuclei of post-mitotic OSNs (Supplementary Fig. 1). Tetraploid embryos injected with P2SN1 and P2SN2 cells resulted in the birth of multiple viable pups that survived to adulthood (Table 2). These mice exhibited no gross anatomic or behavioural abnormalities and were fertile. PCR analyses of non-neuronal tissues revealed recombination at the *Rosa26* reporter allele, indicating that the cloned mice were derived from mature $OMP^{+}$ OSNs (data not shown). In addition, we detected the weak *Rosa26*-driven GFP expression by *in situ* hybridization in both the olfactory epithelium and non-neuronal tissues, demonstrating genetic activation of the reporter (Fig. 3g, h and data not shown).

The expression pattern of P2–IRES-GFP in clones was indistinguishable from that of control donor animals. Approximately 0.1% of the neurons expressed GFP at a high level within zone II of the epithelium (Fig. 3e). The GFP-expressing neurons projected axons to one medial and one lateral glomerulus in the olfactory bulb (Fig. 3f). Moreover, the P2–IRES-GFP allele was expressed at a frequency approximately equal to that of the unmodified P2 allele. *In situ* hybridization was performed on sections through the olfactory epithelium with RNA probes specific to GFP- and P2-coding sequences (Fig. 3g–j). The number of GFP-expressing cells in one section was roughly 50% of the number of P2-expressing cells found in neighbouring sections ($45 \pm 14\%$, s.d.), revealing no preference

## 由表达 P2 受体的神经元克隆小鼠

先前的实验表明，受体基因表达并不需要不可逆的重排。因为这些实验不允许对供体 OSN 细胞核表达的受体基因进行前瞻性鉴定，所以我们无法检查克隆小鼠中的受体表达模式或其 DNA 序列。表达 P2 气味受体的神经元通过在 P2 基因的 3′非翻译区插入 IRES 来标记，该 IRES 可以指导 GFP 翻译（P2–IRES-GFP）[32]。将这些小鼠与同时携带 OMP–IRES-Cre 变异并报告等位基因的小鼠杂交，该报告基因中 *Rosa26* 启动子和弱 GFP 基因通过 LoxP 中间的转录终止子分开 [33]（图 3a）。携带 P2–IRES-GFP，OMP–IRES-Cre 和 *Rosa26*–LoxP-Stop-LoxP– 弱 GFP 报告基因的小鼠在表达 P2–IRES-GFP 等位基因的 OSN 中大约有 0.1%表现出强烈的 GFP 荧光，而其他所有神经元都很暗不发荧光（图 3b）。*Rosa26* 驱动的 GFP 存在于成熟的 OSN 中，但 GFP 信号太弱而不能通过直接荧光检测到。在仅含有 OMP–IRES-Cre 和 *Rosa26*–LoxP-Stop-LoxP– 弱 GFP 基因的动物中未观察到绿色细胞（图 3c）。用 GFP 抗体扩增信号来检测 Cre 介导的转录终止子切除，及其导致的成熟 OSN 中弱 GFP 表达（数据未显示）。

挑选表达 P2–IRES-GFP 等位基因的荧光细胞，并将它们的细胞核注射到去核卵母细胞中（图 3d）。从 141 个重建的卵母细胞发育出 18 个囊胚，产生了两个 ES 细胞系：P2SN1 和 P2SN2（表 1）。PCR 分析显示两种细胞系中均出现 Cre 介导的 *Rosa26* 基因座处的切除，表明它们来自有丝分裂后 OSN 的克隆核（补充图 1）。注射 P2SN1 和 P2SN2 细胞到四倍体囊胚获得存活至成年的多个活动崽（表 2）。这些小鼠没有表现出严重的解剖学异常或行为异常，且可育。非神经元组织的 PCR 分析揭示了 *Rosa26* 报告等位基因的重组，表明克隆的小鼠源自成熟的 $OMP^+$ OSN（数据未显示）。此外，我们通过嗅上皮和非神经组织中的原位杂交检测到弱 *Rosa26* 驱动的 GFP 表达，证明了报告基因的遗传激活（图 3g、3h，未显示数据）。

克隆中 P2–IRES-GFP 的表达模式与对照供体动物的表达模式不可区分。大约 0.1%的神经元在上皮区 II 内以高水平表达 GFP（图 3e）。表达 GFP 的神经元将轴突投射到嗅球中的一个内侧嗅小球和一个旁侧嗅小球（图 3f）。此外，P2–IRES-GFP 等位基因的表达频率大约等于未修饰的 P2 等位基因的表达频率。使用对 GFP 和 P2 编码序列特异的 RNA 探针对嗅上皮切片进行原位杂交（图 3g ~ 3j）。在一个切片中表达 GFP 的细胞数量大约是在相邻切片中发现的表达 P2 的细胞数量的 50%（45%±14%，标准差），表明 P2–IRES-GFP 等位基因并不优先表达。此外，用对另

for the expression of the P2–IRES-GFP allele. Furthermore, *in situ* hybridization with probes specific to six additional olfactory receptors revealed similar expression patterns in control and cloned mice (Fig. 3k, l and data not shown).

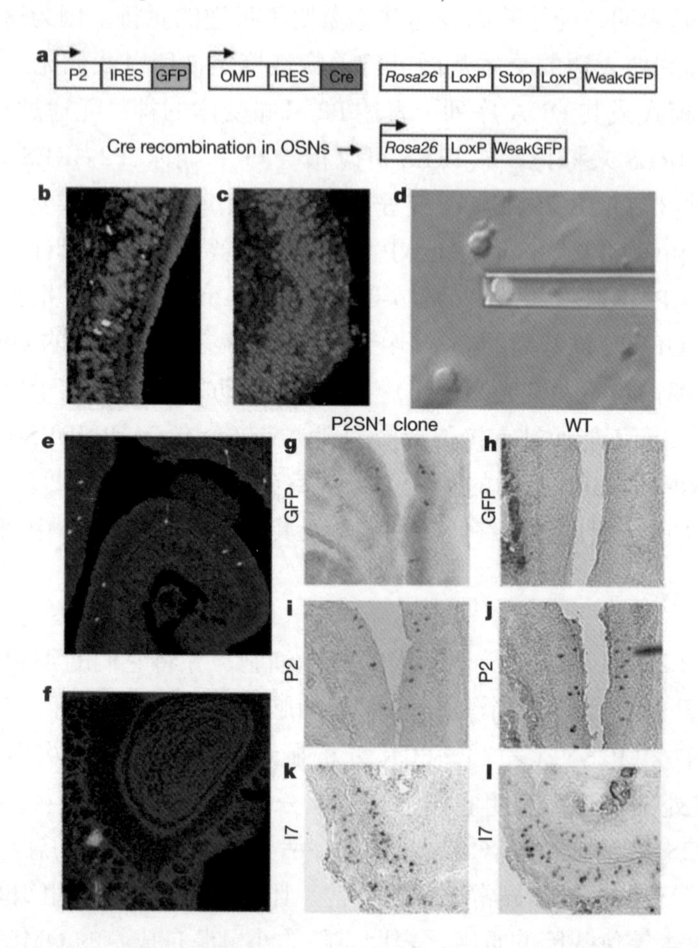


Fig. 3. Mice produced from OSNs expressing the P2 odorant receptor. **a**, Strategy to label P2 sensory neurons with GFP and genetically mark mature OSNs. **b**, **c**, Immunofluorescence staining with nuclear marker TOTO-3 (blue), GFP fluorescence (green) and antibodies to Cre recombinase (red) of the olfactory epithelium of donor animals heterozygous for the P2–IRES-GFP, OMP–IRES-Cre and *Rosa26*–LoxP-Stop-LoxP–weak GFP alleles (**b**) or of control animals heterozygous for the OMP–IRES-Cre and *Rosa26*–LoxP-Stop-LoxP–weak GFP alleles (**c**). **d**, Bright-field and fluorescence merged image of picking a P2–IRES-GFP$^{+}$ neuron from the dissociated olfactory epithelium of a donor for nuclear transfer. **e**, **f**, GFP fluorescence (green) and TOTO-3$^{+}$ nuclei (blue) visualized in sections of the olfactory epithelium (**e**) and olfactory bulb (**f**) of an animal generated by tetraploid complementation with P2SN1 ES cells. **g–l**, *In situ* hybridization on olfactory epithelium sections derived entirely from P2SN2 ES cells (**g**, **i**, **k**) and a wild-type control (**h**, **j**, **l**) with probes for GFP (**g**, **h**), P2 receptor (**i**, **j**) and I7 receptor (**k**, **l**).

We also performed Southern blotting, PCR and genome sequencing to examine the organization of the P2–IRES-GFP allele in the chromosome of cloned mice in an effort to detect potential DNA rearrangement events. If choice of a single receptor gene involved gene conversion into a single active locus, cells expressing the P2–IRES-GFP allele might contain a second copy of this allele at the active locus. Southern blotting to distinguish

外六种气味受体特异的探针进行的原位杂交揭示了对照和克隆小鼠中相似的受体表达模式（图 3k、3l，未显示数据）。

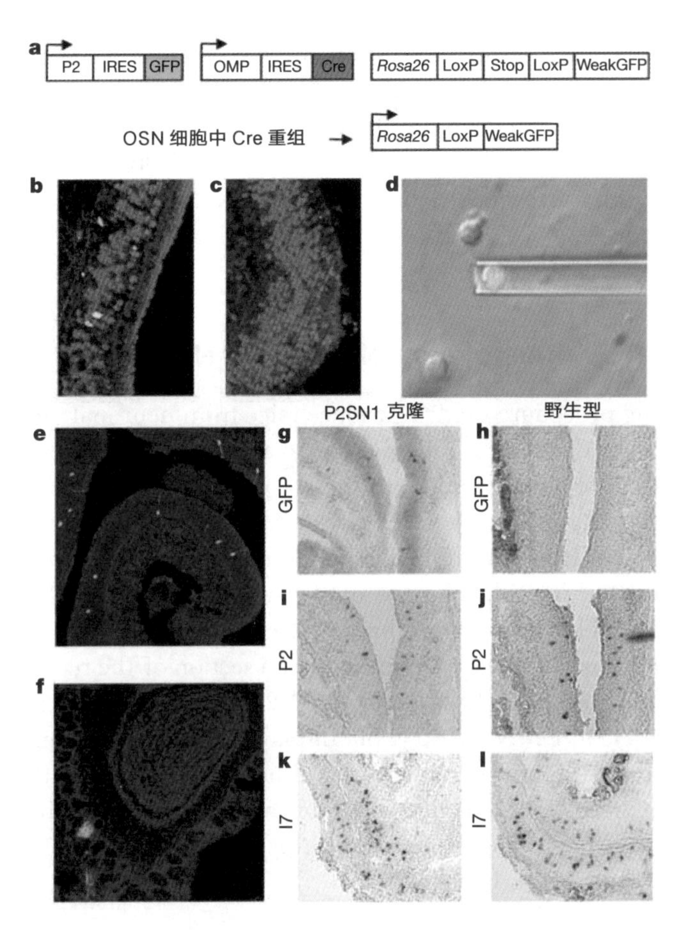


图 3. 由表达 P2 气味受体的 OSN 获得的小鼠。**a**，用 GFP 标记 P2 感觉神经元和遗传标记成熟 OSN 的方法。**b**、**c**，用核标记物 TOTO-3（蓝色）、GFP 荧光（绿色）和 Cre 重组酶抗体（红色）对嗅上皮进行免疫荧光染色。嗅上皮分别来自携带 P2–IRES-GFP、OMP–IRES-Cre 和 *Rosa26*–LoxP-Stop-LoxP– 弱 GFP 等位基因的杂合供体动物（**b**）和携带 OMP–IRES-Cre 和 *Rosa26*–LoxP-Stop-LoxP– 弱 GFP 等位基因的杂合对照动物（**c**）。**d**，从供体裂解的嗅上皮细胞中挑选 P2–IRES-GFP$^+$ 神经元进行核移植的明场和荧光合并图像。**e**、**f**，GFP 荧光（绿色）和 TOTO-3$^+$ 细胞核（蓝色）在通过 P2SN1 ES 细胞与四倍体囊胚补偿得到的动物嗅上皮（**e**）和嗅球（**f**）切片中可视化。**g**～**l**，对完全来源于 P2SN2 ES 细胞的嗅上皮切片（**g**、**i**、**k**）和野生型对照切片（**h**、**j**、**l**）进行原位杂交，探针为 GFP（**g**、**h**）、P2 受体（**i**、**j**）和 I7 受体（**k**、**l**）。

我们还进行了 Southern 印迹、PCR 和基因组测序，以检查克隆小鼠染色体中 P2–IRES-GFP 等位基因的组构，来检测潜在的 DNA 重排事件。如果单个受体基因的选择涉及基因转变成单个活性基因座，则表达 P2–IRES-GFP 等位基因的细胞可能在活性基因座处含有该等位基因的第二个拷贝。用于区分修饰的克隆小鼠 P2–IRES-

the modified P2–IRES-GFP allele from the unmodified endogenous P2 allele in DNA from control and cloned mice revealed only two bands of equal intensity, suggesting that gene conversion into an active locus does not accompany olfactory receptor gene choice (Supplementary Fig. S2). In addition, Southern blot analyses using multiple DNA probes to examine the organization of about 60 kilobases (kb) of DNA 5′ and 3′ of the P2 coding sequence failed to detect any differences between donor, control and cloned animals (Supplementary Fig. S3). Sequencing of 10 kb 3′ of the P2 coding sequence showed that the P2–IRES-GFP allele was identical in donor, control and cloned mice (data not shown). These results suggest that irreversible changes in DNA do not accompany choice of the P2 odorant receptor gene.

## Totipotency of Neuronal Nuclei

The two-step cloning procedure used to produce mice from neuronal nuclei generates mice in which the neuronally derived ES cells give rise to all embryonic tissues, whereas cells from the tetraploid host blastocyst contribute to the embryonic trophectoderm[30]. Thus, neither this work nor the cloning of lymphocytes via an ES cell intermediate[28] demonstrated the totipotency of a nucleus from a terminally differentiated cell[34]. To demonstrate totipotency of mature OSN nuclei, we transplanted nuclei from P2SN1 and P2SN2 ES cells into enucleated oocytes[29]. The resulting embryos were cultured for 24 h and transferred to pseudopregnant recipients (Table 1). Upon caesarean section of the recipients, we recovered full-term pups from both the P2SN1 and P2SN2 cell lines. These pups had enlarged placentas (P2SN1, 0.35 g; P2SN2, 0.40 g) but displayed no overt anatomical or behavioural abnormalities, were fertile and survived to adulthood, consistent with previous cloning experiments[29]. These observations demonstrate that nuclei of terminally differentiated olfactory neurons can be reprogrammed to totipotency, directing development of both embryonic and extraembryonic lineages.

## Discussion

We have asked whether the nucleus of a post-mitotic olfactory sensory neuron can re-enter the cell cycle and undergo reprogramming to direct development of a mouse. ES cell lines were generated from OSN nuclei at frequencies similar to those obtained with differentiated lymphoid cells that can be induced to proliferate under physiological conditions[2]. Thus the mechanisms that lead to the cell-cycle exit and irreversible mitotic arrest that accompany neural differentiation do not result from irreversible epigenetic or genetic events that would interfere with nuclear totipotency.

The differentiation of neurons requires that neural progenitors exit the cell cycle before a restriction point late in G1. This decision is governed by a complex balance of cell-cycle regulators and proneural genes that drive cells into G0 and prevent them from progressing beyond the restriction point[35]. Although most cells that enter G0 can re-enter the cycle on

GFP 等位基因与来自对照小鼠 DNA 中未修饰的内源性 P2 等位基因的 Southern 印迹仅显示两条相等强度的条带，表明基因转变为活性基因座不伴随气味受体基因的选择(补充图 S2)。此外，使用多个 DNA 探针对 P2 编码序列 5′和 3′端约 60 千碱基(kb)的 DNA 进行 Southern 印迹分析，未能检测到供体、对照和克隆动物之间的任何差异(补充图 S3)。对 P2 编码序列 3′端的 10 kb 进行测序，显示 P2–IRES-GFP 等位基因在供体、对照和克隆小鼠中是相同的(数据未显示)。这些结果表明 DNA 的不可逆变化不伴随 P2 气味受体基因的选择。

## 神经元细胞核的全能性

使用两步克隆法从神经元细胞核获得的小鼠中，神经元衍生的 ES 细胞发育为所有胚胎组织，而来自四倍体宿主囊胚的细胞发育成滋养外胚层[30]。因此，这一工作和通过 ES 细胞中间体克隆淋巴细胞[28]都不能证明来自终末分化细胞的细胞核具有全能性[34]。为了证明成熟 OSN 细胞核的全能性，我们将 P2SN1 和 P2SN2 衍生 ES 细胞的细胞核移植到去核卵母细胞中[29]。将得到的胚胎培养 24 小时并转移到假孕雌鼠中(表 1)。在对雌鼠进行剖宫产后，我们获得了来源于 P2SN1 和 P2SN2 细胞系的足月幼崽。这些幼仔具有大于常态的胎盘(P2SN1，0.35 g；P2SN2，0.40 g)，但没有显示出明显的解剖学异常或行为异常，具有生育能力并存活至成年期，与先前的克隆实验结果一致[29]。这些观察结果表明，终末分化的嗅觉神经元的细胞核可以重编程恢复全能性，指导胚胎和胚胎外细胞系的发育。

## 讨　论

我们提出疑问，有丝分裂后嗅觉神经元的细胞核是否可以重新进入细胞周期并进行重编程以指导小鼠的发育。ES 细胞系由 OSN 细胞核产生，该过程的频率与用分化的淋巴细胞获得 ES 细胞系的频率相似，后者可在生理条件下可诱导增殖[2]。因此，伴随着神经分化，导致细胞周期退出和不可逆有丝分裂停滞的机制不是由不可逆的表观遗传或遗传事件引起的，尽管这些事件会干扰核的全能性。

神经元的分化需要神经祖细胞在 G1 晚期的限制点之前退出细胞周期。这一行为受细胞周期调节因子和原神经基因的复杂平衡调控，原神经基因使细胞进入 G0 并阻止它们越过限制点[35]。虽然大多数进入 G0 的细胞可以在适当的刺激下重新进

appropriate stimulation, neurons normally undergo an irreversible mitotic arrest. Whatever mechanisms keep neurons in a post-mitotic state, our experiments demonstrate that they can be overcome in the environment of the egg.

The nervous system contains a diverse array of neural cell types and this diversity is reflected by distinct patterns of gene expression in different neurons. The regulation of gene expression by DNA rearrangements is rare but this mechanism has nonetheless been suggested to explain the diversity inherent in complex nervous systems[36]. DNA recombination events provide *Saccharomyces cerevisiae*, trypanosomes and lymphocytes with a mechanism to stochastically express one member of a set of genes that modulate cellular interactions with the environment. One attractive feature shared by gene rearrangements in trypanosomes and lymphocytes is that gene choice by recombination is a random event. Cells that undergo correct or successful rearrangements are then afforded a selective survival advantage. The stochastic rearrangement of one gene from a gene family and subsequent selection could also provide a mechanism to generate the vast diversity of neuronal cell types. In this manner, neurons with subtly different genotypes would exhibit the array of neuronal phenotypes required for a functioning nervous system.

The olfactory sensory epithelium provides a clear example of neuronal diversity, and it has been suggested that this diversity is generated by stochastic DNA rearrangement events[19,20]. However, efforts to examine the DNA of OSNs expressing a given receptor have been seriously hampered by the inability to obtain homogeneous populations of neurons or clonal cell lines in which each cell expresses the same receptor. We have addressed this problem by generating cloned ES cell lines and mice derived from the nuclei of olfactory sensory neurons expressing the P2 receptor. Analyses of the sequence and organization of the DNA surrounding the expressed P2 allele from cloned ES cells and mice revealed no evidence for either gene conversion or local transpositions at the P2 locus. These results concur with fluorescence *in situ* hybridization studies showing that gene conversion into an active locus is an unlikely mechanism for odorant receptor expression[37]. In addition, the pattern of receptor gene expression in the sensory epithelium of cloned mice was wild type; multiple odorant receptor genes are expressed without preference for the P2 allele expressed in the donor nucleus.

These data demonstrate that the mechanism responsible for the choice of a single odorant receptor gene does not involve irreversible changes in DNA. More dynamic, reversible recombination events might accompany odorant receptor gene choice, but this is unlikely given the current data. Our results, in combination with experiments showing that the expression of various odorant receptor transgenes is influenced by proximal sequence elements, are consistent with epigenetic models of odorant receptor choice[37-42]. In a broader context, the generation of fertile cloned mice that are anatomically and behaviourally indistinguishable from wild-type mice indicates that olfactory sensory neurons do not undergo other irreversible DNA rearrangements that would interfere with either the development or function of the nervous system during adult life.

入细胞周期，但神经元通常会发生不可逆的有丝分裂停滞。无论使神经元保持在有丝分裂后的状态的机制是什么，我们的实验证明神经元可以在卵细胞的环境中重新进入细胞周期。

神经系统包含多种神经细胞类型，这种多样性通过不同神经元中不同的基因表达模式反映出来。通过 DNA 重排调节基因表达是罕见的，但这种机制仍被建议用于解释复杂神经系统固有的多样性[36]。DNA 重组能够使酿酒酵母、锥虫和淋巴细胞随机表达调节细胞与环境相互作用的一组基因中的一个。锥虫和淋巴细胞中基因重排共有一个明显特征，即通过重组导致的基因选择是随机事件。之后，经历正确或成功重排的细胞具有选择性存活优势。来自基因家族的一个基因的随机重排和随后的选择也可以提供产生大量多样的神经元细胞类型的机制。以这种方式，具有微妙不同基因型的神经元将表现出功能性神经系统所需的多种神经元表型。

嗅觉上皮提供了一个神经元多样性的明显例子，并且已经表明这种多样性是由随机 DNA 重排事件造成的[19,20]。然而，由于无法获得其中每个细胞表达相同受体的均质神经元群或克隆细胞系，所以一直难以对表达给定受体的 OSN 的 DNA 进行检测。我们通过用表达 P2 受体的嗅觉神经元细胞核获得克隆的 ES 细胞系和小鼠来解决这个问题。对来自克隆的 ES 细胞和小鼠的表达的 P2 等位基因周围的 DNA 的序列和组构的分析显示，在 P2 基因座处没有发生基因转变或局部转座。这些结果与荧光原位杂交研究一致，表明基因转变为活性基因座不太可能是气味受体表达的机制[37]。此外，克隆小鼠感觉上皮细胞中受体基因表达的模式为野生型；表达多种气味受体基因而并不优先表达在供体细胞核中表达的 P2 等位基因。

这些数据表明，负责选择单一气味受体基因的机制不涉及 DNA 的不可逆变化。更活跃的、可逆的重组事件可能伴随气味受体基因选择，但鉴于目前的数据，这也不太可能。我们的结果与表明各种气味受体转基因的表达受近端序列元件影响的实验相结合，与气味受体选择的表观遗传模型一致[37-42]。在更广泛的背景下，能够产生在解剖学上和行为上与野生型小鼠无显著区别的可育克隆小鼠表明，嗅觉神经元不经历其他干扰神经系统发育或者成年期功能的不可逆的 DNA 重排。

## Methods

### Preparation of donor neurons

Olfactory epithelia were dissected in L15 medium (Gibco) at 4 °C, then chopped into small pieces and incubated with type IV collagenase at 1 μg $ml^{-1}$ at 37 °C for 15 min with occasional vigorous shaking. The collagenase digestion was stopped by addition of DMEM plus 10% FBS. Cells were spun down and resuspended in trituration medium (PBS plus 30% glucose plus 10% FBS plus penicillin/streptomycin) and triturated with several widths of pipette tips to produce single cell suspensions. Cells were pelleted and resuspended in L15/10% FBS before picking.

### ES cell lines and tetraploid embryo complementation

Production of ES cell lines from nuclear transfer embryos was exactly as described[28], and generation of mice by tetraploid embryo complementation was exactly as described[29].

### Generation of cloned embryos and mice

Production of cloned embryos by nuclear transfer was essentially as described[25,29] except that only $GFP^+$ cells from the two donor populations, as identified by epifluorescence, were picked and used for nuclear transfer.

### *In situ* hybridization and immunohistochemistry

Animals were killed by approved methods and tissues of interest were either fresh frozen in OCT or fixed for 2–12 h in 4% PFA/1 × PBS, then washed and equilibrated in 30% sucrose before freezing. Frozen sections (from 20–30 μm) were placed on slides and standard immunohistochemistry and digoxigenin-labelled probe *in situ* hybridization protocols were used[32]. We used the following antibodies: rabbit anti-GFP antibody (Molecular Probes), rabbit anti-Cre recombinase antibody (Novagen), goat anti-Ki67 antibody (Santa Cruz Biotechnology), rat anti-BrdU antibody (abcam) and rabbit anti-MASH-1 antibody (a gift of J. Johnson).

### BrdU labelling and visualization

Mice were injected intraperitoneally with a 5 mg $ml^{-1}$ solution of BrdU in PBS every 2 h for 12 h. Each mouse received BrdU at 100 μg $g^{-1}$ body weight for each injection. Two hours after the last injection mice were killed and tissues were fixed for 2 h in 4% PFA and PBS, washed and sucrose protected, and frozen. To permit simultaneous visualization of BrdU and GFP, sections were first stained with anti-GFP antibody then fixed in 4% PFA/PBS for 15 min, then washed, treated with

# 方　法

## 供体神经元的制备

将嗅上皮于 4℃ L15 培养基（Gibco）中裂解，然后切成小块并与 1μg/ml 的 IV 型胶原酶在 37℃下孵育 15 分钟，偶尔剧烈摇动。通过加入含有 10% FBS 的 DMEM 终止胶原酶消化。将细胞离心并重悬于研磨培养基（PBS 加 30% 葡萄糖加 10% FBS 加青霉素/链霉素）中，并用不同口径的移液管尖端研磨以获得单细胞悬浮液。收集前将细胞沉淀并重新悬浮于 L15/10% FBS 中。

## ES 细胞系和四倍体胚胎补偿法

来自核移植胚胎的 ES 细胞系的获得完全如之前文献描述[28]，且通过四倍体胚胎补偿获得小鼠的步骤也完全如之前文献描述[29]。

## 克隆胚胎和小鼠的获得

通过核移植产生克隆胚胎基本如文献描述[25,29]，唯一不同点是仅有来自两个供体群的 $GFP^+$ 细胞通过荧光挑选出来，并用于核移植。

## 原位杂交和免疫组织化学法

通过经批准的方法处死动物，目标组织放于 OCT 中尽快冷冻，或于 4% PFA/1×PBS 中固定 2~12 小时，然后洗涤并在冷冻前于 30% 蔗糖中平衡。将冷冻切片（20~30 μm 厚）置于载玻片上，并进行标准免疫组织染色和用地高辛标记的探针进行原位杂交[32]。我们使用以下抗体：兔抗 GFP 抗体（Molecular Probes），兔抗 Cre 重组酶抗体（Novagen），山羊抗 Ki67 抗体（Santa Cruz Biotechnology），大鼠抗 BrdU 抗体（abcam）和兔抗 MASH-1 抗体（来自 J.Johnson 的礼物）。

## BrdU 标记和可视化

每 2 小时向小鼠腹膜内注射 5 mg/ml 的 BrdU PBS 溶液，持续进行 12 小时。每次注射时，每只小鼠接受 100 μg/g 体重的 BrdU。在最后一次注射后 2 小时，处死小鼠并将组织在 4% PFA 和 PBS 中固定 2 小时，洗涤并用蔗糖保护，之后冷冻。为了同时观察 BrdU 和 GFP，首先用抗 GFP 抗体对切片染色，然后在 4% PFA/PBS 中固定 15 分钟，之后洗涤，用

4 M HCl/0.1% Triton X-100 for 10 min, washed and stained with rat anti-BrdU followed by the appropriate secondary antibodies.

## Gene targeting and generation of donor mouse lines

The OMP–IRES-Cre mouse line was generated as described previously except that the Cre recombinase gene sequence was inserted in place of the tTA sequence[32]. An IRES directing the translation of Cre recombinase was introduced into the 3′ untranslated region of the OMP locus by homologous recombination in ES cells. Transgenic mice were generated and crossed with a strain bearing the Z/EG reporter transgene. The *Rosa26*–LoxP-Stop-LoxP–GFP line was generated as described previously except that the GFP sequence replaced the CFP and YFP sequences in the *Rosa26* cassette[33].

## Southern blotting, PCR and RT–PCR

Southern blots and PCR screening of ES cells and tails used standard methods[43]. RT–PCR was performed on RNA isolated from mouse olfactory turbinates that had been frozen in Trizol (Invitrogen), and then treated as per the manufacturer's protocol. RNA was reverse transcribed using the Superscript II kit (Invitrogen) and complementary DNA was subjected to PCR using standard conditions.

(**428**, 44-49; 2004)

**Kevin Eggan[1*†], Kristin Baldwin[2*], Michael Tackett[1], Joseph Osborne[2†], Joseph Gogos[2], Andrew Chess[1], Richard Axel[2] & Rudolf Jaenisch[1]**

[1] Whitehead Institute for Biomedical Research and Department of Biology, Massachusetts Institute of Technology, 9 Cambridge Center, Cambridge, Massachusetts 02142, USA

[2] Department of Biochemistry and Molecular Biophysics, Howard Hughes Medical Institute, College of Physicians and Surgeons, Columbia University, 701 West 168th Street, New York, New York 10032, USA

[*] These authors contributed equally to this work

[†] Present addresses: Department of Physiology and Cellular Biophysics, College of Physicians and Surgeons, Columbia University, 630 West 168th Street, New York, New York 10032, USA (J.O.); Department of Molecular and Cellular Biology, Harvard University, 7 Divinity Avenue, Cambridge, Massachusetts 02138, USA (K.E.)

Received 13 November 2003; accepted 14 January 2004; doi:10.1038/nature02375.
Published online 15 February 2004.

References:

1. Rideout, W. M., Eggan, K. & Jaenisch, R. Nuclear cloning and epigenetic reprogramming of the genome. *Science* **293**, 1093-1098 (2001).
2. Hochedlinger, K. & Jaenisch, R. Nuclear transplantation: lessons from frogs and mice. *Curr. Opin. Cell Biol.* **14**, 741-748 (2002).
3. Gurdon, J. B. & Byrne, J. A. The first half-century of nuclear transplantation. *Proc. Natl Acad. Sci. USA* **100**, 8048-8052 (2003).
4. Osada, T., Kusakabe, H., Akutsu, H., Yagi, T. & Yanagimachi, R. Adult murine neurons: their chromatin and chromosome changes and failure to support embryonic development as revealed by nuclear transfer. *Cytogenet. Genome Res.* **97**, 7-12 (2002).
5. Yamazaki, Y. *et al.* Assessment of the developmental totipotency of neural cells in the cerebral cortex of mouse embryo by nuclear transfer. *Proc. Natl Acad. Sci. USA* **98**, 14022-14026 (2001).

4 M HCl/0.1% Triton X-100 处理 10 分钟，再次洗涤并用大鼠抗 BrdU 染色，然后用适当的二抗染色。

## 基因靶向和供体小鼠系的获得

除了插入 Cre 重组酶基因序列以代替 tTA 序列 [32]，OMP–IRES-Cre 小鼠系的获得方法如前所述。通过 ES 细胞中的同源重组，将指导 Cre 重组酶翻译的 IRES 引入 OMP 基因座的 3′ 非翻译区。获得转基因小鼠，并与携带 Z/EG 报告转基因的小鼠杂交。获得 *Rosa26*–LoxP-Stop-LoxP–GFP 系小鼠，除用 GFP 序列替换 *Rosa26* 盒中的 CFP 和 YFP 序列外其余步骤均已在文献中描述 [33]。

## Southern 印迹、PCR 和 RT–PCR

对 ES 细胞和小鼠尾部进行的 Southern 印迹和 PCR 筛选均使用标准方法 [43]。对从 Trizol (Invitrogen) 中冷冻的小鼠嗅鼻甲中分离的 RNA 进行 RT–PCR，然后按照试剂制造商的方案进行处理。使用 Superscript II 试剂盒 (Invitrogen) 逆转录 RNA，并使用标准条件对互补 DNA 进行 PCR。

（任奕 翻译；王宇 审稿）

6. Chien, Y. H., Gascoigne, N. R., Kavaler, J., Lee, N. E. & Davis, M. M. Somatic recombination in a murine T-cell receptor gene. *Nature* **309**, 322-326 (1984).

7. Hozumi, N. & Tonegawa, S. Evidence for somatic rearrangement of immunoglobulin genes coding for variable and constant regions. *Proc. Natl Acad. Sci. USA* **73**, 3628-3632 (1976).

8. Brack, C., Hirama, M., Lenhard-Schuller, R. & Tonegawa, S. A complete immunoglobulin gene is created by somatic recombination. *Cell* **15**, 1-14 (1978).

9. Gao, Y. *et al.* A critical role for DNA end-joining proteins in both lymphogenesis and neurogenesis. *Cell* **95**, 891-902 (1998).

10. Frank, K. M. *et al.* Late embryonic lethality and impaired V(D)J recombination in mice lacking DNA ligase IV. *Nature* **396**, 173-177 (1998).

11. Rehen, S. K. *et al.* Chromosomal variation in neurons of the developing and adult mammalian nervous system. *Proc. Natl Acad. Sci. USA* **98**, 13361-13366 (2001).

12. Kaushal, D. *et al.* Alteration of gene expression by chromosome loss in the postnatal mouse brain. *J. Neurosci.* **23**, 5599-5606 (2003).

13. Chun, J. & Schatz, D. G. Rearranging views on neurogenesis: neuronal death in the absence of DNA end-joining proteins. *Neuron* **22**, 7-10 (1999).

14. Chun, J. Selected comparison of immune and nervous system development. *Adv. Immunol.* 77, 297-322 (2001).

15. Yagi, T. Diversity of the cadherin-related neuronal receptor/protocadherin family and possible DNA rearrangement in the brain. *Genes Cells* **8**, 1-8 (2003).

16. Zhang, X. & Firestein, S. The olfactory receptor gene superfamily of the mouse. *Nature Neurosci.* **5**, 124-133 (2002).

17. Ressler, K. J., Sullivan, S. L. & Buck, L. B. A zonal organization of odorant receptor gene expression in the olfactory epithelium. *Cell* **73**, 597-609 (1993).

18. Vassar, R., Ngai, J. & Axel, R. Spatial segregation of odorant receptor expression in the mammalian olfactory epithelium. *Cell* **74**, 309-318 (1993).

19. Buck, L. & Axel, R. A novel multigene family may encode odorant receptors: a molecular basis for odor recognition. *Cell* **65**, 175-187 (1991).

20. Kratz, E., Dugas, J. C. & Ngai, J. Odorant receptor gene regulation: implications from genomic organization. *Trends Genet.* **18**, 29-34 (2002).

21. Goldstein, B. J. & Schwob, J. E. Analysis of the globose basal cell compartment in rat olfactory epithelium using GBC-1, a new monoclonal antibody against globose basal cells. *J. Neurosci.* **16**, 4005-4016 (1996).

22. Holbrook, E. H., Szumowski, K. E. & Schwob, J. E. An immunochemical, ultrastructural, and developmental characterization of the horizontal basal cells of rat olfactory epithelium. *J. Comp. Neurol.* **363**, 129-146 (1995).

23. Ohta, Y. & Ichimura, K. Proliferation markers, proliferating cell nuclear antigen, Ki67, 5-bromo-2′-deoxyuridine, and cyclin D1 in mouse olfactory epithelium. *Ann. Otol. Rhinol. Laryngol.* **109**, 1046-1048 (2000).

24. Gerdes, J. *et al.* Cell cycle analysis of a cell proliferation-associated human nuclear antigen defined by the monoclonal antibody Ki-67. *J. Immunol.* **133**, 1710-1715 (1984).

25. Wakayama, T., Perry, A. C., Zuccotti, M., Johnson, K. R. & Yanagimachi, R. Full-term development of mice from enucleated oocytes injected with cumulus cell nuclei. *Nature* **394**, 369-374 (1998).

26. Wakayama, T., Rodriguez, I., Perry, A. C., Yanagimachi, R. & Mombaerts, P. Mice cloned from embryonic stem cells. *Proc. Natl Acad. Sci. USA* **96**, 14984-14989 (1999).

27. Rideout, W. M. *et al.* Generation of mice from wild-type and targeted ES cells by nuclear cloning. *Nature Genet.* **24**, 109-110 (2000).

28. Hochedlinger, K. & Jaenisch, R. Monoclonal mice generated by nuclear transfer from mature B and T donor cells. *Nature* **415**, 1035-1038 (2002).

29. Eggan, K. *et al.* Hybrid vigor, fetal overgrowth, and viability of mice derived by nuclear cloning and tetraploid embryo complementation. *Proc. Natl Acad. Sci. USA* **98**, 6209-6214 (2001).

30. Nagy, A. *et al.* Embryonic stem cells alone are able to support fetal development in the mouse. *Development* **110**, 815-821 (1990).

31. Wang, F., Nemes, A., Mendelsohn, M. & Axel, R. Odorant receptors govern the formation of a precise topographic map. *Cell* **93**, 47-60 (1998).

32. Gogos, J. A., Osborne, J., Nemes, A., Mendelsohn, M. & Axel, R. Genetic ablation and restoration of the olfactory topographic map. *Cell* **103**, 609-620 (2000).

33. Srinivas, S. *et al.* Cre reporter strains produced by targeted insertion of EYFP and ECFP into the ROSA26 locus. *BMC Dev. Biol.* **1**, 4 (2001).

34. Rossant, J. A monoclonal mouse? *Nature* **415**, 967-969 (2002).

35. Ohnuma, S. & Harris, W. A. Neurogenesis and the cell cycle. *Neuron* **40**, 199-208 (2003).

36. Edelman, G. M. *Neural Darwinism* (Basic Books, New York, NY, 1987).

37. Ishii, T. *et al.* Monoallelic expression of the odourant receptor gene and axonal projection of olfactory sensory neurones. *Genes Cells* **6**, 71-78 (2001).

38. Serizawa, S. *et al.* Negative feedback regulation ensures the one receptor-one olfactory neuron rule in mouse. *Science* **302**, 2088-2094 (2003).

39. Serizawa, S. *et al.* Mutually exclusive expression of odorant receptor transgenes. *Nature Neurosci.* **3**, 687-693 (2000).

40. Ebrahimi, F. A., Edmondson, J., Rothstein, R. & Chess, A. YAC transgene-mediated olfactory receptor gene choice. *Dev. Dyn.* **217**, 225-231 (2000).

41. Qasba, P. & Reed, R. R. Tissue and zonal-specific expression of an olfactory receptor transgene. *J. Neurosci.* **18**, 227-236 (1998).

42. Vassalli, A., Rothman, A., Feinstein, P., Zapotocky, M. & Mombaerts, P. Minigenes impart odorant receptor-specific axon guidance in the olfactory bulb. *Neuron* **35**, 681-696 (2002).

43. Sambrook, J., Fritsch, E. F. & Maniatis, T. *Molecular Cloning: a Laboratory Manual* 2nd edn (Cold Spring Harbor Laboratory Press, Cold Spring Harbor, NY, 1989).

**Supplementary Information** accompanies the paper on www.nature.com/nature.

**Acknowledgements.** We thank L. Moring, A. Nemes, M. Mendelsohn, J. Loring, J. Dausman, A. Meissner and K. Hochedlinger for assistance during the course of these experiments; T. Cutforth, J. de Nooij, T. Jessell and J. Johnson for sharing reagents necessary for our experiments; and members of the Jaenisch, Axel and Chess laboratories for discussion and assistance during the course of these experiments, especially F. A. Ebrahimi, M. Rios, W. M. R. Rideout, L. Jackson-Grusby and B. Shykind. This research was sponsored by NIH grants to R.J., R.A., A.C. and M.T.. K.B. is an Associate and R.A. is an Investigator of the Howard Hughes Medical

Institute. K.E. is a Junior Fellow in the Harvard Society of Fellows.

**Competing interests statement**. The authors declare that they have no competing financial interests.

Correspondence and requests for materials should be addressed to R.J. (jaenisch@wi.mit.edu) or A.C. (achess@wi.mit.edu).

# A Precision Measurement of the Mass of the Top Quark

DØ Collaboration*

## Editor's Note

**Although the standard model of particle physics is extremely successful, it does not in itself explain the masses of subatomic particles. Instead, they are hypothesized to arise from the interactions of particles with the so-called Higgs boson. Here an international team called the D0 collaboration, working at the Tevatron supercollider at Fermilab in Illinois, report a high-precision measurement of the mass of the top quark, which helps to constrain the range of allowable masses for the Higgs boson. The results suggest that it is more massive than was previously thought, so that more energy will be required to make it. The Higgs boson was finally observed in the Large Hadron Collider at CERN in Geneva in 2012. Its mass is lighter than was expected.**

---

The standard model of particle physics contains parameters—such as particle masses—whose origins are still unknown and which cannot be predicted, but whose values are constrained through their interactions. In particular, the masses of the top quark ($M_t$) and $W$ boson ($M_W$)[1] constrain the mass of the long-hypothesized, but thus far not observed, Higgs boson. A precise measurement of $M_t$ can therefore indicate where to look for the Higgs, and indeed whether the hypothesis of a standard model Higgs is consistent with experimental data. As top quarks are produced in pairs and decay in only about $10^{-24}$ s into various final states, reconstructing their masses from their decay products is very challenging. Here we report a technique that extracts more information from each top-quark event and yields a greatly improved precision (of $\pm 5.3$ GeV/$c^2$) when compared to previous measurements[2]. When our new result is combined with our published measurement in a complementary decay mode[3] and with the only other measurements available[2], the new world average for $M_t$ becomes[4] $178.0 \pm 4.3$ GeV/$c^2$. As a result, the most likely Higgs mass increases from the experimentally excluded[5] value[6] of 96 to 117 GeV/$c^2$, which is beyond current experimental sensitivity. The upper limit on the Higgs mass at the 95% confidence level is raised from 219 to 251 GeV/$c^2$.

---

THE discovery of the top quark in 1995 served as one of the major confirmations of the validity of the standard model (SM)[7,8]. Of its many parameters, the mass of the top quark, in particular, reflects some of the most crucial aspects of the SM. This is because, in principle, the top quark is point-like and should be massless; yet, through

---

* A full list of participants and affiliations is given in the original paper.

# 顶夸克质量的精确测量

DØ 项目合作组 *

## 编者按

尽管粒子物理的标准模型是非常成功的，但模型本身解释不了亚原子粒子的质量问题，亚原子粒子的质量被假定来源于粒子与所谓的希格斯玻色子之间的相互作用。本文中，一个名为 D0 合作组的国际团队通过伊利诺伊州费米实验室的万亿电子伏特加速器 (Tevatron) 得到了顶夸克质量的精确测定，由此有助于限定希格斯玻色子允许质量的范围。结果表明，希格斯玻色子的质量比之前预期的要大，因而需要更大的能量才能产生希格斯玻色子。2012 年，位于瑞士日内瓦的欧洲核子研究组织 (CERN) 的大型强子对撞机终于观测到希格斯玻色子，但其质量比预期的要轻。

---

粒子物理标准模型包含了像粒子质量这一类的参数，这类参数起源不清，无法预言，但是其量值可以通过它们之间的相互作用来限定。特别是一直被预言存在但是从未被观测到的希格斯玻色子的质量，可以由顶夸克的质量 ($M_t$) 和 $W$ 玻色子的质量 ($M_W$) [1] 进行限定。因而，$M_t$ 的精确测量可以更好地指导我们寻找希格斯子，甚至可验证标准模型预言的希格斯子是否与实验数据相符。由于顶夸克成对产生，并在 $10^{-24}$ 秒后就衰变为各种末态，通过衰变的产物来重建其质量是非常困难的。本文提出的技术可以从每一个顶夸克对事例中获取更多信息并得到比之前测量值 [2] 更高的精度 ($\pm 5.3\ \text{GeV}/c^2$)。我们最新的结果结合之前发表的互补衰变模式 (全轻衰变模式) 的测量值 [3] 和目前其余已知的结果 [2]，得到的新的全球 $M_t$ 平均值是 $178.0 \pm 4.3\ \text{GeV}/c^2$ [4]。由此，最有可能的希格斯子质量的值从已经被实验 [5,6] 排除的 $96\ \text{GeV}/c^2$ 提高到 $117\ \text{GeV}/c^2$，这比当前的实验灵敏度要高。在 95% 置信度水平，希格斯子的质量上限从 $219\ \text{GeV}/c^2$ 提高到 $251\ \text{GeV}/c^2$。

---

1995 年顶夸克的发现是对标准模型 (SM) 的重要支持之一 [7,8]。在众多参数中，顶夸克的质量尤其反映了标准模型的一些根本特征。这是因为本质上顶夸克是一个无质量的点粒子，然而通过与预言的希格斯场的相互作用，顶夸克获得的物理质量大致相当于一个金原子核质量。这个质量如此之大，使得顶夸克 (和 $W$ 玻色子一起)

---

* 合作组的参与者及所属单位的名单请参见原文。

its interactions with the hypothesized Higgs field, the physical mass of the top quark appears to be about the mass of a gold nucleus. Because it is so heavy, the top quark (along with the $W$ boson) provides an unusually sensitive tool for investigating the Higgs field. $M_W$ is known to a precision of 0.05%, while the uncertainty on $M_t$ is at the 3% level[1]. Improvements in both measurements are required to restrict further the allowed range of mass for the Higgs; however, given the large uncertainty in $M_t$, an improvement in its precision is particularly important. As has been pointed out recently[9,10], a potential problem for the SM is that, on the basis of the currently accepted value for $M_t$, the most likely value of the Higgs mass[6] lies in a range that has already been excluded by experiment[5]. Precise knowledge of the Higgs mass is crucial for our understanding of the SM and any possible new physics beyond it. For example, in a large class of supersymmetric models (theoretically preferred solutions to the deficiencies of the SM), the Higgs mass has to be less than about 135 GeV/$c^2$. Although, unlike the SM, supersymmetry predicts more than one Higgs boson, the properties of the lightest one are expected to be essentially the same as those for the SM Higgs boson. Thus, if the SM-like Higgs is heavier than about 135 GeV/$c^2$, it would disfavour a large class of supersymmetric models. In addition, some of the current limits on supersymmetric particles from LEP[11] are extremely sensitive to $M_t$. In fact, for $M_t$ greater than 179 GeV/$c^2$, the bounds on one of the major supersymmetry parameters, tan$\beta$, which relates the properties of the SM-like Higgs boson and its heavier partners, would disappear completely[12]. Hence, in addition to the impact on searches for the Higgs boson, other important consequences call for improved precision on $M_t$, and this goal is the main subject of this paper.

The DØ experiment at the Fermilab Tevatron has studied a sample of $t\bar{t}$ events produced in proton–antiproton ($p\bar{p}$) interactions[13]. The total energy of 1.8 TeV released in a head-on collision of a 900-GeV $p$ and a 900-GeV $\bar{p}$ is almost as large as the rest energy of ten gold nuclei. Each top (antitop) quark decays almost immediately into a bottom $b(\bar{b})$ quark and a $W^+$ ($W^-$) boson, and we have reexamined those events in which one of the $W$ bosons decays into a charged lepton (electron or muon) and a neutrino, and the other $W$ into a quark and an antiquark (see Fig. 1). These events and their selection criteria are identical to those used to extract the mass of the top quark in our previous publication, and correspond to an integrated luminosity of 125 events per pb. (That is, given the production cross-section of the $t\bar{t}$ in $p\bar{p}$ collisions at 1.8 TeV of 5.7 pb, as measured by DØ[14], these data correspond to approximately 700 produced $t\bar{t}$ pairs, a fraction of which is fully detected in various possible decay modes. Approximately 30% of these correspond to the lepton + jets topology categorized in Fig. 2, where "jet" refers to products of the fragmentation of a quark into a collimated group of particles that are emitted along the quark's original direction.) The main background processes correspond to multijet production (20%), where one of the jets is reconstructed incorrectly as a lepton, and the $W$ + jets production with leptonic $W$ decays (80%), which has the same topology as the $t\bar{t}$ signal.

成为一个超常灵敏的研究希格斯场的手段。$M_W$ 的测量值仅有 0.05% 的不确定度，而 $M_t$ 的测量值有 3% 的不确定度 [1]。要进一步缩小希格斯子质量的可能范围，需要改进这两个量值的精确度。但因为 $M_t$ 的测量值不确定度大，所以对其精度的改进就格外地重要。正如最近 [9,10] 已指出的，标准模型的一个潜在问题在于，根据当前公认的 $M_t$ 测量值，最有可能的希格斯子质量 [6] 落在已被实验排除的区间 [5]。准确地认识希格斯子的质量，对于我们理解标准模型或者标准模型以外的其他可能的新物理理论，都是非常重要的。比如说，在一大类超对称模型中（解决标准模型缺陷的首选理论），希格斯子质量必须小于约 135 GeV/$c^2$。和标准模型不同的是，超对称性预言的希格斯子不止一个，但是其中最轻的那一个的性质基本上与标准模型中的希格斯玻色子相同。因此，如果类标准模型希格斯玻色子的质量超过 135 GeV/$c^2$，这一大类超对称模型就不再被看好。另外，从大型正负电子对撞机（LEP）[11] 得到的一些超对称粒子当前的限制对 $M_t$ 非常敏感。实际上，当 $M_t$ 大于 179 GeV/$c^2$ 时，一个联系类标准模型希格斯玻色子及其更重的伴子的主要的超对称参数（$\tan\beta$）的限制将完全消失 [12]。除了对希格斯玻色子的寻找有影响，改进 $M_t$ 的测量精度还有其他重要意义，这个目标也是本文的主要研究问题。

DØ 项目在费米实验室的万亿电子伏特加速器（Tevatron）上研究了用质子–反质子（$p\bar{p}$）对撞产生的顶夸克–反顶夸克（$t\bar{t}$）事例 [13]。900 GeV 的质子（$p$）和 900 GeV 的反质子（$\bar{p}$）对撞释放了 1.8 TeV 的总能量，几乎相当于十个金原子核的静止质量。每个顶（反顶）夸克几乎立即衰变为一个底（反底）夸克 $b(\bar{b})$ 和一个 $W^+(W^-)$ 玻色子。我们重新检查了那些一个 $W$ 玻色子衰变为一个带电轻子（电子或者 μ 子）和一个中微子，而另一个 $W$ 玻色子衰变为一个夸克和一个反夸克的事例（图 1）。挑选这些事例所用选择条件及实验数据样本与我们之前发表的获取顶夸克质量的工作中所用的选择条件和数据样本相同，即对应每 pb 有 125 个事例积分亮度的实验数据。（就是说，与 DØ 实验中测得的结果 [14] 一致，假定 $p\bar{p}$ 对撞中的 $t\bar{t}$ 在能量为 1.8 TeV 时产生截面为 5.7 皮靶，这些数据大概对应于 700 个 $t\bar{t}$ 对，其中只有部分通过各种可能的衰变模式检测到了。大概有 30% $t\bar{t}$ 衰变模式对应于图 2 归类的轻子 + 喷注拓扑，其中“喷注”这个词表示夸克分裂产生的成群粒子沿夸克原来的方向平行前进）。主要的背景过程包括多喷注产生（20%，其中的一个喷注被错误地重建为轻子）和与 $t\bar{t}$ 信号有相同拓扑结构的轻子型 $W$ 衰变的 $W$+ 喷注产生（80%）。

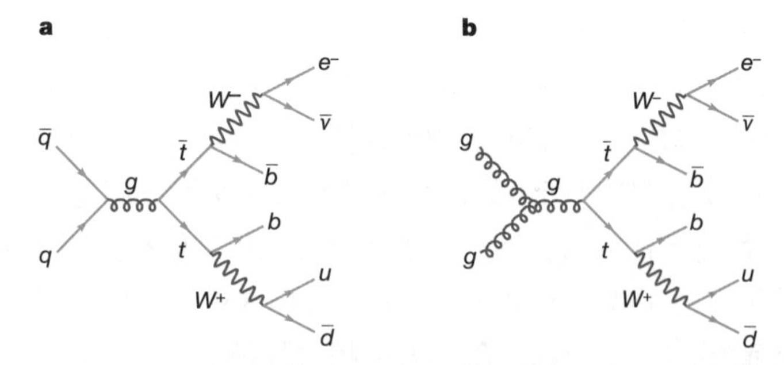


Fig. 1. Feynman diagrams for $t\bar{t}$ production in $p\bar{p}$ collisions, with subsequent decays into an electron, neutrino, and quarks. Quark–antiquark production (**a**) is dominant, but gluon fusion (**b**) contributes ~10% to the cross-section. This particular final state ($e\nu u\bar{d}b\bar{b}$) is one of the channels used in the analysis.

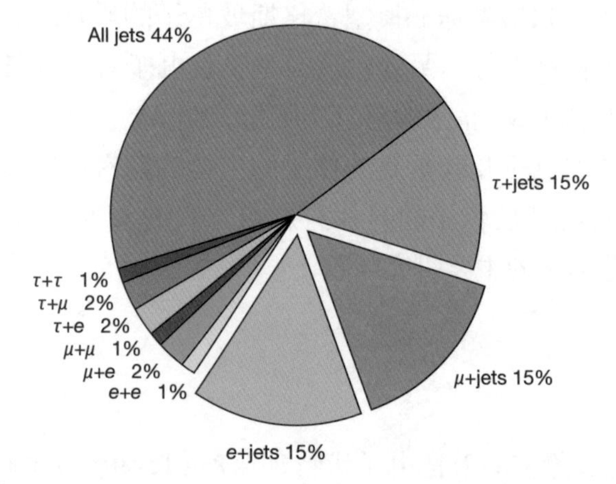


Fig. 2. Relative importance of various $t\bar{t}$ decay modes. The "lepton+jets" channel used in this analysis corresponds to the two offset slices of the pie-chart and amounts to 30% of all the $t\bar{t}$ decays

The previous DØ measurement of $M_t$ in this lepton + jets channel is $M_t = 173.3 \pm 5.6$ (stat) $\pm 5.5$ (syst) GeV/$c^2$, and is based on 91 candidate events. Information pertaining to the older analysis and the DØ detector can be found elsewhere[13,15].

The new method of $M_t$ measurement is similar to one suggested previously (ref. 16 and references therein, and ref. 17) for $t\bar{t}$ dilepton decay channels (where both $W$ bosons decay leptonically), and used in previous mass analyses of dilepton events[3], and akin to an approach suggested for the measurement of the mass of the $W$ boson at LEP[18-20]. The critical differences from previous analyses in the lepton + jets decay channel lie in: (1) the assignment of more weight to events that are well measured or more likely to correspond to $t\bar{t}$ signal, and (2) the handling of the combinations of final-state objects (lepton, jets and imbalance in transverse momentum, the latter being a signature for an undetected neutrino) and their identification with top-quark decay products in an event (such as from ambiguity in choosing jets that correspond to $b$ or $\bar{b}$ quarks from the decays of the $t$ and $\bar{t}$ quarks). Also, because leading-order matrix elements were used to calculate the event weights, only events with exactly four jets are kept in this analysis, resulting in a candidate sample of 71 events. Although we are left with fewer events, the new method for extracting $M_t$ provides

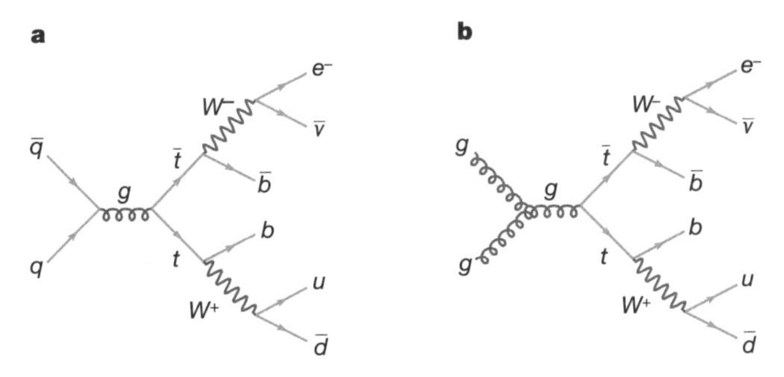


图 1. $p\bar{p}$（正反质子）对撞中 $t\bar{t}$（正反顶夸克）产生的费曼图。产生的顶夸克随即衰变为一个电子、中微子和数个夸克。(a) 图所示的夸克–反夸克产生过程是主要的，(b) 图中的胶子聚变过程贡献大约 10% 的截面。这个末态（$e\nu u\bar{d}b\bar{b}$）是分析中用到的通道之一。

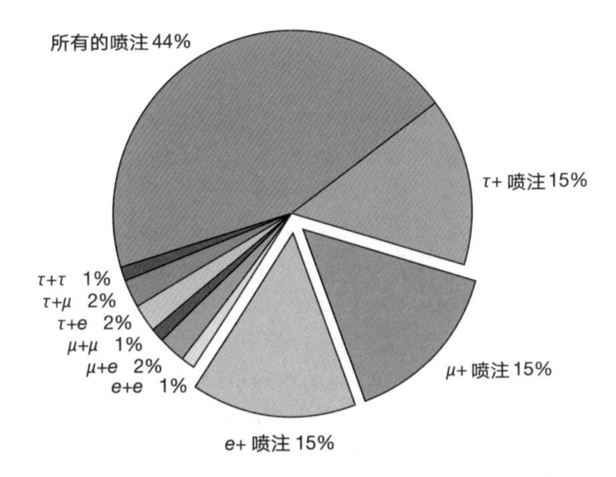


图 2. 各个 $t\bar{t}$ 衰变模式的比例。本分析中用到的“轻子 + 喷注”衰变道对应于扇面图中分割出的两片，总共占 $t\bar{t}$ 衰变的 30%。

以前在通过轻子 + 喷注衰变道测量 $M_t$ 值的 DØ 实验中，根据 91 个候选事例得到的量值是 $M_t = 173.3 \pm 5.6$（统计不确定度）$\pm 5.5$（系统不确定度）GeV/$c^2$。以前的分析以及 DØ 探测器的相关信息可以在相关文献 [13,15] 中找到。

测量 $M_t$ 的新方法类似于之前文献（参考文献 16 以及其中引用的文献，参考文献 17）提出的关于 $t\bar{t}$ 双轻子衰变道（其中两个 $W$ 玻色子的衰变都是轻子型的）中的测量方法，该方法曾被应用于之前双轻子事例的质量分析 [3]，与之前建议的在 LEP 上测量 $W$ 玻色子质量的方法也是相似的 [18-20]。轻子 + 喷注衰变道的分析与之前的分析关键的不同之处在于：(1) 测量得比较精确的事例或者更可能对应于 $t\bar{t}$ 信号的事例这点被赋予了更大的权重，(2) 对末态产物（轻子、喷注和不平衡的横向动量，最后一项表明了一个未探测到的中微子的存在）的各种组合的处理以及在事例中用顶夸克衰变产物对它们的确证（比如解决喷注是对应于 $t$ 夸克衰变中的 $b$ 夸克还是 $\bar{t}$ 夸克衰变中的 $\bar{b}$ 夸克的不确定性问题）。另外，由于使用领头阶矩阵元计算事例权重，分析中只保留了恰好有 4 个喷注的事例，这样得到的候选样本有 71 个事例。虽然事例数

substantial improvement in both statistical and systematic uncertainties.

We calculate as a function of $M_t$ the differential probability that the measured variables in any event correspond to signal. The maximum of the product of these individual event probabilities provides the best estimate of $M_t$ in the data sample. The impact of biases from imperfections in the detector and event-reconstruction algorithms is taken into account in two ways. Geometric acceptance, trigger efficiencies, event selection, and so on enter through a multiplicative acceptance function that is independent of $M_t$. Because the angular directions of all the objects in the event, as well as the electron momentum, are measured with high precision, their measured values are used directly in the calculation of the probability that any event corresponds to $t\bar{t}$ or background production. The known momentum resolution is used to account for uncertainties in measurements of jet energies and muon momenta.

As in the previous analysis[13], momentum conservation in $\gamma$+jet events is used to check that the energies of jets in the experiment agree with Monte Carlo (MC) simulations. This calibration has an uncertainty $\delta E = (0.025E + 0.5\ \text{GeV})$. Consequently, all jet energies in our sample are rescaled by $\pm\delta E$, the analysis redone, and half of the difference in the two rescaled results for $M_t$ ($\delta M_t = 3.3\ \text{GeV}/c^2$) is taken as a systematic uncertainty from this source. All other contributions to systematic uncertainty: MC modelling of signal ($\delta M_t = 1.1\ \text{GeV}/c^2$) and background ($\delta M_t = 1.0\ \text{GeV}/c^2$), effect of calorimeter noise and event pile-up ($\delta M_t = 1.3\ \text{GeV}/c^2$), and other corrections from $M_t$ extraction ($\delta M_t = 0.6\ \text{GeV}/c^2$) are much smaller, and discussed in detail elsewhere[21,22]. It should be noted that the new mass measurement method provides a significant (about 40%, from $\pm 5.5$ to $\pm 3.9\ \text{GeV}/c^2$) reduction in systematic uncertainty, which is ultimately dominated by the measurement of jet energies. For details on the new analysis, see the Methods.

The final result is $M_t = 180.1 \pm 3.6\ (\text{stat}) \pm 3.9\ (\text{syst})\ \text{GeV}/c^2$. The improvement in statistical uncertainty over our previous measurement is equivalent to collecting a factor of 2.4 as much data. Combining the statistical and systematic uncertainties in quadrature, we obtain $M_t = 180.1 \pm 5.3\ \text{GeV}/c^2$, which is consistent with our previous measurement in the same channel (at about 1.4 standard deviations), and has a precision comparable to all previous $M_t$ measurements combined[1].

The new measurement can be combined with that obtained for the dilepton sample that was also collected at DØ during run I (ref. 3) ($M_t = 168.4 \pm 12.3\ (\text{stat}) \pm 3.6\ (\text{syst})\ \text{GeV}/c^2$), to yield the new DØ average for the mass of the top quark:

$$M_t = 179.0 \pm 5.1\ \text{GeV}/c^2 \tag{1}$$

Combining this with measurements from the CDF experiment[2] provides a new "world average" (based on all measurements available) for the mass of the top quark[4]:

$$M_t = 178.0 \pm 4.3\ \text{GeV}/c^2 \tag{2}$$

少了，但是提取 $M_t$ 的新方法在统计不确定度和系统不确定度上都有实质性的改进。

我们计算了以 $M_t$ 及事例观测量作为变量的函数，并将该函数对应为信号的微分概率。这些独立事例概率的最大乘积提供了数据样本中 $M_t$ 的最佳估计值。我们从两个方面考虑了探测器和事例重建算法的不完美引起的偏差影响。几何接收率、触发效率、事例选择等被包含在一个不依赖 $M_t$ 的多元接收函数中。由于事例中所有对象的角度方向和电子动量的测量精度都很高，它们的测量值被直接用来计算任何对应于 $t\bar{t}$ 事例或者背景事例的产生概率。使用已知的动量分辨率来解释测量喷注能量和 μ 子动量时的不确定度。

和以前的分析 [13] 一样，γ+ 喷注事例中的动量守恒被用来检查喷注的能量是否与蒙特卡罗（MC）模拟结果相符合。这个校准的不确定度 $\delta E = (0.025E + 0.5\ \text{GeV})$。因此，我们的样本中所有喷注能量都以 $\pm\delta E$ 重新标度，两次重新标度和分析后的 $M_t$（$\delta M_t = 3.3\ \text{GeV}/c^2$）结果之差的一半被作为喷注能量测量不确定度引起的系统性不确定度。对系统不确定度有贡献的其他因素包括：信号和背景的蒙特卡罗建模（$\delta M_t = 1.1\ \text{GeV}/c^2$ 和 $\delta M_t = 1.0\ \text{GeV}/c^2$），量能器噪声和事例堆积的影响（$\delta M_t = 1.3\ \text{GeV}/c^2$）以及其他源于 $M_t$ 提取的修正（$\delta M_t = 0.6\ \text{GeV}/c^2$）。以上这些因素的贡献要小得多，有另文 [21,22] 具体讨论。这里需要指出，新的质量测量方法明显减小了主要由测量喷注能量导致的系统不确定度（减小了约 40%，从 $\pm 5.5\ \text{GeV}/c^2$ 减到 $\pm 3.9\ \text{GeV}/c^2$）。关于新分析方法的细节，请参看本文后面的**方法**。

最后的结果是 $M_t = 180.1 \pm 3.6$（统计误差）$\pm 3.9$（系统误差）$\text{GeV}/c^2$。和之前的测量结果比较，新方法在统计不确定度上的改进相当于将数据量增加至原来的 2.4 倍。将统计和系统不确定度求均方根，我们得到 $M_t = 180.1 \pm 5.3\ \text{GeV}/c^2$，这与我们之前在相同衰变道的测量结果相符（在大约 1.4 个标准差偏离下），与之前所有的 $M_t$ 测量的联合精度相当 [1]。

将新的测量结果与通过第 I 轮 DØ 实验（参考文献 3）中搜集到的双轻子样本得到的结果（$M_t = 168.4 \pm 12.3$（统计误差）$\pm 3.6$（系统误差）$\text{GeV}/c^2$）相结合，得出顶夸克质量的新的 DØ 平均：

$$M_t = 179.0 \pm 5.1\ \text{GeV}/c^2 \tag{1}$$

把这一结果和 CDF 实验的测量结果 [2] 进一步结合起来，得到以我们最近的结果为主导的顶夸克质量新的“全球平均”（基于目前所有已知的测量结果）[4]：

$$M_t = 178.0 \pm 4.3\ \text{GeV}/c^2 \tag{2}$$

dominated by our new measurement. This new world average shifts the best-fit value of the expected Higgs mass from 96 GeV/$c^2$ to 117 GeV/$c^2$ (see Fig. 3), which is now outside the experimentally excluded region, yet accessible in the current run of the Tevatron and at future runs at the Large Hadron Collider (LHC), at present under construction at CERN. (The upper limit on the Higgs mass at the 95% confidence level changes from 219 GeV/$c^2$ to 251 GeV/$c^2$.) Figure 3 shows the effect of using only the new DØ top mass for fits to the Higgs mass, and indicates a best value of 123 GeV/$c^2$ and the upper limit of 277 GeV/$c^2$ at the 95% confidence level. It should be noted that the horizontal scale in Fig. 3 is logarithmic, and the limits on the Higgs boson mass are therefore asymmetric.

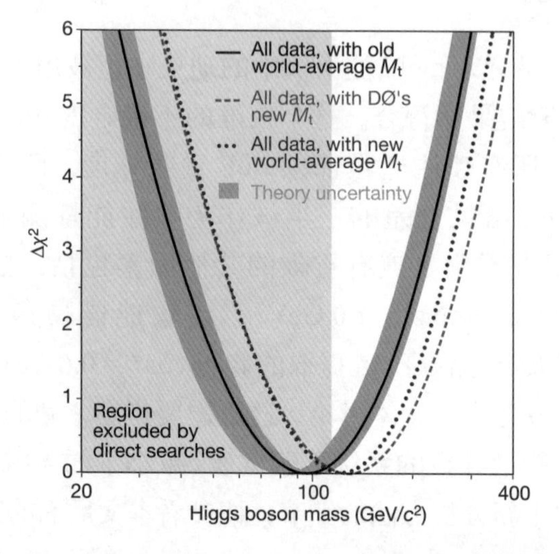


Fig. 3. Current experimental constraints on the mass of the Higgs boson. The $\chi^2$ for a global fit to electroweak data[6] is shown as a function of the Higgs mass. The solid line corresponds to the result for the previous world-averaged $M_t = 174.3 \pm 5.1$ GeV/$c^2$, with the blue band indicating the impact of theoretical uncertainty. The dotted line shows the result for the new world-average $M_t$ of $178.0 \pm 4.3$ GeV/$c^2$, whereas the dashed line corresponds to using only the new DØ average of $179.0 \pm 5.1$ GeV/$c^2$. The yellow-shaded area on the left indicates the region of Higgs masses excluded by experiment (> 114.4 GeV/$c^2$ at the 95% confidence level[5]). The improved $M_t$ measurement shifts the most likely value of the Higgs mass above the experimentally excluded range.

The new method is already being applied to data being collected by the CDF and DØ experiments at the new run of the Fermilab Tevatron and should provide even higher precision on the determination of $M_t$, equivalent to more than a doubling of the data sample, relative to using the conventional method. An ultimate precision of about 2 GeV/$c^2$ on the mass of the top quark is expected to be reached in several years of Tevatron operation. Further improvement may eventually come from the LHC.

## Methods

The probability density as a function of $M_t$ can be written as a convolution of the calculable cross-

这个新的全球平均将预言的希格斯子质量的最佳拟合值从 96 GeV/$c^2$ 提高到 117 GeV/$c^2$(见图 3)，现在这个值已经落在了实验排除的区间之外，但可以在 Tevatron 上目前正运行的实验中以及 CERN 目前正在建设的大型强子对撞机(LHC)上将开展的实验中达到。(在 95% 置信度水平下，希格斯子质量的上限值从 219 GeV/$c^2$ 变为 251 GeV/$c^2$)。图 3 显示如果只采用最新的 DØ 顶夸克质量来拟合希格斯子质量，获得的最佳的结果为 123 GeV/$c^2$，在 95% 置信度水平下最佳上限值为 277 GeV/$c^2$。注意图 3 中水平坐标是对数尺度，因而希格斯玻色子质量的上下限值是不对称的。

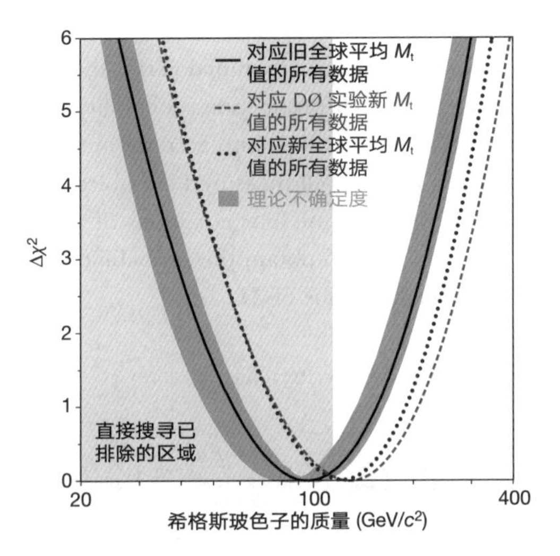


图 3. 当前实验得出的对希格斯玻色子质量的限定。图中显示电弱数据[6]全局拟合的 $\chi^2$ 值随希格斯子质量的变化。实线对应于之前的全球平均 $M_t$ = 174.3 ± 5.1 GeV/$c^2$ 的结果，蓝色带状区域表示理论不确定度的影响。点线表示新的全球平均 $M_t$ = 178.0 ± 4.3 GeV/$c^2$ 的结果，短横线对应只采用 DØ 实验新平均值 179.0 ± 5.1 GeV/$c^2$ 的结果。左边黄色阴影区域表示实验排除的希格斯质量范围( > 114.4 GeV/$c^2$，95% 置信水平下[5])。改进后的 $M_t$ 测量值使得最有可能的希格斯子质量移至实验排除的区间之外。

新方法已经被用来处理费米实验室 Tevatron 上新一轮 CDF 实验和 DØ 实验中所收集到的数据，以期得到更精确的 $M_t$ 值，以达到相当于常规方法两倍多数据样本的精度。可以预期经过未来几年在 Tevatron 上的实验，顶夸克质量的精度最终能够达到 2 GeV/$c^2$。而更进一步的精度提升最终将有赖于 LHC 的投入使用。

## 方　法

作为 $M_t$ 函数的概率密度可以表示为可计算的截面与所有来自测量分辨率的效应的

section and any effects from measurement resolution

$$P(x, M_t) = \frac{1}{\sigma(M_t)} \int d\sigma(y, M_t) dq_1 dq_2 f(q_1) f(q_2) W(y, x) \tag{3}$$

where $W(y, x)$, our general transfer function, is the normalized probability for the measured set of variables $x$ to arise from a set of nascent (partonic) variables $y$, $d\sigma(y, M_t)$ is the partonic theoretical differential cross-section, $f(q)$ are parton distribution functions that reflect the probability of finding any specific interacting quark (antiquark) with momentum $q$ within the proton (antiproton), and $\sigma(M_t)$ is the total cross-section for producing $t\bar{t}$. The integral in equation (3) sums over all possible parton states, leading to what is observed in the detector.

The acceptance of the detector is given in terms of a function $A(x)$ that relates the probability $P_m(x, M_t)$ of measuring the observed variables $x$ to their production probability $P(x, M_t)$: $P_m(x, M_t) = A(x)P(x, M_t)$. Effects from energy resolution, and so on are taken into account in the transfer function $W(y, x)$. The integrations in equation (3) over the eleven well-measured variables (three components of charged-lepton momentum and eight jet angles) and the four equations of energy-momentum conservation leave five integrals that must be performed to obtain the probability that any event represents $t\bar{t}$ (or background) production for some specified value of $M_t$.

The probability for a $t\bar{t}$ interpretation can be written as:

$$P_{t\bar{t}} = \frac{1}{12\sigma_{t\bar{t}}} \int d^5\Omega \sum_{\text{perm.},\nu} |M_{t\bar{t}}|^2 \frac{f(q_1)f(q_2)}{|q_1||q_2|} \Phi_6 W_{\text{jets}}(E_{\text{part}}, E_{\text{jet}})$$

where $\Omega$ represent a set of five integration variables, $M_{t\bar{t}}$ is the leading-order matrix element for $t\bar{t}$ production[23,24], $f(q_1)$ and $f(q_2)$ are the CTEQ4M parton distribution functions for the incident quarks[25], $\Phi_6$ is the phase-space factor for the six-object final state, and the sum is over all 12 permutations of the jets and all possible neutrino solutions. $W_{\text{jets}}(E_{\text{part}}, E_{\text{jet}})$ corresponds to a function that maps parton-level energies $E_{\text{part}}$ to energies measured in the detector $E_{\text{jet}}$ and is based on MC studies. A similar expression, using a matrix element for $W$+jets production (the dominant background source) that is independent of $M_t$, is used to calculate the probability for a background interpretation, $P_{\text{bkg}}$.

Studies of samples of HERWIG (ref. 26; we used version 5.1) MC events indicate that the new method is capable of providing almost a factor-of-two reduction in the statistical uncertainty on the extracted $M_t$. These studies also reveal that there is a systematic shift in the extracted $M_t$ that depends on the amount of background there is in the data. To minimize this effect, a selection is introduced, based on the probability that an event represents background. The specific value of the $P_{\text{bkg}}$ cut-off is based on MC studies carried out before applying the method to data, and, for $M_t = 175$ GeV/$c^2$, retains 71% of the signal and 30% of the background. A total of 22 data events out of our 71 candidates pass this selection.

The final likelihood as a function of $M_t$ is written as:

卷积

$$P(x, M_t) = \frac{1}{\sigma(M_t)}\int d\sigma(y, M_t)dq_1dq_2f(q_1)f(q_2)W(y, x) \tag{3}$$

其中广义传递函数 $W(y, x)$ 表示从一组部分子水平的变量 $y$ 中产生探测器水平变量 $x$ 的归一化概率；$d\sigma(y, M_t)$ 是部分子水平的理论微分截面；$f(q)$ 是部分子的分布函数，表示质子（反质子）内发现处于特定相互作用中、动量为 $q$ 的夸克（反夸克）的概率，$\sigma(M_t)$ 是产生 $t\bar{t}$ 的总截面。式（3）中的积分要对所有可能的部分子态进行求和，从而得到探测器水平的观测结果。

探测器的接受特性由函数 $A(x)$ 表达。这个函数将观测到的探测器水平变量 $x$ 的概率 $P_m(x, M_t)$ 与其产生概率 $P(x, M_t)$ 联系起来：$P_m(x, M_t) = A(x)P(x, M_t)$。能量分辨率等效应包含在传递函数 $W(y, x)$ 中。式（3）中对 11 个精确测量的变量（带电轻子动量的 3 个分量以及 8 个喷注角度）的积分，加上四个能量–动量守恒关系，使得要完成 5 次积分才能得到特定 $M_t$ 质量值下任何可代表 $t\bar{t}$（或者背景）产生事例的概率。

$t\bar{t}$ 事例判读的概率可以表示为

$$P_{t\bar{t}} = \frac{1}{12\sigma_{t\bar{t}}}\int d^5\Omega \sum_{\text{perm.},\nu} |M_{t\bar{t}}|^2 \frac{f(q_1)f(q_2)}{|q_1||q_2|} \Phi_6 W_{\text{jets}}(E_{\text{part}}, E_{\text{jet}})$$

其中 $\Omega$ 代表一组积分变量（5 个），对于 $t\bar{t}$ 产生事例，$M_{t\bar{t}}$ 是其领头阶矩阵元[23,24]。$f(q_1)$ 和 $f(q_2)$ 是入射夸克的 CTEQ4M 部分子分布函数[25]，$\Phi_6$ 是 6 体末态的相空间因子，求和运算针对喷注的全部 12 种排列和所有可能的中微子解进行。$W_{\text{jets}}(E_{\text{part}}, E_{\text{jet}})$ 对应于一个基于蒙特卡罗模拟研究的、将部分子水平的能量 $E_{\text{part}}$ 映射到探测器测量到的能量 $E_{\text{jet}}$ 的函数。利用一个不依赖 $M_t$ 的 $W$+喷注产生（主要的背景来源）的矩阵元，一个类似表达式被用来计算背景判读概率 $P_{\text{bkg}}$。

对 HERWIG（参考文献 26，我们用的是 5.1 版本）蒙特卡罗事例样本的研究表明，新方法能够使得到的 $M_t$ 统计不确定度降低将近一半。这些研究也揭示，获取的 $M_t$ 值存在一个系统漂移，这一漂移依赖于数据中的背景数目。为了尽量减小这一效应，我们根据一个事例代表背景的概率引入了一个遴选程序。应用本方法处理数据之前，根据蒙特卡罗模拟，确定了具体的 $P_{\text{bkg}}$ 截断值。对于 $M_t = 175\ \text{GeV}/c^2$，71% 的信号和 30% 的背景都得到保留。71 个候选事例中，有 22 个通过了遴选。

最终作为 $M_t$ 函数的似然函数表达式如下：

$$\ln L(M_t) = \sum_{i=1}^{n} \ln[c_1 P_{t\bar{t}}(x_i, M_t) + c_2 P_{bkg}(x_i)] - N \int A(x)[c_1 P_{t\bar{t}}(x, M_t) + c_2 P_{bkg}(x)]\mathrm{d}x$$

The integration is performed using MC methods. The best value of $M_t$ (when $L$ is at its maximum $L_{max}$) represents the most likely mass of the top quark in the final $N$-event sample, and the parameters $c_i$ reflect the amounts of signal and background. MC studies show that there is a downward shift of 0.5 GeV/$c^2$ in the extracted mass, and this correction is applied to the result. Reasonable changes in the cut-off on $P_{bkg}$ do not have a significant impact on $M_t$.

Figure 4 shows the value of $L(M_t)/L_{max}$ as a function of $M_t$ for the 22 events that pass all selection criteria, after correction for the 0.5 GeV/$c^2$ bias in mass. The likelihood is maximized with respect to the parameters $c_i$ at each mass point. The gaussian fit in the figure yields $M_t = 180.1$ GeV/$c^2$, with a statistical uncertainty of $\delta M_t = 3.6$ GeV/$c^2$. The systematic uncertainty, dominated by the measurement of jet energies, as discussed above, amounts to $\delta M_t = 3.9$ GeV/$c^2$. When added in quadrature to the statistical uncertainty from the fit, it yields the overall uncertainty on the new $M_t$ measurement of $\pm 5.3$ GeV/$c^2$.

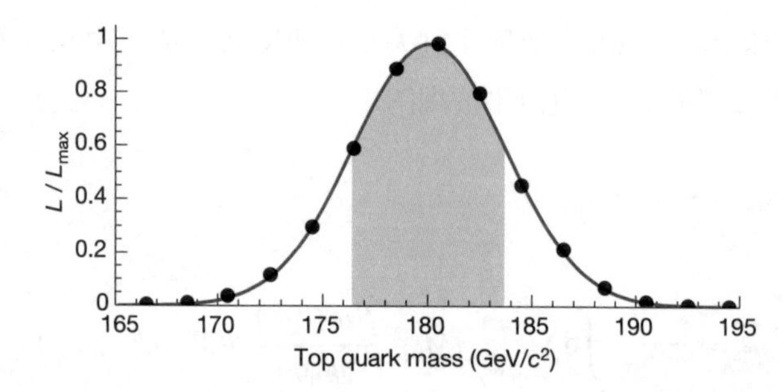


Fig. 4. Determination of the mass of the top quark using the maximum-likelihood method. The points represent the likelihood of the fit used to extract $M_t$ divided by it maximum value, as a function of $M_t$ (after a correction for a −0.5 GeV/$c^2$ mass bias, see text). The solid line shows a gaussian fit to the points. The maximum likelihood corresponds to a mass of 180.1 GeV/$c^2$, which is the new DØ measurement of $M_t$ in the lepton+jets channel. The shaded band corresponds to the range of $\pm 1$ standard deviation, and indicates the $\pm 3.6$ GeV/$c^2$ statistical uncertainty of the fit.

(**429**, 638-642; 2004)

Received 23 January; accepted 21 April 2004; doi:10.1038/nature02589.

References:

1. Hagiwara, K. *et al.* Review of particle physics. *Phys. Rev. D* **66**, 010001 (2002).
2. Affolder, T. *et al.* (CDF Collaboration). Measurement of the top quark mass with the Collider Detector at Fermilab. *Phys. Rev. D* **63**, 032003 (2001).
3. Abbott, B. *et al.* (DØ Collaboration). Measurement of the top quark mass in the dilepton channel. *Phys. Rev. D* **60**, 052001 (1999).
4. The CDF Collaboration, the DØ Collaboration, and the TEVATRON Electro-Weak Working Group. Combination of CDF and DØ Results on the Top-Quark Mass. Preprint at http://www.arXiv.org/hep-ex/0404010 (2004).
5. Barate, R. *et al.* (ALEPH Collaboration, DELPHI Collaboration, L3 Collaboration, OPAL Collaboration, and LEP Working Group for Higgs boson searches). Search for the standard model Higgs boson at LEP. *Phys. Lett. B* **565**, 61-75 (2003).
6. The LEP Collaborations ALEPH, DELPHI, L3, and OPAL, the LEP Electroweak Working Group, and the SLD Heavy Flavour Group. A combination of preliminary electroweak measurements and constraints on the standard model. Preprint at http://www.arXiv.org/hep-ex/0312023 (2003).
7. Abe, F. *et al.* (CDF Collaboration). Observation of top quark production in $p\bar{p}$ collisions with the Collider Detector at Fermilab. *Phys. Rev. Lett.* **74**, 2626-2631 (1995).

$$\ln L(M_t) = \sum_{i=1}^{n} \ln[c_1 P_{t\bar{t}}(x_i, M_t) + c_2 P_{bkg}(x_i)] - N\int A(x)[c_1 P_{t\bar{t}}(x, M_t) + c_2 P_{bkg}(x)]dx$$

积分运算是通过蒙特卡罗方法完成的。$M_t$ 的最佳值（当 $L$ 等于其最大值 $L_{max}$）代表了最终的 $N$ 事例样本中最有可能的顶夸克质量值，参数 $c_i$ 反映信号和背景的数量。蒙特卡罗模拟研究表明获取的质量值有一个向下的 0.5 GeV/$c^2$ 的漂移，由此我们对结果进行了修正。在合理范围内 $P_{bkg}$ 截断值的变动对 $M_t$ 没有显著的影响。

图 4 显示了进行 0.5 GeV/$c^2$ 的质量偏差的修正之后，通过所有遴选标准的 22 个事例的 $L(M_t)/L_{max}$ 比值随 $M_t$ 的变化。在图中每一个质量数据点，对参数 $c_i$ 取最大似然。图中所示数据的高斯拟合得到 $M_t$ = 180.1 GeV/$c^2$，统计不确定度 $\delta M_t$ = 3.6 GeV/$c^2$。如上文所述，系统不确定度主要受喷注能量的测量的影响，大小为 $\delta M_t$ = 3.9 GeV/$c^2$。系统不确定度与拟合得到的统计不确定度求均方根，得到新的 $M_t$ 测量值的总不确定度为 ±5.3 GeV/$c^2$。

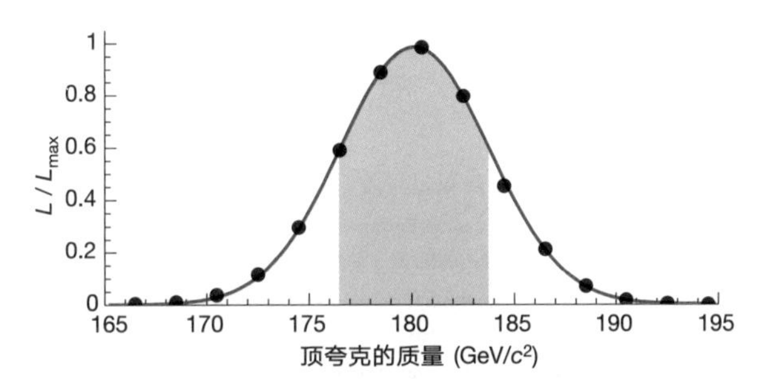


图 4. 用最大似然法确定顶夸克的质量。数据点表示用以获取 $M_t$ 的拟合的似然值与其最大值的比，该比是 $M_t$ 的函数（已进行 0.5 GeV/$c^2$ 的质量偏差的修正，见正文）。实线表示对数据点的高斯拟合结果。最大似然值对应质量为 180.1 GeV/$c^2$，这也是新的轻子 + 喷注衰变道中 $M_t$ 的 DØ 实验测定值。带颜色的区域对应正负一个标准差区间，表示 ±3.6 GeV/$c^2$ 的拟合统计不确度。

（何钧 翻译；曹庆宏 审稿）

8. Abachi, S. *et al.* (DØ Collaboration). Observation of the top quark. *Phys. Rev. Lett.* **74**, 2632-2637 (1995).

9. Ellis, J. The 115 GeV Higgs odyssey. *Comments Nucl. Part. Phys. A* **2**, 89-103 (2002).

10. Chanowitz, M. S. Electroweak data and the Higgs boson mass: a case for new physics. *Phys. Rev. D* **66**, 073002 (2002).

11. The LEP Collaborations ALEPH, DELPHI, L3 & OPAL, the LEP Higgs Working Group. Searches for the neutral Higgs bosons of the MSSM: preliminary combined results using LEP data collected at energies up to 209 GeV. Preprint at http://www.arXiv.org/hep-ex/0107030 (2001).

12. Degrassi, G., Heinemeyer, S., Hollik, W., Slavich, P. & Weiglein, G. Towards high-precision predictions for the MSSM Higgs sector. *Eur. Phys. J. C* **28**, 133-143 (2003).

13. Abbott, B. *et al.* (DØ Collaboration). Direct measurement of the top quark mass by the DØ collaboration. *Phys. Rev. D* **58**, 052001 (1998).

14. Abazov, V. M. *et al.* (DØ Collaboration). $t\bar{t}$ production cross section in $p\bar{p}$ collisions at $\sqrt{s}$ = 1.8 TeV. *Phys. Rev. D* **67**, 012004 (2003).

15. Abachi, S. *et al.* (DØ Collaboration). The DØ detector. *Nucl. Instrum. Methods A* **338**, 185-253 (1994).

16. Dalitz, R. H. & Goldstein, G. R. Test of analysis method for top–antitop production and decay events. *Proc. R. Soc. Lond. A* **445**, 2803-2834 (1999).

17. Kondo, K. *et al.* Dynamical likelihood method for reconstruction of events with missing momentum. 3: Analysis of a CDF high $p_T$ $e\mu$ event as $t\bar{t}$ production. *J. Phys. Soc. Jpn* **62**, 1177-1182(1993).

18. Berends, F. A., Papadopoulos, C. G. & Pittau, R. On the determination of $M_W$ and TGCs in $W$-pair production using the best measured kinematical variables. *Phys. Lett. B* **417**, 385-389 (1998).

19. Abreu, P. *et al.* (DELPHI Collaboration). Measurement of the $W$ pair cross-section and of the $W$ Mass in $e^+e^-$ interactions at 172 GeV. *Eur. Phys. J. C* **2**, 581-595 (1998).

20. Juste, A. Measurement of the $W$ mass in $e^+e^-$ annihilation PhD thesis, 1-160, Univ. Autonoma de Barcelona (1998).

21. Estrada, J. Maximal use of kinematic information for extracting the top quark mass in single-lepton $t\bar{t}$ events PhD thesis, 1-132, Univ. Rochester (2001).

22. Canelli, F. Helicity of the $W$ boson in single-lepton $t\bar{t}$ events PhD thesis. 1-241, Univ. Rochester (2003).

23. Mahlon, G. & Parke, S. Angular correlations in top quark pair production and decay at hadron colliders. *Phys. Rev. D* **53**, 4886-4896 (1996).

24. Mahlon, G. & Parke, S. Maximizing spin correlations in top quark pair production at the Tevatron. *Phys. Lett. B* **411**, 173-179 (1997).

25. Lai, H. L. *et al.* (CTEQ Collaboration). Global QCD analysis and the CTEQ parton distributions. *Phys. Rev. D* **51**, 4763-4782 (1995).

26. Marchesini, G. *et al.* HERWIG: a Monte Carlo event generator for simulating hadron emission reactions with interfering gluons. *Comput. Phys. Commun.* **67**, 465-508 (1992).

**Acknowledgements**. We are grateful to our colleagues A. Quadt and M. Mulders for reading of the manuscript and comments. We thank the staffs at Fermilab and collaborating institutions, and acknowledge support from the Department of Energy and National Science Foundation (USA), Commissariat à L'Energie Atomique and CNRS/Institut National de Physique Nucléaire et de Physique des Particules (France), Ministry for Science and Technology and Ministry for Atomic Energy (Russia), CAPES, CNPq and FAPERJ (Brazil), Departments of Atomic Energy and Science and Education (India), Colciencias (Colombia), CONACyT (Mexico), Ministry of Education and KOSEF (Korea), CONICET and UBACyT (Argentina), The Foundation for Fundamental Research on Matter (The Netherlands), PPARC (UK), Ministry of Education (Czech Republic), the A. P. Sloan Foundation, and the Research Corporation.

**Authors' contributions**. We wish to note the great number of contributions made by the late Harry Melanson to the DØ experiment, through his steady and inspirational leadership of the physics, reconstruction and algorithm efforts.

**Competing interests statement**. The authors declare that they have no competing financial interests.

Correspondence and requests for materials should be addressed to J. Estrada (estrada@fnal.gov).

# Genetic Evidence Supports Demic Diffusion of Han Culture

B. Wen *et al.*

## Editor's Note

**Long-established disciplines situated between the natural sciences and the humanities, such as linguistics, anthropology and archaeology, have been transformed in the past several decades by the availability of genetic data for large numbers of people. The distribution of genes in a population encodes a historical record of cultural movements, interactions and relationships, which can be used to test theories about the dissemination of culture, language and traditions. In this paper by geneticist Li Jin and coworkers, genetic profiles of more than 1,000 individuals in China are used to examine alternative theories of how the Han culture came to be dominant throughout China. One possibility is that the culture was spread by migration (demic diffusion); another is that it happened by cultural exchange without mass movement of people. The Han people of the north are genetically distinct from those of the south, but the results show that southern Han are more closely related to northern Han than they are to southern ethnic minorities, especially through paternal descent—supporting the demic model, driven mostly by migration of northern males. This migration happened largely in three waves within the first and early second millennium, during the Western Jin, Tang and Song dynasties.**

---

The spread of culture and language in human populations is explained by two alternative models: the demic diffusion model, which involves mass movement of people; and the cultural diffusion model, which refers to cultural impact between populations and involves limited genetic exchange between them[1]. The mechanism of the peopling of Europe has long been debated, a key issue being whether the diffusion of agriculture and language from the Near East was concomitant with a large movement of farmers[1-3]. Here we show, by systematically analysing Y-chromosome and mitochondrial DNA variation in Han populations, that the pattern of the southward expansion of Han culture is consistent with the demic diffusion model, and that males played a larger role than females in this expansion. The Han people, who all share the same culture and language, exceed 1.16 billion (2000 census), and are by far the largest ethnic group in the world. The expansion process of Han culture is thus of great interest to researchers in many fields.

---

ACCORDING to the historical records, the Hans were descended from the ancient Huaxia tribes of northern China, and the Han culture (that is, the language and its associated cultures) expanded into southern China—the region originally inhabited by the southern natives, including those speaking Daic, Austro-Asiatic and Hmong-

# 遗传学证据支持汉文化的扩散源于人口扩张

文波等

编者按

在过去的数十年间，科学家获得了大量的人群遗传数据，也因此改变了处在自然科学与人文科学之间的一批学科，诸如语言学、人类学和考古学。一个群体中基因的分布，记录着文化变革、互动和源流关系的历史，可以用来验证关于文化、语言和传统传播的各种假说。遗传学家金力及合作者在这篇论文中，利用中国一千多个个体的遗传学数据，验证了汉文化如何在中国成为主流的两种假说。一种可能性是移民传播了文化（人口扩张说）；也有可能只是文化交流而没有大规模的人口移动。从遗传学看，北方的汉族人群与南方的显然有差异，但是结果却显示：南方汉族人群更接近北方汉族人群，而不是接近南方少数民族，特别是在父系血缘中。这支持了人口扩张说，主要是北方男性移民传播了文化。这种迁移大量发生于公元后第一个千年，以及第二个千年早期，有西晋、唐代和宋代三次高潮。

---

语言和文化在人群间的扩散有两种不同的模式：一种是人口扩张、人群迁徙模式；另一种是文化传播模式，人群之间有文化传播，而基因交流却很有限[1]。同一语系的欧洲人群的形成机制争议颇多，争论的焦点在于来自近东的农业文明和语言的扩散是否伴随着大量的农业人口的迁移[1-3]。本文中，通过对汉族群体的 Y 染色体和线粒体 DNA 多态性系统地进行分析，我们发现汉文化向南扩散的格局符合人口扩张模式，而且在扩张过程中男性占主导地位。汉族有着共同的文化和语言，人口超过了十一亿六千万（2000 年人口统计），无疑是全世界人口最多的民族。因此汉文化的扩散过程广受各领域研究者的关注。

---

史载汉族源于古代中国北方的华夏部族，在过去的两千多年间，汉文化（汉语和相关的文化传统）扩散到了中国南方，而中国南方原住民族则是说侗台语、南亚

Mien languages—in the past two millennia[4,5]. Studies on classical genetic markers and microsatellites show that the Han people, like East Asians, are divided into two genetically differentiated groups, northern Han and southern Han[6,8], separated approximately by the Yangtze river[9]. Differences between these groups in terms of dialect and customs have also been noted[10]. Such observations seem to support a mechanism involving primarily cultural diffusion and assimilation (the cultural diffusion model) in Han expansion towards the south. However, the substantial sharing of Y-chromosome and mitochondrial lineages between the two groups[11,12] and the historical records describing the expansion of Han people[5] contradict the cultural diffusion model hypothesis of Han expansion. In this study, we aim to examine the alternative hypothesis; that is, that substantial population movements occurred during the expansion of Han culture (the demic diffusion model).

To test this hypothesis, we compared the genetic profiles of southern Hans with their two parental population groups: northern Hans and southern natives, which include the samples of Daic, Hmong-Mien and Austro-Asiatic speaking populations currently residing in China, and in some cases its neighbouring countries. Genetic variation in both the non-recombining region of the Y chromosome (NRY) and mitochondrial DNA (mtDNA)[13-16] were surveyed in 28 Han populations from most of the provinces in China (see Fig. 1 and Supplementary Table 1 for details).

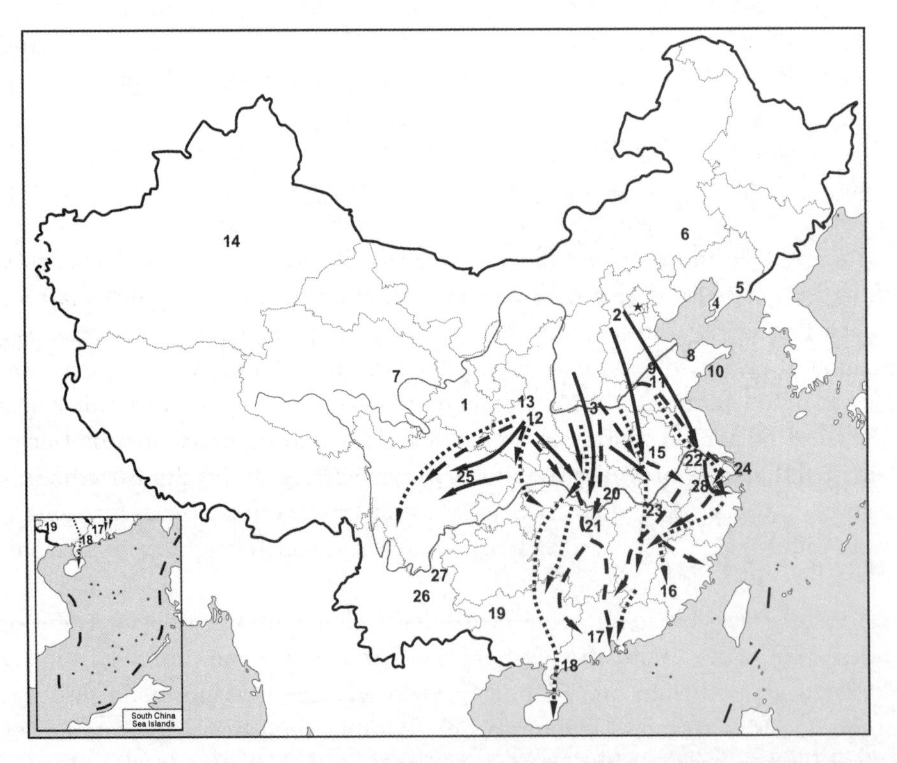


Fig. 1. Geographic distribution of sampled populations. Shown are the three waves of north-to-south migrations according to historical record. The identifications of populations are given in Supplementary Table 1. Populations 1–14 are northern Hans, and 15–28 are southern Hans. The solid, dashed and dotted arrows refer to the first, second and third waves of migrations, respectively. The first wave involving 0.9

语和苗瑶语的人群[4,5]。经典遗传标记和微卫星位点研究显示，汉族和其他东亚人群一样都可以以长江为界[9]分为两个遗传亚群：北方汉族和南方汉族[6,8]。两个亚群之间的方言和习俗差异也很显著[10]。这些现象看似支持文化传播模式，即汉族向南扩张主要是文化传播和同化的结果。然而，两个亚群之间有着许多共同的Y染色体和线粒体类型[11,12]，历史记载的汉族移民史[5]也与汉族的文化传播模式假说相矛盾。本研究对这两种假说进行了检验，证实汉文化的扩散中的确发生了大规模的人群迁徙（人口扩张模式）。

为了验证这些假说，我们把南方汉族的遗传结构与两个亲本群体作比较，其一是北方汉族，其二是南方原住民族，即现居于中国境内和若干邻国的侗台语、苗瑶语和南亚语群体。我们分析了中国汉族28个群体的Y染色体非重组区（NRY）和线粒体DNA(mtDNA)遗传多态[13-16]，这些样本覆盖了中国绝大部分的省份（详见图1和补充信息表1）。

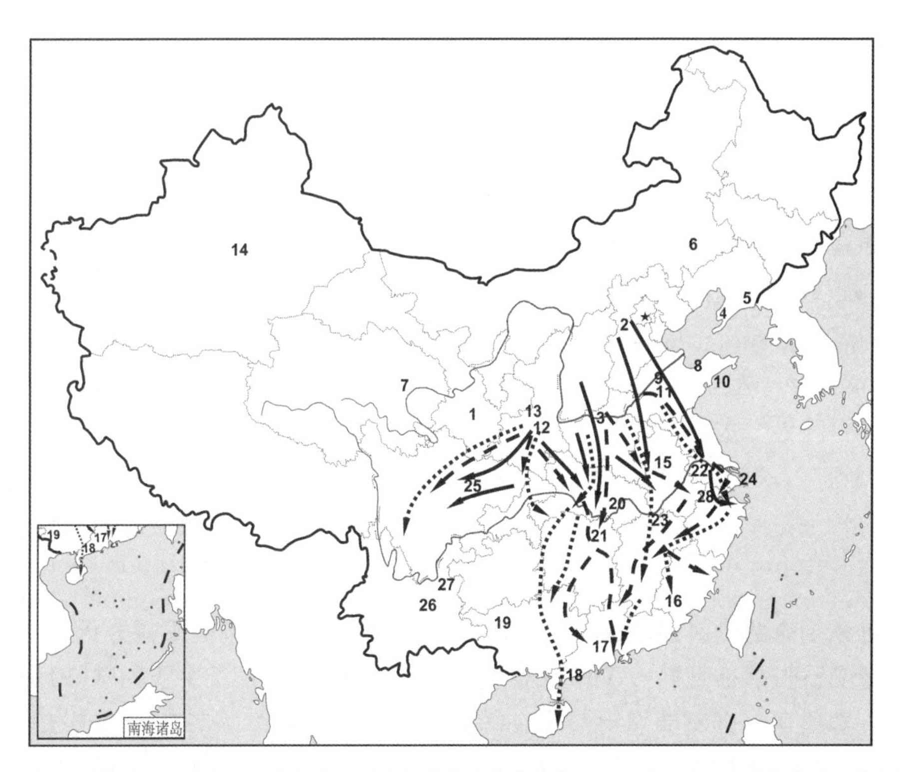


图1. 调查群体的地理分布。图中标出了历史记载中自北而南的三次迁徙浪潮。各群体的详细信息见补充信息表1。群体1～14是北方汉族，15～28是南方汉族。实线、短划线和虚线箭头依次表示三次迁徙浪潮。第一次发生于西晋时期（公元265~316年），迁徙人口约90万（大约是当时南方人口的

million (approximately one-sixth of the southern population at that time) occurred during the Western Jin Dynasty (AD 265–316); the second migration, more extensive than the first, took place during the Tang Dynasty (AD 618–907); and the third wave, including, ~5 million immigrants, occurred during the Southern Song Dynasty (AD 1127–1279).

On the paternal side, southern Hans and northern Hans share similar frequencies of Y-chromosome haplogroups (Supplementary Table 2), which are characterized by two haplogroups carrying the M122-C mutations (O3-M122 and O3e-M134) that are prevalent in almost all Han populations studied (mean and range: 53.8%, 37–71%; 54.2%, 35–74%, for northern and southern Hans, respectively). Haplogroups carrying M119-C (O1* and O1b) and/or M95-T (O2a* and O2a1) (following the nomenclature of the Y Chromosome Consortium) which are prevalent in southern natives, are more frequent in southern Hans (19%, 3–42%) than in northern Hans (5%, 1–10%). In addition, haplogroups O1b-M110, O2a1-M88 and O3d-M7, which are prevalent in southern natives[17], were only observed in some southern Hans (4% on average), but not in northern Hans. Therefore, the contribution of southern natives in southern Hans is limited, if we assume that the frequency distribution of Y lineages in southern natives represents that before the expansion of Han culture that started 2,000 yr ago[5]. The results of analysis of molecular variance (AMOVA) further indicate that northern Hans and southern Hans are not significantly different in their Y haplogroups ($F_{ST} = 0.006$, $P > 0.05$), demonstrating that southern Hans bear a high resemblance to northern Hans in their male lineages.

On the maternal side, however, the mtDNA haplogroup distribution showed substantial differentiation between northern Hans and southern Hans (Supplementary Table 3). The overall frequencies of the northern East Asian-dominating haplogroups (A, C, D, G, M8a, Y and Z) are much higher in northern Hans (55%, 49–64%) than are those in southern Hans (36%, 19–52%). In contrast, the frequency of the haplogroups that are dominant lineages (B, F, R9a, R9b and N9a) in southern natives[12,14,18] is much higher in southern (55%, 36–72%) than it is in northern Hans (33%, 18–42%). Northern and southern Hans are significantly different in their mtDNA lineages ($F_{ST} = 0.006$, $P < 10^{-5}$). Although the $F_{ST}$ values between northern and southern Hans are similar for mtDNA and the Y chromosome, $F_{ST}$ accounts for 56% of the total among-population variation for mtDNA but only accounts for 18% for the Y chromosome.

A principal component analysis is consistent with the observation based on the distribution of the haplogroups in Han populations. For the NRY, almost all Han populations cluster together in the upper right-hand part of Fig. 2a. Northern Hans and southern natives are separated by the second principal component (PC2) and southern Hans' PC2 values lie between northern Hans and southern natives but are much closer to northern Hans (northern Han, $0.58 \pm 0.01$; southern Han, $0.46 \pm 0.03$; southern native, $-0.32 \pm 0.05$), implying that the southern Hans are paternally similar to northern Hans, with limited influence from southern natives. In contrast, for mtDNA, northern Hans and southern natives are distinctly separated by PC2 (Fig. 2b), and southern Hans are located between them but are closer to southern natives (northern Han, $0.56 \pm 0.02$; southern Han, $0.09 \pm 0.06$; southern native,

六分之一)；第二次发生于唐代(公元618~907年)，规模比第一次大得多；第三次发生于南宋(公元1127~1279年)，迁徙人口近500万。

父系方面，南方汉族与北方汉族的Y染色体单倍群频率分布非常相近(见补充信息表2)，尤其是具有M122-C突变的单倍群(O3-M122和O3e-M134)普遍存在于我们研究的汉族群体中(北方汉族在37%~71%之间，平均53.8%；南方汉族在35%~74%之间，平均54.2%)。南方原住民族中普遍出现的单倍群M119-C(O1*和O1b)和(或)M95-T(O2a*和O2a1)(遵循Y染色体委员会的命名法)在南方汉族中的频率(3%~42%，平均19%)高于北方汉族(1%~10%，平均5%)。而且，南方原住民族中普遍存在的单倍群O1b-M110、O2a1-M88和O3d-M7[17]，在南方汉族中低频存在(平均4%)，而北方汉族中却没观察到。如果我们假定在起始于两千多年前的汉文化扩散[5]之前南方原住民族的Y类型频率与现在基本一致的话，南方汉族中南方原住民族的成分应该是不多的。分子方差分析(AMOVA)进一步显示北方汉族和南方汉族的Y染色体单倍群频率分布没有显著差异($F_{ST}=0.006$，$P>0.05$)，说明南方汉族在父系上与北方汉族非常相似。

母系方面，北方汉族与南方汉族的mtDNA单倍群分布非常不同(补充信息表3)。东亚北部的主要单倍群(A，C，D，G，M8a，Y，Z)在北方汉族中的频率(49%~64%，平均55%)比在南方汉族中(19%~52%，平均36%)高得多。另一方面，南方原住民族的主要单倍群(B，F，R9a，R9b，N9a)[12,14,18]在南方汉族中的频率(36%~72%，平均55%)要比在北方汉族中(18%~42%，平均33%)高得多。mtDNA类型的分布在南北汉族之间有极显著差异($F_{ST}=0.006$，$P<10^{-5}$)。虽然南北汉族之间mtDNA和Y染色体的$F_{ST}$值相近，但mtDNA的南北差异$F_{ST}$值占群体间总方差的56%，而Y染色体仅仅占18%。

用汉族群体的单倍群频率数据所做的主成分分析与以上结果相一致。对NRY的分析发现，几乎所有的汉族群体都聚在图2a的右上方。北方汉族和南方原住民族在第2主成分上分离，南方汉族的第2主成分值处于北方汉族和南方原住民族之间，但是更接近于北方汉族(北方汉族$0.58\pm0.01$；南方汉族$0.46\pm0.03$；南方原住民族$-0.32\pm0.05$)，这表明南方汉族在父系上与北方汉族相近，受到南方原住民族的影响很小。就mtDNA而言，北方汉族和南方原住民族仍然被第2主成分分开(图2b)，南方汉族也在两者之间但稍微接近南方原住民族(北方汉族$0.56\pm0.02$；南方汉族$0.09\pm0.06$；南方原住民族$-0.23\pm0.04$)，表明南方汉族的女性基因库比男性

−0.23 ± 0.04), indicating a much more substantial admixture in southern Hans' female gene pool than in its male counterpart.

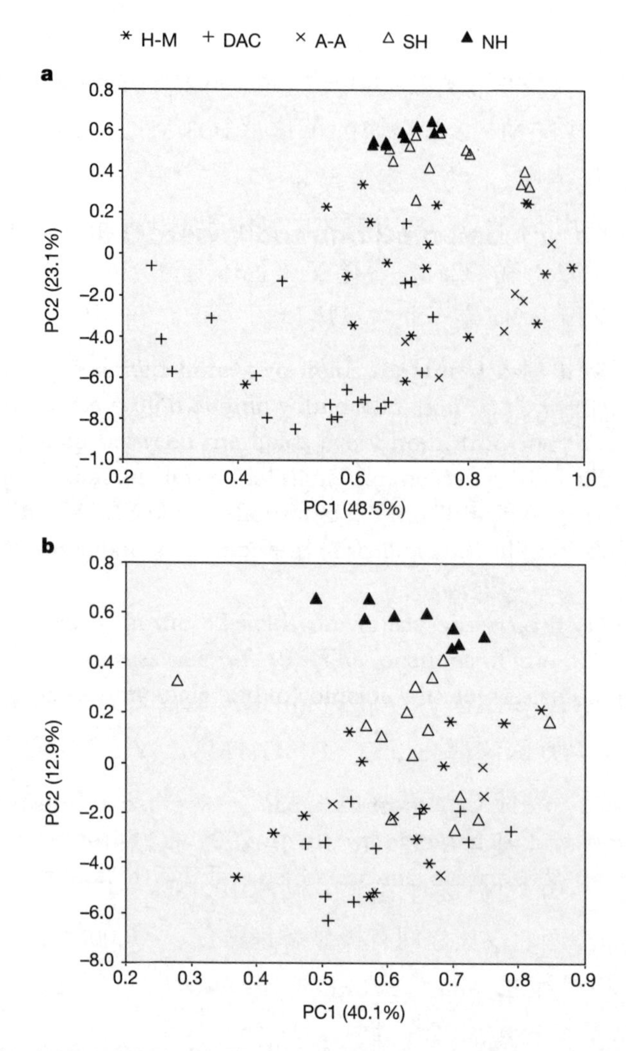


Fig. 2. Principal component plot. **a**, **b**, Plots are of Y-chromosome (**a**) and mtDNA (**b**) haplogroup frequency. Population groups: H-M, Hmong-Mien; DAC, Daic; A-A, Austro-Asiatic; SH, southern Han; NH, northern Han.

The relative contribution of the two parental populations (northern Hans and southern natives) in southern Hans was estimated by two different statistics[19,20], which are less biased than other statistics for single-locus data[21] (Table 1). The estimations of the admixture coefficient ($M$, proportion of northern Han contribution) from the two methods are highly consistent (for the Y chromosome, $r = 0.922$, $P < 0.01$; for mtDNA, $r = 0.970$, $P < 0.01$). For the Y chromosome, all southern Hans showed a high proportion of northern Han contribution ($M_{BE}$: 0.82 ± 0.14, range from 0.54 to 1; $M_{RH}$: 0.82 ± 0.12, range from 0.61 to 0.97) (see refs 20 and 19 for definitions of $M_{BE}$ and $M_{RH}$, respectively) indicating that males from the northern Hans are the primary contributor to the gene pool of the southern

基因库有更多的混合成分。

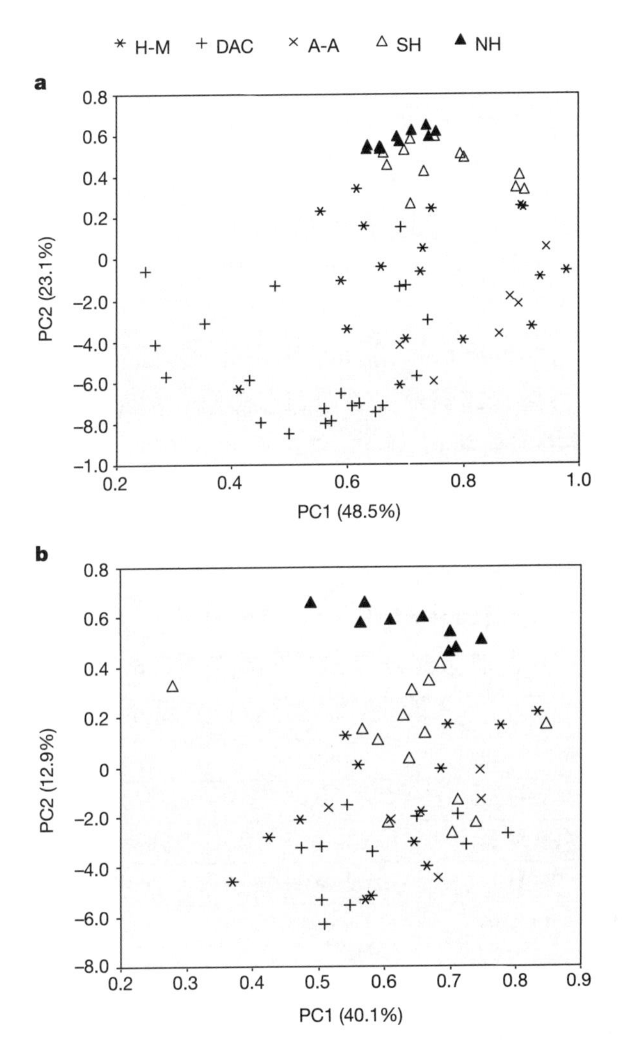


图 2. 主成分散点图。**a** 为 Y 染色体单倍群散点图，**b** 为 mtDNA 单倍群散点图。群体标记：▲北方汉族，△南方汉族，+ 侗台语民族，× 南亚语民族，* 苗瑶语民族。

我们进一步用两种不同的统计方法 [19,20] 来估计两个亲本（北方汉族和南方原住民）对南方汉族基因库的相对贡献（表 1），这两个统计量用于单位点分析时比其他的方法更为准确 [21]。两种方法得到的混合系数估计值（$M$，北方汉族的贡献比例）高度一致（Y 染色体，$r = 0.922$，$P < 0.01$；mtDNA，$r = 0.970$，$P < 0.01$）。就 Y 染色体而言，所有的南方汉族都包含很高比例的北方汉族混合比率（$M_{BE}$:0.82 ± 0.14，范围 0.54 ~ 1；$M_{RH}$:0.82 ± 0.12，范围 0.61 ~ 0.97）（$M_{BE}$ 和 $M_{RH}$ 的定义分别见参考文献 20 和 19），这表明南方汉族男性基因库的主要贡献成分来自北方汉族。相反，南

Hans. In contrast, northern Hans and southern natives contributed almost equally to the southern Hans' mtDNA gene pool ($M_{BE}$: 0.56±0.24 [0.15, 0.95]; $M_{RH}$: 0.50±0.26 [0.07, 0.91]). The contribution of northern Hans to southern Hans is significantly higher in the paternal lineage than in the maternal lineage collectively ($t$-test, $P < 0.01$) or individually (11 out of 13 populations for $M_{BE}$, and 13 out of 13 populations for $M_{RH}$: $P < 0.01$, assuming a null binomial distribution with equal male and female contributions), indicating a strong sex-biased population admixture in southern Hans. The proportions of northern Han contribution ($M$) in southern Hans showed a clinal geographic pattern, which decreases from north to south. The $M$s in southern Hans are positively correlated with latitude ($r^2 = 0.569$, $P < 0.01$) for mtDNA, but are not significant for the Y chromosome ($r^2 = 0.072$, $P > 0.05$), because the difference of $M$s in the paternal lineage among southern Hans is too small to create a statistically significant trend.

Table 1. Northern Han admixture proportion in southern Hans

| Population | Y Chromosome | | mtDNA | |
|---|---|---|---|---|
| | $M_{BE}$ (±s.e.m) | $M_{RH}$ | $M_{BE}$ (±s.e.m) | $M_{RH}$ |
| Anhui | 0.868±0.119 | 0.929 | 0.816±0.214 | 0.755 |
| Fujian | 1 | 0.966 | 0.341±0.206 | 0.248 |
| Guangdong1 | 0.677±0.121 | 0.669 | 0.149±0.181 | 0.068 |
| Guangdong2 | ND | ND | 0.298±0.247 | 0.312 |
| Guangxi | 0.543±0.174 | 0.608 | 0.451±0.263 | 0.249 |
| Hubei | 0.981±0.122 | 0.949 | 0.946±0.261 | 0.907 |
| Hunan | 0.732±0.219 | 0.657 | 0.565±0.297 | 0.490 |
| Jiangsu | 0.789±0.078 | 0.821 | 0.811±0.177 | 0.786 |
| Jiangxi | 0.804±0.113 | 0.829 | 0.374±0.343 | 0.424 |
| Shanghai | 0.819±0.087 | 0.902 | 0.845±0.179 | 0.833 |
| Sichuan | 0.750±0.118 | 0.713 | 0.509±0.166 | 0.498 |
| Yunnan1 | 1 | 0.915 | 0.376±0.221 | 0.245 |
| Yunnan2 | 0.935±0.088 | 0.924 | 0.733±0.192 | 0.645 |
| Zhejiang | 0.751±0.084 | 0.763 | 0.631±0.180 | 0.540 |
| Average | 0.819 | 0.819 | 0.560 | 0.500 |

$M_{BE}$ and $M_{RH}$ refer to the statistics described in refs 20 and 19, respectively. The standard error of $M_{BE}$ was obtained by bootstrap with 1,000 replications. The proportions of contribution from northern Hans were estimated using northern Hans and southern natives as the parental populations of the southern Hans. It was assumed that the allele frequency in the southern natives remained unchanged before and after the admixture, which started about 2,000 yr ago, and the genetic exchange between northern Hans and southern natives has been limited. In fact, the gene flow from northern Hans to southern natives has been larger than that from southern natives to northern Hans; therefore, the level of admixture presented in this table is underestimated and is without proper adjustment. The demic expansion of Han would have been more pronounced than was observed in this study.

We provide two lines of evidence supporting the demic diffusion hypothesis for the expansion of Han culture. First, almost all Han populations bear a high resemblance in

方汉族的 mtDNA 基因库中北方汉族和南方原住民族的贡献比例几乎相等（$M_{BE}$：0.56 ± 0.24 [0.15，0.95]；$M_{RH}$：0.50 ± 0.26 [0.07，0.91]）。总体上北方汉族对南方汉族的遗传贡献父系比母系高得多（$t$ 检验，$P < 0.01$）；各群体分别看也是这样：绝大部分南方汉族群体中北方汉族的贡献在父系上大于母系（$M_{BE}$，11/13，$M_{RH}$，13/13，$P < 0.01$，零假设为男女的贡献相等为二项分布），这表明南方汉族的群体混合过程有很强的性别偏向。南方汉族中北方汉族贡献的比例（$M$）呈现出由北向南递减的梯度地理格局。南方汉族 mtDNA 的 $M$ 值与纬度正相关（$r^2 = 0.569$，$P < 0.01$），但 Y 染色体的相关性不显著（$r^2 = 0.072$，$P > 0.05$），因为南方汉族父系的 $M$ 值差异太小，不足以导致统计上的显著性。

表 1. 南方汉族中的北方汉族混合比例

| 群体 | Y 染色体 | | 线粒体 DNA | |
|---|---|---|---|---|
| | $M_{BE}$（± s.e.m） | $M_{RH}$ | $M_{BE}$（± s.e.m） | $M_{RH}$ |
| 安徽 | 0.868 ± 0.119 | 0.929 | 0.816 ± 0.214 | 0.755 |
| 福建 | 1 | 0.966 | 0.341 ± 0.206 | 0.248 |
| 广东 1 | 0.677 ± 0.121 | 0.669 | 0.149 ± 0.181 | 0.068 |
| 广东 2 | ND | ND | 0.298 ± 0.247 | 0.312 |
| 广西 | 0.543 ± 0.174 | 0.608 | 0.451 ± 0.263 | 0.249 |
| 湖北 | 0.981 ± 0.122 | 0.949 | 0.946 ± 0.261 | 0.907 |
| 湖南 | 0.732 ± 0.219 | 0.657 | 0.565 ± 0.297 | 0.490 |
| 江苏 | 0.789 ± 0.078 | 0.821 | 0.811 ± 0.177 | 0.786 |
| 江西 | 0.804 ± 0.113 | 0.829 | 0.374 ± 0.343 | 0.424 |
| 上海 | 0.819 ± 0.087 | 0.902 | 0.845 ± 0.179 | 0.833 |
| 四川 | 0.750 ± 0.118 | 0.713 | 0.509 ± 0.166 | 0.498 |
| 云南 1 | 1 | 0.915 | 0.376 ± 0.221 | 0.245 |
| 云南 2 | 0.935 ± 0.088 | 0.924 | 0.733 ± 0.192 | 0.645 |
| 浙江 | 0.751 ± 0.084 | 0.763 | 0.631 ± 0.180 | 0.540 |
| 平均 | 0.819 | 0.819 | 0.560 | 0.500 |

$M_{BE}$ 和 $M_{RH}$ 分别为参考文献 20 和 19 所描述的统计量。$M_{BE}$ 的标准误差通过 1,000 次自展获得。把南方原住民族和北方汉族作为南方汉族的亲本群体估计北方汉族的遗传贡献比例，假定 2,000 多年前开始的混合过程前后南方原住民族的等位基因频率基本不变，并且北方汉族和南方原始民族之间的遗传交流不多。实际上，从北方汉族到南方原住民族的基因流动比反向的流动大得多，所以表中的估计值在没有适当调整前是低估的。因而汉族实际的人口扩张程度应该大于本项研究得出的数值。

综上所述，我们提出了两项证据支持汉文化扩散的人口扩张假说。首先，几乎所有的汉族群体的 Y 染色体单倍群分布都极为相似，Y 染色体主成分分析也把几

Y-chromosome haplogroup distribution, and the result of principal component analysis indicated that almost all Han populations form a tight cluster in their Y chromosome. Second, the estimated contribution of northern Hans to southern Hans is substantial in both paternal and maternal lineages and a geographic cline exists for mtDNA. It is noteworthy that the expansion process was dominated by males, as is shown by a greater contribution to the Y-chromosome than the mtDNA from northern Hans to southern Hans. A sex-biased admixture pattern was also observed in Tibeto-Burman-speaking populations[22].

According to the historical records, there were continuous southward movements of Han people due to warfare and famine in the north, as illustrated by three waves of large-scale migrations (Fig. 1). Aside from these three waves, other smaller southward migrations also occurred during almost all periods in the past two millennia. Our genetic observation is thus in line with the historical accounts. The massive movement of the northern immigrants led to a change in genetic makeup in southern China, and resulted in the demographic expansion of Han people as well as their culture. Except for these massive population movements, gene flow between northern Hans, southern Hans and southern natives also contributed to the admixture which shaped the genetic profile of the extant populations.

## Methods

### Samples

Blood samples of 871 unrelated anonymous individuals from 17 Han populations were collected across China. Genomic DNA was extracted by the phenol-chloroform method. By integrating the additional data obtained from the literatures on the Y chromosome and on mtDNA variation, the final sample sizes for analysis expanded to 1,289 individuals (23 Han populations) for the Y chromosome and 1,119 individuals (23 Han populations) for mtDNA. These samples encompass most of the provinces in China (Fig. 1 and Supplementary Table 1).

### Genetic markers

Thirteen bi-allelic Y-chromosome markers, YAP, M15, M130, M89, M9, M122, M134, M119, M110, M95, M88, M45 and M120 were typed by polymerase chain reaction-restriction-fragment length polymorphism methods[11]. These markers are highly informative in East Asians[23] and define 13 haplogroups following the Y Chromosome Consortium nomenclature[24].

The HVS-1 of mtDNA and eight coding region variations, 9-bp deletion, 10397 *Alu*I, 5176 *Alu*I, 4831 *Hha*I, 13259 *Hinc*II, 663 *Hae*III, 12406 *Hpa*I and 9820 *Hinf*I were sequenced and genotyped as in our previous report[22]. Both the HVS-1 motif and the coding region variations were used to infer haplogroups following the phylogeny of East Asian mtDNAs[18].

乎所有的汉族群体都集合成一个紧密的聚类。其二，北方汉族对南方汉族的遗传贡献无论父系方面还是母系方面都是可观的，在线粒体 DNA 分布上也存在地理梯度。北方汉族对南方汉族的遗传贡献在父系（Y 染色体）上远大于母系（mtDNA），表明这一扩张过程中汉族男性处于主导地位。性别偏向的混合格局也同样存在于藏缅语人群中 [22]。

据历史记载，受北方战乱和饥荒的影响，汉人不断南迁，图 1 中画出了三次大规模移民的浪潮。在两千多年间，除了这三次大潮，各个时期几乎都有小规模的南迁。所以，我们的遗传研究也与历史记载相吻合。大量的北方移民改变了中国南方的遗传构成，而汉族人口扩张的同时也带动了汉文化的扩散。除了大规模的人群迁徙，北方汉族、南方汉族和南方原住民族之间的基因交流造成的族群混合也在很大程度上改变了中国人群的遗传结构。

## 方　法

### 样本

采集中国各地的 17 个汉族群体 871 个随机不相关个体的血样。用酚–氯仿法抽提基因组 DNA。结合文献报道的 Y 染色体和 mtDNA 多态性数据，总共分析的样本量是：Y 染色体 23 个群体 1,289 人，mtDNA23 个群体 1,119 人。这些样本涉及了中国的大部分省份（图 1 和补充信息表 1）。

### 遗传标记

通过聚合酶链式反应–限制性片段长度多态性方法 [11] 分型 Y 染色体上的 13 个双等位标记：YAP，M15，M130，M89，M9，M122，M134，M119，M110，M95，M88，M45，M120。根据 Y 染色体委员会的命名系统 [24]，这些标记构成 13 个单倍群，在东亚人群中具有较高的信息量 [23]。

mtDNA 上，对高变 1 区进行测序，对编码区 8 个多态位点（9-bp 缺失，10397 *Alu*I，5176 *Alu*I，4831 *Hha*I，13259 *Hinc*II，663 *Hae*III，12406 *Hpa*I，9820 *Hinf* I）做了分型，有关方法已有报道 [22]。根据东亚 mtDNA 系统树 [18]，用高变 1 区突变结构和编码区多态性构建单倍群。

## Data analysis

Population relationship was investigated by principal component analysis, which was conducted using mtDNA and Y-chromosome haplogroup frequencies and SPSS10.0 software (SPSS Inc.). The genetic difference between northern and southern Hans was tested by AMOVA[25], using ARLEQUIN software[26]. ADMIX 2.0 (ref. 27) and LEADMIX[21] software were used to estimate the level of admixture of the northern Hans and southern natives in the southern Han populations, using two different statistics[19,20]. The selection of parental populations is critical for appropriate estimation of admixture proportion[28,29] and we were careful to minimize bias by using large data sets across East Asia. In this analysis, the average haplogroup frequencies (for Y-chromosome or mtDNA markers, respectively) of northern Hans (arithmetic mean of 10 northern Hans) were taken for the northern parental population. The frequency of southern natives was estimated by the average of three groups including Austro-Asiatic (NRY, 6 populations; mtDNA, 5 populations), Daic (NRY, 22 populations; mtDNA, 11 populations) and Hmong-Mien (NRY, 18 populations; mtDNA, 14 populations). The geographic pattern of Han populations was revealed by the linear regression analysis of admixture proportion against the latitudes of samples[1,3].

(**431,** 302-305, 2004)

**Bo Wen[1,2], Hui Li[1], Daru Lu[1], Xiufeng Song[1], Feng Zhang[1], Yungang He[1], Feng Li[1], Yang Gao[1], Xianyun Mao[1], Liang Zhang[1], Ji Qian[1], Jingze Tan[1], Jianzhong Jin[1], Wei Huang[2], Ranjan Deka[3], Bing Su[1,3,4], Ranajit Chakraborty[3] & Li Jin[1,3]**

[1] State Key Laboratory of Genetic Engineering and Center for Anthropological Studies, School of Life Sciences and Morgan-Tan International Center for Life Sciences, Fudan University, Shanghai 200433, China

[2] Chinese National Human Genome Center, Shanghai 201203, China

[3] Center for Genome Information, Department of Environmental Health, University of Cincinnati, Cincinnati, Ohio 45267, USA

[4] Key Laboratory of Cellular and Molecular Evolution, Kunming Institute of Zoology, the Chinese Academy of Sciences, Kunming 650223, China

Received 28 April; accepted 20 July, 2004; doi: 10.1038/nature02878.

---

References:

1. Cavalli-Sforza, L. L., Menozzi, P. & Piazza, A. *The History and Geography of Human Genes* (Princeton Univ. Press, Princeton, 1994).

2. Sokal, R., Oden, N. L. & Wilson, C. Genetic evidence for the spread of agriculture in Europe by demic diffusion. *Nature* **351**, 143-145 (1991).

3. Chikhi, L. *et al.* Y genetic data support the Neolithic demic diffusion model. *Proc. Natl Acad. Sci. USA* **99**, 11008-11013 (2002).

4. Fei, X. T. *The Pattern of Diversity in Unity of the Chinese Nation* (Central Univ. for Nationalities Press, Beijing, 1999).

5. Ge, J. X., Wu, S. D. & Chao, S. J. *Zhongguo yimin shi (The Migration History of China)* (Fujian People's Publishing House, Fuzhou, China, 1997).

6. Zhao, T. M. & Lee, T. D. Gm and Km allotypes in 74 Chinese populations: a hypothesis of the origin of the Chinese nation. *Hum. Genet.* **83**, 101-110 (1989).

7. Du, R. F., Xiao, C. J. & Cavalli-Sforza, L. L. Genetic distances calculated on gene frequencies of 38 loci. *Sci. China* **40**, 613 (1997).

8. Chu, J. Y. *et al.* Genetic relationship of populations in China. *Proc. Natl Acad. Sci. USA* **95**, 11763-11768 (1998).

9. Xiao, C. J. *et al.* Principal component analysis of gene frequencies of Chinese populations. *Sci. China* **43**, 472-481 (2000).

10. Xu, Y. T. A brief study on the origin of Han nationality. *J. Centr. Univ. Natl* **30**, 59-64 (2003).

11. Su, B. *et al.* Y chromosome haplotypes reveal prehistorical migrations to the Himalayas. *Hum. Genet.* **107**, 582-590 (2000).

12. Yao, Y. G. *et al.* Phylogeographic differentiation of mitochondrial DNA in Han Chinese. *Am. J. Hum. Genet.* **70**, 635-651 (2002).

13. Cavalli-Sforza, L. L. & Feldman, M. W. The application of molecular genetic approaches to the study of human evolution. *Nature Genet.* **33**, 266-275 (2003).

14. Wallace, D. C., Brown, M. D. & Lott, M. T. Nucleotide mitochondrial DNA variation in human evolution and disease. *Gene* **238**, 211-230 (1999).

15. Underhill, P. A. *et al.* Y chromosome sequence variation and the history of human populations. *Nature Genet.* **26**, 358-361 (2000).

## 数据分析

根据mtDNA和Y染色体单倍群频率，用SPSS10.0软件(SPSS公司)作主成分分析，研究群体间关系。南北汉族的遗传差异用ARLEQUIN软件[26]做AMOVA检验[25]。南方汉族中北方汉族和南方原住民族的混合比例估计用两种不同的统计方法[19,20]：ADMIX 2.0[27]和LEADMIX[21]软件。亲本群体的选择对混合比例的适当估计很重要[28,29]，我们通过扩大东亚的参考数据来减小偏差。分析中，10个北方汉族群体的各单倍群频率(Y染色体和mtNDA标记分别分析)的算术平均作为北方亲本群体。南方原住民族的频率是三个族群的平均：侗台语群(NRY，22群体；mtNDA，11群体)，南亚语群(NRY，6群体；mtNDA，5群体)，苗瑶语群(NRY，18群体；mtNDA，14群体)。通过样本的混合比例与纬度[1,3]的线性回归分析揭示汉族群体的地理格局。

(李辉 翻译；徐文堪 审稿)

16. Jobling, M. A. & Tyler-Smith, C. The human Y chromosome: an evolutionary marker comes of age. *Nature Rev. Genet.* **4**, 598-612 (2003).

17. Su, B. *et al.* Y-chromosome evidence for a northward migration of modern humans into eastern Asia during the last ice age. *Am. J. Hum. Genet.* **65**, 1718-1724 (1999).

18. Kivisild, T. *et al.* The emerging limbs and twigs of the East Asian mtDNA tree. *Mol. Biol. Evol.* **19**, 1737-1751 (2002).

19. Roberts, D. F. & Hiorns, R. W. Methods of analysis of the genetic composition of a hybrid population. *Hum. Biol.* **37**, 38-43 (1965).

20. Bertorelle, G. & Excoffier, L. Inferring admixture proportions from molecular data. *Mol. Biol. Evol.* **15**, 1298-1311 (1998).

21. Wang, J. Maximum-likelihood estimation of admixture proportions from genetic data. *Genetics* **164**, 747-765 (2003).

22. Wen, B. *et al.* Analyses of genetic structure of Tibeto-Burman populations revealed a gender-biased admixture in southern Tibeto-Burmans. *Am. J. Hum. Genet.* **74**, 856-865 (2004).

23. Jin, L. & Su, B. Natives or immigrants: modern human origin in East Asia. *Nature Rev. Genet.* **1**, 126-133 (2000).

24. The Y Chromosome Consortium, A nomenclature system for the tree of human Y-chromosomal binary haplogroups. *Genome Res.* **12**, 339-348 (2002).

25. Excoffier, L., Smouse, P. E. & Quattro, J. M. Analysis of molecular variance inferred from metric distances among DNA haplotypes: application to human mitochondrial DNA restriction data. *Genetics* **131**, 479-491 (1992).

26. Schneider, S., *et al.* Arlequin: Ver. 2.000. A software for population genetic analysis. (Genetics and Biometry Laboratory, Univ. of Geneva, Geneva, 2000).

27. Dupanloup, I. & Bertorelle, G. Inferring admixture proportions from molecular data: extension to any number of parental populations. *Mol. Biol. Evol.* **18**, 672-675 (2001).

28. Chakraborty, R. Gene admixture in human populations: Models and predictions. *Yb. Phys. Anthropol.* **29**, 1-43 (1986).

29. Sans, M. *et al.* Unequal contributions of male and female gene pools from parental populations in the African descendants of the city of Melo, Uruguay. *Am. J. Phys. Anthropol.* **118**, 33-44 (2002).

**Supplementary Information** accompanies the paper on www.nature.com/nature.

**Acknowledgements.** We thank all of the donors for making this work possible. The data collection was supported by NSFC and STCSM to Fudan and a NSF grant to L.J. L.J., R.D. and R.C. are supported by NIH.

**Competing interests statement.** The authors declare that they have no competing financial interests.

Correspondence and requests for materials should be addressed to L.J. (lijin@fudan.edu.cn or li.jin@uc.edu). The mtDNA HVS-1 sequences of 711 individuals from 15 Han populations were submitted to GenBank with accession numbers AY594701–AY595411.

# A New Small-bodied Hominin from the Late Pleistocene of Flores, Indonesia

P. Brown *et al.*

## Editor's Note

**When archaeologist Michael Morwood set out to trace how early modern humans got to Australia, the last thing he and his colleagues expected to unearth in Liang Bua cave on Flores was a creature out of folk-tale. But that's what they discovered. "The Hobbit": a human-like creature that stood barely a metre tall, had a very small brain and a body reminiscent of hominids that lived in Africa, three or more million years ago—but which survived until at least 18,000 years ago. *Homo floresiensis* challenged conventional wisdom about the adaptability of the human form, as well as the assumption that *Homo sapiens* reigned alone on Earth. The sensation created by this strangest of fossil humans has yet to fade.**

---

Currently, it is widely accepted that only one hominin genus, *Homo*, was present in Pleistocene Asia, represented by two species, *Homo erectus* and *Homo sapiens*. Both species are characterized by greater brain size, increased body height and smaller teeth relative to Pliocene *Australopithecus* in Africa. Here we report the discovery, from the Late Pleistocene of Flores, Indonesia, of an adult hominin with stature and endocranial volume approximating 1 m and 380 $cm^3$, respectively—equal to the smallest-known australopithecines. The combination of primitive and derived features assigns this hominin to a new species, *Homo floresiensis*. The most likely explanation for its existence on Flores is long-term isolation, with subsequent endemic dwarfing, of an ancestral *H. erectus* population. Importantly, *H. floresiensis* shows that the genus *Homo* is morphologically more varied and flexible in its adaptive responses than previously thought.

---

THE LB1 skeleton was recovered in September 2003 during archaeological excavation at Liang Bua, Flores[1]. Most of the skeletal elements for LB1 were found in a small area, approximately 500 $cm^2$, with parts of the skeleton still articulated and the tibiae flexed under the femora. Orientation of the skeleton in relation to site stratigraphy suggests that the body had moved slightly down slope before being covered with sediment. The skeleton is extremely fragile and not fossilized or covered with calcium carbonate. Recovered elements include a fairly complete cranium and mandible, right leg and left innominate. Bones of the left leg, hands and feet are less complete, while the vertebral column, sacrum, scapulae, clavicles and ribs are only represented by fragments. The position of the skeleton suggests that the arms are still in the wall of the excavation, and may be recovered in the future. Tooth eruption, epiphyseal union and tooth wear indicate an adult, and pelvic

# 印度尼西亚弗洛勒斯晚更新世一身材矮小人族新成员

布朗等

## 编者按

当考古学家迈克尔·莫伍德开始研究早期现代人是如何到达澳大利亚的时候，他和他的同事们最不希望在弗洛勒斯的利昂布阿洞穴发现的是一种民间传说中的生物，但这就是他们的发现。“霍比特人”：这是一种类似人类的生物，身高只有一米，大脑非常小，身体让人想起300万年前或更早以前生活在非洲的原始人，但这种原始人至少存活到18,000年前。弗洛勒斯人挑战了关于人类形态适应性的传统观点，也挑战了智人独自统治地球的假设。这种由最奇怪的人类化石产生的轰动还没有消退。

---

目前，人们普遍认为，更新世人族在亚洲仅存在一个属，即人属，以直立人和智人两个物种为代表。相对于非洲上新世的南方古猿属，这两个种的特征是脑容量更大，身高更高，牙齿更小。在此我们报道，在印度尼西亚弗洛勒斯岛晚更新世发现的一个人族成员成年个体化石，其身高和颅腔容量分别约为1米和380立方厘米，相当于已知最小的南方古猿。其原始与衍生特征的组合，将这一人族成员定为一个新的物种——弗洛勒斯人。对其为何存在于弗洛勒斯岛上，最可能的解释是一个祖先直立人种群的长期隔离，并在随后出现了地方性的矮化。重要的是，弗洛勒斯人显示出人属在形态上的适应性反应比以前所想象的更多样化，也更灵活。

---

LB1骨架是2003年9月在弗洛勒斯岛的利昂布阿进行考古挖掘期间出土的[1]。LB1的大部分骨骼部位都是在大约500平方厘米的一个很小的区域内被发现的，骨架不少部位仍然互相关节，胫骨弯曲在股骨下方。骨架方位与遗址地层的相对关系表明躯体在被沉积物掩埋之前，沿坡轻微向下移动了一些。骨骼非常脆弱，没有石化，也未被碳酸钙覆盖。出土的部位包括相当完整的颅骨与下颌骨，右腿和左髋骨。左腿、手和脚的骨头的完整性稍差，而脊柱、荐椎、两侧肩胛骨、两侧锁骨和肋骨仅有一些碎片保存。骨架的姿态表明其手臂还埋在发掘探方的侧壁里，将来可能会被发现。牙齿的萌发状况、骨骺的愈合程度及牙齿的磨损程度表明这是一个成年个

anatomy strongly supports the skeleton being that of a female. On the basis of its unique combination of primitive and derived features we assign this skeleton to a new species, *Homo floresiensis*.

## Description of *Homo floresiensis*

Order Primates Linnaeus, 1758

Suborder Anthropoidea Mivart, 1864

Superfamily Hominoidea Gray, 1825

Family Hominidae Gray, 1825

Tribe Hominini Gray, 1825

Genus *Homo* Linnaeus, 1758

***Homo floresiensis*** sp. nov.

**Etymology.** Recognizing that this species has only been identified on the island of Flores, and a prolonged period of isolation may have resulted in the evolution of an island endemic form.

**Holotype.** LB1 partial adult skeleton excavated in September 2003. Recovered skeletal elements include the cranium and mandible, femora, tibiae, fibulae and patellae, partial pelvis, incomplete hands and feet, and fragments of vertebrae, sacrum, ribs, scapulae and clavicles. The repository is the Centre for Archaeology, Jakarta, Indonesia.

**Referred material.** LB2 isolated left mandibular $P_3$. The repository is the Centre for Archaeology, Jakarta, Indonesia.

**Localities.** Liang Bua is a limestone cave on Flores, in eastern Indonesia. The cave is located 14 km north of Ruteng, the provincial capital of Manggarai Province, at an altitude of 500 m above sea level and 25 km from the north coast. It occurs at the base of a limestone hill, on the southern edge of the Wae Racang river valley. The type locality is at 08° 31′ 50.4″ south latitude 120° 26′ 36.9″ east longitude.

**Horizon.** The type specimen LB1 was found at a depth of 5.9 m in Sector VII of the excavation at Liang Bua. It is associated with calibrated accelerator mass spectrometry (AMS) dates of approximately 18 kyr and bracketed by luminescence dates of $35 \pm 4$ kyr and $14 \pm 2$ kyr. The referred isolated left $P_3$ (LB2) was recovered just below a discomformity

体，骨盆的解剖学结构特征有力地支持这是一个女性个体的骨架。根据其原始与衍生特征的独特组合，我们将这具骨架定为一新种——弗洛勒斯人。

## 弗洛勒斯人的描述

灵长目　Primates Linnaeus，1758

类人猿亚目　Anthropoidea Mivart，1864

人猿超科　Hominoidea Gray，1825

人科　Hominidae Gray，1825

人族　Hominini Gray，1825

人属　*Homo* Linnaeus，1758

弗洛勒斯人（新种）*Homo floresiensis* sp. nov.

**词源**　考虑到该种仅发现于弗洛勒斯岛，并且可能由于长期的隔离才造成了一个岛屿特有类型的演化。

**正型标本**　2003 年 9 月发掘出土的不完整成年个体骨架 LB1。出土的骨骼部位包括颅骨与下颌骨，两侧股骨，胫骨，腓骨和髌骨，不完整骨盆，不完全的双手和双脚，以及椎体、荐椎、肋骨、两侧肩胛骨和两侧锁骨的碎片。存放在印度尼西亚雅加达考古中心。

**归入材料**　LB2 游离左下第 3 前臼齿。存放在印度尼西亚雅加达考古中心。

**产地**　利昂布阿是印度尼西亚东部弗洛勒斯岛上的一个石灰岩洞穴。该洞穴位于芒加莱省会城市鲁滕以北 14 千米处，其海拔 500 米，距北海岸 25 千米。它位于威拉肯河谷南缘的石灰岩山丘底部。模式产地位于南纬 08° 31′ 50.4″ 东经 120° 26′ 36.9″。

**层位**　模式标本 LB1 发现于利昂布阿发掘 VII 区 5.9 米深处。与其直接相关的校准加速器质谱法（AMS）测定年代大约为 18,000 年，释光法把年代限定在 35,000 ± 4,000 年与 14,000 ± 2,000 年之间。归入材料中的游离左下第 3 前臼齿（LB2）发现于 IV 区 4.7 米处不整合面正下方，其年代限定范围在流石 U 系测定年代

at 4.7 m in Sector IV, and bracketed by a U-series date of 37.7 ± 0.2 kyr on flowstone, and 20 cm above an electron-spin resonance (ESR)/U-series date of $74^{+14}_{-12}$ kyr on a *Stegodon* molar.

**Diagnosis.** Small-bodied bipedal hominin with endocranial volume and stature (body height) similar to, or smaller than, *Australopithecus afarensis*. Lacks masticatory adaptations present in *Australopithecus* and *Paranthropus*, with substantially reduced facial height and prognathism, smaller postcanine teeth, and posteriorly orientated infraorbital region. Cranial base flexed. Prominent maxillary canine juga form prominent pillars, laterally separated from nasal aperture. Petrous pyramid smooth, tubular and with low relief, styloid process absent, and without vaginal crest. Superior cranial vault bone thicker than *Australopithecus* and similar to *H. erectus* and *H. sapiens*. Supraorbital torus arches over each orbit and does not form a flat bar as in Javan *H. erectus*. Mandibular $P_3$ with relatively large occlusal surface area, with prominent protoconid and broad talonid, and either bifurcated roots or a mesiodistally compressed Tomes root. Mandibular $P_4$ also with Tomes root. First and second molar teeth of similar size. Mandibular coronoid process higher than condyle, and the ramus has a posterior orientation. Mandibular symphysis without chin and with a posterior inclination of the symphysial axis. Posteriorly inclined alveolar planum with superior and inferior transverse tori. Ilium with marked lateral flare. Femur neck long relative to head diameter, the shaft circular and without pilaster, and there is a high bicondylar angle. Long axis of tibia curved and the midshaft has an oval cross-section.

## Description and Comparison of the Cranial and Postcranial Elements

Apart from the right zygomatic arch, the cranium is free of substantial distortion (Figs 1 and 2). Unfortunately, the bregmatic region, right frontal, supraorbital, nasal and subnasal regions were damaged when the skeleton was discovered. To repair post-mortem pressure cracks, and stabilize the vault, the calvarium was dismantled and cleaned endocranially before reconstruction. With the exception of the squamous suture, most of the cranial vault sutures are difficult to locate and this problem persists in computed tomography (CT) scans. As a result it is not possible to locate most of the standard craniometric landmarks with great precision.

37,700±200 年以及 20 厘米之上的一个剑齿象臼齿电子自旋共振(ESR)/U 系测定年代 $74,000^{+14,000}_{-12,000}$ 年之间。

**特征** 身材矮小两足行走的人族成员，其颅腔容积和身材(身高)相当于或小于南方古猿阿法种。缺少南方古猿与傍人所表现出的咀嚼适应性变化，面部高度与凸颌程度明显降低，犬后齿较小，眶下区朝后。颅底角度收缩。突出的上犬齿隆突呈突出柱状，在侧面与鼻孔分离。岩锥平滑、呈管状且凸出程度低，茎突缺失，且没有茎突鞘脊。上部颅顶骨厚于南方古猿而与直立人及智人相当。眶上圆枕在每一个眼眶上呈拱形，且不像爪哇直立人那样呈水平条状。下第 3 前臼齿有相对较大的咀嚼面表面积，下原尖突出，下跟座宽阔，而且牙根或者完全分叉或者为前后压扁的托姆斯式齿根。下第 4 前臼齿也具有托姆斯式齿根。第一和第二臼齿大小相当。颌骨冠状突高于髁突，且上升支向后倾。下颌联合部没有颏，联合部轴线向后倾斜。齿槽面向后倾斜，并且有上下圆枕。髂骨具明显侧向展开。股骨颈部相对于股骨头直径相对显长，骨干截面呈圆形且没有粗线嵴发育，双髁角角度大。胫骨长轴弯曲且骨干中部的横截面为椭圆形。

## 颅骨与颅后骨骼部位的描述与对比

除了右侧颧弓，颅骨基本没有变形(图 1、图 2)。不幸的是，骨架发现时，其前囟区、右侧额骨、眶上、鼻骨及鼻下区域被损坏。为了修复死后遭受压力产生的裂缝、稳固穹隆，在重建前拆开了颅顶穹隆并从颅内进行了清理。除鳞骨骨缝以外，大部分颅骨穹隆骨缝难以定位，这一问题在计算机断层扫描(CT)中也存在。因此大多数标准颅骨测量标志点都不可能高精度定位。

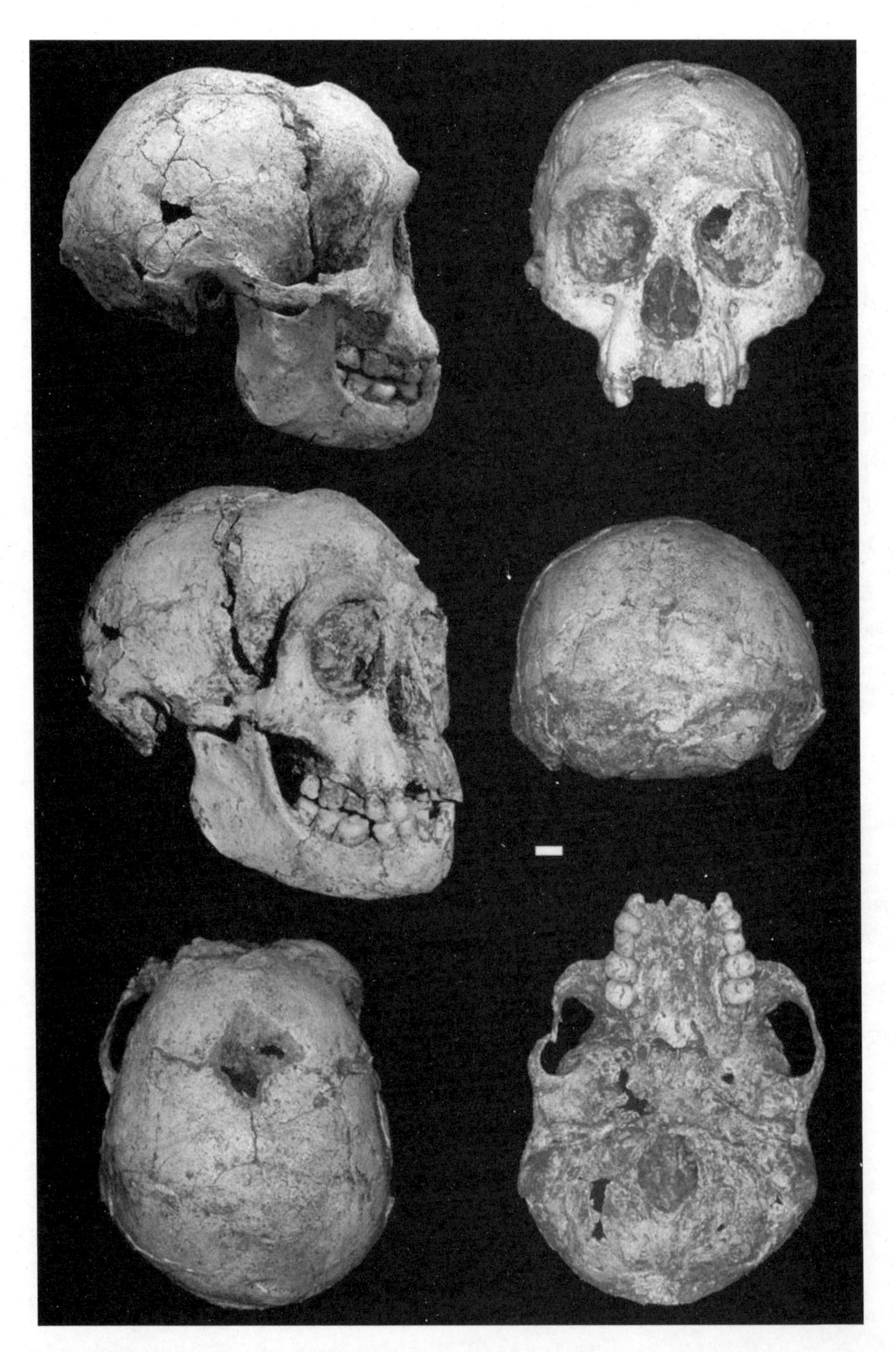

Fig. 1. The LB1 cranium and mandible in lateral and three-quarter views, and cranium in frontal, posterior, superior and inferior views. Scale bar, 1 cm.

The LB1 cranial vault is long and low. In comparison with adult *H. erectus* (including

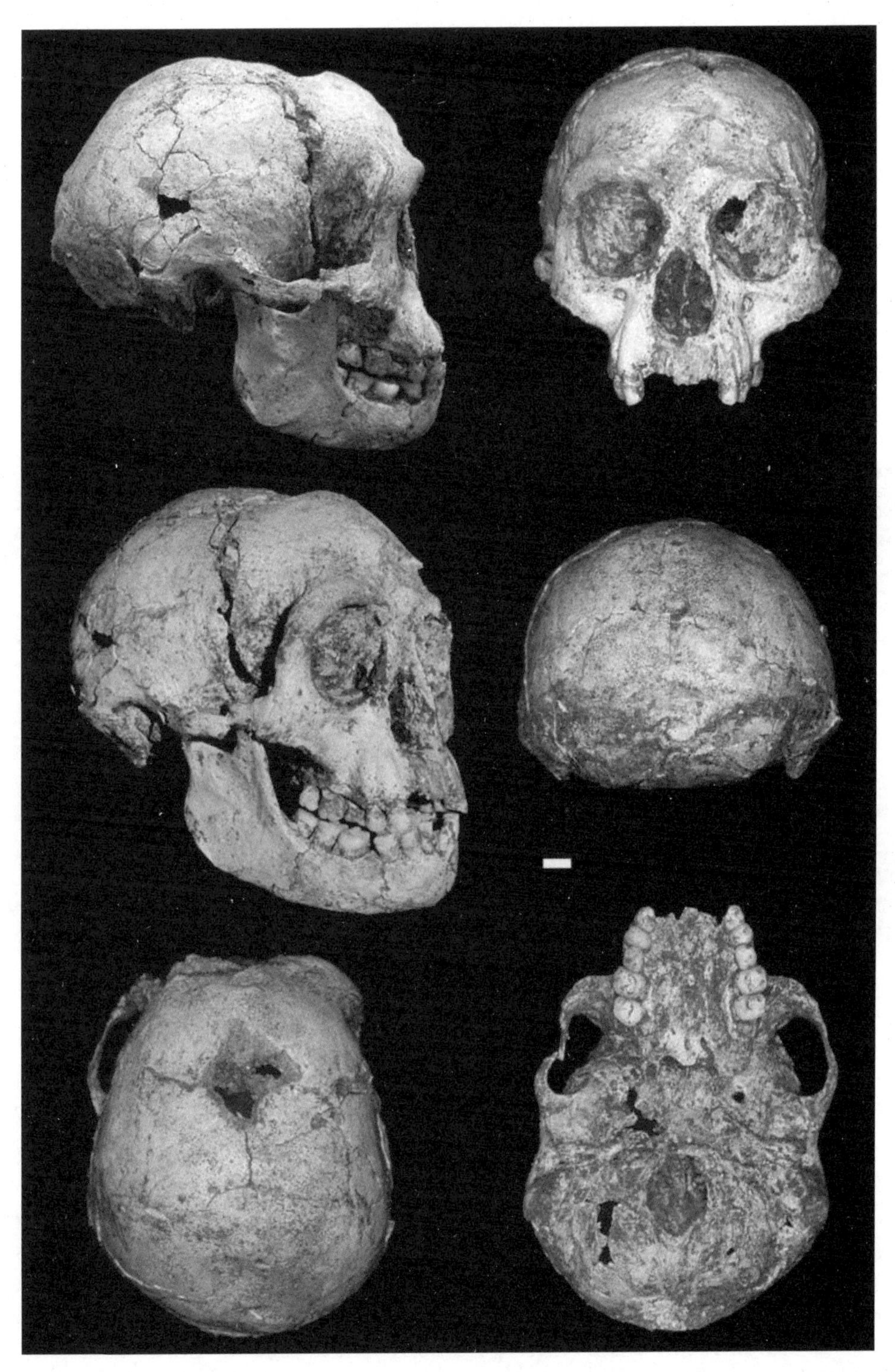

图 1. LB1 颅骨与下颌骨的侧视图与四分之三视图。颅骨的正视、后视、顶视及底视图。比例尺，1 厘米。

LB1 颅骨穹隆长而低。与成年直立人（包括被归属为匠人与格鲁吉亚人的标本）

specimens referred to as *Homo ergaster* and *Homo georgicus*) and *H. sapiens* the calvarium of LB1 is extremely small. Indices of cranial shape closely follow the pattern in *H. erectus* (Supplementary Table 1). For instance, maximum cranial breadth is in the inflated supramastoid region, and the vault is broad relative to its height. In posterior view the parietal contour is similar to *H. erectus* but with reduced cranial height[2,3]. Internal examination of the neurocranium, directly and with CT scan data, indicates that the brain of LB1 had a flattened platycephalic shape, with greatest breadth across the temporal lobes and reduced parietal lobe development compared with *H. sapiens*. The cranial base angle (basion–sella–foramen caecum) of 130° is relatively flexed in comparison with both *H. sapiens* (mean 137°–138° (refs 4, 5)) and Indonesian *H. erectus* (Sambungmacan 4 141° (ref. 6)). Other small-brained hominins, for instance STS 5 *Australopithecus africanus*, have the primitive less-flexed condition.

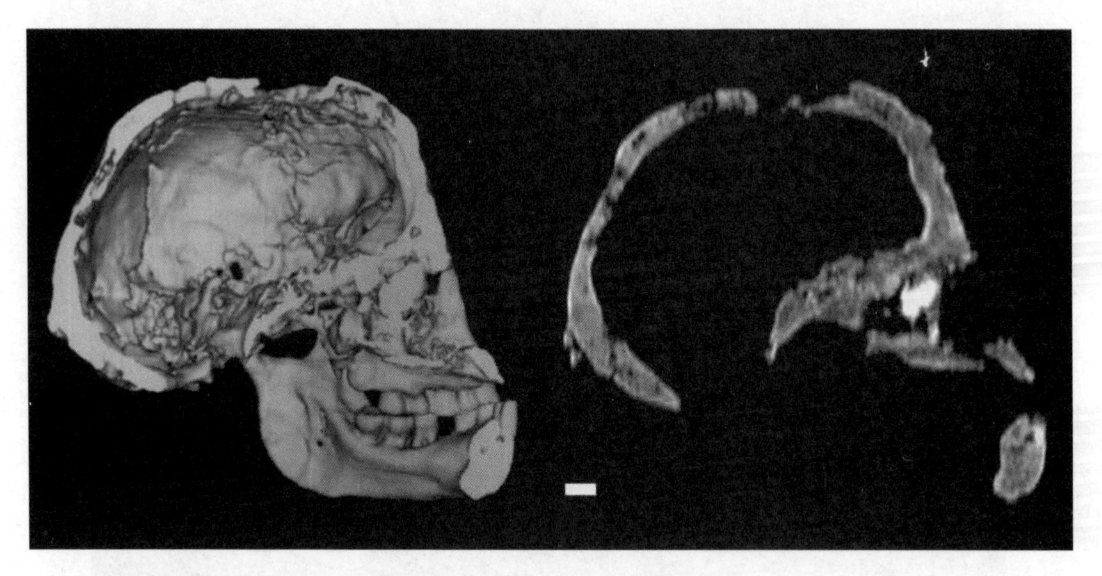

Fig. 2. Rendered three-dimensional and individual midsagittal CT section views of the LB1 cranium and mandible. Scale bar, 1 cm.

The endocranial volume, measured with mustard seed, is 380 $cm^3$, well below the previously accepted range for the genus *Homo*[7] and equal to the minimum estimates for *Australopithecus*[8]. The endocranial volume, relative to an indicator of body height (maximum femur length 280 mm), is outside the recorded hominin normal range (Fig. 3). Medially, laterally and basally, the cranial vault bone is thick and lies within the range of *H. erectus* and *H. sapiens*[9,10] (Supplementary Table 1 and Fig. 2). Reconstruction of the cranial vault, and CT scans, indicated that for most of the cranial vault the relative thickness of the tabular bone and diploë are similar to the normal range in *H. erectus* and *H. sapiens*. In common with *H. erectus* the vault in LB1 is relatively thickened posteriorly and in areas of pneumatization in the lateral cranial base. Thickened vault bone in LB1, relative to that in *Australopithecus* and early *Homo*[2], results in a substantially reduced endocranial volume in comparison to Plio-Pleistocene hominins with similar external vault dimensions.

及智人相比，LB1 的颅骨穹隆部极小。颅骨形状指数非常符合直立人的模式（补充表 1）。例如，颅骨宽度最大在膨胀的乳突上方区域，穹隆相对于其高度更显宽阔。在后视图中，顶骨轮廓与直立人相似，但是颅骨高度减小[2,3]。对脑颅内部的直接观察与 CT 扫描数据观察表明 LB1 的大脑具有变平了的扁头形状，最大宽度横过颞叶；与智人相比，顶叶发育减弱。颅底角（颅底点–蝶鞍点–孔盲）为 130°，它与智人（平均为 137° ~ 138°（参考文献 4、5））及印度尼西亚直立人（萨姆伯戈默卡 4 141°（参考文献 6））相比，相对屈曲。其他脑容量小的人族成员，例如，STS 5 南方古猿非洲种，拥有屈曲程度较低的原始状态。

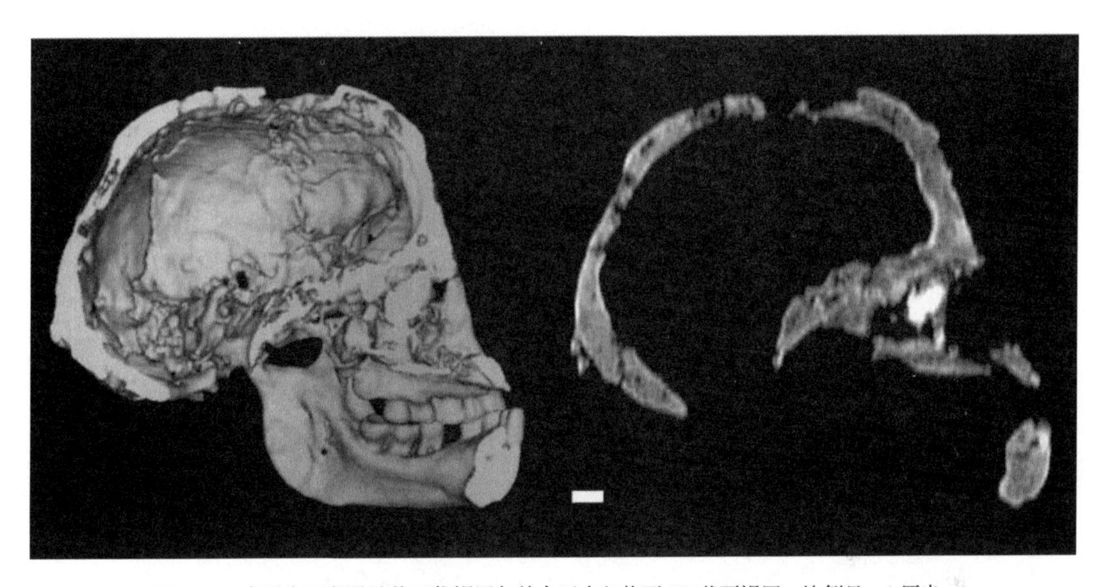

图 2. LB1 颅骨与下颌骨渲染三维视图与单个正中矢状面 CT 截面视图。比例尺，1 厘米。

用芥菜籽来测量的颅腔容积是 380 立方厘米，远低于以前公认的人属[7]的范围，而与南方古猿[8]的最小估计值相当。相对于身高指标（最大股骨长度 280 毫米），颅腔容积位于已知人族成员的正常范围之外（图 3）。颅骨穹隆部骨骼在中间、侧面及底部都厚且在直立人与智人[9,10]的范围之内（补充表 1 和图 2）。颅骨穹隆的重建及 CT 扫描表明，对于颅骨穹隆的大部分来讲，板状骨与板障骨的相对厚度与直立人及智人的正常范围相当。与直立人相同，在 LB1 穹隆后部与颅底侧面气腔区域相对变厚。相对于南方古猿和早期人属成员[2]，LB1 增厚的穹隆骨骼，导致与外部穹隆尺寸相似的上新世 – 更新世人族成员相比，颅腔容积大幅减少。

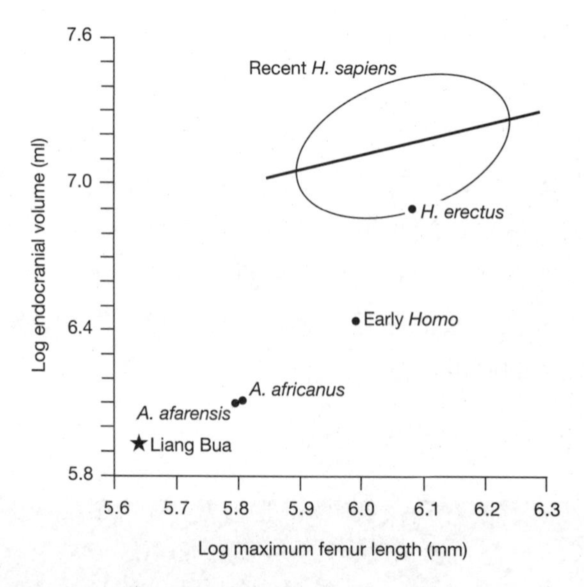


Fig. 3. Relationship between endocranial volume and femur length in LB1, *A. afarensis*, *A. africanus*, early *Homo* sp., *H. erectus* and modern *H. sapiens*. Modern human data, with least squares regression line and 95% confidence ellipse, from a global sample of 155 individuals collected by P.B. Details of the hominin samples are in the Supplementary Information.

The occipital of LB1 is strongly flexed, with an occipital curvature angle of 101° (Supplementary Information), and the length of the nuchal plane dominates over the occipital segment. The occipital torus forms a low extended mound, the occipital protuberance is not particularly prominent compared with Indonesian *H. erectus* and there is a shallow supratoral sulcus. The endinion is positioned 12 mm inferior to the inion, which is within the range of *H. erectus* and *Australopithecus*[10]. Compared with *Australopithecus* and early *Homo*[2] the foramen magnum is narrow (21 mm) relative to its length (28 mm), and mastoid processes are thickened mediolaterally and are relatively deep (20.5 mm). In common with Asian, and some African, *H. erectus* a deep fissure separates the mastoid process from the petrous crest of the tympanic[10,11]. Bilaterally there is a recess between the tympanic plate and the entoglenoid pyramid. These two traits are not seen in modern humans, and show varied levels of development in Asian and African *H. erectus* and Pliocene hominins[10]. The depth and breadth of the glenoid fossae and angulation of the articular eminence are within the range of variation in *H. sapiens*. The inferior surface of the petrous pyramid has numerous similarities with Zhoukoudian *H. erectus*[12], with a smooth tubular external surface as in chimpanzees, and a constricted foramen lacerum. Styloid processes and vaginal crests are not present.

The temporal lines approach to within 33 mm of the coronal suture and have a marked posterior extension. There are no raised angular tori as is common in *H. erectus*[10] and some terminal Pleistocene Australians, and no evidence of parietal keeling. Posteriorly there is some asymmetrical obelionic flattening and CT scans indicate that the parietals reduce in thickness in this slightly depressed area (Fig. 2). A principal component analysis (PCA) of five cranial vault measurements separates LB1, STS5 (*A. africanus*) and KNM-ER 1813

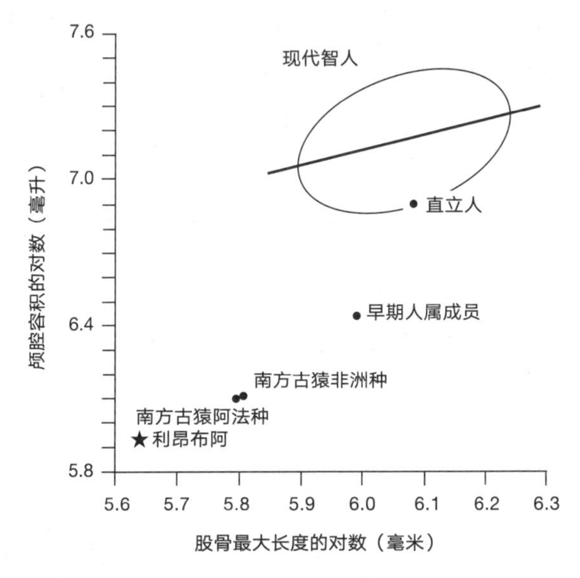


图 3. 在 LB1、南方古猿阿法种、南方古猿非洲种、早期人属未定种、直立人及现代智人中，颅腔容积与股骨长度之间的关系。现代人类的数据带有最小二乘回归线与 95% 置信椭圆，来自布朗收集的全球 155 个个体样本。人族成员样本的详细情况见补充信息。

LB1 的枕骨强烈向下屈曲，枕骨弯曲角 101°(补充信息)，而且项平面长度占了枕节的大部分。枕骨圆枕形成了一个较低的延伸小丘，与印度尼西亚直立人相比，枕骨隆突不是特别突出，并且有一浅上圆枕沟。枕内点位置低于枕外隆凸点 12 毫米，在直立人与南方古猿的范围之内 [10]。与南方古猿及早期人属成员 [2] 相比，枕骨大孔相对于其长度(28 毫米)显狭窄(21 毫米)，乳突由中线向侧面增厚，相对较深(20.5 毫米)。与亚洲和某些非洲直立人相同，有一条很深的裂缝将乳突与鼓室的岩骨脊分开 [10,11]。在鼓板与内颞颌关节锥之间有一两侧凹陷。这两个特征在现代人中是看不到的，显示出亚洲及非洲直立人和上新世人族成员的多样的发育水平 [10]。关节窝的深度和宽度以及关节隆起的角度在智人的变异范围之内。岩锥的下表面与周口店直立人 [12] 有许多相似之处，拥有黑猩猩那样光滑管状的外表面，以及收缩的破裂孔。无茎突与茎突鞘脊突。

颞线与冠状缝相交的位置，距离冠状缝起点在 33 毫米之内，且颞线有明显的向后延伸。不具有直立人 [10] 与一些更新世末尾的澳大利亚人中常见的升高的角圆枕，也没有顶骨矢状隆起的证据。在后方矢状缝的顶骨孔之间区域有一不对称变平，CT 扫描表明顶骨在这个稍微凹陷的区域厚度减少(图 2)。5 个颅骨穹隆测量值的主成分分析(PCA)，将 LB1、STS5(南方古猿非洲种)与 KNM-ER 1813(早期人属成员)

(early *Homo*) from other hominin calvaria in size and shape. Shape, particularly height and breadth relationships, placed LB1 closest to ER-3883, ER-3733 and Sangiran 2 *H. erectus* (Supplementary Fig. 1).

The face of LB1 lacks most of the masticatory adaptations evident in *Australopithecus* and its overall morphology is similar to members of the genus *Homo*[2,3]. In comparison with *Australopithecus*, tooth dimensions and the alveolar segment of the maxillae are greatly reduced, as are facial height and prognathism. The facial skeleton is dominated by pronounced canine juga, which form prominent pillars lateral to the nasal aperture. However, these are distinct from the anterior pillars adjacent to the nasal aperture in *A. africanus*[2,3]. The infraorbital fossae are deep with large infraorbital foramina, the orbits have a particularly arched superior border and a volume of 15.5 $cm^3$ (ref. 13). On the better preserved right-hand side, the supraorbital torus arches over the orbit and does not form a straight bar, with bulbous laterally projecting trigones, as in Indonesian *H. erectus*[11]. The preserved section of the right torus only extends medially slightly past mid-orbit, and the morphology of the glabella region and medial torus is unknown. In facial view the zygo-maxillary region is medially deep relative to facial height, and the inferior border of the malars are angled at 55° relative to the coronal plane. In lateral view the infraorbital region is orientated posteriorly as in other members of the genus *Homo*, rather than the more vertical orientation in *A. africanus*[2,3]. The root of the maxillary zygomatic process is centred above the first molar, and the incisive canal is relatively large and has an anterior location, contrasting with African and Javan *H. erectus*. In lateral view, curvature of the frontal squama is more similar to African early *Homo* and Dmanisi *H. ergaster*[3,14] than it is to the Javan hominins. The frontal squama is separated from the supraorbital torus by a supraorbital sulcus. In the middle third of the frontal there is a slight sagittal keel, extending into the remains of a low, broad prebregmatic eminence. On the midfrontal squama there is a circular healed lesion, probably the remains of a depressed fracture, which is about 15 mm across.

The mandible is complete, apart from some damage to the right condyle (Fig. 4) and combines features present in a variety of Pliocene and Pleistocene hominins. Post-mortem breaks through the corpus at the right $P_3$ and $M_2$, and the left canine have resulted in some lateral distortion of the right ramus. There is a strong Curve of Spee. The ramus root inserts on the corpus above the lateral prominence, and in lateral aspect obscures the distal $M_3$. The ramus is broadest inferiorly, slopes slightly posteriorly and is thickened medio-laterally, and the coronoid process is higher than the condyle. The right condyle has a maximum breadth of 18 mm. There is a narrow and shallow extramolar sulcus and moderate lateral prominence. The anterior portion of the corpus is rounded and bulbous and without a chin. In the posterior symphyseal region the alveolar planum inclines postero-inferiorly, there is a moderate superior torus, deep and broad digastric fossa, and the inferior transverse torus is low and rounded rather than shelf-like (Fig. 4). There is a strong posterior angulation of the symphyseal axis, and the overall morphology of the symphysis is very similar to LH4 *A. afarensis* and unlike Zhoukoudian and Sangiran *H. erectus*. There are bilaterally double mental foramina, with the posterior foramina smaller and located more inferiorly. Double

在大小与形状上与其他人族成员颅顶分开。形状，特别是高度和宽度之间的关系，把 LB1 放到最靠近 ER-3883、ER-3733 及桑吉兰 2 直立人的位置（补充图 1）。

LB1 面部缺少绝大多数南方古猿中明显的与咀嚼相关的适应特征，且其总体形态与人属成员相似[2,3]。与南方古猿对比，其牙齿大小与上颌骨的齿槽部分都明显减小，面部高度和凸颌程度也是同样的情况。犬齿隆凸突出，在面部骨骼上非常显著，在鼻孔侧面形成突出圆柱。然而，这与南方古猿非洲种中邻近鼻孔的前部圆柱截然不同[2,3]。眶下窝深，眶下孔大，眼窝具有一个特别拱曲的上缘，其容积为 15.5 立方厘米（参考文献 13）。在保存较好的右侧，眶上圆枕在眼眶上成拱形，并不是形成平直条状，具有向侧面突出的球状三角区，正如印度尼西亚直立人中那样[11]。保存的圆枕右侧仅向中间延伸稍微超过眼眶中部，眉间区与中间圆枕形态未知。在正视图中，颧骨上颌骨区域，相对于面高，至面部中线的深度显大，颧骨下缘相对于冠状面所成角度为 55°。侧视图中，眶下区域朝向后方，如人属其他成员一样，而不像南方古猿非洲种那样朝向更加垂直[2,3]。上颌骨颧突根部以第一臼齿上方为中心，门齿管相当大且位置靠前，与非洲和爪哇直立人不同。在侧面图中，额骨鳞部的弯曲程度较爪哇人族成员与非洲的早期人属与德马尼西匠人[3,14]更接近。一眶上沟把额骨鳞部与眶上圆枕分开。在额骨中部三分之一，有一轻微矢状隆起，延伸成为低而宽阔的前囟前部隆起的残留。在中额骨鳞部，有一圆形已愈合的损伤，可能是凹陷骨折的痕迹，大约 15 毫米宽。

下颌骨完整，仅右侧髁突部分损坏（图 4），且复合了多种上新世与更新世人族成员的特征。右下第 3 前臼齿和第 2 臼齿以及左犬齿处的死后破裂贯穿了下颌体，使得右侧上升枝有些侧向变形。下颌齿列表现出强烈的施佩曲线。上升枝根部在侧面下颌突之上嵌入下颌体，从侧面遮挡下第 3 臼齿远中部。上升枝下部最宽，向后稍有倾斜，中侧向变厚，冠状突高于髁突。右侧髁突最大宽度 18 毫米。外臼齿沟窄而浅，下颌突中等程度发育。下颌体前部浑圆肥硕且没有颏。在下颌联合后部区域，齿槽面向后向下倾斜，上圆枕中等程度发育，二腹肌窝深且宽阔，下圆枕低而圆，不呈搁架状（图 4）。下颌联合轴线有一个强烈向后的角度，下颌联合的总体形态与 LH4 南方古猿阿法种很相似，而与周口店及桑吉兰直立人不相像。在两侧都有两个颏孔，后一个孔较小，位置更靠下。双颏孔在印度尼西亚直立人[15]中普遍存在。而

mental foramina are common in Indonesian *H. erectus*[15]. While the mandibular dental arch is narrow anteriorly, and long relative to its breadth, the axis of $P_3$–$M_3$ is laterally convex rather than straight (Fig. 4).

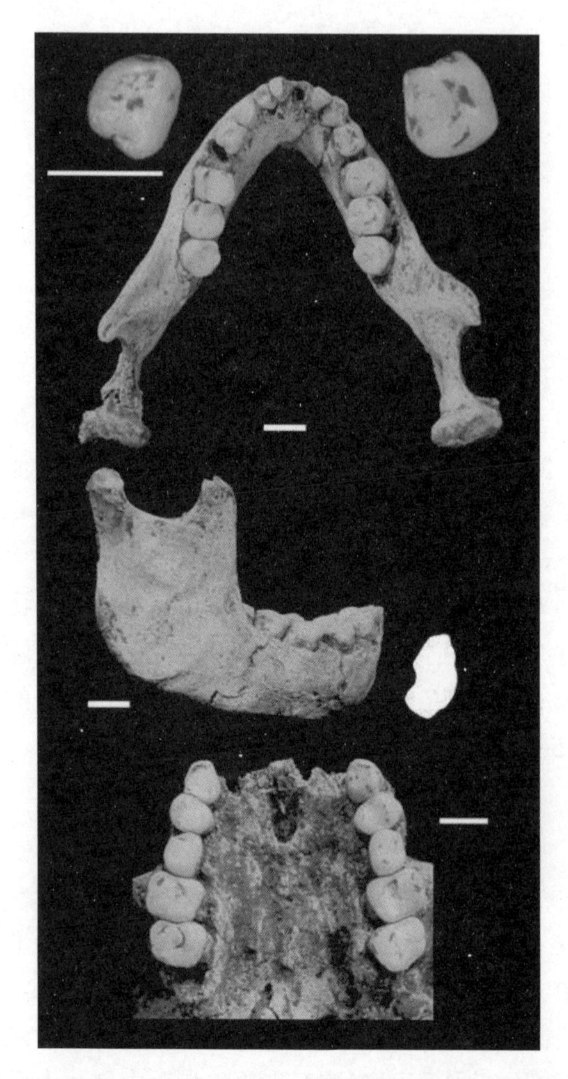

Fig. 4. Right lateral and occlusal views of the LB1 mandible, sagittal profile of the symphysis, occlusal view of the mandibular dentition and occlusal views of the mandibular premolars. Scale bars, 1 cm.

The right $P_4$ is absent and the alveolus completely fused, the left $P_4$ was lost after death, and CT scans indicate that the maxillary right $M^3$ was congenitally absent. The relatively small and conical alveolus for the missing left $M^3$ suggests that it had a much smaller crown than $M^1$ and $M^2$. Size, spacing and angulation of the maxillary incisor alveoli, and absence of a mesial facet on the canines suggest that incisor $I^2$ was much smaller than $I^1$, and there may have been a diastema. Occlusal wear has removed details of cusp and fissure morphology from most of the maxillary and mandibular teeth. The canines have worn down to a relatively flat surface and there would have been an edge-to-edge bite anteriorly.

下颌骨齿弓前面窄，相对于宽度显得长，下第 3 前臼齿至第 3 臼齿轴线向侧面凸出而不是直的(图 4)。

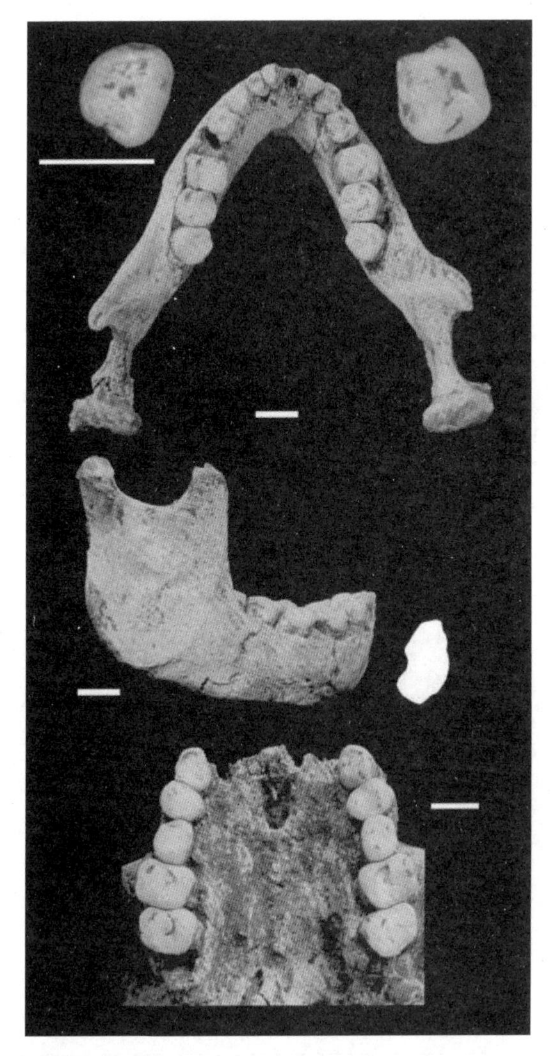

图 4. LB1 下颌骨的右侧视图与咀嚼面视图、联合部矢状面轮廓、下齿列咀嚼面视图及下前臼齿咀嚼面视图。比例尺，1 厘米。

右下第 4 前臼齿缺失且齿槽完全合并，左下第 4 前臼齿为死后丢失，CT 扫描显示右上第 3 臼齿先天缺失。丢失的左上第 3 臼齿齿槽相对小且呈圆锥形，这表明其齿冠要远小于上第 1 臼齿和上第 2 臼齿。上门齿齿槽的大小、间距和角度，以及犬齿近中接触面缺失表明，上第 2 门牙比上第 1 门牙小得多，并且可能存在齿隙。上下颌牙齿咀嚼面的磨耗已经把大部分上下颌牙齿上的齿尖及裂隙形态的细节都磨掉了。犬齿已经磨耗到一个相当平坦的平面，在牙齿前部可能曾有边缘对边缘的咬合。

Interproximal wear is pronounced and in combination with the loss of crown height means that mesio-distal crown dimensions convey little phylogenetic information. With the exception of $P_3$ the size and morphology of the mandibular teeth follow the pattern in *H. erectus* and *H. sapiens* (Fig. 5, Supplementary Table 2). There is not a great deal of difference between the size of the molar teeth in each quadrant, and the size sequence for both mandibular and maxillary teeth is $M1 \geqslant M2 > M3$. Using the megadontial quotient as a measure of relative tooth size[16], and substituting $P_3$ crown area for the missing $P_4$s, LB1 is megadont (1.8) relative to *H. sapiens* (0.9) and *H. ergaster* (0.9), but not *H. habilis* (1.9) (ref. 8) (Supplementary Information). The $P_3$s have a relatively great occlusal surface area (molariform) and when unworn had a prominent protoconid and broad talonid. Both $P_3$s have bifurcated roots and the alveolus for the left $P_4$ indicates a mesiodistally compressed, broad Tomes' root. A larger, less worn, isolated left $P_3$ from the deposit (LB2) has a more triangular occlusal outline, and a Tomes' root (Supplementary Fig. 2). Mandibular $P_3$s and $P_4$s with similar crown and root morphology have been recorded for *Australopithecus* and early *Homo*[17,18], and some Indonesian *H. erectus* mandibular premolars also have bifurcated or Tomes' roots[15]. Unusually, both maxillary $P^4$s are rotated parallel to the tooth row, a trait that seems to be unrecorded in any other hominin. Maxillary canines and $P^3$s have long roots and very prominent juga. The $P^3$ juga are emphasized by the rotation of the adjacent $P^4$ roots.

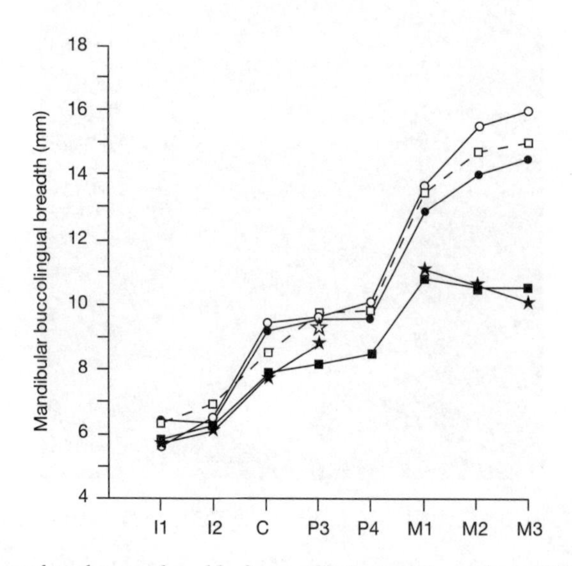


Fig. 5. Mean buccolingual tooth crown breadths for mandibular teeth in *A. afarensis* (filled circles), *A. africanus* (open circles), early *Homo* sp. (open squares), modern *H. sapiens* (filled squares), LB1 (filled stars) and LB2 (open stars). There are no mandibular $P_4$s preserved for LB1. Data for *Australopithecus* and early *Homo* are from ref. 49. Modern human data from a global sample of 1,199 individuals collected by P.B.

The pelvic girdle is represented by a right innominate, with damage to the iliac crest and pubic region, and fragments of the sacrum and left innominate. The right innominate, which is undistorted, has a broad greater sciatic notch suggesting that LB1 is a female (Fig. 6). In common with all bipedal hominins, the iliac blade is relatively short and wide[19]; however, the ischial spine is not particularly pronounced. Compared with modern humans the LB1

齿间磨耗显著，加上齿冠高度的磨损，意味着齿冠近远中尺寸所含有的系统发育信息很少。除了下第 3 前臼齿，下颌牙齿的大小和形态遵循直立人与智人的模式（图 5 和补充表 2）。在每个四分之一齿枝上，臼齿的大小没有很大差别，上下颌臼齿的大小次序是 M1 ≥ M2 ＞ M3。如果以牙齿巨大系数作为牙齿相对尺寸的量度 [16]，用下第 3 前臼齿齿冠面积替代失去的下第 4 前臼齿，相对于智人（0.9）与匠人（0.9），LB1 属于牙齿巨大（1.8），但相对于能人（1.9）则不是如此（参考文献 8）（补充信息）。下第 3 前臼齿拥有相对大的咀嚼面表面积（臼齿形），未磨耗时下原尖突出，下跟座宽阔。两侧下第 3 前臼齿均拥有二分叉齿根，左下第 4 前臼齿的齿槽孔表明其拥有近远中向压扁且宽阔的托姆斯式齿根。1 枚来自沉积物的更大且磨耗程度较低的左下第 3 前臼齿（LB2）拥有更趋于三角形的咀嚼面轮廓和托姆斯式齿根（补充图 2）。下颌第 3 和第 4 前臼齿具有相似的齿根和齿冠形态，这在南方古猿与早期人属中已有记载 [17,18]，而且一些印度尼西亚直立人下颌前臼齿也有二分叉或托姆斯式齿根 [15]。不同寻常的是，两侧的上颌上第 4 前臼齿均旋转并平行于齿列，这个特征似乎在任何其他人族成员中都没有记载过。上颌犬齿与上第 3 前臼齿牙根长且隆凸十分突出。上第 3 前臼齿隆凸由于相邻上第 4 前臼齿齿根的旋转而显加强。

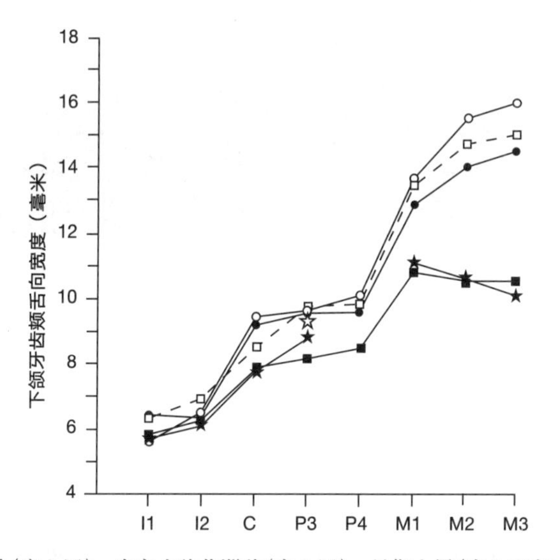


图 5. 南方古猿阿法种（实心圆）、南方古猿非洲种（空心圆）、早期人属（空心正方形）、现代智人（实心正方形）、LB1（实心五角形）及 LB2（空心五角形）的下颌牙齿齿冠颊舌向的平均宽度。LB1 的两侧下第 4 前臼齿未保存。南方古猿与早期人属数据来自参考文献 49。现代人类数据来自布朗收集的一个包括 1,199 个个体的全球样本。

腰带部分保存了髂嵴与耻骨区域破损的右侧髋骨和骶骨及左侧髋骨碎片。右侧髋骨，未变形，坐骨大切迹宽阔，表明 LB1 为一女性（图 6）。和所有两足直立行走的人族成员一样，髂骨刃相对短而宽 [19]；不过，坐骨棘不是特别明显。与现代人相

ilium has marked lateral flare, and the blade would have projected more laterally from the body, relative to the plane of the acetabulum. The left acetabulum is of circular shape, and has a maximum width of 36 mm.

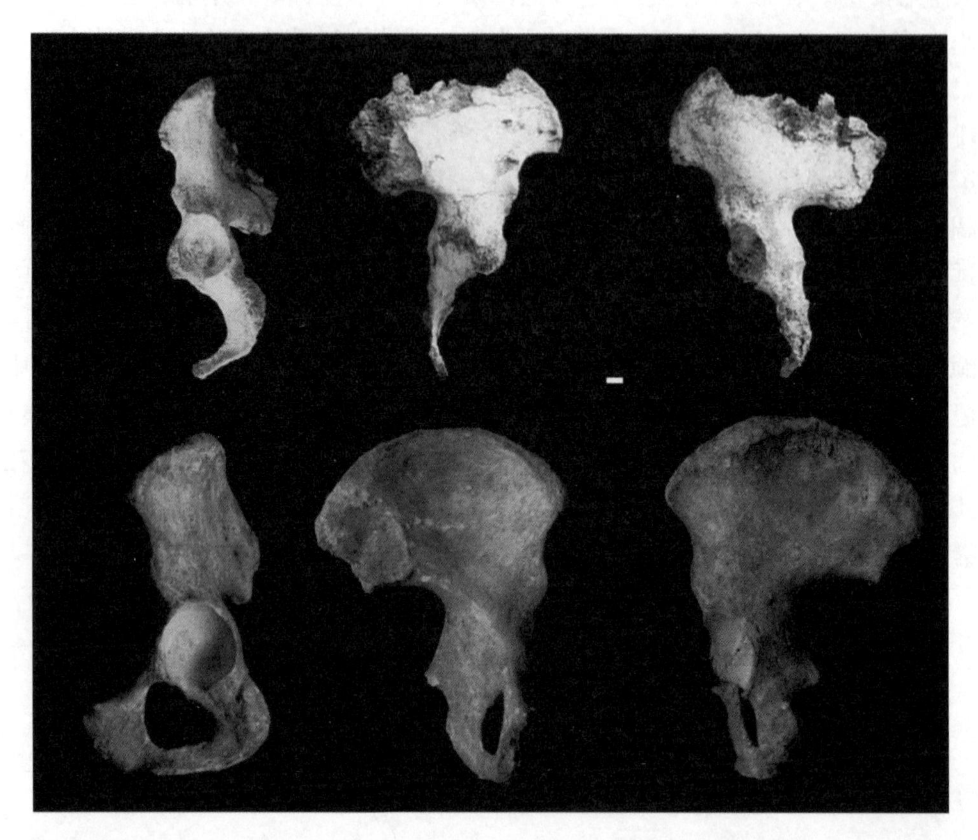

Fig. 6. Comparison of the left innominate from LB1 with a modern adult female *H. sapiens*. Lateral (external), and medial and lateral views of maximum iliac breadth. The pubic region of LB1 is not preserved and the iliac crest is incomplete. Scale bar, 1 cm.

Apart from damage to the lateral condyle and distal shaft, the right femur is complete and undistorted (Fig. 7). The overall anatomy of the femur is most consistent with the broad range of variation in *H. sapiens*, with some departures that may be the result of the allometric effects of very small body size. The femur shaft is relatively straight, and areas of muscle attachment, including the linea aspera, are not well developed. In contrast with some examples of Asian and African *H. erectus*, the femora do not have reduced medullary canals[20]. On the proximal end, the lesser trochanter is extremely prominent and the strong development of the intertrochanteric crest is similar to *H. sapiens* rather than the flattened intertrochanteric area in *Australopithecus* and *H. erectus* (KNM-ER 1481A, KNM-WT 15000). The biomechanical neck length is 55.5 mm and the neck is long relative to the femoral head diameter (31.5 mm), as is common to both *Australopithecus* and early *Homo*[19]. The neck–head junction is 31.5 mm long, with a shaft–neck angle of 130°, and the femur neck is compressed anteroposteriorly (Fig. 7). Several indices of femoral size and shape, for example the relationship between femoral head size and midshaft circumference (66 mm), and femur

比，LB1 髂骨侧面显著变宽，这表明，相对于髋臼平面，髂骨刃会向身体侧面凸出更多。左侧髋臼呈圆形，最大宽 36 毫米。

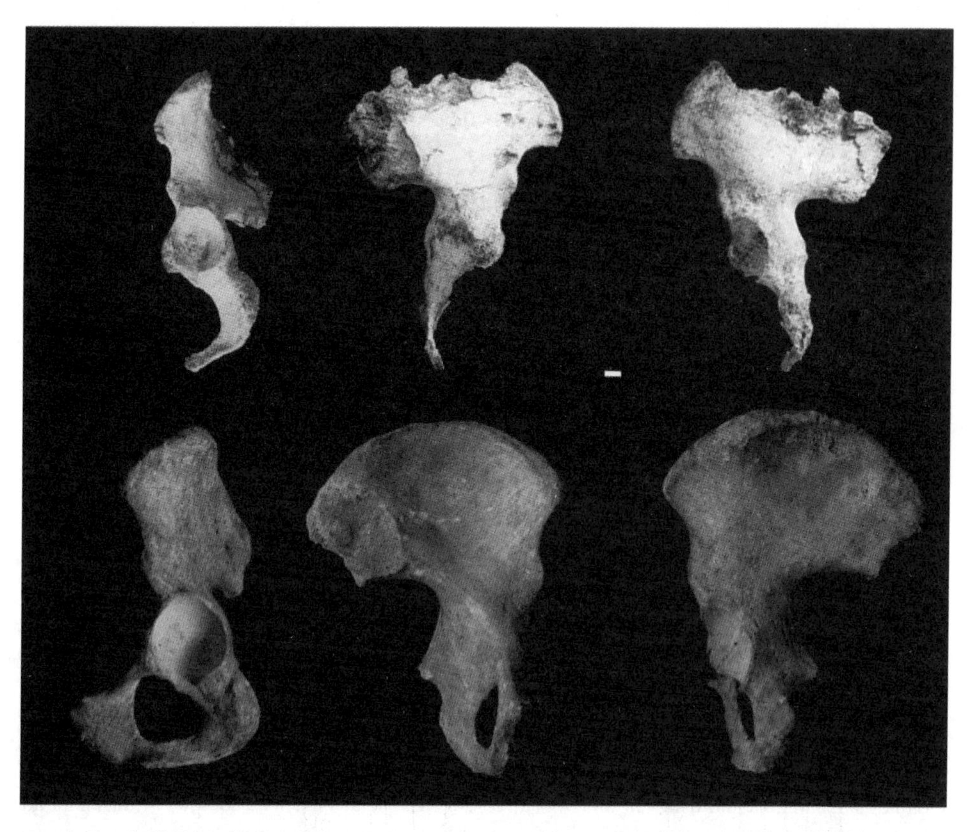

图 6. LB1 左侧髋骨与一现代女性智人的对比。髂骨最大宽度侧（外）视图、内侧视图及侧视图。LB1 的耻骨区域未保存，且髂嵴不完整。比例尺，1 厘米。

除了外侧髁突和远端骨干有破损，右侧股骨完整且未变形（图 7）。股骨的整体解剖学特征与智人宽泛变异范围是最为一致的，一些偏差可能是由体型矮小的异速生长效应导致。股骨干相对较直，包括股骨粗线在内的肌肉附着区不很发育。与亚洲及非洲直立人的一些例子不同，其两侧股骨没有减弱的髓腔 [20]。在近中末端，小转子极其突出，转子间嵴的强烈发育类似于智人，而与南方古猿属和直立人（KNM-ER 1481A，KNM-WT 15000）的平坦的转子间区不同。股骨颈的生物力学长度为 55.5 毫米，且相对于股骨头直径（31.5 毫米）显长，这在南方古猿属及早期人属中也普遍存在 [19]。股骨颈–股骨头联合处长 31.5 毫米，股骨干–股骨颈成 130° 角，股骨颈前后压缩（图 7）。股骨大小与形状的几个指标，例如股骨头大小和股骨干中部周长（66 毫米）间及股骨长度和转子下方骨干大小间的关系 [21]，都落入黑猩猩与南方古猿的变异范围之内。股骨干没有粗线嵴，其横截面呈圆形，骨干中部横截面面积

length and sub-trochanteric shaft size[21], fall within the chimpanzee and australopithecine range of variation. The femur shaft does not have a pilaster, is circular in cross-section, and has cross-sectional areas of 370 $mm^2$ at the midshaft and 359 $mm^2$ at the midneck. It is therefore slightly more robust than the best-preserved small-bodied hominin femur of similar length (AL288-1; ref. 21). Distally there is a relatively high bicondylar angle of 14°, which overlaps with that found in *Australopithecus*[22].

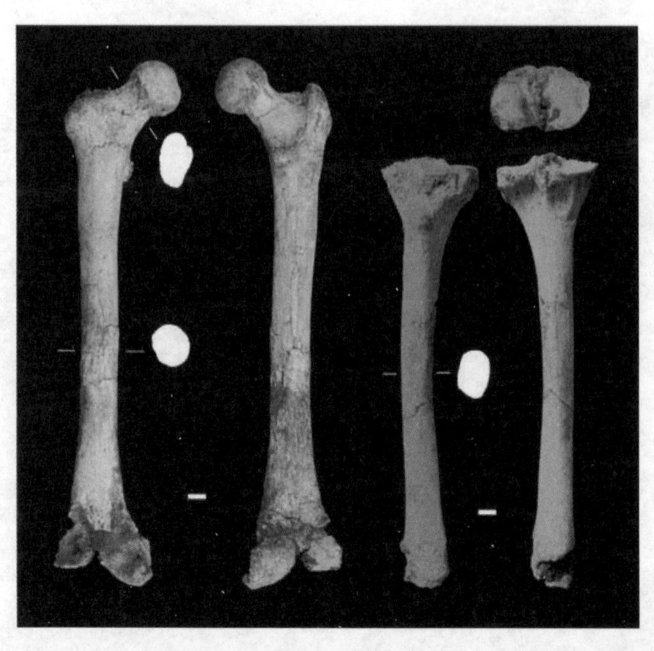

Fig. 7. Anterior and posterior views of the LB1 right femur and tibia, with cross-sections of the femur neck and midshaft, and tibia midshaft. The anterior surfaces of the medial and lateral condyles of the femur are not preserved. With the exception of the medial malleolus, the tibia is complete and undistorted. Scale bar, 1 cm.

The right tibia is complete apart from the tip of the medial malleolus (Fig. 7). Its most distinctive feature, apart from its small size (estimated maximum length 235 mm, bicondylar breadth 51.5 mm) and the slight curvature in the long axis, is a shaft that is oval in cross-section (midshaft 347 $mm^2$), without a sharp anterior border, and relatively thickened medio-laterally in the distal half. The relationship between the midshaft circumference and the length of the tibia is in the chimpanzee range of variation and distinct from *Homo*[21].

Additional evidence of a small-bodied adult hominin is provided by an unassociated left radius shaft, without the articular ends, from an older section of the deposit (74–95 kyr). The estimated maximum length of this radius when complete is approximately 210 mm. Although the arms of LB1 have not been recovered, the dimensions of this radius are compatible with a hominin of LB1 proportions.

Although there is considerable interspecific variation, stature has been shown to have phylogenetic and adaptive significance among hominins[23]. Broadly speaking, *Australopithecus*

为370平方毫米，股骨颈中部横截面面积为359平方毫米。因此，它比保存最完好且具有相似长度的身材矮小的人族成员的股骨略显更加粗壮(AL288-1；参考文献21)。在远端双髁角相对显大，为14°，与南方古猿属重叠[22]。

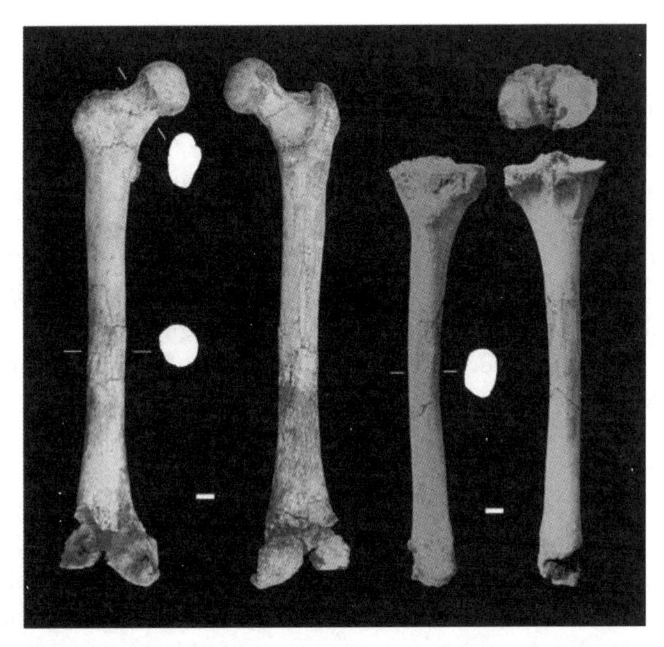

图7. LB1右侧股骨和胫骨的前视图与后视图，及股骨颈、股骨干与胫骨干的横截面。股骨的中髁突与侧髁突前方未保存。除了内踝，胫骨完整且未变形。比例尺，1厘米。

除了内踝末端破损，右胫骨完整(图7)。除了其尺寸较小(估计最大长为235毫米，双髁宽为51.5毫米)以及长轴微弯外，它最与众不同的特征是其骨干的横截面呈椭圆形(骨干中部横截面面积为347平方毫米)，无明显的前嵴，远端半部的内–侧向加厚。骨干中部周长和胫骨长度之间的关系在黑猩猩的变异范围之内，而与人属截然不同[21]。

另外一根采自更老堆积物(74,000 ~ 95,000年)且与之不相关的缺失关节部位的左侧桡骨骨干，为身材矮小的成年人族成员提供了附加的证据。该桡骨完整状态下推测最大长度约为210毫米。虽然LB1的双臂还没有被发现，但该桡骨的尺寸与LB1这一人族成员的比例一致。

虽然种间变异相当大，但身高已经在人族之中被证明具有系统发育和适应方面

and the earliest members of the genus *Homo* are shorter than *H. erectus* and more recent hominins[8]. The maximum femur length of LB1 (280 mm) is just below the smallest recorded for *A. afarensis* (AL-288-1, 281 mm[24]) and equal to the smallest estimate for the OH 62 *H. habilis* femur (280–404 mm)[21]. Applying stature estimation formulae developed from human pygmies[25] gives a stature estimate of 106 cm for LB1 (Supplementary Information). This is likely to be an overestimation owing to LB1's relatively small cranial height.

A stature estimate for LB1 of 106 cm gives a body mass of 16 to 28.7 kg, and a femur cross-sectional area of 525 $mm^2$ gives a mass of 36 kg (Supplementary Information). The brain mass for LB1, calculated from its volume[26], is 433.2 g; this gives an encephalization quotient (EQ)[27] range of 2.5–4.6, which compares with 5.8–8.1 for *H. sapiens*, 3.3–4.4 for *H. erectus/ergaster* and 3.6–4.3 for *H. habilis*, and overlaps with the australopithecine range of variation[28,29]. If LB1 shared the lean and relatively narrow body shape typical of Old World tropical modern humans then the smallest body weight estimate, based on Jamaican school children data[19], is probably most appropriate. This would support the higher EQ estimate and place LB1 within the *Homo* range of variation. Although neurological organization is at least as important as EQ in determining behavioural complexity, these data are consistent with *H. floresiensis* being the Pleistocene toolmaker at Liang Bua.

## Origins and Evolution

The LB1 skeleton was recovered from Flores, an island of 14,000 $km^2$ east of the Wallace Line, in Indonesia. It combines extremely small stature and an endocranial volume in the early australopithecine range, with a unique mosaic of primitive and derived traits in the cranium, mandible and postcranial skeleton. Both its geographic location and comparatively recent date suggest models that differ to those for more expected geological contexts, such as Pliocene eastern Africa. Among modern humans, populations of extremely small average stature were historically found in predominantly rainforest habitat in the equatorial zone of Africa, Asia and Melanesia[30,31]. Explanations for the small body size of these people generally focus on the thermoregulatory advantages for life in a hot and humid forest, either through evaporative cooling[32] or reduced rates of internal heat production[30]. For African pygmies, smaller body size is the result of reduced levels of insulin-like growth factor 1 (IGF-1) throughout the growth period[33], or reduced receptivity to IGF-1 (ref. 34). Although adult stature is reduced, cranio-facial proportions remain within the range of adjacent larger-bodied populations, as does brain size[35,36]. The combination of small stature and brain size in LB1 is not consistent with IGF-related postnatal growth retardation. Similarly, neither pituitary dwarfism, nor primordial microcephalic dwarfism (PMD) in modern humans replicates the skeletal features present in LB1 (refs 37–40).

Other mechanisms must have been responsible for the small body size of these hominins, with insular dwarfing being the strongest candidate. Although small body size was an attribute of Pliocene australopithecines, the facial and dental characteristics of LB1 link it with larger-bodied Pleistocene *Homo*. In this instance, body size is not a direct expression

的意义[23]。宽泛地讲，南方古猿属与人属最早成员矮于直立人与更晚的人族成员[8]。LB1 的股骨最大长度（280 毫米）刚好低于南方古猿阿法种（AL-288-1，281 毫米[24]）的最小记录值，而等于 OH 62 能人股骨长度（280 ~ 404 毫米）的最小估计值[21]。应用基于俾格米人[25]开发的身高估算公式，估算出 LB1 身高为 106 厘米（补充信息）。由于 LB1 的颅骨高度相对较小，这一身高可能是过高估值。

根据 LB1 106 厘米的身高估值，得出其体重为 16 至 28.7 千克，而根据股骨横截面面积为 525 平方毫米，得出体重为 36 千克（补充信息）。根据 LB1 的脑容量来计算[26]，其大脑重量为 433.2 克，从而得到脑系数（EQ）[27]的范围为 2.5 ~ 4.6，而智人为 5.8 ~ 8.1，直立人/匠人为 3.3 ~ 4.4，能人为 3.6 ~ 4.3，与南方古猿的变异范围重叠[28,29]。如果 LB1 享有旧世界热带现代人典型的清瘦和相对狭窄的体型，那么，基于牙买加入学儿童的数据[19]，体重的最小估值或许是最适当的。这会支持较高的 EQ 估值，而把 LB1 置于人属变异范围之内。虽然在决定行为的复杂性上，神经结构至少也与 EQ 一样重要，但这些数据与利昂布阿的弗洛勒斯人是更新世工具制造者是一致的。

## 起源与演化

LB1 骨架发现于印度尼西亚华莱士线以东方圆 14,000 平方千米的弗洛勒斯岛。它兼具落入早期南方古猿范围的极其矮小的身材和颅内容积，及在颅骨、下颌骨和颅后骨骼中独特的原始和衍生特征的镶嵌。其地理位置和相当晚的年代都表明了与更加期望的地质背景（如上新世非洲东部）不同的模型。在现代人中，平均身高极其矮小的种群在历史时期主要发现于非洲赤道地区、亚洲及美拉尼西亚的热带雨林栖息地[30,31]。对这些人体型小的解释，一般聚焦于在炎热潮湿的森林里生活的热调节优势，无论是通过蒸发降温[32]还是降低体内热量产生的速度[30]。对非洲俾格米人而言，体型较小是由于在整个生长期类胰岛素生长因子 1（IGF-1）分泌水平的降低[33]，或对 IGF-1 的接受能力的降低（参考文献 34）。虽然成年身高降低，但像大脑大小一样，其颅–面比例仍然保持在邻近大体型种群的范围之内[35,36]。LB1 中身材矮小与脑容量小兼备，与 IGF 相关的后天发育延缓不一致。同样，现代人中无论是垂体性侏儒症还是先天性小头侏儒症（PMD）都不能再现 LB1 中出现的骨骼特征（参考文献 37 ~ 40）。

一定是其他机制造成了这些人族成员的身材矮小，而岛屿侏儒化是最佳候选。虽然体型小是上新世南方古猿的特性，但 LB1 的面部与牙齿的特征将其与体型较

of phylogeny. The location of these small hominins on Flores makes it far more likely that they are the end product of a long period of evolution on a comparatively small island, where environmental conditions placed small body size at a selective advantage. Insular dwarfing, in response to the specific ecological conditions that are found on some small islands, is well documented for animals larger than a rabbit[41,42]. Explanations of the island rule have primarily focused on resource availability, reduced levels of interspecific competition within relatively impoverished faunal communities and absence of predators. It has been argued that, in the absence of agriculture, tropical rainforests offer a very limited supply of calories for hominins[43]. Under these conditions selection should favour the reduced energy requirements of smaller individuals. Although the details of the Pleistocene palaeoenvironment on Flores are still being documented, it is clear that until the arrival of Mesolithic humans the faunal suit was relatively impoverished, and the only large predators were the Komodo dragon and another larger varanid. Dwarfing in LB1 may have been the end product of selection for small body size in a low calorific environment, either after isolation on Flores, or another insular environment in southeastern Asia.

Anatomical and physiological changes associated with insular dwarfing can be extensive, with dramatic modification of sensory systems and brain size[44], and certainly exceed what might be predicted by the allometric effects of body size reduction alone. Evidence of insular dwarfing in extinct lineages, or the evolution of island endemic forms, is most often provided by the fossil record. Whereas there is archaeological evidence of hominins being on Flores by approximately 840 kyr[45], there is no associated hominin skeletal material, and the currently limited evidence from Liang Bua is restricted to the Late Pleistocene. The first hominin immigrants may have had a similar body size to *H. erectus* and early *Homo*[21,46], with subsequent dwarfing; or, an unknown small-bodied and small-brained hominin may have arrived on Flores from the Sunda Shelf.

## Discussion

When considered as a whole, the cranial and postcranial skeleton of LB1 combines a mosaic of primitive, unique and derived features not recorded for any other hominin. Although LB1 has the small endocranial volume and stature evident in early australopithecines, it does not have the great postcanine tooth size, deep and prognathic facial skeleton, and masticatory adaptations common to members of this genus[2,47]. Instead, the facial and dental proportions, postcranial anatomy consistent with human-like obligate bipedalism[48], and a masticatory apparatus most similar in relative size and function to modern humans[48] all support assignment to the genus *Homo*—as does the inferred phylogenetic history, which includes endemic dwarfing of *H. erectus*. For these reasons, we argue that LB1 is best placed in this genus and have named it accordingly.

On a related point, the survival of *H. floresiensis* into the Late Pleistocene shows that the genus *Homo* is morphologically more varied and flexible in its adaptive responses than is generally recognized. It is possible that the evolutionary history of *H. floresiensis* is unique,

大的更新世人属联系起来。在这种情况下，体型大小不是系统发育的直接表现。弗洛勒斯岛上这些矮小人族成员的发现位置，使其更有可能是在一个相对较小的岛屿上经过长期演化的最终产物，那里的环境条件使得小体型更具有选择优势。岛屿侏儒化，作为对一些小岛上特殊生态条件的响应，在比兔子大的动物中有很详细的记载[41,42]。对岛屿规则的解释主要聚焦于资源的可获得性、相对贫困的动物群落中种间竞争的减少以及捕食者的消失。有人提出，在农业出现前，热带雨林提供给人族成员的卡路里十分有限[43]。在这样的条件下，选择应该偏向于能量需求降低的较矮小个体。虽然弗洛勒斯的更新世古环境资料正处于记载中，但很明显，在中石器时代人类到来前，整个动物群还比较单薄，唯一的大型捕食者是科莫多巨蜥和另一种体型较大的巨蜥。无论是在弗洛勒斯岛上被隔离后，还是在东南亚的另一个孤立环境中，LB1 的侏儒化可能是在低卡路里环境下选择偏向小体型的最终产物。

与岛屿侏儒化相关的解剖学和生理学变化非常之多，伴随对感觉系统和大脑尺寸的改变[44]，并且肯定超过了仅由身体体型减小的异速生长效应所能预测出的程度。已经绝灭的谱系中岛屿侏儒化，或岛屿特有类型的演化的证据，最常由化石记录提供。虽然有考古证据表明，人族成员在弗洛勒斯岛上大约有 840,000 年了[45]，但并没有相关的人族成员的骨骼材料，而现在来自利昂布阿的有限证据仅限于晚更新世。第一批人族成员移民可能有类似于直立人和早期人属成员的身体大小[21,46]，但随后矮化；或者，一种未知的身材矮小、脑容量小的人族成员从巽他大陆架来到了弗洛勒斯。

## 讨　论

当作为一个整体来看时，LB1 颅骨和颅后骨骼兼具原始和独特的衍生特征的镶嵌，这些衍生特征在其他任何人族成员中都没有记载过。虽然 LB1 像早期南方古猿那样颅内容积及身高小，但其没有硕大的犬后齿尺寸、深而前突的面部骨骼以及这一属的成员所共有的咀嚼适应特征[2,47]。相反，面部和牙齿的比例、颅后骨骼的解剖特征与似人的习惯性两足行走一致[48]，而且其咀嚼器官与现代人在相对大小与功能上最相似[48]，这一切都支持将其归入人属，包括直立人的地方性矮化在内的系统发育史同样支持将其归入人属。基于这些原因，我们认为 LB1 最适合放在这个属，并因此而给她命名。

关于另外相关的一点，弗洛勒斯人能幸存下来延续到晚更新世，显示了人属在其适应性响应中，形态上比人们一般认为的更加多样化和灵活。弗洛勒斯人的演化历史可能是独一无二的，但我们认为更有可能的情况是，随着人属走出非洲，该属

but we consider it more likely that, following the dispersal of *Homo* out of Africa, there arose much greater variation in the morphological attributes of this genus than has hitherto been documented. We anticipate further discoveries of highly endemic, hominin species in locations similarly affected by long-term genetic isolation, including other Wallacean islands.

(**431**, 1055-1061; 2004)

**P. Brown[1], T. Sutikna[2], M. J. Morwood[1], R. P. Soejono[2], Jatmiko[2], E. Wayhu Saptomo[2] & Rokus Awe Due[2]**

[1] Archaeology & Palaeoanthropology, School of Human & Environmental Studies, University of New England, Armidale, New South Wales 2351, Australia

[2] Indonesian Centre for Archaeology, Jl. Raya Condet Pejaten No. 4, Jakarta 12001, Indonesia

Received 3 March; accepted 8 September 2004.

---

References:

1. Morwood, M. J. *et al.* Archaeology and age of a new hominin from Flores in eastern Indonesia. *Nature* doi:10.1038/nature02956 **431**, 1087-1091 (2004).

2. Wood, B. A. *Koobi Fora Research Project, Vol. 4: Hominid Cranial Remains* (Clarendon, Oxford, 1991).

3. Vekua, A. K. *et al.* A new skull of early *Homo* from Dmanisi, Georgia. *Science* **297**, 85-89 (2002).

4. Spoor, C. F. Basicranial architecture and relative brain size of STS 5 (*Australopithecus africanus*) and other Plio-Pleistocene hominids. *S. Afr. J. Sci.* **93**, 182-186 (1997).

5. Lieberman, D., Ross, C. F. & Ravosa, M. J. The primate cranial base: ontogeny, function, and integration. *Yearb. Phys. Anthropol.* **43**, 117-169 (2000).

6. Baba, H. *et al. Homo erectus* calvarium from the Pleistocene of Java. *Science* **299**, 1384-1388 (2003).

7. Tobias, P. V. *The Skulls, Endocasts and Teeth of* Homo habilis (Cambridge Univ. Press, Cambridge, 1991).

8. McHenry, H. M. & Coffing, K. E. *Australopithecus* to *Homo*: Transformations of body and mind. *Annu. Rev. Anthropol.* **29**, 125-166 (2000).

9. Brown, P. Vault thickness in Asian *Homo erectus* and modern *Homo sapiens*. *Courier Forschungs-Institut Senckenberg* **171**, 33-46 (1994).

10. Bräuer, G. & Mbua, E. *Homo erectus* features used in cladistics and their variability in Asian and African hominids. *J. Hum. Evol.* **22**, 79-108 (1992).

11. Santa Luca, A. P. *The Ngandong Fossil Hominids* (Department of Anthropology Yale Univ., New Haven, 1980).

12. Weidenreich, F. The skull of *Sinanthropus pekinensis*: a comparative study of a primitive hominid skull. *Palaeontol. Sin.* **D10**, 1-485 (1943).

13. Brown, P. & Maeda, T. Post-Pleistocene diachronic change in East Asian facial skeletons: the size, shape and volume of the orbits. *Anthropol. Sci.* **112**, 29-40 (2004).

14. Gabunia, L. K. *et al.* Earliest Pleistocene hominid cranial remains from Dmanisi, Republic of Georgia: taxonomy, geological setting, and age. *Science* **288**, 1019-1025 (2000).

15. Kaifu, Y. *et al.* Taxonomic affinities and evolutionary history of the Early Pleistocene hominids of Java: dento-gnathic evidence. *Am. J. Phys. Anthropol.* (in the press).

16. McHenry, H. M. in *Evolutionary History of the "Robust" Australopithecines* (ed. Grine, F. E.) 133-148 (Aldine de Gruyter, New York, 1988).

17. Wood, B. A. & Uytterschaut, H. Analysis of the dental morphology of the Plio-Pleistocene hominids. III. Mandibular premolar crowns. *J. Anat.* **154**, 121-156 (1987).

18. Wood, B. A., Abbott, S. A. & Uytterschaut, H. Analysis of the dental morphology of Plio-Pleistocene hominids. IV. Mandibular postcanine root morphology. *J. Anat.* **156**, 107-139 (1988).

19. Aiello, A. & Dean, C. *An Introduction to Human Evolutionary Anatomy* (Academic, London, 1990).

20. Kennedy, G. E. Some aspects of femoral morphology in *Homo erectus*. *J. Hum. Evol.* **12**, 587-616 (1983).

21. Haeusler, M. & McHenry, H. M. Body proportions of *Homo habilis* reviewed. *J. Hum. Evol.* **46**, 433-465 (2004).

22. Stern, J. T. J. & Susman, R. L. The locomotor anatomy of *Australopithecus afarensis*. *Am. J. Phys. Anthropol.* **60**, 279-317 (1983).

23. Ruff, C. B. Morphological adaptation to climate in modern and fossil hominids. *Yearb. Phys. Anthropol.* **37**, 65-107 (1994).

24. Jungers, W. L. Lucy's limbs: skeletal allometry and locomotion in *Australopithecus afarensis*. *Nature* **297**, 676-678 (1982).

25. Jungers, W. L. Lucy's length: stature reconstruction in *Australopithecus afarensis* (A.L.288−1) with implications for other small-bodied hominids. *Am. J. Phys. Anthropol.* **76**, 227-231 (1988).

26. Count, E. W. Brain and body weight in man: their antecendants in growth and evolution. *Ann. NY Acad. Sci.* **46**, 993-1101 (1947).

27. Martin, R. D. Relative brain size and basal metabolic rate in terrestrial vertebrates. *Nature* **293**, 57-60 (1981).

28. Jerison, H. J. *Evolution of the Brain and Intelligence* (Academic, New York, 1973).

29. McHenry, H. M. in *The Primate Fossil Record* (ed. Hartwig, C. H.) 401-406 (Cambridge Univ. Press, Cambridge, 2002).

30. Cavalli-Sforza, L. L. (ed.) *African Pygmies* (Academic, Orlando, 1986).

31. Shea, B. T. & Bailey, R. C. Allometry and adaptation of body proportions and stature in African Pygmies. *Am. J. Phys. Anthropol.* **100**, 311-340 (1996).

32. Roberts, D. F. *Climate and Human Variability* (Cummings Publishing Co., Menlo Park, 1978).

的形态特性产生了比截止目前已记载的更多的变异。我们预期会有更多高度特化的人族成员物种在同样受到长期遗传隔离的地方被发现，包括其他华莱士区的岛屿。

（田晓阳 翻译；张颖奇 审稿）

33. Merimee, T. J., Zapf, J., Hewlett, B. & Cavalli-Sforza, L. L. Insulin-like growth factors in pygmies. *N.* Engl. *J. Med.* **15**, 906-911 (1987).

34. Geffner, M. E., Bersch, N., Bailey, R. C. & Golde, D. W. Insulin-like growth factor I resistance in immortalized T cell lines from African Efe Pygmies. *J. Clin. Endocrinol. Metab.* **80**, 3732-3738 (1995).

35. Hiernaux, J. *The People of Africa* (Charles Scribner's Sons, New York, 1974).

36. Beals, K. L., Smith, C. L. & Dodd, S. M. Brain size, cranial morphology, climate and time machines. *Current Anthropology* **25**, 301-330 (1984).

37. Rimoin, D. L., Merimee, T. J. & McKusick, V. A. Growth-hormone deficiency in man: an isolated, recessively inherited defect. *Science* **152**, 1635-1637 (1966).

38. Jaffe, H. L. *Metabolic, Degenerative and Inflammatory Disease of Bones and Joints* (Lea and Febiger, Philadelphia, 1972).

39. Seckel, H. P. G. *Bird-Headed Dwarfs* (Karger, Basel, 1960).

40. Jeffery, N. & Berkovitz, B. K. B. Morphometric appraisal of the skull of Caroline Crachami, the Sicilian "Dwarf" 1815?–1824: A contribution to the study of primordial microcephalic dwarfism. *Am. J. Med. Genet.* **11**, 260-270 (2002).

41. Sondaar, P. Y. in *Major Patterns in Vertebrate Evolution* (eds Hecht, M. K., Goody, P. C. & Hecht, B. M.) 671-707 (Plenum, New York, 1977).

42. Lomolino, M. V. Body size of *mammals* on islands: The island rule re-examined. *Am. Nat.* **125**, 310-316 (1985).

43. Bailey, R. C. & Headland, T. The tropical rainforest: Is it a productive habitat for human foragers? *Hum. Ecol.* **19**, 261-285 (1991).

44. Köhler, M. & Moyà-Solà, S. Reduction of brain and sense organs in the fossil insular bovid *Myotragus*. *Brain Behav. Evol.* **63**, 125-140 (2004).

45. Morwood, M. J., O'Sullivan, P. B., Aziz, F. & Raza, A. Fission-track ages of stone tools and fossils on the east Indonesian island of Flores. *Nature* **392**, 173-176 (1998).

46. Walker, A. C. & Leakey, R. (eds) *The Nariokotome* Homo erectus *skeleton* (Harvard Univ. Press, Cambridge, 1993).

47. Rak, Y. The *Australopithecine Face* (Academic, New York, 1983).

48. Wood, B. A. & Collard, M. The human genus. *Science* **284**, 65-71 (1999).

49. Johanson, D. C. & White, T. D. A systematic assessment of early African Hominids. *Science* **202**, 321-330 (1979).

**Supplementary Information** accompanies the paper on www.nature.com/nature.

**Acknowledgements**. We would like to thank F. Spoor and L. Aiello for data and discussion. Comments by F. Spoor and D. Lieberman greatly improved aspects of the original manuscript. Conversation with S. Collier, C. Groves, T. White and P. Grave helped clarify some issues. CTscans were produced by CT-Scan KSU, Medical Diagnostic Nusantara, Jakarta. S. Wasisto completed complex section drawings and assisted with the excavation of Sector VII. The 2003 excavations at Liang Bua, undertaken under Indonesian Centre for Archaeology Permit Number 1178/SB/PUS/ BD/24.VI/2003, were funded by a Discovery Grant to M.J.M. from the Australian Research Council. UNE Faculty of Arts, and M. Macklin, helped fund the manufacture of stereolithographic models of LB1.

**Authors contributions**. P.B. reconstructed the LB1 cranium and was responsible for researching and writing this article, with M.J.M. T.S. directed many aspects of the Liang Bua excavations, including the recovery of the hominin skeleton. M.J.M. and R.P.S. are Principal Investigators and Institutional Counterparts in the ARC project, as well as Co-Directors of the Liang Bua excavations. E.W.S. and Jatmiko assisted T.S., and had prime responsibility for the work in Sector VII. R.A.D. did all of the initial faunal identifications at Liang Bua, including hominin material, and helped clean and conserve it.

**Competing interests statement**. The authors declare that they have no competing financial interests.

**Correspondence** and requests for materials should be addressed to P.B. (pbrown3@pobox.une.edu.au).

# Unusual Activity of the Sun during Recent Decades Compared to the Previous 11,000 Years

S. K. Solanki *et al.*

## Editor's Note

**In the debate over whether human activities are the main cause of global warming during the past century, one of the main arguments for this trend being instead of natural origin is that it might be due to enhanced solar activity—the Sun giving out more heat. That idea is examined in this paper, in which the authors reconstruct solar activity over the past 11,400 years (since the last ice age ended) by using levels of atmospheric carbon-14 as a proxy. $^{14}C$ is produced by cosmic rays impinging on the atmosphere—a process affected by changes in solar activity—and its accumulation in tree rings provided an annual, dateable record over many millennia. The authors conclude that there has been exceptional solar activity over the past 70 years, but that this does not suffice to explain recent global warming.**

---

Direct observations of sunspot numbers are available for the past four centuries[1,2], but longer time series are required, for example, for the identification of a possible solar influence on climate and for testing models of the solar dynamo. Here we report a reconstruction of the sunspot number covering the past 11,400 years, based on dendrochronologically dated radiocarbon concentrations. We combine physics-based models for each of the processes connecting the radiocarbon concentration with sunspot number. According to our reconstruction, the level of solar activity during the past 70 years is exceptional, and the previous period of equally high activity occurred more than 8,000 years ago. We find that during the past 11,400 years the Sun spent only of the order of 10% of the time at a similarly high level of magnetic activity and almost all of the earlier high-activity periods were shorter than the present episode. Although the rarity of the current episode of high average sunspot numbers may indicate that the Sun has contributed to the unusual climate change during the twentieth century, we point out that solar variability is unlikely to have been the dominant cause of the strong warming during the past three decades[3].

---

SUNSPOTS—strong concentrations of magnetic flux at the solar surface—are the longest-studied direct tracers of solar activity. Regular telescopic observations are available after AD 1610. In addition to the roughly 11-year solar cycle, the number of sunspots, formalized in the group sunspot number[1] (GSN), exhibits prominent fluctuations on longer timescales. Notable are an extended period in the seventeenth century called the Maunder minimum, during which practically no sunspots were present[2], and the period of high solar activity since about AD 1940 with average sunspot numbers above 70.

# 最近几十年太阳活动的异动：与过去的 11,000 年相比

索兰基等

## 编者按

在关于人类活动是否是上个世纪全球变暖主因的辩论中，认为变暖乃是自然发生一派的一个主要观点是说太阳活动的加强导致了这种趋势——即太阳释放了更多的热量。本文作者检验了上述观点，他们利用大气中碳-14 活度作为代用指标，重建了过去 11,400 年（即从末次冰期结束时开始）的太阳活动。碳-14 由宇宙射线与大气层中的物质碰撞而产生，这一过程会受到太阳活动变化的影响，而碳-14 则会在树木年轮中积累，进而提供了年度的、可用于定年的千年纪记录。作者的结论是，尽管最近 70 年太阳活动出现异常，但这不足以解释近年来全球变暖的现象。

---

太阳黑子数目的直接观测数据只包含过去四个世纪[1,2]，但我们需要更长时间序列的数据以完成特定的研究。譬如，识别太阳活动对气候可能的影响以及检验太阳发电机模型都需要长序列的太阳活动数据。这里，我们基于树轮年代学所确定的放射性碳元素丰度，重建了过去 11,400 年的太阳黑子数目。我们将每个基于物理学的过程模型进行了整合，这些模型将放射性碳元素丰度与太阳黑子联系起来。根据我们的重建结果，过去 70 年太阳活动比较异常，而上一次太阳有如此高的活动性要追溯到 8,000 多年前。我们发现在过去的 11,400 年里，太阳只有 10% 的时间处在相似的磁场活动水平上，而且几乎所有早期的高活动性时段都短于当今阶段。需要指出的是，尽管当前高平均太阳黑子数目可能说明太阳对于二十世纪不寻常的气候变化存在影响，但太阳的变化不太可能是过去三十年间全球剧烈变暖的主导因素[3]。

---

太阳黑子——太阳表面磁场流量的强聚集区——是太阳活动的直接示踪物中被研究最久的一种。早在公元 1610 年就开始了利用常规望远镜对其的观测。除了大致上为 11 年的太阳周期，太阳黑子的数目，即正式文献中所谓的“太阳黑子团数”（GSN）[1]，在长时间尺度下具有显著地波动。延续到十七世纪的“蒙德极小期”尤为明显，在此期间几乎没有任何太阳黑子出现[2]；而在大约始于 1940 年的高太阳活动期当中，平均太阳黑子数目在 70 个以上。

A physical approach to reconstruction of the sunspot number back in time is based on archival proxies, such as the concentration of the cosmogenic isotopes $^{14}C$ in tree rings[4-6] or $^{10}Be$ in ice cores[7,8]. This approach has recently been strengthened by the development of physics-based models describing each link in the chain of processes connecting the concentration of cosmogenic isotopes with the sunspot number[9-12]. This advance allowed a reconstruction of the sunspot number since AD 850 based on $^{10}Be$ records from Antarctica and Greenland[13,14]. The current period of high solar activity is unique within this interval, but the covered time span is too short to judge just how unusual the current state of solar activity is.

Here we present a reconstruction of the sunspot number covering the Holocene epoch, the modern period of relatively warm climate that superseded the glacial period about 11,000 years ago. The reconstruction is based on $\Delta^{14}C$, the $^{14}C$ activity in the atmosphere[15] obtained from high-precision $^{14}C$ analyses on decadal samples of mid-latitude tree-ring chronologies. The data set has been created in an international collaboration of dendrochronologists and radiocarbon laboratories[16]. The absolutely and precisely dated original data set used for the sunspot number reconstruction is represented by the black line in Fig. 1. Starting at a level 15% higher than the reference level of AD 1950, the atmospheric $^{14}C$ shows a long-term trend (indicated by the red line), which is mainly the result of changes in the intensity of the geomagnetic dipole field before and during the Holocene epoch. The fluctuations on shorter timescales predominantly result from variations of the $^{14}C$ production rate due to heliomagnetic variability, which modulates the cosmic ray flux.

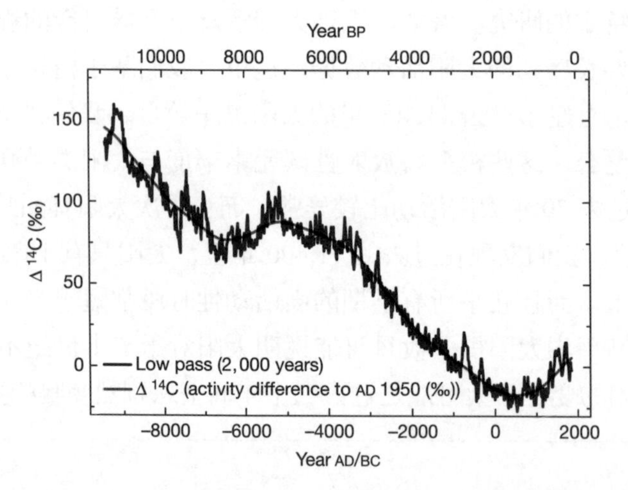


Fig. 1. Atmospheric radiocarbon level $\Delta^{14}C$ (expressed as deviation, in ‰, from the AD 1950 standard level[15]) derived from mostly decadal samples of absolutely dated tree-ring chronologies (INTCAL98 data set)[16]. The $\Delta^{14}C$ measurement precision is generally 2–3‰, although in the earlier part of the time series it can reach up to 4–5‰. The INTCAL98 data for times earlier than 11,400 BP are not directly employed for the reconstruction because of larger errors and uncertainties in the carbon cycle acting at that time. See Supplementary Information for more information on the data set, initial conditions used for the reconstruction, and error estimates. The long-term decline (indicated by the red curve) is caused by a reduction in $^{14}C$ production rate due mainly to an increase in the geomagnetic shielding of the cosmic ray flux. The short-term fluctuations (duration one to two centuries) reflect changes of the production rate due to solar variability. Years BC are shown negative here and in other figures.

重建太阳黑子的数目，物理方法是利用档案代用指标，譬如树木年轮中宇宙成因同位素 $^{14}C$[4-6] 或冰芯中的 $^{10}Be$。近来，这种方法借助于物理学模型的发展而得以强化，模型描述了宇宙成因同位素丰度与太阳黑子数目之间联系的每一环节。这一最新进展，使得利用南极和格陵兰 $^{10}Be$ 记录来重建自公元 850 年以来的太阳黑子数目成为可能 [13,14]。当前的高太阳活动期在此区间内是独一无二的，但由于涵盖的时间太短，尚不足以对现阶段的异常程度做出判断。

这里，我们重建了涵盖整个全新世及 11,000 年前由冰期更替为气候温暖的现代时期之后的太阳黑子数目。重建工作基于 $\Delta^{14}C$，大气中的 $^{14}C$ 活度是基于中纬度树木年轮几十年的样本的高精度分析而获得的 [15]。所用的数据集是多个树轮年代学和放射性碳实验室间国际合作的成果 [16]。用于太阳黑子数重建且具有被绝对而精确定年的原始数据在图 1 用黑色实线表示。以公元 950 年的数据为参照点，起始数据与之相比高出 15%，且大气中的 $^{14}C$ 表现出一种长期的变化趋势（以红线标出），这主要是地磁偶极场在全新世之前和期间的强度变化的结果。更短时间的波动主要是太阳磁场变化引起宇宙射线通量的变化，进而对 $^{14}C$ 的产生速率产生影响的结果。

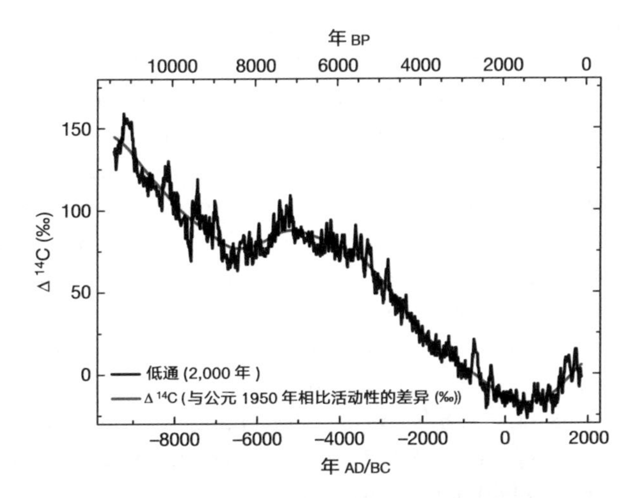


图 1. 基于具有绝对年代的树木年轮的十年际样本（INTCAL98 数据集）[16] 获得的大气放射性碳水平 $\Delta^{14}C$（以相对公元 1950 年标准水平的偏差表示，单位为千分之一 [15]）。尽管早期 $\Delta^{14}C$ 时间序列的误差部分会达到 4‰ ~ 5‰，但总体为 2‰ ~ 3‰。早于 11,400 BP 的 INTCAL98 数据不被直接用于重建，因为当时的碳循环存在较大的误差与不确定性。更多关于数据、重建初始条件，以及误差估计的信息参见补充信息。长时间尺度的下降趋势（图中红色曲线）是 $^{14}C$ 产率下降的结果，主要是地磁场对宇宙射线屏蔽增强造成的。短期的波动（为期 1 ~ 2 个世纪）反映了 $^{14}C$ 产率随太阳活动的变化。图中公元前（BC）在数轴上记为负数。

The atmospheric $^{14}C$ level may also be affected by changes in the partition of carbon between the major reservoirs, that is, deep ocean, ocean mixed layer, biosphere and atmosphere. Variations in ocean circulation[17] could influence $^{14}C$ via a variable uptake of $CO_2$ into the ocean or by the exchange of $^{14}C$-depleted carbon from the deep ocean, but, owing to the rather small $^{14}C$ gradients among the reservoirs, strong changes in these processes need to be invoked. For the Holocene, there is no evidence of considerable oceanic variability, so we can assume that the short- and mid-term fluctuations of $^{14}C$ predominantly reflect solar variability. This is supported by the strong similarity of the fluctuations of $^{10}Be$ in polar ice cores compared to $^{14}C$, despite their completely different geochemical history[18-20].

We first determine the $^{14}C$ production rate in the Earth's atmosphere following Usoskin and Kromer[21]. They used two distinct methods, which take into account carbon cycle effects in different ways. Both methods give similar results when applied to the tree-ring $\Delta^{14}C$ data set described above. For the current reconstruction we use the average of the $^{14}C$ production rate deduced using both techniques. In accordance with the decadal sampling of the $^{14}C$ data we reconstruct the 10-year averaged sunspot number. Because the $\Delta^{14}C$ data are contaminated by extensive burning of $^{14}C$-free fossil fuel since the late nineteenth century[22] and later by atmospheric nuclear bomb tests, we use $^{14}C$ data before AD 1900 only and take the historical sunspot number record for the most recent period.

From the $^{14}C$ production rate we obtain the sunspot number in multiple steps, each substantiated by a physics-based model. A model describing the transport and modulation of galactic cosmic rays within the heliosphere[11] is inverted to find the cosmic ray flux corresponding to the determined $^{14}C$ production rate. The transport of galactic cosmic rays in the heliosphere is affected by the Sun's open magnetic flux, that is, the fraction of the Sun's total magnetic flux that reaches out into interplanetary space[12]. The open flux is linked with the sunspot number by inverting a model describing the evolution of the open magnetic flux for given sunspot number[9,10]. All adjustable parameters entering this chain of models were fixed using independent data prior to the current reconstruction, so that no free parameter remains when reconstructing the sunspot number from $^{14}C$ data (see Supplementary Table S1). This reconstruction method was previously applied to $^{10}Be$ data from Greenland and Antarctica. Only the first step changes when using $\Delta^{14}C$ instead of $^{10}Be$ data to reconstruct the sunspot number. Hence, possible errors and uncertainties in the later steps are similar to those studied in our earlier papers[13,14].

Applying our reconstruction method to $\Delta^{14}C$, we first determine the sunspot number since AD 850 in order to compare these values with the historical record of GSNs since 1610 and with the reconstruction on the basis of $^{10}Be$ data[14]. Figure 2 shows that the reconstructed average sunspot number from $\Delta^{14}C$ is remarkably similar to the 10-year averaged GSN series (correlation coefficient $0.925^{+0.02}_{-0.03}$ with a false alarm probability $< 10^{-6}$). The difference between the reconstructed and measured sunspot number is nearly gaussian with a standard deviation of 5.8, which is smaller than the theoretical estimate of the reconstructed sunspot number uncertainty (about 8 for the last millennium, see Supplementary Information),

大气 $^{14}C$ 水平可能也会受到 $^{14}C$ 主要碳库——深海、海洋混合层、生物圈、大气圈——之间碳配比变化的影响。洋流的变化 [17] 可能会通过改变海洋吸收的 $CO_2$，或者通过与深海 $^{14}C$ 贫化的碳之间的交换来影响 $^{14}C$。但是，由于不同碳库间的 $^{14}C$ 梯度非常的小，需要考虑上述过程中的剧烈变化。对于全新世来说，没有海洋碳库剧烈变化的明显证据。因此，我们可以假设 $^{14}C$ 短期和中期变化主要反映了太阳活动的变化。上述假设得到了极地冰芯 $^{10}Be$ 与 $^{14}C$ 对比分析结果的支持。尽管具有不同的地球化学历史，但两者的波动变化极为相似 [18-20]。

我们根据乌索斯金和克罗默的方法，首先来确定地球大气中 $^{14}C$ 的产生速率 [21]。他们采用了两种不同的办法，即以不同的方式考虑了碳循环的影响。将两者应用于前述树木年轮的 $\Delta^{14}C$ 数据集时，两种方法给出了相似的结果。对当前生产速率的重建工作，我们采用了两种方法所得结果的平均值。与 $^{14}C$ 十年期采样一致，我们重建了十年平均的太阳黑子数。因为 $\Delta^{14}C$ 数据受到 19 世纪晚期大量的无 $^{14}C$ 化石燃料燃烧及后期大气核弹试验的污染 [22]，我们仅采用公元 1900 年以前的数据，对于最近的一段时期，我们只使用历史记录的太阳黑子数数据。

从 $^{14}C$ 的产生速率，我们分多步得到了太阳黑子数目，每一步都有物理模型证实。利用限定太阳风层内宇宙射线的传输和调节模型 [11] 的反演，来寻找已确定 $^{14}C$ 产率对应的宇宙射线通量。太阳风层内宇宙射线的传输受太阳开放磁通量的影响。开放通量是指太阳总磁通量中到达行星际空间的组分 [12]。通过给定太阳黑子数目下的开放磁通量演化模型的反演，将开放通量与太阳黑子的数目联系起来 [9,10]。所有进入模型链的可变参数都用以前重建工作的独立数据来确定，所以在利用 $^{14}C$ 数据重建太阳黑子数目时已经没有自由参数(见补充表格 S1)。这种重建方法曾被用于格陵兰和南极 $^{10}Be$ 的数据重建。在用 $\Delta^{14}C$ 数据取代 $^{10}Be$ 数据重建太阳黑子数时，仅第一步发生了改变。因此，后续步骤中可能出现的误差和不确定性都与我们先前的研究相似 [13,14]。

将我们的重建方法应用于 $\Delta^{14}C$。为了将这些值与自 1610 年以来 GSN 的历史记录和基于 $^{10}Be$ 数据的重建结果进行对比 [14]，我们首先确定了自公元 850 年至今的太阳黑子数目。图 2 表明由 $\Delta^{14}C$ 重建的平均黑子数目与 10 年平均的 GNS 序列十分相似(相关系数 $0.925^{+0.02}_{-0.03}$，误报可能性 $<10^{-6}$)。重建数据和观测到的黑子数目的误差基本符合标准差为 5.8 的高斯分布，这个标准差小于重建黑子数目的理论估计(对于上个千年大约为 8，见参考信息)，这表明了后者的保守特性。基于 $^{10}Be$ 重建的两个

indicating the conservative nature of the latter. Two $^{10}$Be-based sunspot number reconstructions are plotted in Fig. 2, which correspond to extreme assumptions about the geographic area of $^{10}$Be production relevant for its deposition in polar ice. The local polar production model (green curve) provides an upper limit to the sunspot number[14], while the global production model (magenta dashed curve) gives a lower limit[13]. The sunspot number time series obtained from $\Delta^{14}$C lies between the two $^{10}$Be-based curves, and for the period after AD 1200 is closer to the $^{10}$Be-based reconstruction under the assumption of global production.

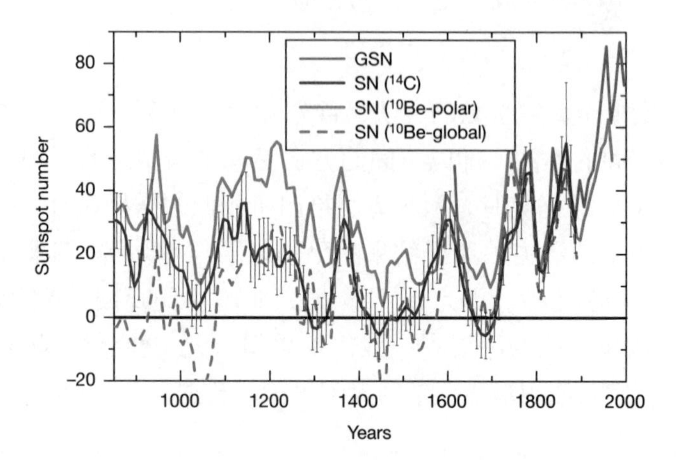


Fig. 2. Comparison between directly measured sunspot number (SN) and SN reconstructed from different cosmogenic isotopes. Plotted are SN reconstructed from $\Delta^{14}$C (blue), the 10-year averaged group sunspot number[1] (GSN, red) since 1610 and the SN reconstruction[14] from $^{10}$Be under the two extreme assumptions of local (green) and global (magenta, dashed) production, respectively. The slightly negative values of the reconstructed SN during the grand minima are an artefact; they are compatible with SN = 0 within the uncertainty of these reconstructions as indicated by the error bars. $\Delta^{14}$C is connected with the $^{14}$C production rate via a carbon cycle model[21]. The connection between the $^{14}$C production rate, $R$, and the cosmic ray flux is given by $R = \int_{\theta=0}^{\pi}\int_{P_C(\theta,M)}^{\infty} X(P,\Phi)Y(P)\mathrm{d}P\sin\theta\mathrm{d}\theta$ , where $\theta$ is the colatitude relative to the geomagnetic dipole axis, and $P_C(\theta, M)$ is the local cosmic ray rigidity cutoff (which depends on $\theta$ and the virtual geomagnetic dipole moment, $M$)[23]. $X(P, \Phi)$ is the differential cosmic ray rigidity spectrum near Earth, $\Phi$ is the modulation strength describing the average rigidity losses of cosmic rays inside the heliosphere, $Y(P)$ is the differential yield function[24] of $^{14}$C, and $P$ is the rigidity of the primary cosmic rays. For studies of long-term changes of the cosmic ray flux, the parameter $\Phi$ alone adequately describes the modulation of the cosmic ray spectrum $X(P)$[11,24]. The two most abundant cosmic ray species, protons and $\alpha$-particles, are taken into account in the model[13]. The cosmic ray transport model relates $R$ to $\Phi$, which in turn depends on the Sun's open magnetic flux[12]. The open flux is linked with the magnetic flux in sunspots (and thus with the SN) via the source term in a system of differential equations[9,10]. The value of $R$ is obtained from $\Delta^{14}$C and $M$ is known for the whole interval of interest[25,26], so that $\Phi$ can be obtained from the inversion of the equation given above. Error bars depict the 68% confidence interval for the reconstructed SN, which takes into account both random and systematic uncertainties (see Supplementary Information).

Figure 3a shows the reconstruction based on the 11,400-year set of $\Delta^{14}$C data. Clearly, the level of activity has remained variable, with episodes of particularly low numbers of sunspots (grand minima) distributed over the whole record. Episodes of high activity are also present. These are mostly concentrated in the earliest three millennia (before 6000

黑子数目序列见图 2，它们对应于对极地冰中与 $^{10}$Be 沉积有关的产率的地理区域的极端假设。局地极地产生模型（绿色曲线）为太阳黑子数提供了上限，而全球产生模型（品红色曲线）则提供了下限。基于 $\Delta^{14}$C 重建的太阳黑子时间序列介于基于 $^{10}$Be 重建的两条曲线之间。对于公元 1200 年之后的时期来说，该曲线要更靠近于 $^{10}$Be 全球产生模型下的重建。

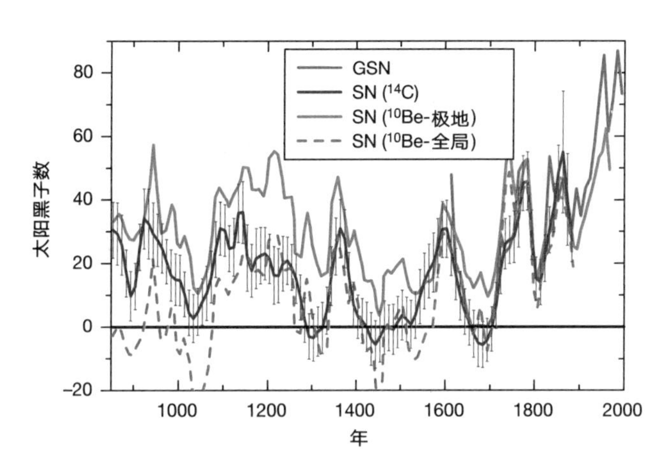


图 2. 直接测量获得的太阳黑子数（SN）与基于宇宙成因同位素重建的太阳黑子数的比较。图中 SN 分别通过以下数据重建：$\Delta^{14}$C（蓝色）、1610 年来十年期平均太阳黑子群数 [1]（GSN，红色），以及在两个极端假设条件（局域，绿色实线；全局，洋红色虚线）下的 $^{10}$Be[14]。在大的极小期 SN 的偏负数值是人为造成的，在重建的误差范围内（图中误差棒所示）SN 近似为零。$\Delta^{14}$C 通过碳循环模型与 $^{14}$C 的产率联系起来 [21]，而 $^{14}$C 的生成率（$R$）与宇宙射线通量之间的关系式为：$R = \int_{\theta=0}^{\pi}\int_{P_C(\theta,M)}^{\infty} X(P, \Phi)Y(P)\mathrm{d}P\sin\theta\mathrm{d}\theta$，其中 $\theta$ 是相对于地磁偶极子轴的磁余纬，$P_C(\theta, M)$ 是局域宇宙射线刚性截止（值取决于 $\theta$ 以及虚拟地磁偶极子矩 $M$）[23]。$X(P, \Phi)$ 是靠近地球处宇宙射线刚性谱的微分，$\Phi$ 是调制强度，用来限定太阳风层内宇宙射线的平均刚性损失，$Y(P)$ 是 $^{14}$C 产率函数的微分，$P$ 是原始宇宙射线的刚性。对于宇宙射线通量变化的长期研究，只利用参数 $\Phi$ 就足以描述宇宙射线谱的调制 $X(P)$[11,24]。模型也考虑了两种最丰富的宇宙射线，即质子和 $\alpha$ 粒子 [13]。宇宙射线传输模型将 $R$ 和 $\Phi$ 相联系，它们转而依赖于太阳开放磁场通量 [12]。在太阳黑子中，开放通量通过系统微分方程组中的源项与磁通量联系起来（因而也与 SN 相关联）[9,10]。$R$ 的值由 $\Delta^{14}$C 获得，$M$ 在整个研究时段是已知的 [25,26]，故 $\Phi$ 可以通过上面给出的方程反演得到。误差棒表示考虑了随机和系统误差的重建 SN 的 68% 置信区间（详见补充信息）。

图 3a 显示了基于 11,400 年 $\Delta^{14}$C 数据集的太阳黑子重建情况。显然，活动水平持续变化，太阳黑子数很少的时期（长极小期）在整个记录中都有分布。高活动性的时期也有，主要集中在最早的三个千年（公元前 6000 年之前），这些时期也表现为

BC), which also exhibit a high average sunspot number (35.6 compared to 25.6 after 6000 BC). During the last eight millennia, the episode with the highest average sunspot number is the ongoing one that started about 60 years ago. The sunspot number averaged over the whole period is 28.7 with a standard deviation of 16.2. The average number of 75 since 1940 thus lies 2.85 standard deviations above this long-term average.

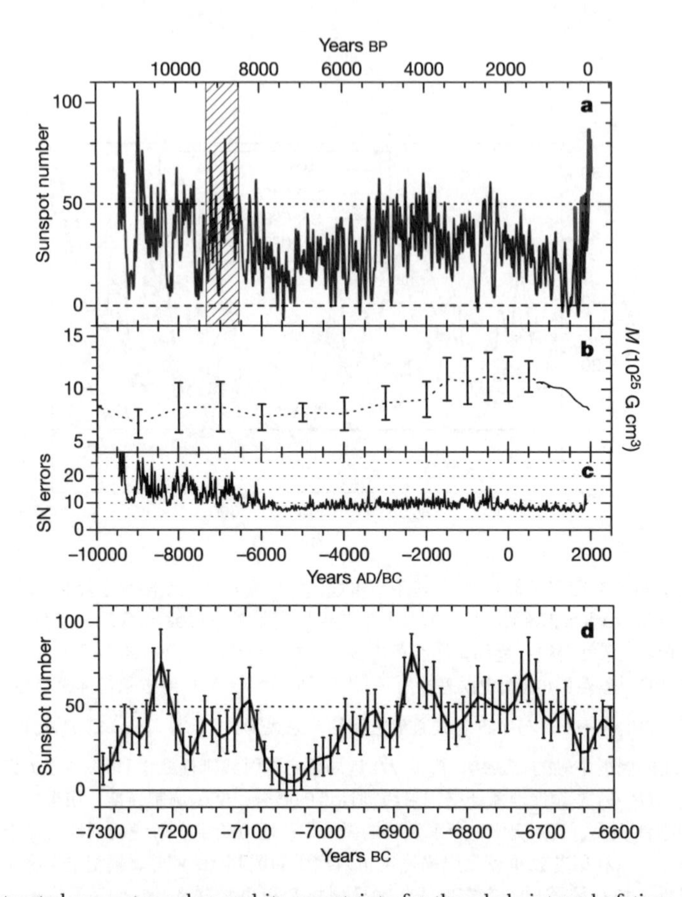


Fig. 3. Reconstructed sunspot number and its uncertainty for the whole interval of time considered. **a**, 10-year averaged SN reconstructed from $\Delta^{14}C$ data since 9500 BC (blue curve) and 10-year averaged group sunspot number[1] (GSN) obtained from telescopic observations since 1610 (red curve). The horizontal dotted line marks the threshold above which we consider the Sun to be exceptionally active. It corresponds to 1.3 standard deviations above the mean. **b**, Evolution of the virtual geomagnetic dipole moment[26] with error bars that take into account the scatter between different palaeomagnetic reconstructions. (The error bars give the s.d. in the reconstructed virtual geomagnetic dipole moment.) The geomagnetic field data of ref. 25 are given by the dotted line. **c**, Uncertainty in the reconstructed SN. It includes errors introduced at each step of the reconstruction process. The largest sources of random errors are the uncertainty in the knowledge of the geomagnetic dipole moment and in the $^{14}C$ production rate. We also consider systematic errors—for example, due to uncertainties in the $^{14}C$ production rate prior to the considered period of time. A discussion of how these uncertainties are estimated is given in Supplementary Information. Clearly, the uncertainties are sufficiently small that they do not affect the presence or absence of grand minima or of episodes of high activity, except in already marginal cases. **d**, A detail from the full time series of reconstructed SN with expanded temporal scale. The chosen interval (corresponding to the shaded part of **a**) exhibits three episodes of high solar activity and a grand minimum. The error bars indicate the total uncertainty, $\sigma$, in the reconstruction. (They depict the 68% confidence interval for the reconstructed SN,

高平均太阳黑子数目（35.6 个，而公元前 6000 年之后仅为 25.6 个）。在过去的八个千年之中，具有最高平均太阳黑子数目的时期就是开始于 60 年前且当前还在进行的时段。太阳黑子数目在整个时期的平均值是 28.7 个，标准差为 16.2。而自 1940 年以来的平均数达 75，高于 2.85 个标准差的平均变化。

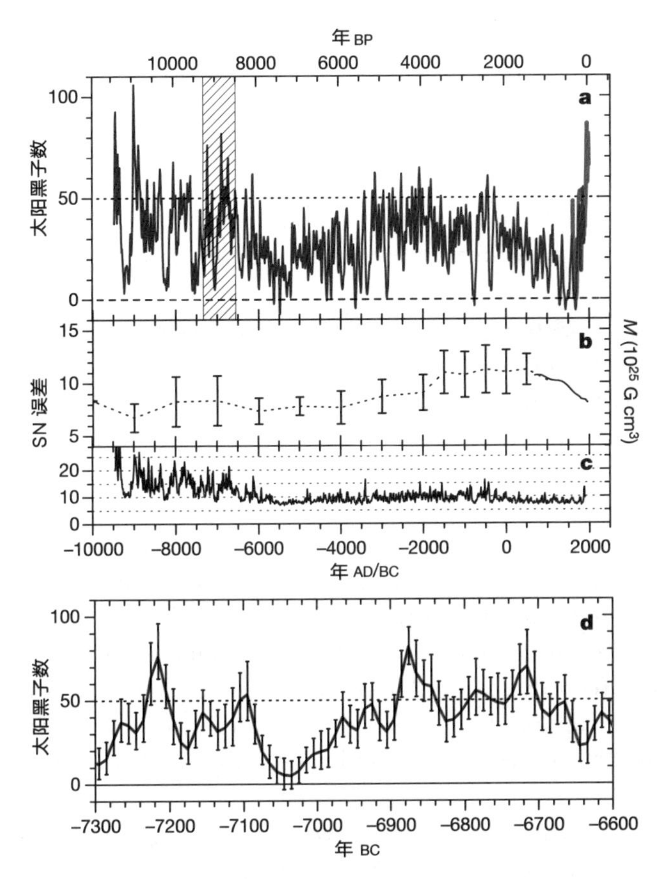


图 3. 重建的太阳黑子数与整个研究时段的不确定性。**a**. 公元前 9500 年以来基于 $\Delta^{14}C$ 数据重建的 10 年平均 SN（蓝色曲线）和由 1610 年来望远镜观测数据获得的 10 年平均太阳黑子群数（GSN）[1]（红色曲线）。水平虚线表示我们认为太阳活动异常活跃的阈值，其对应了平均值之上 1.3 个标准差。**b.** 虚拟地磁偶极矩的演化 [26]，其中误差棒考虑了不同的古地磁重建之间的离散（误差棒给出了重建虚拟地磁偶极矩的标准差）。来自文献 25 的地磁场数据用虚线表示。**c.** 重建太阳黑子数的不确定性，包括了重建过程中每一步的误差。最大的随机误差源是对于地磁偶极矩和对研究期之前的 $^{14}C$ 产率了解不足。我们也考虑了系统误差，如早于研究时段的 $^{14}C$ 产率的不确定性。关于误差估计的详细内容参见补充信息。显然，这些不确定性因素较小，除了边际情况，不足以影响大的极小期和高活动性事件的出现和缺失。**d.** 时间尺度扩展到整个时段的 SN 序列细节图。选取时段（对应于图 **a** 的阴影部分）展示了三个高太阳活动事件和一个大的极小期。误差棒指示重建过程中总误差 $\sigma$（包括随机误差和系统误差的 68% 置信区间，详见补充信息）。两个最强的极大值分别位于高活动性阈值（50）的 $2.1\sigma$ 和 $3.0\sigma$ 处。因而，与这些误差相关的统计波动导致误差的概率分别为 3% 和 0.2%。整个高活动期（比如持续 50 年）是由统计波动引起的误差的概率相当小。

which takes into account both random and systematic uncertainties (see Supplementary Information).) The two strongest maxima lie 2.1σ and 3.0σ, respectively, above the high-activity threshold of 50. Hence the probability that they are due to statistical fluctuations related to these errors is 3% and 0.2%, respectively. The probability that a whole episode of high activity (lasting, say, 50 years) is due to a statistical fluctuation is significantly smaller.

A major uncertainty in the reconstructed sunspot number is related to the evolution of the geomagnetic field, which is represented in Fig. 3b. A weaker geomagnetic field leads to an increased cosmic ray flux impinging on the terrestrial atmosphere and thus to a higher $^{14}C$ production rate, mimicking a lower value of sunspot number if not properly taken into account. The uncertainties in the reconstructed sunspot numbers are discussed in Supplementary Information and enter into Fig. 3c, where the sum of the random and systematic uncertainties affecting the reconstruction is given. For the whole time interval, the mean uncertainty of the reconstructed sunspot number is about 10.

In Fig. 3d we show a sub-interval of 700 years duration from our reconstruction, which exhibits three prominent periods of high sunspot number (average values exceeding 50) and a grand minimum. Within the 95% confidence interval, the Sun spent in total between 780 and 1,060 years in a high-activity state (sunspot number > 50), which corresponds to 6.9–9.3% of the total duration of our reconstruction. The most probable values are 950 years and 8.3%, respectively. Although the rarity of the current episode of high average sunspot number may be taken as an indication that the Sun has contributed to the unusual degree of climate change during the twentieth century, we stress that solar variability is unlikely to be the prime cause of the strong warming during the last three decades[3]. In ref. 3, reconstructions of solar total and spectral irradiance as well as of cosmic ray flux were compared with surface temperature records covering approximately 150 years. It was shown that even under the extreme assumption that the Sun was responsible for all the global warming prior to 1970, at the most 30% of the strong warming since then can be of solar origin.

There are 31 periods during which the 10-year averaged sunspot number consistently exceeds a level of 50. The average length of such episodes is about 30 years, the longest being 90 years (around 9000 BC). The distribution of the durations of such episodes is given in Fig. 4a. The number of high-activity periods decreases exponentially with increasing duration. The current level of high solar activity has now already lasted close to 65 years and is marked by the arrow on the figure. This implies that not only is the current state of solar activity unusually high, but also this high level of activity has lasted unusually long. Assuming the previous episodes of high activity to be typical, we can estimate the probability with which the solar activity level will remain above a sunspot number of 50 over the next decades. The result is given in Fig. 4b, which shows that there is only a probability of $8\%^{+3\%}_{-4\%}$ that the current high-activity episode will last another 50 years (and thus reach a total duration of 115 years), while the probability that it will continue until the end of the twenty-first century is below 1%.

在重建黑子数目的过程中，一个主要的不确定性因素与地磁场的演化相关，见图 3b。较弱的地磁场导致射入地球大气中的宇宙射线流通量增加，$^{14}C$ 产率增高。如果无法对这一影响进行正确处理，其产生的结果与低太阳黑子产生的结果类似。关于重建黑子数目过程中的不确定性的讨论见补充信息以及图 3c。在图中，我们给出了重建的随机以及系统误差总和。对整个时间区间而言，太阳黑子数目重建的平均误差大约为 10。

图 3d 是我们重建结果中为期 700 年的一个子区间，上面展示了三个明显的高太阳黑子数时期（平均值超过 50）和一个大的极小期。在 95% 置信区间内，太阳高活动期（太阳黑子数 > 50）持续时间在 780 到 1,060 年之间，对应于总重建时间的 6.9% ~ 9.3%。最有可能的时间跨度为 950 年，占总重建时间的 8.3%。尽管当代罕见的高平均太阳黑子数或许可以作为太阳对二十世纪异常气候变化影响的标示，但我们仍要强调太阳活动的变化不可能是最近三十年气候强烈变暖的主要原因 [3]。参考文献 3 重建了过去 150 年来的太阳总辐照度和光谱辐照度，以及宇宙射线通量，并与地表温度记录进行对比。结果显示，即使在最极端的假设下，即太阳对 1970 年前所有的全球变暖现象皆有影响，最多也只有 30% 的强气候变暖源于太阳活动。

太阳黑子数十年平均值持续高于 50 的有 31 个时段，这些时段的平均长度大约为 30 年，最长达 90 年（约在公元前 9000 年）。这些事件的持续时间分布见图 4a。随持续时间的增加，高活跃度时段的数量呈指数下降。现今的高太阳活动已经持续了将近 65 年，在图中以箭头标示。这说明不仅当下太阳活动异常活跃，而且持续时间也异常长。假设过去的高活跃事件具有代表性，我们可以估计未来数十年太阳黑子数保持大于 50 的概率。结果见图 4b，太阳现有高活动水平再持续 50 年（持续总时间达到 115 年）的概率只有 $8\%^{+3\%}_{-4\%}$，而持续到二十一世纪末的概率小于 1%。

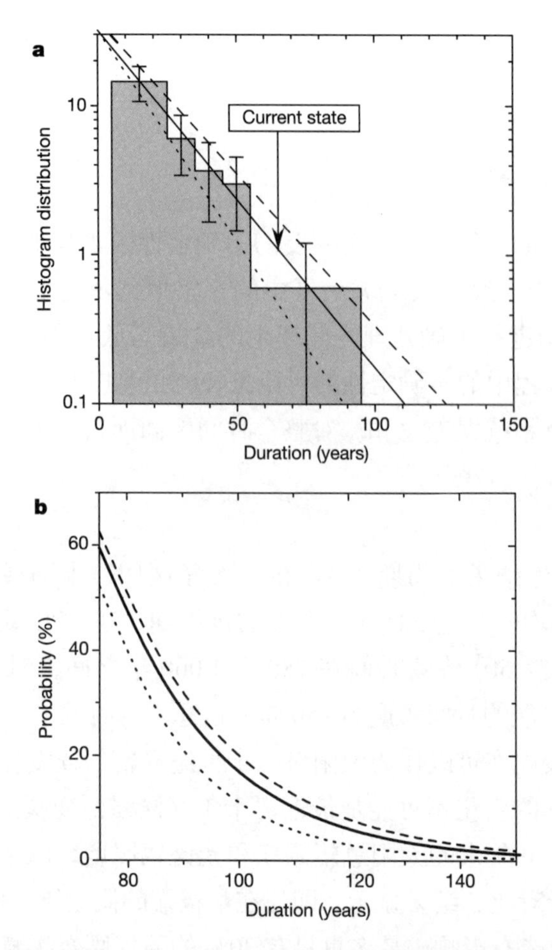


Fig. 4. Distribution of the duration of episodes of high solar activity and the probability that the current episode will reach a given duration. **a**, Histogram of the distribution function of the duration of episodes of high solar activity during which the 10-year averaged SN exceeds 50. Some bins have been enlarged in order to improve the statistics; in such cases the average number per bin is given. The vertical error bars correspond to $\sqrt{N}$, where $N$ is the number of events combined in one data point. The length of the current period of high activity is marked by the arrow. The solid line is a least-squares exponential fit to the plotted points. The dashed and dotted lines represent exponential fits to the distributions obtained from extreme SN reconstructions including the influence of random and systematic errors as given in Fig. 3 (see also the discussion in Supplementary Information). **b**, The probability of the total duration of a state of high activity (SN level exceeding 50). For the current episode, which started in AD 1940 (~65 years ago), the start of the diagram corresponds to the year AD 2015. Each curve is based upon the corresponding fit shown in **a**. The probability (for the reference curve) that the high activity continues for another 5 decades for a total duration of 115 years is only 8%.

(**431**, 1084-1087, 2004)

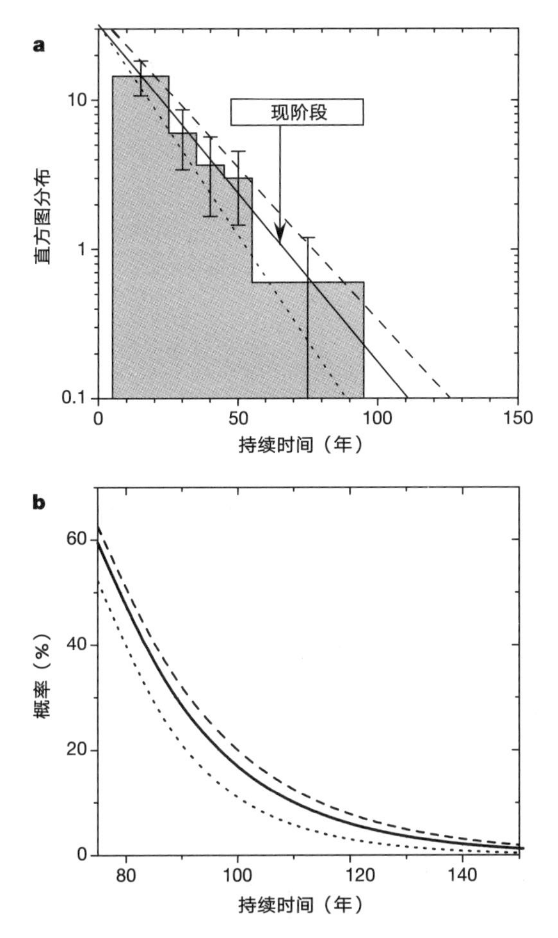


图 4. 高太阳活动持续时间的分布与当代事件达到给定时长的概率。**a.** 10 年平均 SN 超过 50 的高太阳活动事件持续时间分布函数的直方图。有些组被放大以方便统计，在这些情况下，给出的是每组的平均数。垂直误差棒对应于$\sqrt{N}$，其中 $N$ 代表组合成一个数据点的事件数目。当代高活动期的长度在图中用箭头标示。实线代表基于标出点并利用最小二乘法指数拟合的曲线。虚线和点线表示基于极端 SN 重建分布（包括随机和系统误差的影响，见图 3）的指数拟合（参见补充信息中的讨论）。**b.** 高活动性（SN 水平高于 50）总持续时间的概率。当今时段始于 1940 年，即大约 65 年前，图中开始于 2015 年。每条曲线都是基于图 **a** 所示对应状态的拟合。未来高活动性再持续 50 年，总时长达 115 年的概率仅为 8%。

（汪浩 翻译；许冰 审稿）

**S. K. Solanki[1], I. G. Usoskin[2], B. Kromer[3], M. Schüssler[1] & J. Beer[4]**

[1] Max-Planck-Institut für Sonnensystemforschung (formerly the Max-Planck-Institut für Aeronomie), 37191 Katlenburg-Lindau, Germany

[2] Sodankylä Geophysical Observatory (Oulu unit), University of Oulu, 90014 Oulu, Finland

[3] Heidelberger Akademie der Wissenschaften, Institut für Umweltphysik, Neuenheimer Feld 229, 69120 Heidelberg, Germany

[4] Department of Surface Waters, EAWAG, 8600 Dübendorf, Switzerland

Received 20 February; accepted 1 September 2004; doi 10.1038/nature02995.

---

References:

1. Hoyt, D. V. & Schatten, K. H. Group sunspot numbers: A new solar activity reconstruction. *Sol. Phys.* **179**, 189-219 (1998).
2. Eddy, J. A. The Maunder minimum. *Science* **192**, 1189-1202 (1976).
3. Solanki, S. K. & Krivova, N. Can solar variability explain global warming since 1970? *J. Geophys. Res.* **108**, doi: 10.1029/2002JA009753 (2003).
4. Stuiver, M. & Braziunas, T. F. Atmospheric $^{14}C$ and century-scale solar oscillations. *Nature* **338**, 405-408 (1989).
5. Stuiver, M. & Braziunas, T. F. Sun, ocean, climate and atmospheric $^{14}CO_2$: an evaluation of causal and spectral relationships. *Holocene* **3**, 289-305 (1993).
6. Damon, P. E. & Sonett, C. P. in *The Sun in Time* (eds Sonnet, C. P., Giampapa, M. S. & Matthews, M. S.) 360-388 (Univ. Arizona, Tucson, 1991).
7. Beer, J. *et al.* Use of $^{10}Be$ in polar ice to trace the 11-year cycle of solar activity. *Nature* **347**, 164-166 (1990).
8. Beer, J. Long-term indirect indices of solar variability. *Space Sci. Rev.* **94**, 53-66 (2000).
9. Solanki, S. K., Schüssler, M. & Fligge, M. Evolution of the Sun's large-scale magnetic field since the Maunder minimum. *Nature* **408**, 445-447 (2000).
10. Solanki, S. K., Schüssler, M. & Fligge, M. Secular variation of the Sun's magnetic flux. *Astron. Astrophys.* **383**, 706-712 (2002).
11. Usoskin, I. G., Alanko, K., Mursula, K. & Kovaltsov, G. A. Heliospheric modulation strength during the neutron monitor era. *Sol. Phys.* **207**, 389-399 (2002).
12. Usoskin, I. G., Mursula, K., Solanki, S. K., Schüssler, M. & Kovaltsov, G. A. A physical reconstruction of cosmic ray intensity since 1610. *J. Geophys. Res.* **107**, doi:10.1029/2002JA009343 (2002).
13. Usoskin, I. G., Mursula, K., Solanki, S. K., Schüssler, M. & Alanko, K. Reconstruction of solar activity for the last millenium using $^{10}Be$ data. *Astron. Astrophys.* **413**, 745-751 (2004).
14. Usoskin, I. G., Solanki, S. K., Schüssler, M., Mursula, K. & Alanko, K. A millenium scale sunspot number reconstruction: evidence for an unusually active Sun since the 1940s. *Phys. Rev. Lett.* **91**, 211101 (2003).
15. Stuiver, M. & Pollach, P. Discussion: reporting of $^{14}C$ data. *Radiocarbon* **19**, 355-363 (1977).
16. Stuiver, M. *et al.* INTCAL98 Radiocarbon age calibration. *Radiocarbon* **40**, 1041-1083 (1998).
17. Broecker, W. S. An unstable superconveyor. *Nature* **367**, 414-415 (1994).
18. Bond, G. *et al.* Persistent solar influence on North Atlantic surface circulation during the Holocene. *Science* **294**, 2130-2136 (2001).
19. Muscheler, R., Beer, J. & Kromer, B. *Solar Variability as an Input to the Earth's Environment* 305-316 (ESA SP-535, European Space Agency, Noordwijk, 2003).
20. Bard, E., Raisbeck, G. M., Yiou, F. & Jouzel, J. Solar modulation of cosmogenic nuclide production over the last millennium: comparison between $^{14}C$ and $^{10}Be$ records. *Earth Planet. Sci. Lett.* **150**, 453-462 (1997).
21. Usoskin, I. G. & Kromer, B. Reconstruction of the $^{14}C$ production rate from measured relative abundance. *Radiocarbon* (in the press).
22. Suess, H. E. Radiocarbon content in modern wood. *Science* **122**, 415-417 (1955).
23. Elsasser, W., Ney, E. P. & Winckler, J. R. Cosmic-ray intensity and geomagnetism. *Nature* **178**, 1226-1227 (1956).
24. Castagnoli, G. & Lal, D. Solar modulation effects in terrestrial production of carbon-14. *Radiocarbon* **22**, 133-158 (1980).
25. Hongre, L., Hulot, G. & Khokhlov, A. An analysis of the geomagnetic field over the past 2000 years. *Phys. Earth Planet. Inter.* **106**, 311-335 (1998).
26. Yang, S., Odah, H. & Shaw, J. Variations in the geomagnetic dipole moment over the last 12000 years. *Geophys. J. Int.* **140**, 158-162 (2000).

**Supplementary Information** accompanies the paper on www.nature.com/nature.

**Competing interests statement.** The authors declare that they have no competing financial interests.

Correspondence and requests for materials should be addressed to S.K.S. (solanki@mps.mpg.de).

# Tomographic Imaging of Molecular Orbitals

J. Itatani *et al.*

## Editor's Note

**In molecules, the electron orbitals of the constituent atoms are considered to overlap and reconfigure into molecular orbitals that determine the molecule's structure and chemical behaviour. The shape and nature of these molecular orbitals can be calculated from quantum theory, but they have been largely treated as a mere mathematical formalism that permits a quantum description of the chemical bond. Here, however, David Villeneuve in Canada and his coworkers show that molecular orbitals can actually be imaged in a slice-wise tomographic fashion by looking at the way the electron clouds of molecules interact with extremely short pulses of laser light. They present a map of the quantum-mechanical wavefunction for a molecular orbital of the nitrogen molecule.**

---

Single-electron wavefunctions, or orbitals, are the mathematical constructs used to describe the multi-electron wavefunction of molecules. Because the highest-lying orbitals are responsible for chemical properties, they are of particular interest. To observe these orbitals change as bonds are formed and broken is to observe the essence of chemistry. Yet single orbitals are difficult to observe experimentally, and until now, this has been impossible on the timescale of chemical reactions. Here we demonstrate that the full three-dimensional structure of a single orbital can be imaged by a seemingly unlikely technique, using high harmonics generated from intense femtosecond laser pulses focused on aligned molecules. Applying this approach to a series of molecular alignments, we accomplish a tomographic reconstruction of the highest occupied molecular orbital of $N_2$. The method also allows us to follow the attosecond dynamics of an electron wave packet.

---

THE electrons that make up molecules are organized by energy in orbitals[1,2]. Although total electron density in molecules is routinely measured by X-ray diffraction or electron scattering, only two methods are able to "see" the highest occupied molecular orbitals (HOMOs)—electron momentum spectroscopy and scanning tunnelling microscopy. Electron momentum spectroscopy[3] is an (*e*, 2*e*) scattering technique that can determine the radially averaged density of the outermost valence electrons. Scanning tunnelling microscopy[4] gives the electron density, distorted by surface states. We show that high harmonic emission from molecules[5-11] allows the three-dimensional shape of the highest electronic orbital to be measured, including the relative phase of components of the wavefunction. Our results may be compared with predictions of the various models that are used to describe many-electron systems[12], such as the Hartree-Fock, the Kohn-Sham and the Dyson approaches.

# 分子轨道的层析成像

板谷治郎等

## 编者按

*在分子中，大家认为构成原子的电子轨道通过重叠和重新组合成为分子轨道，这种分子轨道可决定其分子结构及化学行为。通过量子论可以估算出这些分子轨道的形状和性质，但是，在很大程度上，它们只不过被视为数学形式体系，该数学形式可以量子化地描述化学键。然而，加拿大的戴维·维尔纳夫和他的合作者们在本文中表明，通过观察分子的电子云与激光的极短脉冲之间相互作用的方式，证明了分子轨道实际上可以以切片层析的方式成像。他们提供了一张氮分子的分子轨道的量子力学波函数的图像。*

---

单电子波函数（或称轨道）是用来描述分子的多电子波函数的数学构造。由于最外层轨道决定分子的化学性质，因而值得我们格外关注。观察这些轨道在化学键形成和断裂时的变化就是观察化学反应的本质。不过单轨道很难用实验方法加以观测，而且到目前为止，在化学反应的时间尺度上是不可能做到这种观测的。我们要在本文中说明的是，单一轨道的完整三维结构可以通过一种看似不大可能的技术来成像，即采用高次谐波，它是通过高强度的飞秒激光脉冲聚焦于排列的分子而产生的。将这种方法应用于一系列分子排列，我们实现了对于 $N_2$ 中最高占据分子轨道的层析重建。这种方法还使我们得以探索一个电子波包的阿秒动力学。

---

构成分子的电子依其能量而分列于轨道之中 [1,2]。尽管分子中总的电子密度可用 X 射线衍射或电子散射等常规的方式测得，但只有两种方法能够“看到”最高占据分子轨道（HOMOs）——电子动量谱学和扫描隧道显微镜。电子动量谱学 [3] 是一种 $(e, 2e)$ 散射技术，它能测定最外层价电子的径向平均密度。扫描隧道显微镜 [4] 则给出被表面状态扭曲的电子密度。我们要说明的是，来自分子的高次谐波发射 [5-11] 使我们得以测定最外层电子轨道的三维形状，包括波函数分量的相对相位。可以将我们的结果与用来描述多电子体系的各种模型 [12]——例如哈特里–福克、科恩–沙姆以及戴森方法——的预测结果加以比较。

If we can measure orbitals with femtosecond laser technology, we can also observe orbital changes occurring on the timescale of chemical reactions. The next challenge will be to measure attosecond bound-state electron dynamics. We show that imaging orbitals can be extended to imaging coherent electron motion in atoms or molecules. Quantum mechanically, electron motion is described by a coherent superposition of electronic states. This is the closest that quantum mechanics allows to imaging the "Bohr orbital motion" of electrons.

Tomographic imaging of a molecular orbital is achieved in three crucial steps. (1) Alignment of the molecular axis in the laboratory frame. (2) Selective ionization of the orbital. (3) Projection of the state onto a coherent set of plane waves.

## Alignment of Gas-phase Molecules

Computed tomography refers to the technology of retrieving sectional images of an object, such as a human body, from a series of one-dimensional projections[13]. For a projection direction fixed in the laboratory frame, the object must be rotated. Thus, we begin by discussing how molecules can be aligned[14].

A relatively intense, non-resonant laser pulse induces a Stark shift of the ground state of molecules that depends on the angle of the molecular axis to the laser polarization[15]. It has been demonstrated that gas-phase molecules can be aligned adiabatically[16], aligned in three dimensions[17] or oriented[18]. We use non-adiabatic alignment[19,20]. It results in field-free alignment of linear molecules well after the aligning pulse has terminated. A 60-fs laser pulse produces a rotational wave packet in $N_2$. This wave packet periodically rephases, giving periods when the molecular axes are aligned in space. The probe pulse coincides with the first half-revival at 4.1 ps (ref. 20). The degree of angular alignment is good enough to see a difference in the harmonic spectra for a $5^\circ$ rotation of the alignment direction.

## High Harmonic Generation from Molecules

High harmonics are produced by ionizing atoms or molecules. We choose to operate in the tunnel ionization regime[21] because the rate of tunnel ionization depends exponentially on the ionization potential. Hence the orbital with the lowest ionization potential (the HOMO in an unexcited molecule) is preferentially selected. This is the second requirement for tomographic imaging of a single molecular orbital.

The third requirement for tomography is that we can take a series of projected images of the molecular orbital. For this we record high harmonic emission generated from the molecule itself. High harmonics have been extensively studied from atoms, whereas harmonics from molecules have been studied with isotropic distributions[5,6] or with weak alignment[7]. It has

如果能够用飞秒激光技术测定轨道，那么我们也能观测到在化学反应时间尺度上发生的轨道变化。下一个挑战将会是测定阿秒束缚态电子动力学。我们将会说明，轨道成像技术的适用范围可以扩展至原子或分子中相干电子运动的成像。在量子力学中，电子运动是用电子态的相干叠加来描述的，就量子力学所允许的意义而言，这是最接近于电子"玻尔轨道运动"的成像。

分子轨道的层析成像是通过三个重要的步骤来实现的。(1)在实验室坐标系中分子轴的排列。(2)轨道的选择性电离。(3)该状态在一组相干平面波上的投影。

## 气相分子的排列

计算机层析成像术指的是从物体(诸如人体)的一系列一维投影中恢复剖视图像的技术[13]。对于实验室坐标系中的一个固定的投影方向，必然需要将该物体旋转。因此，我们从分子可以被如何排列[14]开始讨论。

一个相当强度的且非共振的激光脉冲会激发分子基态的斯塔克位移，该位移取决于分子轴与激光偏振方向之间的夹角[15]。实验已经证明，气相分子可以在绝热的条件下排列[16]，即在三维空间中排列[17]或者定向排列[18]。我们使用非绝热的排列[19,20]。这导致线性脉冲结束后线型分子在无外场的作用下实现排列。一个 60 fs 激光脉冲在 $N_2$ 中产生了一个旋转的波包。这个波包周期性地重现，使分子轴在空间中排列时呈现周期性。探测脉冲的作用时间与位于 4.1 ps 的前半个重现相一致(参考文献 20)。角度排列的程度足够好，使得当分子的排列方向旋转了 5° 以后就可以从谐波谱中观察到它们的差别了。

## 分子高次谐波的产生

通过电离原子或分子可以产生高次谐波。我们选择在隧道电离区域[21]进行操作，因为隧道电离速率以指数方式随电离势而变化。因此，具有最低电离势的轨道(未激发分子中的 HOMO)便是优选的。这是一个单分子轨道层析成像所需要的第二个条件。

层析需要的第三个条件是，我们能够获得一系列分子轨道的投影图像。为此我们记录下分子自身所产生的高次谐波发射。源于原子的高次谐波已被人们广泛地研究，而源于分子的谐波则是通过各向同性分布[5,6]或弱的排列[7]进行研究。采用实

been observed[8-11] both experimentally and numerically, that harmonic emission from aligned molecules is sensitive to both the alignment angle and to the spatial structure of the electronic wavefunction. These observations provide the basis for our analysis.

High harmonic generation is understood using the three-step quasi-static model[22]—(1) tunnel ionization by an intense low-frequency laser field, (2) acceleration of the free electron, and (3) re-collision. Tunnel ionization coherently transfers a part of the bound-state electron wavefunction to the continuum. Once free, the strong laser field dominates the motion of the continuum electron wave packet. First it propagates away from the molecule (Fig. 1a) and is then driven back after the field reverses its direction. Figure 1b shows the re-colliding wave packet expanding laterally. By the time of re-collision, the electron wave packet has a typical transverse $1/e$ width of 9 Å for 800-nm laser fields[23,24]. Because the wave packet is much larger than the size of small molecules (typically ~1 Å, shown by the lines in Fig. 1b), the molecule sees essentially a plane electron wave re-colliding. We will see that the planar nature of the re-collision electron allows us to experimentally retrieve a one-dimensional projection of the ground-state wavefunction.

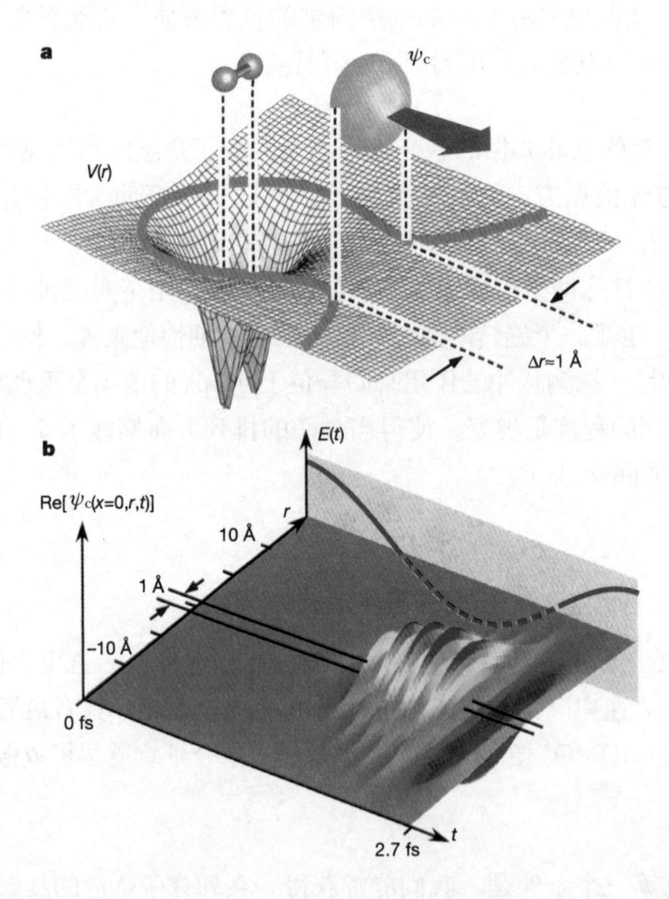


Fig. 1. Illustration of the tunnel ionization process from an aligned molecule. **a**, The orange line on the potential surface is an isopotential contour slightly above the energy level of the bound state at the peak of the field amplitude. The opening of the contour shows the saddle point region where the bound-state

验方法和数值方法进行的研究[8-11]均表明，源于排列分子的谐波发射对于排列角度和电子波函数的空间结构都很敏感。这些结果为我们的分析提供了研究的基础。

利用三步准静态模型[22]可以理解高次谐波生成——(1)利用强低频激光场进行隧道电离，(2)自由电子的加速，(3)再碰撞。隧道电离以相干方式将一部分束缚态电子波函数转化成连续态。一旦不受约束，强激光场控制连续电子波包的运动。它先是从分子传播开去(如图1a所示)，接着在场反转方向后返回。图1b显示了再碰撞波包横向扩展。再碰撞时，800 nm激光场条件下电子波包的一个典型的横向 $1/e$ 宽度为9 Å[23,24]。因为波包的尺寸比小分子(典型情况约为1 Å，如图1b中直线所示)要大得多，所以在分子看来实际上是一个平面电子波的再碰撞。我们将会看到，再碰撞电子的平面属性使我们得以通过实验恢复基态波函数的一维投影。

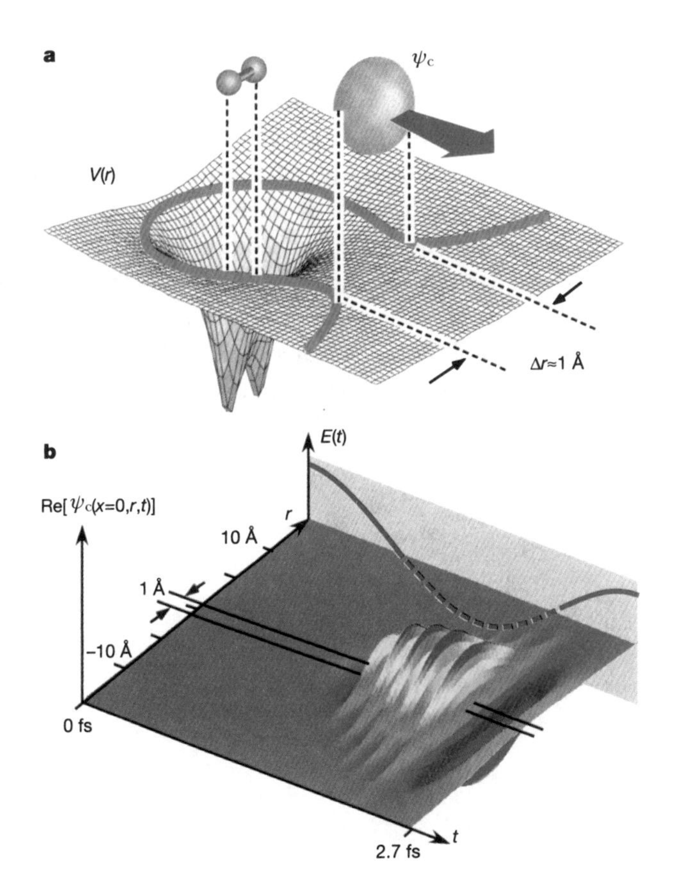


图1. 排列分子隧道电离过程的图示。**a**，势能面上的橙色线是略高于电场振幅峰值处束缚态能级的等势线。等势线的开口部分显示鞍点区域，束缚态电子波函数会从这里隧穿为连续态。电子波包 $\psi_c$ 的横向伸展是由鞍点区域的宽度 $\Delta r$ 决定的，后者取决于分子取向。在连续态中传播时，波包会根据不确定性

electron wavefunction will tunnel through to the continuum. The lateral spread of the electron wave packet $\psi_c$ is determined by the width of the saddle point region $\Delta r$ that depends on the molecular alignment. The wave packet expands in the lateral direction during propagation in the continuum because of the initial momentum spread $\Delta p$ given by the uncertainty principle, $\Delta p \approx h/\Delta r$. **b**, Illustration of the re-colliding wave packet seen by a molecule (the real part of the wave packet $\mathrm{Re}[\psi_c(x=0, t)]$ is shown). The kinetic energy at the time of re-collision is taken from the classical trajectory, and determines the instantaneous frequency of the wave packet as seen by the molecule. The lateral spread is calculated by the free expansion of a gaussian wave packet with an initial $1/e$ full width of 1 Å.

The laser field shears the initial continuum wave packet in such a way that the low kinetic energy (long de Broglie wavelength) part of the wave packet re-collides first. Over the next half laser period, it chirps to high kinetic energy (short wavelength) and then back to low energy[25]. In our experiment, the highest-energy electron has a wavelength of ~1.5 Å. This wavelength is related to the high harmonic cut-off. The broad range of electron wavelengths allows us to experimentally retrieve the spatial shape of the bound-state wavefunction.

Figure 2 illustrates the final step in the harmonic generation process. Figure 2a shows a bound-state wavefunction, $\psi_g$. A fraction of the electron wavefunction tunnels from this orbital. The re-colliding electron wave packet, $\psi_c$, shown as a plane wave in Fig. 2b, overlaps the remaining portion of the initial wavefunction. The coherent addition of the two wavefunctions induces a dipole, seen as the asymmetric localization of electron density in Fig. 2c. The induced dipole oscillates as the continuum wavefunction propagates. It is this oscillating dipole that emits high harmonic radiation. The instantaneous frequency of the oscillating dipole is related to the electron kinetic energy by $\hbar\omega = E_k$ (refs 22, 25). In other words, the electron wave packet and the emitted photons are related by energy conservation, and are mutually coherent. We do not use the more usual expression[22], $\hbar\omega = E_k + I_p$, where $I_p$ is the ionization potential and $E_k$ is the instantaneous kinetic energy of the electron in the continuum at the time of re-collision calculated for a delta function potential, because here we are concerned with the electron kinetic energy seen by the bound-state electron wavefunction.

原理 $\Delta p \approx h/\Delta r$，由于初始动量的伸展 $\Delta p$ 而发生横向扩展。**b**，一个分子的再碰撞波包的图示（所示为波包的实部 $\mathrm{Re}[\psi_c(x=0,t)]$）。再碰撞时刻的动能是根据经典轨迹得到的，它决定了分子“看到”的波包瞬时频率。横向扩展是利用具有 1 Å 初始 $1/e$ 全宽的高斯波包的自由展开而计算出来的。

激光场以特定的方式剪切初始连续波包，以使波包的低动能（长的德布罗意波长）部分首先发生再碰撞。在下面的半个激光周期中，高动能的部分继而发生再碰撞，接着再变回低能[25]。在我们的实验中，最高能量的电子具有约为 1.5 Å 的波长。该波长与高次谐波的截止区有关。电子波长的宽广范围使我们能够通过实验恢复束缚态波函数的空间形状。

图 2 描绘了谐波生成过程中的最后一步。图 2a 显示了束缚态波函数 $\psi_g$。电子波函数的一部分从这一轨道隧穿。图 2b 显示为平面波的再碰撞电子波包 $\psi_c$ 与初始波函数的剩余部分重叠。两个波函数的相干叠加产生一个偶极，看起来如同图 2c 中电子密度的不对称分布。当连续波函数传播时，诱发偶极振荡。正是这个振荡偶极发射出高次谐波辐射。振荡偶极的瞬时频率与电子动能的关系是 $\hbar\omega = E_k$（参考文献 22 和 25）。换句话说，电子波包和发射的光子通过能量守恒而联系起来，因而是彼此相干的。我们没有使用更常见的表达式[22]，$\hbar\omega = E_k + I_p$，式中 $I_p$ 为电离势，$E_k$ 为连续态中再碰撞时刻的电子瞬时动能，是对于 δ 函数势能计算得到的；因为在本文中我们所考虑的是束缚态电子波函数所观察到的电子动能。

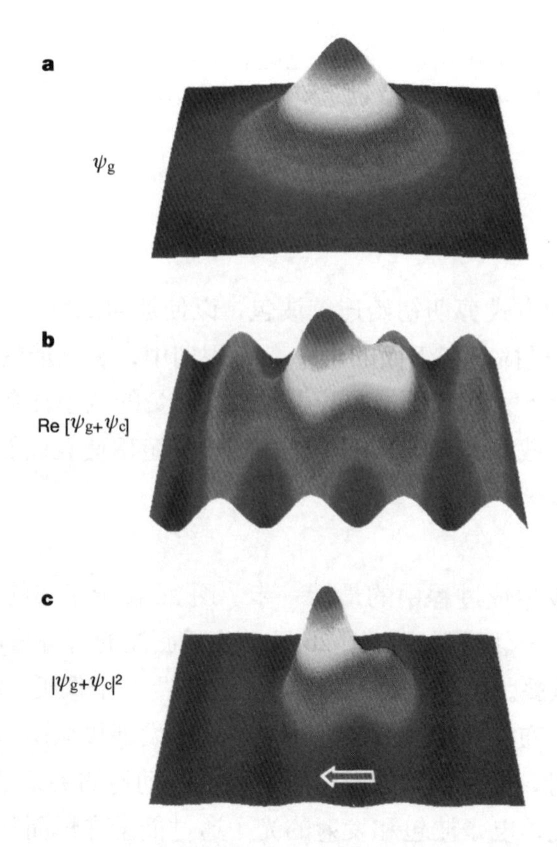


Fig. 2. Illustration of a dipole induced by the superposition of a ground-state wavefunction $\psi_g$ and a re-colliding plane wave packet $\psi_c$. **a**, Bound-state wavefunction (for example, H atom, 1 *s*). **b**, The real part of the superposition of the bound-state wavefunction $\psi_g$ and the continuum plane wave $\psi_c$. **c**, The total electron density distribution $|\psi_g+\psi_c|^2$. The superposition of the two wavefunctions induces a dipole $d(t)$ as shown by the red arrow. As the wave packet propagates, the induced dipole oscillates back and forth and leads to the emission of harmonics.

Formally, we are measuring the transition dipole moment from the highest occupied state to a set of continuum wavefunctions. Mathematically, the harmonic spectrum from a single atom or molecule is given by the Fourier transform of the dipole acceleration[26]. We relate the dipole acceleration spectrum to the dipole moment using the slowly varying envelope approximation (see Methods section). The spectral intensity ($I$) and phase of harmonics ($\phi$) at frequency $\omega$ can respectively be written as $I(\omega) \propto \omega^4 |d(\omega)|^2$ and $\phi(\omega) = \arg[d(\omega)]$, where the transition dipole moment in the spectral domain is $\mathbf{d}(\omega) = a[k(\omega)] \int \psi_g(\mathbf{r})(e\mathbf{r})\exp[ik(\omega)x]\mathrm{d}\mathbf{r}$. Here the electron wave packet seen by the molecule is expanded in a superposition of plane waves as $\psi_c = \int a(k)\exp[ik(\omega)x]\mathrm{d}k$, $k(\omega)$ is the wavenumber (momentum) of the electron corresponding to harmonic frequency $\omega$, and $a[k(\omega)]$ is its complex amplitude[25]. A similar approach was taken by Sanpera *et al.*[27], who divided the electron wavefunction into bound and continuum parts and modelled the continuum part as a gaussian wave packet. Provided that $a[k(\omega)]$ is known, then measuring the harmonic spectrum—its amplitude, phase and polarization—is equivalent to measuring this integral, evaluated for different $k(\omega)$. This is an experimental determination of the one-dimensional spatial

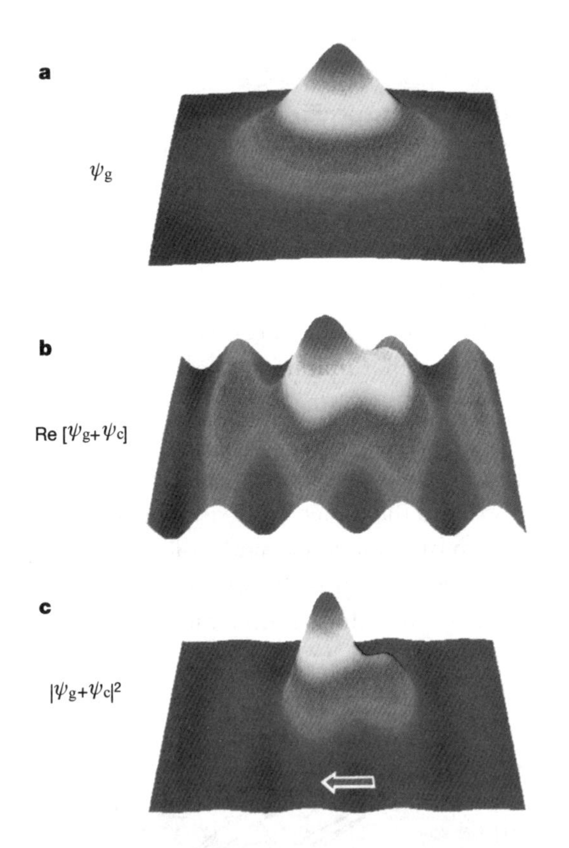


图 2. 由基态波函数 $\psi_g$ 与再碰撞平面波包 $\psi_c$ 间叠加而诱导出的偶极的图像。**a**，束缚态波函数（例如 H 原子的 1 *s* 轨道）。**b**，束缚态波函数 $\psi_g$ 与连续平面波 $\psi_c$ 叠加的实部。**c**，总电子密度分布 $|\psi_g+\psi_c|^2$。两个波函数的叠加诱导出一个偶极 $d(t)$，如图中红色箭头所示。随着波包的传播，诱导偶极来回振动从而导致谐波发射。

从形式上来讲，我们将要测定从最高占据态到一组连续波函数的跃迁偶极矩。用数学语言来讲，单个原子或分子的谐波谱是通过偶极加速度的傅里叶变换而给出的[26]。我们利用缓变包络近似（参见介绍方法的小节）将偶极加速度谱和偶极矩联系起来。光谱强度（$I$）与频率为 $\omega$ 的谐波相位（$\phi$）可以分别写成 $I(\omega)\propto\omega^4|d(\omega)|^2$ 和 $\phi(\omega)=\arg[d(\omega)]$，式中光谱域中的跃迁偶极矩为 $\mathbf{d}(\omega)=a[k(\omega)]\int\psi_g(\mathbf{r})(e\mathbf{r})\exp[ik(\omega)x]\mathrm{d}\mathbf{r}$。这里，分子所“看到”的电子波包扩展为平面波的叠加，$\psi_c=\int a(k)\exp[ik(\omega)x]\mathrm{d}k$，式中 $k(\omega)$ 为对应于谐波频率 $\omega$ 的电子波数（动量），$a[k(\omega)]$ 则是其复振幅[25]。桑佩拉等人[27]使用了一种类似的方法，将电子波函数划分成束缚态部分和连续态部分，并将连续态部分按照高斯波包来建模。假设 $a[k(\omega)]$ 为已知的，那么测定谐波谱——它的振幅、相位和偏振——就等价于测定对于不同 $k(\omega)$ 值

Fourier transform of $\mathbf{r}\psi_g(\mathbf{r})$.

Experimentally, we concentrate on a specific molecule, $N_2$. High harmonics are generated with a 30-fs, $2 \times 10^{14}$ W cm$^{-2}$, 800-nm, horizontally polarized probe pulse in a $\sim 10^{17}$ cm$^{-3}$ molecular gas jet described in the Methods section. The molecular alignment axis is rotated relative to the polarization of the probe pulse.

The probe pulse intensity was kept as low as possible to avoid saturation or depletion of the ground state. The molecular beam thickness was less than 1 mm and the laser focus was before the beam—conditions that minimize phase mismatch and select only the short electron trajectories[28].

Figure 3 shows the intensities of each harmonic as the laser polarization is rotated from 0 to 90° (in steps of 5°) with respect to the molecular axis. The intensity contrast between 0 and 90° exceeds an order of magnitude for some harmonics. We use these data to reconstruct the molecular orbital of $N_2$. Also shown is the spectrum from the reference atom, argon, recorded under conditions identical to those for the nitrogen.

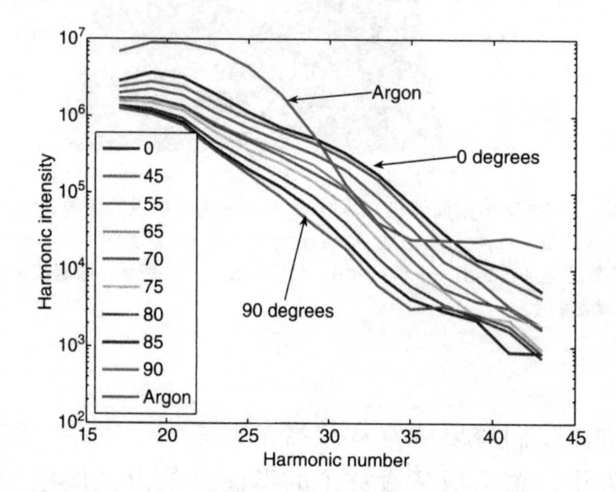


Fig. 3. High harmonic spectra were recorded for $N_2$ molecules aligned at 19 different angles between 0 and 90° relative to the polarization axis of the laser. For clarity, only some of the angles have been plotted above. The high harmonic spectrum from argon is also shown; argon is used as the reference atom. Clearly the spectra depend on both the alignment angle and shape of the molecular orbital.

## Calibrating the Continuum Wave Packet

As discussed above, the harmonic spectrum is an experimental evaluation of the dipole, $\mathbf{d}(\omega)$. If we could evaluate the plane-wave amplitude $a[k(\omega)]$ independently, then our measurement would determine $\int \psi_g(\mathbf{r})(e\mathbf{r})\exp[ik(\omega)x]\,d\mathbf{r}$—that is, the spatial Fourier components of $\mathbf{r}\psi_g(\mathbf{r})$. One way to do this is to perform the same experiment with a reference atom.

的积分。这是对于 $\mathbf{r}\psi_g(\mathbf{r})$ 的一维空间傅里叶变换的一种实验测定方法。

在实验方面，我们重点关注一种特定的分子，$N_2$。在下面方法的小节中我们会谈到，利用分子密度约为 $10^{17}\ \mathrm{cm}^{-3}$ 的气体喷嘴，30 fs、$2\times10^{14}\ \mathrm{W\cdot cm^{-2}}$、800 nm、水平偏振的探测脉冲，可以产生高次谐波。分子的排列轴相对于探测脉冲的偏振方向旋转。

探测脉冲强度保持尽可能低以免基态饱和或耗散。分子束粗细小于 1 mm，并且激光要在分子束前面聚焦——这些条件能使相位失配最小化从而只选取电子短轨道 [28]。

图 3 显示了随着激光偏振方向相对于分子轴从 0° 到 90° 旋转（步长为 5°）时每个谐波的强度。对于 0° 和 90° 之间的某些谐波，强度对比超出一个数量级。我们利用这些数据来重建 $N_2$ 的分子轨道。同时显示出来的还有参考原子氩的光谱，这是在与氮相同的条件下记录的。

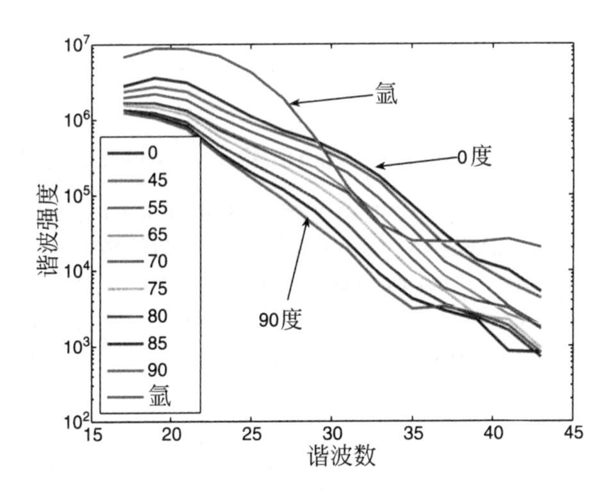


图 3. 在相对于激光偏振轴呈 0° 到 90° 之间的 19 个不同的指定夹角处记录的 $N_2$ 分子的高次谐波谱。为清晰起见，上面只画出了某些角度的情况。同时显示的还有氩的高次谐波谱；氩用作参照原子。很明显，光谱同时决定于所排列的角度和分子轨道的形状。

## 校准连续波包

如同上面已讨论过的，谐波谱是偶极 $\mathbf{d}(\omega)$ 的实验估算值。如果我们能够独立地估算出平面波振幅 $a[k(\omega)]$ 的数值，那么我们的测量结果就能确定 $\int\psi_g(\mathbf{r})(e\mathbf{r})\exp[ik(\omega)x]\mathrm{d}\mathbf{r}$，也就是 $\mathbf{r}\psi_g(\mathbf{r})$ 的空间傅里叶分量。做到这一点的一种方法，就是用参考原子进行同样的实验。

Argon is very similar to $N_2$ in its response to strong laser fields, having nearly the same ionization potential and intensity-dependent ionization probability[6]. This is confirmed by the dependence of the instantaneous ionization rates[29] for atoms, and for different orientations of $N_2$ (ref. 30). That means that the first, critical, step in the three-step high harmonic generation process is the same. Because the laser field dominates wave packet motion in the direction of the laser field, the second step, which determines the chirp of the re-colliding wave packets seen by Ar or $N_2$, will be the same. Thus, $a[k(\omega)]$ will be the same.

The continuum wave packet will also be similar for Ar and $N_2$. The narrow saddle point through which the electron tunnels acts as a spatial filter that removes much of the structure of the orbital from the continuum wave packet. This can be seen in numerical simulations[31]. By measuring the ellipticity dependence of the high harmonic signal[24] produced by $N_2$ and argon, we confirmed that the lateral spread of the wave packets is similar. The ionization rate of $N_2$ is angle-dependent[30,32], but is readily measured from the ion yield, and varies only by 25% for $N_2$ (ref. 33). This variation is almost cancelled by the angular dependence of the wave-packet spread in the lateral direction[23]. Thus we use the harmonic spectrum from argon to determine $|a[k(\omega)]|$ without including the angular dependence of tunnel ionization rate and wave-packet spread in the continuum.

## Tomographic Reconstruction of the Orbital

We have shown that (1) tunnel ionization, owing to its nonlinearity, is extremely selective[29,30]. In $N_2$, it ionizes only the highest electronic state—the HOMO. And (2), the harmonic spectra contain a range of spatial Fourier components of the shape of this single orbital. We have recorded the harmonic spectra at a series of angles between the molecular axis and the re-colliding electron. We now show that we can invert this information to obtain the shape of the orbital.

The Fourier slice theorem[13] proves that the Fourier transform of a projection $P_\theta$ is equal to a cut at angle $\theta$ through the two-dimensional Fourier transform $F$ of the object. This is the basis of computed tomography based on the inverse Radon transform. Our dipole is the Fourier transform of a projection of the wavefunction, and so can be inverted.

We describe the mathematical details of the tomographic reconstruction in the Methods section. This procedure can reconstruct orbital shapes with symmetries such as $\sigma_g$, $\pi_g$ and $\pi_u$, using harmonics 17–51 of an 800-nm laser field and 25 angles from 0° to 180° (fewer angles are needed for symmetric molecules).

A complete inversion of a general orbital requires knowledge of the relative phase and amplitude of each harmonic for two orthogonal polarizations. Although we measured only the amplitude (the phase of each harmonic can be directly measured[28]) the relative phase of the harmonics is known from first principles. As with any classical or quantum resonance,

氩与 $N_2$ 在强激光场的作用非常相似，具有几乎相同的电离势和依赖于强度的电离概率 [6]。瞬时电离速率 [29] 对于原子以及 $N_2$ 不同取向的依赖性确证了这一点（参考文献 30）。这意味着，在分三步进行的高次谐波生成过程中，第一步（关键的步骤）是相同的。由于激光场主导着波包在激光场方向上的运动，因此第二个步骤——它决定了 Ar 或 $N_2$ 所"看"到的再碰撞波包的啁啾——也会是相同的。于是，$a[k(\omega)]$ 就应该相同。

Ar 和 $N_2$ 的连续波包同样会是相似的。电子隧穿所经的狭窄的鞍点起到空间滤波器的作用，从连续波包中移除了大部分的轨道结构。在许多数值模拟中可以看到这一点 [31]。通过测定 $N_2$ 和 Ar 产生的高次谐波信号对于椭偏率的依赖性 [24]，我们证实了波包的横向伸展是类似的。虽然 $N_2$ 的电离速率是依赖于角度的 [30,32]，但是很容易利用离子产量测得，而且对于 $N_2$ 只变化了 25%（参考文献 33）。横向上的伸展的波包的角度依赖性基本上消除了这一差异 [23]。因此我们可以利用来自氩的谐波谱来确定 $|a[k(\omega)]|$，而不必考虑隧道电离速率和连续态中的波包伸展对于角度的依赖性。

## 轨道的层析成像重建

我们已经表明：(1) 隧道电离，由于其非线性，是极具有选择性的 [29,30]。对于 $N_2$，它只电离最高电子态——HOMO。(2) 谐波谱中包含这一单个轨道形状的一部分空间傅里叶分量。我们已经在分子轴与再碰撞电子之间的一系列角度处记录了谐波谱。现在，我们来展示能够将这一信息逆推以得知轨道的形状。

傅里叶切片定理 [13] 证明，投影 $P_\theta$ 的傅里叶变换等于物体的二维傅里叶变换 $F$ 在角度 $\theta$ 处的切片。这是基于逆向拉东变换的计算机层析成像术的基础。我们的偶极是波函数投影的傅里叶变换，因而可以被逆推出来。

我们在关于方法的小节中描述层析重建方法的细节。利用 800 nm 激光场的 17 ~ 51 阶谐波，以及从 0° 到 180° 中的 25 个角度（对称分子需要更少的角度），这个过程能够重建具有诸如 $\sigma_g$、$\pi_g$ 和 $\pi_u$ 等对称性的轨道形状。

对于一个一般轨道的全反演，需要了解对于两个正交偏振的每一个谐波的相对相位和振幅。尽管我们仅仅测定了振幅（每一个谐波的相位可以被直接测得 [28]），但是利用第一性原理即可知道谐波的相对相位。如同具有任何经典的或量子共振的时

a phase jump of $\pi$ occurs as the driving (electron) frequency moves across resonance. Calculations by Hay *et al.*[9] show minimum harmonic signal and a $\pi$ phase jump at the 25th harmonic for 0° when the re-collision electron wavelength resonates with the projected molecular dimension. We assume a phase jump of $\pi$ where the measured dipole of $N_2$ goes through a minimum, corresponding to a change of sign of the dipole. This occurs at the 25th harmonic.

The reconstructed molecular orbital of $N_2$ is shown in Fig. 4a, based upon the 19 projections. It can be compared with the *ab initio* orbital calculation of the $N_2$ $2p\ \sigma_g$ orbital. Note that the recovered wavefunction has both positive and negative lobes, so it is not the square of the wavefunction, but the wavefunction itself, up to an arbitrary phase.

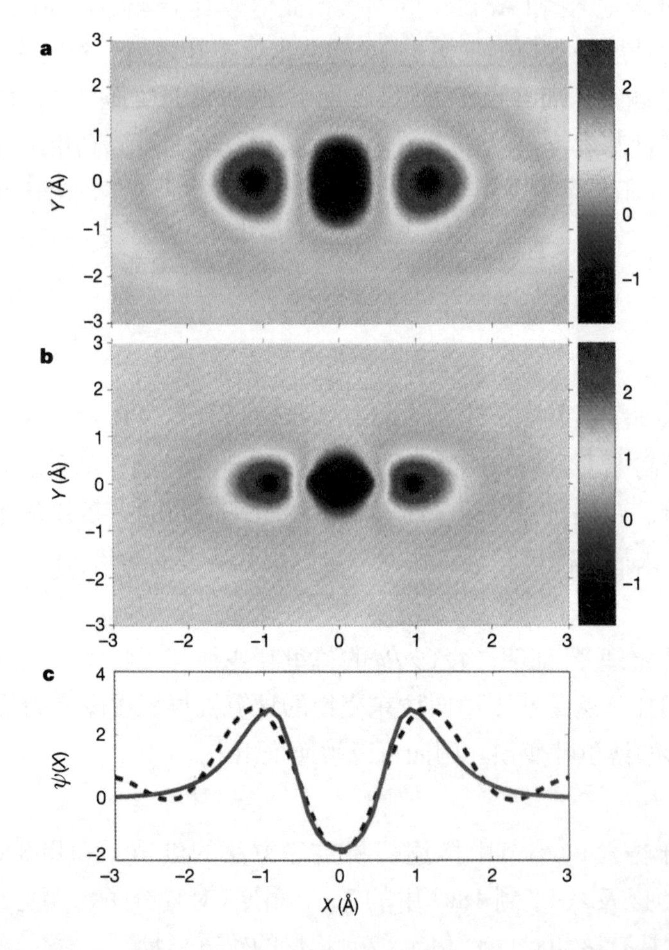


Fig. 4. Molecular orbital wavefunction of $N_2$. **a**, Reconstructed wavefunction of the HOMO of $N_2$. The reconstruction is from a tomographic inversion of the high harmonic spectra taken at 19 projection angles. Both positive and negative values are present, so this is a wavefunction, not the square of the wavefunction, up to an arbitrary phase. **b**, The shape of the $N_2$ $2p\ \sigma_g$ orbital from an ab initio calculation. The colour scales are the same for both images. **c**, Cuts along the internuclear axis for the reconstructed (dashed) and *ab initio* (solid) wavefunctions.

候一样，π 的相位阶跃随驱动（电子）频率穿过共振而发生。通过海等人的计算[9]表明，当再碰撞电子波长与分子尺度的投影发生共振时，对于 0° 的第 25 阶谐波处有极小谐波信号和 π 相位阶跃。我们假定 π 相位阶跃——此时所测得的 $N_2$ 偶极经历极小值——对应于偶极符号的变化。它在第 25 阶谐波处发生。

图 4a 中显示了基于 19 个投影重建的 $N_2$ 的分子轨道。可以将它与第一性原理轨道计算所得的 $N_2$ 的 $2p\ \sigma_g$ 轨道相比较。需要我们注意的是重新获得的波函数兼有正负的两片，因此它不是波函数的平方而是波函数本身，相对于任意相位。

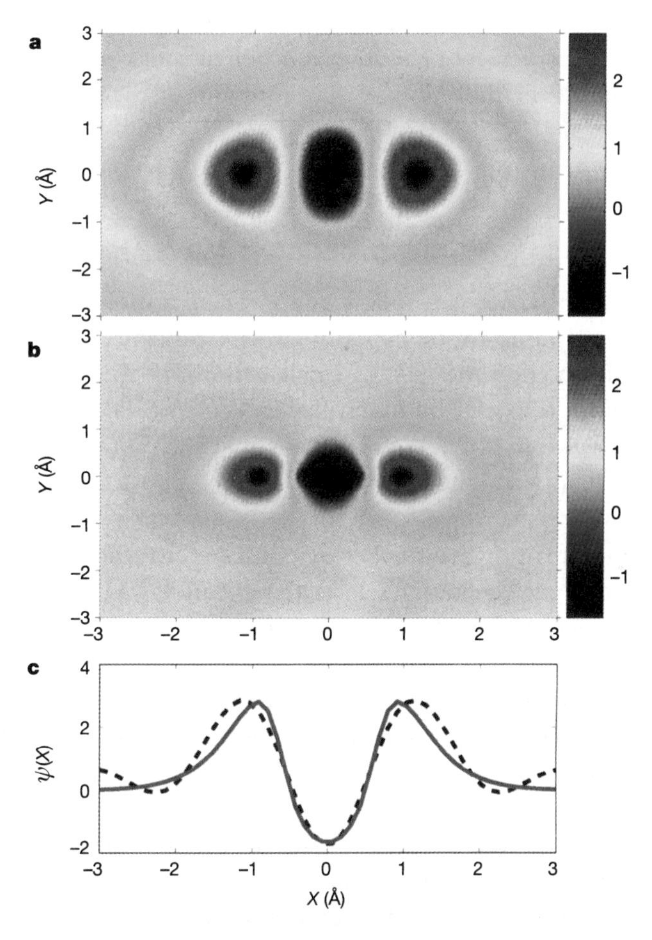


图 4. $N_2$ 的分子轨道波函数。**a**，重建的 $N_2$ 的 HOMO 波函数。重建结果来自对从 19 个投影角度取得的高次谐波谱的层析成像反演。相对于任意相位，正值和负值都包括在内，因此这是一个波函数而不是波函数的平方。**b**，利用第一性原理计算得到的 $N_2$ 的 $2p\ \sigma_g$ 轨道形状。两幅图像的彩色标度是一样的。**c**，重建的（短划线）和第一性原理计算的（实线）波函数沿核间轴的截面。

Our orbital reconstruction method is similar to medical tomography. However, we image a single orbital among many orbitals. The coherence of the re-collision provides this specificity. Of all the occupied or unoccupied states of the molecule, the re-colliding electron wave packet is only coherent with the state from which it tunnelled (or, as we will show below, with any other coherently related state). Measurable high harmonic emission requires a macroscopic number of ionizing molecules, so only the coherent emission interferes constructively and is observed experimentally. This coherent filtering is conceptually similar to the homodyne detection technique and is naturally implemented in harmonic generation.

Tomographic approaches have previously been used for molecular measurements, but in a very different way[34,35]. Although tomographic reconstruction algorithms are used, they are based on "rotations" in the position–momentum phase-space distributions, not physical rotations of an object. These techniques yield the one-dimensional Wigner distribution $W(x,p)$ of the phase-space density of a vibrating[34] or dissociating[35] molecule. In contrast, we measure the actual three-dimensional wavefunction of a single electron orbital of a molecule.

## Watching Electrons Move

So far, we have shown that we can take a snapshot of a molecular orbital using 30-fs laser pulses. Thus, in a pump-probe experiment, it is possible to image (probe) changes to molecular orbitals on the timescale of nuclear motion (the femtosecond timescale). However, electronic motion can be much faster. Can orbital imaging be extended to measuring electronic wave packets? We now show that the motion of this electronic wave packet can indeed be mapped to the harmonic spectrum.

It may seem that measuring wave-packet dynamics contradicts our assertion that the coherence of high harmonic generation selects a single orbital. However, an electronic wave packet is a coherent superposition of electronic states. Therefore, all states forming the wave packet are coherent with the returning electron provided it tunnelled from one or more of the states. They all contribute to the induced dipole that emits harmonics, even though only the highest-energy state contributed to the ionization. Therefore the relative intensity of each harmonic will depend on the relative phases between these electronic states.

To demonstrate how the bound-state wave-packet motion appears in harmonic spectra, we use the one-dimensional time-dependent Schrödinger equation to simulate an electron in a diatomic potential in the presence of an intense laser field. The field is chosen at a wavelength of 1.6 μm and duration of 8 fs to induce a single re-collision. To form a fast-moving bound-state wave packet, two electronic states with energy difference of $\Delta E = 9.5$ eV are equally populated before the arrival of the laser pulse. This energy difference corresponds to wave-packet motion with a period of 435 as. By varying the initial phase of the two states, we perform a numerical pump–probe measurement of the wave-packet motion.

我们的轨道重建方法与医学中的层析成像相类似。然而，我们描绘的是很多轨道中的一个。再碰撞的相干性提供了这种特征。在分子所有占据态或未占据态中，再碰撞电子波包都只与它隧穿时所来自的那个态相干(或者，如同我们下面将要说明的那样，与任意其他相关态相干)。可观测的高次谐波发射需要大量的电离分子，因此只有相干发射才能有效地干涉从而能用实验方法观测到。从概念上说，这种相干滤波类似于零差检测技术，因而能够自然地被应用于谐波产生中。

以前曾将层析成像方法用于分子测量，不过是以一种极为不同的方式[34,35]。尽管使用了层析成像的重建算法，但是它们所依据的是位置–动量相空间分布中的“旋转”，而不是物体的物理旋转。这些技术产生出一个振动分子[34]或离解分子[35]的相空间密度的一维维格纳分布 $W(x, p)$。与之相比，我们测得的是一个分子中单独一个电子轨道的实际的三维波函数。

## 观看电子运动

目前为止，实验表明，我们能够利用 30 fs 激光脉冲获得一个分子的快照。因此，在一次泵浦–探测实验中，有可能在核运动的时间尺度上(飞秒时间尺度)描绘(探测)出分子轨道的变化。然而，电子运动可能还要快很多。轨道成像能够扩展以观测电子波包吗？现在，我们来说明这种电子波包的运动确实能够映射到谐波谱上。

看来有可能的是，对波包动力学的观测与我们的下列断言相矛盾：高次谐波生成的相干性选择了单一轨道。然而，一个电子波包是电子态的相干叠加。因此，如果该电子从一个或多个态进行隧穿的话，那么构成波包的所有态都是与该返回电子相干的。虽然只有最高能量态才对电离有贡献，但所有态都对发射谐波的诱导偶极有贡献。因此，每一个谐波的相对强度都会依赖于这些电子态之间的相对相位。

为了说明束缚态波包的运动如何出现在谐波谱中，我们利用一维含时薛定谔方程来模拟在有强激光场存在时双原子势场中的一个电子。选取该场波长为 1.6 μm，持续时间为 8 fs，以诱导出单独一次再碰撞。为形成一个快速运动的束缚态波包，要在激光脉冲到达之前等量地占据在能量差为 $\Delta E = 9.5\ \text{eV}$ 的两个电子态。这一能量差对应于具有 435 as 周期的波包运动。通过改变两种态的初始相位，我们对波包运动进行了数值的泵浦–探测测量。

Figure 5 shows the calculated harmonic spectrum versus the initial phase of the two states. The structure in the harmonic spectrum synchronizes with the bound-state wave-packet motion. Both the modulation in the single time delay spectrum and the movement of the modulation in the pump–probe spectrum measure wave-packet dynamics.

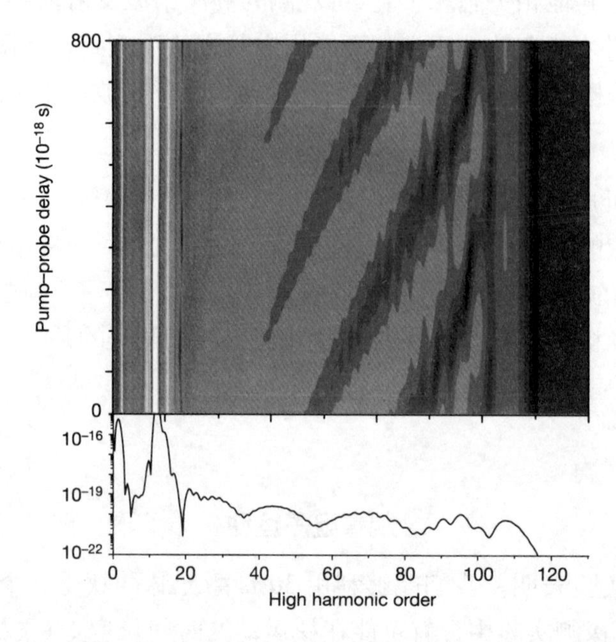


Fig. 5. A one-dimensional Schrödinger calculation shows that attosecond electronic wave-packet motion is resolved in the high harmonic spectra. The bottom curve shows a spectrum at a particular pump–probe delay time; the minima are due to interference caused by the wave-packet motion. The top picture shows the spectra at a range of time delays, showing that the minima move with pump–probe time delay. The simulation populated the ground and first excited state (9.5 eV above ground) of a model atom to create an electronic wave packet.

## Promises and Challenges

Our experimental condition—$N_2$ illuminated by an 800-nm pulse at an intensity of $2 \times 10^{14}$ W cm$^{-2}$—is not special. Ionization intensities of $\sim 10^{14}$ W cm$^{-2}$ are typical for small- to intermediate-sized molecules[36]. For these laser parameters a re-collision electron has a wavelength of 1–2 Å—the correct range for measuring orbital structure. Thus, it appears feasible to extend tomography to many other molecules, provided there is a general method for determining $a[k(\omega)]$. This seems very promising. The phase dependence of the ionization rate determines $a[k(\omega)]$ up to a global constant describing the total ionization probability. Theories of time-dependent ionization of atoms[29] and small molecules[30] shows that the phase-dependent ionization rate is determined mainly by the ionization potential. If this remains true for more-complex molecules, then a reference atom (or molecule) can always be found to experimentally determine $a[k(\omega)]$. If a reference atom is not possible then $a[k(\omega)]$ can be characterized experimentally through the ellipticity dependence of the harmonics[24], and by directly measuring the electron spectrum in elliptically or circularly polarized light.

图 5 显示出计算所得的谐波谱与两种态初始相位的关系。谐波谱的结构与束缚态波包同步运动。单时间延迟光谱中的调制和泵浦-探测光谱中调制的运动都量度了波包动力学。

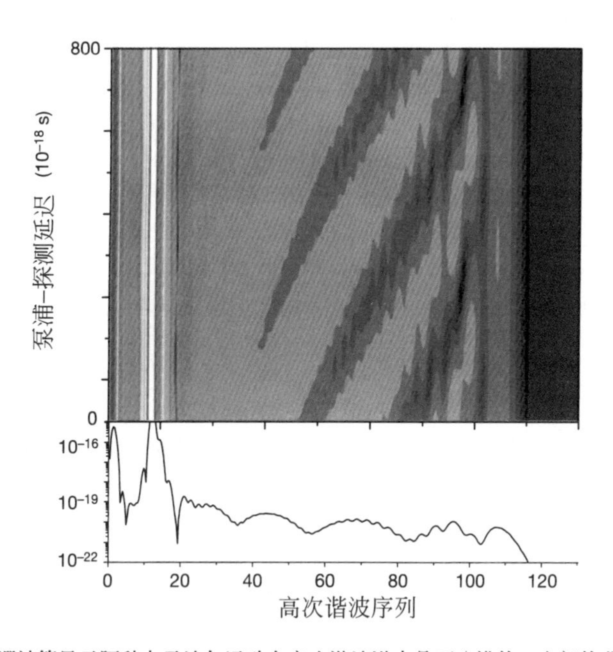


图 5. 一维的薛定谔计算显示阿秒电子波包运动在高次谐波谱中是可分辨的。底部的曲线显示出特定的泵浦-探测延迟时间时的光谱；极小值源于波包运动所导致的干涉。顶部的图像显示出一个时间延迟范围内的光谱，表明极小值会随着泵浦-探测时间延迟的改变而运动。模拟构造了一个模型原子的基态和第一激发态（高于基态 9.5 eV）以制造出一个电子波包。

## 前景与挑战

我们的实验条件——用强度为 $2\times10^{14}$ W · cm$^{-2}$ 的 800 nm 脉冲照射 $N_2$——并不是特定的。约为 $10^{14}$ W · cm$^{-2}$ 的激光诱发的电离强度对于小型和中型分子来说是典型的[36]。对于上述这些激光参数而言，再碰撞电子具有 1 ~ 2 Å 的波长，这是测定轨道结构的合理范围。因此，假定存在一种确定 $a[k(\omega)]$ 的通用方法，那么将层析方法拓展用于很多其他分子看来是可行的。这似乎是大有前景的。电离速率的相位依赖性决定了 $a[k(\omega)]$ 是一个描述总电离概率的全局常数。原子[29] 和小分子[30] 含时电离理论指出，相位相关电离速率主要是由电离势决定的。如果这一点对于更复杂的分子仍然是正确的话，那么就总可以找到一个参考原子（或分子）以便用实验方法测定 $a[k(\omega)]$。如果不存在参考原子的话，那么借助于谐波对椭偏率的依赖性[24]，并且通过在椭圆或圆偏振光中直接测定电子光谱，也能够用实验方法来表征 $a[k(\omega)]$。

But does the strong electric field of the laser modify the orbital? In small molecules the modification is small, because the re-colliding electron returns near the zero-crossing of the optical field. However, it may not be small in large molecules whose bound electrons can move more freely: that is, where the molecular polarizability is larger. Thus, it may be necessary to "shield" the orbital from the intense field. One method of shielding is to exploit traditional spectroscopy. A state, first excited by resonant light, can provide the continuum electron needed for tomography as long as it remains coherent with the ground state. Because the excited state will ionize much more easily than the ground state, tomography can be performed at field strengths in which the ground-state wavefunction is unperturbed. (Longer laser wavelengths will be required to keep the electron kinetic energy high in order to maintain sufficient spatial resolution.) Exploiting excited states has two additional advantages. First, the direction of the transition moment selects the alignment of those molecules that will ultimately ionize. Second, any orbital that is optically accessible can be imaged, not just the HOMO.

The ability of harmonics to measure the three-dimensional structure of orbitals is most important for molecular dynamics. It should be possible to observe an orbital while it changes its symmetry as a result of a curve-crossing, or to observe a chemical bond being broken—the very foundations of chemistry. With our wave-packet approach to align molecules, the only restriction is that the molecular dynamics must be faster than the duration of a rotational revival. To observe electronic rearrangement during chemical dynamics we would need to add an additional pulse to our pulse sequence[37]. Its function would be to induce dynamics. The first pulse aligns the molecule; the second one initiates the vibrational (or electronic) dynamics that we wish to study; and the final pulse would produce the harmonic radiation that could probe the evolution of the wavefunction.

## Methods

### Experimental details

For all measurements, we used a pulsed gas jet with a density of $\sim 10^{17}$ $cm^{-3}$ in the interaction region. The laser beams were focused ~1 mm below the orifice of the gas nozzle. The intensity of the pump pulse was set at $\sim 4 \times 10^{13}$ W $cm^{-2}$. This intensity is below the ionization threshold, where no ions are detected by a d.c.-biased electrode placed near the gas jet, and no harmonic emission is detected. The intensity of the probe pulse was estimated to be $\sim 2 \times 10^{14}$ W $cm^{-2}$ from the cut-off of high harmonics. This laser intensity is well below the saturation of ionization and harmonic generation, ensuring that high harmonics are emitted mostly at the peak of the laser pulse without depleting the ground state.

### Tomographic procedure

The angular frequency $\omega$ of the radiated extreme-ultraviolet field, and the wavenumber $k$

但是，激光的强电场是否会改变轨道？在小分子中，这种改变是较小的，因为再碰撞电子在光场过零点附近返回。然而，在大分子中，这种改变就可能不小了，因为大分子的束缚电子能够更为自由地运动，也就是说，分子的极化性更大了。于是，就必然需要对轨道“屏蔽”强场。一种屏蔽方法是开发传统的光谱学。一个态在最初被共振光激发后，只要它保持与基态相干，那么就能够为层析方法提供连续电子。由于激发态会比基态更容易电离，因此在基态波函数未受扰动的场强中能够使用层析方法(需要较长的激光波长来维持电子的高动能，以保证充分的空间分辨率)。对激发态的利用还有两个附加优点：第一，跃迁偶极矩的方向选择了最终将要电离的那些分子的排列。第二，任一光学意义上可接触的轨道都能成像，而不仅仅是 HOMO。

谐波的这种检测轨道三维结构的能力对于分子动力学而言是极为重要的。通过这个可能观测到一个轨道由于曲线交叉而改变其对称性，或者观测到一个化学键正在断裂——化学的基础所在。使用我们的波包方法排列分子时，唯一的限制是，分子动力学必须要比旋转复原的时间更快。要观测到化学动力学过程中的电子重排，需要在我们的脉冲序列中增添一个附加的脉冲[37]。它的功能是诱导动力学过程。第一个脉冲将分子排列；第二个脉冲引发我们希望研究的振动(或电子)动力学过程；而最后一个脉冲则产生谐波辐射，此谐波辐射能够检测出波函数的演化。

## 方　法

### 实验细节

在所有的观测中，我们使用了在相互作用区域中密度约为 $10^{17}\ cm^{-3}$ 的脉动式气体喷嘴。在气体喷嘴口的下方，激光束聚焦至约 1 mm。泵浦脉冲的强度设置约为 $4\times10^{13}\ W\cdot cm^{-2}$。这一强度低于电离阈值，此处用置于气体喷嘴附近的直流偏压电极没有检测到任何离子，也没有检测到任何谐波发射。通过高次谐波的截止位置，估计探测脉冲的强度约为 $2\times10^{14}\ W\cdot cm^{-2}$。这一激光强度大大低于电离和谐波产生的饱和值，以确保高次谐波几乎都是在激光脉冲峰值处发射的，没有基态耗散。

### 层析过程

辐射出的极紫外场的角频率 $\omega$，以及对应于在再碰撞时刻生成它的电子波的波数 $k$，通

corresponding to the electron wave that produced it at the time of recollision, are related by $k(\omega) = (2\omega)^{1/2}$ in atomic units. The transition dipole moment between the orbital wavefunction and the continuum electron is $\mathbf{d}(\omega;\theta) = \langle \psi(\mathbf{r};\theta) | \mathbf{r} | \exp[ik(\omega)x] \rangle$. This is a complex vector, and $\psi(\mathbf{r};\theta)$ represents the orbital wavefunction rotated by Euler angle $\theta$. Assuming perfect phase-matching, the extreme-ultraviolet signal that is emitted is given as $S(\omega;\theta) = N^2(\theta)\, \omega^4 | a[k(\omega)]\mathbf{d}(\omega;\theta)|^2$. Here $a[k(\omega)]$ is the complex amplitude of component $k$ of the continuum wave packet, and $N(\theta)$ is the number of ions produced.

The value of $a[k(\omega)]$ is determined experimentally by recording $S(\omega)$ for a reference argon atom whose orbital ($2p_x$), and hence $d(\omega)$, is known. Hence, $a[k(\omega)] = \omega^{-2}[S_{\text{ref}}(\omega)]^{1/2}[| \langle \psi_{\text{ref}} | r | k \rangle |^2]^{-1/2}$. This calibration not only characterizes the continuum wave packet, it also includes experimental factors such as detector efficiency.

Then $S(\omega;\theta)$ is recorded for an unknown molecule, yielding its transition dipole moment $d(\omega;\theta)$. The definition of $d(\omega;\theta)$ is a spatial Fourier transform of the orbital in the $x$ direction. We apply the Fourier slice theorem[13] to do the inversion, with one important modification. The Fourier transform contains the factor $\mathbf{r}$, which, being defined in the laboratory frame, does not rotate with the molecular frame, $\theta$.

The tomographic inversion requires two intermediate functions,

$$f_x(x,y) = \sum_{\theta}\sum_{\omega}[d_x(\omega;\theta)\cos\theta + d_y(\omega;\theta)\sin\theta]\exp[ik(\omega)(x\cos\theta + y\sin\theta)]$$

$$f_y(x,y) = \sum_{\theta}\sum_{\omega}[-d_x(\omega;\theta)\sin\theta + d_y(\omega;\theta)\cos\theta]\exp[ik(\omega)(x\cos\theta + y\sin\theta)]$$

Then the wavefunction is given as:

$$\psi(x,y) = \mathrm{Re}(f_x/x + f_y/y).$$

We must also include the angle dependence of the ionization rate to normalize each spectrum. The ionization rate is known for simple molecules such as $N_2$ (refs 30, 32) but can be measured at the same time as the harmonic spectrum is being recorded[33] simply by measuring the total ion yield, $N(\theta)$.

(**432**, 867-871; 2004)

**J. Itatani[1,2], J. Levesque[1,3], D. Zeidler[1], Hiromichi Niikura[1,4], H. Pépin[3], J. C. Kieffer[3], P. B. Corkum[1] & D. M. Villeneuve[1]**
[1] National Research Council of Canada, 100 Sussex Drive, Ottawa, Ontario K1A 0R6, Canada
[2] University of Ottawa, 150 Louis Pasteur, Ottawa, Ontario K1N 6N5, Canada
[3] INRS- Energie et Materiaux, 1650 boulevard Lionel-Boulet, CP 1020, Varennes, Québec J3X 1S2, Canada
[4] PRESTO, Japan Science and Technology Agency, 4-1-8 Honcho Kawaguchi Saitama, 332-0012, Japan

Received 8 June; accepted 9 November 2004; doi:10.1038/nature03183.

---

References:

1. Mulliken, R. S. Electronic structures of polyatomic molecules and valence. II. General considerations. *Phys. Rev.* **41**, 49-71(1932).

过用原子单位表示的 $k(\omega) = (2\omega)^{1/2}$ 联系起来。轨道波函数与连续电子之间的跃迁偶极矩为 $\mathbf{d}(\omega;\theta) = <\psi(\mathbf{r};\theta)|\mathbf{r}|\exp[ik(\omega)x]>$。这是一个复向量，$\psi(\mathbf{r};\theta)$ 表示旋转了欧拉角 θ 的轨道波函数。假定相位匹配是完美的，则所发射的极紫外信号由 $S(\omega;\theta) = N^2(\theta)\omega^4|a[k(\omega)]\mathbf{d}(\omega;\theta)|^2$ 给出。这里，$a[k(\omega)]$ 是连续波包中成分 $k$ 的复振幅，而 $N(\theta)$ 则是所产生的离子数。

通过记录参考原子氩的 $S(\omega)$——氩原子的轨道 ($2p_x$) 以及 $d(\omega)$ 是已知的——可以用实验方法测定 $a[k(\omega)]$ 的数值。因此，$a[k(\omega)] = \omega^{-2}[S_{ref}(\omega)]^{1/2}[|<\psi_{ref}|r|k>|^2]^{-1/2}$。这一校准不仅表征了连续波包，而且包含了诸如检测器效率等实验因素。

接着对于一种未知分子记录其 $S(\omega;\theta)$，得出跃迁偶极矩 $d(\omega;\theta)$。$d(\omega;\theta)$ 界定为轨道 $x$ 方向上的空间傅里叶变换。我们利用傅里叶切片定理[13]来进行逆变换，并进行了一个重要的修正。傅里叶变换包含因子 $\mathbf{r}$，它是在实验室坐标系中定义的，不随着分子结构 $\theta$ 旋转。

层析逆变换需要两个中间函数，

$$f_x(x, y) = \sum_{\theta}\sum_{\omega} [d_x(\omega;\theta)\cos\theta + d_y(\omega;\theta)\sin\theta]\exp[ik(\omega)(x\cos\theta + y\sin\theta)]$$

$$f_y(x, y) = \sum_{\theta}\sum_{\omega} [-d_x(\omega;\theta)\sin\theta + d_y(\omega;\theta)\cos\theta]\exp[ik(\omega)(x\cos\theta + y\sin\theta)]$$

接着波函数由下式给出：

$$\psi(x, y) = \mathrm{Re}(f_x/x+f_y/y)$$

我们还必须考虑电离速率的角度依赖性以将每一谱图归一化。诸如 $N_2$ 等简单分子的电离速率是已知的（参考文献 30 和 32），但是也可以在记录谐波谱[33]的同时，简单地通过测量总的离子产量 $N(\theta)$ 而得到。

（王耀杨 翻译；陆培祥 审稿）

2. Linus, C. P. *The Nature of the Chemical Bond and the Structure of Molecules and Crystals* (Cornell Univ. Press, Ithaca, New York, 1960).
3. Brion, C. E., Cooper, G., Zheng, Y., Litvinyuk, I. V. & McCarthy, I. E. Imaging of orbital electron densities by electron momentum spectroscopy—a chemical interpretation of the binary (e, 2e) reaction. *Chem. Phys.* **270**, 13-30 (2001).
4. Binning, G., Rohrer, H., Gerber, Ch. & Weibel, E. Surface studies by scanning tunneling microscopy. *Phys. Rev. Lett.* **49**, 57-61 (1982).
5. Sakai, H. & Miyazaki, K. High-order harmonic generation in nitrogen molecules with subpicosecond visible dye-laser pulses. *Appl. Phys. B* **61**, 493-498 (1995).
6. Liang, Y., Augst, A., Chin, S. L., Beaudoin, Y. & Chaker, M. High harmonic generation in atomic and diatomic molecular gases using intense picosecond laser pulses—a comparison. *J. Phys. B* **27**, 5119-5130 (1994).
7. Velotta, R., Hay, N., Manson, M. B., Castillejo, M. & Marangos, J. P. High-order harmonic generation in aligned molecules. *Phys. Rev. Lett.* **87**, 183901 (2001).
8. de Nalda, R. *et al.* Role of orbital symmetry in high-order harmonic generation from aligned molecules. *Phys. Rev. A* **69**, 031804(R) (2004).
9. Hay, N. *et al.* Investigations of electron wave-packet dynamics and high-order harmonic generation in laser-aligned molecules. *J. Mod. Opt.* **50**, 561-571 (2003).
10. Hay, N. *et al.* High-order harmonic generation in laser-aligned molecules. *Phys. Rev.* A **65**, 053805 (2002).
11. Lein, M., Corso, P. P., Marangos, J. P. & Knight, P. L. Orientation dependence of high-order harmonic generation in molecules. *Phys. Rev. A* **67**, 023819 (2003).
12. Parr, R. G. & Yang, W. *Density-functional Theory of Atoms and Molecules* (Oxford Univ. Press, New York, 1989).
13. Kak, A. C. & Slaney, M. *Principles of Computerized Tomographic Imaging* (Society for Industrial and Applied Mathematics, New York, 2001).
14. Stapelfeldt, H. & Seideman, T. Aligning molecules with strong laser pulses. *Rev. Mod. Phys.* **75**, 543-557 (2003).
15. Stapelfeldt, H., Sakai, H., Constant, E. & Corkum, P. B. Deflection of neutral molecules using the nonresonant dipole force. *Phys. Rev. Lett.* **79**, 2787-2790 (1997).
16. Sakai, H. *et al.* Controlling the alignment of neutral molecules by a strong laser field. *J. Chem. Phys.* **110**, 10235-10238 (1999).
17. Larsen, J. J., Hald, K., Bjerre, N. & Stapelfeldt, H. Three dimensional alignment of molecules using elliptically polarized laser fields, *Phys. Rev. Lett.* **85**, 2470-2473 (2000).
18. Sakai, H., Minemoto, S., Nanjo, H., Tanji, H. & Suzuki, T. Controlling the orientation of polar molecules with combined electrostatic and pulsed nonresonant laser fields. *Phys. Rev. Lett.* **90**, 083001 (2003).
19. Rosca-Pruna, F. & Vrakking, M. J. J. Experimental observation of revival structures in picosecond laser-induced alignment of $I_2$. *Phys. Rev. Lett.* **87**, 153902 (2001).
20. Dooley, P. W. *et al.* Direct imaging of rotational wave-packet dynamics of diatomic molecules. *Phys. Rev. A* **68**, 023406 (2003).
21. Delone, N. B. & Krainov, V. P. *Multiphoton Processes in Atoms* (Springer, Heidelberg, 2000).
22. Corkum, P. B. Plasma perspective on strong-field multiphoton ionization. *Phys. Rev. Lett.* **71**, 1994-1997 (1993).
23. Niikura, H. *et al.* Sub-laser-cycle electron pulses for probing molecular dynamics. *Nature* **417**, 917-922 (2002).
24. Dietrich, P., Burnett, N. H., Ivanov, M. & Corkum, P. B. High-harmonic generation and correlated two-electron multiphoton ionization with elliptically polarized light. *Phys. Rev. A* **50**, R3585-R3588 (1994).
25. Lewenstein, M., Balcou, Ph., Ivanov, M. Yu., L'Huillier, A. & Corkum, P. B. Theory of high-harmonic generation by low-frequency laser fields. *Phys. Rev. A.* **49**, 2117-2132 (1994).
26. Burnett, K., Reed, V. C., Cooper, J. & Knight, P. L. Calculation of the background emitted during high-harmonic generation. *Phys. Rev. A* **45**, 3347-3349 (1992).
27. Sanpera, A. *et al.* Can harmonic generation cause non-sequential ionization? *J. Phys. B* **31**, L841-L848 (1998).
28. Mairesse, Y. *et al.* Attosecond synchronization of high-harmonic soft x-rays. *Science* **302**, 1540-1543 (2003).
29. Yudin, G. & Ivanov, M. Yu. Nonadiabatic tunnel ionization: Looking inside a laser cycle. *Phys. Rev. A* **64**, 013409 (2001).
30. Tong, X. M., Zhao, Z. X. & Lin, C. D. Theory of molecular tunneling ionization. *Phys. Rev. A* **66**, 033402 (2002).
31. Spanner, M., Smirnova, O., Corkum, P. B. & Ivanov, M. Yu. Reading diffraction images in strong field ionization of diatomic molecules. *J. Phys. B* **37**, L243-L250 (2004).
32. Otobe, T., Yabana, K. & Iwata, J.-I. First-principles calculations for the tunnel ionization rate of atoms and molecules. *Phys. Rev. A* **69**, 053404 (2004).
33. Litvinyuk, I. V. *et al.* Alignment-dependent strong field ionization of molecules. *Phys. Rev. Lett.* **90**, 233003 (2003).
34. Dunn, T. J., Walmsley, I. A. & Mukamel, S. Experimental determination of the quantum mechanical state of a molecular vibrational mode using fluorescence tomography. *Phys. Rev. Lett.* **74**, 884-887 (1995).
35. Skovsen, E., Stapelfeldt, H., Juhl, S. & Mølmer, K. Quantum state tomography of dissociating molecules. *Phys. Rev. Lett.* **91**, 090406 (2003).
36. Hankin, S. M., Villeneuve, D. M., Corkum, P. B., & Rayner, D. M. Nonlinear ionization of organic molecules in high intensity laser fields. *Phys. Rev. Lett.* **84**, 5082-5085 (2000).
37. Lee, K. F. *et al.* Two-pulse alignment of molecules. *J. Phys. B* **37**, L43-L48 (2004).


**Acknowledgements.** In addition to the NRC, we acknowledge financial support from the National Science and Engineering Research Council, Photonic Research Ontario, the Canadian Institute for Photonic Innovation, the Alexander von Humboldt-Stiftung and the Japan Society for the Promotion of Science. We thank M. Yu. Ivanov, M. Spanner, J. P. Marangos, M. Lein, P. H. Bucksbaum, I. A. Walmsley, D. Jonas, J. Tse and J. G. Underwood for discussions.

**Competing interests statement.** The authors declare that they have no competing financial interests.

**Correspondence** and requests for materials should be addressed to D.M.V. (david.villeneuve@nrc.ca).

# Stratigraphic Placement and Age of Modern Humans from Kibish, Ethiopia

I. McDougall *et al.*

## Editor's Note

**Where did our own species come from, and when? Fossils discovered in 1967 in the Kibish Formation in southern Ethiopia yielded hominid skull remains identifiable as anatomically modern humans, but their antiquity had been much debated. Geochronologists McDougall and colleagues sought to settle the matter with accurate argon-argon dates on volcanic layers in the Formation, together with work on the history of the Omo river valley compared with that of the Nile. The final date—195,000 years old, plus or minus 5,000—made the Omo Kibish hominids the oldest examples of *Homo sapiens* from anywhere in the world. Ethiopia was the birthplace of humankind, as it had been for so many other species in the wider human family.**

---

In 1967 the Kibish Formation in southern Ethiopia yielded hominid cranial remains identified as early anatomically modern humans, assigned to *Homo sapiens*[1-4]. However, the provenance and age of the fossils have been much debated[5,6]. Here we confirm that the Omo I and Omo II hominid fossils are from similar stratigraphic levels in Member I of the Kibish Formation, despite the view that Omo I is more modern in appearance than Omo II[1-3]. $^{40}$Ar/$^{39}$Ar ages on feldspar crystals from pumice clasts within a tuff in Member I below the hominid levels place an older limit of $198 \pm 14$ kyr (weighted mean age $196 \pm 2$ kyr) on the hominids. A younger age limit of $104 \pm 7$ kyr is provided by feldspars from pumice clasts in a Member III tuff. Geological evidence indicates rapid deposition of each member of the Kibish Formation. Isotopic ages on the Kibish Formation correspond to ages of Mediterranean sapropels, which reflect increased flow of the Nile River, and necessarily increased flow of the Omo River. Thus the $^{40}$Ar/$^{39}$Ar age measurements, together with the sapropel correlations, indicate that the hominid fossils have an age close to the older limit. Our preferred estimate of the age of the Kibish hominids is $195 \pm 5$ kyr, making them the earliest well-dated anatomically modern humans yet described.

---

THE principal outcrops of the Kibish Formation are along the Omo River where it skirts the Nkalabong Range (Fig. 1), with the highest outcrops close in elevation to that of the watershed between the Omo and the Nile rivers[7,8]. Former hydrographic links are apparent from Nilotic fauna in the Turkana Basin sequence[8-10].

# 来自埃塞俄比亚基比什的现代人的地层位置和年代

麦克杜格尔等

## 编者按

我们自己的物种从何而来，何时而来？1967 年，科学家在埃塞俄比亚南部的基比什组中发现了古人类头骨化石，这些化石被认为是解剖学意义上的早期现代人，被归入智人这一类群。但这些化石的出土地点和年代一直备受争议。地质年表学家麦克杜格尔及其同事们试图用地层中火山层的氩精确年代，以及奥莫河谷与尼罗河的历史对比来解决这个问题。最终的年代——19.5 万 ± 0.5 万年——使得奥莫基比什古人类成为目前发现的世界上最古老的智人。埃塞俄比亚是人类的诞生地，同人类大家庭中许多其他物种一样。

---


1967 年，在埃塞俄比亚南部的基比什组挖掘出来了古人类头盖骨遗迹，该头盖骨被鉴定为解剖学上的早期现代人，被划分到了智人 [1-4]。然而，这些化石的出土地点和年代却饱受争议 [5,6]。尽管有观点认为奥莫 I 号化石的形态比奥莫 II 号更加接近现代 [1-3]，但是我们确证了奥莫 I 号和奥莫 II 号古人类化石是来自于基比什组 I 段的相似层位。人们采集了古人类化石出土层位下方的 I 段的凝灰岩内的浮石碎屑，从中得到的长石晶体确定了 $^{40}Ar/^{39}Ar$ 年代，将古人类化石的年代下限确定在 198 ± 14 kyr(kyr 为千年)(加权均值年代为 196 ± 2 kyr)。III 段凝灰岩内的浮石碎屑中的长石则提供了古人类化石的年代上限，为 104 ± 7 kyr。地质学证据表明基比什组的每一段都经历了快速的沉积作用。基比什组的同位素年龄与地中海腐泥的年代一致，这反映了尼罗河流量的增加，以及由此引起的奥莫河流量的必然增加。因此 $^{40}Ar/^{39}Ar$ 方法得出的年代值，以及它与地中海腐泥形成时间的相关关系，共同暗示了古人类化石的年代接近于上述测年结果的下限。我们更倾向于估计基比什古人类的年代约为 195 ± 5 kyr，这样他们就成为迄今描述过的最早的可以清楚追溯到其年代的解剖学上的现代人。


---

基比什组的主要露头地是沿着奥莫河一带，那里位于纳卡拉邦山脉的边缘(图 1)，其中位置最高的露头，在海拔上接近奥莫河与尼罗河的分水岭 [7,8]。在图尔卡纳盆地的沉积序列里，出现了尼罗河动物群，显然，从前这两条河流有十分接近的水文关系 [8-10]。

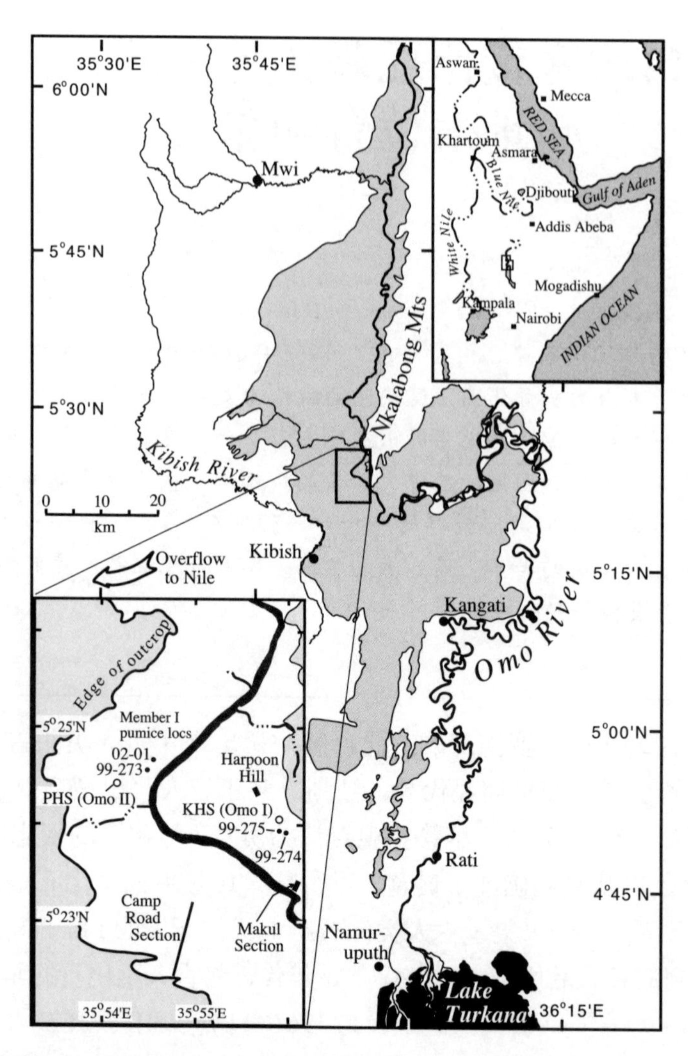


Fig. 1. Map showing the distribution of the Kibish Formation (shaded) in the lower Omo Valley, southern Ethiopia, after Davidson[29]. Inset on lower left, locations of Omo I, Omo II, measured sections, and dated samples.

The Kibish Formation (about 100 m thick) consists of flat-lying, tectonically undisturbed, unconsolidated sediments deposited mainly in deltaic environments over brief periods. It comprises the youngest exposed sedimentary sequence in the Omo Basin, and lies disconformably upon the Nkalabong Formation[11,12] or on the underlying Mursi Formation[12]. Strata are composed principally of claystone and siltstone, with subordinate fine sandstone, conglomerate and tuffs (Fig. 2).

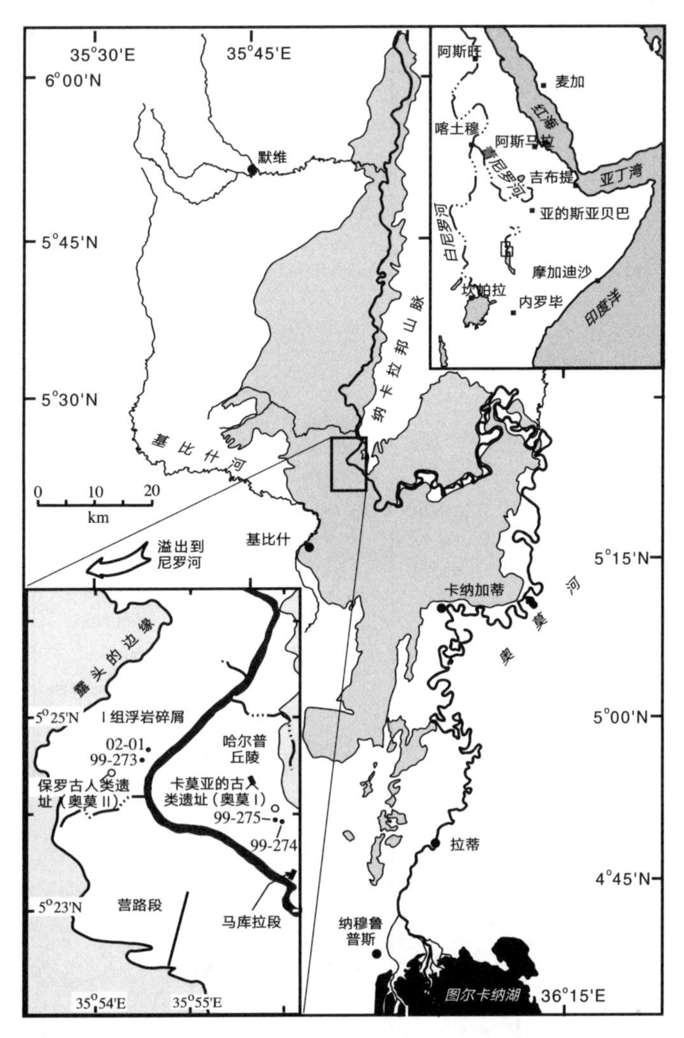


图 1. 埃塞俄比亚南部下奥莫河谷的基比什组（阴影部分）的分布地图（戴维森指定的[29]）。左下方的插图表示的是奥莫 I 号和奥莫 II 号的位置、测量的剖面以及确定年代的样本。

基比什组（约 100 米厚）主要由短时间内堆积在三角洲环境的平整且未受到干扰的、松散的沉积物组成。其包含奥莫盆地最晚近的暴露沉积序列，以不整合的方式位于纳卡拉邦组之上[11,12]或在下方的穆尔斯组上[12]。地层主要由黏土岩和粉砂岩构成，附着有细砂岩、砾岩和凝灰岩（图 2）。

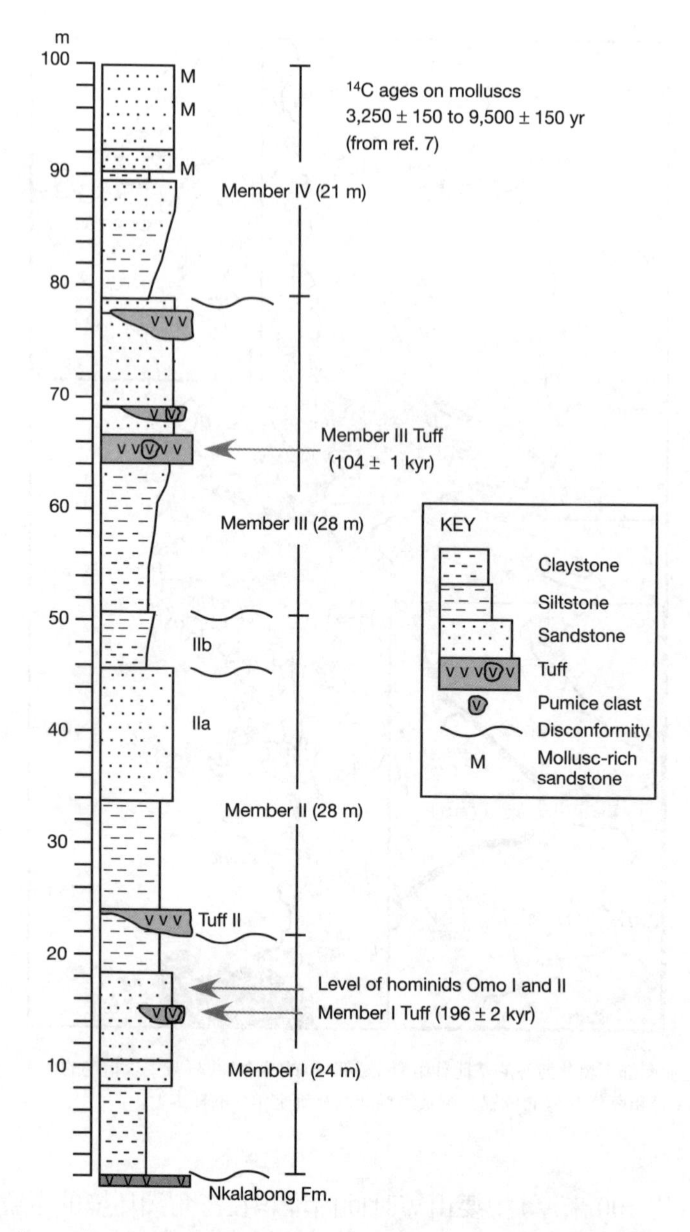


Fig. 2. Composite stratigraphy of the Kibish Formation. Member I was measured near the type section at Makul; Member II was measured near Harpoon Hill, and Members III and IV were measured along Camp Road (see Fig. 1 for locations).

Butzer *et al.*[13] and Butzer[14] divided the Kibish Formation into Members I to IV on the basis of disconformities with up to 30 m relief (Fig. 2). The members record discrete times of deposition when the northern margin of Lake Turkana and the Omo delta lay about 100 km north of their current positions. As the higher lake levels reflect significantly higher precipitation in the region, such periods should be recognizable at least regionally, and perhaps globally.

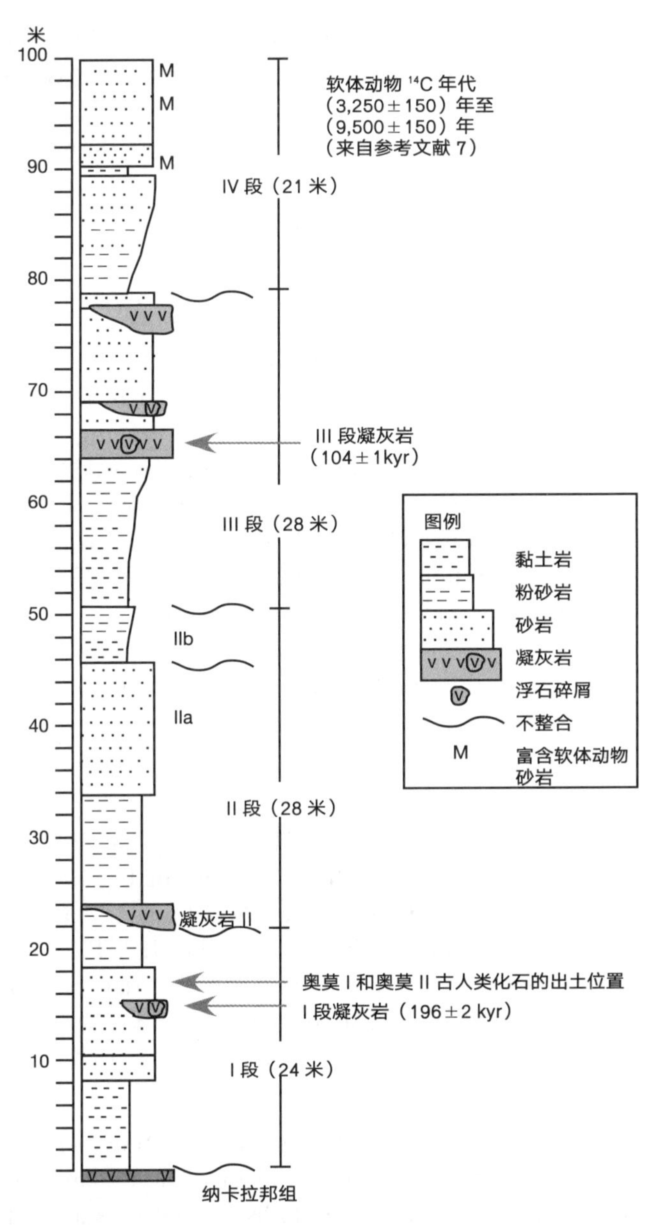


图 2. 基比什组的复合地层学。经测量，I 段位于马库拉的典型剖面附近，II 段位于哈尔普丘陵附近，III 段和 IV 段沿着营路测量的(具体位置见图 1)。

布策等人[13]和布策[14]根据基比什组高达 30 米的地层不整合现象，将其划分成了 I 到 IV 段(图 2)。在那时，图尔卡纳湖的北部边界和奥莫三角洲，在它们现在位置的北边 100 千米，基比什组的四个段精确记录了那时沉积过程的不连续时间。由于湖泊水位的升高反映了该地区降水的显著增加，所以至少在区域上，甚至在全球范围内，都应该可以辨别出这些时期。

Member I (26 m thick; Fig. 2) was deposited disconformably on the Nkalabong Formation in a deltaic environment. Small (less than 30 mm) rounded pumice clasts occur in an impure tuff in Member I, 7 m below the base of Member II. This tuff lies near, but probably slightly below, the levels from which Omo I and Omo II were derived. Glass shards from the tuff are very similar in composition to the glass from three pumice clasts enclosed in the tuff (Supplementary Table 1), indicating a common origin. Further, this composition is distinct from the glass composition of any other tuff in the Kibish Formation. Member II, about 28 m thick, was deposited on a topographic surface developed on Member I with at least 19 m of relief. Member II contains two discrete sequences, separated by an internal disconformity, designated Members IIa and IIb on Fig. 2. Member II was incised by as much as 25 m before the deposition of Member III[11,14], which begins with thin siltstone and claystone beds that drape topography. These beds, averaging about 3 cm thick, fine upward internally, and represent annual flooding. Thus, deposition of the lower part of Member III may record less than 1 kyr. A tuff 3.5–12 m thick, 18 m above the base of Member III, locally contains pumice clasts with alkali feldspar phenocrysts. Glass of the pumices normally differs compositionally from the glass of the tuff (Supplementary Table 1). However, one pumice sample yielded two contrasting sets of analyses: one corresponding to the tuff, the other to the dominant pumices. The youngest unit of the Kibish Formation, Member IV, was deposited on the underlying sediments after up to 30 m of dissection. Member IV comprises at most 21 m of strata deposited between about 9.5 and 3.3 kyr ago, on the basis of $^{14}$C dating of mollusc shell[7,14,15].

Omo I was found at Kamoya's Hominid Site (KHS; 5° 24.15′ N, 35° 55.81′ E), which was identified from contemporary photographs, from evidence of the 1967 excavations and by additional hominid bone material conjoining the 1967 finds[16]. The hominid fossils were recovered from a siltstone 2.4 m below the base of Member II[14]. Thus, Omo I derives from near the top of Member I. Butzer *et al.*[13] reported a $^{230}$Th/$^{234}$U date of 130 ± 5 kyr on *Etheria* from essentially the same level as the hominid, but they and others have questioned the reliability of the age[5,6].

Omo II was found on the surface at PHS (Paul's Hominid Site), which Butzer[14] mapped about 2.6 km northwest of KHS. A map drawn at the time by Paul Abell (discoverer of Omo II), together with contemporary photographs of the site supplied by K.W. Butzer, have positively identified PHS at 5° 24.55′ N, 35° 54.07′ E, about 3.3 km west by north of KHS (Fig. 1). Although PHS was mislocated on the published map, Butzer's[14] stratigraphic description at the site is correct. The base of Member II lies about 3 m above the approximate level from which Omo II was recovered. The basal tuff of Member II is of unique composition, and correlates from KHS to PHS (Supplementary Table 1).

Ages on alkali feldspars separated from pumice clasts from tuffs in Members I and III reflect the time of their eruption, and provide maximum and minimum ages for the hominid fossils, respectively. $^{40}$Ar/$^{39}$Ar results are summarized in Table 1, with details listed in Supplementary Tables 2–4.

I段(26米厚；图2)是在三角洲环境中不整合地沉积在纳卡拉邦组上的。小型(不足30毫米)的圆形浮岩出现于I段的掺杂的凝灰岩中，位于II段基底之下7米处。这种凝灰岩位于奥莫I号和奥莫II号的层位附近，但可能略低于该水平。凝灰岩中的玻璃碎片与凝灰岩中附着的三种浮岩碎屑的玻璃在构成上非常相似(附表1)，这暗示着它们具有共同的来源。此外，这种成分与基比什组的其他凝灰岩中的玻璃成分都不一样。II段约有28米厚，沉积在一个I段上发展起来的表面上，具有至少19米的起伏地势。II段内部有一个不整合面，将这段地层分成了两部分，图2中将它们指定为IIa和IIb段。III段的沉积过程是从覆盖地形的薄的粉砂岩和黏土岩床开始的，但在沉积开始之前，II段被侵入了25米之多[11,14]。这些平均约3厘米厚的河床，内部向上部位的颗粒都很精细，代表了每年一次的洪灾。因此，III段的下半部分的沉积岩，记录的时代可能少于1,000年。一层3.5米到12米厚的凝灰岩，位于III段的基底之上18米处，局部含有浮岩碎屑和碱性长石斑晶。浮岩的玻璃质通常与凝灰岩玻璃质在成分上有所不同(附表1)，但是，对一例浮岩样本的分析，产生了两个截然不同的结果，一套对应凝灰岩，另一套则对应于占优势的浮岩。基比什组中最年轻的单元——IV段是在高达30米的不连续层之后沉积在下面的沉积物上的。根据软体动物外壳进行的$^{14}C$年代测定发现，IV段沉积形成的时代为9,500年到3,300年前，地层厚度约为21米[7,14,15]。

奥莫I号是在卡莫亚古人类遗址(KHS；北纬5°24.15′，东经35°55.81′)发现的，可以根据1967年挖掘的证据结合1967年发现的其他古人类化石以及同时期的照片上鉴别出来[16]。古人类化石从II段的基底之下的2.4米处的粉砂岩处发掘出来[14]。因此，奥莫I号出土位置接近I段的顶部。布策等[13]使用$^{230}Th/^{234}U$测年法，测定了一例与古人类化石来自同一层位的艾特利亚人的年代为130±5 kyr，但是他们和其他人都对这一年代的可靠性表示质疑[5,6]。

奥莫II号发现于保罗古人类遗址(PHS)的地表，根据布策[14]的标记，这一地点位于KHS西北约2.6千米处。根据保罗·埃布尔(奥莫II号的发现者)当时绘制的地图与布策提供的同一时期的照片，最终确定PHS位于KHS北部约3.3千米处，地理坐标为北纬5°24.55′东经35°54.07′(图1)。尽管在已发表的地图上将PHS的位置搞错了，但是布策[14]对该遗址进行的地层学描述是正确的。II段的基底在奥莫II号出土位置之上约3米处。II段的基底凝灰岩具有独特的成分，从KHS到PHS都与其具有相关性(附表1)。

从I段和III段的凝灰岩中的浮岩碎屑分离得到的碱性长石的年代反映了它们爆发的时间，并且分别提供了古人类化石的最大和最小年代。$^{40}Ar/^{39}Ar$的结果在表1中进行了总结，详细情况在附表2～4中列出。

Table 1. Summary of $^{40}Ar/^{39}Ar$ alkali feldspar laser fusion ages from pumice clasts in tuffs of the Kibish area, Turkana Basin, Ethiopia

| Sample no. | Tuff | Locality | Irradiation | *n* | *n* used | Simple mean age (kyr) | Weighted mean age (kyr) | Isochron age (kyr) | MSWD | $(^{40}Ar/^{39}Ar)_i$ |
|---|---|---|---|---|---|---|---|---|---|---|
| Kibish Formation, Member III | | | | | | | | | | |
| 99-275A | Member III | 0.4 km SSE of KHS | ANU58/L10 | 10 | 10 | 105.4 ± 5.0 | 106.0 ± 1.6 | 108.0 ± 3.1 | 1.03 | 292.2 ± 4.1 |
| 99-275C | Member III | 0.4 km SSE of KHS | ANU58/L1 | 13 | 11 | 107.5 ± 7.1 | 105.4 ± 1.8 | 101.1 ± 8.4 | 2.04 | 303.6 ± 13.8 |
| 99-274A | Member III | 0.5 km SE of KHS | ANU58/L6 | 13 | 13 | 98.1 ± 5.3 | 98.2 ± 1.1 | 97.4 ± 2.2 | 2.47 | 298.9 ± 6.8 |
| 99-274B | Member III | 0.5 km SE of KHS | ANU58/L7 | 14 | 13 | 105.0 ± 8.3 | 109.6 ± 1.3 | 110.2 ± 4.4 | 3.27 | 292.9 ± 16.2 |
| Kibish Formation, Member I, near Omo II site, just west of Omo River, Nakaa'kire, 5° 24.6′N, 35° 54.5′E | | | | | | | | | | |
| 99-273A | Member I | Nakaa'kire | ANU58/L3 | 9 | 9 | 319.8 ± 18.6 | 315.3 ± 3.6 | 318.2 ± 16.6 | 3.55 | 287.3 ± 41.5 |
| 99-273B | Member I | Nakaa'kire | ANU58/L4 | 8 | 7 | 220.8 ± 13.3 | 210.9 ± 2.1 | 209.6 ± 2.0 | 0.55 | 301.4 ± 3.1 |
| 02-01A | Member I | Nakaa'kire | ANU98/L9 | 16 | 16 | 204.2 ± 14.9 | 205.3 ± 2.2 | 194.1 ± 5.4 | 2.66 | 315.8 ± 7.3 |
| 02-01B | Member I | Nakaa'kire | ANU98/L10 | 15 | 14 | 194.7 ± 10.8 | 191.5 ± 1.9 | 184.9 ± 3.8 | 2.13 | 304.0 ± 3.5 |
| 02-01C | Member I | Nakaa'kire | ANU98/L3 | 14 | 14 | 195.6 ± 15.0 | 192.7 ± 2.3 | 192.5 ± 6.7 | 3.85 | 295.7 ± 5.8 |

$^{40}K$ decay constant $\lambda = 5.543 \times 10^{-10}$ $yr^{-1}$. Fluence monitor: Fish Canyon Tuff sanidine 92-176 of reference age 28.1 Myr. Samples 99-275A and 99-275C (laboratory sample numbers at ANU) = KIB99-47 (field sample number); 99-274A and 99-274B = KIB99-41; 99-273A and 99-273B = KIB99-19. Results with errors are means ± s.d. MSWD, mean square of weighted deviates.

The five pumice clasts measured from Member I yielded three different ages (Table 1), with at least the oldest age interpreted as evidence for reworking of that particular pumice clast. Nine analyses on 99-273A gave a mean age of 320 ± 19 kyr. Seven analyses on 99-273B yielded a mean age of 221 ± 13 kyr after rejecting one outlier (269 ± 18 kyr). Single feldspar crystals from 02-01B yielded a mean age of 195 ± 11 kyr ($n = 14$), forming a concordant data set after one rejection (333 ± 6 kyr). Multiple crystals ($n \leqslant 3$) were analysed in 12 of 16 measurements for 02-01A, and in 8 of 14 analyses for 02-01C ($n \leqslant 4$). No ages were rejected on 02-01A, whose mean age is 204 ± 15 kyr, or on 02-01C, whose mean age is 196 ± 15 kyr. It is probable that these latter three pumice clasts are products of the same volcanic eruption, as shown by their concordant ages and similar mean K/Ca ratios. When combined, all 44 analyses provide an arithmetic mean age of 198 ± 14 kyr and a weighted mean age of 195.8 ± 1.6 kyr. The mean age calculated for pumice 99-273B is statistically older than the ages determined on the three pumice clasts 02-01, so that 99-273B might also be a reworked pumice clast. Clearly, the age of deposition of a tuff must be younger than the youngest igneous component found within it. Thus, this tuffaceous level of Member I of the Kibish Formation was deposited after 196 kyr ago, on the basis of the mean age determined on the three 02-01 pumice clasts.

表 1. 对埃塞俄比亚图尔卡纳盆地基比什地区的凝灰岩中的浮岩碎屑进行的 $^{40}Ar/^{39}Ar$ 碱性长石激光聚变测年概况

| 样品编号 | 凝灰岩 | 产地 | 放射 | 取样量 | 实际使用数目 | 简单平均年代(kyr) | 加权平均年代(kyr) | 等龄线法年代(kyr) | 加权均方差 | $(^{40}Ar/^{39}Ar)_i$ |
|---|---|---|---|---|---|---|---|---|---|---|
| 基比什组，III 段 | | | | | | | | | | |
| 99-275A | III 段 | KHS 东南偏南 0.4 千米处 | ANU58/L10 | 10 | 10 | 105.4 ± 5.0 | 106.0 ± 1.6 | 108.0 ± 3.1 | 1.03 | 292.2 ± 4.1 |
| 99-275C | III 段 | KHS 东南偏南 0.4 千米处 | ANU58/L1 | 13 | 11 | 107.5 ± 7.1 | 105.4 ± 1.8 | 101.1 ± 8.4 | 2.04 | 303.6 ± 13.8 |
| 99-274A | III 段 | KHS 东南 0.5 千米处 | ANU58/L6 | 13 | 13 | 98.1 ± 5.3 | 98.2 ± 1.1 | 97.4 ± 2.2 | 2.47 | 298.9 ± 6.8 |
| 99-274B | III 段 | KHS 东南 0.5 千米处 | ANU58/L7 | 14 | 13 | 105.0 ± 8.3 | 109.6 ± 1.3 | 110.2 ± 4.4 | 3.27 | 292.9 ± 16.2 |
| 基比什组，I 段，靠近奥莫 II 遗址，就在奥莫河以西，Nakaa'kire，5° 24.6' N, 35° 54.5' E | | | | | | | | | | |
| 99-273A | I 段 | Nakaa'kire | ANU58/L3 | 9 | 9 | 319.8 ± 18.6 | 315.3 ± 3.6 | 318.2 ± 16.6 | 3.55 | 287.3 ± 41.5 |
| 99-273B | I 段 | Nakaa'kire | ANU58/L4 | 8 | 7 | 220.8 ± 13.3 | 210.9 ± 2.1 | 209.6 ± 2.0 | 0.55 | 301.4 ± 3.1 |
| 02-01A | I 段 | Nakaa'kire | ANU98/L9 | 16 | 16 | 204.2 ± 14.9 | 205.3 ± 2.2 | 194.1 ± 5.4 | 2.66 | 315.8 ± 7.3 |
| 02-01B | I 段 | Nakaa'kire | ANU98/L10 | 15 | 14 | 194.7 ± 10.8 | 191.5 ± 1.9 | 184.9 ± 3.8 | 2.13 | 304.0 ± 3.5 |
| 02-01C | I 段 | Nakaa'kire | ANU98/L3 | 14 | 14 | 195.6 ± 15.0 | 192.7 ± 2.3 | 192.5 ± 6.7 | 3.85 | 295.7 ± 5.8 |

$^{40}K$ 衰变常数 $\lambda = 5.543 \times 10^{-10}\ yr^{-1}$。流量监控器：参考年代为 2,810 万年的菲什峡谷凝灰岩透长石 92-176。样本 99-275A 和 99-275C(ANU 的实验室样本编号) = KIB99-47(野外样本编号)；99-274A 和 99-274B = KIB99-41；99-273A 和 99-273B = KIB99-19。结果与误差用平均值 ± 标准差表示。

对 I 段得到的五份浮岩碎屑的测年产生了三个不同的结果(表 1)，至少最古老的年代被解读为这一例浮岩碎屑曾经被搬运的证据。对 99-273A 进行的九项分析给出的平均年代为 320 ± 19 kyr。对 99-273B 进行的七项分析在排除一个离群值(269 ± 18 kyr)之后，测得的平均年代为 221 ± 13 kyr。来自 02-01B 的唯一长石晶体测得的平均年代为 195 ± 11 kyr($n = 14$)，在剔除一个离群值(333 ± 6 kyr)之后，形成了一套整合的数据集。对 02-01A 的 16 个测量值中的 12 个进行了多晶体($n \leqslant 3$)分析，对 02-01C($n \leqslant 4$)的 14 个测量值中的 8 个进行了分析。得到的 02-01A 的所有年代数值中没有离群值，其平均年代为 204 ± 15 kyr，02-01C 的也没有离群值，其平均年代为 196 ± 15 kyr。有可能后面这三种浮岩碎屑是同一次火山喷发的产物，正如它们的一致年代和相似的平均 K/Ca 比例反映的一样。综合分析，所有 44 个结果提供的算术平均年代是 198 ± 14 kyr，加权平均年代是 195.8 ± 1.6 kyr。统计学上看，浮岩 99-273B 计算出来的平均年代比三种 02-01 浮岩碎屑确定的年代更古老，所以 99-273B 可能也是一种二次搬运的浮岩碎屑。显然，凝灰岩层的年代肯定比其中最年轻的火成组分更年轻。因此，根据这三种 02-01 浮岩碎屑确定的平均年代，基比什组 I 段的凝灰岩层是在 196 kyr 之后沉积下来的。

Feldspars from four pumice clasts from the Member III tuff were analysed. Two were *in situ* (99-275A, C), and two (99-274A, B) had weathered out of the same unit about 150 m farther east (Fig. 1). Ages on 99-275A range from 98.0 ± 7.7 kyr to 114.6 ± 4.3 kyr. Five single-crystal measurements have an arithmetic mean age of 105.5 ± 7.0 kyr, identical to five measurements on groups of two or three crystals (mean age 105.4 ± 2.7 kyr), indicating a homogeneous population. The overall mean age is 105.4 ± 5.0 kyr, with a weighted mean age of 106.0 ± 1.6 kyr. Eleven analyses on sample 99-275C yield a mean age of 107.5 ± 7.1 kyr and a weighted mean of 105.4 ± 1.8 kyr, after rejection of two outliers (142.1 ± 3.3 and 128.8 ± 8.3 kyr). Concordant results from 99-274A on seven single crystals and six pairs of crystals gave an overall arithmetic mean age of 98.1 ± 5.3 kyr. The companion clast 99-274B gave a mean age of 105.0 ± 8.3 kyr, after elimination of one outlier (Supplementary Table 4). Again these results reflect a single age population, as shown by the similar mean ages of the individual pumices, and the overlapping average K/Ca ratios. Combining all these results (except outliers), the overall arithmetic mean age is 103.7 ± 7.4 kyr ($n = 47$) and the weighted mean age is 103.7 ± 0.9 kyr. Thus, the depositional age must be equal to or younger than 104 kyr, providing evidence that Members I and II of the Kibish Formation are older than 104 kyr.

Each of the members of the Kibish Formation was deposited during intervals when Lake Turkana was at a much higher level than at present, and Member II has an internal disconformity. The upper part of Member I was being deposited at or after 196 kyr ago, and the upper part of Member III was being deposited at or after 104 kyr ago; $^{14}C$ ages on Member IV correspond to deposition between 9.5 and 3.3 kyr ago. These ages are remarkably similar to ages of Mediterranean sapropels S7, S4 and S1. Sapropels S1–S7 have the following estimated ages: 195 kyr (S7), 172 kyr (S6), 124 kyr (S5), 102 kyr (S4), 81 kyr (S3), 55 kyr (S2) and 8 kyr (S1)[17]. In many cases sapropels are related to a greatly increased flow of the Nile River into the Mediterranean Sea as a consequence of intensification of the African monsoon[18], recorded in more negative $\delta^{18}O$ in planktonic foraminifera. As the Omo River shares a divide with the Blue Nile and with tributaries of the White Nile, the Nile and the Omo must be affected similarly. As noted, deposition of each of the members of the Kibish Formation was probably very rapid. Thus, the close correspondence between the ages of Member I (196 kyr) and of sapropel S7 (195 kyr), of Member III (104 kyr) and of sapropel S4 (102 kyr), and of Member IV (3.3–9.5 kyr) and of sapropel S1 (8 kyr) is probably causally related. Sapropel S2 (55 kyr) is absent from or poorly represented in many Mediterranean sedimentary cores and has a very small $\delta^{18}O$ residual; thus, it is not surprising that no deposits of this age have been identified in the Kibish Formation. Sapropel S6, deposited during a European glacial period[19], might also be absent from the Kibish Formation, because it too has a small $\delta^{18}O$ anomaly. The two parts of Member II may be accommodated by sapropels S5 and S6, or they may correspond to the two phases identified in sapropel S5 (119–124 kyr), which are separated by 700–900 yr (ref. 20). This link between sapropel formation in the Mediterranean and very high levels of Lake Turkana is a particularly notable finding. In contrast, S3 is very well represented in many Mediterranean sedimentary cores and is therefore expected to be recorded in the Kibish Formation, but has not been recognized. Given the large expanse of the plain in the

对来自III段凝灰岩的四例浮岩碎屑中的长石进行了分析。两例是在原位(99-275A，C)采集的，另外两例(99-274A，B)也来自同一地层单元，但它们被风化搬运到了东部约150米远的地方(图1)。99-275A的年代从98.0±7.7 kyr到114.6±4.3 kyr不等。五个单晶体测量结果的算术平均值是105.5±7.0 kyr，与两三种晶体一组的五个结果是一致的(均值为105.4±2.7 kyr)，显示它们是同一群体。总平均年代为105.4±5.0 kyr，加权平均年代为106.0±1.6 kyr。对样本99-275C进行的十一项分析在剔除掉两个离群值(142.1±3.3和128.8±8.3 kyr)之后产生的平均年代为107.5±7.1 kyr，加权均值为105.4±1.8 kyr。对99-274A进行的七种单晶体和六对晶体分析得到的一致结果给出的总算术平均值是98.1±5.3 kyr。在剔除掉一个离群值后，伴随碎屑99-274B给出的平均年代为105.0±8.3 kyr(附表4)。这些结果再次反映了这些样本来源于同一次沉积，另外，单个浮岩样本有相似的平均年代，它们的K/Ca比值和滑动平均K/Ca比所表明的一样。将所有这些结果(除了离群值)综合起来，总算术平均年代为103.7±7.4 kyr($n=47$)，加权平均年代为103.7±0.9 kyr。因此，沉积年代必须等于或小于104 kyr，这为基比什组的I段和II段的年代要早于104 kyr提供了证据。

基比什组的每个地层段都是在图尔卡纳湖处于比现在更高的水位上的间隔期间沉积下来的，II段地层具有内部不整合性。I段地层的上半部分是在196 kyr前或晚些时候沉积下来的，III段的上半部分是在104 kyr前或晚些时候沉积下来的；IV段的$^{14}C$年代分析表明其对应的沉积作用发生于距今9.5 kyr到3.3 kyr前。这些年代与地中海腐泥S7、S4和S1阶段年代非常相似。腐泥S1~S7的估计年代如下：195 kyr(S7)、172 kyr(S6)、124 kyr(S5)、102 kyr(S4)、81 kyr(S3)、55 kyr(S2)和8 kyr(S1)[17]。浮游有孔虫中$\delta^{18}O$的负值显示非洲季风的加剧，尼罗河流入地中海的水量大幅增加，很多情况下这与腐泥形成有关[18]。由于奥莫河与青尼罗河以及白尼罗河支流共有分水岭，所以尼罗河和奥莫河也必然会受到相似的影响。如前所述，基比什组的各成员层的沉积作用可能非常迅速。因此，I段地层(196 kyr)和腐泥S7(195 kyr)的年代之间、III段地层(104 kyr)和腐泥S4(102 kyr)的年代之间、IV段地层(3.3~9.5 kyr)和腐泥S1(8 kyr)的年代之间可能是有因果关系的。腐泥S2层(55 kyr)在许多地中海沉积物岩芯中都不存在或者含量很少，$\delta^{18}O$残值也很小；因此，在基比什组没有测定出该年代的堆积物也不足为奇。腐泥S6是在欧洲冰期[19]堆积下来的，可能在基比什组也不存在，因为它也有一个小的$\delta^{18}O$异常。II段地层存在两个不同的部分，二者可能对应腐泥层S5和S6，或者它们可能对应于腐泥S5层(119~124 kyr)所代表的两个时期，即被700~900年分开的两个时期(参考文献20)。地中海的腐泥地层和图尔卡纳湖的高水位之间的这种联系是一项非常值得注意的发现。相反，很多地中海沉积物岩芯中都能发现S3腐泥层，因此也希望基比什组

region underlain by the Kibish Formation, it is quite possible that deposits correlative with sapropel S3 are present but are not exposed in the immediate Kibish region that we have studied.

Our palaeontological and stratigraphic studies support the original report[14] that Omo I and Omo II are derived from comparable stratigraphic levels within Member I of the Kibish Formation despite their morphological differences[1-3,21,22]. Morphological diversity among fossil hominids from the Middle and Late Pleistocene of Africa is of major importance in understanding the tempo and mode of modern human origins[21,23].

$^{40}Ar/^{39}Ar$ dating of feldspars from tuffs in Member I and Member III of the Kibish Formation shows that its hominid fossils are younger than $195.8 \pm 1.6$ kyr and older than $103.7 \pm 0.9$ kyr. Direct isotopic dating of volcanic eruptions recorded in the Kibish Formation does not enable us to place narrower limits on the age. However, the suggested correlations of Member IIa and Member IIb with either the two identified phases of sapropel S5 or sapropels S6 and S5, respectively, indicate that deposition of Member I of the Kibish Formation occurred earlier than about 125 kyr ago or earlier than 172 kyr ago. The geological evidence for rapid deposition of Member I and the remarkably close correspondence of the isotopic ages on the youngest pumice clasts in the tuff of Member I at 196 kyr with the estimated age of sapropel S7 is regarded as strongly supporting the view that Member I was deposited close to $196 \pm 2$ kyr ago. On this basis we suggest that hominid fossils Omo I and Omo II are relatively securely dated to $195 \pm 5$ kyr old, somewhat older than the age of between 154 and 160 kyr assigned to the hominid fossils from Herto, Ethiopia[24], making Omo I and Omo II the oldest anatomically modern human fossils yet recovered.

## Methods

### Age measurements

Alkali feldspar crystals were separated from pumice clasts and cleaned ultrasonically in 7% HF for 5–10 min to remove adhering volcanic glass and surface alteration. Although separations were performed on the coarsest crystals present, in several cases crystals were less than 1 mm, with masses less than 0.8 mg. In such cases several crystals were used for each analysis (see Supplementary Tables 2–4).

Samples were irradiated in facilities X33 or X34 of the High Flux Australian Reactor (Lucas Heights, Sydney, Australia) for 6 h as described in ref. 25. Cadmium shielding 0.2 mm thick was used to reduce the $(^{40}Ar/^{39}Ar)_K$ correction factor, which was measured by analysis of zero-aged synthetic potassium silicate glass co-irradiated with the unknowns. This correction is particularly important because of the young age of the samples, so interpolated values for each sample are given in the Supplementary Tables 2–4. The fluence monitor employed was sanidine 92-176, separated from the Fish Canyon Tuff, with a reference age of 28.1 Myr (ref. 26).

能够有相应的记录，但尚未得到确认。鉴于该地区位于基比什组之下的平原如此之大，与腐泥 S3 相关的堆积物非常可能存在，只是在我们研究过的基比什地区周围没有暴露出来而已。

尽管奥莫 I 号和奥莫 II 号古人类化石存在形态学差异 [1-3,21,22]，但我们的古生物和地层学研究支持原先的报道 [14]，即奥莫 I 号和奥莫 II 号化石都来自基比什组的 I 段地层，其出土层位比较接近。来自非洲中、晚更新世时期的古人类化石之间存在的形态学多样性，对于理解现代人起源的进度和模式具有非常重要的意义 [21,23]。

对基比什组的 I 段和 III 段地层中的凝灰岩里的长石进行 $^{40}Ar/^{39}Ar$ 年代测定，表明古人类化石的年代区间为 195.8 ± 1.6 kyr 至 103.7 ± 0.9 kyr。对基比什组所记录的火山爆发的同位素直接测年并不能帮助我们将年代范围缩小。然而，IIa 组和 IIb 组分别与地中海腐泥 S5 或腐泥 S6 和 S5 两个已知时期的相关性表明，基比什组的 I 段地层的堆积作用发生的年代早于约 125 kyr 前，或者早于 172 kyr 前。地质学证据表明，I 段存在着快速堆积作用，另外，I 段地层的凝灰岩中的最晚的浮岩碎屑的同位素年代（196 kyr）与腐泥 S7 的年代的非常密切的对应性强烈支持了如下观点，即 I 段是在接近 196 ± 2 kyr 前堆积起来的。以此为基础，我们认为奥莫 I 和奥莫 II 的古人类化石比较肯定的年代是 195 ± 5 kyr，比埃塞俄比亚赫托 [24] 的古人类（154 kyr 至 160 kyr）的年代早一些，因此奥莫 I 和奥莫 II 也成为迄今挖掘出来的年代最古老的解剖学上的现代人化石。

## 方　　法

### 年代测定

碱性长石晶体是从浮岩碎屑中分离出来的，然后用 7% 的氢氟酸超声波洗涤 5 至 10 分钟以去除黏附的火山玻璃和表面蚀变。尽管是在目前最粗糙的晶体上进行分离操作的，但是有些情况下，晶体还不足 1 毫米，质量也不到 0.8 毫克。这些情况下，每个分析都要使用几种晶体（见附表 2 ~ 4）。

样本使用澳大利亚高通量反应器（卢卡斯高地，悉尼，澳大利亚）的 X33 或 X34 仪器照射 6 小时，具体方法见参考文献 25 中的描述。用 0.2 毫米厚的镉屏蔽减小 $(^{40}Ar/^{39}Ar)_K$ 校正因子，该因子通过分析与未知因素共同照射的年代为零的合成硅酸钾玻璃来进行测定。由于样本的年代较近，所以这一修正就格外重要，附表 2~4 中给出了按照此法得到的每个样本的内插值。采用的流量监控器是透长石 92-176，是从菲什峡谷凝灰岩上分离下来的，其参考年代为 2,810 万年（参考文献 26）。

After irradiation, feldspar crystals were loaded into wells in a copper sample tray, installed in the vacuum system and baked overnight. Samples were fused with a focused argon-ion laser beam with up to 10 W of power. After purification of the gases released during fusion, the argon was analysed isotopically in a VG3600 mass spectrometer, using a Daly collector. The overall sensitivity of the system was about $2.5 \times 10^{-17}$ mol $mV^{-1}$. Mass discrimination was monitored through regular measurements of atmospheric argon. The irradiation parameter, $J$, for each unknown was derived by interpolation from the measurements made on the fluence monitor crystals, with at least five analyses per level; the precision generally was in the range 0.3–0.75%, standard deviation of the population. Calcium correction factors[27] used in all calculations were $(^{36}Ar/^{37}Ar)_{Ca} = 3.49 \times 10^{-4}$ and $(^{39}Ar/^{37}Ar)_{Ca} = 7.86 \times 10^{-4}$.

## Data handling

In calculating the arithmetic mean age for each pumice clast, any result more than two standard deviations from the initial mean of each sample was rejected iteratively until no further outliers were identified. No more than two measurements were rejected in any group of analyses, and no results were rejected for about half the groups. The error quoted in Table 1 is the standard deviation of the population, but because uncertainties on individual ages are variable, a weighted mean age and error is also given, weighting each age by the inverse of the variance. Differences in these mean ages are small (Table 1), but the error of the weighted mean age is usually much lower. In addition, data from each pumice clast, after exclusion of outliers, were plotted in an isotope correlation diagram ($^{36}Ar/^{40}Ar$ versus $^{39}Ar/^{40}Ar$), using the York[28] procedure. The derived ages are quite close to the mean ages (Table 1), and the calculated trapped argon composition generally has the atmospheric argon ratio of 295.5, within uncertainty.

(**433**, 733-736; 2005)

**Ian McDougall[1], Francis H. Brown[2] & John G. Fleagle[3]**

[1] Research School of Earth Sciences, Australian National University, Canberra, ACT 0200, Australia

[2] Department of Geology and Geophysics, University of Utah, Salt Lake City, Utah 84112, USA

[3] Department of Anatomical Science, Stony Brook University, Stony Brook, New York 11794, USA

Received 22 September; accepted 8 December 2004; doi:10.1038/nature03258.

References:

1. Day, M. H. Omo human skeletal remains. *Nature* **222**, 1135-1138 (1969).
2. Day, M. H. & Stringer, C. B. in *Congrès International de Paléontologie Humaine I, Nice* Vol. 2, 814-846 (Colloque International du CNRS, 1982).
3. Day, M. H. & Stringer, C. B. Les restes crâniens d'Omo-Kibish et leur classification à l'intérieur du genre *Homo. Anthropologie* **95**, 573-594 (1991).
4. Day, M. H., Twist, M. H. C. & Ward, S. Les vestiges post-crâniens d'Omo I (Kibish). *Anthropologie* **95**, 595-610 (1991).
5. Howell, F. C. in *Evolution of African Mammals* (eds Maglio, V. J. & Cooke, H. B. S.) 154-248 (Harvard Univ. Press, Cambridge, Massachusetts, 1978).
6. Smith, F. H., Falsetti, A. B. & Donnelly, S. M. Modern human origins. *Yb Phys. Anthropol.* **32**, 35-68 (1989).
7. Butzer, K. W., Isaac, G. Ll., Richardson, J. L. & Washbourn-Kamau, C. Radiocarbon dating of East African lake levels. *Science* **175**, 1069-1076 (1972).
8. Fuchs, V. E. The geological history of the Lake Rudolf Basin, Kenya Colony. *Phil. Trans. R. Soc. Lond. B* **229**, 219-274 (1939).
9. Arambourg, C. *Mission Scientifique de l'Omo 1932–1933. Geologie–Anthropologie–Paleontologie* (Muséum National d'Histoire Naturelle, Paris, 1935-1947).
10. Butzer, K. W. The Lower Omo Basin: Geology, fauna and hominids of Plio-Pleistocene formations. *Naturwissenschaften* **58**, 7-16 (1971).

照射之后，将长石晶体加样到一个铜质样品托盘的加样孔中，安装到真空系统后烘烤过夜。使用聚焦氩离子激光束以高达 10 W 的功率使样本熔化。对熔化期间释放出来的气体进行纯化后，使用戴利收集器在 VG3600 质谱仪中对氩进行同位素分析。该系统的综合灵敏度约为 $2.5 \times 10^{-17}$ mol · $mV^{-1}$。质量甄别通过大气中的氩的常规含量进行监控。每个未知的照射参数 $J$ 都是通过在流量监控晶体上进行的测量得到的插入值派生出来的，每个水平至少进行五次分析；通常的精确性范围是 0.3% ~ 0.75%，群体的标准差。在所有计算中使用的钙修正因子[27]都是 $(^{36}Ar/^{37}Ar)_{Ca} = 3.49 \times 10^{-4}$ 和 $(^{39}Ar/^{37}Ar)_{Ca} = 7.86 \times 10^{-4}$。

## 数据处理

在计算每种浮岩碎屑的算术平均年代时，每个样本的最初均值中存在两个以上标准差的任何结果都被反复检查剔除掉了，直到不再出现离群值为止。任何一组分析中，都没有剔除掉两个以上的测量值，而且约有一半的组中都没有剔除掉任何结果。表 1 中引用的误差是群体的标准差，但是由于个体年代的不确定性是不同的，所以给出了加权平均年代和误差，即对变异反向的每个年代进行了加权。这些平均年代的差异很小（表 1），但是加权平均年代的误差通常更低。此外，每种浮岩碎屑的数据在剔除掉离群值之后，都使用 York[28] 程序在同位素相关性表格（$^{36}Ar/^{40}Ar$ 对 $^{39}Ar/^{40}Ar$）中绘图。派生的年代与平均年代非常接近（表 1），而计算出来的捕获氩成分一般具有大气中的氩比值——295.5，属误差之内。

（刘皓芳 翻译；潘雷 审稿）

11. Butzer, K. W. in *Earliest Man and Environments in the Lake Rudolf Basin* (eds Coppens, Y., Howell, F. C., Isaac, G. Ll. & Leakey, R. E. F.) 12-23 (Univ. of Chicago Press, Chicago, 1976).

12. Butzer, K. W. & Thurber, D. L. Some late Cenozoic sedimentary formations of the Lower Omo Basin. *Nature* **222**, 1138-1143 (1969).

13. Butzer, K. W., Brown, F. H. & Thurber, D. L. Horizontal sediments of the lower Omo Valley: the Kibish Formation. *Quaternaria* **11**, 15-29 (1969).

14. Butzer, K. W. Geological interpretation of two Pleistocene hominid sites in the Lower Omo Basin. *Nature* **222**, 1133-1135 (1969).

15. Owen, R. B., Barthelme, J. W., Renaut, R. W. & Vincens, A. Palaeolimnology and archaeology of Holocene deposits north-east of Lake Turkana, Kenya. *Nature* **298**, 523-529 (1982).

16. Fleagle, J. *et al.* The Omo I partial skeleton from the Kibish Formation. *Am. J. Phys. Anthropol. Suppl.* **36**, 95 (2003).

17. Lourens, L. J. *et al.* Evaluation of the Plio-Pleistocene astronomical timescale. *Paleoceanography* **11**, 391-413 (1996).

18. Rossignol-Strick, M., Nesteroff, W., Olive, P. & Vergnaud-Grazzini, C. After the deluge: Mediterranean stagnation and sapropel formation. *Nature* **295**, 05-110 (1982).

19. Rossignol-Strick, M. & Paterne, M. A synthetic pollen record of the eastern Mediterranean sapropels of the last 1 Ma: implications for the time-scale and formation of sapropels. *Mar. Geol.* **153**, 221-237 (1999).

20. Rohling, E. J. *et al.* African monsoon variability during the previous interglacial maximum. *Earth Planet. Sci. Lett.* **202**, 61-75 (2002).

21. Haile-Selassie, Y., Asfaw, B. & White, T. D. Hominid cranial remains from Upper Pleistocene deposits at Aduma, Middle Awash, Ethiopia. *Am. J. Phys. Anthropol.* **123**, 1-10 (2004).

22. Rightmire, G. P. in *The Origins of Modern Humans: A World Survey of the Fossil Evidence* (eds Smith, F. H. & Spencer, F.) 295-325 (Alan R. Liss, New York, 1984).

23. Howell, F. C. in *Origins of Anatomically Modern Humans* (eds Nitecki, M. H. & Nitecki, D. V.) 253-319 (Plenum, New York, 1994).

24. Clark, J. D. *et al.* Stratigraphic, chronological and behavioural contexts of Pleistocene *Homo sapiens* from Middle Awash, Ethiopia. *Nature* **423**, 747-752 (2003).

25. McDougall, I. & Feibel, C. S. Numerical age control for the Miocene–Pliocene succession at Lothagam, a hominoid-bearing sequence in the northern Kenya Rift. *J. Geol. Soc. Lond.* **156**, 731-745 (1999).

26. Spell, T. L. & McDougall, I. Characterization and calibration of $^{40}Ar/^{39}Ar$ dating standards. *Chem. Geol.* **198**, 189-211 (2003).

27. Spell, T. L., McDougall, I. & Doulgeris, A. P. The Cerro Toledo Rhyolite, Jemez Volcanic Field, New Mexico: $^{40}Ar/^{39}Ar$ geochronology of eruptions between two caldera-forming events. *Bull. Geol. Soc. Am.* **108**, 1549-1566 (1996).

28. York, D. Least squares fitting of a straight line with correlated errors. *Earth Planet. Sci. Lett.* **5**, 320-324 (1969).

29. Davidson, A. *The Omo River Project* (Bulletin 2, Ethiopian Institute of Geological Surveys, Addis Ababa, 1983).

**Supplementary Information** accompanies the paper on www.nature.com/nature.

**Acknowledgements.** We thank J. Mya, R. Maier and X. Zhang for technical support for the geochronology; participants in the Kibish expeditions between 1999 and 2003, including Z. Assefa, J. Shea, S. Yirga, J. Trapani and especially C. Feibel, B. Passey and C. Fuller for their geological contributions; and R. Leakey, K. Butzer and especially P. Abell for providing us with information and documents about the 1967 expedition to the Kibish area. We thank the Government of Ethiopia, the Ministry of Youth, Sports and Culture, the Authority for Research and Conservation of Cultural Heritage, and the National Museum of Ethiopia for permission to study the Kibish Formation. Support was provided by the National Science Foundation, the Leakey Foundation, the National Geographic Society and the Australian National University. Neutron irradiations were facilitated by the Australian Institute of Nuclear Science and Engineering and the Australian Nuclear Science and Technology Organization.

**Competing interests statement.** The authors declare that they have no competing financial interests.

**Correspondence** and requests for materials should be addressed to I.McD (ian.mcdougall@anu.edu.au).

# The DNA Sequence of the Human X Chromosome

M. T. Ross *et al.*

## Editor's Note

**The enigmatic X chromosome reveals its genetic secrets in this paper detailing its DNA sequence. Female mammals carry two X chromosomes, whist males carry one X and one Y. Geneticist Mark Ross and colleagues determined the sequence of over 99% of the gene-containing region of the human X chromosome. The data reveal how X and Y chromosomes evolved from a pair of regular chromosomes around 300 million years ago. It also suggests that nearly 10% of the chromosome's 1,098 genes belong to a group that is upregulated in testicular and other cancers, and links a particular type of repetitive sequence with X-inactivation—the process that silences one of the two copies of the X chromosome in females to avoid a double dose of X chromosome genes.**

---

The human X chromosome has a unique biology that was shaped by its evolution as the sex chromosome shared by males and females. We have determined 99.3% of the euchromatic sequence of the X chromosome. Our analysis illustrates the autosomal origin of the mammalian sex chromosomes, the stepwise process that led to the progressive loss of recombination between X and Y, and the extent of subsequent degradation of the Y chromosome. LINE1 repeat elements cover one-third of the X chromosome, with a distribution that is consistent with their proposed role as way stations in the process of X-chromosome inactivation. We found 1,098 genes in the sequence, of which 99 encode proteins expressed in testis and in various tumour types. A disproportionately high number of mendelian diseases are documented for the X chromosome. Of this number, 168 have been explained by mutations in 113 X-linked genes, which in many cases were characterized with the aid of the DNA sequence.

---

THE X chromosome has many features that are unique in the human genome. Females inherit an X chromosome from each parent, but males inherit a single, maternal X chromosome. Gene expression on one of the female X chromosomes is silenced early in development by the process of X-chromosome inactivation (XCI), and this chromosome remains inactive in somatic tissues thereafter. In the female germ line, the inactive chromosome is reactivated and undergoes meiotic recombination with the second X chromosome. The male X chromosome fails to recombine along virtually its entire length during meiosis: instead, recombination is restricted to short regions at the tips of the X chromosome arms that recombine with equivalent segments on the Y chromosome.

# 人类 X 染色体的 DNA 序列

罗斯等

编者按

这篇论文通过详细介绍神秘的 X 染色体的 DNA 序列，揭示了它的遗传秘密。雌性哺乳动物有两条 X 染色体，雄性哺乳动物有一条 X 染色体和一条 Y 染色体。遗传学家马克·罗斯和他的同事们确定了人类 X 染色体中 99% 以上基因区域的序列，这些数据揭示了大约 3 亿年前 X 和 Y 染色体是如何从一对普通染色体进化而来的。它也表明染色体的 1,098 个基因的近 10% 属于一个群体，该群体调节睾丸和其他癌症；并且将一个特定类型的重复序列与 X 染色体失活联系起来，这个过程使得女性当中一个 X 染色体的两个副本之一沉默，来避免 X 染色体基因剂量加倍。

---

人类 X 染色体有一种独特的生物学特性，这种特性是由其进化形成的，即男性和女性共有的性染色体。我们已经确定了 X 染色体上 99.3%的常染色质序列。我们的分析表明了哺乳动物性染色体起源于常染色体，X 和 Y 之间重组逐渐丢失的过程，以及 Y 染色体随后的退化程度。LINE1 重复元件占 X 染色体的三分之一，其分布与它们在 X 染色体失活过程中作为中转站的作用一致。我们在这个序列中发现了 1,098 个基因，其中有 99 个编码在睾丸和各种肿瘤类型中表达的蛋白质。X 染色体记录了大量不成比例的孟德尔病。在这一数字中，有 168 个可以用 113 个 X 连锁基因的突变解释，这些突变在许多情况下是借助 DNA 序列来表征的。

---

X 染色体上有许多人类基因组中独特的特征。女性从父母双方各继承一条 X 染色体，但男性只继承一条母亲的 X 染色体。其中一条女性 X 染色体的基因表达在发育早期被 X 染色体失活（XCI）过程所抑制，此后该染色体在身体组织中仍旧保持失活。在女性生殖系中，失活的染色体被重新激活，经过减数分裂后与第二条 X 染色体进行重组。在减数分裂过程中，男性的 X 染色体不能沿着其整个长度进行重组，相反，重组仅限于与 Y 染色体上的等值段进行重组的 X 染色体臂端的短区域。这些区域内的基因在性染色体间共享，因此它们的行为被称为“假常染色体”。X 染色体

Genes inside these regions are shared between the sex chromosomes, and their behaviour is therefore described as "pseudoautosomal". Genes outside these regions of the X chromosome are strictly X-linked, and the vast majority are present in a single copy in the male genome.

The unique properties of the X chromosome are a consequence of the evolution of sex chromosomes in mammals. The sex chromosomes have evolved from a pair of autosomes within the last 300 million years (Myr)[1]. In the process, the original, functional elements have been conserved on the X chromosome, but the Y chromosome has lost almost all traces of the ancestral autosome, including the genes that were once shared with the X chromosome. The hemizygosity of males for almost all X chromosome genes exposes recessive phenotypes, thus accounting for the large number of diseases that have been associated with the X chromosome[2]. The characteristic pattern of X-linked inheritance (affected males and no male-to-male transmission) was recognized by the eighteenth century for some cases of haemophilia, and gave impetus in the 1980s to the earliest successes in positional cloning—of the genes for chronic granulomatous disease[3] and Duchenne muscular dystrophy[4]. For females, the major consequence of the loss of genes from the Y chromosome is XCI, which equalizes the dosage of X-linked gene products between the sexes.

The biological consequences of sex chromosome evolution account for the intense interest in the human X chromosome in recent decades. However, evolutionary processes are likely to have shaped the behaviour and structure of the X chromosome in many other ways, influencing features such as repeat content, mutation rate, gene content and haplotype structure. The availability of the finished sequence of the human X chromosome, described here, now allows us to explore its evolution and unique properties at a new level.

## The X Chromosome Sequence

We constructed a map of the X chromosome using predominantly P1-artificial chromosome (PAC) and bacterial artificial chromosome (BAC) clones (Supplementary Table 1), which were assembled into contigs using restriction-enzyme fingerprinting and integrated with earlier maps using sequence-tagged site (STS) content analysis[5]. Gaps were closed by targeted screening of clone libraries in bacteria or yeast, and by assessing BAC and fosmid end-sequence data for evidence of spanning clones. Fourteen euchromatic gaps remain intractable, despite using libraries with a combined 80-fold chromosome coverage. Five of these gaps are within the 2.7 megabase (Mb) pseudoautosomal region at the tip of the chromosome short arm (PAR1). This is reminiscent of the situation in other human subtelomeric regions[6], and might reflect cloning difficulties in an area with a high content of (G+C) nucleotides and minisatellite repeats.

We selected 1,832 clones from the map for shotgun sequencing and directed finishing using established procedures[7]. Finished sequences were estimated to be more than 99.99% accurate by independent assessment[8]. The sequence of the X chromosome has been

上这些区域外的基因是严格的 X 连锁的，并且绝大多数都存在于男性基因组的单拷贝中。

X 染色体独特的特性是哺乳动物性染色体进化的结果。性染色体在过去的 3 亿年内从一对常染色体进化而来[1]。在这个过程中，原有的功能元件保留在 X 染色体上，而 Y 染色体已经失去了几乎所有祖先常染色体的痕迹，包括曾经与 X 染色体共享的基因。男性的半杂合性使几乎所有 X 染色体上的基因暴露隐性表型，从而产生大量 X 染色体连锁的疾病[2]。X 连锁遗传的特征模式（影响男性，没有男性之间的传递）因十八世纪血友病的一些案例而被识别出，并推动了 20 世纪 80 年代定位克隆的初步成功——对慢性肉芽肿病[3]和进行性假肥大性肌营养不良[4]的克隆。对女性来说，Y 染色体基因丢失的主要后果是 XCI，它可以使 X 连锁基因产物在两性之间的剂量均衡。

性染色体进化的生物学影响解释了近几十年人们为什么对人类 X 染色体有如此强烈的兴趣。然而，进化过程可能以许多其他方式塑造了 X 染色体的行为和结构，影响如重复序列的含量、突变率、基因含量和单体型结构等特征。这里所描述的已经完成的人类 X 染色体序列让我们在一个新的水平上探索其进化和特性。

## X 染色体序列

我们主要通过限制性内切酶指纹法将 P1 人工染色体（PAC）和细菌人工染色体（BAC）克隆（补充表 1）组装成重叠群，并使用序列标签位点（STS）含量分析与早先图谱整合，构建了一个 X 染色体图谱[5]。通过有针对性的筛选细菌或酵母菌克隆文库，并通过评估 BAC 和 F 黏粒末端序列数据来寻找跨域克隆的证据，以封闭缺口。即使使用了染色体覆盖率达 80 倍的文库，仍有 14 个常染色质缺口无法解决。其中五个缺口在染色体短臂末端 2.7 Mb 的假常染色体区域（PAR1）。这使人想起在人类其他亚端粒区域的情况[6]，并可能反映在（G+C）核苷酸和小卫星重复含量高的区域的克隆难度。

我们从图谱中选择 1,832 个克隆进行鸟枪法测序并用已建立的程序进行定向指导[7]。通过独立评价，完成的序列准确率超过 99.99%[8]。该 X 染色体的序列是

assembled from the individual clone sequences and comprises 16 contigs. These extend into the telomeric (TTAGGG)*n* repeat arrays at the ends of the chromosome arms, and include both pseudoautosomal regions (PARs). The data were frozen for the analyses described below, at which point we had determined 150,396,262 base pairs (bp) of sequence (Supplementary Table 2). Subsequently, we obtained a further 609,664 bp of sequence. The 14 euchromatic gaps are estimated to have a combined size of less than 1 Mb (see Methods and Supplementary Table 2), and the sequence therefore covers at least 99.3% of the X chromosome euchromatin. There is also a single heterochromatic gap corresponding to the polymorphic 3.0 ($\pm$0.4) Mb array[9] of alpha satellite DNA at the centromere. On this basis, we conclude that the X chromosome is approximately 155 Mb in length.

The coverage and quality of the finished sequence have been assessed using independent data. All markers from the deCODE genetic map[10] are found in the sequence and the concordance of marker orders is excellent with only one discrepancy. DXS6807 is the most distal Xp marker on the deCODE map (4.39 cM), but in the sequence this marker is proximal to three others with genetic locations of 9–11 cM on the deCODE map. Out of 788 X chromosomal RefSeq[11] messenger RNAs that were assessed, 783 were found completely in the sequence, and parts of four others are also present (T. Furey, personal communication). The missing segments of *GTPBP6*, *CRLF2*, *DHRSX* and *FGF16* lie within gaps 1, 4, 5 and 10, respectively, and the *GAGE3* gene is in gap 7 (Supplementary Table 2). The sequence assembly was assessed using fosmid end-sequence pairs that match the X chromosome sequence. The orientation and separation of end-pairs of more than 17,000 fosmids were consistent with the sequence assembly. In two cases, sequences had been misassembled owing to long and highly similar repeats. There were six instances of large deletions in sequenced clones, which were resolved by determining fosmid sequences through the deleted regions. Finally, there were two cases of apparent length variation between the reference sequence and the DNA used for the fosmid library.

## Features of the X Chromosome Sequence

The annotated sequence of the X chromosome is presented in Supplementary Fig. 1, and updates are contained in the Vertebrate Genome Annotation (VEGA) database (http://vega.sanger.ac.uk/Homo_sapiens/). The distribution of a number of sequence features on the chromosome is shown in Fig. 1 (Editorial note: this figure was originally included as a fold-out insert. For details, see *Nature*'s web site). Analysis of the sequence reveals a gene-poor chromosome that is highly enriched in interspersed repeats and has a low (G+C) content (39%) compared with the genome average (41%).

Fig. 1. Features of the X chromosome sequence. **a**, X chromosome ideogram according to Francke[65]. **b**, Evolutionary domains of the X chromosome: the X-added region (XAR), the X-conserved region (XCR; dotted region in proximal Xp does not appear to be part of the XCR), the pseudoautosomal region PAR1, and evolutionary strata S5–S1. **c**, Sequence scale in intervals of 1 Mb. Note that correlation between cytogenetic band positions and physical distance is imprecise, owing to varying levels of condensation of different Giemsa bands. **d**, (G+C) content of 100-kb sequence windows. **e**, Number of genes in 1-Mb

由单个克隆序列组装而成，包括 16 个重叠群。它们延伸到染色体臂末端的端粒(TTAGGG)*n* 重复序列，并包括两个假常染色体区域(PAR)。这些数据被冻结用于下述分析，在此，我们已确定 150,396,262 bp 的序列(补充表 2)。随后，我们进一步获得了长度为 609,664 bp 的序列。估计 14 个常染色质缺口合并大小小于 1 Mb(见方法和补充表 2)，因此，该序列至少覆盖 99.3%的 X 染色体常染色质。还有一个异染色质缺口对应着丝粒上 α 卫星 DNA 的多态性 3.0(±0.4) Mb 阵列[9]。在此基础上，我们得出结论：X 染色体长度大约 155 Mb。

我们用独立的数据对已经完成序列的覆盖度和质量进行了评估，在序列中发现了 deCODE 遗传图谱[10]的所有标记，并且标记顺序一致性很高，只有一个除外。DXS6807 是 deCODE 图上最远端 Xp 标记(4.39 cM)，但在序列中这个标记与其他三个在 deCODE 图上遗传位置距离 9～11 cM 的标记最接近。我们评估了 788 个 X 染色体的信使 RNA 的参考序列[11]，其中 783 个在序列中被完全发现，其他 4 个也部分出现在序列中(富里，个人交流)。*GTPBP6*、*CRLF2*、*DHRSX* 和 *FGF16* 丢失的片段分别位于缺口 1、4、5 和 10，*GAGE3* 基因在缺口 7 中(补充表 2)。序列组装使用匹配 X 染色体序列的 F 黏粒末端序列对进行评估。超过 1.7 万 F 黏粒的末端对方向和间距与序列组装一致。序列长度长且高度相似的重复序列导致有两例组装错误。在测序的克隆中有六个大的缺失实例，这些实例已通过确定缺失区域的 F 黏粒序列来解决。最后，有两例在参考序列和用于 F 黏粒文库的 DNA 之间有明显的长度差异。

## X 染色体序列的特征

X 染色体的注解序列见补充图 1，在脊椎动物基因组注释(VEGA)数据库(http://vega.sanger.ac.uk/Homo_sapiens/)里进行更新。此染色体上若干序列特征的分布如图 1(编者注：在原文中，本图是一个折叠插入图，更多细节请看《自然》官网)所示。序列分析揭示了一个基因贫乏的染色体，其高度富集散在重复序列，与基因组的平均水平(41%)相比(G+C)含量(39%)较低。

图 1. X 染色体序列特征。**a**，根据弗兰克的 X 染色体表意图[65]。**b**，X 染色体进化域：X 增加区(XAR)，X 保守区(XCR；Xp 近端虚线区域不属于 XCR)，假常染色体区域 PAR1 和进化层 S5～S1。**c**，序列刻度间隔为 1 Mb。请注意，由于不同吉姆萨带缩合程度不同，细胞遗传学带位置和物理距离间的相关性并不精确，**d**，100 kb 序列窗口的(G+C)含量。**e**，1 Mb 序列窗口中的基因数量(不包括假基因)。**f**～**k**，100 kb 窗口的序列被散在重复序列部分覆盖：L1 重复(**f**)，L1 重复亚族 L1M(**g**)、L1 重复

sequence windows (pseudogenes not included). **f–k**, Fractional coverage of 100-kb sequence windows by interspersed repeats: L1 repeats (**f**), L1M subfamilies of L1 repeats (**g**), L1P subfamilies of L1 repeats (**h**), *Alu* repeats (**i**), L2 repeats (**j**), MIR repeats (**k**). Vertical grey lines in **d–k** represent gaps in the euchromatic sequence of the chromosome. Grey bar centred at approximately 60 Mb shows the position of the centromere. **l**, A selection of landmark genes on the chromosome. *OPN* refers to the three opsin genes in the reference sequence, which are organized as follows: cen-*OPN1LW-OPN1MW-OPN1MW*-tel. **m**, Genes that escape from X-chromosome inactivation as previously identified[48]. **n**, Cancer-testis antigen genes, belonging to the *MAGE* (light green), *GAGE* (dark green), *SSX* (magenta), *SPANX* (orange) or other (grey) CT gene families. For the genes in **l–n**, arrows indicate the direction of transcription.

## Genes

Based on a manual assessment of all publicly available human expressed sequences and genes from other organisms, we have annotated 1,098 genes (7.1 genes per Mb) across four different categories (see Methods): known genes (699), novel coding sequences (132), novel transcripts (166), and putative transcripts (101). We have also identified 700 pseudogenes in the sequence (4.6 pseudogenes per Mb), of which 644 are classified as processed and 56 as non-processed. The gene density (excluding pseudogenes) on the X chromosome is among the lowest for the chromosomes that have been annotated to date. This might simply reflect a low gene density on the ancestral autosomes. Alternatively, selection may have favoured transposition of particular classes of gene from the X chromosome to the autosomes during mammalian evolution. These could include developmental genes for which the protein products are required in double dose in males (or in females after XCI has occurred), or genes for which mutation in male somatic tissues is lethal.

Physical characteristics of the genes and pseudogenes are summarized in Supplementary Table 3. Exons of the 1,098 genes account for only 1.7% of the X chromosome sequence. On the basis of the lengths of these gene loci, 33% of the chromosome is transcribed. This is considerably below the recent estimates for chromosomes 6 (ref. 12), 9 (ref. 6), 10 (ref. 13) and 13 (ref. 14), to which the equivalent gene annotation procedure was applied (Supplementary Table 4), and is a reflection not just of low gene density on chromosome X but also of low gene length. For example, mean gene length is 49 kilobases (kb) on chromosome X compared with 57 kb on chromosome 13. Nevertheless, the X chromosome contains the largest known gene in the human genome, the dystrophin (*DMD*) locus in Xp21.1, which spans 2,220,223 bp. Consistent with its low gene density, the frequency of predicted CpG islands on the X chromosome is only 5.25 per Mb, which is exactly half of the estimated genome average[7]. There is an association with a CpG island for 49% of the known genes, the category for which the most complete gene structures are expected in the current annotation.

We identified evolutionarily conserved regions (ECRs) by comparing the X chromosome sequence to the genomes of mouse, rat, zebrafish and the pufferfishes *Tetraodon nigroviridis* and *Fugu rubripes* (Supplementary Table 5). There are 4,493 ECRs that are conserved between the X chromosome and all of the other species. Of these, 4,393 overlap with

亚族 L1P(**h**)、*Alu* 重复(**i**)、L2 重复(**j**)、MIR 重复(**k**)。**d** ~ **k** 中的垂直灰线代表染色体常染色质序列中的缺口。集中在大约 60 Mb 处的灰色条块代表着丝粒的位置。**l**，染色体上标志性基因的选择。*OPN* 是指参考序列中的三个视蛋白基因，排列如下：着丝粒 -*OPN1LW-OPN1MW-OPN1MW*- 端粒。**m**，如前所述，从 X 染色体失活中逃逸的基因 [48]。**n**，癌/睾丸抗原基因，属于 *MAGE*(浅绿色)、*GAGE*(深绿色)、*SSX*(品红色)、*SPANX*(橙色)或其他(灰色)CT 基因家族。对于 **l** ~ **n** 中的基因，箭头指示转录的方向。

## 基　因

在人工评估所有已公布的人类表达序列和其他生物基因的基础上，我们已注释了四个不同类别的 1,098 个基因(每 Mb 有 7.1 个基因)(见方法)：已知基因(699)、新的编码序列(132)、新转录本(166)、假定转录本(101)。我们还在序列中确定了 700 个假基因(每 Mb 有 4.6 个假基因)，其中 644 个为已加工的基因，56 个为未加工的基因。X 染色体上的基因密度(不包括假基因)是迄今为止已注释的染色体中最低的。这可能只是反映了祖先染色体上的低基因密度。或者，在哺乳动物进化中选择可能有利于从 X 染色体向常染色转移某些类别的基因。这些基因可能包括男性(或 XCI 发生后的女性)需要双倍剂量的蛋白质产物发育基因，或男性体细胞组织中突变致死的基因。

补充表 3 总结了这些基因和假基因的物理特性。1,098 个基因的外显子只占 X 染色体序列的 1.7%。根据这些基因座的长度，有 33%的染色体被转录。这大大低于最近采用等效的基因注释程序(补充表 4)对染色体 6(参考文献 12)、9(参考文献 6)、10(参考文献 13)和 13(参考文献 14)的估计，不仅是 X 染色体上低基因密度的反映，同时也是低基因长度的反映。例如，X 染色体上平均基因长度为 49 千碱基(kb)，相比之下 13 号染色体上的为 57 kb。然而，X 染色体包含人类基因组中已知最大的基因——在 Xp21.1 上的肌萎缩蛋白基因座 *DMD*，它跨越 2,220,223 个碱基对。与其低基因密度相符的是，预测 X 染色体上每 Mb 只有 5.25 个 CpG 岛，这正好是基因组平均估计值的一半 [7]。在已知基因中有 49% 与 CpG 岛相关，预计在目前的注释中这类基因结构最完整。

我们通过将 X 染色体序列与小鼠、大鼠、斑马鱼、黑青斑河鲀和红鳍东方鲀的基因组序列进行比较，确定了进化上的保守区域(ECR)(补充表 5)。在 X 染色体和所有其他物种间有 4,493 ECR 是保守的。其中，4,393 个与 4,373 个已注释的外显子

4,373 annotated exons. The remaining 100 ECRs are most likely to be unannotated exons, although some could be highly conserved control or structural elements. From these data we conclude that we have annotated at least 97.8% of the protein-coding exons on the X chromosome ([4,373/(4,373+100)] × 100).

## Non-coding RNA Genes

The gene set described above includes non-coding RNA (ncRNA) genes only when there is supporting evidence of expression from complementary DNA or expressed-sequence-tag (EST) sources. Using a complementary approach, we analysed the X chromosome sequence using the Rfam[15] database of structural RNA alignments, and predicted 173 ncRNA genes and/or pseudogenes (Supplementary Fig. 1 and Supplementary Table 6). These are physically separate from the genes described in the preceding section and are not included in the total gene count, owing to the difficulty in discriminating between genes and pseudogenes for these ncRNA predictions. Using tRNAscan-SE[16], we predicted only two transfer RNA genes on the X chromosome (Supplementary Table 6), out of the several hundred predicted in the human genome[7]. Thirteen microRNAs from the microRNA registry[17] have also been mapped onto the sequence (Supplementary Table 7).

The most prominent of the ncRNA genes on the X chromosome is *XIST* (X (inactive)-specific transcript)[18], which is critical for XCI. The *XIST* locus spans 32,103 bp in Xq13, and its untranslated transcript coats and transcriptionally silences one X chromosome in *cis*. The RefSeq[11] transcript of *XIST* is an RNA of 19,275 bases, which includes the largest exon on the chromosome (exon 1: 11,372 bp). There is also evidence for shorter *XIST* transcripts generated by alternative splicing, particularly in the 3′ region of the gene[19]. In the mouse, *Tsix* is antisense to *Xist*[20], and its transcript (or the process of its transcription) is believed to repress the accumulation of *Xist* RNA. There is evidence for transcription antisense to *XIST* in human[21,22], but we have been unable to annotate the human *TSIX* gene as there are no corresponding expressed sequences in the public databases, and because there is a lack of primary sequence conservation between the human and mouse regions. In the human sequence, two other ncRNA genes are annotated in the 400 kb region distal to *XIST*, which are orthologues of the mouse genes described previously as *Jpx* and *Ftx* (ref. 23). In the mouse, *Xist*, *Jpx* and *Ftx* are located within a smaller area of approximately 200 kb[23].

## The Cancer-testis Antigen Genes

On assessing the predicted proteome of the X chromosome for Pfam[24] domains, our most prominent finding was the presence of the MAGE domain (IPR002190) in 32 genes (Supplementary Table 8). In comparison, only four other MAGE genes are reported in the rest of the genome: *MAGEF1* on chromosome 3, and *MAGEL2*, *NDN* and *NDNL2* on chromosome 15. The *MAGE* gene products are members of the cancer-testis (CT) antigen group, which are characterized by their expression in a number of cancer types, while their

重叠。其余 100 个 ECR 最可能是未注释的外显子，尽管有一些可能是高度保守的控制元件或结构元件。从这些数据中我们得出结论：我们至少已经注释了 X 染色体 97.8%的编码蛋白质的外显子（$[4,373/(4,373+100)]\times 100$）。

## 非编码 RNA 基因

上文所述基因集合只有存在互补 DNA 表达或表达序列标签（EST）表达的支持性证据时才包含非编码 RNA（ncRNA）基因。利用互补性方法，我们用 Rfam[15] 数据库的结构 RNA 比对分析了 X 染色体序列，并预测了 173 个 ncRNA 基因和（或）假基因（补充图 1 和补充表 6）。由于这些 ncRNA 预测难以区分基因和假基因，因此它们与前面章节所描述的基因有实质区别，不包括在基因总数中。使用 tRNAscan-SE[16]，我们预测 X 染色体上只有两个转移 RNA 基因（补充表 6），而在人类基因组中预测有几百个[7]。13 个来自微 RNA 注册表的微 RNA[17] 也绘制在了序列中（补充表 7）。

X 染色体上最明显的 ncRNA 基因是 *XIST*（X（失活）特异转录本）[18]，它对 X 染色体失活至关重要。*XIST* 基因座在 Xq13 中跨越 32,103 bp，其未翻译的转录本覆盖并转录沉默一条 X 染色体。*XIST* 的 RefSeq[11] 转录本是一个包含 19,275 个碱基的 RNA，它包括此染色体上最大的外显子（外显子 1：11,372 bp）。还有证据表明较短的 *XIST* 转录本通过可变剪接产生，特别是在基因 3′ 区[19]。在小鼠中，*Tsix* 与 *Xist*[20] 是反义的，其转录本（或转录过程）被认为抑制 *Xist* RNA 的积累。有证据证明转录本与人类 *XIST* 是反义的[21,22]，由于在公共数据库中没有相应的表达序列，且人类和小鼠序列区域缺乏基本的保守性，因此我们无法对人类 *TSIX* 基因进行注释。在人类序列中，其他两个注释的 ncRNA 基因在 *XIST* 远端 400 kb 区域，它们是之前描述的小鼠 *Jpx* 和 *Ftx* 基因的直系同源基因（参考文献 23）。在小鼠中，*Xist*、*Jpx* 和 *Ftx* 都位于一个大约 200 kb 的小区域内[23]。

## 癌/睾丸抗原基因

在评估预测 X 染色体蛋白质组的 Pfam[24] 结构域时，我们最突出的发现是在 32 个基因中存在 MAGE 抗原结构域（IPR002190）（补充表 8）。相比之下，在剩余基因组中只报道了另外四个 MAGE 基因：3 号染色体上的 *MAGEF1* 和 15 号染色体上的 *MAGEL2*、*NDN* 和 *NDNL2*。*MAGE* 基因产物是癌/睾丸（CT）抗原组的成员，其特点是在大量癌症类型中表达，而它们在正常组织中的表达仅在或主要是在睾丸中。这

expression in normal tissues is solely or predominantly in testis. This expression profile has led to the suggestion that the CT antigens are potential targets for tumour immunotherapy. A recent report listed 84 CT antigen genes for the human genome[25]. The X chromosome gene set we describe above contains 99 CT antigen genes and includes novel members of the *MAGE*, *GAGE*, *SSX*, *LAGE*, *CSAGE* and *NXF* families (Supplementary Table 9). Assessment of the most recent RefSeq[11] information shows that this set does not include two known *MAGE* genes (*MAGEA5* and *MAGEA7*) and seven *GAGE* genes (*GAGE3–7*, *7B* and *8*), which are expected to lie in gaps 14 and 7, respectively (Supplementary Table 2). Furthermore, gaps 6 and 9 are also within regions of CT antigen gene duplication. Therefore, we predict that approximately 10% of the genes on the X chromosome are of the CT antigen type.

Conclusive data on the normal functions of the CT antigens, or their involvement in disease conditions, are very limited. However, the remarkable enrichment for CT antigen genes on the X chromosome relative to the rest of the genome might be indicative of a male advantage associated with these genes. Recessive alleles that are beneficial to males are expected to become fixed more rapidly on the X chromosome than on an autosome[26]. If these alleles are detrimental to females, their expression could become restricted to male tissues as they rise to fixation. Both the concentration of the CT antigen genes on the X chromosome and their expression profiles are consistent with this model of male benefit. The CT antigen genes on the X chromosome are also notable for the expansion of various gene families by duplication. This degree of duplication is perhaps an indication of selection in males for increased copy number. In this context, it is of interest that the *MAGE* family has independently expanded on the X chromosome in both the human and mouse lineages[27].

## Repetitive Sequences

Interspersed repeats account for 56% of the euchromatic X chromosome sequence, compared with a genome average of 45% (Supplementary Table 10). Within this, the *Alu* family of short interspersed nuclear elements (SINEs) is below average, in keeping with the gene-poor nature of the chromosome. Conversely, long terminal repeat (LTR) retroposon coverage is above average; but the most remarkable enrichment is for long interspersed nuclear elements (LINEs) of the L1 family, which account for 29% of the X chromosome sequence compared to a genome average of only 17%. The possible significance of this enrichment for XCI is discussed later.

Applying the criterion of at least 90% sequence identity over at least 5 kb (ref. 28), we estimate that intrachromosomal segmental duplications account for 2.59% of the X chromosome (Supplementary Table 11 and Supplementary Fig. 2). In contrast, interchromosomal segmental duplications indicated by sequence matches to the autosomes account for a very small fraction (0.24%) of the X chromosome (Supplementary Table 12). Six gaps in the X chromosome map are either flanked by or contained within

种基因表达谱表明 CT 抗原是肿瘤免疫治疗的潜在靶标。最近的一份报告列出了人类基因组中 84 个 CT 抗原基因[25]。上述 X 染色体基因集包含 99 个 CT 抗原基因，包括 *MAGE*、*GAGE*、*SSX*、*LAGE*、*CSAGE* 和 *NXF* 家族的新成员（补充表 9）。对最新 RefSeq[11] 信息的评估表明，这个集合不包括两个已知 *MAGE* 基因（*MAGEA5* 和 *MAGEA7*）和 7 个 *GAGE* 基因（*GAGE3*～*7*，*7B* 和 *8*），预计分别位于缺口 14 和 7（补充表 2）。此外，缺口 6 和 9 也在 CT 抗原基因复制区域内。因此，我们预测，X 染色体上大约 10%的基因是 CT 抗原类型。

关于 CT 抗原的正常功能，或它们在疾病中的作用的结论性数据非常有限。然而，相比于基因组的其他部分，X 染色体上的 CT 抗原基因显著富集可能表明男性优势与这些基因相关。预计对男性有益的隐性等位基因在 X 染色体上比在常染色体上固定得更快[26]。如果这些等位基因对女性有害，当它们上升到固定状态时表达可能会局限于男性组织。CT 抗原基因在 X 染色体上的浓度和它们的基因表达谱都符合男性受益模式。X 染色体上的 CT 抗原基因也通过复制显著扩大了各种基因家族。复制的程度可能是男性选择增加拷贝数量的标志。在此背景下，有意思的是，在人类和小鼠的谱系中，*MAGE* 家族都独立地扩展到了 X 染色体上[27]。

## 重复序列

散在重复序列占 X 染色体常染色质序列的 56%，而基因组的平均水平为 45%（补充表 10）。在这个范围内，*Alu* 家族短散在重复序列（SINE）低于平均水平，这与染色体的基因贫乏性质一致。相反，长末端重复序列（LTR）逆转座子覆盖率高于平均水平；L1 家族长散在重复序列（LINE）很多，占 X 染色体序列的 29%，而基因组平均水平只有 17%。稍后将讨论富含 XCI 可能的意义。

应用在至少 5 kb 上至少 90%以上序列同源这一标准（参考文献 28），我们估计，染色体内片段性重复占 X 染色体的 2.59%（补充表 11 和补充图 2）。与此相反，染色体间片段重复的序列与常染色体的匹配只占 X 染色体很小一部分（0.24%）（补充表 12）。X 染色体图上六个缺口或在染色体内重复片段的两侧或包含在其中（见补充表

intrachromosomally-duplicated segments (gaps 2, 3, 6, 7, 9 and 14 in Supplementary Table 2), which might produce instability of clones or otherwise confound mapping progress. The intrachromosomal duplicates are striking in their proximity. Apart from the two segments containing *SSX* gene copies, which are separated by 4.5 Mb, only six of 229 matches are separated by more than 1 Mb. Among these duplications are well-described cases that are associated with genomic disorders[29]. In Xp22.32, deletions of the steroid sulphatase (*STS*) gene, causing X-linked ichthyosis (Online Mendelian Inheritance in Man (OMIM)[2] entry number 308100), result from recombination between flanking duplications that contain copies of the *VCX* gene. Also, some instances of Hunter syndrome (OMIM 309900), red-green colour blindness (OMIM 303800), Emery-Dreifuss muscular dystrophy (OMIM 310300), incontinentia pigmenti (OMIM 308300) and haemophilia A (OMIM 306700) result from rearrangements involving duplicated sequences in Xq28. In haemophilia A, mutations are frequently the result of inversions between a sequence in intron 22 of the *F8* gene and one of two more distally located copies. A novel finding from our analysis of the X chromosome reference sequence is that the two distal copies are in opposite orientations. Therefore, a large deletion involving *F8* and several more distal genes could be an alternative to the inversion rearrangement. A deletion consistent with this prediction has been reported in a family in which carrier females are affected by a high spontaneous-abortion rate in pregnancy[30].

## The X Chromosome Centromere

The X chromosome sequence extends from both arms into centromeric, higher-order repeat sequences, which are known to be associated functionally with the X centromere[31-33]. The most proximal 494 kb and 360 kb of the Xp and Xq sequences, respectively, consist of extensive regions of satellite DNA, adjacent to euchromatin of the chromosome arms that is exceptionally high in L1 content (Fig. 2). The satellite region on Xp contains small amounts of other satellite families[31], whereas that on Xq consists entirely of alpha satellite. Similar to all other human chromosome arms that have been examined[33,34], these transition regions consist of monomeric alpha satellite that is not associated with centromere function. Both the Xp and Xq contigs reported here, though, extend more proximally and reach into highly homogeneous, higher-order repeat alpha satellite (DXZ1). Critically, the Xp and Xq contig copies of the DXZ1 repeat are themselves 98–100% identical in sequence, and are oriented in the same direction along the chromosome (Fig. 2). On this basis, the two contigs reach the "end" of each chromosome arm and thus also reach the centromeric locus from either side. This represents a logical endpoint for efforts to complete the sequence of chromosome arms in the human genome, and the first demonstration of this endpoint is provided by the X chromosome sequence.

2 中的缺口 2、3、6、7、9 和 14)，这可能会导致克隆的不稳定或混淆绘图过程。染色体内的重复惊人的接近。除了两个含有 *SSX* 基因拷贝的片段被 4.5 Mb 分开，229 个匹配中只有 6 个匹配之间的距离大于 1 Mb。在这些重复中有与基因组无序相联系的并且充分描述的病例 [29]。在 Xp22.32 中，类固醇硫酸酯酶(*STS*)基因缺失导致了 X 连锁鱼鳞病(在线人类孟德尔遗传数据库(OMIM)[2] 索引号 308100)，这是由包含 *VCX* 基因拷贝的侧翼重复间的重组造成的。还有一些列子：亨特综合征(OMIM 309900)，红绿色盲(OMIM 303800)，埃德二氏肌营养不良症(OMIM 310300)，色素失调症(OMIM 308300)和血友病 A(OMIM 306700)都是 Xq28 重复序列重排的结果。在血友病 A 中，突变通常是在 *F8* 基因内含子 22 中的序列和两个以上远端拷贝中的一个之间倒位的结果。从对 X 染色体参考序列的分析中我们有了一个新发现：这两个远端的拷贝方向相反。因此，涉及 *F8* 和多个远端基因的大片段缺失可能是反向重排的替代方法。符合这一预测的缺失已在一个妊娠期有高自然流产率的女性携带者家族中报道过 [30]。

## X 染色体着丝粒

X 染色体序列从两个臂向着丝粒延伸，分布了许多功能上与 X 着丝粒有联系的高度有序的重复序列 [31-33]。Xp 序列近端的 494 kb 和 Xq 序列近端的 360 kb，由广泛的卫星 DNA 区域组成，毗邻 L1 含量非常高的由常染色质组成的染色体臂(见图 2)。Xp 上卫星区域包含少量其他卫星家族 [31]，而 Xq 上的卫星区域包含全部的 α 卫星。与所有其他已检测的人类染色体臂相似 [33,34]，这些过渡区域包括与着丝粒功能无关的单体 α 卫星。但这里报道的 Xp 和 Xq 重叠群以更接近的方式延伸并成为了高度均匀的、更高阶重复的 α 卫星(DXZ1)。更为重要的是，DXZ1 重复的 Xp 和 Xq 重叠序列本身序列一致性达 98% ~ 100%，并且沿染色体的方向相同(见图 2)。在此基础上，这两个重叠群到达各自染色体臂的“末端”，从而也从两侧到达着丝粒基因座。这代表了努力完成人类基因组染色体臂序列的一个合乎逻辑的终点，而这个终点是由 X 染色体序列证实的。

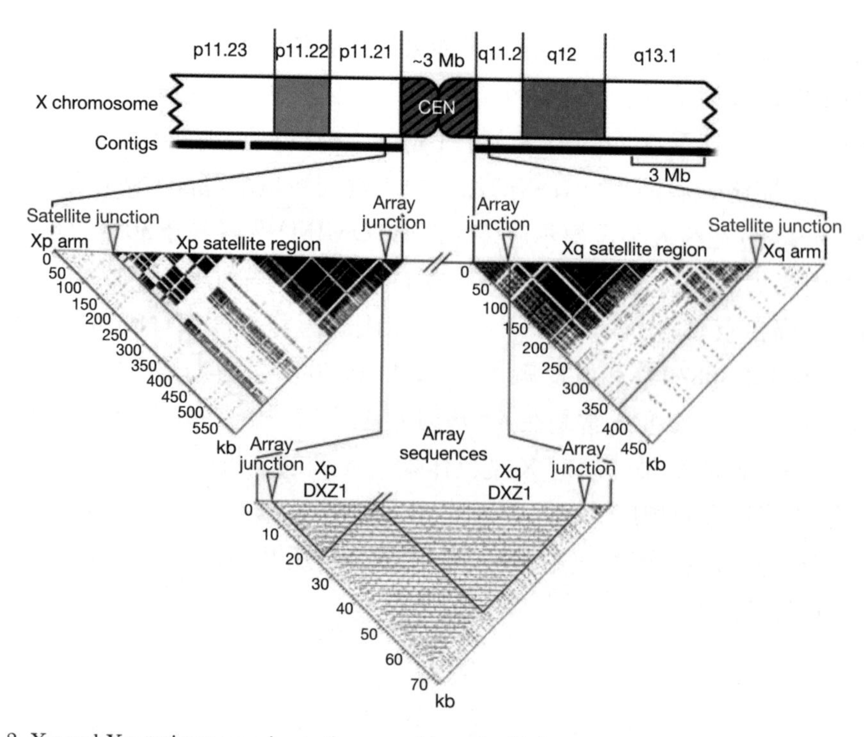


Fig. 2. Xp and Xq pericentromeric contigs extend into the X-chromosome-specific higher-order alpha satellite, DXZ1. The pericentromeric region of the X chromosome is shown as a truncated ideogram. Self-self alignments of proximal sequences from each arm are illustrated by dotter plots below the ideogram. On each plot, the junction between the arm sequence and the arm-specific satellite region is marked by a red arrow, and the junction between the arm-specific satellite region and the X-chromosome-specific alpha satellite array (DXZ1) is marked with a blue arrow. Approximately 594 kb of sequence were analysed from Xp, including ~21 kb of DXZ1 sequence. The ~454 kb of sequence analysed from Xq included ~44 kb of DXZ1 sequence. In each case, ~100 kb of arm sequence were included. The highly repetitive structure of pericentromeric satellites is in stark contrast to the near absence of repetitive structure in the arm sequences, despite an unusually high density of LINE repeats in these regions. Gaps in the dark satellite regions occur where interspersed elements (LINEs, SINEs and LTRs) interrupt the satellite sequences. In the Array Sequences dotter plot, the most proximal ~21 kb of the Xp sequence is joined to the most proximal ~44 kb of the Xq sequence. The periodic nature of the centromeric, higher-order alpha satellite array is evident. Black horizontal lines on the plot reveal near identity of sequences spaced at ~2 kb intervals. This DXZ1 sample represents ~65 kb of the 3 (±0.4) Mb alpha satellite array. The regions outlined in blue are self-self alignments ("Xp DXZ1" and "Xq DXZ1"), and the remaining rectangular region of the plot is an alignment of Xp versus Xq DXZ1, which reveals the close relationship between DXZ1 sequences from each arm.

## Single-nucleotide Polymorphisms

A total of 153,146 candidate single-nucleotide polymorphisms (SNPs) have been mapped onto the X chromosome sequence and are displayed in the VEGA database. These include 901 SNPs that result in non-synonymous changes in protein-coding regions, and are therefore candidate functional protein variants. The heterozygosity level on the X chromosome is known to be well below that of the autosomes, and this difference can be explained partly or entirely by population genetic factors[35]. Included in the mapped

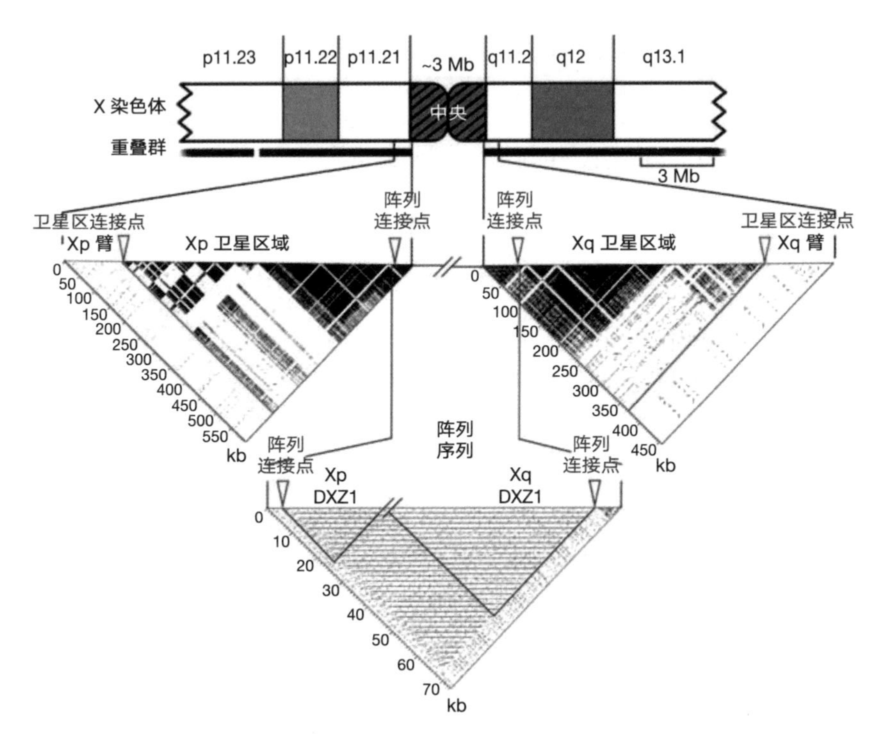


图 2. Xp 和 Xq 着丝粒周边重叠群延伸到 X 染色体特异高阶 α 卫星 DXZ1。X 染色体着丝粒周边区域显示为一个截断表意图。每个臂近端序列自身比对通过表意图下方的带点小块说明。每个小块上，臂序列和臂特异卫星区域间的连接点用红色箭头标记，臂特异卫星区域和 X 染色体特异 α 卫星阵列（DXZ1）的连接点用蓝色箭头标记。从 Xp 中分析了大约 594 kb 的序列，其中包括 ~21 kb 的 DXZ1 序列。从 Xq 中分析了 ~454 kb 的序列，包括 ~44 kb 的 DXZ1 序列。在每一种情况下，都包括 ~100 kb 的臂序列。着丝粒周边卫星序列的高度重复结构与臂序列中几乎没有重复结构形成鲜明对照，尽管这些区域 LINE 重复的密度非常高。黑色卫星区域的缺口发生在散在元件（LINE、SINE 和 LTR）中断卫星序列的地方。在阵列的序列带点小块中，Xp 序列最近端 ~21 kb 与 Xq 序列最近端 ~44 kb 相连接。着丝粒恒定，高阶 α 卫星阵列的周期性是显而易见的。小块上的黑色水平线显示间隔为 ~2 kb 的序列几乎是一致的。这个 DXZ1 样本代表 3（±0.4）Mb 阿尔法卫星阵列的 ~65 kb。蓝色轮廓区域是自身比对（“Xp DXZ1”和“Xq DXZ1”），小块的其余矩形区域是 Xp 与 Xq 的 DXZ1 的比对，它揭示了每个臂的 DXZ1 序列间密切的关系。

## 单核苷酸多态性

共有 153,146 个候选单核苷酸多态性（SNP）绘制到 X 染色体序列并显示在 VEGA 数据库中。其中包括导致蛋白质编码区域非同义改变的 901 个 SNP，因此，是候选功能蛋白变异。X 染色体上的杂合水平远低于常染色体，这种差异可以部分或者全部由群体遗传因素解释[35]。被绘制到图谱里的 62,334 个 SNP 是通过将流式

SNPs are 62,334 that were identified by alignment of flow-sorted X chromosome shotgun sequence reads to the X chromosome reference sequence. Using comparable sequence data for chromosome 20, we calculated that the heterozygosity level on the X chromosome is approximately 57% of that observed for the autosome.

## Evolution of the Human X Chromosome

Males of the three mammalian groups—Eutheria ("placental" mammals), Metatheria (marsupials) and Prototheria (egg-laying mammals)—have X and Y sex chromosomes. Ohno proposed in 1967 that the mammalian sex chromosomes evolved from an autosome pair following their recruitment into a chromosomal system for sex determination[1]. A barrier to recombination developed between these "proto" sex chromosomes, isolating the sex-determining regions and eventually spreading throughout the two homologues. In the absence of recombination, the accumulation of mutation events subsequently led to the degeneration of the Y chromosome. The sex chromosomes of birds are not homologous to those of the mammals. The sex chromosome system of birds evolved independently during the last 300 Myr, giving rise to homogametic (ZZ) male birds and heterogametic (ZW) female birds, in contrast to the mammalian system of XY males and XX females.

The autosomal origin of the mammalian sex chromosomes is vividly illustrated by alignment of the human X and chicken whole genome sequences (Fig. 3a). Orthologues of some human X chromosome genes were previously mapped to chicken chromosomes 1q13-q21 and 4p11-p14 (ref. 36). Using genomic sequence alignment, we identified approximately 30 regions of homology that together cover most of human Xq and are confined to a single section of approximately 20 Mb at the end of chicken chromosome 4p (Fig. 3a). In contrast, most of the short arm (Xp11.3–pter), including the pseudoautosomal region PAR1, matches a single block of chicken chromosome 1q. No clear picture emerges regarding the origin of the remainder of the short arm (Xcen–p11.3). We were unable to detect large regions of conserved synteny using sequence alignment, and genes from this region have orthologues on several chicken autosomes, including chromosomes 12, 1 and 4 (ref. 37). This region is also characterized by the expansion of several families of CT antigen genes (Fig. 1), which have no readily detectable orthologues in chicken. The present analysis supports the notion of a mammalian "X-conserved region" (XCR)[38], which includes the long arm and is descended from the proto-X chromosome. It also supports a separate, large addition ("X-added region" or XAR[38]) to the established X chromosome by translocation from a second autosome, which occurred in the eutherian mammals before their radiation (~105 Myr ago). In contrast to earlier hypotheses, however, it appears that much of the proximal short arm (Xcen–p11.3) should no longer be considered part of an XCR.

X 染色体鸟枪序列数据与 X 染色体参考序列比对后确定的。比较 20 号染色体的序列数据后，我们计算出 X 染色体的杂合水平大约是常染色体的 57%。

## 人类 X 染色体的进化

三个哺乳动物群体——真兽次亚纲（有“胎盘”的哺乳动物），后兽次亚纲（有袋的哺乳动物）和原兽亚纲（产卵哺乳动物），它们的雄性体内含有 X 和 Y 性染色体。1967 年大野提出哺乳动物性染色体是从一对常染色体进化而来的，它们进入染色体系统之后用来决定性别 [1]。这些“原始”性染色体之间形成了重组障碍，分离出决定性别的区域并最终在这两个同源物中传播。如果没有重组，随后的基因突变事件积累导致 Y 染色体退化。鸟的性染色体与哺乳动物的性染色体没有同源性。鸟类性染色体在过去的 3 亿年独立进化，从而形成同型配子（ZZ）的雄鸟和异型配子（ZW）的雌鸟，与哺乳动物系统的 XY 雄性和 XX 雌性相反。

人类 X 染色体和鸡全基因组序列的比对，生动地说明哺乳动物性染色体起源于常染色体（图 3a）。人类的一些 X 染色体基因直系同源物以前被绘制到鸡染色体 1q13-q21 和 4p11-p14 中（参考文献 36）。利用基因组序列比对，我们确定了大约 30 个覆盖了人类 Xq 大部分的同源区域，这些区域都限制在鸡染色体 4p 末端大约 20 Mb 的单一区域内（图 3a）。相比之下，短臂（Xp11.3–pter）的大部分区域，包括假常染色体区域 PARl，与鸡染色体 1q 的单一区块相匹配。关于短臂（Xcen–p11.3）剩余部分的起源没有清楚的图片展示。我们用序列比对无法检测到大的同线性保守区域，并且这些区域的基因在鸡的一些常染色体上有直系同源物，这些常染色体包括 12 号、1 号和 4 号染色体（参考文献 37）。这一区域还表现出若干 CT 抗原基因家族（图 1）扩展的特点，而这在鸡中并没有检测到同源物。这个分析支持哺乳动物“X 保守区”（XCR）的概念 [38]，X 保守区包括长臂，它是从原始 X 染色体留下来的。它还支持从第二个常染色体易位到已有的 X 染色体上从而形成的独立的、大型增加片段（“X 补充区域”或 XAR[38]）的概念，这发生在真兽类哺乳动物适应辐射前（~ 1 亿 500 万年前）。但是，与早期假设相反，好像大部分的近端短臂（Xcen–p11.3）不应该被认为是 XCR 的一部分。

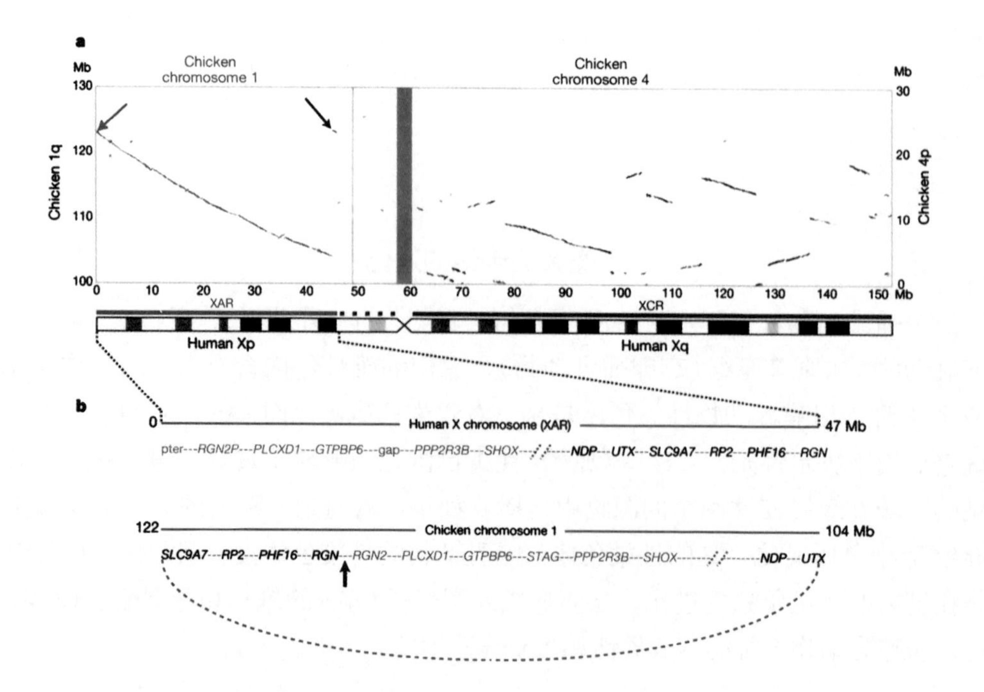


Fig. 3. Homologies between the human X chromosome and chicken autosomes. **a**, Plot of BLASTZ sequence alignments between the X chromosome and chicken chromosomes 1 (red) and 4 (blue). Grey bar centred at approximately 60 Mb shows the position of the X centromere. Only the relevant section of each chicken chromosome is shown (see Mb scale at left for chromosome 1 and at right for chromosome 4). A schematic interpretation of the homologies shows the XAR and XCR as red and blue bars, respectively (see Fig. 1). Homologies at the ends of the XAR are indicated with arrows and are expanded in **b**. **b**, (Top) Genes at the ends of the human XAR. Genes from distal Xp (magenta arrow in **a**) are in magenta and genes from Xp11.3 (black arrow in **a**) are shown in black. (Bottom) Arrangement of the orthologous genes on chicken chromosome 1. A hypothetical ring chromosome, with the equivalent gene order to that observed in the chicken, is indicated by the curved, dotted red line. Recombination between one end of the established X chromosome and the ring chromosome at the arrowed position could, in a single step, have added the XAR and created the gene order observed on the human X chromosome.

The precise location of genes that demarcate the XAR suggests a possible mechanism for the addition. The annotated genes at the extreme ends of the 47 Mb XAR are *PLCXD1* (cU136G2.1 in Supplementary Fig. 1) near Xpter, and *RGN* in Xp11.3. We also found an unprocessed *RGN* pseudogene (*RGN2P*) at Xpter, distal to *PLCXD1*. The orthologues for these three loci are adjacent on chicken chromosome 1, in the order (tel)–*RGN*–*RGN2*–*PLCXD1*–(cen) (Fig. 3b). The generation of these two different gene orders from a common ancestral sequence would require a minimum of two rearrangements as well as the translocation that added the XAR. A more parsimonious model suggested by these data, however, is that the XAR was acquired by recombination between the X chromosome and a ring chromosome in which the ancestral *PLCXD1*, *RGN* and *RGN2* sequences were neighbours (Fig. 3b).

In order to examine more recent patterns of evolution, we compared the human X chromosome with other mammalian sequences. We saw nine major blocks of sequence homology between

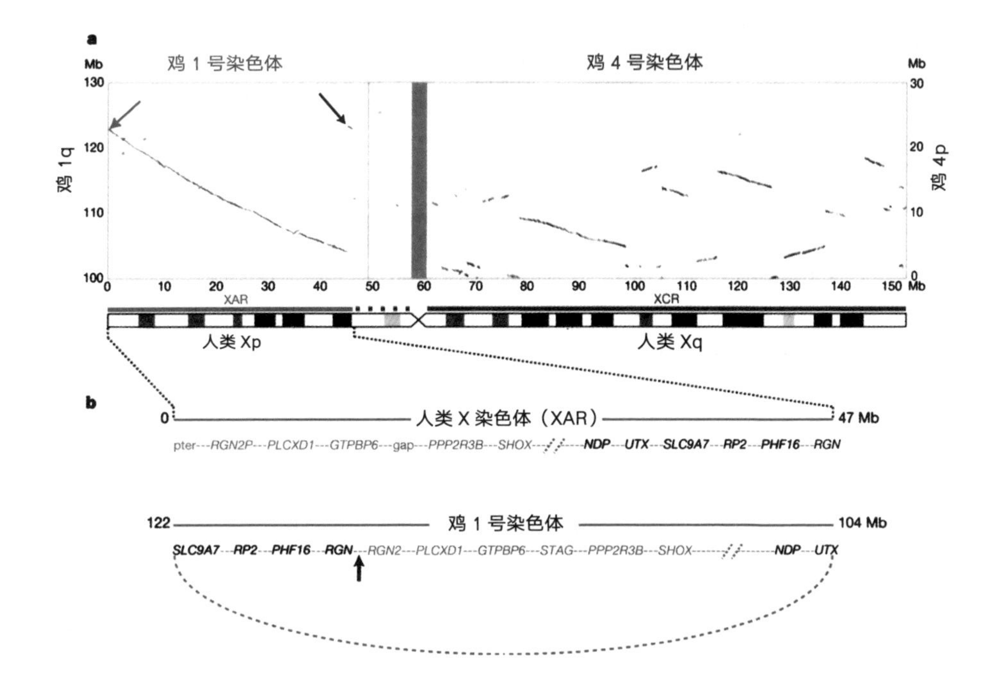


图 3. 人类 X 染色体和鸡常染色体的同源性。**a**，X 染色体和鸡 1 号（红色）和 4 号（蓝色）染色体间 BLASTZ 序列比对。集中在约 60 Mb 的灰色条块显示的是 X 染色体着丝粒的位置。鸡的每条染色体只显示有关的片段（1 号染色体看左边的 Mb 比例尺，4 号染色体看右边的 Mb 比例尺）。一个解释同源性的示意图分别显示了红色条 XAR 和蓝色条 XCR（见图 1）。XAR 末端的同源性用箭头指示，并在 **b** 中扩大显示。**b**，（上）人 XAR 末端的基因。Xp 远端的基因（**a** 中品红色箭头）是品红色，Xp11.3 的基因（**a** 中黑色箭头）显示为黑色。（下）鸡的 1 号染色体上直系同源基因的排列。由弯曲的红色虚线表示的一个假设的环状染色体，其基因顺序与在鸡染色体中观察到的一致。已建立的 X 染色体一端与箭头所指位置的环状染色体之间的重组可以一步完成 XAR 的添加，重组还能创建在人类 X 染色体上观察到的基因顺序。

划分 XAR 的基因的确切位置表明序列增加的可能机制。在 47 Mb XAR 末端的注释基因是接近 Xpter 的 *PLCXD1*（cU136G2.1 见补充图 1）以及在 Xp11.3 中的 *RGN*。我们还在 *PLCXD1* 远端的 Xpter 上发现一个未处理的 *RGN* 假基因（*RGN2P*）。这三个基因座的直系同源物在鸡的 1 号染色体上相邻，顺序为（端粒）–*RGN*–*RGN2*–*PLCXD1*–（着丝粒）（图 3b）。从一个共同祖先的序列产生这两个不同的基因顺序最少需要两个重排，以及增加 XAR 的易位。然而，这些数据暗示了一个更为简洁的模型，即 XAR 是通过 X 染色体与祖先 *PLCXD1*、*RGN* 和 *RGN2* 序列相邻的环状染色体重组获得的（图 3b）。

为了研究更近期的进化模式，我们将人类 X 染色体与其他哺乳动物序列进行了比对。我们发现人类和小鼠 X 染色体之间有 9 个主要的序列同源性区块，人类与

human and mouse X chromosomes, and eleven between human and rat (Fig. 4). The homology blocks occupy almost the entirety of each X chromosome, confirming the remarkable degree of conserved synteny of this chromosome within the eutherian mammalian lineage. This is consistent with Ohno's law, which predicts that the establishment of a dosage compensation mechanism had a stabilizing effect on the gene content of the mammalian X chromosome[1]. On the long arm, just two blocks of homology account for the entire alignment of the human and corresponding mouse sequences, but the mouse homologous regions are punctuated with three additional segments, each containing long and very similar repeats (arrowed in Fig. 4). Alignment of human Xq with the rat sequence reveals four discrete homology blocks; the greater fragmentation compared with the mouse alignment would be explained by a minimum of two rearrangements, one in each of the two mouse–human homology blocks, specifically on the rat lineage. The mouse-specific repeat segments are not detected in the current version of the rat genome sequence. On the short arm of the human X chromosome, seven major blocks of homology with each rodent account for most of the human sequence (Fig. 4). Using the dog as an outgroup, we established that the human and dog X chromosome sequences are essentially collinear (K. Lindblad-Toh, personal communication). Therefore, all of the rearrangements indicated in Fig. 4 occurred in the rodent lineage, and the human X chromosome appears to have been remarkably stable in its organization since the radiation of eutherian mammals. This is consistent with the recent prediction, derived from a comparison of human, rodent and chicken chromosomes, that the human X chromosome is identical to the putative ancestral (eutherian) mammalian X chromosome[39].

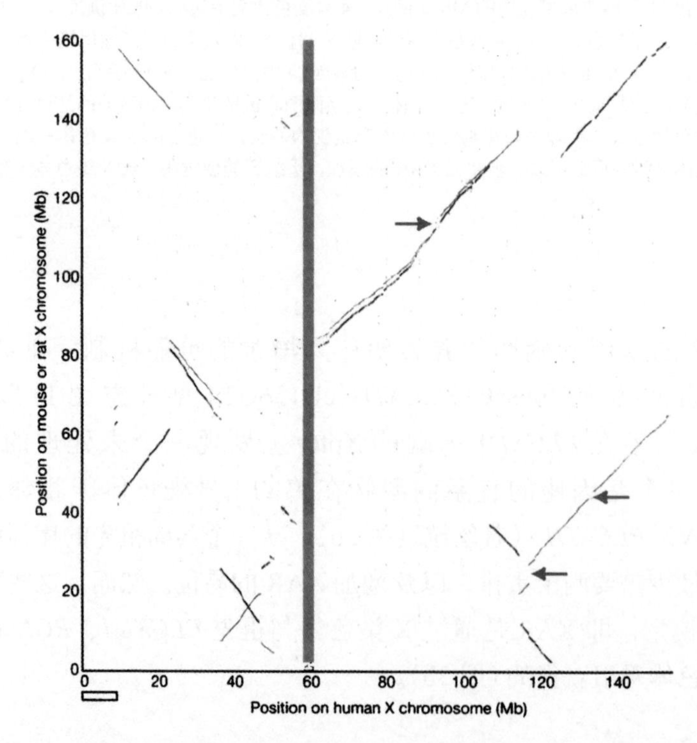


Fig. 4. Conservation of the X chromosome in eutherian mammals. Plot of BLASTZ sequence alignments between the human X chromosome and the mouse (red) and rat (blue) X chromosomes. The rodent

大鼠之间有 11 个（图 4）。这些同源区块几乎占据整个 X 染色体，证实了这条染色体在真兽类哺乳动物谱系内具有显著的保守同线性。这与大野的定律一致，该定律预测剂量补偿机制的建立对哺乳动物 X 染色体的基因含量有稳定作用[1]。在染色体长臂上，人类和小鼠相应序列整体比对后只有两个同源性区块，但小鼠同源区域被另外三个额外的片段隔开，每个片段包含长的、非常相似的重复序列（图 4 中箭头所示）。人类 Xq 与大鼠序列比对揭示四个分离的同源区块；与小鼠比对相比，将通过至少两个重排来解释更大的片段化，即在两个鼠–人同源性区块中分别有一个重排，特别是在大鼠谱系中。小鼠特异性重复片段在目前的大鼠基因组序列中未检测到。人类 X 染色体短臂上与每个啮齿动物同源的 7 个主要区块占了人类序列的大部分（图 4）。使用狗作为外类群，我们确定人和狗 X 染色体的 DNA 序列本质上是共线性的（林德布拉德–托，个人交流）。因此，图 4 所显示的所有重排发生在啮齿动物谱系中，从真兽类哺乳动物适应辐射以来，人类 X 染色体在其系统内似乎非常稳定。这与最近比较人类、啮齿动物和鸡染色体得出的预测一致，即人类 X 染色体与假定哺乳动物祖先（真兽类）X 染色体一致[39]。

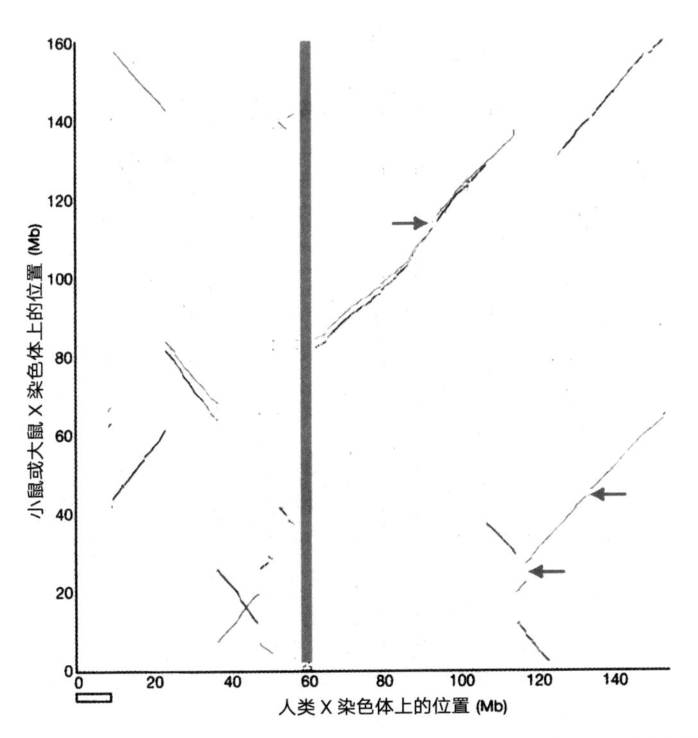


图 4. 在真兽类哺乳动物中 X 染色体的保守性。人类 X 染色体与小鼠（红色）和大鼠（蓝色）X 染色体之间 BLASTZ 序列比对图。啮齿类动物的染色体着丝粒是向下的。箭头指示的区域在小鼠序列中是长度

chromosomes are oriented with their centromeres pointing downwards. Regions indicated with arrows are long, highly similar repeats in the mouse sequence that are absent from the human and rat sequences. These repeats were apparently collapsed in an earlier analysed version of the mouse sequence, which also had a large inversion with respect to the mouse assembly used here (NCBI32)[66]. The NCBI32 assembly has a gap from 0–3 Mb, which explains the absence of homology to the human X sequence in this part of the plot. The open horizontal bar shows the terminal section of human Xp, which is not conserved on the rodent X chromosomes.

The most notable difference we found between the human and rodent X chromosomes is the existence of 9 Mb of sequence at the tip of the human short arm (including human PAR1) that is apparently missing from the rodent X chromosomes (Fig. 4). There are 34 known and novel protein-coding genes in this segment of the human X chromosome (Supplementary Fig. 1), enabling us to investigate how this difference arose. A comprehensive database search of the rodent genome sequences revealed convincing orthologues for only thirteen of these genes in rat and five in mouse. Most of the rat orthologues are located in two groups on chromosome 12, and the only genes for which X-linked orthologues could be found in both rodents were *PRKX* and *STS*. In contrast, we found 24 of these 34 genes on chicken chromosome 1, and the order of these genes is perfectly conserved between the two genomes. Therefore, we conclude that this large terminal segment was present in the XAR and was subsequently removed from the X chromosome in a common murid ancestor of mouse and rat. The relative paucity of rodent ECRs in this segment of the X chromosome sequence (Supplementary Fig. 1) suggests that much of the region may be absent altogether from the genomes of *Mus musculus* and *Rattus norvegicus*.

## Comparison of the Human X and Y Chromosomes

The evolutionary process has eradicated most traces of the ancestral relationship between the human X and Y chromosomes. At the cytogenetic level, the Y chromosome has a large and variably sized heterochromatic block and is considerably smaller than the X chromosome, and the euchromatic part of the X chromosome is six times longer than that of Y. Few genes on human chromosome X have an active counterpart on the Y chromosome, and the majority of these are contained in regions where XY homology is of relatively recent origin.

A detailed comparison of the human X and Y chromosome sequences reveals the extent of Y chromosome decay in non-recombining regions. All of the large homologous blocks visible in Fig. 5 (and represented schematically in Fig. 6) are descended from material that was added to the established sex chromosomes. The tip of the short arm of the X and Y chromosomes comprises the 2.7 Mb pseudoautosomal region PAR1. Homology between the X and Y chromosomes in PAR1 is maintained by an obligatory recombination in male meiosis; gene loci in this region are present in two copies in both males and females and are not subject to dosage compensation by XCI. At the tip of the long arm of X and Y is a second pseudoautosomal region, the 330 kb PAR2, which was created by duplication

很长且高度相似的重复序列，在人类和大鼠序列中缺失。这些重复序列在较早分析版本的小鼠序列中显然是塌陷的，该版本的小鼠序列与此处使用的小鼠序列（NCBI32）[66] 相比也有较大的不同。NCBI32 序列有一个 0～3 Mb 的缺口，这就解释了在图的这一部分中缺少与人类 X 序列的同源性。空心的横杠显示人类 Xp 末端部分，它在啮齿动物 X 染色体上并不保守。

我们在人类和啮齿动物的 X 染色体间发现的最显著的差异是人类染色体短臂末端（包括人类 PAR1）存在的 9 Mb 序列在啮齿动物 X 染色体中明显缺失（图 4）。在人类 X 染色体的这一段序列中，有 34 个已知的、新的蛋白编码基因（补充图 1），使我们能够研究这种差异是如何产生的。对啮齿动物基因组序列数据库进行全面搜索显示，大鼠中只有 13 个基因有高可信度直系同源物，而小鼠只有 5 个。大多数大鼠直系同源物位于 12 号染色体的两个组上，在两种啮齿动物中发现的 X 连锁同源基因只有 *PRKX* 和 *STS*。相比之下，我们在鸡的 1 号染色体上发现了这 34 个基因中的 24 个，并且这些基因的顺序在两个基因组间非常保守。因此，我们得出这样的结论：这个大型末端序列曾出现在 XAR 中，后来从小鼠和大鼠共同的鼠科祖先的 X 染色体中被移除。啮齿动物 ECR 在 X 染色体序列的这一段相对缺失（补充图 1）表明，该区域的大部分序列可能在小家鼠和褐家鼠的基因组中共同缺失。

## 比较人类 X 和 Y 染色体

进化过程已消除人类 X 染色体和 Y 染色体之间祖先关系的大部分痕迹。在细胞遗传学水平上，Y 染色体有一个大的大小可变的异染色质块，但这个异染色质块比 X 染色体小得多，X 染色体常染色质部分是 Y 的 6 倍。人类 X 染色体上很少有基因在 Y 染色体上有活跃的对应基因，其中大多数包含在 XY 同源性起源较晚的区域。

对人类 X 染色体和 Y 染色体序列的详细比较揭示了 Y 染色体在非重组区域衰减的程度。在图 5 中可看到的所有大型同源区块（并在图 6 中示意出）都来自于增加到既定的性染色体的物质。X 染色体和 Y 染色体短臂末端包含 2.7 Mb 的假常染色体区域 PAR1。PAR1 中的 X 染色体和 Y 染色体之间的同源性是通过雄性减数分裂中必要的重组来维持的；这个区域的基因座在男性和女性中都有两个拷贝，不受 XCI 的剂量补偿影响。在 X 染色体和 Y 染色体长臂末端是第二个假常染色体区域，即 330 kb 的 PAR2，它是自人类和黑猩猩谱系分化以来，通过复制从 X 到 Y 的物质

of material from X to Y since the divergence of human and chimpanzee lineages[40]. Some genes in PAR2 are subject to XCI, presumably reflecting their status on the X chromosome before the duplication event. Outside the PARs, homologies between the X and Y chromosomes are in non-recombining regions, predominantly in other parts of the XAR, together with a large "X-transposed region" (XTR)[41] in Xq21.3 and Yp11.2–p11.3 (see below). It is thought that the XAR originally formed a large pseudoautosomal region with an equivalent YAR, which is now largely eroded. At a gross level, the homology between the XAR and YAR is continuous for 6 Mb proximal to the pseudoautosomal boundary on the X (PABX), but is considerably more fragmented on the Y chromosome (Figs 5b and 6). Beyond this, the remaining 38.5 Mb of the XAR detects few other remnants of the YAR. Homologies are mostly in small islands around genes with functional orthologues on both sex chromosomes (for example, *AMELX*/*AMELY*, *ZFX*/*ZFY*, see Table 1).

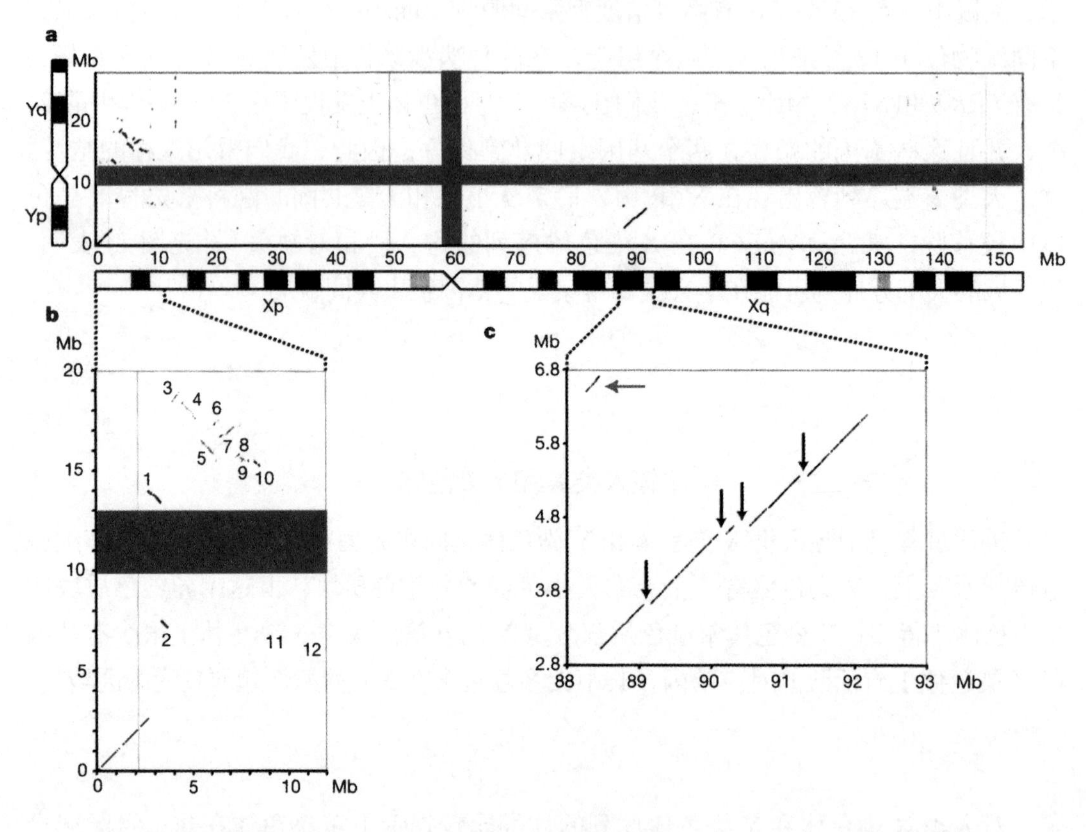


Fig. 5. Limited homology between the human sex chromosomes illustrates the extent of Y chromosome erosion in non-recombining regions. **a**, BLASTN alignments (length ⩾ 80 bp, sequence identity ⩾ 70%) between the finished sequences of the X and Y chromosomes. The centromere positions are represented by grey bars. The analysed Y chromosome sequence ends at the large, heterochromatic segment on Yq, which is indicated by the black bar on the truncated Y chromosome ideogram. **b**, Major blocks of homology remaining between the XAR and the YAR. Expansion of the BLASTN plot from 0–12 Mb on the X chromosome and 0–20 Mb on the Y chromosome. On the X chromosome, the major homologies lie in the terminal 8.5 Mb of Xp: PAR1 (magenta line) and numbered blocks 1–10. Lesser homologies 11 and 12 contain the *TBL1X*/*TBL1Y* and *AMELX*/*AMELY* genes, respectively. **c**, The XTR region in detail (88–93 Mb on X and 2.8–6.8 Mb on Y). Black arrows show large segments deleted from the Y chromosome

产生的[40]。PAR2 中有些基因受 XCI 影响，这很可能反映了它们在复制事件之前在 X 染色体上的状态。PAR 以外，X 染色体和 Y 染色体之间的同源性在非重组区域，主要位于 XAR 的其他区域，以及在 Xq21.3 和 Yp11.2–p11.3 中的大型"X 转置区"(XTR)[41](见下文)。我们认为，XAR 最初形成与 YAR 等同的大型假常染色体区域，现在 YAR 在很大程度上消退了。总的来说，XAR 和 YAR 之间的同源性有 X 假常染色体边界近端(PABX)的连续 6 Mb，但在 Y 染色体上是相当多的分散片段(图 5b 和 6)。除此之外，XAR 其余的 38.5 Mb 检测到其他几个残余的 YAR。同源性大多在两条性染色体上有功能性直系同源物的基因周围(例如，*AMELX*/*AMELY*、*ZFX*/*ZFY*，见表 1)。

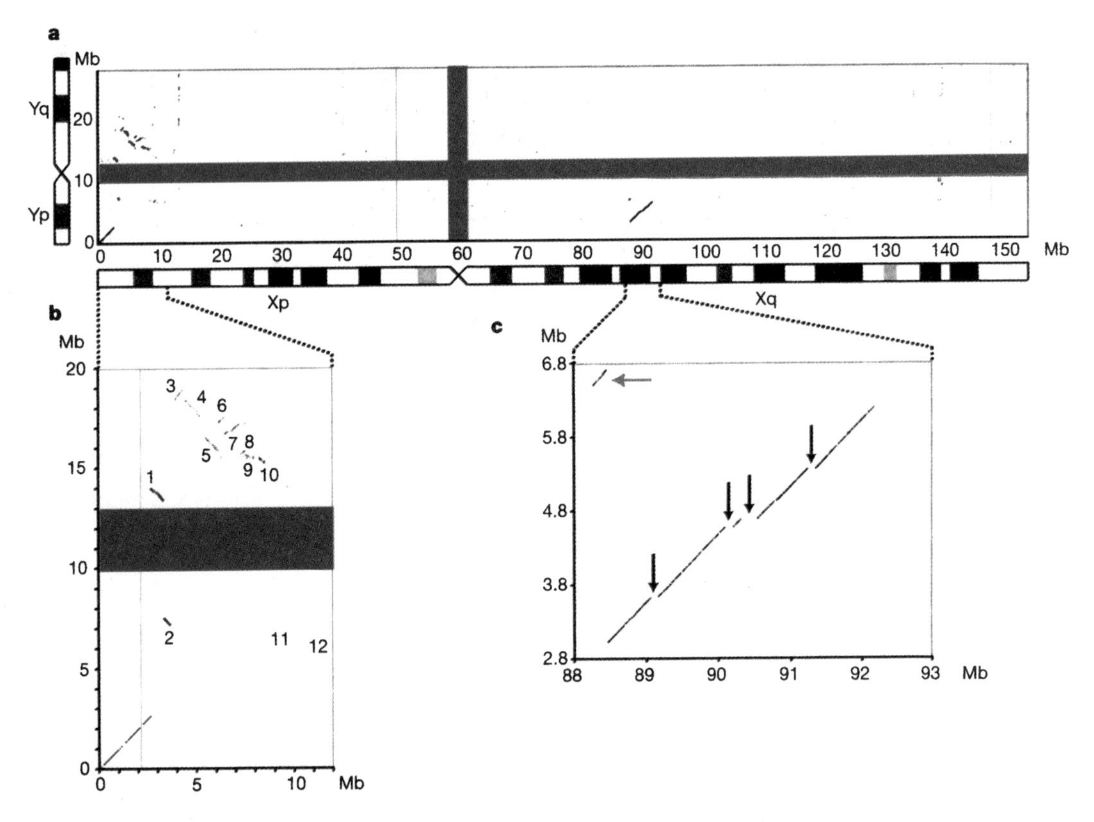


图 5. 人类性染色体之间有限的同源性说明 Y 染色体在非重组区域消退的程度。**a**，X 染色体和 Y 染色体完成的序列之间 BLASTN 比对(长度 ≥ 80 bp，序列一致性 ≥ 70%)。灰色条框指示着丝粒的位置。所分析的 Y 染色体序列末端在 Yq 上大的异染色质片段处，通过图上截断 Y 染色体的黑色条块表示。**b**，XAR 和 YAR 之间主要同源性区块。X 染色体上从 0 ~ 12 Mb，Y 染色体从 0 ~ 20 Mb 的 BLASTN 扩展图。在 X 染色体上，主要的同源性位于 8.5 Mb 的 Xp 末端：PAR1(品红色线)和编号区块 1 ~ 10。较小的同源性片段 11 和 12 分别包含 *TBL1X*/*TBL1Y* 和 *AMELX*/*AMELY* 基因。**c**，XTR 区域详细信息(在 X 染色体上 88 ~ 93 Mb 和在 Y 染色体上 2.8 ~ 6.8 Mb)。黑色箭头显示从 XTR 的 Y 染色体拷贝中删除的大片段。品红色箭头表示通过 Y 染色体臂内倒位后从 XTR 剩余部分分离出的短片段。人类群体中

copy of the XTR. The magenta arrow indicates the short segment that is separated from the rest of the XTR by a paracentric inversion on the Y chromosome. An independent inversion polymorphism on Yp in human populations encompasses this small segment. The position and orientation of the segment shows that the Y chromosome reference sequence is of the less common, derived Y chromosome.

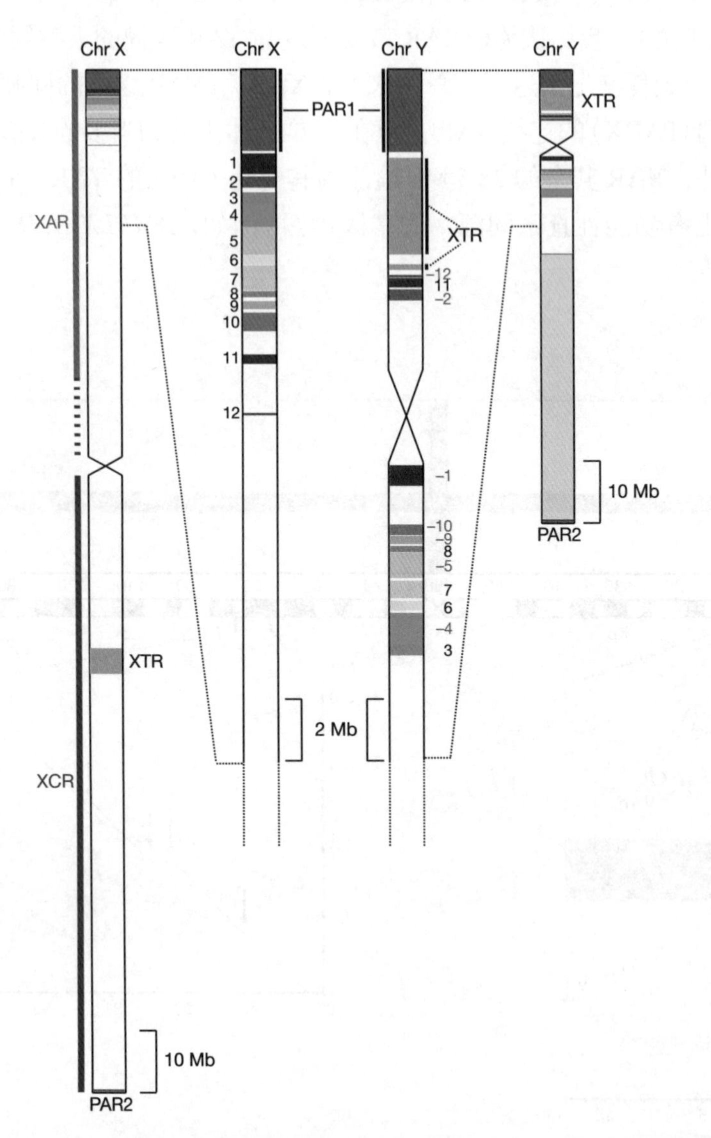


Fig. 6. Schematic representation of major homologies between the human sex chromosomes. The entire X and Y chromosomes are shown using the same scale on the left and right sides of the figure, respectively. The major heterochromatic region on Yq is indicated by the pale grey box proximal to PAR2. Expanded sections of X and Y are shown in the centre of the figure. Homologies coloured in the figure are either part of the XAR (PAR1 and blocks 1–12), or were duplicated from the X chromosome to the Y chromosome since the divergence of human and chimpanzee lineages (XTR and PAR2). The numbering of XAR-YAR blocks follows that in Fig. 5b. Blocks inverted on the Y chromosome relative to the X chromosome are assigned red, negative numbers.

Yp 上的一个独立的倒位多态现象包括这个小片段。这个片段的位置和方向表明，Y 染色体参考序列是不太常见的衍生 Y 染色体。

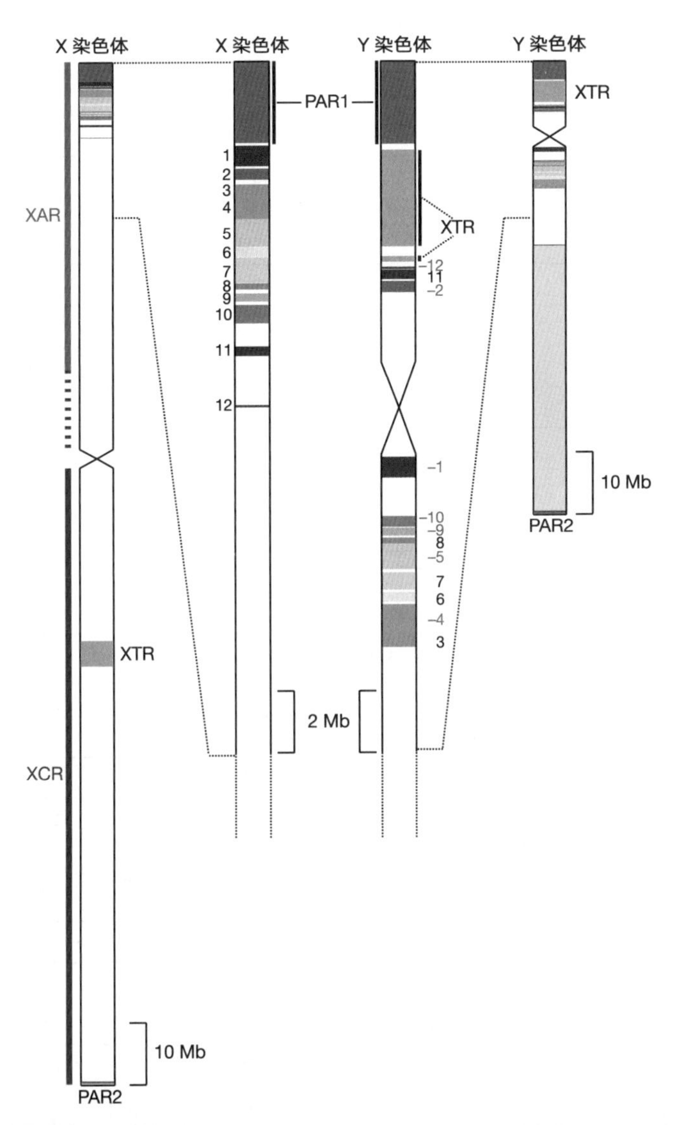


图 6. 人类性染色体之间主要的同源性示意图。整个 X 染色体和 Y 染色体分别以相同的比例显示在图左侧和右侧。Yq 上主要的异染色质区域用接近 PAR2 的浅灰色框指示。图中央显示了 X 和 Y 的展开部分。图中着色的同源物或者是 XAR 的一部分（PAR1 和 1 ~ 12 区块），或者是自人类和黑猩猩谱系分化以来（XTR 和 PAR2）从 X 染色体到 Y 染色体的物质所复制的。XAR-YAR 的编号按照图 5b 所示。Y 染色体相对 X 染色体反转的区块用红色负数标记。

Table 1. Homologous genes on the human X and Y chromosomes

| Region | Distance from Xpter (Mb) | X gene* | Y gene | Distance from Ypter (Mb)† | XY homology block‡ |
|---|---|---|---|---|---|
| Pseudoautosomal region PAR1 (XAR) | 0.15 | cU136G2.1 (*PLCXD1*) | cU136G2.1 (*PLCXD1*) | 0.15 | PAR1 |
| | 0.17 | cU136G2.2 (*GTPBP6*) | cU136G2.2 (*GTPBP6*) | 0.17 | PAR1 |
| | 0.25 | cM56G10.2§ | cM56G10.2§ | 0.25 | PAR1 |
| | 0.29 | cM56G10.1 (*PPP2R3B*) | cM56G10.1 (*PPP2R3B*) | 0.29 | PAR1 |
| | 0.57 | *SHOX* | *SHOX* | 0.57 | PAR1 |
| | 0.92 | bA309M23.1§ | bA309M23.1§ | 0.92 | PAR1 |
| | 1.31 | *CRLF2* | *CRLF2* | 1.31 | PAR1 |
| | 1.38 | *CSF2RA* | *CSF2RA* | 1.38 | PAR1 |
| | 1.52 | *IL3RA* | *IL3RA* | 1.52 | PAR1 |
| | 1.55 | *SLC25A6* | *SLC25A6* | 1.55 | PAR1 |
| | 1.56 | bA261P4.5§ | bA261P4.5§ | 1.56 | PAR1 |
| | 1.57 | bA261P4.6 (*CXYorf2*) | bA261P4.6 (*CXYorf2*) | 1.57 | PAR1 |
| | 1.59 | *ASMTL* | *ASMTL* | 1.59 | PAR1 |
| | 1.66 | bA261P4.4 (*P2RY8*) | bA261P4.4 (*P2RY8*) | 1.66 | PAR1 |
| | 1.76 | *DXYS155E* (*CXYorf3*) | *DXYS155E* (*CXYorf3*) | 1.76 | PAR1 |
| | 1.79 | *ASMT* | *ASMT* | 1.79 | PAR1 |
| | 1.79 | bB297E16.3§ | bB297E16.3§ | 1.79 | PAR1 |
| | 1.91 | bB297E16.4§ | bB297E16.4§ | 1.91 | PAR1 |
| | 1.93 | bB297E16.5§ | bB297E16.5§ | 1.93 | PAR1 |
| | 2.37 | *DHRSX* | *DHRSX* | 2.37 | PAR1 |
| | 2.41 | *ALTE* (*ZBED1*) | *ALTE* (*ZBED1*) | 2.41 | PAR1 |
| | 2.54 | Em:AC097314.2§ | Em:AC097314.2§ | 2.54 | PAR1 |
| | 2.53 | Em:AC097314.3§ | Em:AC097314.3§ | 2.53 | PAR1 |
| | 2.63 | *CD99* | *CD99* | 2.63 | PAR1 |
| X-added region (XAR) | 3.57 | *PRKX* | *PRKY* | 7.23 | 2 |
| | 5.81 | *NLGN4X* | *NLGN4Y* | 15.23 | 5 |
| | 6.31 | Em:AC108684.1 (*VCX3A*) | *VCY*, *VCY1B* | 14.54, 14.6 | 6 |
| | 7.62 | *VCX* | *VCY*, *VCY1B* | 14.54, 14.6 | 9 |
| | 7.95 | Em:AC097626.1 (*VCX2*) | *VCY*, *VCY1B* | 14.54, 14.6 | 10 |
| | 8.24 | Em:AC006062.2 (*VCX3B*) | *VCY*, *VCY1B* | 14.54, 14.6 | 10 |
| | 9.37 | *TBL1X* | *TBL1Y* | 6.97 | 11 |
| | 11.07 | *AMELX* | *AMELY* | 6.78 | 12 |
| | 12.75 | *TMSB4X* | *TMSB4Y* | 14.25 | |

表 1. 人类 X 和 Y 染色体上的同源基因

| 区域 | 与 Xpter 的距离 (Mb) | X 基因 * | Y 基因 | 与 Ypter 的距离 (Mb)† | XY 同源性区块 ‡ |
|---|---|---|---|---|---|
| 假常染色体区 PAR1 (XAR) | 0.15 | cU136G2.1 (*PLCXD1*) | cU136G2.1 (*PLCXD1*) | 0.15 | PAR1 |
| | 0.17 | cU136G2.2 (*GTPBP6*) | cU136G2.2 (*GTPBP6*) | 0.17 | PAR1 |
| | 0.25 | cM56G10.2§ | cM56G10.2§ | 0.25 | PAR1 |
| | 0.29 | cM56G10.1 (*PPP2R3B*) | cM56G10.1 (*PPP2R3B*) | 0.29 | PAR1 |
| | 0.57 | *SHOX* | *SHOX* | 0.57 | PAR1 |
| | 0.92 | bA309M23.1§ | bA309M23.1§ | 0.92 | PAR1 |
| | 1.31 | *CRLF2* | *CRLF2* | 1.31 | PAR1 |
| | 1.38 | *CSF2RA* | *CSF2RA* | 1.38 | PAR1 |
| | 1.52 | *IL3RA* | *IL3RA* | 1.52 | PAR1 |
| | 1.55 | *SLC25A6* | *SLC25A6* | 1.55 | PAR1 |
| | 1.56 | bA261P4.5§ | bA261P4.5§ | 1.56 | PAR1 |
| | 1.57 | bA261P4.6 (*CXYorf2*) | bA261P4.6 (*CXYorf2*) | 1.57 | PAR1 |
| | 1.59 | *ASMTL* | *ASMTL* | 1.59 | PAR1 |
| | 1.66 | bA261P4.4 (*P2RY8*) | bA261P4.4 (*P2RY8*) | 1.66 | PAR1 |
| | 1.76 | *DXYS155E* (*CXYorf3*) | *DXYS155E* (*CXYorf3*) | 1.76 | PAR1 |
| | 1.79 | *ASMT* | *ASMT* | 1.79 | PAR1 |
| | 1.79 | bB297E16.3§ | bB297E16.3§ | 1.79 | PAR1 |
| | 1.91 | bB297E16.4§ | bB297E16.4§ | 1.91 | PAR1 |
| | 1.93 | bB297E16.5§ | bB297E16.5§ | 1.93 | PAR1 |
| | 2.37 | *DHRSX* | *DHRSX* | 2.37 | PAR1 |
| | 2.41 | *ALTE* (*ZBED1*) | *ALTE* (*ZBED1*) | 2.41 | PAR1 |
| | 2.54 | Em:AC097314.2§ | Em:AC097314.2§ | 2.54 | PAR1 |
| | 2.53 | Em:AC097314.3§ | Em:AC097314.3§ | 2.53 | PAR1 |
| | 2.63 | *CD99* | *CD99* | 2.63 | PAR1 |
| X 增加区 (XAR) | 3.57 | *PRKX* | *PRKY* | 7.23 | 2 |
| | 5.81 | *NLGN4X* | *NLGN4Y* | 15.23 | 5 |
| | 6.31 | Em:AC108684.1 (*VCX3A*) | *VCY*, *VCY1B* | 14.54, 14.6 | 6 |
| | 7.62 | *VCX* | *VCY*, *VCY1B* | 14.54, 14.6 | 9 |
| | 7.95 | Em:AC097626.1 (*VCX2*) | *VCY*, *VCY1B* | 14.54, 14.6 | 10 |
| | 8.24 | Em:AC006062.2 (*VCX3B*) | *VCY*, *VCY1B* | 14.54, 14.6 | 10 |
| | 9.37 | *TBL1X* | *TBL1Y* | 6.97 | 11 |
| | 11.07 | *AMELX* | *AMELY* | 6.78 | 12 |
| | 12.75 | *TMSB4X* | *TMSB4Y* | 14.25 | |

*Continued*

| Region | Distance from Xpter (Mb) | X gene* | Y gene | Distance from Ypter (Mb)† | XY homology block‡ |
|---|---|---|---|---|---|
| X-added region (XAR) | 16.59 | *CXorf15* | *CYorf15A, CYorf15B* | 20.13, 20.15 | |
| | 19.91 | *EIF1AX* | *EIF1AY* | 21.08 | |
| | 23.96 | *ZFX* | *ZFY* | 2.87 | |
| | 40.78 | *USP9X* | *USP9Y* | 13.33 | |
| | 40.96 | *DDX3X* | *DDX3Y* | 13.46 | |
| | 44.61 | *UTX* | *UTY* | 13.91 | |
| X-conserved region (XCR) | 53.00 | dJ290F12.2 (*TSPYL2*) | *TSPY* (~35) | 9.50 | |
| | 53.12 | *SMCX* | *SMCY* | 20.27 | |
| | 71.27 | *RPS4X* | *RPS4Y1, RPS4Y2* | 2.77, 21.27 | |
| X-transposed region (XTR) | 88.50 | bB348B13.2§ | n/a | 2.96 | XTR |
| | 88.99 | *TGIF2LX* | *TGIF2LY* | 3.49 | XTR |
| | 91.26 | *PCDH11X* | *PCDH11Y* | 5.28 | XTR |
| X-conserved region (XCR) | 135.68 | *RNMX* (*RBMX*) | *RBMY* (6) | 22.02, 22.04, 22.37, 22.41, 22.66, 22.85 | |
| | 139.31 | *SOX3* | *SRY* | 2.70 | |
| | 148.38 | Em:AC016940.3 (*HSFX2*)§ | *HSFY1, HSFY2* | 19.3, 19.12 | |
| | 148.56 | Em:AC016939.4 (*HSFX1*)§ | *HSFY1, HSFY2* | 19.3, 19.12 | |
| Pseudoautosomal region PAR2 | 154.57 | *SPRY3* | *SPRY3* | 57.44 | PAR2 |
| | 154.71 | *SYBL1* | *SYBL1* | 57.58 | PAR2 |
| | 154.81 | *IL9R* | *IL9R* | 57.67 | PAR2 |
| | 154.81 | Em:AJ271736.5§ | Em:AJ271736.5§ | 57.69 | PAR2 |
| | 154.82 | Em:AJ271736.6 (*FAM39A*)§ | Em:AJ271736.6 (*FAM39A*)§ | 57.69 | PAR2 |

Pseudogenes are not included in the table.

* Gene names as shown in Supplementary Fig. 1. HUGO name is in parentheses when the two names differ. Em, EMBL entry.

† Distances refer to Y chromosome sequence assembly NCBI35. Where multiple Y chromosome orthologues exist, the locations of all copies are shown on the Y chromosome. The exception is TSPY, which has ~35 copies in an array centred at approximately 9.5 Mb on the Y chromosome[41].

‡ Major homology blocks as shown in Figs 5 and 6.

§ Novel cases of X genes with Y homologues assigned to these categories.

The XTR arose by duplication of material from X to Y since the divergence of the human and chimpanzee lineages[42]. The duplicated region spans 3.91 Mb on X, but the corresponding region is only 3.38 Mb on the Y chromosome (Fig. 5c). We have aligned the entire X and Y copies of this region. Excluding insertions and deletions, sequence identity between the copies is 98.78%. We estimate that the transposition event occurred

续表

| 区域 | 与 Xpter 的距离 (Mb) | X 基因 * | Y 基因 | 与 Ypter 的距离 (Mb)† | XY 同源性区块 ‡ |
|---|---|---|---|---|---|
| X 增加区 (XAR) | 16.59 | *CXorf15* | *CYorf15A*, *CYorf15B* | 20.13, 20.15 | |
| | 19.91 | *EIF1AX* | *EIF1AY* | 21.08 | |
| | 23.96 | *ZFX* | *ZFY* | 2.87 | |
| | 40.78 | *USP9X* | *USP9Y* | 13.33 | |
| | 40.96 | *DDX3X* | *DDX3Y* | 13.46 | |
| | 44.61 | *UTX* | *UTY* | 13.91 | |
| X 保守区 (XCR) | 53.00 | dJ290F12.2 (*TSPYL2*) | *TSPY* (~35) | 9.50 | |
| | 53.12 | *SMCX* | *SMCY* | 20.27 | |
| | 71.27 | *RPS4X* | *RPS4Y1*, *RPS4Y2* | 2.77, 21.27 | |
| X 转置区 (XTR) | 88.50 | bB348B13.2§ | n/a | 2.96 | XTR |
| | 88.99 | *TGIF2LX* | *TGIF2LY* | 3.49 | XTR |
| | 91.26 | *PCDH11X* | *PCDH11Y* | 5.28 | XTR |
| X 保守区 (XCR) | 135.68 | *RNMX* (*RBMX*) | *RBMY* (6) | 22.02, 22.04, 22.37, 22.41, 22.66, 22.85 | |
| | 139.31 | *SOX3* | *SRY* | 2.70 | |
| | 148.38 | Em:AC016940.3 (*HSFX2*)§ | *HSFY1*, *HSFY2* | 19.3, 19.12 | |
| | 148.56 | Em:AC016939.4 (*HSFX1*)§ | *HSFY1*, *HSFY2* | 19.3, 19.12 | |
| 假常染色体区 PAR2 | 154.57 | *SPRY3* | *SPRY3* | 57.44 | PAR2 |
| | 154.71 | *SYBL1* | *SYBL1* | 57.58 | PAR2 |
| | 154.81 | *IL9R* | *IL9R* | 57.67 | PAR2 |
| | 154.81 | Em:AJ271736.5§ | Em:AJ271736.5§ | 57.69 | PAR2 |
| | 154.82 | Em:AJ271736.6 (*FAM39A*)§ | Em:AJ271736.6 (*FAM39A*)§ | 57.69 | PAR2 |

假基因不列入表中。

* 基因名称显示在补充图 1 中。当两个名字不同时，HUGO 用括号括起来。Em，EMBL 索取号。

† 距离参考 Y 染色体序列组装 NCBI35。存在多个 Y 染色体直系同源物，所有拷贝的位置显示在 Y 染色体上。唯一的例外是 TSPY，它在 Y 染色体上有 ~35 个拷贝，位于以 9.5 Mb 为中心的阵列中[41]。

‡ 主要同源区块如图 5 和图 6 所示。

§ 与 Y 染色体同源的 X 基因的新实例纳入这些类别。

自人类和黑猩猩谱系分化以来，XTR 通过从 X 染色体到 Y 染色体的物质复制产生[42]。复制区域横跨 X 染色体上的 3.91 Mb，但相应的区域在 Y 染色体上只有 3.38 Mb(图 5c)。我们比对了这一区域整个 X 和 Y 拷贝。不包括插入和缺失，拷贝之间序列一致性达到 98.78%。我们估计，转座事件发生在大约 470 万年前(补充讨

approximately 4.7 Myr ago (Supplementary Discussion 1), which is close to the suggested date of the speciation event that led to humans and chimpanzees, assumed here to be 6 Myr ago. The sequence alignment demonstrates the substantial changes to the XTR on the Y chromosome since the transposition. An inversion is known to have separated a 200-kb section from the rest of the XTR[43] (Fig. 5c). Also, the main block of homology is 540 kb shorter on Y than X, owing in particular to the absence of four large regions from the Y chromosome (Fig. 5c). The detection of these sequences at the expected positions on the chimpanzee X chromosome confirms that they were deleted from the Y chromosome after the transposition.

We found that only 54 of the 1,098 genes annotated on the X chromosome have functional homologues on the Y chromosome (Table 1). We obtained direct evidence for 24 genes in PAR1. Twenty-three of them are annotated (Supplementary Fig. 1), and the location of the 5′ end of *CRLF2* indicates that the rest of this gene is in gap 4 of the human X sequence (see the VEGA database). On the basis of the excellent conservation of synteny between human PAR1 and the chicken sequence, we infer that a stromal antigen gene (orthologue of chicken Ensembl gene ENSGALG00000016716) lies in gap 1 (see Fig. 3b). As the annotated putative transcript cM56G10.2 might represent the 3′ end of this gene, we conclude that PAR1 contains at least 24 genes. Together with the five annotated genes in PAR2, 29 genes lie entirely within the recombining regions of the sex chromosomes. Additionally, the *XG* locus spans the boundary between PAR1 and X-specific DNA, but has been disrupted by rearrangement on the Y chromosome.

Outside the XY-recombining regions of the X chromosome, we observed 25 genes that have functional homologues on the Y chromosome (Table 1). Fifteen of these are within the XAR, and a further three genes are shared by the X and Y copies of the XTR. The seven other XY gene pairs are believed to have descended from the proto-sex chromosomes. Only five cases have been described previously[44,45]: the X chromosome genes are *SOX3*, *SMCX*, *RPS4X*, *RBMX* and *TSPYL2*, which are located on the long arm and proximal short arm (Table 1). The two additional cases we report here involve heat-shock transcription factor genes, designated *HSFX1* and *HSFX2*. They are assigned to the category of XCR genes on the basis of a high degree of divergence from their Y chromosome homologues and their location distal to *SOX3* within the XCR. *HSFX1* and *HSFX2* lie within the separate copies of a palindromic repeat in Xq28 and are identical to each other. By analogy, their Y chromosome homologues (*HSFY1* and *HSFY2*) lie within the arms of a Y chromosome palindrome, the similarity of which is thought to be maintained by gene conversion[41].

On the basis of this and previously published information[41], we can conclude that approximately 15 protein-coding genes on the Y chromosome have no detectable X chromosome homologue.

论 1)，这与人类和黑猩猩物种分化的日期接近，分化时间假设是 600 万年前。序列比对显示，自转座以来，Y 染色体上的 XTR 发生了实质性的变化。已知的倒位将一个 200 kb 的片段从 XTR 其余的部分分离出来 [43](图 5c)。此外，Y 染色体上的主要同源性区块比 X 染色体的短 540 kb，尤其是因为 Y 染色体缺失了四个大型区域(图 5c)。对黑猩猩 X 染色体上预期位置序列的检测证实，它们是在转座后从 Y 染色体上删除的。

我们发现，X 染色体注释的 1,098 个基因中只有 54 个在 Y 染色体上有功能同源基因(表 1)。我们在 PAR1 中获得了 24 个基因的直接证据。其中 23 个是有注释的(补充图 1)，*CRLF2* 的 5′ 端的位置表明该基因的其余部分在人类 X 染色体序列缺口 4 中(见 VEGA 数据库)。根据人类 PAR1 和鸡序列间良好的同线性保守关系，我们推断，一个基质抗原基因(鸡 Ensembl 基因 ENSGALG00000016716 的直系同源物)位于缺口 1 中(见图 3b)。因为注释的假定转录本 cM56G10.2 可能代表这个基因的 3′ 末端，我们推断 PAR1 至少包含 24 个基因。连同 PAR2 中 5 个注释基因，29 个基因完全在性染色体重组区域内部。此外，*XG* 基因座跨越 PAR1 和 X 特异性 DNA 边界，但被 Y 染色体上的重排打乱。

在 X 染色体上 XY 重组区域之外，我们观察到 25 个基因在 Y 染色体上有功能同源基因(表 1)。其中 15 个基因在 XAR 内，另外三个基因由 XTR 的 X 和 Y 拷贝共享，其他 7 个 XY 基因对被认为是原始性染色体的后代。这其中只有 5 个实例是之前报道过的 [44,45]：在 X 染色体上的基因是 *SOX3*、*SMCX*、*RPS4X*、*RBMX* 和 *TSPYL2*，它们位于长臂和近端短臂上(表 1)。这里我们报道另外两个涉及热激转录因子基因的实例，这两个基因分别为 *HSFX1* 和 *HSFX2*。根据它们的 Y 染色体同源物与 XCR 远端 *SOX3* 位置的高度分化，我们将其归为 XCR 基因。*HSFX1* 和 *HSFX2* 位于 Xq28 一个回文重复的单独拷贝中并且完全一样。通过类推，其 Y 染色体同源物(*HSFY1* 和 *HSFY2*)位于 Y 染色体回文臂内，其相似性被认为是通过基因转变来维持的 [41]。

基于此信息以及以前公布的信息 [41]，我们可以得出结论，Y 染色体上大约 15 个蛋白编码基因没有可检测的 X 染色体同源物。

## The Progressive Loss of XY Recombination

The barrier to recombination between the proto-X and Y chromosomes initially encompassed the sex-determining locus on the Y (*SRY*) and possibly other loci affecting male fitness. It is proposed that rearrangement of the Y chromosome led to the development of this barrier. Thereafter, successive rearrangements that encompassed parts of the pseudoautosomal region resulted in segments of Y-linked DNA that could no longer recombine and consequently degenerated over time. Evidence for the role of Y-specific (as opposed to X-specific) rearrangement in this phenomenon is most clearly illustrated by our analysis of the XAR, which shows very little rearrangement between human and avian lineages (Fig. 3a).

In a previous study[46], four broad physical and evolutionary regions were defined on the X chromosome. The X chromosome genes within a given region all showed a similar level of divergence from their Y chromosome counterparts. However, between regions, levels of divergence were very different, presumably reflecting the stepwise loss of recombination between the X and Y chromosomes.

The physical order of the four regions on the X chromosome was seen to parallel their evolutionary ages, and therefore the chromosome was described as having four "evolutionary strata"[46]. In general, gene pairs were found to be less divergent moving through the strata from Xqter to Xpter. The first two strata (S1 and S2) encompass the long arm and proximal short arm, respectively, and were defined by the genes that survive from the proto-sex chromosomes. Gene pairs were found to be increasingly similar moving through strata 3 and 4, which occupy the proximal and distal sections of the XAR, respectively.

We re-evaluated XY homology in S4 and S3 using finished, genomic sequences from the two chromosomes. For S4 in particular, substantial blocks of homology exist between the chromosomes (blocks 1–10 in Fig. 5b and Fig. 6). Aligning the X and Y chromosome sequences across this region, we observed a bipartite organization, with markedly greater XY identity in the distal 1.0 Mb compared with the proximal 4.5 Mb (Fig. 7a). On this basis, the distal portion containing the *GYG2*, *ARSD*, *ARSE*, *ARSF*, *ADLICAN* and *PRKX* genes can be redefined as a new, fifth stratum, S5 (Figs 1 and 7a). A most parsimonious series of inversions, from the current arrangement of homologous blocks on X to that on Y, is consistent with the proposed strata (Fig. 7b). These data refine the picture of loss of XY recombination during evolution, which occurred by migration of the PABX in a stepwise manner distally through the XAR. The available evidence now suggests that there have been at least four PABX positions within the XAR, which are at the S2/S3, S3/S4 and S4/S5 boundaries (~47 Mb, ~8.5 Mb and ~4 Mb from Xpter, respectively), and at the current position (2.7 Mb from Xpter). We estimate that the two most recent PABX movements, which created first S4 and then S5, occurred 38–44 Myr ago and 29–32 Myr ago, respectively (Supplementary Discussion 2).

## XY 重组逐步丧失

原始 X 染色体和 Y 染色体之间重组的障碍最初包括 Y 染色体上的性别决定基因（*SRY*）和其他可能影响男性适合度的基因座。有人提出 Y 染色体的重排导致了这一障碍的发展。之后，包括部分假常染色体区域的连续重排导致 Y 连锁的 DNA 片段不能再重组，因此随着时间的推移这些片段逐渐退化。我们对 XAR 的分析清楚地说明了 Y 染色体特异性（与 X 特异性相反）重排在这一现象中所起的作用。XAR 显示人类和鸟类谱系之间几乎没有重排（图 3a）。

在先前的研究中[46]，在 X 染色体上定义了四个广泛的物理和进化区域。给定区域内 X 染色体基因都表现出与 Y 染色体对应物相似程度的分化。然而，在不同的区域之间，分化程度有很大的不同，这大概反映了 X 和 Y 染色体间重组的逐步丧失。

可以看出 X 染色体上四个区域的物理顺序与它们的进化年龄平行，因此，染色体被描述为有四个"进化阶段"[46]。一般情况下，在从 Xqter 到 Xpter 的基因阶段中，基因对的差异很小。前两个阶段（S1 和 S2）分别包含长臂和近端短臂，这两个阶段通过原始性染色体存留下来的基因确定。位于 XAR 近端和远端的基因对在第 3 阶段和第 4 阶段越来越相似。

我们使用来自两条染色体的完整基因组序列重新评估了 S4 和 S3 中的 XY 同源性。尤其是 S4，大量同源区块存在于染色体之间（在图 5b 和图 6 中的区块 1 ~ 10）。比对这一区域的 X 染色体和 Y 染色体序列，我们观察到了一个由两部分构成的组织。与近端 4.5 Mb 相比，远端 1.0 Mb 的 XY 特性更显著（图 7a）。在此基础上，我们将包含 *GYG2*、*ARSD*、*ARSE*、*ARSF*、*ADLICAN* 和 *PRKX* 基因的远端部分重新定义为一个新的第五阶段——S5（图 1 和 7a）。从目前的 X 染色体上同源区块的排列，到 Y 染色体上同源区块的排列这一系列最简约的倒位与之前提出的阶段一致（图 7b）。这些数据完善了进化过程中 XY 重组缺失的图像，这种缺失是由 PABX 在远端通过 XAR 的逐步迁移造成的。现有的证据表明，在 XAR 内至少有四个 PABX 位置，分别位于 S2/S3、S3/S4 和 S4/S5 边界（分别位于 Xpter 的 ~47 Mb，~8.5 Mb 和 ~4 Mb）和当前位置（位于 Xpter 的 2.7 Mb）。我们估计最近的两次 PABX 运动，先产生了 S4，之后产生了 S5，分别发生在 3,800 万 ~ 4,400 万年前和 2,900 万 ~ 3,200 万年前（补充讨论 2）。

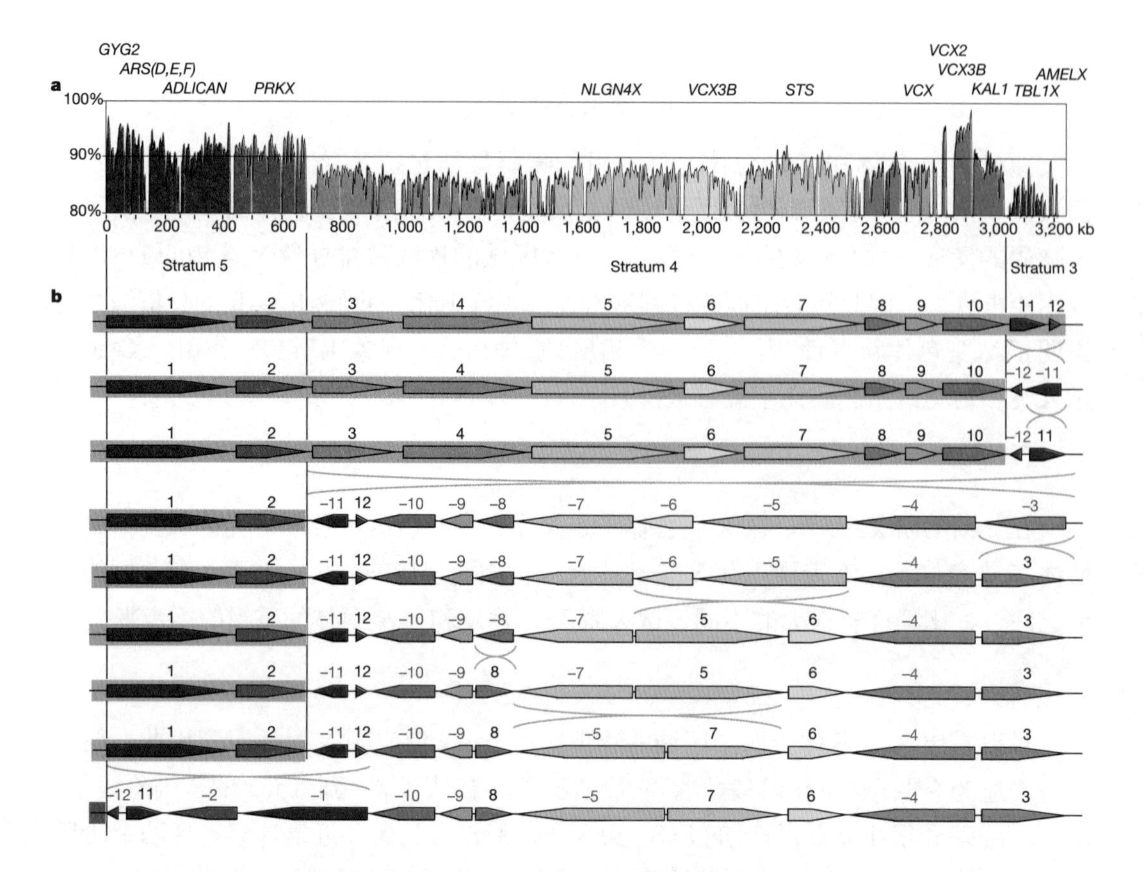


Fig. 7. Evidence for a fifth evolutionary stratum on the X chromosome. **a**, Sequence identity between the X and Y homology blocks 1–12 (see Figs 5b and 6) plotted in 5-kb windows. The scale shows the total amount of sequence aligned, excluding insertions and deletions (see Methods). A 10-kb spacer is placed between each consecutive block of homology. Segments of the plot are coloured according to the system used in Figs 6 and 7b. On the basis of this plot, a new evolutionary stratum S5 is defined, which includes homology blocks 1 and 2. **b**, A most parsimonious series of inversion events from the arrangement of homology blocks 1–12 on the X chromosome (top) to the Y chromosome (bottom), calculated using GRIMM[64]. The grey boxes show the suggested extents of former pseudoautosomal regions within the distal part of the XAR, and the magenta box (bottom row) shows the position of the current pseudoautosomal region. This inversion sequence provides independent support for the proposed pseudoautosomal boundary movements and evolutionary strata. It was previously suggested that *AMELX* (in block 12) is in S4 (ref. 46), or possibly at the boundary between S3 and S4 (ref. 67). However, the more distal location of block 11, which contains *TBL1X* (an S3 gene[46]), is not consistent with these suggestions. The two regions of increased sequence identity within block 10 contain the *VCX2* and *VCX3B* genes on the X chromosome and the *VCY1B* and *VCY* genes on the Y chromosome. This gene family might have arisen *de novo* in the simian lineage[68], which could account for the unusual characteristics of this part of the alignment.

In addition to the varied degree of XY sequence identity within S3, S4, S5 and PAR1, we found marked differences in their sequence composition, which were presumably also caused by the loss of recombination in each region during evolution. Specifically, we observed that L1, L2 and mammalian interspersed repeat (MIR) coverage decrease with each more distal stratum and PAR1 (Table 2 and Fig. 1), but (G+C) levels and *Alu* repeat content increase abruptly at the boundary between S4 and S5 (Table 2 and Fig. 8); variations in the incidence of different *Alu* subfamilies (Y, S and J) also contribute to the

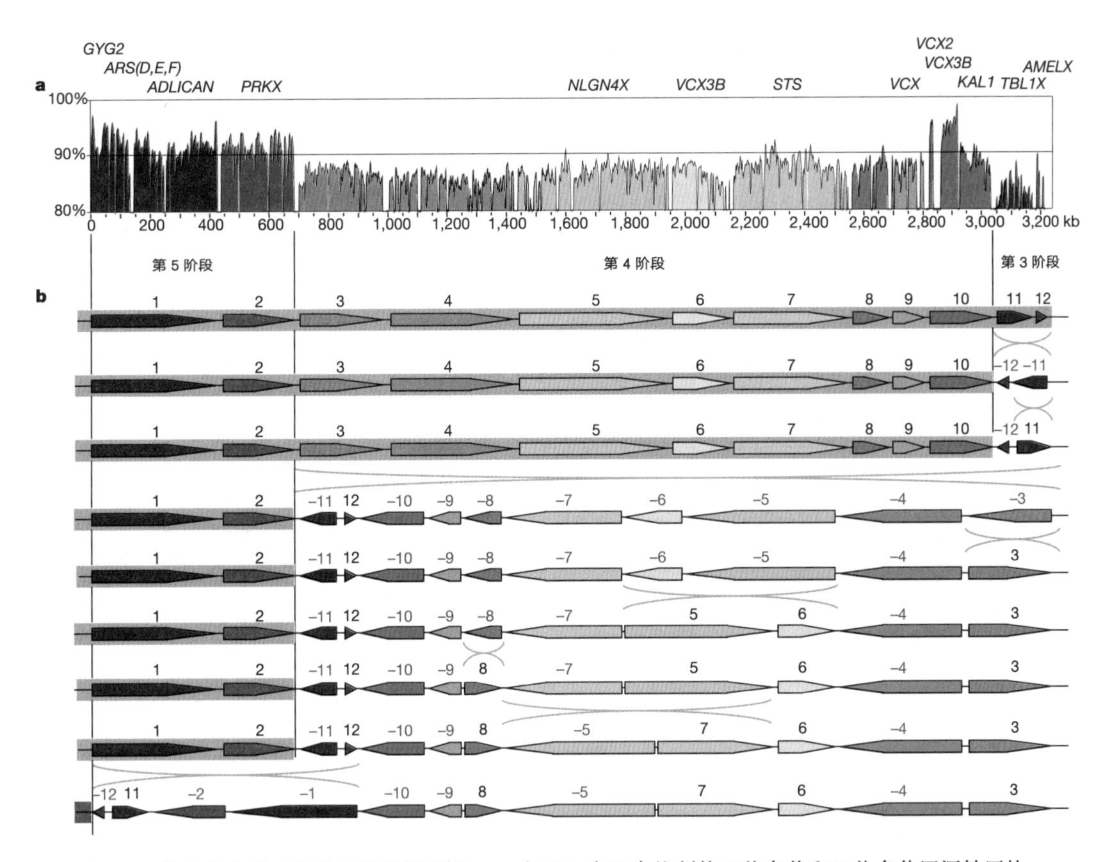


图 7. X 染色体上第五个进化阶段的证据。**a**，在 5 kb 窗口中绘制的 X 染色体和 Y 染色体同源性区块 1 ~ 12 的序列一致性(见图 5b 和 6)。该比例显示序列比对的总量，但不包括插入和缺失(见方法)。在每个同源性连续区块之间有 10 kb 的间隔。根据图 6 和 7b 中用的系统对每个片段着色。在此划分的基础上，我们确定了一个新的进化阶段 S5，包括同源块 1 和 2。**b**，用 GRIMM 计算，X 染色体(上)到 Y 染色体(下)同源区块 1 ~ 12 排列的一系列最简约的倒位事件[64]。灰色框显示 XAR 远端部分内之前的假常染色体区域的建议范围，品红色方块(下排)表明目前的假常染色体区域的位置。倒位序列为假常染色体边界运动和进化阶段的观点提供了独立的支持。早前有人提出，*AMELX*(在区块 12)位于 S4 中(参考文献 46)，也可能位于 S3 和 S4 间的边界(参考文献 67)。然而，包含 *TBL1X*(一个 S3 基因[46])的区块 11 的较远端的位置与这些观点不符。区块 10 内序列一致性增加的两个区域包含 X 染色体上的 *VCX2* 和 *VCX3B* 基因以及 Y 染色体上的 *VCY1B* 和 *VCY* 基因。这一基因家族可能是在类人猿谱系中新产生的[68]，这可以解释这部分排列的不寻常特征。

除了 S3、S4、S5 和 PAR1 内 XY 序列同源程度不同外，我们还发现它们的序列组成存在显著差异，这也可能是进化时每个区域内重组缺失造成的。具体来说，我们观察到 L1、L2 和哺乳动物散在重复(MIR)覆盖度随每个更远端阶段和 PAR1 减少(表 2 和图 1)，但(G+C)水平和 *Alu* 重复含量在 S4 和 S5 边界突然增加(表 2 和图 8)；不同 *Alu* 亚家族(Y、S 和 J)发病率的变化促成了每个阶段和 PAR1 的不同

distinct character of each stratum and PAR1 (Supplementary Table 13). The compositional differences between S4 and S5 provide additional support for the subdivision of the original stratum 4 (Fig. 8).

Table 2. Sequence characteristics of evolutionary domains of the X chromosome

| Region | (G+C) (%) | L1 (%) | L1P (%) | L1M (%) | *Alu* (%) | L2 (%) | MIR+MIR3 (%) |
|---|---|---|---|---|---|---|---|
| X chromosome | 39.46 | 28.87 | 13.39 | 15.21 | 8.23 | 2.98 | 2.07 |
| XAR | 39.87 | 17.89 | 6.60 | 11.23 | 10.28 | 2.63 | 1.76 |
| XCR | 39.28 | 33.50 | 16.38 | 16.97 | 7.28 | 3.12 | 2.19 |
| PAR1 | 48.11 | 6.97 | 2.64 | 4.38 | 28.88 | 0.24 | 0.21 |
| S5 | 42.86 | 8.89 | 4.36 | 4.59 | 18.72 | 0.66 | 0.31 |
| S4 | 38.87 | 11.10 | 4.31 | 6.80 | 8.60 | 1.59 | 0.95 |
| S3 | 39.46 | 19.55 | 7.14 | 12.34 | 9.24 | 2.94 | 1.98 |

See Supplementary Table 13 for additional repeat element data.

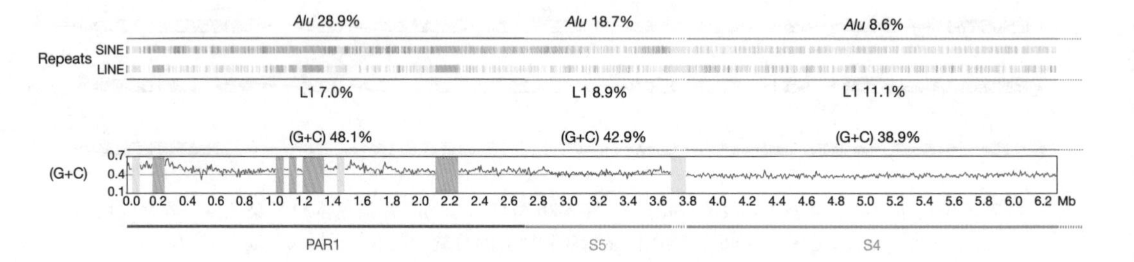


Fig. 8. Sequence compositional changes in the distal evolutionary strata of the X chromosome. Shown are the positions of SINE and LINE repeats and (G+C) content within PAR1, S5 and the distal half of S4. The percentage of *Alu*, L1 and (G+C) are shown for each region (including the whole of S4). There is an abrupt increase in *Alu* repeat levels and (G+C) content from S4 to S5. The five euchromatic gaps in PAR1 are shown as light brown bars. Pale blue bars represent clones for which the sequences were unfinished at the time of the sequence assembly.

## X-chromosome Inactivation

XCI in mammals achieves dosage compensation between males and females for X-linked gene products. Inactivation of one X chromosome occurs early in female development and is initiated from the X-inactivation centre (XIC). The *XIST* transcript is expressed initially on both X chromosomes, but later the transcript from the chromosome that is destined for inactivation becomes more stable than the other. Finally, the transcript is expressed only from the inactive X chromosome ($X_i$). Coating with the *XIST* transcript is the earliest of many chromatin modifications on $X_i$.

XCI was first proposed based partly on the study of X: autosome translocations in female mice[47]. Studies of derivative chromosomes containing inactivated X chromosome segments

特性(补充表 13)。S4 和 S5 之间的成分差异为原始阶段 4 的细分提供了额外的支持(图 8)。

表 2. X 染色体进化区域的序列特征

| 区域 | (G+C) (%) | L1 (%) | L1P (%) | L1M (%) | *Alu* (%) | L2 (%) | MIR+MIR3 (%) |
|---|---|---|---|---|---|---|---|
| X 染色体 | 39.46 | 28.87 | 13.39 | 15.21 | 8.23 | 2.98 | 2.07 |
| XAR | 39.87 | 17.89 | 6.60 | 11.23 | 10.28 | 2.63 | 1.76 |
| XCR | 39.28 | 33.50 | 16.38 | 16.97 | 7.28 | 3.12 | 2.19 |
| PAR1 | 48.11 | 6.97 | 2.64 | 4.38 | 28.88 | 0.24 | 0.21 |
| S5 | 42.86 | 8.89 | 4.36 | 4.59 | 18.72 | 0.66 | 0.31 |
| S4 | 38.87 | 11.10 | 4.31 | 6.80 | 8.60 | 1.59 | 0.95 |
| S3 | 39.46 | 19.55 | 7.14 | 12.34 | 9.24 | 2.94 | 1.98 |

其他重复元件数据见补充表 13。

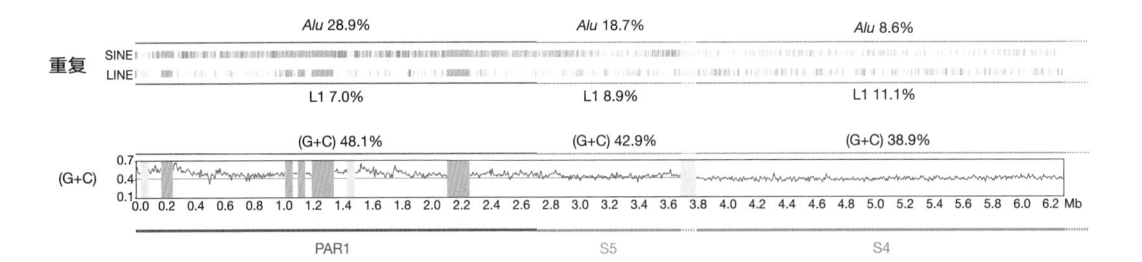


图 8. 在 X 染色体进化最远阶段序列成分变化。图中所示的是 SINE 和 LINE 重复的位置以及 PAR1、S5 和 S4 远端一半的(G+C)含量。每个区域显示了 *Alu*、L1 和(G+C)的百分比(包括整个 S4)。*Alu* 重复水平和(G+C)含量从 S4 到 S5 突然增加。PAR1 中五个常染色质缺口显示为浅褐色方框。淡蓝色方框代表序列组装时未完成序列的克隆。

## X 染色体失活

在哺乳动物中，XCI 实现了雄性与雌性之间 X 连锁基因产物的剂量补偿效应。一条 X 染色体失活发生在雌性发育早期，而且是从 X 失活中心(XIC)开始。*XIST* 转录本最初在两条 X 染色体上都有表达，但后来在注定要失活的染色体上的转录本变得比另一条更稳定。最后，转录本只在失活的 X 染色体($X_i$)上表达。在 $X_i$ 上的许多染色质修饰中，包覆 *XIST* 转录本是最早的。

XCI 的首次提出部分是基于对 X 染色体的研究：雌性老鼠中常染色体易位[47]。我们通过对含有失活 X 染色体片段的衍生染色体进行研究后得出结论，失活可以跨

later concluded that the inactivation could spread across the translocation boundary to the autosomal segment, but that inactivation of this segment was incomplete. More recently, it has become clear that more than 15% of the genes on the human X chromosome, including many without functional equivalents on the Y, escape from XCI, as presented in detail elsewhere[48]. The majority of the genes that escape XCI lie within the distal regions of the XAR (Fig. 1): all genes studied in PAR1, S5 and S4 were found to escape from XCI, but there is a lower proportion of escapees in S3, and very few examples in the XCR[48]. This observation correlates with our picture of X chromosome evolution: XCI follows Y chromosome attrition[49], which is less advanced in the distal strata of the XAR.

Inefficient inactivation of the autosomal segment in $X_i$: autosome translocations led to the proposal that "way stations" on the X chromosome boost the spread of XCI. According to this model, way stations are present throughout the genome but are enriched on the X chromosome, particularly in the region of the XIC[50]. Lyon suggested that L1 elements are good candidates for acting as way stations on account of their enrichment on the mammalian X chromosome[51]. We observe a distribution of L1 elements on the chromosome that is consistent with both the way station and the Lyon hypotheses (Fig. 1 and Table 2). The coverage of L1 repeats is very high in the XCR, especially around the XIC. As noted previously[52], this enrichment in L1 levels is accounted for particularly by elements that were active more recently in mammalian evolution[53] (L1P in Fig. 1). In the XAR, L1 coverage is close to autosome levels, whereas L1 levels are particularly low in the distal evolutionary strata of the XAR, where genes consistently escape inactivation. The *XIST* locus itself lies in a 60 kb region that is virtually devoid of L1 elements, whereas L1 levels are extremely high in the adjacent regions. Based on their distributions, other interspersed repeats are not strong candidates for way stations. For example, although L2 and MIR elements are reduced in S4, S5 and particularly PAR1 relative to the rest of the chromosome, their overall levels on the X chromosome are not enriched relative to the autosomes but are slightly reduced. Furthermore, L2 and MIR levels are low in the region distal to the XIC. These characteristics do not preclude an involvement in XCI, but are not consistent with a role as way stations.

The possible causal relationship of L1 elements to the spread of XCI remains a subject of debate. Some studies have reported significant associations between L1 coverage and inactivation[52], and others have refuted this[54]. Our observations on regional differences in composition emphasize that such studies should compare active and inactivated genes (or domains) from the same evolutionary stratum, in order to avoid correlations that are unrelated to XCI.

## Medical Genetics and the X Chromosome Sequence

The X chromosome holds a unique place in the history of medical genetics. Ascertainment of X-linked diseases is enhanced by the relative ease of recognizing this mode of inheritance. More important, however, is the fact that a disproportionately large number of

越易位边界扩散到常染色体片段中，但该片段失活尚不完全。最近，人们已经清楚地看到：人类 X 染色体上超过 15%的基因，包括许多在 Y 染色体上没有功能等同物的基因，没有发生 XCI，这在其他地方也有详细说明 [48]。大多数没发生 XCI 的基因位于 XAR 远端区域(图 1)：在 PAR1、S5 和 S4 中研究的所有基因都没发生 XCI，但 S3 中没发生 XCI 的比例很低，在 XCR 中实例也很少 [48]。这一观察结果与我们的 X 染色体进化情况有关：XCI 发生在 Y 染色体削弱之后 [49]，而 Y 染色体的削弱在 XAR 较远的阶段几乎不发生。

$X_i$ 中常染色体片段低效失活：常染色体易位导致提出 X 染色体上的“中转站”促进 XCI 的扩散这一观点。根据这一模型，中转站出现在整个基因组中，但在 X 染色体上富集，特别是在 XIC 区域 [50]。莱昂认为由于 L1 元件在哺乳动物 X 染色体上富集，所以它是作为中转站的理想选择 [51]。我们观察到 L1 元件在染色体上的分布符合中转站和莱昂假设(图 1 和表 2)。在 XCR 中，L1 重复的覆盖率非常高，特别是 XIC 周围。正如之前所述 [52]，最近在哺乳动物进化中活跃的元件解释了 L1 水平的富集 [53](图 1 中的 L1P)。在 XAR 中，L1 覆盖率接近常染色体水平，而 L1 水平在 XAR 远端进化阶段特别低，在该阶段基因始终没有发生 XCI。*XIST* 基因座本身位于一个实际上缺少 L1 元件的 60 kb 区域，而在邻近区域 L1 水平极其高。根据它们的分布，其他散在重复中没有强有力的中转站候选。例如，虽然 L2 和 MIR 元件在 S4、S5，尤其是 PAR1 中相对于染色体的其余部分有所减少，但它们在 X 染色体上的总体水平相对于常染色体并没有富集，而是略有降低。此外，L2 和 MIR 水平在 XIC 远端较低。这些特征不排除参与 XCI，但不符合中转站的功能。

L1 元件与 XCI 扩展可能的因果关系仍然是一个有争议的话题。一些研究报告了 L1 的覆盖率和失活之间的显著联系 [52]，而另一些研究则反驳了这一说法 [54]。我们经过对区域组成差异的观察后，强调了这种研究应比较同一进化阶段的活性和失活基因(或域)，以避免与 XCI 无关的相关性。

## 医学遗传学和 X 染色体序列

X 染色体在医学遗传学历史上占有独特的地位。识别这种遗传模式相对容易，于是增强了对 X 连锁疾病的确定。然而更重要的是，实际上大量的疾病条件都与 X 染色体有关，因为任何在 Y 染色体上没有活性对应物的基因，其隐性突变的表型结

disease conditions have been associated with the X chromosome because the phenotypic consequence of a recessive mutation is revealed directly in males for any gene that has no active counterpart on the Y chromosome. Thus, although the X chromosome contains only 4% of all human genes, almost 10% of diseases with a mendelian pattern of inheritance have been assigned to the X chromosome (307 out of 3,199; information obtained from OMIM[2]). These two aspects of the medical genetics of the X chromosome have greatly stimulated progress in the positional cloning of many genes associated with human disease. To date, the molecular basis for 168 X-linked phenotypes has been determined, and the X chromosome sequence has aided this process for 43 of them, by providing positional candidate genes or a reference sequence for comparison to patient samples (Supplementary Table 14).

Identifying genes involved in rare conditions yields important biological insights. For example, discovery of mutations in the *SH2D1A* gene[55] (involved in X-linked lymphoproliferative disease (XLP, OMIM 308240)) led to identification of a new mediator of signal transduction between T and NK cells, and a novel family of proteins involved in the regulation of the immune response. Mental retardation is one of the most common problems in clinical genetics, and affects significantly more males than females. To date, 16 genes from the X chromosome have been associated with cases of non-syndromic X-linked mental retardation (NS-XLMR), in which mental retardation is the only phenotypic feature. These genes encode a range of protein types, and some are also involved in syndromic forms of mental retardation. For example, the *ARX* gene encodes an aristaless-related homeobox transcription factor and is linked to NS-XLMR cases, as well as to syndromic mental retardation associated with epilepsy (infantile spasm syndrome, ISSX, OMIM 308350) or with dystonic hand movements (Partington syndrome, PRTS, OMIM 309510)[56]. The *MECP2* gene, which encodes a methyl-CpG-binding protein, was initially linked to cases of Rett syndrome in girls[57] (RTT, OMIM 312750) but was later also seen to be mutated in males or females with NS-XLMR[58]. The molecular defect has been determined in only a minority of families affected by NS-XLMR, which has led to speculation that there could be as many as 100 genes on the X chromosome that are associated with NS-XLMR[59]. Discovering the genes for these and other rare, monogenic disorders is of critical value in extending our understanding of fundamental new processes in human biology, and the annotated X chromosome will further facilitate this process.

## Concluding Remarks

The completion of the X chromosome sequencing project is an essential component of the goal of obtaining a high-quality, annotated human genome sequence for use in studies of gene function, sequence variation, disease and evolution. It also means that for the first time, we now have the finished sex chromosome sequences of an organism. The study of these sequences gives a greater insight into mammalian sex chromosome evolution and its consequences. As these analyses are extended to other genomes, we will gain a greater appreciation of the different evolutionary forces that shape sex chromosome and autosome

果直接在男性中显示。因此，尽管 X 染色体只含有 4%的人类基因，但几乎 10%的孟德尔遗传模式的疾病与 X 染色体有关（3,199 种疾病中有 307 种；信息从 OMIM[2] 获得）。X 染色体的医学遗传学的这两个方面极大地促进了与人类疾病相关的许多基因定位克隆的发展。迄今为止，我们已经确定了 168 个 X 连锁表型的分子基础，通过为患者样本的比较提供位置候选基因或参考序列，X 染色体序列为其中 43 个的分子基础的确定提供了帮助（补充表 14）。

识别罕见情况下涉及的基因可以获得重要的生物学见解。例如，*SH2D1A* 基因突变的发现[55]（涉及 X 连锁淋巴组织增生性疾病（XLP，OMIM 308240））使得我们在 T 细胞和 NK 细胞之间鉴定出一个新的信号转导调节物，以及一个涉及免疫反应调节的新的蛋白家族。精神发育迟缓是临床遗传学最常见问题之一，并且它对男性的影响比对女性的大。迄今为止，已有 16 个来自 X 染色体的基因与非综合征 X 连锁智力低下（NS-XLMR）病例相关，其中智力低下是唯一的表型特征。这些基因编码一系列蛋白类型，其中一些基因还与精神发育迟缓的综合征形式有关。例如，*ARX* 基因编码一个无芒样同源框转录因子，并与 NS-XLMR 病例以及与癫痫（婴儿痉挛综合征，ISSX，OMIM 308350）或手部运动异常（帕廷顿综合征，PRTS，OMIM 309510）有关的精神发育迟缓的综合征相关[56]。编码甲基化 CpG 结合蛋白的 *MECP2* 基因，最初与女孩中的雷特综合征的病例有关[57]（RTT，OMIM 312750），但后来也被发现在患有 NS-XLMR 的男性或女性中发生突变[58]。这种分子缺陷只在少数受 NS-XLMR 影响的家族中被发现，人们推测 X 染色体上可能有多达 100 个基因与 NS-XLMR 相关[59]。发现上述基因和其他罕见的单基因疾病对于扩展我们对人类生物学新的基本过程的理解具有重要价值，而标注的 X 染色体将进一步促进这一过程。

## 结 束 语

X 染色体测序项目的完成是获得用于基因功能、序列变异、疾病和进化研究的高质量、带注释的人类基因组序列的重要组成部分。这也意味着，我们第一次拥有了一个有机体的已完成的性染色体序列。对这些序列的研究使我们对哺乳动物性染色体进化及其结果有了更深入的了解。当这些分析扩大到其他基因组时，我们将对形成性染色体和常染色体的不同进化力量有更深入的了解。非常重要的是研究突变过程速率的差异，并考虑不同寻常的雄性重组模式对这些过程的影响。显然，这种

evolution. It will be important to study differences in the rates of mutational processes, and to consider the influence of the unusual pattern of male recombination on these processes. Clearly, this analysis should not be restricted to a consideration of mammalian sex chromosomes, and it will be of great interest to make comparisons with non-mammalian systems that arose independently in evolution.

## Methods

The approach used to establish a bacterial clone map of the X chromosome has been previously described[5]. 13,264 clones were identified using 4,363 STS markers derived from published genetic or physical maps, from shotgun sequencing of flow-sorted X chromosomes, or from end-sequences of clones at contig ends. Clones were assembled into contigs using restriction-enzyme fingerprinting, and were integrated with the Washington University Genome Sequencing Center whole genome BAC map[60] in order to identify additional clones. Nine euchromatic gaps were measured using fluorescent *in situ* hybridization of clones to extended DNA fibres, and a tenth gap was estimated on the basis of end-sequence data from spanning, unstable BAC clones (Supplementary Table 2). On the basis of pulsed-field gel electrophoresis experiments, we expect the sizes of the other four euchromatic gaps to have a combined size of less than 400 kb.

Finished sequences of individual clones were determined using procedures described in ref. 7. For the analyses described above, the sequence was frozen in March 2004, at which point 150,396,262 bp of sequence had been determined from a minimal tiling path of 1,832 clones (1,616 sequence accessions). This sequence is available at http://www.sanger.ac.uk/HGP/ChrX/, and its annotation is represented in Supplementary Fig. 1. Updates to the sequence and annotation can be obtained from the VEGA database.

Manual annotation of gene structures has been described elsewhere[14], and used guidelines agreed at the human annotation workshop (HAWK; http://www.sanger.ac.uk/HGP/havana/hawk.shtml). Genes were assigned to one of four groups: (1) known genes that are identical to human cDNAs or protein sequences and have a RefSeq RNA (and RefSeq protein, if the gene encodes a protein); (2) novel coding sequences, which have an open reading frame (ORF) and are identical to spliced ESTs, or have similarity to other genes/proteins (any species); (3) novel transcripts, which are similar to novel coding sequences, except that no ORF can be determined with confidence; and (4) putative transcripts, which are identical to splicing human ESTs but have no ORF. Gene symbols were approved by the HUGO Gene Nomenclature Committee wherever possible. Predicted protein translations were analysed for Pfam domains using InterProScan (http://www.ebi.ac.uk/InterProScan/). CpG islands were predicted using the program GpG (G. Micklem, personal communication).

Interspersed repeats were identified and classified using RepeatMasker (http://repeatmasker.genome.washington.edu). In order to search for segmental duplications, WU-BLASTN (http://blast.wustl.edu) was used to align the current X chromosome sequence to itself or to the NCBI34 autosome assemblies. Duplicated blocks at least 5 kb in length were defined as described in ref. 28.

分析不应局限于对哺乳动物性染色体的考虑，将哺乳动物系统与在进化中独立产生的非哺乳动物系统进行比较将是非常有趣的。

## 方　法

建立 X 染色体细菌克隆图谱的方法以前已描述过[5]。13,264 个克隆通过 4,363 个 STS 标记被鉴定出来，这些 STS 标记来自已发表的遗传或物理图谱、流式 X 染色体鸟枪法测序、重叠群末端克隆的末端序列。用限制酶指纹法将克隆组装到重叠群，并与华盛顿大学基因组测序中心全基因组 BAC 图谱[60]进行整合，用以确定其他的克隆。使用荧光原位杂交克隆来测量 9 条常染色体缺口以扩展 DNA 纤维，基于不稳定 BAC 克隆末端序列数据，我们对第十个缺口进行了评估(补充表 2)。根据脉冲场凝胶电泳实验，我们预计其他四个常染色体缺口的合并大小小于 400 kb。

使用参考文献 7 中描述的步骤确定单个克隆的完成序列。在上述的分析中，该序列在 2004 年 3 月被冻结，那时从 1,832 个克隆的最小叠瓦式中确定了 150,396,262 bp 的序列(1,616 个序列登记入册)。该序列可在 http://www.sanger.ac.uk/HGP/ChrX/获得，其注解如补充图 1 所示。更新的序列和注解可从 VEGA 数据库获取。

人工注释基因结构在其他地方也有描述[14]，并且使用的准则经过了人类注解研讨会的同意(HAWK；http://www.sanger.ac.uk/HGP/havana/hawk.shtml)。基因被分为四组：(1) 与人类 cDNA 或蛋白质序列一致且具有 RNA 参考序列(以及蛋白质参考序列，如果该基因编码了一种蛋白质)的已知基因；(2) 有开放阅读框且与剪切表达序列标签一致或与其他基因/蛋白(任何物种)相似的新编码序列；(3) 与新编码序列一致的新转录本，但排除了确定没有开放阅读框的情况；(4) 与剪切人类表达序列标签一致但没有开放阅读框的假定转录本。基因符号尽可能由 HUGO 基因命名委员会批准。用 InterProScan 分析 Pfam 域的预测蛋白翻译(http://www.ebi.ac.uk/InterProScan/)。使用 CpG 程序分析 CpG 岛(米克勒姆，个人交流)。

使用 RepeatMasker(http://repeatmasker.genome.washington.edu)对散在重复序列进行识别和分类。为了寻找片段重复，我们使用 WU-BLASTN(http://blast.wustl.edu)将当前的 X 染色体序列与自身或与 NCBI34 常染色体组进行比对。如参考文献 28 所述，确定的重复区块长度至少有 5 kb。

SNPs (dbSNP release 119) were mapped onto the X chromosome sequence using first SSAHA[61] and then Cross-match (http://www.phrap.org/phredphrapconsed.html).

## Comparative analysis

The genome assemblies used for comparative analyses were: *Gallus gallus* WASHUC1 (Washington University Genome Sequencing Center, http://www.genome.wustl.edu/projects/chicken), *Rattus norvegicus* RGSC3.1 (Rat Genome Sequencing Consortium http://www.hgsc.bcm.tmc.edu/projects/rat/), *Mus musculus* NCBI32 (Mouse Genome Sequencing Consortium, http://www.ncbi.nlm.nih.gov/genome/seq/NCBIContigInfo.html), *Danio rerio* version 3 (Sanger Institute, http://www.sanger.ac.uk/Projects/D_rerio), *T. nigroviridis* version 6 (Genoscope and the Broad Institute, http://www.genoscope.cns.fr/externe/tetraodon/Ressource.html), and *F. rubripes* version 2 (International Fugu Genome Consortium, http://www.fugu-sg.org/project/info.html). ECRs between the X chromosome and the rodent and fish genomes were obtained as described elsewhere[13]. In order to visualize regions of conserved synteny, the X chromosome sequence was aligned to the chicken and rodent genome sequences using BLASTZ (with default parameters), and matches were plotted by chromosome position. Matches to the rodent genomes were filtered to include only those with a sequence identity of at least 70% to the human sequence. The Ensembl database (http://www.ensembl.org) was used to search for orthologous gene pairs between the X chromosome and the other three genomes.

Genomic sequence homologies between the X and Y chromosomes were identified by aligning the two finished chromosome sequences using WU-BLASTN, and then filtering the alignments to include only those of at least 70% sequence identity and 80 bp length. In order to calculate the sequence identity between large, XY-homologous regions, a global alignment of unmasked sequence was generated using LAGAN[62]. Gapped regions, which result from insertions or deletions, were removed from the alignment, and then the nucleotide sequence identity was calculated for the remainder. Sequence identity plots were produced by parsing the LAGAN output into VISTA[63]. GRIMM[64] was used to calculate a most parsimonious series of inversions that would account for differences in homology block order and orientation between the X and Y chromosomes. Homologous protein-coding gene pairs between the X and Y chromosomes were identified by TBLASTN searching with the coding sequences of annotated coding genes on the Y chromosome against the X chromosome genomic sequence.

(**434**, 325-337; 2005)

**Mark T. Ross[1], Darren V. Grafham[1], Alison J. Coffey[1], Steven Scherer[2], Kirsten McLay[1], Donna Muzny[2], Matthias Platzer[3], Gareth R. Howell[1], Christine Burrows[1], Christine P. Bird[1], Adam Frankish[1], Frances L. Lovell[1], Kevin L. Howe[1], Jennifer L. Ashurst[1], Robert S. Fulton[4], Ralf Sudbrak[5,6], Gaiping Wen[3], Matthew C. Jones[1], Matthew E. Hurles[1], T. Daniel Andrews[1], Carol E. Scott[1], Stephen Searle[1], Juliane Ramser[7], Adam Whittaker[1], Rebecca Deadman[1], Nigel P. Carter[1], Sarah E. Hunt[1], Rui Chen[2], Andrew Cree[2], Preethi Gunaratne[2], Paul Havlak[2], Anne Hodgson[2], Michael L. Metzker[2], Stephen Richards[2], Graham Scott[2], David Steffen[2], Erica Sodergren[2], David A. Wheeler[2], Kim C. Worley[2], Rachael Ainscough[1], Kerrie D. Ambrose[1], M. Ali Ansari-Lari[2], Swaroop Aradhya[2], Robert I. S. Ashwell[1], Anne K. Babbage[1], Claire L. Bagguley[1], Andrea Ballabio[2], Ruby Banerjee[1], Gary E. Barker[1], Karen F. Barlow[1], Ian P. Barrett[1], Karen N. Bates[1], David M. Beare[1],**

首先使用 SSAHA[61] 将 SNP(dbSNP 发布版 119)映射到 X 染色体序列上，然后进行交叉匹配(http://www.phrap.org/phredphrapconsed.html)。

## 对比分析

用于比较分析的基因组组合是：原鸡 WASHUC1(华盛顿大学基因组测序中心，http://www.genome.wustl.edu/project/chicken)，褐家鼠 RGSC3.1(大鼠基因组测序协作组 http://www.hgsc.bcm.tmc.edu/projects/rat/)，小家鼠 NCBI32(小鼠基因组测序协作组，http://www.ncbi.nlm.nih.gov/genome/seq/NCBIContigInfo.html)，斑马鱼第 3 版(桑格研究所，http://www.sanger.ac.uk/Project/D_rerio)，黑青斑河鲀第 6 版(Genoscope 和布罗德研究所，http://www.genoscope.cns.fr/externe/tetraodon/Ressource.html)和红鳍东方鲀第 2 版(国际河豚基因协作组 http://www.fugu-sg.org/project/info.html)。X 染色体与啮齿动物和鱼类基因组之间的 ECR 如其他地方所述 [13]。为了可视化保守的同源区域，X 染色体序列使用 BLASTZ(默认参数)与鸡和啮齿类动物基因组序列进行比对，并通过染色体位置绘制匹配。与啮齿动物基因组匹配的基因经过筛选，只包括那些至少与人类序列一致性达 70%的基因。使用 Ensembl 数据库(http://www.ensembl.org)来寻找 X 染色体和其他三个基因组间直系同源基因对。

通过用 WU-BLASTN 对两个已完成的染色体序列进行比对，确定了 X 染色体和 Y 染色体间的基因组序列同源性，然后对比对结果进行筛选，只包含序列一致性不低于 70%、长度不小于 80 bp 的染色体序列。为了计算大的 XY 同源性区域间的序列一致性，使用 LAGAN[62] 生成了未掩蔽全序列比对。将插入或删除导致的缺口区域从比对中移除，然后计算剩余序列的核苷酸序列一致性。通过将 LAGAN 输出解析为 VISTA[63] 来生成序列标识图。用 GRIMM[64] 计算出一系列最简单的逆序序列，这些逆序序列可以解释 X 染色体和 Y 染色体在同源块顺序和方向上的差异。利用带注释的编码序列进行 TBLASTN 搜索，确定 X 染色体和 Y 染色体之间的同源蛋白编码基因对。

(李梅 翻译；胡松年 审稿)

Helen Beasley[1], Oliver Beasley[1], Alfred Beck[5], Graeme Bethel[1], Karin Blechschmidt[3], Nicola Brady[1], Sarah Bray-Allen[1], Anne M. Bridgeman[1], Andrew J. Brown[1], Mary J. Brown[2], David Bonnin[2], Elspeth A. Bruford[8], Christian Buhay[2], Paula Burch[2], Deborah Burford[1], Joanne Burgess[1], Wayne Burrill[1], John Burton[1], Jackie M. Bye[1], Carol Carder[1], Laura Carrel[9], Joseph Chako[2], Joanne C. Chapman[1], Dean Chavez[2], Ellson Chen[10], Guan Chen[2], Yuan Chen[11], Zhijian Chen[2], Craig Chinault[2], Alfredo Ciccodicola[12], Sue Y. Clark[1], Graham Clarke[1], Chris M. Clee[1], Sheila Clegg[1], Kerstin Clerc-Blankenburg[2], Karen Clifford[1], Vicky Cobley[1], Charlotte G. Cole[1], Jen S. Conquer[1], Nicole Corby[1], Richard E. Connor[1], Robert David[2], Joy Davies[1], Clay Davis[2], John Davis[1], Oliver Delgado[2], Denise DeShazo[2], Pawandeep Dhami[1], Yan Ding[2], Huyen Dinh[2], Steve Dodsworth[1], Heather Draper[2], Shannon Dugan-Rocha[2], Andrew Dunham[1], Matthew Dunn[1], K. James Durbin[2], Ireena Dutta[1], Tamsin Eades[1], Matthew Ellwood[1], Alexandra Emery-Cohen[2], Helen Errington[1], Kathryn L. Evans[13], Louisa Faulkner[1], Fiona Francis[14], John Frankland[1], Audrey E. Fraser[1], Petra Galgoczy[3], James Gilbert[1], Rachel Gill[2], Gernot Glöckner[3], Simon G. Gregory[1], Susan Gribble[1], Coline Griffiths[1], Russell Grocock[1], Yanghong Gu[2], Rhian Gwilliam[1], Cerissa Hamilton[2], Elizabeth A. Hart[1], Alicia Hawes[2], Paul D. Heath[1], Katja Heitmann[5], Steffen Hennig[5], Judith Hernandez[2], Bernd Hinzmann[3], Sarah Ho[1], Michael Hoffs[1], Phillip J. Howden[1], Elizabeth J. Huckle[1], Jennifer Hume[2], Paul J. Hunt[1], Adrienne R. Hunt[1], Judith Isherwood[1], Leni Jacob[2], David Johnson[1], Sally Jones[2], Pieter J. de Jong[15], Shirin S. Joseph[1], Stephen Keenan[1], Susan Kelly[2], Joanne K. Kershaw[1], Ziad Khan[2], Petra Kioschis[16], Sven Klages[5], Andrew J. Knights[1], Anna Kosiura[5], Christie Kovar-Smith[2], Gavin K. Laird[1], Cordelia Langford[1], Stephanie Lawlor[1], Margaret Leversha[1], Lora Lewis[2], Wen Liu[2], Christine Lloyd[1], David M. Lloyd[1], Hermela Loulseged[2], Jane E. Loveland[1], Jamieson D. Lovell[1], Ryan Lozado[2], Jing Lu[2], Rachael Lyne[1], Jie Ma[2], Manjula Maheshwari[2], Lucy H. Matthews[1], Jennifer McDowall[1], Stuart McLaren[1], Amanda McMurray[1], Patrick Meidl[1], Thomas Meitinger[17], Sarah Milne[1], George Miner[2], Shailesh L. Mistry[1], Margaret Morgan[2], Sidney Morris[2], Ines Mü ller[5,18], James C. Mullikin[19], Ngoc Nguyen[2], Gabriele Nordsiek[3], Gerald Nyakatura[3], Christopher N. O'Dell[1], Geoffery Okwuonu[2], Sophie Palmer[1], Richard Pandian[1], David Parker[2], Julia Parrish[2], Shiran Pasternak[2], Dina Patel[1], Alex V. Pearce[1], Danita M. Pearson[1], Sarah E. Pelan[1], Lesette Perez[2], Keith M. Porter[1], Yvonne Ramsey[1], Kathrin Reichwald[3], Susan Rhodes[1], Kerry A. Ridler[1], David Schlessinger[20], Mary G. Schueler[19], Harminder K. Sehra[1], Charles Shaw-Smith[1], Hua Shen[2], Elizabeth M. Sheridan[1], Ratna Shownkeen[1], Carl D. Skuce[1], Michelle L. Smith[1], Elizabeth C. Sotheran[1], Helen E. Steingruber[1], Charles A. Steward[1], Roy Storey[1], R. Mark Swann[1], David Swarbreck[1], Paul E. Tabor[2], Stefan Taudien[3], Tineace Taylor[2], Brian Teague[2], Karen Thomas[1], Andrea Thorpe[1], Kirsten Timms[2], Alan Tracey[1], Steve Trevanion[1], Anthony C. Tromans[1], Michele d'Urso[12], Daniel Verduzco[2], Donna Villasana[2], Lenee Waldron[2], Melanie Wall[1], Qiaoyan Wang[2], James Warren[2], Georgina L. Warry[1], Xuehong Wei[2], Anthony West[1], Siobhan L. Whitehead[1], Mathew N. Whiteley[1], Jane E. Wilkinson[1], David L. Willey[1], Gabrielle Williams[2], Leanne Williams[1], Angela Williamson[2], Helen Williamson[1], Laurens Wilming[1], Rebecca L. Woodmansey[1], Paul W. Wray[1], Jennifer Yen[2], Jingkun Zhang[2], Jianling Zhou[2], Huda Zoghbi[2], Sara Zorilla[2], David Buck[1], Richard Reinhardt[5], Annemarie Poustka[16], André Rosenthal[3], Hans Lehrach[5], Alfons Meindl[7], Patrick J. Minx[4], LaDeana W. Hillier[4], Huntington F. Willard[21], Richard K. Wilson[4], Robert H. Waterston[4], Catherine M. Rice[1], Mark Vaudin[1], Alan Coulson[1], David L. Nelson[2], George Weinstock[2], John E. Sulston[1], Richard Durbin[1], Tim Hubbard[1], Richard A. Gibbs[2], Stephan Beck[1], Jane Rogers[1] & David R. Bentley[1]

[1] The Wellcome Trust Sanger Institute, Wellcome Trust Genome Campus, Hinxton, Cambridge CB10 1SA, UK

[2] Baylor College of Medicine Human Genome Sequencing Center, Department of Molecular and Human Genetics, One Baylor Plaza, Houston, Texas 77030, USA

[3] Genomanalyse, Institut für Molekulare Biotechnologie, Beutenbergstr. 11, 07745 Jena, Germany

[4] Washington University Genome Sequencing Center, Box 8501, 4444 Forest Park Avenue, St. Louis, Missouri 63108, USA

[5] Max Planck Institute for Molecular Genetics, Ihnestrasse 73, 14195 Berlin, Germany

[6] Institute for Clinical Molecular Biology, Christian-Albrechts-University, 24105 Kiel, Germany

[7] Medizinische Genetik, Ludwig-Maximilian-Universität, Goethestr. 29, 80336 München, Germany

[8] HUGO Gene Nomenclature Committee, The Galton Laboratory, Department of Biology, University College London, Wolfson House, 4 Stephenson Way, London NW1 2HE, UK

[9] Department of Biochemistry and Molecular Biology, Pennsylvania State College of Medicine, Hershey, Pennsylvania 17033, USA

[10] Advanced Center for Genetic Technology, PE-Applied Biosystems, Foster City, California 94404, USA

[11] European Bioinformatics Institute, Wellcome Trust Genome Campus, Hinxton, Cambridge CB10 1SD, UK

[12] Institute of Genetics and Biophysics, Adriano Buzzati-Traverso, Via Marconi 12, 80100 Naples, Italy

[13] Medical Genetics Section, University of Edinburgh, Western General Hospital, Edinburgh EH4 2XU, UK

[14] Laboratoire de Génétique et de Physiopathologie des Retards Mentaux, Institut Cochin. Inserm U567, Université Paris V., 24 rue du Faubourg Saint Jacques, 75014 Paris, France

[15] BACPAC Resources, Children's Hospital Oakland Research Institute, 747 52nd Street, Oakland, California 94609, USA

[16] Molekulare Genomanalyse, Deutsches Krebsforschungszentrum, Im Neuenheimer Feld 580, 69120 Heidelberg, Germany
[17] Institute of Human Genetics, GSF National Research Center for Environment and Health, Ingolstädter Landstr. 1, 85764 Neuherberg, Germany
[18] RZPD Resource Center for Genome Research, 14059 Berlin, Germany
[19] National Human Genome Research Institute, National Institutes of Health, Bethesda, Maryland 20892, USA
[20] Laboratory of Genetics, National Institute on Aging, 333 Cassell Drive, Baltimore, Maryland 21224, USA
[21] Institute for Genome Sciences & Policy, Duke University, Durham, North Carolina 27708, USA

Received 1 February; accepted 7 February 2005; doi:10.1038/nature03440.

---

References:

1. Ohno, S. *Sex Chromosomes and Sex-linked Genes* (Springer, Berlin, 1967).
2. McKusick-Nathans Institute for Genetic Medicine, Johns Hopkins University and National Center for Biotechnology Information, National Library of Medicine. *OMIM: Online Mendelian Inheritance in Man.* 〈http://www.ncbi.nlm.nih.gov/omim/〉 (2000).
3. Royer-Pokora, B. *et al.* Cloning the gene for an inherited human disorder—chronic granulomatous disease—on the basis of its chromosomal location. *Nature* **322**, 32-38 (1986).
4. Monaco, A. P. *et al.* Isolation of candidate cDNAs for portions of the Duchenne muscular dystrophy gene. *Nature* **323**, 646-650 (1986).
5. Bentley, D. R. *et al.* The physical maps for sequencing human chromosomes 1, 6, 9, 10, 13, 20 and X. *Nature* **409**, 942-943 (2001).
6. Humphray, S. J. *et al.* DNA sequence and analysis of human chromosome 9. *Nature* **429**, 369-374 (2004).
7. International Human Genome Sequencing Consortium. Initial sequencing and analysis of the human genome. *Nature* **409**, 860-921 (2001).
8. Schmutz, J. *et al.* Quality assessment of the human genome sequence. *Nature* **429**, 365-368 (2004).
9. Mahtani, M. M. & Willard, H. F. Physical and genetic mapping of the human X chromosome centromere: repression of recombination. *Genome Res.* **8**, 100-110 (1998).
10. Kong, A. *et al.* A high-resolution recombination map of the human genome. *Nature Genet.* **31**, 241-247 (2002).
11. Pruitt, K. D., Katz, K. S., Sicotte, H. & Maglott, D. R. Introducing RefSeq and LocusLink: curated human genome resources at the NCBI. *Trends Genet.* **16**, 44-47 (2000).
12. Mungall, A. J. *et al.* The DNA sequence and analysis of human chromosome 6. *Nature* **425**, 805-811 (2003).
13. Deloukas, P. *et al.* The DNA sequence and comparative analysis of human chromosome **10**. *Nature* 429, 375-381 (2004).
14. Dunham, A. *et al.* The DNA sequence and analysis of human chromosome 13. *Nature* **428**, 522-528 (2004).
15. Griffiths-Jones, S., Bateman, A., Marshall, M., Khanna, A. & Eddy, S. R. Rfam: an RNA family database. *Nucleic Acids Res.* **31**, 439-441 (2003).
16. Lowe, T. M. & Eddy, S. R. tRNAscan-SE: a program for improved detection of transfer RNA genes in genomic sequence. *Nucleic Acids Res.* **25**, 955-964 (1997).
17. Griffiths-Jones, S. The microRNA Registry. *Nucleic Acids Res.* **32** (Database issue), D109-111 (2004).
18. Brown, C. J. *et al.* A gene from the region of the human X inactivation centre is expressed exclusively from the inactive X chromosome. *Nature* 349, 38-44 (1991).
19. Brown, C. J. *et al.* The human XIST gene: analysis of a 17 kb inactive X-specific RNA that contains conserved repeats and is highly localized within the nucleus. *Cell* **71**, 527-542 (1992).
20. Lee, J. T., Davidow, L. S. & Warshawsky, D. Tsix, a gene antisense to Xist at the X-inactivation centre. *Nature Genet.* **21**, 400-404 (1999).
21. Migeon, B. R., Chowdhury, A. K., Dunston, J. A. & McIntosh, I. Identification of TSIX, encoding an RNA antisense to human XIST, reveals differences from its murine counterpart: implications for X inactivation. *Am. J. Hum. Genet.* **69**, 951-960 (2001).
22. Chow, J. C., Hall, L. L., Clemson, C. M., Lawrence, J. B. & Brown, C. J. Characterization of expression at the human XIST locus in somatic, embryonal carcinoma, and transgenic cell lines. *Genomics* **82**, 309-322 (2003).
23. Chureau, C. *et al.* Comparative sequence analysis of the X-inactivation center region in mouse, human, and bovine. *Genome Res.* **12**, 894-908 (2002).
24. Bateman, A. *et al.* The Pfam protein families database. *Nucleic Acids Res.* **32** (Database issue), D138-141 (2004).
25. Scanlan, M. J., Simpson, A. J. & Old, L. J. The cancer/testis genes: review, standardization, and commentary. *Cancer Immun.* [online] **4**, 1 (2004).
26. Hurst, L. D. Evolutionary genomics: Sex and the X. *Nature* **411**, 149-150 (2001).
27. Chomez, P. *et al.* An overview of the MAGE gene family with the identification of all human members of the family. *Cancer Res.* **61**, 5544-5551 (2001).
28. Cheung, J. *et al.* Genome-wide detection of segmental duplications and potential assembly errors in the human genome sequence. *Genome Biol.* **4**, R25 (2003).
29. Stankiewicz, P. & Lupski, J. R. Genome architecture, rearrangements and genomic disorders. *Trends Genet.* **18**, 74-82 (2002).
30. Pegoraro, E. *et al.* Familial skewed X inactivation: a molecular trait associated with high spontaneous- abortion rate maps to Xq28. *Am. J. Hum. Genet.* **61**, 160-170 (1997).
31. Schueler, M. G., Higgins, A. W., Rudd, M. K., Gustashaw, K. & Willard, H. F. Genomic and genetic definition of a functional human centromere. *Science* **294**, 109-115 (2001).
32. Spence, J. M. *et al.* Co-localization of centromere activity, proteins and topoisomerase II within a subdomain of the major human X alpha-satellite array. *EMBO J.* **21**, 5269-5280 (2002).
33. Rudd, M. K. & Willard, H. F. Analysis of the centromeric regions of the human genome assembly. *Trends Genet.* **20**, 529-533 (2004).

34. She, X. *et al.* The structure and evolution of centromeric transition regions within the human genome. *Nature* **430**, 857-864 (2004).

35. The International SNP Map Working Group. A map of human genome sequence variation containing 1.42 million single nucleotide polymorphisms. *Nature* **409**, 928-933 (2001).

36. Schmid, M. *et al.* First report on chicken genes and chromosomes 2000. *Cytogenet. Cell Genet.* **90**, 169-218 (2000).

37. Kohn, M., Kehrer-Sawatzki, H., Vogel, W., Graves, J. A. & Hameister, H. Wide genome comparisons reveal the origins of the human X chromosome. *Trends Genet.* **20**, 598-603 (2004).

38. Graves, J. A. The origin and function of the mammalian Y chromosome and Y-borne genes—an evolving understanding. *Bioessays* **17**, 311-320 (1995).

39. Hillier, L. W. *et al.* Sequence and comparative analysis of the chicken genome provide unique perspectives on vertebrate evolution. *Nature* **432**, 695-716 (2004).

40. Freije, D., Helms, C., Watson, M. S. & Donis-Keller, H. Identification of a second pseudoautosomal region near the Xq and Yq telomeres. *Science* **258**, 1784-1787 (1992).

41. Skaletsky, H. *et al.* The male-specific region of the human Y chromosome is a mosaic of discrete sequence classes. *Nature* **423**, 825-837 (2003).

42. Page, D. C., Harper, M. E., Love, J. & Botstein, D. Occurrence of a transposition from the X-chromosome long arm to the Y-chromosome short arm during human evolution. *Nature* **311**, 119-123 (1984).

43. Sargent, C. A. *et al.* The sequence organization of Yp/proximal Xq homologous regions of the human sex chromosomes is highly conserved. *Genomics* **32**, 200-209 (1996).

44. Toder, R., Wakefield, M. J. & Graves, J. A. The minimal mammalian Y chromosome - the marsupial Y as a model system. *Cytogenet. Cell Genet.* **91**, 285-292 (2000).

45. Delbridge, M. L. *et al.* TSPY, the candidate gonadoblastoma gene on the human Y chromosome, has a widely expressed homologue on the X - implications for Y chromosome evolution. *Chromosome Res.* **12**, 345-356 (2004).

46. Lahn, B. T. & Page, D. C. Four evolutionary strata on the human X chromosome. *Science* **286**, 964-967 (1999).

47. Lyon, M. F. Gene action in the X-chromosome of the mouse (*Mus musculus* L.). *Nature* **190**, 372-373 (1961).

48. Carrel, L. & Willard, H. F. X-inactivation profile reveals extensive variability in X-linked gene expression in females. *Nature* doi:10.1038/nature03479 (this issue).

49. Jegalian, K. & Page, D. C. A proposed path by which genes common to mammalian X and Y chromosomes evolve to become X inactivated. *Nature* **394**, 776-780 (1998).

50. Gartler, S. M. & Riggs, A. D. Mammalian X-chromosome inactivation. *Annu. Rev. Genet.* **17**, 155-190 (1983).

51. Lyon, M. F. X-chromosome inactivation: a repeat hypothesis. *Cytogenet. Cell Genet.* **80**, 133-137 (1998).

52. Bailey, J. A., Carrel, L., Chakravarti, A. & Eichler, E. E. Molecular evidence for a relationship between LINE-1 elements and X chromosome inactivation: the Lyon repeat hypothesis. *Proc. Natl Acad. Sci. USA* **97**, 6634-6639 (2000).

53. Smit, A. F., Toth, G., Riggs, A. D. & Jurka, J. Ancestral, mammalian-wide subfamilies of LINE-1 repetitive sequences. *J. Mol. Biol.* **246**, 401-417 (1995).

54. Ke, X. & Collins, A. CpG islands in human X-inactivation. *Ann. Hum. Genet.* **67**, 242-249 (2003).

55. Coffey, A. J. *et al.* Host response to EBV infection in X-linked lymphoproliferative disease results from mutations in an SH2-domain encoding gene. *Nature Genet.* **20**, 129-135 (1998).

56. Stromme, P. *et al.* Mutations in the human ortholog of *Aristaless* cause X-linked mental retardation and epilepsy. *Nature Genet.* **30**, 441-445 (2002).

57. Amir, R. E. *et al.* Rett syndrome is caused by mutations in X-linked MECP2, encoding methyl-CpG-binding protein 2. *Nature Genet.* **23**, 185-188 (1999).

58. Orrico, A. *et al.* MECP2 mutation in male patients with non-specific X-linked mental retardation.*FEBS Lett.* **481**, 285-288 (2000).

59. Ropers, H. H. *et al.* Nonsyndromic X-linked mental retardation: where are the missing mutations?*Trends Genet.* **19**, 316-320 (2003).

60. The International Human Genome Mapping Consortium. A physical map of the human genome. *Nature* **409**, 934-941 (2001).

61. Ning, Z., Cox, A. J. & Mullikin, J. C. SSAHA: a fast search method for large DNA databases. *Genome Res.* **11**, 1725-1729 (2001).

62. Brudno, M. *et al.* LAGAN and Multi-LAGAN: efficient tools for large-scale multiple alignment of genomic DNA. *Genome Res.* **13**, 721-731 (2003).

63. Frazer, K. A., Pachter, L., Poliakov, A., Rubin, E. M. & Dubchak, I. VISTA: computational tools for comparative genomics. *Nucleic Acids Res.* **32**, W273–W279 (2004).

64. Bourque, G. & Pevzner, P. A. Genome-scale evolution: reconstructing gene orders in the ancestral species. *Genome Res.* **12**, 26-36 (2002).

65. Francke, U. Digitized and differentially shaded human chromosome ideograms for genomic applications. *Cytogenet. Cell Genet.* **65**, 206-218 (1994).

66. Gibbs, R. A. *et al.* Genome sequence of the Brown Norway rat yields insights into mammalian evolution. *Nature* **428**, 493-521 (2004).

67. Iwase, M. *et al.* The amelogenin loci span an ancient pseudoautosomal boundary in diverse mammalian species. *Proc. Natl Acad. Sci. USA* **100**, 5258-5263 (2003).

68. Lahn, B. T. & Page, D. C. A human sex-chromosomal gene family expressed in male germ cells and encoding variably charged proteins. *Hum. Mol. Genet.* **9**, 311-319 (2000).

**Supplementary Information** accompanies the paper on www.nature.com/nature.

**Acknowledgements**. We thank the Washington University Genome Sequencing Center for access to chicken and chimpanzee genome sequence data before publication; the Broad Institute for access to dog and chimpanzee genome sequence data before publication, for *T. nigroviridis* genome data, and for fosmid end-sequence data and clones; members of the Sanger Institute zebrafish genome project; the mouse, rat and *F. rubripes* sequencing consortia; Genoscope for *T. nigroviridis* genome data; the Ensembl, UCSC, EMBL and GenBank database groups; G. Schuler for information on sequence overlaps; T. Furey for information on RefSeq RNA coverage; D. Jaffe for data on fosmid end-sequence matches; D. Vetrie, E. Kendall, D. Stephan, J. Trent, A. P. Monaco, J. Chelly, D. Thiselton, A. Hardcastle, G. Rappold and the Resource Centre of the German Human Genome Project (RZPD) for the provision of clones and mapping data; the HUGO Gene Nomenclature

Committee (S. Povey (chair), M. W. Wright, M. J. Lush, R. C. Lovering, V. K. Khodiyar, H. M. Wain and C. C. Talbot Jr) for assigning official gene symbols; C. Rees for assistance with the manuscript; C. Tyler-Smith for critical reading of the manuscript; and the Wellcome Trust, the NHGRI, and the Ministry of Education and Research (Germany) for financial support.

**Competing interests statement.** The authors declare that they have no competing financial interests.

**Correspondence** and requests for materials should be addressed to M.T.R. (mtr@sanger.ac.uk). All DNA sequences reported in this study have been deposited in the EMBL or GenBank databases, and accession numbers are given in Supplementary Fig. 1.